A Century of Honors

Past to Present

A
Century
of
Honors

A Century of Honors

The First One-Hundred Years of Award Winners, Honorary Members, Past Presidents, and Fellows of the Institute

IEEE
PRESS

The Institute of Electrical and Electronics Engineers, Inc., New York

---

Library of Congress Cataloging in Publication Data:

Institute of Electrical and Electronics Engineers.
  A century of honors.

  1. Institute or Electrical and Electronics Engineers.
  2. Electric engineering—Awards.   3. Electronics—Awards.
  I. Title.
  TK1.I422   621.3'.079   84-6712
  ISBN 0-87942-177-0

PRINTED IN THE UNITED STATES OF AMERICA
IEEE Order Number PC017119

A CENTURY OF ELECTRICAL PROGRESS

HONORARY CENTENNIAL COMMITTEE

Richard J. Gowen, *1984 IEEE President*

John Bardeen
William R. Hewlett
William C. Norris
Robert N. Noyce
Simon Ramo

Ian M. Ross
Roland W. Schmitt
Mark Shepard, Jr.
John W. Simpson
Charles H. Townes

CENTENNIAL TASK FORCE

John D. Ryder, *Chairman*
Donald S. Brereton, *Vice Chairman*

Nathan Cohn
Robert F. Cotellessa
Lawrence P. Grayson

Donald T. Michael
William W. Middleton
Mac E. Van Valkenburg

Table of Contents

Part I
Award Winners, Honorary Members, and Institute Presidents

Section I-A: *IEEE Medal of Honor*

Section I-B: *Major Annual Medals*

Section I-C: *Field Awards*

A Legacy of Excellence

MAY 13, 1984 marks the centennial of the world's largest engineering society. That key event, the founding of the IEEE under the banner of the American Institute of Electrical Engineers in 1884, also signaled the emergence of electrical engineering as a recognized profession. Therefore, 1984 is truly a significant milestone for electrical engineering.

At this special moment when we can look back with pride on a century of electrical progress, it is appropriate that we give particular recognition to those whose contributions were most responsible for that progress. Towards that end, the following pages bring together the names of a select group of individuals who have been the most highly honored by the IEEE and its predecessors, the American Institute of Electrical Engineers and the Institute of Radio Engineers, during these past 100 years.

Whether as President of the Institute, as recipient of a major award, or as Fellow of the Institute, collectively these individuals have succeeded in building an esteemed profession, a vigorous Institute to serve it, and a rich base of accomplishments for others to build upon. Together they form the honor roll of our profession.

As our first century ends, we express our grateful appreciation to those listed in this special volume for their outstanding legacy of engineering excellence. As our second century begins, it becomes our opportunity and obligation to begin building our own legacy of excellence and of even greater service to mankind.

Richard J. Gowen
IEEE President, 1984

Introduction

IN 1984, the Institute of Electrical and Electronics Engineers (IEEE) celebrates an important milestone—its one-hundredth anniversary. From the vantage point of the IEEE centennial, the Institute's members can look back on a century of outstanding progress and achievements and can count as colleagues, past and present, the giants of electrical technology. This book is a compilation of the honors bestowed upon those great men and women by their fellow engineers. Thanks to the contributions of these and others, our society is, in part, the product of electrical engineering. Indeed, the history of the IEEE and its predecessors, the American Institute of Electrical Engineers (AIEE) and the Institute of Radio Engineers (IRE), is part of the record of the impact of electrical science and technology on the shaping of the twentieth century.

The AIEE was born during a period of optimism and enthusiasm. By 1884, applications for electricity were rapidly increasing, progress in electrical theory and practice were accelerating, and scientists and electricians, as well as entrepreneurs and investors, saw only greater growth ahead. With such growth, electrical technology was becoming more complex, and practitioners began to feel the need for a national forum to exchange ideas and experiences and an organization to define and guide a new profession.

In the spring of 1884, a call was issued for a meeting to form a national electrical society, and after some preliminary gatherings, the American Institute of Electrical Engineers was established in New York City on May 13. Impetus had been given to the new organization by the planning for an International Electrical Exhibition to be held by the Franklin Institute in Philadelphia later that year, and the AIEE quickly gained recognition as a spokesman for American electrical engineers.

From the beginning, wire communications and light and power systems were the major interests of the AIEE. An early and active participant in the development of standards for the electrical industry, the Institute became the primary arbiter for electrical standards in the United States. During the first three decades of its existence, the AIEE confronted and resolved such internal concerns as locating permanent headquarters for the organization; providing mechanisms for contact with a far-flung membership and with students; and fostering new technical interests through committees that were established to meet the challenge of increasing specialization.

By 1912, however, the interests and needs of those specializing in the expanding field of radio could no longer be satisfied by a technical committee meeting two or three times a year. In that year, two largely local radio organizations—The Society of Wireless Telegraph Engineers and the Wireless Institute—merged to form a national society for scientists and engineers involved in the development of wireless communication—the Institute of Radio Engineers. Many of the original members of the IRE were members of the AIEE and both organizations continued to have members in common until they merged in 1963.

The structural development and general activities of the IRE were similar to those of the AIEE. Specialized segments were gathered into professional groups under a central governing body; geographical units and student branches were formed; the creation of an extensive literature and the exchange of knowledge was facilitated through meetings and publications; membership grades were established; and standards were, from the beginning, a major concern.

The nature of radio technology meant that the interests of the IRE went beyond national boundaries. Therefore, the new organization sought and attracted members from many countries and eventually estabished units in several areas throughout the world.

In the 1930's, the word "electronics" became part of the vocabulary of electrical engineering. Electronics engineers tended to become members of the IRE, but the applications of electron tube technology became so extensive that the technical boundaries differentiating the IRE and the AIEE became difficult to distinguish. After World War II, the two organizations became increasingly competitive. Problems of overlap and duplication of effort arose, only partially resolved by joint committees and meetings.

Finally, in 1961, the leadership of both the IRE and the AIEE resolved to seek an end to these difficulties through consolidation. The next year a merger plan was formulated and approved, becoming effective on January 1, 1963. Plans were made for melding the technical activities and geographical units of the two societies and for establishing a unified publications program for the new organization, the Institute of Electrical and Electronics Engineers.

Throughout the past one-hundred years the IEEE and its predecessors developed ways to honor the achievements of outstanding electrical engineers. These accolades include awards, Honorary Membership, and the grade of Fellow.

The Edison Medal, the oldest of the awards, was conceived by a group of Thomas Edison's associates in 1904 and first awarded in 1909 to Elihu Thomson, under the auspices of the AIEE. The Medal of Honor was the IRE's first medal; it was initially given to Edwin Armstong in 1917 and is now the highest award given by the IEEE. In addition to these, the Institute presently awards six Major Annual Medals, fourteen Field Awards for outstanding achievement in specific fields of electrical and electronics engineering, a Service Award, and three Prize Paper Awards for outstanding papers in IEEE publications. Part I of *A Century of Honors* explains the purpose and origin of each of these awards, and lists the past and present winners of each.

Part I also contains a list of Honorary Members. The AIEE established Honorary Membership at its founding in 1884, and first awarded the grade to Sir William Preece that same year. While the AIEE did not initially define the qualifications for Honorary Membership, the grade was clearly reserved for those with truly great accomplishments. In its first ten years the honor went to such men as Cyrus Field, William Thomson (Lord Kelvin), and Hermann von Helmholz. In 1901 the Institute stipulated in its constitution that Honorary Members were to be chosen "from among those who have rendered acknowledged eminent service to electrical engineering or its allied sciences." In 1957 it broadened the definition by making eligible those who had "rendered meritorious service to mankind in engineering or other allied fields." During the seventy-nine years that the AIEE existed as an independent organization, it elected a total of 49 men as Honorary Members, the last two being Philip Sporn and Allen B. DuMont in 1961.

The first constitution of the Institute of Radio Engineers provided that persons who had "rendered acknowledged eminent service to the art or science of radio transmission" could be elected Honorary Members. The IRE never awarded the grade, however, and all reference to Honorary Members was dropped from its constitution in 1931.

When the AIEE and IRE merged in 1963, provision for Honorary Members was made in the bylaws of the new Institute of Electrical and Electronics Engineers, but no new Honorary Members were elected until 1981. In that year the requirements were changed so that only non-IEEE members were eligible, and the Institute also began to issue citations describing the accomplishments which resulted in the Honorary Membership. The complete list of all Honorary Members, including citations where appropriate, follows the award winners and is, in turn, followed by a list of all Presidents of the AIEE, IRE, and IEEE.

Parts II and III of *A Century of Honors* are devoted to Fellows of the Institute. The grade of Fellow first appeared in the AIEE constitution of 1912. In that year the AIEE revised its membership structure and established the grade of Fellow for those engineers who had demonstrated outstanding proficency and had achieved distinction in their profession. Potential Fellows had to be at least thirty-two years of age, with a minimum of ten years experience. When the IRE established its Fellow grade in 1914 the requirements were clearly modeled after those of the AIEE. Much of the wording in the relevant sections of the IRE constitution is identical to the corresponding wording in the AIEE constitution.

For the first several years after the establishment of the Fellow grade, both the AIEE and the IRE allowed Members to make direct application for transfer to Fellow. In both cases applications had to be accompanied by references from five existing Fellows and required the approval of the Board of Directors. In 1939 the IRE modified its procedure to make admission or transfer to Fellow grade possible only by direct invitation of the Board of Directors, a policy it maintained until the merger in 1963. In 1938 the AIEE modified its constitution to provide that "Applications to the grade of Fellow shall result only from a proposal of five Members or Fellows." In 1951 the AIEE prohibited applications for Fellow grade altogether and adopted a policy of direct invitation similar to that of the IRE.

As noted above, numerous electrical engineers were members of both the AIEE and IRE, and many of these became Fellows of both organizations. When the two institutes merged in 1963 all AIEE and IRE Fellows automatically became Fellows of the IEEE. In 1942 the IRE had begun to issue citations to new Fellows, briefly describing their accomplishments. The AIEE followed suit in 1952, and the IEEE continued the practice after the merger.

Parts II and III of this book are a listing of all members who are now or have even been Fellows of the AIEE, IRE, or IEEE. Part II lists Fellows who are no longer members of the IEEE or are deceased.* Part III consists of Fellows currently active in the IEEE. In Parts II and III the entry for each individual includes the year the member became a Fellow, the organization that originally bestowed the fellowship, and the citation, where appropriate. Also, in Part III, a brief biographical sketch is given.

A Century of Honors is the first comprehensive listing of Fellows ever compiled by the IEEE or its predecessors. While every effort has been made to make this directory as accurate as possible, it, like all historical works, is subject to the limitations imposed by the sources from which it is drawn.

The IEEE has maintained complete records of all Fellows elected since 1963 but, unfortunately, similar records are not available for the IRE before 1942 or the AIEE before 1952. The names of Fellows elected prior to those years are scattered among several sources. Fellows were sometimes listed in publications such as the *AIEE Journal* and *IRE Proceedings,* or in the minutes of the Board of Directors of the IRE. Such listings were not consistently maintained, however; the most reliable source turned out to be membership directories. Since most directories do not contain separate listings of Fellows, it was necessary to search through the entire membership listing of each book, extracting the names of those listed as Fellows. Needless to say, this was a laborious task that was subject to error. A further problem arose from the fact that membership directories were not issued for every year. The IRE, for instance, did not publish books in 1915, 1917–1925, 1933–1936, and 1938–1942; the AIEE did not publish in 1932, 1933, 1935, 1945–1947, and 1952. Thus it is possible (indeed probable) that some members were designated as Fellows but died before their names could appear in a directory.

The IEEE Centennial Task Force encourages those who note errors or omissions to notify the Center for the History of Electrical Engineering. In this way *A Century of Honors* can serve not only as a permanent registry of the honors bestowed by the electrical engineering profession upon its most eminent members, but can also help to close some of the gaps in the historical record.

---

*In addition to the Fellows listed in Part II, a special honor of *Life Fellow Emeritus* was given to *Emily L. Sirjane* in 1979. This distinguished title was created solely for Ms. Sirjane as she had "provided the Institute of Electrical and Electronics Engineers with the benefit of her leadership, wisdom, and dedication and constant support as a member of the staff of the Institute of Radio Engineers and the Institute of Electrical and Electronics Engineers for 31 years."

Part I
Award Winners, Honorary Members, and Institute Presidents

Section I-A
IEEE Medal of Honor

Medal of Honor

Five years after the formation of the Institute of Radio Engineers, the Medal of Honor was established as its first award, to recognize distinguished service in the then fledgling art of radio communication. The first Medal of Honor was bestowed on Major Edwin H. Armstrong in 1917.

In the years that followed, the Medal of Honor became the highest award in the radio engineering profession, given only to those who had attained preeminence in the field through outstanding technical contributions.

Following the merger of the Institute of Radio Engineers and the American Institute of Electrical Engineers in 1963, the Medal of Honor became the highest award of the Institute of Electrical and Electronics Engineers, and its scope was correspondingly broadened. The award consists of a gold medal, a bronze replica, a certificate, and ten thousand dollars, and receives support from the Frank A. Cowan Fund. The recipient need not be a member of the IEEE, and the medal is not necessarily awarded on an annual basis. It is presented only when a candidate is identified as having made a particular contribution which forms a clearly exceptional addition to the science and technology of concern to the Institute.

Medal of Honor Recipients

1917   **E. H. Armstrong**   "In recognition of his work and publications dealing with the action of the oscillating and non-oscillating audion."

1919   **E. F. W. Alexanderson**   "In recognition of his pioneer accomplishments in the field of long distance radio communication, including his development of the radio frequency alternator which bears his name, a magnetic amplifier permitting effective modulation of the output of such an alternator, and a cascade radio frequency vacuum tube amplifier yielding exceptional total amplification."

1920   **Guglielmo Marconi**   "In recognition of his pioneer work in radio telegraphy."

1921   **R. A. Fessenden**   (No citation)

1922   **Lee de Forest**   "For his major contributions to the communications arts and sciences, as particularly exemplified by his invention of that outstandingly significant device: the three-electrode vacuum tube, and his work in the fields of radio telephonic transmission and reception."

1923   **John Stone-Stone**   "For his valuable pioneer contributions to the radio art."

1924   **M. I. Pupin**   "In recognition of his fundamental contributions in the field of electrical tuning and the rectification of alternating currents used for signalling purposes."

1926   **G. W. Pickard**   "For his contributions as to crystal detectors, coil antennas, wave propagation and atmospheric disturbances."

1927   **L. W. Austin**   "For his pioneer work in quantitive measurement and correlation of factors involved in radio wave transmission."

1928   **Jonathan Zenneck**   "For his contribution to original researches in radio circuit performance and to the scientific and educational contributions to the literature of the pioneer radio art."

1929   **G. W. Pierce**   "For his major contributions in the theory and operation of crystal detectors, piezoelectric-crystals and magnetostriction frequency controls and magnetostriction devices for the production of sound; and for his instructional leadership as a teacher and as a writer of important texts in the electric wave fields."

1930   **P. O. Pedersen**   (No citation)

1931   **G. A. Ferrié**   "For his pioneer work in the up building of radio communication in France and in the world, his long continued leadership in the communication field, and his outstanding contributions to the organization of international cooperation in radio."

1932   **A. E. Kennelly**   "For his studies of radio propagation phenomena and his contributions to the theory and measurement methods in the alternating current circuit field which now have extensive radio application."

1933   **J. A. Fleming**   "For the conspicuous part he played in introducing physical and engineering principles into the radio art."

1934   **S. C. Hooper**   "For the orderly planning and systematic organization of radio communication in the Government Service with which he is associated, and the concomitant and resulting advances in the development of radio equipment and procedure."

1935   **Balth. van der Pol**   "For his fundamental studies and contributions in the field of circuit theory and electromagnetic wave propagation phenomena."

1936   **G. A. Campbell**   "For his contributions to the theory of electrical network."

1937   **Melville Eastham**   "For his pioneer work in the field of radio measurements, his constructive influence on laboratory practice in communication engineering, and his unfailing support of the aims and ideals of the Institute."

1938   **J. H. Dellinger**   "For his contributions to the development of radio measurements and standards, his researches and discoveries of the relation between radio wave propagation and other natural phenomena, and his leadership in international conferences contributing to the world wide cooperation in telecommunications."

1939   **A. G. Lee**   "For his accomplishments in promoting international radio services and in fostering advances in the art and science of radio communication."

1940   **Lloyd Espenschied**   "For his accomplishments as an engineer, as an inventor, as a pioneer in the development of radio telephony, and for his effective contributions to the progress of international radio coordination."

1941   **A. N. Goldsmith**   "For his contributions to radio research, engineering, and commercial development, his leadership in standardization, and his unceasing devotion to the establishment and upbuilding of the Institute and its PROCEEDINGS."

1942   **A. H. Taylor**   "For his contributions to radio communication as an engineer and organizer, including pioneering work in the practical application of piezoelectric control to radio transmitters, early recognition and investigation of skip distances and other high-frequency wave-propagation problems, and many years of service to the government of the United States as an engineering executive of outstanding ability in directing the Radio Division of the Naval Research Laboratory."

1943   **William Wilson**   "For his achievements in the development of modern electronics, including its application to radio-telephony, and for his contributions to the welfare and work of the Institute."

1944   **Haraden Pratt**   "In recognition of his engineering contributions to the development of radio, of his work in the extension of communication facilities to distant lands, and of his constructive leadership in Institute affairs."

1945   **H. H. Beverage**   "In recognition of his achievements in radio research and invention, of his practical applications of engineering developments that greatly extended and increased the efficiency of domestic and world-wide radio communications and of his devotion to the affairs of the Institute of Radio Engineers."

1946   **R. V. L. Hartley**   "For his early work on oscillating circuits employing triode tubes and likewise for his early recognition and clear exposition of the fundamental relationship between the total amount of information which may be transmitted over a transmission system of limited bandwidth and the time required."

1948   **L. C. F. Horle**   "For his contributions to the radio industry in standardization work, both in peace and war, particularly in the field of electron tubes, and for his guidance of a multiplicity of technical committees into effective action."

1949   **Ralph Bown**   "For his extensive contributions to the field of radio and for his leadership in Institute affairs."

1950   **F. E. Terman**   "For his many contributions to the radio and electronic industry as teacher, author, scientist and administrator."

1951   **V. K. Zworykin**   "For his outstanding contributions to the concept and development of electronic apparatus basic to modern television, and his scientific achievements that led to fundamental advances in the application of electronics to communications, to industry and to national security."

1952　**W. R. G. Baker**　"In recognition of his outstanding direction of scientific and engineering projects; for his statesmanship in reconciling conflicting viewpoints and obtaining cooperative effort; and for his service to the Institute."

1953　**J. M. Miller**　"In recognition of his pioneering contributions to the fundamentals of electron tube theory and measurements, to crystal controlled oscillators and to receiver development."

1954　**W. L. Everitt**　"For his distinguished career as author, educator and scientist; for his contributions in establishing electronics and communications as a major branch of electrical engineering; for his unselfish service to his country; for his leadership in the affairs of The Institute of Radio Engineers."

1955　**H. T. Friis**　"For his outstanding technical contributions in the expansion of the useful spectrum of radio frequencies, and for the inspiration and leadership he has given to young engineers."

1956　**J. V. L. Hogan**　"For his contributions to the electronic field as a founder and builder of The Institute of Radio Engineers, for the long sequence of his inventions, and for his continuing activity in the development of devices and systems useful in the communications art."

1957　**J. A. Stratton**　"For his inspiring leadership and outstanding contributions to the development of radio engineering, as teacher, physicist, engineer, author and administrator."

1958　**A. W. Hull**　"For outstanding scientific achievement and pioneering inventions and development in the field of electron tubes."

1959　**E. L. Chaffee**　"For his outstanding research contributions and his dedication to training for leadership in radio engineering."

1960　**Harry Nyquist**　"For fundamental contributions to a quantitative understanding of thermal noise, data transmission and negative feedback."

1961　**Ernst A. Guillemin**　"For outstanding scientific and engineering achievements."

1962　**Edward V. Appleton**　"For his distinguished pioneer work in investigating the ionosphere by means of radio waves."

1963　**John H. Hammond, Jr.**　"For pioneering contributions to circuit theory and practice, to the radio control of missiles and to basic communication methods."

1963　**George C. Southworth**　"For pioneering contributions to microwave radio physics, to radio astronomy, and to waveguide transmission."

1964　**Harold A. Wheeler**　"For his analyses of the fundamental limitations on the resolution in television systems and on wideband amplifiers, and for his basic contributions to the theory and development of antennas, microwave elements, circuits, and receivers."

1966   **Claude E. Shannon**   "For his development of a mathematical theory of communication which unified and significantly advanced the state of the art."

1967   **Charles H. Townes**   "For his significant contributions in the field of quantum electronics which have led to the maser and the laser."

1968   **Gordon K. Teal**   "For his contributions to single crystal germanium and silicon technology and the single crystal grown junction transistor."

1969   **Edward L. Ginzton**   "For his outstanding contributions in advancing the technology of high power klystrons and their application, especially to linear particle accelerators."

1970   **Dennis Gabor**   "For his ingenious and exciting discovery and verification of the principles of holography."

1971   **John Bardeen**   "For his profound contributions to the understanding of the conductivity of solids, to the invention of the transistor, and to the microscopic theory of superconductivity."

1972   **Jay W. Forrester**   "For exceptional advances in the digital computer through his invention and application of the magnetic-core random-access memory, employing coincident current addressing."

1973   **Rudolf Kompfner**   "For a major contribution to world-wide communication through the conception of the traveling wave tube embodying a new principle of amplification."

1974   **Rudolf Emil Kalman**   "For pioneering modern methods in system theory, including concepts of controllability, observability, filtering, and algebriac structures."

1975   **John R. Pierce**   "For his pioneering proposals and the leadership of communication satellite experiments, and for contributions in theory and design of electron beam devices essential to their success."

1977   **H. Earle Vaughan**   "For his vision, technical contributions and leadership in the development of the first high-capacity pulse-code-modulation time-division telephone switching system."

1978   **Robert N. Noyce**   "For his contributions to the silicon integrated circuit, a cornerstone of modern electronics."

1979   **Richard Bellman**   "For contributions to decision processes and control system theory, particularly the creation and application of dynamic programming."

1980   **William Shockley**   "For the invention of the junction transistor, the analog and the junction field-effect transistor, and the theory underlying their operation."

1981   **Sidney Darlington**   "For fundamental contributions to filtering and signal processing leading to chirp radar."

1982   **John Wilder Tukey** "For his contributions to the spectral analysis of random processes and the fast Fourier transform algorithm."

1983   **Nicolaas Bloembergen** "For pioneering contributions to Quantum Electronics including the invention of the three-level solid state maser."

1984   **Norman F. Ramsey** "For fundamental contributions to very high accuracy time and frequency standards exemplified by the cesium atomic clock and hydrogen maser oscillator."

Section I-B
Major Annual Medals
Alexander Graham Bell Medal

The telecommunications industry, spawned by Alexander Graham Bell's invention of the telephone in 1876, has played a major role in developing America's economic, social, and political strength during the last 100 years. Bell himself exemplified the contributions scientists and engineers have made to the betterment of mankind. Therefore, to commemorate the centennial of the telephone's invention, the Alexander Graham Bell Medal was established in 1976 through agreement between the American Telephone and Telegraph Company and the IEEE Foundation, Incorporated. The award consists of a gold medal, a bronze replica, a certificate, and ten thousand dollars. The award may be made each year to an individual or group of individuals, up to three in number, "for exceptional contributions to the advancement of telecommunications."

Alexander Graham Bell Medal Recipients

1976   **Amos E. Joel, Jr., William Keister, and Raymond W. Ketchledge**   "For conception and development of Electronic Switching Systems and their effective introduction into a nationwide telephone system."

1977   **Eberhardt Rechtin**   "For pioneering and lasting contributions to deep-space-vehicle communications technology and for leadership in defense telecommunications."

1978   **M. Robert Aaron, John S. Mayo, and Eric E. Sumner**   "For personal contributions to, and leadership in, the practical realization of high-speed digital communications."

1979   **Christian Jacobaeus**   "For pioneering work in the theory of switching systems and technical leadership in the development of telecommunication systems."

1980   **Richard R. Hough**   "For his contributions to the nationwide and international telephone network and, particularly, the introduction of electronic switching therein."

1981   **David Slepian**   "For fundamental contributions to communication theory."

1982   **Harold A. Rosen**   "For pioneering contributions to, and leadership in, geostationary communication satellites."

1983   **Stephen O. Rice**   "For his contributions to the fundamental understanding of communications systems and to the underlying mathematics, and for inspiring younger scientists and engineers."

1984   **Andrew James Viterbi**   "For fundamental contributions to telecommunication theory and practice and for leadership in teaching."

Edison Medal

In 1904, twenty-five years after Thomas Edison produced the first practical incandescent electric light, a group of Mr. Edison's friends and associates created a medal in his honor, to commemorate the achievements of a quarter of a century in the art of electric lighting. In their words "The Edison Medal should, during the centuries to come, serve as an honorable incentive to scientists, engineers, and artisans to maintain by their works the high standard of accomplishment set by the illustrious man whose name and feats shall live while human intelligence continues to inhabit the world."

The American Institute of Electrical Engineers, some four years later, entered into an agreement with the founders to award the medal, with the responsibilities of the Institute carried out through the Edison Medal Committee. The Committee currently functions under the auspices of the IEEE.

The gold Edison Medal, together with a small gold replica, a certificate, and ten thousand dollars, may be awarded annually to some one person for "a career of meritorious achievements in electrical science or electrical engineering or electrical arts."

Edison Medal Recipients

1909   **Elihu Thomson**   "For meritorious achievement in electrical science, engineering and arts as exemplified in his contributions thereto during the past thirty years."

1910   **Frank J. Sprague**   "For meritorious achievement in electrical science, engineering and arts as exemplified in his contributions thereto."

1911   **George Westinghouse**   "For meritorious achievement in connection with the development of the alternating current system for light and power."

1912   **William Stanley**   "For meritorious achievement in invention and development of alternating current systems and apparatus."

1913   **Charles F. Brush**   "For meritorious achievement in the invention and development of the series arc lighting systems."

1914   **Alexander Graham Bell**   "For meritorious achievement in the invention of the telephone."

1916   **Nikola Tesla**   "For meritorious achievement in his early original work in polyphase and high-frequency electrical currents."

1917   **John J. Carty**   "For his work in the science and art of telephone engineering."

1918   **Benjamin G. Lamme**   "For invention and development of electrical machinery."

1919   **W. L. R. Emmet**   "For inventions and developments of electrical apparatus and primo movers."

1920   **Michael I. Pupin** "For his work in mathematical physics and its application to the electric transmission of intelligence."

1921   **Cummings C. Chesney** "For early developments in alternating current transmission."

1922   **Robert A. Millikan** "For his experimental work in electrical science."

1923   **John W. Lieb** "For the development and operation of electric central stations for illumination and power."

1924   **John W. Howell** "For his contributions toward the development of the incandescent lamp."

1925   **Harris J. Ryan** "For his contributions to the science and the art of high-tension transmission of power."

1927   **William D. Coolidge** "For his contributions to the incandescent electric lighting and the X-ray arts."

1928   **Frank B. Jewett** "For his contributions to the art of electric communication."

1929   **Charles F. Scott** "For his contributions to the science and art of polyphase transmission of electrical energy."

1930   **Frank Conrad** "For his contributions to radio broadcasting and short wave radio transmission."

1931   **E. W. Rice, Jr.** "For his contributions to the development of electrical systems and apparatus and his encouragement of scientific research in industry."

1932   **Bancroft Gherardi** "For his contributions to the art of telephone engineering and the development of electrical communication."

1933   **Arthur E. Kennelly** "For meritorious achievements in electrical science, electrical engineering and the electrical arts as exemplified by his contributions to the theory of electrical transmission and to the development of international electrical standards."

1934   **Willis R. Whitney** "For his contributions to electrical science, his pioneer inventions, and his inspiring leadership in research."

1935   **Lewis B. Stillwell** "For his distinguished engineering achievements and his pioneer work in the generation, distribution, and utilization of electric energy."

1936   **Alex Dow** "For outstanding leadership in the development of the central station industry and its service to the public."

1937   **Gano Dunn**   "For distinguished contributions in extending the science and art of electrical engineering, in the administration of great engineering works, and for inspiring leadership in the profession."

1938   **Dugald C. Jackson**   "For outstanding and inspiring leadership in engineering education and in the field of geneation and distribution of electric power."

1939   **Philip Torchio**   "For distinguished contributions to the art of central station engineering and for achievement in the production distribution, and utilization of electrical energy."

1940   **George Ashley Campbell**   "In recognition of his distinction as scientist and inventor and for his outstanding original contributions to the theory and application of electric circuits and apparatus."

1941   **John B. Whitehead**   "For his contributions to the field of electrical engineering, his pioneering and development in the field of dielectric research, and his achievements in the advancement of engineering education."

1942   **Edwin H. Armstrong**   "For distinguished contributions to the art of electric communciation, notably the regenerative circuit, the superheterodyne, and frequency modulation."

1943   **Vannevar Bush**   "For his contribution to the advancement of electrical engineering, particularly through the development of new applications of mathematics to engineering problems, and for his eminent service to the nation in guiding the war research program."

1944   **E. F. W. Alexanderson**   "For his outstanding inventions and developments in the radio, transportation, marine and power fields."

1945   **Philip Sporn**   "For his contributions to the art of economical and dependable power generation and transmission."

1946   **Lee de Forest**   "For pioneering achievements in radio and for the invention of the grid-controlled vacuum tube with its profound technical and social conequences."

1947   **Joseph Slepian**   "For his theoretical and practical contributions to power systems through circuit analysis, arc control, and current interruption."

1948   **Morris E. Leeds**   "For his contributions to industry through development and production of electrical precision measuring devices and controls."

1949   **Karl B. McEachron**   "For his contributions to the advancement of electrical science in the field of lightning and other high voltage phenomena and for the application of this knowledge to the design and protection of electric apparatus systems."

1950   **Otto B. Blackwell**   "For his pioneer contributions to the art of telephone transmission."

1951   **Charles F. Wagner**   "For his distinguished contributions in the field of power system engineering."

1952   **Vladimir K. Zworykin**   "For outstanding contributions to concept and design of electronic components and systems."

1953   **John F. Peters**   "For his contributions to the fundamentals of transformer design, his invention of the Klydonograph, his contributions to Military Computers and for his sympathetic understanding in the training of young engineers."

1954   **Oliver E. Buckley**   "For his personal contributions to the science and art which have made possible a transatlantic telephone cable; for wise leadership of a great industrial laboratory; for outstanding services to the government of his country."

1955   **Leonid A. Umansky**   "For his outstanding contribution to the electrification of industry through the application of electrical machines, devices and systems to automatic process machinery; and for his inspiration, leadership, and teaching of men in this work."

1956   **Comfort A. Adams**   "For pioneering achievements in the development of alternating current electric machinery and in electric welding; for vision and initiative in the formation of an engineering standards organization and for eminence as an educator and consulting engineer."

1957   **John K. Hodnette**   "For his significant contributions to the electrical industry through creative design and development of transformer apparatus which marked new advances in protection, performance and service. For his vision, judgment and management skill which fostered and achieved the practical application of his ideas with resulting advancements in the electrical industry."

1958   **Charles F. Kettering**   "For invention, research and development in the broad fields of industry, engineering, transportation, medicine, education, energy and power resulting in service to all mankind."

1959   **James F. Fairman**   "For outstanding performance in improving the design of large electric power systems; for far-sighted leadership in atomic power development; and for unremitting efforts to improve the engineering profession."

1960   **Harold S. Osborne**   "For his contributions to the art of telecommunication and his leadership and vision in extending its application; for his achievements in the coordination of international communication and in national and international standardization; and for his advancement of the engineering profession."

1961   **William B. Kouwenhoven**   "For his inspiring leadership in education, for his contributions in the fields of electrical insulation, electrical measurements, and electrical science applied to medicine, and especially for his investigations of the effects of electricity on the human body with the successful development of countershock for the cure of fibrillation of the heart."

1962   **Alexander C. Monteith**   "For meritorious achievement in engineering, education, management, and development of young engineers."

1963   **John R. Pierce**   "For his pioneer work and leadership in satellite communications and for his stimulus and contributions to electron optics, travelling wave tube theory and the control of noise in electron streams."

1964   **Schedule Revised***

1965   **Walker Lee Cisler**   "For his achievements in the power industry, including the development of nuclear power; for his services to his country and to international understanding, including the application of electric power to economic growth among all nations; and for his broad contributions to the profession of engineering."

1966   **Wilmer L. Barrow**   "For a career of meritorious achievement—innovating, teaching a developing means for transmission of electromagnetic energy at microwave frequencies."

1967   **George H. Brown**   "For a meritorious career distinguished by significant engineering contributions to antenna development, electromagnetic propagation, the broadcast industry, the art of radio frequency heating, and color television."

1968   **Charles F. Avila**   "For his early contribution to underground transmission, for his continuing guidance in the field of electrical research and for his positive leadership in the development of the electrical utility industry."

1969   **Hendrik W. Bode**   "For fundamental contributions to the arts of communication, computation, and control; for leadership in bringing mathematical science to bear on engineering problems; and for guidance and creative counsel in systems engineering."

1970   **Howard H. Aiken**   "For a meritorious career of pioneering contributions to the development and application of large-scale digital computers and important contributions to education in the digital computer field."

1971   **John W. Simpson**   "For sustained contributions to society through the development and engineering design of nuclear power systems."

1972   **William H. Pickering**   "For contributions to telecommunications, rocket guidance and spacecraft control, and for inspiring leadership in unmanned exploration of the solar system."

1973   **B. D. H. Tellegen**   "For a creative career of significant achievement in electrical circuit theory, including the gyrator."

1974   **Jan A. Rajchman**   "For a creative career in the development of electronic devices and for pioneering work in computer memory systems."

*Date of selection changed to date of presentation.

A CENTURY OF HONORS

1975   **Sidney Darlington** "For basic contributions to network theory and for important inventions in radar systems and electronic circuits."

1976   **Murray Joslin** "For his leadership in overcoming technical and financial obstacles to nuclear power generation and for managerial guidance and foresight in the planning, building and operating the early Dresden Nuclear Power Station."

1977   **Henri Busignies** "For technical contributions and leadership in the field of radar, radio communication and radio navigation."

1978   **Daniel E. Noble** "For leadership and innovation in meeting important public needs, especially in developing mobile communications and solid-state electronics."

1979   **Albert Rose** "For basic inventions in television camera tubes and fundamental contributions to the understanding of photoconductivity, insulators, and human and electronic vision."

1980   **Robert Adler** "For many inventions in the fields of electronic beam tubes and ultrasonic devices, and for leadership in innovative research and development."

1981   **C. Chapin Cutler** "For his creative contributions to microwave electronics, space communications, and technology of communication systems."

1982   **Nathan Cohn** "For a career of creative contributions and leadership in the instrument, control and process computer industry, in the control and economic dispatch of power in large interconnected electrical systems."

1983   **Herman P. Schwan** "For a career of creative endeavor by which engineering, physics, biology, and medicine have been amalgamated into a coherent field of electromagnetic bioengineering."

1984   **Eugene I. Gordon** "For a singular career of invention, development, and leadership in electron devices."

Founders Medal

The Founders Medal was established by the Institute of Radio Engineers in 1952 and is now supported by the IEEE Life Member Fund. It derives its character and purpose from the inspirational example of leadership and service set by the three founders of the IRE: the late Alfred N. Goldsmith, who was Director Emeritus and Editor Emeritus of the IEEE, and the late Messrs. John V. L. Hogan and Robert Marriott. The medal is given for major contributions in the "leadership, planning, or administration of affairs of great value to the electrical and electronics engineering profession," and consists of a gold medal, a bronze replica, and a certificate.

Founders Medal Recipients

1953 **David Sarnoff** "For outstanding contributions to the radio engineering profession through wise and courageous leadership in the planning and administration of technical developments which greatly increased the impact of electronics on the public welfare."

1954 **Alfred N. Goldsmith** "For outstanding contributions to the radio engineering profession through wise and courageous leadership in the planning and administration of technical developments which have greatly increased the impact of electronics on the public welfare."

1957 **Raymond A. Heising** "For his leadership in Institute affairs, for his contributions to the establishment of the permanent IRE Headquarters and for originating the Professional Group system."

1958 **W. R. G. Baker** "For outstanding contributions to the radio engineering profession through wise and courageous leadership in the planning and administration of technical developments which have greatly increased the impact of electronics on the public welfare."

1960 **Haraden Pratt** "For outstanding contributions to the radio engineering profession and to The Institute of Radio Engineers through wise and courageous leadership in the planning and administration of technical developments which have greatly increased the impact of Electronics on the public welfare."

1961 **Ralph Bown** "For outstanding service to the IRE and for outstanding contributions to the radio engineering profession through wise and courageous leadership in the planning and administration of technical developments which have greatly increased the impact of electronics on the public welfare."

1963 **Frederick E. Terman** "For distinguished leadership in the organization and administration of, and contributions to, scientific research and education."

1964 **Andrew G. L. McNaughton** "For his inspiring leadership and his personal contributions in the field of electrical engineering and radio communications."

1966 **Elmer W. Engstrom** "For his leadership in management and integration of research and development programs and for his foresighted application of the systems engineering concept in bringing television to the public."

1967   **Harvey Fletcher** "For his creative contributions to the science of physical acoustics, electrical engineering, and for his management skills in the operation of a leading research laboratory."

1968   **Patrick E. Haggerty** "For outstanding contributions to the leadership of the electrical and electronics engineering profession, with special reference to the development of the worldwide semiconductor industry and service to the profession through his contributions leading to the creation of the IEEE."

1969   **E. Finley Carter** "For outstanding contributions to the electrical engineering profession and to the Institute of Electrical and Electronics Engineers through wise and imaginative leadership in the planning and administration of technical developments in electronics and telecommunications."

1970   **Morris D. Hooven** "In recognition of many years of creative leadership in the advancement of electric power systems engineering, and for contributions to the community in engineering education and in the effective uses of water resources."

1971   **Ernst Weber** "For leadership in the advancement of the electrical and electronics engineering profession in the fields of education, engineering societies, industry and government."

1972   **Masaru Ibuka** "For outstanding administrative leadership in applying solid-state devices in consumer electronics, thereby enhancing industry growth and bringing distinction to the profession."

1973   **William R. Hewlett and David Packard** "For leadership in the development of electronic instruments, for creative management of an industrial activity, and for their unselfish public service."

1974   **Lawrence A. Hyland** "For leadership and management in the field of electronics."

1975   **John G. Brainerd** "For his leadership in electronics in fields encompassing computer technology, high frequency techniques, engineering education, and national and international electrical standardization."

1976   **Edward W. Herold** "For his outstanding contributions to the electrical engineering profession at large, and in particular his insight and leadership in the development of color television."

1977   **Jerome B. Weisner** "For leadership and service to the national and the engineering and scientific professions in matters of technical developments, public policy and education."

1978   **Donald G. Fink** "For contributions and leadership in television, radar and technical journalism and service to the electrical and electronics engineering profession."

1979   **Hanzo Omi** "For pioneering leadership on computer technology, the promotion of international cooperation in research and development in electronics and communications, and for devoted service to the electrical profession."

1980   **Simon Ramo**   ''For contributions and leadership in the development, application, and management of systems engineering in the field of electronics.''

1981   **James Hillier**   ''For original contributions in electron microscopy and leadership in fostering a creative laboratory environment.''

1982   **Shigeru Yonezawa**   ''For technical contributions in the development of VHF multichannel telephone systems, and for leadership in promoting international cooperation between developing and industrialized nations in telecommunications technologies.''

1983   **Joseph M. Pettit**   ''For contributions in electronic and engineering education; for leadership in engineering organizations; and for service to the world as an advisor to government and industry.''

1984   **Kojo Kobayashi**   ''For leadership in the development of computer and communications technologies, their integration into modern networks, and the worldwide expansion of electronics.''

Lamme Medal

The Lamme Medal was established in 1928 through a bequest made by the late Benjamin Garver Lamme, Chief Engineer of the Westinghouse Electric and Manufacturing Company to recognize members of the American Institute of Electrical Engineers who had shown "meritorious achievement in the development of electrical apparatus or machinery." The award now carries the designation of the IEEE and is supported by the Westinghouse Foundation. Its scope has been broadened and it is presently awarded for "meritorious achievement in the development of electrical or electronic apparatus or systems." The award consists of a gold medal, a bronze replica, and a certificate.

Lamme Medal Recipients

1928   **Allan B. Field**   "For the mathematical and experimental investigation of eddy current losses in large slot-wound conductors in electrical machinery."

1929   **Rudolf B. Hellmund**   "For his contributions to the design and development of rotating electrical machinery."

1930   **William J. Foster**   "For his contributions to the design of rotating alternating current machinery."

1931   **Guiseppi Faccioli**   "For his contributions to the development and standardization of high-voltage, oil-filled bushings, capacitors, lightning arresters, and numerous features in high voltage transformers and power transmission."

1932   **Edward Weston**   "For his achievements in the development of electrical apparatus, especially in connection with precision measuring instruments."

1933   **Lewis B. Stillwell**   "For his distinguished career in connection with the design, installation, and operation of electrical machinery and equipment."

1934   **Henry E. Warren**   "For outstanding contributions to the development of electric clocks and means for controlling central station frequencies."

1935   **Vannevar Bush**   "For his development of methods and devices for application of mathematical analysis to problems of electrical engineering."

1936   **Frank Conrad**   "For his pioneering and basic developments in the fields of electric metering and protective systems."

1937   **Robert E. Doherty**   "For his extension of the theory of alternating current machinery, his skill in introducing that theory into practice and his encouragement of young men to aspire to excellence in this field."

1938   **Marion A. Savage**   "For able and original work in the development and improvement of mechanical construction and the efficiency of large high speed turbine alternators."

1939   **Norman W. Storer**   "For pioneering development and application of equipment of electrical traction."

1940   **Comfort A. Adams**   "For his contributions to the theory and design of alternating current machinery and his work in the field of electric welding."

1941   **Forrest E. Ricketts**   "For his contributions to the high reliability of power-supply systems, especially in the design of apparatus for selective relaying and circuit reclosure."

1942   **Joseph Slepian**   "For his contribution to the development of circuit interrupting and current rectifying apparatus."

1943   **Arthur H. Kohoe**   "For his pioneer work in the development of alternating current networks and associated apparatus for power distribution."

1944   **S. H. Mortensen**   "For his pioneer work in the development of self-starting synchronous motors and for his contributions to the development of large hydraulic and steam turbine driven generators."

1945   **David C. Prince**   "For his distinguished work in the development of high voltage switching equipment and electronic converters."

1946   **J. B. MacNeill**   "For his foresight, leadership and creative contribution in the development of switching equipment."

1947   **A. M. MacCutcheon**   "For his distinguished accomplishments in the development of motors for industrial needs, notably in the steel industry."

1948   **V. K. Zworykin**   "For his outstanding contribution to the concept and design of electronic apparatus basic to modern television."

1949   **C. M. Laffoon**   "For outstanding contributions to the design of electrical machines, particularly large turbine generators and high frequency generators."

1950   **Donald I. Bohn**   "For his pioneering development and application of electrical equipment for controlling rectifying systems in the production of aluminum."

1951   **Arthur E. Silver**   "For his pioneering electrification by designing the simplified farm-type transformer combined with high-voltage, lone span, and common neutral line construction."

1952   **I. F. Kinnard**   "For his outstanding contributions in design and developments in instrumentation and measurements."

1953   **F. A. Cowan**   "For his outstanding contributions to long distance communication and the development of modulating and transmission measuring apparatus of original design and wide application."

1954   **A. M. deBellis** "For his contributions to the design and development of power station equipment, especially air-insulated phase-isolated metal-clad high-voltage bus structures and disconnecting switches."

1955   **C. R. Hanna** "For his fundamental calculations and developments in the field of electrodynamics, and particularly for his achievements in the design of generator voltage regulators, automatic rolling mill controls, and tank gun stabilizers."

1956   **H. H. Beverage** "For his pioneering and outstanding engineering achievements in the conception and application of principles basic to progress in national and world-wide radio communications."

1957   **H. S. Black** "For his many outstanding contributions to telecommunication and allied electronic arts, especially the invention of the negative feedback amplifier and the successful development and application of the negative feedback amplification principle."

1958   **P. L. Alger and Sterling Beckwith** "In recognition of his contributions to the art and science of design and application of rotating electric machines."

1959   **Lee A. Kilgore** "For meritorious achievements in the design of electrical machinery; more specifically, for analyses of synchronous machine reactances; for inventions of special armature windings; and for inventions and designs related to large adjustable speed alternating current motors."

1960   **John G. Trump** "For meritorious achievements in the design of particle accelerators and x-ray generators; more specifically for invention and design related both to multi-electrode acceleration tubes and to Van de Graaff generators; and also for exceptional contributions which led to applications in treatment of malignant disease."

1961   **Charles Concordia** "For meritorious achievements in the design of electrical machinery; more specifically, for analyses of synchronous machine characteristics leading to improved designs and for exceptional contributions to the application and control of machines used in electric power systems."

1962   **Edwin L. Harder** "For meritorious achievements in the design, understanding and application of electric apparatus; more specifically for analyses of complex problems involved in rotating machinery, relays, regulators, ground detectors, saturable reactors, industrial control, magnetic amplifiers and computers; and solving these problems by the invention of new and novel forms of such apparatus as well as conceiving new combinations thereof."

1963   **Loyal V. Bewley** "For meritorious achievement in the theoretical analysis of high voltage surges resulting in an advancement of insulation design and improvement of protection for machines, transformers, station apparatus and transmission lines."

1964   **Scheduled Revised***

*Date of selection changed to date of presentation.

1965   **A. Uno Lamm**   ''For outstanding achievement in developing the high-power high-voltage mercury arc valve and a unique system of control and protection for its application as a rectifier and inverter in high-voltage d-c power transmission.''

1966   **René Andre Baudry**   ''For his significant contributions to the design of large electric generators.''

1967   **Warren P. Mason**   ''For outstanding contributions in the fields of sonics and ultrasonics and for his original work in designs of and applications for electro-mechanical transducers.''

1968   **Nathan Cohn**   ''For meritorious achievement in the field of automatic control of power generation and frequency for interconnected electric power systems.''

1969   **James D. Cobine**   ''For his contribution to the knowledge and development of gaseous discharge devices and their adaptation to the development of high-power vacuum interrupters.''

1970   **Harry F. Olson**   ''For his pioneering and continuing leadership in the field of electroacoustics, notably the invention and development of the velocity microphone.''

1971   **Winthrop M. Leeds**   ''For contributions to the development of high voltage, high power circuit breakers, specifically using SF6 gas, and for his effective exposition of the theory of arc interruption.''

1972   **Yu H. Ku and Robert H. Park**   ''In recognition of his outstanding contributions to analysis of the transient behavior of a-c machines and systems.''

1973   **Charles Stark Draper**   ''For outstanding contributions to vehicle guidance, control, and instrumentation through his pioneering development of inertial navigation systems.''

1974   **Seymour B. Cohn**   ''For outstanding contributions to the theory and practice of microwave component design.''

1975   **Harold B. Law**   ''For outstanding contribution in developing color picture tubes, including the fabrication technique which made color television practical.''

1976   **C. Kumar N. Patel**   ''For the invention and development of the carbon dioxide and spin-flip Raman lasers and for contributions to infrared spectroscopy of gases and solids.''

1977   **Bernard M. Oliver**   ''For his contributions to the theory and practice of electronic instrumentation and measurements.''

1978   **Harry Winston Mergler**   ''For pioneering research and creative industrial application of digital technology to machine tool and industrial control systems.''

1979   **James M. Lafferty**   ''For contributions to thermionic emitters and to high-vacuum technology as applied to high-power vacuum switches.''

1980   **Eugene C. Starr**   "For outstanding contributions in the field of long-distance high-voltage electric power transmission systems."

1981   **George B. Litchford**   "For outstanding contributions in the development of electronic systems for air navigation and air traffic control."

1982   **Marvin Chodorow**   "For contributions to the theory and design of high power klystrons and traveling wave tubes."

1983   **Marion E. Hines**   "For sustained, innovative contributions to microwave device applications of semiconductor diodes."

1984   **William McMurray**   "For meritorious achievement in the development of forced-commutation thyristor circuitry and its application to alternating-current adjustable-speed drive systems."

Education Medal

The medal in engineering education was established by the American Institute of Electrical Engineers in 1956. The Education Medal is supported by the IEEE Life Member Fund which is sustained principally by continuing voluntary payments of dues by Life Members after they qualify for exemption from dues.

The award consists of a gold medal, a bronze replica, a certificate, and five thousand dollars. It is awarded annually to recognize "outstanding contributions to education for excellence in teaching and ability to inspire students; leadership in electrical engineering education through publication of textbooks and writings on engineering education; innovations in curricula and teaching methodology; and contributions to the teaching and engineering profession through research, engineering achievements, technical papers, and participation in the education activities of professional societies."

Education Medal Recipients

1956   **F. E. Terman**   (No citation)

1957   **W. L. Everitt**   (No citation)

1958   **J. F. Calvert**   (No citation)

1959   **Gordon S. Brown**   (No citation)

1960   **Ernst Weber**   "For excellence as a teacher in science and electrical engineering, for creative contributions in research and development, for broad professional and administrative leadership and in all for a considerate approach to human relations."

1961   **George F. Corcoran**   "For his contributions to electrical engineering education both in classroom teaching and in the creating of a series of undergraduate text books which have had far reaching influence on the education of electrical engineers."

1962   **Ernst A. Guillemin**   "For inspirational and intellectual leadership in the revolutionary changes in engineering education."

1963   **William G. Dow**   "For his outstanding contributions to teaching, to educational theory and to research, and his inspiring leadership in faculty development."

1964   **B. Richard Teare, Jr.**   "In recognition of his dedication to excellence in the field of teaching, his singular devotion to improvement of this art, his outstanding contributions to the technology of his profession, and his ability to combine a keen sense of practical values with a thorough grounding in science to solve general problems of staff and curricula development in engineering."

1965   **Hugh Hildreth Skilling**   "For leadership in undergraduate education and, in particualr, for innovation and lucid exposition of complex ideas in his textbooks."

1966   **William H. Huggins**   "For his creative approach to and extraordinary effectiveness in teaching; and for his abiding interest in and rapport with students."

1967   **John R. Whinnery**   "For his outstanding teaching, his inspired leadership in educational administration, and his excellent pioneering textbooks."

1968   **Edward C. Jordan**   "For leadership in bringing new technological developments into electrical engineering education, and in creating an integration of research with education in a major department."

1969   **Donald O. Pederson**   "For inspiring teaching and for educational leadership in electronics, notably integrated circuits."

1970   **Jacob Millman**   "For his impact in the areas of electronic devices and circuits through his outstanding textbooks and his stimulating teaching."

1971   **Franz Ollendorff**   "For contributions to the teaching of electrical engineering, especially the preparation of classic texts on electromagnetic fields, and for leadership in building distinguished program in a new institute."

1972   **Mac Elwyn Van Valkenburg**   "For outstanding textbooks in circuit theory, innovations in undergraduate teaching, inspired guidance of students, and professional leadership in electrical engineering."

1973   **Lotfi A. Zadeh**   "For inspired and dedicated teaching, and distinguished leadership in system theory."

1974   **John G. Truxal**   "For pioneering the introduction of engineering concepts into the high schools and for his outstanding textbooks and leadership in engineering education."

1975   **Charles A. Desoer**   "For innovative teaching through publication of standards-setting textbooks and through dedicated and inspired supervision of a generation of highly successful students."

1976   **John G. Linvill**   "For leadership as a teacher, author and administrator, and for contributions in solid-state electronics and technology."

1977   **Robert M. Fano**   "For leadership in engineering education through teaching and outstanding research in computer science, information theory, and electromagnetic theory."

1978   **Harold A. Peterson**   "In recognition of contributions to engineering education through teaching and research in power systems."

1979   **John R. Ragazzini**   "For leadership in engineering education and pioneering research in sampled-data control systems."

1980   **Aldert van der Ziel**   "For leadership in engineering education, and for contributions to noise theory in electronic devices."

1981   **Ernest S. Kuh**   "For leadership in engineering education and contributions to research in circuits and systems."

1982   **King-Sun Fu**   "For contributions to engineering education through inspired teaching and research in computer engineering, system theory and pattern recognition."

1983   **Mischa Schwartz**   "For leadership in engineering education through devoted teaching and publication of outstanding textbooks."

1984   **Athanasios Papoulis**   "For inspirational leadership in teaching through thought-provoking lectures, research, and creative textbooks in signal analysis, stochastic processes, and systems."

Simon Ramo Medal

The Simon Ramo Medal, sponsored by TRW, Inc., was established in 1980 in recognition of the distinguished engineering contributions of Dr. Simon Ramo, Vice Chairman of the Board and Chairman of the Executive Committee of TRW, Inc. It is awarded for "significant achievement in systems engineering and systems science as evidenced by some major engineering contribution or for technical leadership in a major innovative engineering project within the scope of the IEEE."

If the candidate has not previously published an IEEE paper on the subject of the award, the recipient may be requested to present a Simon Ramo lecture on an appropriate subject at a designated IEEE meeting. The award consists of a gold medal, a certificate, five thousand dollars, and a travel allowance not to exceed one half of the honorarium.

Simon Ramo Medal Recipients

1984   **Samuel C. Phillips**   "For technical leadership in system engineering leading to the success of the Apollo lunar program and of various space and missile programs enhancing national security and expanding technical frontiers."

Section I-C
Field Awards

Cledo Brunetti Award

The Cledo Brunetti Award was established in 1975 through a bequest made by the late Cledo Brunetti, an executive of the FMC Corporation. The award is made for "outstanding contributions in the field of miniaturization in the electronic arts," and consists of a certificate and one thousand dollars.

Cledo Brunett Award
Recipients

1978   **Jack S. Kilby and Robert N. Noyce**   "For contributions to miniaturization through invention and the development of integrated circuits."

1979   **Geoffrey W. A. Dummer and Philip J. Franklin**   "For contributions to materials development and fabrication techniques for miniature passive electronic components and assemblies."

1980   **Marcian E. Hoff, Jr.**   "For the conception and development of the microprocessor."

1981   **Donald R. Herriott**   "For key contributions to the development of a practical electron beam system for fabrication of integrated circuit masks and to other aspects of microlithography."

1982   **Robert H. Dennard**   "For the invention of the one-transistor dynamic random access memory cell and for contributions to scaling of MOS devices."

1983   **Abe Offner**   "For the invention and design of the optics which made possible the projection lithography systems that were key to advancing integrated circuit manufacture."

1984   **Harry W. Rubinstein**   "For early key contributions to the development of printed components and conductors on a common insulating substrate."

Control Systems Science and Engineering Award

The award was established in 1980 and is sponsored by Systems Control, Inc. It is awarded to an individual for "meritorious achievement in contributions to theory, design, or techniques, as evidence by publications or patents in the area of control systems science and engineering." The award consists of a certificate and one thousand dollars.

Control Systems Science and Engineering Award Recipients

1982   **Howard H. Rosenbrock**   "For contributions to multivariable control theory and design methods."

1984   **Arthur E. Bryson, Jr.**   "For pioneering contributions to optimal control and estimation and their applications."

Harry Diamond Memorial Award

The award was established in 1949 by friends of the late Harry Diamond, Chief of the Electronics Division of the National Bureau of Standards. Mr. Diamond's associates felt that his professional life exemplified the highest degree of scientific effort in government service. The award was originally sponsored by the Institute of Radio Engineers and consists of a certificate and two thousand dollars. In honors "outstanding technical contributions in the field of government service in any country, as evidence by publication in professional society journals."

Harry Diamond Memorial Award Recipients

1950 **Andrew V. Haeff** "For his contribution to the study of the interaction of electrons and radiation, and for his contribution to the storage tube art."

1951 **Marcel J. E. Golay** "For his many contributions in the overall Signal Corps research and development program and particularly for his accomplishments leading toward a reduction in the infrared-radio gap."

1952 **Newbern Smith** "For his fundamental work during a period of many years on radio wave propagation, this work being the basis for the practical use of ionospheric observations in the operation of worldwide communication systems."

1953 **Robert M. Page** "For outstanding contributions to the development of radar through pioneering work and through sustained efforts over the years."

1954 **Harold A. Zahl** "For his technical contributions, his long service, and his leadership in the U.S. Army Signal Corps research program."

1955 **Bernard Salzberg** "For his contributions in the fields of electron tubes, circuits and military electronics."

1956 **Wilbur S. Hinman, Jr.** "For his contributions to the electronic art in the fields of meteorology and proximity fuzes."

1957 **Georg Goubau** "For his many contributions in ionospheric research and circuit theory, and for his discovery of the surface wave transmission principle."

1958 **Edward W. Allen, Jr.** "For his technical and administrative contributions in the field of radio spectrum utilization."

1959 **Jack W. Herbstreit** "For original research and leadership in radio-wave propagation."

1960 **K. A. Norton** "For contributions to the understanding of radio wave propagation."

1961   **Helmut L. Brueckmann**   "For outstanding contributions to the theory and technology of antennas."

1962   **William Culshaw**   "For outstanding accomplishments in the field of microwave optics and interferometry."

1963   **Allen H. Schooley**   "For contributions in government service to radar and electronic research."

1964   **James R. Wait**   "For outstanding contributions in the field of electromagnetic wave theory."

1965   **George J. Thaler**   "For his outstanding contributions to the published literature in the area of linear and non-linear feedback control theory."

1966   **John J. Egli**   "For outstanding contribution in govenment service in the fields of wave propagation, electromagnetic compatibility and advanced radio communications."

1967   **Rudolf A. Stampfl**   "For is outstanding technical contribution and his able direction of a highly complex engineering organization that has contributed greatly to the exploration of space."

1968   **Harry I. Davis**   "For his outstanding contributions in the conception and development of innovations in military electronics."

1969   **Maurice Apstein**   "For contributions to ordnance electronics and inspiring leadership in the work of government laboratories."

1970   **Allen V. Astin**   "For outstanding contributions and for inspiring technical leadership in instrument and measurement technology."

1971   **Arthur H. Guenther**   "For his contributions to high power pulse techniques to simulate the environmental effects of nuclear weapons explosions, a vital part of the safeguards to the nuclear test ban treaty."

1972   **William B. McLean**   "For outstanding leadership of developments in guided missiles and undersea exploration and transport."

1973   **Harold Jacobs**   "For identification of new bulk semiconductor effects at millimeter waves, with application to the fields of imaging and surveillance."

1974   **Chester H. Page**   "For outstanding technical leadership and contributions to ordnance electronics, metrology and standards, and especially in domestic and international standardization of electrical quantities and symbols."

1975   **Louis Costrell**   "For outstanding achievements in nuclear radiation measurement techniques and related instrumentation standardization."

1976   **Maxime A. Faget**   "For contributions to the design and development of Gemini, Apollo and Space Shuttle manned space-craft."

1977   **Jacob Rabinow**   "For important inventions in ordnance, computers, and post office automation."

1978   **David M. Kerns**   "For contributions to microwave theory basic to power, impedance and antenna standards and measurements."

1979   **Henry P. Kalmus**   "For contributions to electronic ordnance systems."

1980   **Martin Greenspan**   "For contributions to the fields of acoustics and elasticity."

1981   **George Abraham**   "For development of multistable semicondcutor devices and integrated circuits and their application to military systems."

1982   **Jules Aarons**   "For contributions to the understanding of trans-ionospheric wave propagation and its application to satellite communications and navigation."

1983   **Merrill I. Skolnik**   "For fundamental contributions to radar system engineering including antennas, propagation, clutter, accuracy and bistatic techniques."

1984   **Sydney R. Parker**   "For research contributions in sensitivity analysis, digital filters, and discrete time modeling."

William M. Habirshaw Award

The William M. Habirshaw Award was established in 1958 through agreement between Phelps Dodge Foundation and the American Institute of Electrical Engineers. Mr. Habirshaw was a distinguished chemist who pioneered in the first ac high voltage underground cables used in America. The award consists of a bronze medal, a certificate, and one thousand dollars. It may be awarded each year to an individual or group of individuals who have made "outstanding contributions to the field of electrical transmission and distribution."

William M. Habirshaw Award Recipients

1959   **William A. Del Mar**

1960   **Selden B. Crary**   "For contributions to the stability, reliability, and economy, of bulk power transmission systems, and international leadership in the field."

1961   **Samuel B. Griscom**   "For outstanding contributions to the development and application of improved lightning and short-circuit protection of electric power systems, and high voltages for commercial and residential area distribution."

1962   **Herman Halperin**   "For his contributions in the field of high-voltage cable insulation design, development of sheath materials, and allowable economic loading of cables and electrical equipment under varying environments."

1963   **Lawrence M. Robertson**   "For his vigorous leadership and unique contributions in the field of high voltage transmission, particularly including lightning, corona and radio influence studies at high altitudes, and extra-high transmission voltages."

1964   **Clement S. Schifreen**   "In recognition of his important and continuing contributions in the field of cable technology and practice for the transmission and distribution of electric power."

1965   **Wilfred F. Skeats**   "For his outstanding contributions to the understanding of the transient behavior of electric power systems under switching conditions and to the development and testing of high voltage switchgear."

1966   **I. Birger Johnson**   "For his contribution to the reliability and economy of electric power transmission particularly as affected by insulation requirements, and to the understanding of surge voltage phenomena."

1967   **Robert J. Wiseman**   "For his many contributions to power cable technology, with particular emphasis on the development of high-voltage pipe-type cable."

1968   **Eugene C. Starr**   "For outstanding contributions to the development of more economical and reliable ac and dc transmission."

1969   **James A. Rawls** "For outstanding leadership in extending the application of EHV transmission."

1970   **Fred J. Vogel** "For outstanding contributions to electrical equipment development, electrical insulation technology, and industry standards resulting in increased reliability and major economies in ac transmission and distribution systems."

1971   **Gunnar Jancke** "For leadership in the creation of the world's first 400 kV extra high voltage transmission system, including the application of series capacitors at that voltage."

1972   **Jean-Jacques Archambault and Lionel Cahill** "For the pioneering and achievements in the successful planning, engineering and design of the Hydro-Quebec 735 kV bulk power supply transmission system."

1973   **Eugene W. Boehne** "For his studies and evaluation of the performance of EHV transmission facilities under varying conditions and for his contributions to studies and analysis of system behavior."

1974   **Herbert R. Stewart** "For significant contributions to transmission system analysis and design, particularly protection, stability, and control."

1975   **Everett J. Harrington** "For contributions in the application of high-voltage power circuit breakers, high-voltage series and shunt capacitors, and high-voltage direct-current transmission technology."

1976   **Francis John Lane** "For his international leadership in development and application of high-voltage ac and dc transmision."

1977   **Schedule Revised***

1978   **Martin H. McGrath** "For contributions to the theory, design, manufacture, testing and application of underground power cables."

1979   **Howard C. Barnes and Theodore H. Nagel** "For planning, design, and advancement of 765 kV transmission systems."

1980   **Edward W. Kimbark** "For advancement of electric power transmission through innovative research, classic textbooks, and inspirational teaching."

1981   **William R. Johnson** "For contributions to electric power technology by combining high-voltage ac and dc transmission power."

1982   **Peter L. Bellaschi** "For contributions in the field of transmission and distribution of electric power and to the development of extra-high voltage apparatus and systems."

*Date of selection changed to date of presentation.

1983   **Andrew. F. Corry** "For outstanding technical acomplishments in designing and operating underground cable systems and in leadership in organizing and directing high-voltage underground electric power transmission research and development."

1984   **Ralph S. Gens**   "For contributions to the advancement of electric power transmission technology in extra high-voltage ac and dc system research and development."

IEEE Award in International Communication in Honor of Hernand and Sosthènes Behn

The award was established in 1966 through agreement with the International Telephone and Telegraph Corporation, which was founded by Hernand and Sosthènes Behn. It consists of a plaque, a certificate, and two thousand dollars and may be awarded each year to an individual or team of not more than three individuals for "outstanding achievement in the field of international communication."

IEEE Award in International Communication Recipients

1966   **E. Maurice Deloraine**   "In consideration of his outstanding technical and scientific contributions in the field of international communication extending over a period of more than 45 years and particularly for his active leadership in many fields of communication in Europe and in the U.S.A.."

1967   **Leonard Jaffe**   "For his outstanding foresight, initiative and competence in developing and executing an experimental program in communications satellites, and in organizaing and guiding international participation therein."

1968   **Edward W. Allen**   "For work in the development of international communication systems, particularly satellite communications."

1969   **Henri G. Busignies**   "For his outstanding leadership and technical contributions in the fields of electronics technology and communication techniques."

1970   **Herre Rinia**   "For outstanding contributions as a distinguished research organizer and leader in the technology of international communications."

1971   **Eugene F. O'Neill**   "For outstanding technical innovations and management in the development of many key technologies underlying the present day international communications art, especially TELSTAR, the first operational telecommunications satellite, as well as his earlier contributions to trans-oceanic cable telephony."

1972   **Frank de Jager and Johannes A. Greefkes**   "For their contributions to communication systems research, in particular for their inventions in the delta-modulation area."

1973   **Vladimir A. Kotelnikov**   "For fundamental contributions to communication theory and practice, and for pioneering research and leadership in radar astronomy."

1974   **Leslie H. Bedford**   "For his outstanding pioneering and development work in the fields of telecommunications and radar."

1975   **John G. Puente**   "For leadership in and contributions to the development of digital techniques and multiple access methods for satellite communications."

1976 **Sidney Metzger** "For his contribution to the successful development and placing in service of the Intelsat System."

1977 **Schedule Revised***

1978 **F. Louis H. M. Stumpers** "For contribution to the theory and application of frequency modulation and for the promotion of radio sciences on an international level."

1979 **A. Nejat Ince** "For contributions to satellite communications systems and the planning and design of automatically switched international communications."

1980 **Armig B. Kandoian** "For pioneering contributions to international communications, television broadcasting and air navigation."

1981 **Richard Cyril Kirby** "For sustained leadership in the development and management of international radio communications."

1982 **Hiroshi Inose** "For pioneering work in digital switching and modulation and for leadership in international professional activities."

1983 **Lynn W. Ellis** "For improving international communications through advanced planning, cost-effective designs and technology transfer."

*Date of selection changed to date of presentation.

Mervin J. Kelly Award

The Mervin J. Kelly Award was established in 1959 through agreement between the Bell Telephone Laboratories and the American Institute of Electrical Engineers. Dr. Kelly was the President of Bell Laboratories from 1951–1959. The award consisted of a bronze medal, a certificate, and one thousand dollars. It was made each year to an individual or group of individuals for "outstanding contributions to the advancement of the art of telecommunication." It was discontinued in 1976.

Mervin J. Kelly Award Recipients

1960   **Mervin J. Kelly**   "For outstanding contributions in the technology of telecommunication; as a distinguished organizer and eminent leader."

1961   **Harry Nyquist**   "For his fundamental role in the evolution of modern communication and control theories."

1962   **Claude E. Shannon**   "For his fundamental role in the evolution of modern communication, information and switching circuit theories."

1963   **William L. Everitt**   "For his outstanding leadership and many contributions in the field of telecommunication."

1964   **Harald Trap Friis**   "For his many contributions to the development of radio systems from high frequency to microwaves, particularly in the fields of radio measurements and antennas."

1965   **Stephen O. Rice**   "For his outstanding and fundamental contributions in the field of communications, particularly in the understanding of the effects of noise."

1966   **Murray Grinshaw Crosby**   "For his pioneering work in the circuit technology, propagation characteristics, and noise suppression theory of frequency and phase modulation."

1967   **Harry F. Olson**   "For his important and continuing contributions in the fields of electroacoustics, speech processing, and wideband signal recording."

1968   **William R. Bennett**   "For contributions to telecommunication theory in the fields of modulation, noise, and pulse transmission."

1969   **Hendrik C. A. Van Duuren**   "For fundamental contributions to data communications."

1970   **David A. McLean**   "For his pioneering contributions to telecommunication components and particularly for originating and perfecting tantalum thin film technology."

1971   **Harold A. Rosen**   "For his outstanding contributions to telecommunications through his pioneering inventions, developments and leadership in stationary communication satellites."

1972   **Roger A. Sykes**   "For outstanding contributions to the understanding and application of frequency control and filtering."

1973   **Leon S. Nergaard**   "For outstanding contributions and leadership in the introduction of very high and ultra high frequencies for telecommunications."

1974   **Harold M. Barlow**   "For outstanding work in the measurement and properties of radio frequency waves, and their application to telecommunications."

1975   **Alton C. Dickieson**   "For contributions in the field of telecommunications encompassing voice frequency, cable and radio systems for domestic and international services."

Morris E. Leeds Award

The Morris E. Leeds Award was established in 1958 through agreement between the Leeds & Northrup Foundation and the American Institute of Electrical Engineers. Mr. Leeds was the founder of Leeds & Northrup Company, manufacturers of electric instruments and automatic control systems. The award consists of a certificate and one thousand dollars and may be awarded once each year to an individual or group of individuals "making an outstanding contribution to the field of electrical measurement." Special consideration is given to the value of contributions made before the candidate has reached his thirty-sixth birthday.

Morris E. Leeds Award Recipients

1959   **Herbert B. Brooks**   (No citation)

1960   **Perry A. Borden**   "For outstanding contributions to electrical measurements in the field of recording, control and telemetering; and helpful encouragement to younger colleagues."

1961   **Theodore A. Rich**   "For outstandingly creative contributions to the field of measurement through exceptional awareness of new measurement needs and timely invention to accomplish the measurement."

1962   **Bernard E. Lenehan**   "For his keen understanding of electrical measurement phenomena and his outstanding ability to apply fundamental analysis and conceive basically new and sound ideas for solving difficult instrumentation problems."

1963   **Francis B. Silsbee**   "For his contributions to the international standardization of basis electrical units and their measurement."

1964   **John Gilbert Ferguson**   "For his contributions in expanding the knowledge of the measurement of fundamental electrical units."

1965   **Harold E. Edgerton**   "For his contributions to the field of measurements particularly through the development and application of high speed, high intensity, precisely timed light sources."

1966   **William W. Mumford**   "For his outstanding contribution to the theory and technique of microwave measurements, including his invention and application of standard noise sources and directional couplers."

1967   **Henry R. Chope**   "For accomplishment in the field of industrial process measurement through the use of nucleonic and radio-frequency technique."

1968   **Albert J. Williams, Jr.**   "For outstanding contributions to electrical instrumentation and measurement."

1969   **Harry William Houck**   "For outstanding contributions to the field of RF instrumentation."

1970   **Harold I. Ewen**   "For contributions to the design of sensitive radiometric systems, and for the co-discovery of the 21 cm spectral line of interstellar hydrogen."

1971   **Martin E. Packard**   "For his pioneering research leading to the practical use of nuclear magnetic resonance for the accurate measurement of magnetic fields, and for his contributions to the spectrometry of complex molecules."

1972   **Forest K. Harris**   "For a lifetime of making outstanding advances in the science of high-accuracy electrical measurements, and of stimulating further advances through his teaching, authorship, and committee activity."

1973   **Charles Howard Vollum**   "For outstanding contributions in the field of electrical measurements through many developments of the oscilloscope."

1974   **Norbert L. Kusters**   "For inspiring leadership in the field of electrical measurements, in particular for the development of the current comparator and its associated applications."

1976   **Francis Lewis Hermach**   "For research and development of extremely accurate ac–dc transfer standards and for outstanding service on standards committees."

1977   **Arthur Melville Thompson**   "For outstanding advances in absolute electrical measurements, particularly capacitance and resistance."

1978   **Thomas M. Dauphinee**   "For the application of physical principles to electrical instrumentation and measurements."

1979   **Robert D. Cutkosky**   "For internationally significant contributions to the field of precision impedance measurements and standards."

1980   **Wallace H. Coulter**   "For his invention and development of basic electrical measurement methods for the counting, sizing and identification of biological cells and inorganic particles."

1981   **Frank C. Creed**   "For contributions to the theory and practice of high-voltage impulse measurements."

1982   **Lothar Rohde**   "For contributions to the field of precise electronic measurement equipment for very high frequencies."

1983   **Erich P. Ippen and Charles V. Shank**   "For their pioneering contributions to extending the techniques of time-domain measurement into the subpicosecond range."

1984   **Leonard Samuel Cutler**   "For outstanding contributions to the development of advanced time standards, in particular to the cesium beam standard."

Morris N. Liebmann Memorial Award

The Morris N. Liebmann Award was established in 1919 by the Institute of Radio Engineers using funds donated by Emil J. Simon, a charter member of the IRE. The award perpetuates the memory of Colonel Morris N. Liebmann, and associate of Mr. Simon's who was killed during World War I. It is made for the "most important contribution to emerging technologies recognized within recent years," and consists of a certificate and two thousand dollars.

Morris N. Liebmann Award Recipients

1919   **L. F. Fuller** "For his contributions to long distance radio communication."

1920   **R. A. Weagant** "For his experimental researchers and results in the field of the reduction of atmospheric disturbances in radio reception."

1921   **R. A. Heising** "In recognition of the publication of his basic work in the field of the signal modulation of an oscillator out put, and in particular of his invention and development of constant current modulations as first applied to radiotelephony."

1922   **C. S. Franklin** "For his investigations of short wave directional transmission and reception."

1923   **H. H. Deverage** "For his work on directional antennas."

1924   **J. R. Carson** "In recognition of his valuable contributions to alternating current circuit theory and, in particular, to his investigation of filter systems and of single side band telephony."

1925   **Frank Conrad** "For his research work in the short wave transmitting and receiving field."

1926   **Ralph Bown** "For researches and investigations into the more difficult element of wave transmission phenomena which have resulted in extensive and useful additions to existing knowledge."

1927   **A. H. Taylor** "For his work in connection with the investigation of radio transmission phenomena."

1928   **W. G. Cady** "For his fundamental investigation in piezo-electric phenomena and their application to radio technique."

1929   **E. V. Appleton** "For his investigations in the field of wave propagation."

1930   **A. W. Hull** "In recognition of the many advances in vacuum tube development which were due to his fundamental researches in the field of electronics."

1931   **Stuart Ballantine**   "For his outstanding theoretical and experimental investigations of numerous radio and acoustic devices."

1932   **Edmond Bruce**   "For his theoretical investigations and field developments in the domain of directional antennas."

1933   **Heinrich Barkhausen**   "For his work on oscillation circuits and particularly on that type of oscillator which now bears his name."

1934   **V. K. Zworykin**   "For his contributions to the development of television."

1935   **Frederick B. Llewellyn**   "For his analysis and disclosures of the effects and reactions within vacuum tubes at ultra high frequencies."

1936   **B. J. Thompson**   "For his contribution to the vacuum tube art in the field of very high frequencies."

1937   **W. H. Doherty**   "For his improvement in the efficiency of radio frequency power amplifiers."

1938   **G. C. Southworth**   "For his theoretical and experimental investigations of the propagation of ultra high frequency waves through confined dielectric channels and the development of a technique for the generation and measurement of such waves."

1939   **Harald T. Friis**   "For his investigations in radio transmission including the development of method of measuring signals and noise and the creation of a receiving system for mitigating selective fading and noise interference."

1940   **Harold A. Wheeler**   "For his contributions to the analysis of wide band high frequency circuits particularly suitable for television."

1941   **P. T. Farnsworth**   "For his contributions in the field of applied electronics."

1942   **S. A. Schelkunoff**   "For his contributions to the theory of electromagnetic fields in wave transmission and radiation."

1943   **Wilmer L. Barrow**   "For his theoretical and experimental investigations of ultra high frequency propagation in wave guides and radiation from horns, and the application of these principles to engineering practice."

1944   **W. W. Hansesn**   "For his application of electromagnetic theory in radiation antennas, resonators, and electron bunching, and for the development of practical equipment and measurement techniques in the microwave field."

1945   **P. C. Goldmark**   "For his contributions to the development of television systems, particularly in the field of color."

1946   **Albert Rose**   "For his contributions to the art of converting optical images to electrical signals, particularly the image orthicon."

1947   **John R. Pierce**   "For his development of a traveling wave tube having both high gain and very great bandwidth."

1948   **S. W. Seeley**   "For his development of ingenious circuits related to frequency modulation."

1949   **Claude E. Shannon**   "For his original and important contributions to the theory of the transmission of information in the presence of noise."

1950   **O. H. Schade**   "For outstanding contributions to analysis, measurement technique, and system development in the field of television and related optics."

1951   **R. B. Dome**   "For many technical contributions to the profession, but notably his contributions to the inter carrier sound system of television reception, wide band phase shift networks and various simplifying innovations in FM receiver circuits."

1952   **William Shockley**   "In recognition of his contributions to the creation and development of the transistor."

1953   **J. A. Pierce**   "For his pioneering and sustained outstanding contributions to radio navigation, and his related fundamental studies of radio wave propagation."

1954   **R. R. Warnecke**   "For his many valuable contributions and scientific advancements in the field of electron tubes, and in particular, the magnetron class of traveling wave tubes."

1955   **A. V. Loughren**   "For his leadership and technical contributions in the formulation of the signal specification for compatible color television."

1956   **Kenneth Bullington**   "For his contributions to the knowledge of tropospheric transmission beyond the horizon, and to the application of the principles of such transmission to practical communications systems."

1957   **O. G. Villard, Jr.**   "For his contributions in the field of meteor astronomy and ionosphere physics which led to the solution of outstanding problems in radio propagation."

1958   **E. L. Ginzton**   "For his creative contribution to the generation and useful application of high enegy at microwave frequencies."

1959   **Nicolaas Bloembergen and C. H. Townes**   "For fundamental and original contributions to the maser."

1960   **J. A. Rajchman**   "For contributions to the development of magnetic devices for information processing."

1961   **Leo Esaki**   "For important contributions to the theory and technology of solid state devices, particularly as embodied in the tunnel diode."

1962   **Victor H. Rumsey**   "For basic contributions to the development of frequent independent antennas."

1963   **Ian Munro Ross**   "For contributions to the development of the epitaxial transistor and other semiconductor devices."

1964   **Arthur L. Schawlow**   "For his pioneering and continuing contributions in the field of optical masers."

1965   **William R. Bennett, Jr.**   "For basic contributions to the field of radio in the theoretical understanding and physical realization of continuously operated oscillators at optical frequencies."

1966   **Paul K. Weimer**   "For invention, development and applications of the thin film transistor."

1968   **Emmett N. Leith**   "For establishing the pace of coherent optics in radar and communications system for major advances in modern holography."

1969   **John B. Gunn**   "For contributions to solid state microwave power generation."

1970   **John A. Copeland**   "For the discovery of the limited space-charge-accumulation mode of oscillation."

1971   **Martin Ryle**   "For his contributions in applying aperture synthesis to extend the capabilities of radio telescopes, thereby increasing man's knowledge of the Universe."

1972   **Stewart E. Miller**   "For pioneering research in guided millimeter wave and optical transmission systems."

1973   **Nick Holonyak, Jr.**   "For outstanding contributions to the field of visible light emitting diodes and diode lasers."

1974   **Willard S. Boyle and George E. Smith**   "For the invention of the charge-coupled device and leadership in the field of MOS device physics."

1975   **A. H. Bobeck, P. C. Michaelis, and H. E. D. Scovil**   "For the concept and development of single-walled magnetic domains (magnetic bubbles), and for recognition of their importance to memory technology."

1976   **Herbert John Shaw**   "For contributions to the development of acoustics surface wave devices."

1977   **Horst H. Berger and Siegfried K. Wiedmann**   "For the invention and exploration of the Merged Transistor Logic, MTL."

1978   **Kuen C. Kao, John B. MacChesney, and Robert D. Maurer**   "For making communication at optical frequencies practical by discovering, inventing and developing the materials, techniques and configurations for glass fiber waveguides."

1979   **Ping King Tien**   "For contributions to integrated optics technology."

1980   **A. J. DeMaria**   "For contributions to the initiation and demonstration of the first picosecond optical pulse generator."

1981   **Calvin F. Quate**   "For development of an acoustic microscope capable of submicron resolution."

1982   **John R. Arthur, Jr., and Alfred Y. Cho**   "For the development and application of molecular beam epitaxy technology."

1983   **Robert W. Brodersen, Paul R. Gray, and David A. Hodges**   "For pioneering contributions and leadership in research on switched-capacitor circuits for analog-digital conversion and filtering."

1984   **David E. Carlson and Christopher R. Wronski**   "For crucial contributions to the use of amorphous silicon in low-cost, high-performance photovoltaic solar cells."

Jack A. Morton Award

The award was established in 1974 in memory of Dr. Jack A. Morton, Past Vice President of Electronics Technology at Bell Telephone Laboratories, who was known for his outstanding contributions to the development of solid-state electron devices. It is sponsored by twenty semiconductor organizations in the United States, Europe, and Japan.

The award is made to an individual or group of individuals for "outstanding contributions in the field of solid-state devices." It consists of a bronze medal, a certificate, and two thousand dollars.

Jack A. Morton Award Recipients

1976   **Robert N. Hall**   "For outstanding achievement in solid-state physics and chemistry and the invention and development of semiconductor devices."

1977   **Morgan Sparks**   "For contributions to solid state device technology and the management of research and development."

1978   **Juri Matisoo**   "For pioneering the Josephson computer technology."

1979   **Martin P. Lepselter**   "For invention of the beam-lead structure and metallurgy used in silicon integrated circuits."

1980   **James F. Gibbons**   "For pioneering contributions to the use of ion implantation in the fabrication of semiconductor devices."

1981   **Nick Holonyak, Jr.**   "For pioneering work in quantum well lasers and contributions to visible semiconductor lasers and light-emitting diodes."

1982   **Dove Frohman-Bentchkowsky**   "For contributions to non-volatile semiconductor memories."

1983   **Jun-Ichi Nishizawa**   "For invention and development of the class of static induction transistors (SIT) and for advances in optoelectronic devices."

1984   **Hans S. Ruprecht and Jerry M. Woodall**   "For pioneering work in gallium aluminum arsenide heterojunctions and high efficiency light emitting diodes and injection lasers prepared by liquid phase epitaxy."

Frederik Philips Award

The Frederik Philips Award was established in 1971 through agreement with the N. V. Philips' Gloeilampenfabrieken. In that year Mr. Philips retired as President of the Dutch electrical and electronics firm. The award is made each year to an individual or group of individuals up to three in number for "outstanding accomplishments in the management of research and development resulting in effective innovation in the electrical and electronics industry." It consists of a gold medal, a certificate, and two thousand dollars.

Frederik Philips Award Recipients

1971   **Frederik J. Philips**   "For fostering management of research and development that has been effective in the creation of numerous new porducts."

1972   **William O. Baker**   "For leadership of an industrial laboratory where pioneering research was used to create a wide spectrum of new technology throughout the telecommunication industry."

1973   **John H. Dessauer**   "For his pioneering research and development efforts and outstanding accomplishments in the field of xerography."

1974   **Chauncey Guy Suits**   "For his leadership in guiding the research efforts of a major technologically diversified industrial company."

1975   **C. Lester Hogan**   "For outstanding accomplishments in the management of research and develoment resulting in important innovation in the semiconductor device industry."

1976   **Koji Kobayashi**   "For leadership in the management of research, development, and the production of telecommunication and electronic systems."

1977   **Schedule Revised***

1978   **William E. Shoupp**   "For organizational ability in managing electronics and nuclear energy research and development, particularly in development of the first submarine atomic power plant."

1979   **Gordon E. Moore**   "For effective direction of research and development in silicon integrated circuit technology."

1980   **William M. Webster**   "For sustained leadership in the management of research and development."

1981   **Dean A. Watkins**   "For managerial leadership in the research and development of traveling-wave tubes."

*Date of selection changed to date of presentation.

1982   **Werner J. Kleen** "For technical and managerial roles in the development and manufacture of microwave tubes in Europe."

1983   **Allen E. Puckett** "For foresight and judgment in the management of research and development leading to significant advances in communication satellite systems."

1984   **John K. Galt** "For managerial leadership that resulted in the understanding of the Physics and Chemistry of III–V compounds associated with the concepts of luminescence and of heterostructure devices and of the molecular beam epitaxial process."

Emanuel R. Piore Award

The award was established in 1976 through agreement with the International Business Machines Corporation, and honors Emanuel R. Piore, longtime IBM executive and former chief scientist. It consists of a gold plated bronze medal, a certificate, two thousand dollars, and a twenty five-hundred dollar international travel grant. The award may be made each year to an individual or team of individuals for "outstanding achievement in the field of information processing, in relation to computer science, deemed to have contributed significantly to the advancement of science and to the betterment of society."

Emanuel R. Piore Award Recipients

1977   **George R. Stibitz** "For pioneering contributions to the development of computers, utilizing binary and floating-point arithmetic, memory indexing, operation from a remote console, and program-controlled computations."

1978   **J. Presper Eckert, Jr., and John W. Mauchly** "For the design and construction of electronic digital computers which stimulated the development of the computer industry."

1979   **Richard W. Hamming** "For introduction of error correcting codes, pioneering work in operating systems and programming languages, and the advancement of numerical computation."

1980   **Lawrence R. Rabiner and Ronald W. Schafer** "For their contributions to digital speech processing and digital filter design."

1982   **Kenneth L. Thompson and Dennis M. Ritchie** "For the creation and development of an operating system of high utility, availability, and instructive value, as embodied in UNIX and its related facilities."

1983   **Niklaus Wirth** "For creative contribution to programming language and design methodology as exemplified by his development of the Pascal language."

1984   **Harvey G. Cragon** "For creative contributions and leadership in uniting computer architecture with the inherent capabilities of the integrated circuit."

David Sarnoff Award

The David Sarnoff Award was established in 1959 through agreement between the RCA Corporation and the American Institute of Electrical Engineers. At that time Mr. Sarnoff was Chairman of the Board and Chief Executive Officer of RCA. The award consists of a gold medal, a bronze replica, a certificate, and one thousand dollars. The award may be made each year to an individual or group of individuals up to three in number for an "outstanding contribution to the field of electronics."

David Sarnoff Award Recipients

1959   **David Sarnoff**   (No citation)

1960   **Rudolf Kompfner**   "For creative achievements in research and development in the field of electronics and for his leadership in this field."

1961   **Charles H. Townes**   "For research in resonance physics leading to major advances in communication technology."

1962   **Harry B. Smith**   "For his contributions to the field of Doppler radar and other areas of applied electronics."

1963   **Robert N. Hall**   "For his outstanding contributions to the scientific understanding of semiconductors and for their applications to electronics."

1964   **Henri Busignies**   "For outstanding contributions to electronic systems especially in the fields of direction finding and air navigation."

1965   **Jack A. Morton**   "For his outstanding leadership and contributions to the development and understanding of solid-state electron devices."

1966   **Jack S. Kilby**   "For his outstanding creative contribution in the field of monolithic integrated circuits, for his key inventions and for his team leadership in the application of integrated circuits to large scale systems."

1967   **James Hillier**   "For his pioneering research on the electron microscope, including both electron optics and biological microscopy, and for his accomplishments as a research director and his inspiration to scientists—young and old."

1968   **Walter P. Dyke**   "For pioneering contributions to the development of field emission in X-ray technology."

1969   **Robert H. Rediker**   "For contributions to semiconductor device research and injection lasers."

1970   **John B. Johnson**   "For fundamental contributions of major importance to electronics and communications."

1971   **Alan L. McWhorter**   "For outstanding contributions leading to a better understanding of semiconductor devices."

1972   **Edward G. Ramberg**   "For outstanding contributions to electron physics, electron optics and television."

1973   **Max Vernon Mathews**   "For leadership in applying electronics to art and for his contribution to the production of musical sounds by computer."

1974   **Frederik J. L. Sangster**   "For the invention of the integrated bucket-brigade delay line and ingenuity in finding new realizations and applications of this principle."

1975   **Bernard C. De Loach, Jr.**   "For contributions to and leadership in the development of the impact avalanche and transit time (IMPATT) device."

1976   **George Harry Heilmeier**   "For combining science with engineering to create the dynamic scattering liquid crystal display."

1977   **Jack M. Manley and Harrison E. Rowe**   "For their work on the properties of nonlinear devices resulting in the well-known Manley–Rowe Relations."

1978   **Stephen E. Harris**   "For scientific discoveries and device inventions in the fields of lasers, quantum electronics and non-linear optics."

1979   **A. Gardner Fox and Tingye Li**   "For the discovery of modes in open structures and their applications to laser resonators."

1980   **Marshall I. Nathan**   "For his role in the discovery of the injection laser, and other contributions to the physics of semiconductor devices."

1981   **Cyril Hilsum**   "For contributions in the fields of III–V compound semiconductors, solid state microwave components and display devices."

1982   **Nobutoshi Kihara**   "For major contributions to magnetic video tape recording."

1983   **Hermann K. Gummel**   "For contributions and leadership in device analysis and development of computer-aided design tools for semiconductor devices and circuits."

1984   **Alan D. White and Jameson D. Rigden**   "For invention of the visible light helium neon laser."

Charles Proteus Steinmetz Award

Dr. Steinmetz, whose researches into the mathematical bases of electrical phenomena form the foundation of much of modern electrical engineering, was also an active participant in American Institute of Electrical Engineers standardization efforts. The Charles Proteus Steinmetz Award, established in 1979, is presented annually to an individual for "major contributions to the development of standards in the field of electrical and electronics engineering." The award is sponsored by the IEEE Standards Board and consists of a certificate and one thousand dollars.

Charles Proteus Steinmetz Award Recipients

1980   **Leon Podolsky**   "For distinguished leadership in assuring that standards will facilitate international trade."

1982   **Ralph M. Showers**   "For distinguished achievement as engineer, entrepreneur, elightened manager of people, industrialist, and stateman in the application of advanced technology."

1983   **William A. McAdams**   "For outstanding leadership in national and international electrical and electronic standardization activities."

1984   **H. Baron Whitaker**   "For technical and administrative contributions to the development of safety standards for electrical equipment and systems."

Nikola Tesla Award

The award was established in 1975 and is sponsored by the IEEE Power Engineering Society. Nikola Tesla was the inventor of the induction motor and developed the elements of the system that made possible large scale production and distribution of polyphase ac power. The Nikola Tesla Award consists of a plaque and one thousand dollars. It may be awarded each year to an individual or group of individuals who have made "outstanding contributions to the field of generation and utilization of electric power."

Nikola Tesla Award Recipients

1976   **Leon T. Rosenberg**   "For his half-century of development and design of large steam turbine driven generators and his important contributions to literature."

1977   **Cyril G. Veinott**   "For his leadership in development and application of small induction motors."

1978   **Charles H. Holley**   "For contributions to the evolution of turbine generator designs with achievement in performance and reliability."

1979   **John W. Batchelor**   "For contributions to the design of large turbine driven generators and the development of related international standards."

1980   **Philip H. Trickey**   "For advancement in the development and application of Tesla's theories through precise designs of small induction machines."

1981   **Dean B. Harrington**   "For contributions to the design, development and performance analysis of large steam turbine-generators."

1982   **Sakae Yamamura**   "For contributions to the theory of linear induction motors and the development of magnetic levitation of track vehicles."

1984   **Herbert H. Woodson**   "For contributions to power generation technology particularly in superconducting generators and magnetohydrodynamic generators."

Vladimir K. Zworykin Award

The award was established in 1950 under the auspices of the Institute of Radio Engineers, by Dr. Zworykin, one of the seminal figures in the development of electronic television. It is supported by the RCA Corporation and consists of a certificate and one thousand dollars. It is awarded annually when the Institute determines that there is a qualifying candidate for the "most important technical contribution to electronic television."

Vladimir K. Zworykin Award Recipients

1952 **B. D. Loughlin** "For his outstanding contributions to the theory, the understanding and the practice of color television."

1953 **Frank Gray** "For recognition of the fundamental importance to color television of his early studies of the television signal spectrum."

1954 **A. V. Bedford** "For his contributions to the principle of mixed highs and its application to color television."

1955 **H. B. Law** "For development of techniques and processes resulting in a practical form of shadow-mask tri-color kinescope."

1956 **F. J. Bingley** "For contributions to colorimetric science as applied to television."

1957 **Donald Richman** "For contributions to the theory of synchronization, particularly that of color subcarrier reference oscillator synchronization in color television."

1958 **C. P. Ginsburg** "For pioneering contributions to the development of video magnetic recording."

1959 **Paul K. Weimer** "For contributions to photoconductive-type pickup tubes."

1961 **P. C. Goldmark** "For important contributions to the development and utilization of electronic television in military reconnaissance and in medical education."

1962 **G. A. Morton** "For his contribution to electronic television through the development of camera and imaging tubes."

1963 **P. G. Rice, Jr., and W. E. Evans, Jr.** "For the development of techniques and equipment for fixing televised images on paper."

1965 **Norman F. Fyler** "For key contributions to the basic design of color television tubes."

1966 **Ray D. Kell** "For his extensive and significant contributions, papers, and inventions which have been fundamental in the development of both black-and-white and color TV."

1967   **Keiji Suzuki**   "For his outstanding technical contribution in the field of television and engineering leadership in the application of the relaying of television signals via satellites."

1968   **Kurt Schlesinger**   "For sustained and pioneering contributions to television çircuitry and electron optical devices."

1969   **Otto H. Schade, Sr.**   "For broad technical contributions to the electronics and optics of television."

1970   **Charles H. Coleman**   "For highly significant contributions to the technology of recording monochrome and color television signals."

1971   **Alfred C. Schroeder**   "For his outstanding technical contributions to television and particularly his leadership in the development of color television."

1972   **Robin E. Davies**   "For his development of the field-store standards converter which permits U.S. colour television (NTSC) signals and those from the European (PAL) system to be exchanged directly."

1973   **Albert Macovski**   "For contribution to single tube encoded color cameras and color television receiving circuits."

1974   **Senri Miyaoka**   "For his contributions to the development of new concepts in color TV display tubes."

1975   **Eugene I. Gordon and Ralph E. Simon**   "For the invention and leadership in the development of the silicon target camera tube, and in the extension of electronic television into new applications."

1977   **Dalton H. Pritchard**   "For significant contributions to color television technology."

1978   **Sam H. Kaplan**   "For contributions to the theory and development of shadow mask tubes."

1979   **Albert M. Morrell**   "For contributions to shadow mask color picture tube development and design."

1980   **Walter Bruch**   "For the development of the Phase Alternating Line (PAL) color television system."

1981   **Naohiro Goto**   "For the development of the photosensitive structure of the SATICON TV camera tube."

1983   **Jon Clemens and Eugene O. Keizer**   "For outstanding contributions to the development of an electronic disc system for recorded television programs."

1984   **Takehiro Kakizaki and Yasuharu Kubota**   "For the invention and development of a unique single-tube color TV camera using a single carrier and electronic indexing."

Section I-D
IEEE Service Award

Haraden Pratt Award

The Haraden Pratt Award was established in 1971 in honor of Haraden Pratt, who served as an IEEE Officer, Director, and Director Emeritus and who gave dedicated and distinguished service to the Institute. The purpose of the award is to recognize annually individuals who have conferred "outstanding service to the Institute." It consists of an illuminated certificate.

Haraden Pratt Award Recipients

1972   **Alfred N. Goldsmith**   "In recognition of Dr. Goldsmith's dedicated service to the Institute of Electrical and Electronics Engineers and the Institute of Radio Engineers over a span of sixty years, and of his leadership in establishing the high standards for which the Institute is respected throughout the World, the Board of Directors confers this first Award, established through the bequest of his friend, and close associate, Haraden Pratt. On the honored list of those who have rendered outstanding service to the Institue, no one has performed so well, for so long a time, with such unswerving dedication to truth and excellence."

1973   **Elgin B. Robertson, Sr.**   "In recognition of Elgin B. Robertson's half century of dedicated service to the Institute of Electrical and Electronics Engineers and the American Institute of Electrical Engineers, which has included his firm leadership as President of AIEE and his wise counsel as Director of AIEE and Director Emeritus of IEEE, and for his part in founding the IEEE through the union of AIEE and IRE, the Board of Directors confers this Award, established through the bequest of his friend and associate, Haraden Pratt. Attesting to the esteem with which he is held by the Institute and its members is the fact that he is one of four living members on whom the grade of Honorary Member has been conferred. Long devoted to the cause of unity among engineers, he has truly enriched the profession he has served."

1974   **James H. Mulligan, Jr.**   "In recognition of James H. Mulligan's dedicated service to the engineering profession through his wise counsel and action in the IEEE both as Director and President, particularly to the younger members, and of his excellent service in promoting coordination of IEEE with other technical and professional organizations, thereby elevating the professional status of all engineers, the Board of Directors confers this Award established by Haraden Pratt with the expectation that his accomplishments will act as a catalyst to future leaders."

1975   **Walter J. Barrett**   "In recognition of Walter J. Barrett's long and dedicated service to the engineering profession and the Institute of Electrical and Electronics Engineers and the American Institute of Electrical Engineers as Director and President of both AIEE and the United Engineering Trustees, and particularly for his dynamic leadership in planning and in organizing financial support for the new United Engineering Center and in locating its distinguished site, the Board of Directors confers this award which appropriately recognizes a man who served so well and ardently at all levels in his professional society."

1976   **Clarence H. Linder**   "In recognition of Clarence H. Linder's dedicated service to the engineering profession, to the American Institute of Electrical and Electronics Engineers as Director and President of both AIEE and IEEE, as President of the United Engineering Trustees, for his service to the profession as President of the National Academy of Engineering, for his strong support in the establishment of the United Engineering Center and particularly for his leadership in planning and perfecting the merger of AIEE and IRE, the Board of Directors confers this Award in deep appreciation of his many contributions to the Institute."

1977   **Schedule Revised***

1978   **Ivan S. Coggeshall**   "In recognition of Ivan S. Coggeshall's dedicated contribution over a span of fifty years to the engineering profession through his service to the Institute of Radio Engineers and the Institute of Electrical and Electronics Engineers and his leadership in integration of wire and radio media through his wise counsel and action as Officer and Staff member of technical and professional organizations, the Board of Directors confers this Award."

1979   **John D. Ryder**   "In recognition of John D. Ryder's dedicated contribution over a span of fifty years to electrical engineering education as teacher, author of textbooks and engineering dean and through his service and wise counsel to the Institute of Radio Engieers and the Institute of Electrical and Electronics Engineers as Editor of the PROCEEDINGS of the IRE and the PROCEEDINGS of the IEEE, Director and President of the IRE, Chairman of the History Committee and Chairman of the IEEE Centennial Committee, the Board of Directors confers this Award."

1980   **Raymond W. Sears**   "For long-standing, resourceful, and unselfish contributions to the welfare of the Institute, particularly as Institute Secretary, Treasurer, and IEEE representative in various intersociety affairs."

1981   **Robert H. Tanner**   "For contributions toward professionalism and dedicated service to the Canadian Region, to the IEEE, and to the profession over many years."

1983   **Thomas H. Lee**   "For meritorious service to the Institute, for the development of the IEEE Energy Committee, and for promoting public understanding of energy issues."

1984   **William W. Middleton**   "For metitorious service to the Institute and dedicated service to the Communications Society and to Region 2."

*Date of selection changed to date of presentation.

Section I-E
Prize Paper Awards

W. R. G. Baker Award

W. R. G. Baker, Vice-President of General Electric, and later Vice-President for Research at Syracuse University, New York, established this award under the auspices of the Institute of Radio Engineers in 1956. It consists of a certificate and one thousand dollars and is awarded for the "most outstanding paper reporting original work in the *Tranactions, Journals,* and *Magazines* of IEEE Societies or in the *Proceedings of the IEEE*" issued between January 1 and December 31 of the preceding year.

W. R. G. Baker Award Recipients

1957   **D. R. Fewer**   "Design Principles of Junction Transistor Audio Power Amplifiers," *IRE Transactions on Audio,* November–December 1955.

1957   **R. J. Kircher**   "Properties of Junction Transistors," *IRE Transactions on Audio,* July–August 1955.

1957   **R. L. Trent**   "Design Principles of Junction Transistor Audio Amplifiers," *IRE Transactions on Audio,* September–October 1955.

1958   **R. L. Kyhl and H. F. Webster**   "Breakup of Hollow Cylindrical Electron Beams," *IRE Transactions on Electron Devices,* October 1956.

1959   **R. D. Thornton**   "Active RC Networks," *IRE Transactions on Circuit Theory,* September 1957.

1960   **E. J. Nalos**   "A Hybrid Type Traveling-Wave Tube for High-Power Pulsed Amplification," *IRE Transactions on Electron Devices,* July 1958.

1961   **Manfred Clynes**   "Respiratory Control of Heart Rate: Laws Derived from Analog Computer Simulation,"*IRE Transactions on Medical Electronics,* January 1960.

1962   **Marvin Chodorow and Tore Wessel-Berg**   "A High Efficiency Klystron with Distributed Interaction," *IRE Transactions on Electron Devices,* January 1961.

1963   **Leonard Lewin**   "On the Resolution of a Class of Waveguide Discontinuity Problems by the Use of Singular Integral Equations," *IRE Transactions on Microwave Theory and Techniques,* July 1961.

1964   **Donald L. White**   "The Depletion Layer Transducer," *IRE Transactions on Ultrasonics Engineering,* July 1962.

1965   **D. C. Youla**   "A New Theory of Broad-Band Matching," *IEEE Transactions on Circuit Theory*, March 1964.

1966   **Robert G. Gallager**   "A Simple Derivation of the Coding Theorem and Some Applications," *IEEE Transactions on Information Theory*, January 1965.

1967   **Dean E. McCumber and Allan G. Chynoweth**   "Theory of Negative-Conductance Amplification and of Gunn Instabilities in 'Two Valley' Semiconductors," *IEEE Transactions on Electron Devices*, January 1966.

1968   **J. Andersen and H. B. Lee**   "Network Synthesis Using Lossy Reactances," *IEEE Transactions on Circuit Theory*, September 1966.

1969   **Tosiro Koga**   "Synthesis of Finite Passive *n*-Ports with Prescribed Positive Real Matrices of Several Variables," *IEEE Transactions on Circuit Theory*, March 1968.

1970   **George J. Friedman and Cornelius T. Leondes**   "Constraint Theory—Parts I, II, and III," *IEEE Transactions on Systems Science and Cybernetics*, January, April, and July 1969.

1971   **Andrew H. Bobeck, Robert F. Fischer, Anthony J. Perneski, J. P. Remeika, and L. G. Van Uitert**   "Application of Orthoferrites to Domain-Wall Devices," *IEEE Transactions on Magnetics*, September 1969.

1972   **Dirk J. Kuizenga and A. E. Siegman**   "FM and AM Mode Locking of the Homogeneous Laser—Parts I and II," *IEEE Journal of Quantum Electronics*, November 1970.

1973   **Leon O. Chua**   "Memristor—The Missing Circuit Element," *IEEE Transactions on Circuit Theory*, September 1971.

1974   **David B. Large, Lawrence Ball, and Arnold J. Farstad**   "Radio Transmission to and from Underground Coal Mines—Theory and Measurement," *IEEE Transactions on Communications*, March 1973.

1975   **Tingye Li, Stewart E. Miller, and Enrique A. J. Marcatili**   "Research Toward Optical-Fiber Transmission Systems," *Proceedings of the IEEE*, May 1975.

1976   **Robert W. Keyes**   "Physical Limits in Digital Electronics," *Proceedings of the IEEE*, May 1975.

1977   **Manfred R. Schroeder**   "Models of Hearing," *Proceedings of the IEEE*, September 1975.

1978   **E. C. Sakshaug, J. S. Kresge, and Stanley A. Miske, Jr.**   "A New Concept in Station Arrester Design," *IEEE Transactions on Power Apparatus and Systems*, March–April 1977.

1979   **Stephen W. Director and Gary D. Hachtel**   "The Simplicial Approximation Approach to Design Centering," *IEEE Transactions on Circuits and Systems*, July 1977.

1980   **Gordon M. Jacobs, David J. Allstot, Robert W. Brodersen, and Paul R. Gray**   ''Design Techniques for MOS Switched Capacitor Ladder Filters,'' *IEEE Transactions on Circuits and Systems*, December 1978.

1981   **Timothy C. May and Murray H. Woods**   ''Alpha-Particle-Induced Soft Errors in Dynamic Memories,'' *IEEE Transactions on Electron Devices*, January 1979.

1982   **Carl O. Bozler and Gary D. Alley**   ''Fabrication and Numerical Simulation of the Permeable Base Transistor,'' *IEEE Transactions on Electron Devices*, June 1980.

1983   **Ryszard Malewski, Chinh T. Nguyen, Kurt Feser, and Nils Hylten-Cavallius**   ''Elimination of the Skin Effect Error in Heavy-Current Shunts,'' *IEEE Transactions on Power Apparatus and Systems*, March 1981.

1984   **Yannis Tsividis**   ''Signal Processors with Transfer Function Coefficients Determined by Timing,'' *IEEE Transactions on Circuits and Systems*, December 1982.

Donald G. Fink Prize Award

The award was established in 1979, in honor of Donald G. Fink, eminent electrical engineer, editor of numerous technical publications, and first General Manager of the IEEE. Supported by the IEEE Life Member Fund Committee, it honors the "outstanding survey, review, or tutorial paper in any of the *Transactions, Journals, Magazines,* or *Proceedings* of the IEEE" issued between January 1 and December 31 of the preceding year. The award consists of a certificate and one thousand dollars.

Donald G. Fink Prize Award Recipients

1981　**Whitfield Diffie and Martin E. Hellman**　"Privacy and Authentication: An Introduction to Cryptography," *Proceedings of the IEEE*, March 1979.

1982　**Arun N. Netravali and John O. Limb**　"Picture Coding: A Review," *Proceedings of the IEEE*, March 1980.

1983　**Anil K. Jain**　"Image Data Compression: A Review," *Proceedings of the IEEE*, March 1981.

1984　**Robert A. Scholtz**　"The Origins of Spread-Spectrum Communications," *IEEE Transactions on Communication*, May 1982.

1984　**Enders A. Robinson**　"A Historical Perspective of Spectrum Estimation," *Proceedings of the IEEE*, September 1982.

Browder J. Thompson Memorial Prize Award

The award was established by the Institute of Radio Engineers in 1945 in memory of Browder J. Thompson, who gave his life in the service of his country while prosecuting technical work with the United States Armed Forces in Italy. The award is made for the "most outstanding paper in any IEEE publication" by an author or authors who have not reached the age of thirty years at the date of submission of the original manuscript. The award consists of a certificate and one thousand dollars.

Browder J. Thompson Memorial Prize Award Recipients

1946　**Gordon M. Lee**　"A Three Beam Oscillograph for Recording at Frequencies up to 10,000 megacycles," *IRE Proceedings-Waves and Electronc,* March 1946.

1947　**C. L. Dolph**　"A Current Distribution for Broadside Arrays which Optimizes the Relationship between Beam Width and Side-Lobe Level," *Proceedings of the IRE,* June 1946.

1948　**W. H. Huggins**　"Broadband Noncontacting Short Circuits for Coaxial Lines," *Proceedings of the IRE,* September, October, and November 1947.

1949　**R. V. Pound**　"Frequency Stabilization of Microwave Oscillators," *Proceedings of the IRE,* December 1947.

1950　**Joseph F. Hull and Arthur W. Randals**　"High-Power Interdigital Magnetrons," *Proceedings of the IRE,* 1948.

1951　**A. B. Macnee**　"An Electronic Differential Analyzer," *Proceedings of the IRE,* November 1949.

1952　**H. W. Welch, Jr.**　"Effects of Space Charge on Frequency Characteristics of Magnetrons," *Proceedings of the IRE,* December 1950.

1953　**Richard C. Booton, Jr.**　"An Optimization Theory for Time-Varying Linear Systems with Nonstationary Statistical Inputs," *Proceedings of the IRE,* August 1952.

1954　**R. L. Petritz**　"On the Theory of Noise in *P-N* Junctions and Related Devices," *Proceedings of the IRE,* November 1952.

1955　**B. D. Smith, Jr.**　"Coding by Feedback Methods," *Proceedings of the IRE,* August 1953.

1956　**Jack E. Bridges**　"Detection of Television Signals in Thermal Noise," *Proceedings of the IRE,* September 1954.

1957　**D. A. Buck**　"The Cryotron—A Superconductive Computer Component," *Proceedings of the IRE,* April 1956.

1958   **Arthur Karp** "Backward-Wave Oscillator Experiments at 100 to 200 Kilomegacycles," *Proceedings of the IRE*, April 1957.

1959   **F. H. Blecher** "Design Principles for Single Loop Transistor Feedback Amplifiers," *IRE Transactions on Circuit Theory*, September 1957.

1960   **J. W. Gewartowski** "Velocity and Current Distributions in the Spent Beam of the Backward-Wave Oscillator," *IRE Transactions on Electron Devices*, October 1958.

1961   **Eiichi Goto** "The Parametron, a Digital Computing Element which Utilizes Parametric Oscillation," *Proceedings of the IRE*, August 1959.

1962   **Henri B. Smets** "Analysis & Synthesis of Nonlinear Systems," *IRE Transactions on Circuit Theory*, December 1960.

1963   **Chih-Tang Sah** "Effect of Surface Recombination and Channel on *P-N* Junction and Transistor Characteristics," *IRE Transactions on Electron Devices*, January 1962.

1964   **H. B. Lee** "A New Canonic Realization Procedure," *IEEE Transactions on Circuit Theory*, March 1963.

1965   **S. R. Hofstein and F. P. Heiman** "The Silicon Insulated-Gate Field-Effect Transistor," *Proceedings of the IRE*, September 1963.

1966   **Kenneth M. Johnson** "High-Speed Photodiode Signal Enhancement at Avalanche Breakdown Voltage," *IEEE Transactions on ELectron Devices*, February 1965.

1967   **L. O. Chua and R. A. Rohrer** "On the Dynamic Equations of a Class of Nonlinear RLC Networks," *IEEE Transactions on Circuit Theory*, December 1965.

1968   **Michael L. Dertouzos** "Phaseplot: An On-Line Graphical Display Technique," *IEEE Transactions on Electronic Computers*, April 1967.

1969   **Malvin C. Teich** "Infrared Heterodyne Detection," *Proceedings of the IRE*, January 1968.

1970   **John David Rhodes** "The Stepped Digital Elliptic Filter," *IEEE Transactions on Microwave Theory and Techniques*, April 1969.

1971   **L. J. Griffiths** "A Simple Adaptive Algorithm for Real Time Processing of Antenna Arrays," *Proceedings of the IEEE*, October 1969.

1972   **G. David Forney, Jr.** "Convolutional Codes I: Algebraic Structure," *IEEE Transactions on Information Theory*, November 1970.

1973   **Jerry Mar** "A Two-Terminal Transistor Memory Cell Using Breakdown," *IEEE Journal of Solid-State Circuits*, October 1971.

A CENTURY OF HONORS

1974   **Jörn Justesen** "A Class of Constructive Asymptotically Good Algebraic Codes," *IEEE Transactions on Information Theory*, September 1972.

1975   **Nuggehally S. Jayant** "Digital Coding of Speech Waveforms: PCM, DPCM, and DM Quantizers," *Proceedings of the IEEE*, May 1974.

1976   **Russell M. Mersereau and Dan E. Dudgeon** "Two-Dimensional Digital Filtering," *Proceedings of the IEEE*, April 1975.

1977   **Michael R. Portnoff** "Implementation of the Digital Phase Vocoder Using the Fast Fourier Transform," *IEEE Transactions on Acoustics, Speech, and Signal Processings*, June 1976.

1978   **David A. Hounshell** "Bell and Gray: Contrasts in Style, Politics, and Etiquette," *Proceedings of the IEEE*, September 1976.

1979   **Marvin B. Lieberman** "A Literature Citation Study of Science-Technology Coupling in Electronics," *Proceedings of the IEEE*, January 1978.

1980   **Alan S. Willsky** "Relationship Between Digital Signal Processing and Control and Estimation Theory," *Proceedings of the IEEE*, September 1978.

1981   **Lawrence H. Goldstein** "Controllability/Observability Analysis of Digital Circuits," *IEEE Transactions on Circuits and Systems*, September 1979.

1982   **Stig Skelboe** "Computation of the Periodic Steady-State Response of Nonlinear Networks by Extrapolation Methods," *IEEE Transactions on Circuits and Systems*, March 1980.

1983   **Daniel S. Kimes and Julie A. Kirchner** "Modeling the Effects of Various Radiant Transfers in Mountainous Terrain on Sensor Response," *IEEE Transactions on Geoscience and Remote Sensing*, April 1981.

1984   **John C. Curlander** "Location of Spaceborn SAR Imagery," *IEEE Transactions on Geoscience and Remote Sensing*, July 1982.

Armstrong, Edwin Howard 1953 AIEE

Arnold, Bion Joseph 1937 AIEE

Bardeen, John 1981 IEEE
"For significant contributions to solid state physics leading to the invention of the TRANSISTOR and to the theoretical understanding of superconductivity."

Blondel, Andre E. 1912 AIEE

Brattain, Walter H. 1981 IEEE
"For significant contributions to fundamental semiconductor physics leading to the invention of the TRANSISTOR."

Brown, Charles Eugene Lancelot 1912 AIEE

Brush, Charles F. 1929 AIEE

Budde, Emil A. 1912 AIEE

Bush, Vannevar 1949 AIEE

Carty, John J. 1928 AIEE

Chesney, Cummings C. 1938 AIEE

Danzig, George B. 1983 IEEE
"For the development of linear programming and its numerous extensions, which have led to the widespread use of optimization techniques in many branches of electrical engineering design."

Dow, Alex 1937 AIEE

DuMont, Allen B. 1961 AIEE

Dunn, Gano 1945 AIEE

Edison, Thomas A. 1928 AIEE

Emanueli, Luigi 1958 AIEE

Emmet, William LeRoy 1933 AIEE

Fairman, James F. 1955 AIEE

Farmer, Prof. Moses G. 1890 AIEE

de Ferranti, Sebastian Ziani 1912 AIEE

Field, Cyrus W. 1892 AIEE

Foch, Ferdinand 1921 AIEE

Green, Cecil H. 1983 IEEE
"For creative philanthropy fostering collaboration of educational institutions and industry in the advancement of electrical science."

Green, Dr. Norvin 1889 AIEE

Hamilton, George A. 1933 AIEE

Heaviside, Oliver 1918 AIEE

Helmholtz, Herman Ludwig Ferdinand von 1894 AIEE

Hoover, Herbert 1929 AIEE

Jackson, Dugald C. 1944 AIEE

Jewett, Frank Baldwin 1945 AIEE

Josephson, Brian D. 1982 IEEE
"For his prediction (discovery) of pair tunneling between superconductors which constitutes the basis of a Josephson junction technology for high speed and low power computing elements and memories."

Kelvin, Lord 1892 AIEE

Kennelly, Arthur E. 1933 AIEE

Knuth, Donald E. 1982 IEEE
"For his contributions to the field of computer science, through research and education."

Land, Edwin H. 1981 IEEE
"For his pioneering work in science and technology of photography and light."

Marconi, Guglielmo 1917 AIEE

McNaughton, Andrew George Latta 1942 IEEE

Menon, M. G. K. 1984 IEEE
"For contributions to the planning and growth of science and technology, particularly electronics, and its application to development; promotion of international scientific cooperation; experimental research in particle physics and cosmic rays."

Millikan, Robert Andrews 1933 AIEE

Pacinotti, Antonio 1912 AIEE

Panofsky, Wolfgang K. H. 1982 IEEE
"For extensive pioneering contributions to the development of linear particle accelerators, advancement of high power microwave technology, and high energy particle physics."

Preece, Sir William H. 1884 AIEE

Pupin, Michael I. 1928 AIEE

Quarles, Donald Aubrey 1956 AIEE

Ramsey, Norman 1983 IEEE
"For the invention of the hydrogen maser, the development of high precision atomic clocks and contributions to the standardization of the second in terms of the cesium resonance."

Rice, Edwin Wilbur, Jr. 1933 AIEE

Robertson, Elgin Barnett 1959 AIEE

Scott, Charles F. 1929 AIEE

Shibusawa, Motoji 1929 AIEE

Siemens, Dr. Werner von 1892 AIEE

Simon, Herbert A. 1984 IEEE
"For his contributions in the field of complex information processing."

Skinner, Charles Edward 1945 AIEE

Sorensen, Royal Wasson 1954 AIEE

Sprague, Frank J. 1932 AIEE

Sporn, Philip 1961 AIEE

Swasey, Ambrose 1928 AIEE

Thompson, Silvanus P. 1914 AIEE

Thomson, Elihu 1928 AIEE

Weston, Edward 1933 AIEE

Yamashita, Hideo 1984 IEEE
"For his contributions to research in computers, international cooperation in the science and technology of information processing, and for his leadership in electrical engineering education."

Past Presidents of AIEE, IRE, and IEEE

The American Institute of Electrical Engineers
(AIEE)

Norvin Green	1884–85–86	Cummings C. Chesney	1926–27
Franklin L. Pope	1886–87	Bancroft Gherardi	1927–28
T. Commerford Martin	1887–88	Rudolph F. Schuchardt	1928–29
Edward Weston	1888–89	Harold B. Smith	1929–30
Elihu Thomson	1889–90	William S. Lee	1930–31
William A. Anthony	1890–91	Charles E. Skinner	1931–32
Alexander Graham Bell	1891–92	Harry P. Charlesworth	1932–33
Frank Julian Sprague	1892–93	John B. Whitehead	1933–34
Edwin J. Houston	1893–94–95	J. Allen Johnson	1934–35
Louis Duncan	1895–96–97	Edward B. Meyer	1935–36
Francis B. Crocker	1897–98	Alexander M. MacCutcheon	1936–37
Arthur E. Kennelly	1898–1900	William H. Harrison	1937–38
Carl Hering	1900–01	John Castlereagh Parker	1938–39
Charles P. Steinmetz	1901–02	F. Malcolm Farmer	1939–40
Charles F. Scott	1902–03	Royal W. Sorensen	1940–41
Bion J. Arnold	1903–04	David C. Prince	1941–42
John W. Lieb	1904–05	Harold S. Osborne	1942–43
Schuyler Skaats Wheeler	1905–06	Nevin E. Funk	1943–44
Samuel Sheldon	1906–07	Charles A. Powel	1944–45
Henry G. Stott	1907–08	William E. Wickenden	1945–46
Louis A. Ferguson	1908–09	J. Elmer Housley	1946–47
Lewis B. Stillwell	1909–10	Blake D. Hull	1947–48
Dugald C. Jackson	1910–11	Everett S. Lee	1948–49
Gano Dunn	1911–12	James F. Fairman	1949–50
Ralph D. Mershon	1912–13	Titus G. LeClair	1950–51
C. O. Mailloux	1913–14	Fred O. McMillan	1951–52
Paul M. Lincoln	1914–15	Donald A. Quarles	1952–53
John J. Carty	1915–16	Elgin B. Robertson	1953–54
Harold W. Buck	1916–17	Alexander C. Monteith	1954–55
Edwin W. Rice, Jr.	1917–18	Morris D. Hooven	1955–56
Comfort A. Adams	1918–19	Mervin S. Coover	1956–57
Calvert Townley	1919–20	Walter J. Barrett	1957–58
Arthur W. Berresford	1920–21	L. F. Hickernell	1958–59
William McClellan	1921–22	James H. Foote	1959–60
Frank B. Jewett	1922–23	Warren J. Chase	1961–62
Harris J. Ryan	1923–24	B. Richard Teare, Jr.	1962–63
Farley Osgood	1924–25	Clarence H. Linder	1960–61
Michael I. Pupin	1925–26		

The Institute of Radio Engineers
(IRE)

Robert H. Marriott	1912	Haraden Pratt	1938
Greenleaf W. Pickard	1913	Raymond A. Heising	1939
Loius W. Austin	1914	Laurence C. F. Horle	1940
John Stone-Stone	1915	Frederick E. Terman	1941
Arthur E. Kennelly	1916	Arthur F. Van Dyck	1942
Michael I. Pupin	1917	Lynde P. Wheeler	1943
George W. Pierce	1918–19	Hubert M. Turner	1944
John V. L. Hogan	1920	William L. Everitt	1945
E. F. W. Alexanderson	1921	Frederick B. Llewellyn	1946
Fulton Cutting	1922	Walter R. G. Baker	1947
Irving Langmuir	1923	Benjamin E. Shackelford	1948
John Harold Morecroft	1924	Stuart L. Bailey	1949
John H. Dellinger	1925	Raymond F. Guy	1950
Donald M. Nicol	1926	Ivan S. Coggeshall	1951
Ralph Bown	1927	Donald B. Sinclair	1952
Alfred N. Goldsmith	1928	James W. McRae	1953
Albert Hoyt Taylor	1929	William R. Hewlett	1954
Lee de Forest	1930	John D. Ryder	1955
Ray H. Manson	1931	Arthur V. Loughren	1956
Walter G. Cady	1932	John T. Henderson	1957
Lewis M. Hull	1933	Donald G. Fink	1958
C. M. Jansky, Jr.	1934	Ernst Weber	1959
Chas. S. Ballantine	1935	Ronald L. McFarlan	1960
Alan Hazeltine	1936	Lloyd V. Berkner	1961
Harold H. Beverage	1937	Partick E. Haggerty	1962

The Institute of Electrical and Electronics Engineers, Inc.
(IEEE)

Ernst Weber	1963	John J. Guarrera	1974
Clarence H. Linder	1964	Arthur P. Stern	1975
Bernard M. Oliver	1965	Joseph K. Dillard	1976
William G. Shepherd	1966	Robert M. Saunders	1977
Walter K. MacAdam	1967	Ivan A. Getting	1978
Seymour W. Herwald	1968	Jerome J. Suran	1979
F. Karl Willenbrock	1969	Leo Young	1980
John V. N. Granger	1970	Richard W. Damon	1981
James H. Mulligan, Jr.	1971	Robert E. Larson	1982
Robert H. Tanner	1972	James B. Owens	1983
Harold Chestnut	1973		

Part II
Past Fellows

A

ABBOTT, ARTHUR LAURIE, AIEE Fellow 1944.

ABBOTT, W.L., AIEE Fellow 1913.

ABRAHAM, LEONARD GLADSTONE, IRE Fellow 1957 "For contributions to the engineering of broad-band coaxial transmission systems." AIEE Fellow 1959 "For contributions to the engineering and development of long distance telephone and transmission systems."

ACHESON, MARCUS A., IRE Fellow 1941.

ACKER, ALBERT, AIEE Fellow 1921.

ADAMS, COMFORT A., AIEE Fellow 1913.

ADAMS, FRANCIS JOSEPH, AIEE Fellow 1951.

ADAMS, LEE FRANCIS, AIEE Fellow 1942.

ADAMS, P.R., IRE Fellow 1961 "For contribution in the development of radio aids to navigation."

ADAMS, PATRICK HENRY, AIEE Fellow 1930.

ADAMS, ROBERT T., IRE Fellow 1962 "For contributions to over-the-horizon communications."

ADAMS, ROBERT WINTHROP, AIEE Fellow 1944.

ADAMSON, COLIN, IEEE Fellow 1974 "For contributions to engineering education and to protection of high-voltage direct-current transmission systems."

ADDEY, FREDERICK, IRE Fellow 1920.

ADDICKS, LAURENCE, AIEE Fellow 1919.

ADLER, HANS ANTON, AIEE Fellow 1962 "For contributions to insulation research and the application of probability mathematics to power system problems."

ADOLPH, JOHANNES, AIEE Fellow 1929.

ADORIAN, PAUL, IRE Fellow 1961 "For development of electronic distribution networks used in broadcasting and television."

ADSIT, CHARLES G., AIEE Fellow 1913.

AFFEL, HERMAN ANDREW, AIEE Fellow 1941. IRE Fellow 1949 "For his contributions to the communications art, and his guidance of important developments in carrier systems for multiplex telephone and television transmission."

AHMED, WAQUER, IRE Fellow 1962 "For contributions to engineering education in Pakistan."

AIKEN, HOWARD H., IRE Fellow 1960 "For contributions to development of computer science and technology."

AISBERG, EUGENE, IEEE Fellow 1968 "For his outstanding work, for more than forty years, as editor of technical magazines and author of educational books successfully known on a world-wide basis."

ALBERSHEIM, W.J., IRE Fellow 1957 "For contributions in the fields of sound reproduction and military electronics."

ALBRIGHT, ARTHUR S., AIEE Fellow 1943.

ALBRIGHT, H. FLEETWOOD, AIEE Fellow 1912.

ALDRIDGE, THOMAS H.U., AIEE Fellow 1918.

ALEXANDER, BEN, IEEE Fellow 1966 "For inventive and analytical contributions in the fields of electronic and inertial navigation."

ALEXANDER, EVERETT H., AIEE Fellow 1949.

ALEXANDER, HARRY, AIEE Fellow 1920.

ALEXANDER, S.N., IRE Fellow 1956 "For contributions to the development and application of digital computers."

ALEXANDERSON, E.F.W., IRE Fellow 1915. AIEE Fellow 1920.

ALGER, PHILIP L., AIEE Fellow 1930.

ALLEN, ALBERT J., AIEE Fellow 1950.

ALLEN, ASA EDWARD, AIEE Fellow 1942.

ALLEN, C.E., AIEE Fellow 1914.

ALLEN, EDWARD W., JR., IRE Fellow 1953 "For his technical and administrative contributions in the fields of radio wave propagation and radio noise."

ALLEN, EDWIN W., AIEE Fellow 1922.

ALLEN, ELBERT G., AIEE Fellow 1922.

ALLEN, JOHN E., AIEE Fellow 1947.

ALLEN, OLIVER F., AIEE Fellow 1924.

ALLEN, THOMAS HAMPTON, AIEE Fellow 1944.

ALLIBONE, THOMAS EDWARD, AIEE Fellow 1949.

ALLISON, FRED, AIEE Fellow 1917.

ALMQUIST, CARL T., AIEE Fellow 1951.

ALMQUIST, MILTON LEROY, AIEE Fellow 1951. IRE Fellow 1958 "For his contributions to the planning of telephone and radio systems."

AMES, NORMAN B., AIEE Fellow 1948.

AMSTUZ, J. OSCAR, AIEE Fellow 1948.

ANDEREGG, GUSTAVUS A., AIEE Fellow 1930.

ANDERSON, A. EUGENE, IRE Fellow 1959 "For contributions to transistor switching circuits."

ANDERSON, ALBERT S., AIEE Fellow 1951.

ANDERSON, BURTON E., AIEE Fellow 1945.

ANDERSON, STEWART WISE, AIEE Fellow 1931.

ANDERTON, JOHN H., AIEE Fellow 1918.

ANDREW, V.J., IRE Fellow 1955 "For his contributions to radio antennas and transmission lines."

ANDREWS, FRANCIS E., AIEE Fellow 1950.

ANDREWS, W.S., AIEE Fellow 1912.

ANDRUS, LUCIUS B., AIEE Fellow 1919.

ANGOT, ANDRE M., IEEE Fellow 1966 "For achievement leadership in military telecommunications and engineering education."

ANGUS, DONALD JAMES, AIEE Fellow 1945.

ANKENBRANDT, F.L., IRE Fellow 1957 "For leadership in areas of military air navigation and communications."

ANSON, EDWARD H., AIEE Fellow 1957 "For contributions to railway electrification, power supply systems and engineering management."

APPLEGARTH, ALEXANDER R., JR., IEEE Fellow 1967 "For contributions to the development of airborne electronic equipment."

APPLETON, EDWARD V., IRE Fellow 1931.

ARENBERG, DAVID L., IEEE Fellow 1968 "For basic inventions and other contributions to the design and the manufacture of ultrasonic delay lines."

ARENDT, MORTON, AIEE Fellow 1913.

ARGERSINGER, ROY EDWIN, AIEE Fellow 1936.

ARMAN, ADRIAN NEVILLE, AIEE Fellow 1949.

ARMBRUST, GEORGE MAURICE, AIEE Fellow 1933.

ARMOR, JAMES COE, AIEE Fellow 1942.

ARMSTRONG, EDWIN H., IRE Fellow 1927.

ARMSTRONG, HARRISON R., AIEE Fellow 1961 "For contributions to insulation coordination and transient over-voltage protection."

ARMSTRONG, J.R.C., AIEE Fellow 1913.

ARMSTRONG, RALPH W., AIEE Fellow 1913.

ARNOLD, BION JOSEPH, AIEE Fellow 1912.

ARNOLD, HAROLD DEFOREST, AIEE Fellow 1929.

ARNOLD, J. LORING, AIEE Fellow 1922.

ASHBRIDGE, NOEL, IRE Fellow 1938.

ATKINSON, RALPH W., AIEE Fellow 1928.

ATTWOOD, S.S., IRE Fellow 1961 "For contributions to the understanding of electromagnetic field theory."

ATTWOOD, STEPHEN S., AIEE Fellow 1943.

AUBERT, ROGER A., IEEE Fellow 1969 "For contributions to the development of radiocommunication systems,

and for management of research laboratories and production facilities."

AUSTIN, ARTHUR O., AIEE Fellow 1925. IRE Fellow 1930.

AUSTIN, BASCUM OTTO, AIEE Fellow 1951.

AUSTIN, LOUIS W., IRE Fellow 1915.

AYDELOTT, JOHN CHARLES, AIEE Fellow 1948.

B

BABCOCK, ALLEN H., AIEE Fellow 1912.

BABITS, VICTOR A., IRE Fellow 1959 "For contributions to engineering education and pioneering in television."

BACK, G.I., IRE Fellow 1955 "For his leadership in the field of military communications and communication systems."

BAGNALL, VERNON BARNARD, AIEE Fellow 1948.

BAGNO, SAMUEL, IRE Fellow 1961 "For creative contributions in the fields of instrumentation, medical electonics, and electronic aids to law enforcement."

BAILEY, ARNOLD B., IEEE Fellow 1965 "For contributions in the field of radio navigation and communications."

BAILEY, AUSTIN, IRE Fellow 1936. AIEE Fellow 1950.

BAILEY, BENJAMIN F., AIEE Fellow 1921.

BAILEY, EDGAR L., AIEE Fellow 1943.

BAILEY, G.W., IRE Fellow 1956 "For sustaining service to amateur radio, and administrative leadership."

BAILEY, WILLIAM M., IRE Fellow 1960 "For contributions to capacitor technology."

BAKER, BENJAMIN PIFER, AIEE Fellow 1948.

BAKER, ERNEST WELLINGTON, AIEE Fellow 1951.

BAKER, MARVIN A., AIEE Fellow 1960 "For contributions to large rotating amplifying generators, DC Motors and generators."

BAKER, ROBERT A., IEEE Fellow 1968 "For outstanding contributions in the design, construction, and management of a major electric-power system."

BAKER, ROBERT M., IEEE Fellow 1966 "For contributions to the theory and practice of high frequency heating, and for his successful calorimetry of laser power."

BAKER, WALTER R.G., IRE Fellow 1928. AIEE Fellow 1947.

BALCH, ELWYN C., AIEE Fellow 1943.

BALCH, JOHN A., IRE Fellow 1940.

BALDWIN, M.W., IRE Fellow 1948 "For his fundamental investigations of the quality of television pictures."

BALDWIN, ROBERT LEE, AIEE Fellow 1937.

BALL, JOHN DUDLEY, AIEE Fellow 1920.

BALL, REGINALD DONAHOE, AIEE Fellow 1947.

BALL, THOMAS FAUNTLEROY, AIEE Fellow 1938.

BALLARD, B. GUY, AIEE Fellow 1955 "For his contribution in applying electrical engineering technology to problems of national defense." IRE Fellow 1955 "For his direction of radar and electronic research in Canada."

BALLARD, WILLIAM CYRUS, JR., AIEE Fellow 1951.

BALLENTINE, STUART, IRE Fellow 1928.

BALLMAN, EDWIN CHARLES, AIEE Fellow 1959 "For contributions to design and fabrication of small electric motors."

BALSBAUGH, JAYSON CLAIR, AIEE Fellow 1951.

BALTZLY, CLIFFORD C., AIEE Fellow 1943.

BANCROFT, CHARLES F., AIEE Fellow 1912.

BANKUS, JOHN, AIEE Fellow 1949.

BARCLAY, ROBERT HAMILTON, AIEE Fellow 1928.

BARCO, A.A., IRE Fellow 1957 "For contributions to radio and television receiver circuits."

BARCUS, E. DALE, AIEE Fellow 1961 "For contributions in the field of noise, crosstalk, and electrical protection and in the design of special communication transmission facilities."

BARFIELD, JOE W., AIEE Fellow 1956 "For contributions to the development of an electrical utility system."

BARING, J.W., AIEE Fellow 1953 "In recognition of contributions to fundamental date in the field of power equipment design and operation."

BARKER, HARRY, AIEE Fellow 1955 "For over a half a century of active service to the Electrical Engineering Profession as editor, educator and consultant."

BARKER, JOSEPH WARREN, AIEE Fellow 1930.

BARKHAUSEN, HEINRICH, IRE Fellow 1930.

BARKLEY, W.J., IRE Fellow 1956 "For pioneering and management in the field of electronic communication."

BARLOW, STUART L.M., IEEE Fellow 1972 "For contributions to the development of international goodwill and collaboration in the electrical engineering industry."

BARNES, CHARLES C., IEEE Fellow 1967 "For contributions to power transmission cable systems."

BARNES, GEORGE CARROLL, JR., AIEE Fellow 1962 "For contribtuions to electrical engineering education."

BARNES, HOWELL H., AIEE Fellow 1913.

BARNES, JAMES P., AIEE Fellow 1927.

BARNES, JOHN L., IRE Fellow 1952 "For contributions to mathematical theory and exposition in the field of transients in linear electrical and mechanical systems."

BARNHOLDT, HAROLD L., AIEE Fellow 1919.

BARNUM, THOMAS EDISON, AIEE Fellow 1912.

BARRETO, MARIO DE BARROS, IRE Fellow 1924.

BARRETT, WALTER J., AIEE Fellow 1950.

BARRON, JACOB T., AIEE Fellow 1927.

BARROW, WILMER L., IRE Fellow 1942.

BARSTOW, WILLIAM S., AIEE Fellow 1912.

BARTHOLOMEW, FRANCIS JOHN, AIEE Fellow 1952.

BARTLETT, LAWRENCE, AIEE Fellow 1938.

BARTON, FREDERICK S., IRE Fellow 1935.

BARTON, PHILLIP PRIZE, AIEE Fellow 1912.

BARY, CONSTANTINE W., AIEE Fellow 1956 "For his work in engineering research and development of methods for dealing with present and future economics of electric utility service."

BASCOM, HENRY MELVIN, AIEE Fellow 1930.

BATCHER, RALPH R., IRE Fellow 1950 "For his pioneering work with cathode ray tube instruments and for his development of variable frequency standards and instruments."

BATEMAN, IVAN LESTER, AIEE Fellow 1959 "For contributions to the development and operation of a major metropolitan utility."

BATES, LOUIS W., AIEE Fellow 1947.

BATSEL, MAX C., IRE Fellow 1927.

BATTEY, WALTER RAYMOND, AIEE Fellow 1943.

BAUER, BENJAMIN B., IRE Fellow 1953 "For his important contributions to the development of microphone and other audio devices."

BAUGH, CLIVE EVERETT, AIEE Fellow 1948.

BAUGHN, EARL, AIEE Fellow 1950.

BAUHAN, OSCAR, AIEE Fellow 1951.

BAUMANN, ERNST, IEEE Fellow 1971 "For contributions to semiconductor technology and manufacturing processes."

BAUMANN, FREDERICK W., IEEE Fellow 1976 "For improvements to alternating current motors and generators."

BEACH, ROBIN, AIEE Fellow 1935.

BEAMES, CLARE F., AIEE Fellow 1913.

BEAN, GEORGE W., AIEE Fellow 1957 "For engineering achievements and administrative ability in the design and building of a rapidly expanding far flung power system."

BEAR, FREDERICK THOMAS, AIEE Fellow 1958 "For contributions to increased capability of electric power systems."

BEARD, JAMES ROBERT, AIEE Fellow 1941.

BEARDSLEY, CLIFFORD RAY, AIEE Fellow 1930.

BEATTY, ROBERT W., IEEE Fellow 1967 "For contributions to microwave measurements and techniques."

BEAVER, J. LYNFORD, AIEE Fellow 1926.

BEAVERS, MARTIN F., AIEE Fellow 1955 "For his contributions to the understanding and better design of inductive apparatus, particularly in the field of dielectrics and heat transfer."

BECHMANN, RUDOLF, IRE Fellow 1960 "For contributions in the field of piezoelectric crystals."

BECHTEL, ERNEST J., AIEE Fellow 1913.

BECK, E.F.W., AIEE Fellow 1953 "For outstanding contributions in the development of lightning arrestors and in the science of protecting electric apparatus from damage by lightning voltage."

BECK, ROBERT WILLIAM, AIEE Fellow 1960 "For contributions to organization and operation of multiplicity of power systems."

BECKER, JOSEPH ADAM, AIEE Fellow 1943.

BEDELL, FREDERICK, AIEE Fellow 1926.

BEDFORD, L.H., IRE Fellow 1948 "For his development of special circuits, in particular those used for scanning purposes, in television."

BEEBE, MURRAY C., AIEE Fellow 1913.

BEEKMAN, ROYCE ALLEN, AIEE Fellow 1949.

BEEUWKES, REINIER, AIEE Fellow 1920.

BEEZER, JOSEPH E., IRE Fellow 1928.

BEGGS, GEORGE E., JR., IEEE Fellow 1965 "For leadership in the development and manufacture of electronic instruments for measurement, control and display."

BEGGS, J.E., IRE Fellow 1956 "For contributions to the development of new designs of radio tubes."

BEHR, FRANCIS J., IRE Fellow 1920.

BEHREND, BERNARD A., AIEE Fellow 1912.

BELKNAP, J. HARRISON, AIEE Fellow 1949.

BELTZ, W.H., IRE Fellow 1956 "For leadership in improving the reliability of military electronic systems."

BENECKE, Q.O., AIEE Fellow 1913.

BENKESSER, GRANT E., AIEE Fellow 1948.

BENNETT, CHARLES E., AIEE Fellow 1925.

BENNETT, CLAUDIUS EDMUND, AIEE Fellow 1937.

BENNETT, EDWARD, IRE Fellow 1918. AIEE Fellow 1918.

BENNETT, JOHN WILLIAM, AIEE Fellow 1955 "For contributions to the design and construction of electric distribution and generating systems."

BENNETT, RAWSON, II, IRE Fellow 1950 "For his contributions to the development of sonar systems and for effective administration of military electronics laboratories." AIEE Fellow 1959 "For contributions to accelerated and coordinated naval research, particularly in electronics."

BENNETT, REGINALD SAMUEL, AIEE Fellow 1946.

BENNETT, W. R., IRE Fellow 1956 "For contributions in the fields of circuit and transmission theory."

BENNION, HOWARD SHARP, AIEE Fellow 1944.

BENOIT, R.C., JR., IEEE Fellow 1964 "For contributions in the field of military electronics."

BENSON, ERICK ALFRED, AIEE Fellow 1959 "For contributions of engineering innovations to construction and operation of a large hydro-project."

BENSON, FRANCIS SCHAAR, AIEE Fellow 1950.

BERG, ERNST J., AIEE Fellow 1913.

BERGER, F.B., IRE Fellow 1961 "For fundamental contributions to the theory and development of doppler navigation."

BERGVALL, ROYAL C., AIEE Fellow 1950.

BERKNER, LLOYD VIEL, AIEE Fellow 1947. IRE Fellow 1947 "For his investigations of ionospheric phenomena and his contributions to airborne radar development."

BERRESFORD, ARTHUR W., AIEE Fellow 1914.

BERRY, EDWARD ROBIE, AIEE Fellow 1929.

BERRY, JOEL HALBERT, AIEE Fellow 1946.

BETTANNIER, EUGENE L., AIEE Fellow 1946.

BETTIS, ALEXANDER E., AIEE Fellow 1926.

BETTS, PHILANDER, AIEE Fellow 1913.

BETZ, PAUL L., AIEE Fellow 1962 "For achievements in research in a large electric and gas utility."

BETZER, CECIL EVERETT, AIEE Fellow 1962 "For contributions to the design and operating characteristics of underground power cables."

BIBBONS, JAMES R., AIEE Fellow 1920.

BICKELHAUPT, CARROLL OWEN, AIEE Fellow 1928.

BIEGLER, PHILIP SHERIDAN, AIEE Fellow 1929.

BIGELOW, WALTER B., AIEE Fellow 1947.

BIJUR, JOSEPH, AIEE Fellow 1913.

BILLAU, LEWIS SCOVILLE, AIEE Fellow 1950.

BILLINGS, ASA WHITE KENNY, AIEE Fellow 1913.

BINGLEY, FRANK J., IRE Fellow 1950 "In recognition of his contributions in the field of television broadcast engineering."

BINYON, BASIL, IEEE Fellow 1967 "For his outstanding work as a pioneer in international electronic communications."

BIRCKHEAD, LENNOX, AIEE Fellow 1949.

BIRD, WILLIAM L., AIEE Fellow 1920.

BISHOP, HAROLD, IRE Fellow 1938.

BJORQUIST, CARL HENRY, AIEE Fellow 1958 "For his contributions to standards for and tests of high voltage power system equipment."

BLACK, K.C., IRE Fellow 1949 "For his outstanding wartime work on radio countermeasures and his many contributions to the design of coaxial cable transmission systems."

BLACK, LEONARD J., IRE Fellow 1953 "For research in the field of electro-magnetic radiation, and a distinguished record in teaching radio engineering."

BLACKMON, HENDLEY, AIEE Fellow 1949.

BLACKWELL, OTTO B., AIEE Fellow 1917. IRE Fellow 1928.

BLAISDELL, LEONARD T., AIEE Fellow 1939.

BLAKE, S.H., AIEE Fellow 1917.

BLANKENBUEHER, JOHN H., AIEE Fellow 1948.

BLATHY, OTTO T., AIEE Fellow 1913.

BLISS, LOUIS D., AIEE Fellow 1920.

BLISS, WILLIAM LARD, AIEE Fellow 1917.

BLOMBERG, K.H., IRE Fellow 1950 "In recognition of his many contributions to development and engineering in the field of communications in Sweden."

BLONDEL, ANDRE, IRE Fellow 1917.

BLOOD, JOHN BALCH, AIEE Fellow 1912.

BLOOD, W.H., JR., AIEE Fellow 1913.

BLUME, LOUIS FREDERICK, AIEE Fellow 1939.

BLYE, PAUL W., AIEE Fellow 1951.

BOARD, V.L., AIEE Fellow 1938.

BOCK, CHARLES D., IEEE Fellow 1971 "For contributions to analog computer components and application, and to inertial navigation."

BODE, HENDRICK W., AIEE Fellow 1951, IRE Fellow 1952 "In recognition of his many contributions in the field of circuit and wave-filter theory."

BODICKY, ANDREW, AIEE Fellow 1953 "For long and outstanding work on important submarine power cable installations and authoritative contributions to the literature of cable engineering."

BODLER, OSCAR W., AIEE Fellow 1919.

BOGEN, LOUIS E., AIEE Fellow 1912.

BOHN, DONALD IVAN, AIEE Fellow 1949.

BOHNERT, JOHN I., IRE Fellow 1960 "For contribution to the field of microwave antennas."

BOLLINGER, HOWARD MOORE, AIEE Fellow 1942.

BOLLJAHN, JOHN T., IRE Fellow 1959 "For contributions to the fundamental theory and design of antennas."

BOLSER, M.O., AIEE Fellow 1926.

BOLT, R.H., IEEE Fellow 1964 "For contributions to the field of acoustics through research and teaching."

BOLTON, FRANK CLEVELAND, AIEE Fellow 1942.

BONNEY, ROBERT BRIDGE, AIEE Fellow 1937.

BOORZHINSKY, NICHOLAS P., AIEE Fellow 1950.

BOOTH, JESSE J., AIEE Fellow 1939.

BOOTH, RALPH D., AIEE Fellow 1953 "For contributions as to the successful solution of problems arising in

the development of electric power systems, and in the application of electric power to industry."

BOOTHROYD, W.P., IRE Fellow 1956 "For contributions to the development of microwave relays, multiplex equipment, and television receivers."

BORCH, FREDERICK, AIEE Fellow 1939.

BORDEN, PERRY A., AIEE Fellow 1944.

BORGNIS, FRITZ E., IRE Fellow 1962 "For leadership in engineering education and research and contribution in the fields of electronics and acoustics."

BORN, ARTHUR S., IEEE Fellow 1965 "For contributions in the field of air navigational aids."

BOSCH, LESTER LOUIS, AIEE Fellow 1945.

BOSE, SURENDRA NATH, AIEE Fellow 1948.

BOSOMWORTH, G.P., IRE Fellow 1957 "For contributions to the use of electronics in the rubber industry."

BOSSARD, GISBERT L., AIEE Fellow 1937.

BOSSON, FREDERICK N., AIEE Fellow 1916.

BOSWELL, CLAY CARLTON, AIEE Fellow 1962 "For contributions to the engineering, construction and administration of an electric power system."

BOURLAND, LANGFORD TODD, AIEE Fellow 1947.

BOUTHILLON, LEON, IRE Fellow 1917.

BOUTRY, GEORGES A., XXXX Fellow 1962 "For contributions to electro-optical science, color television, and European television standards."

BOVERI, THEODORE, AIEE Fellow 1947.

BOWDEN, B. VIVIAN IEEE Fellow 1970 "For contributions in the fields of radar and digital computers."

BOWMAN, KENNETH KARL, AIEE Fellow 1951.

BOWN, RALPH, IRE Fellow 1925. AIEE Fellow 1941.

BOYAJIAN, ARAM, AIEE Fellow 1926.

BOYCE, BENJAMIN KNOWLTON, AIEE Fellow 1941.

BOYCE, FRANK GEORGE, AIEE Fellow 1945.

BOYER, ELMER E., AIEE Fellow 1912.

BOYRER, WILLIAM CHARLES, AIEE Fellow 1924.

BOYSE, CYRIL OLIVER, AIEE Fellow 1950.

BOZELL, HAROLD V., AIEE Fellow 1923.

BRACKETT, BYRON BRIGGS, AIEE Fellow 1935.

BRACKETT, HAROLD H., AIEE Fellow 1944.

BRADFIELD, WILLIAM W., IRE Fellow 1915.

BRADSTRUM, ROY E., AIEE Fellow 1961 "For contributions to operation, refinement and expansion of a major telephone utility."

BRADT, A.W., AIEE Fellow 1945.

BRAILLARD, RAYMOND, IRE Fellow 1935.

BRAINARD, FRANK K., AIEE Fellow 1938.

BRAMHALL, FAY BEAUMONT, AIEE Fellow 1944. IRE Fellow 1961 "For contributions to the theory and practice of electrical communications."

BRAND, FREDERICK FERMOR, AIEE Fellow 1939.

BRANDON, EDGAR T.J., AIEE Fellow 1920.

BRANIN, FRANKLIN H., IEEE Fellow 1974 "For contributions to circuit theory and the use of computers in circuit analysis."

BRAUN, FERDINAND, IRE Fellow 1915.

BRAUNIG, VICTOR H., AIEE Fellow 1957 "For his leadership in design and operation of thermo-electric generating stations and in the management of a large electric utility."

BRAZIER, LESLIE GIDDENS, AIEE Fellow 1949.

BRECHT, WINSTON ALLEN, AIEE Fellow 1948.

BREIT, GREGORY, IRE Fellow 1945 "For pioneering in the experimental probing of the ionosphere and giving to the world the first publication of the experimental proof of the existence of the ionospere; and for having initiated at an early date the pulse method of probing by reflection which is the basis of modern radar."

BRENNECKE, CORNELIUS GODFREY, AIEE Fellow 1951. IRE Fellow 1951 "Teacher, engineer and physicist, in recognition of his excellent work in basic research."

BRENTON, WALTER, AIEE Fellow 1940.

BRETCH, EDWARD, AIEE Fellow 1948.

BRICE, NEIL M., IEEE Fellow 1974 "For contributions to the understanding of the interactions between the magnetosphere and the ionosphere, especially in the polar regions."

BRIDGWATER, MALCOLM MACFARLAN, AIEE Fellow 1960 "For contributions to electric power system development and interconnected operation."

BRIGGS, ALBERT F., AIEE Fellow 1954 "For his contributions to the application of electric power and automatic controls in the oil industry, including pump line pumping, gathering systems and loading installations."

BRIGHT, GRAHAM, AIEE Fellow 1922.

BRIGHT, R. LOUIS, IRE Fellow 1962 "For contributions to transistor and computer circuits."

BRILLOUIN, LEON N., IRE Fellow 1952 "For his many teaching accomplishments and his contributions to the literature of physics and radio."

BROOKES, A. SIDNEY, IEEE Fellow 1965 "For contributions in the fields of lightning protection and power cable design."

BROOKS, ALLERTON FRANK, AIEE Fellow 1950.

BROOKS, FORREST EDMUND, AIEE Fellow 1948.

BROOKS, HERBERT BARTON, AIEE Fellow 1931.

BROOKS, MORGAN, AIEE Fellow 1913.

BROSNAN, THOMAS JOSEPH, AIEE Fellow 1950 "For contributions and leadership in the field of electronic components."

BROWN, A.S., IRE Fellow 1956 "For contributions to, and leadership in, military electronics."

BROWN, BURTON P., IEEE Fellow 1967 "For contributions to radar antennas and electronic defensive systems."

BROWN, CHARLES D., AIEE Fellow 1939.

BROWN, ELBERT C., AIEE Fellow 1962 "For contributions to the economic planning and operation of interconnected power systems."

BROWN, FREDERIC STREETER, AIEE Fellow 1945.

BROWN, FREDERICK W., IRE Fellow 1962 "For direction and administration of research in the field of radio wave propagation."

BROWN, HAROLD HASLEY, AIEE Fellow 1951.

BROWN, HARRY FARNSWORTH, AIEE Fellow 1947.

BROWN, J. STANLEY, AIEE Fellow 1950.

BROWN, KENNETH R., AIEE Fellow 1955 "For contributions to the development and technical integration of electric utilities and to protection of transformers on rural systems."

BROWN, MELVIN KEELER, AIEE Fellow 1950.

BROWN, ROY L., AIEE Fellow 1955 "For contributions to the theory, design and technical advancement of transformers in the power field."

BROWN, THERON ADELBERT, AIEE Fellow 1961 "For contributions to the design of distibution systems, substations and power plants."

BROWN, WALTER J., AIEE Fellow 1950.

BROWN, WILLIAM W., IEEE Fellow 1969 "For contributions to the development and design of low-frequency antenna systems and stations."

BROWNE, RALPH COWAN, AIEE Fellow 1928.

BROWNING, G.H., IRE Fellow 1955 "For his early contributions and his inspirational leadership in the electronics field."

BRUECKMANN, HELMUT L., IRE Fellow 1958 "For his research on antennas."

BRUNETTI, CLEDO, AIEE Fellow 1948. IRE Fellow 1949 "In recognition of his pioneering work on printed circuits."

BRUSTMAN, JOSEPH A., IRE Fellow 1959 "For contributions in the development of digital computing systems."

BRUUN, GEORG THEODOR, AIEE Fellow 1942.

BRYANT, JOHN M., AIEE Fellow 1943.

BUCHANAN, OMAR BAILEY, AIEE Fellow 1930.

BUCHER, E.E., IRE Fellow 1929.

BUCHHOLD, THEODORE A., IEEE Fellow 1965 "For basic contributions to guidance systems and superconductivity equipment."

BUCK, A. MORRIS, AIEE Fellow 1923.

BUCK, H.W., AIEE Fellow 1912.

BUCKLEY, OLIVER E., AIEE Fellow 1929.

BUDENBOM, H.T., IRE Fellow 1956 "For contributions to electronic navigation and to precision military radar systems."

BUEHNER, ROBERT OTJEN, AIEE Fellow 1943.

BUELL, ROY C., AIEE Fellow 1951.

BUILDER, GEOFFREY, IRE Fellow 1942.

BULLARD, WILLIAM RALPH, AIEE Fellow 1951.

BULLEN, CHARLES VICTOR, AIEE Fellow 1962 "For contributions to engineering education."

BULLER, FRANCIS H., AIEE Fellow 1953 "For contributions to the knowledge of electrical, thermal and mechanical characteristics of electric power cable."

BUNDY, EDWIN S., AIEE Fellow 1933.

BURCH, EDWARD P., AIEE Fellow 1913.

BURGER, EDWARD J., AIEE Fellow 1950.

BURGER, RUDOLPH E., AIEE Fellow 1930.

BURGESS, CHARLES F., AIEE Fellow 1913.

BURKE, JAMES, AIEE Fellow 1913.

BURKEHOLDER, CHARLES IRVINE, AIEE Fellow 1931.

BURKETT, CHARLES W., AIEE Fellow 1921.

BURLEIGH, CHARLES B., AIEE Fellow 1912.

BURNAP, ROBERT SAMUEL, AIEE Fellow 1951.

BURNHAM, WALTER W., IRE Fellow 1923.

BURNS, HARRY ROBERT, AIEE Fellow 1950.

BURROWS, CHARLES RUSSELL, IRE Fellow 1943 "For his contributions in the field of radio wave propagation, particularly for his investigations of propagation along the ground at ultra-high frequency modulation." AIEE Fellow 1945.

BURROWS, CHARLES W., AIEE Fellow 1918.

BURT, AUSTIN, AIEE Fellow 1912.

BURT, BYRON T., AIEE Fellow 1913.

BURT, ROBERT C., AIEE Fellow 1943.

BURTON, PAUL GIBSON, AIEE Fellow 1916.

BUSCH, ALOYSIUS JOHN, AIEE Fellow 1951.

BUSH, VANNEVAR, AIEE Fellow 1924.

BUSHMAN, ANDREW KIDD, AIEE Fellow 1947.

BUSIGNIES, HENRI G., IRE Fellow 1945 "For his accomplishments in the field of radio direction finders, particularly pioneering work on instruments having automatic indication features." AIEE Fellow 1959 "For contributions in the fields of electronic direction finding, air navigation, radar and radio communications."

BUSSEY, HENSON ESTES, AIEE Fellow 1918.

BUTCHER, CHARLES ALBERT, AIEE Fellow 1943.

BUTLER, JOHN WENDELL, AIEE Fellow 1946.

BUTTNER, H.H., IRE Fellow 1944 "In recognition of his direction of radio communication activities in the international field."

BUTTOLPH, LEROY JAMES, AIEE Fellow 1942.

BUTTON, CHARLES T., AIEE Fellow 1956 "For his contributions to the theory and design of small meters and aircraft inverters."

BYLLESBY, HENRY M., AIEE Fellow 1920.

BYNG, E.S., AIEE Fellow 1929.

BYRNE, JOHN F., IRE Fellow 1950 "For his development of a system of polyphase broadcasting and for effective wartime engineering administration."

BYRNES, I.F., IRE Fellow 1954 "For his contributions to development and design of equipment in the field of marine communication and navigation."

C

CADWALLADER, JAMES ALBERT, AIEE Fellow 1939.

CADY, WALTER G., IRE Fellow 1927.

CAHAN, FRANCOIS MICHEL, AIEE Fellow 1958 "For contributions to stable, secured operations of an extra high voltage continental interconnected system."

CAHOON, R.D., IRE Fellow 1956 "For contributions to international shortwave services and to Canadian broadcasting."

CALABRESE, GIUSEPPE, AIEE Fellow 1951.

CALDERWOOD, EVERETT M., AIEE Fellow 1951.

CALDWELL, BASCOM H., JR., AIEE Fellow 1949.

CALDWELL, EUGENE, AIEE Fellow 1947.

CALDWELL, F.C., AIEE Fellow 1913.

CALDWELL, HARRY L., AIEE Fellow 1920.

CALDWELL, ORESTES H., IRE Fellow 1944 "For his contribution in broadening the horizon of the engineer by his long continued effort to increase the use of electronic principles in industrial operations." AIEE Fellow 1954 "For his services to electronic engineering in the fields of publication and public service."

CALLAHAN, JOHN L., AIEE Fellow 1948. IRE Fellow 1953 "For his contributions to international radio communications and especially to radiophoto and multiplex techniques."

CALLANDER, DELMER W., AIEE Fellow 1955 "For contributions to the development of transformers, particularly large power units."

CALVERT, JOHN F., IRE Fellow 1960 "For contributions to electrical engineering and education."

CALVERT, JOHN FYRER, AIEE Fellow 1945.

CAMBELL, GEORGE A., AIEE Fellow 1913.

CAMILLI, GUGLIELMO, AIEE Fellow 1943.

CAMPAIGNE, HOWARD H., IEEE Fellow 1969 "For pioneering contributions in computer design, and leadership in the advancement of computer technology for scientific and military applications."

CAMPBELL, ALLAN B., AIEE Fellow 1938.

CAMPBELL, ARTHUR BALLARD, AIEE Fellow 1935.

CAMPBELL, RICHARD D., IRE Fellow 1951 "For his work in the internationally important field of radiofrequency allocation."

CAMPBELL, W. LEE, AIEE Fellow 1913.

CANN, JOHN O.G., IRE Fellow 1921.

CANNADY, NATHANIEL ELLIS, AIEE Fellow 1947.

CANNON, WILLIAM D., AIEE Fellow 1954 "For developments in instrumentation in the communications field."

CAPARO, JOSE ANGEL, AIEE Fellow 1932.

CARHART, FRANK MILTON, AIEE Fellow 1932.

CARLE, N.A., AIEE Fellow 1913.

CARLTON, WILLARD G., AIEE Fellow 1912.

CARNAHAN, CHALON W., IRE Fellow 1952 "For original contributions in the fields of frequency modulation, television and electronic systems engineering."

CARPENTER, CHARLES BENJAMIN, AIEE Fellow 1944.

CARPENTER, HENRY C., AIEE Fellow 1922.

CARPENTER, HUBERT VINTON, AIEE Fellow 1918.

CARPENTER, JOHN WILLIAM, AIEE Fellow 1946.

CARPENTER, WILLIAM PHILIP, AIEE Fellow 1958 "For contributions to techniques of controlling electric illumination."

CARROLL, DOUGLAS V., IEEE Fellow 1971 "For contributions to naval and industrial electronics."

CARROLL, JOSEPH SNYDER, AIEE Fellow 1947.

CARSON, JOHN RENSHAW, AIEE Fellow 1933.

CARTER, EMMETT F., IRE Fellow 1936.

CARTER, HERBERT G., AIEE Fellow 1930.

CARTER, LEONARD LEROY, AIEE Fellow 1938.

CARTER, P.S., IRE Fellow 1949 "For his many contributions in the fields of radio transmission and communication systems."

CARTER, THOMAS, AIEE Fellow 1920.

CARTY, JOHN J., AIEE Fellow 1913.

CARUTHERS, ROBERT S., IRE Fellow 1962 "For contributions in the fields of multiplexing and long-distance telephony."

CASE, HERBERT MONROE, AIEE Fellow 1938.

CASKEY, ARTHUR DAVID, AIEE Fellow 1960 "For contributions to planning and design of a large power system."

CASSELL, WALLACE LEWIS, AIEE Fellow 1958 "For contributions to electrical curricula and inspirational teaching."

CASTLE, SCAUEL N., AIEE Fellow 1918.

CHACE, WILLIAM G., AIEE Fellow 1913.

CHAFFEE, E. LEON, IRE Fellow 1921.

CHAIT, HERMAN N., IEEE Fellow 1970 "For development and invention in microwave ferrites and antennas."

CHAMBERLAIN, ADOLPH B., IRE Fellow 1942.

CHANDLER, RALPH BORTHWICK, AIEE Fellow 1944.

CHANG, HSU, IEEE Fellow 1973 "For contributions to magnetic memory research."

CHAPMAN, J.H., IEEE Fellow 1964 "For leadership in space research and scientific achievement in upper atmospheric radio physics."

CHARLESWORTH, HARRY PRESCOTT, AIEE Fellow 1928.

CHASE, DELMAR DUDLEY, AIEE Fellow 1959 "For contributions to the planning and performance of a large power system and to the design of large two-winding turbine generators."

CHASE, PHILIP HARTLEY, AIEE Fellow 1943.

CHASE, WARREN H., AIEE Fellow 1951.

CHASE, WILLIAM GREGORY, AIEE Fellow 1924.

CHATELAIN, MICHAEL A., AIEE Fellow 1926.

CHENEY, H.W., AIEE Fellow 1913.

CHERNYSHOFF, ALEXANDER, AIEE Fellow 1926.

CHERRY, LLOYD BENJAMIN, AIEE Fellow 1962 "For contributions to electrical engineering education."

CHESNEY, C.C., AIEE Fellow 1913.

CHESTERMAN, FRANCIS J., AIEE Fellow 1922.

CHEYNEY, ALGERNON R., AIEE Fellow 1913.

CHI, ANDREW R., IEEE Fellow 1973 "For contributions in worldwide precise time and frequency synchronization, and of frequency control."

CHIEN, ROBERT T., IEEE Fellow 1972 "For contributions to coding theory, graph theory, and engineering education."

CHILDERHOSE, ERWIN ALFRED, AIEE Fellow 1937.

CHIPP, R.D., IRE Fellow 1955 "For his contributions to the development of radar and television apparatus for the Navy."

CHRISTENSEN, J.W., IRE Fellow 1961 "For contributions to photo transmission and color television systems."

CHRISTIE, CLARENCE VICTOR, AIEE Fellow 1932.

CHU, LAN J., IRE Fellow 1952 "In recognition of his contributions to the theory and practice of wave guides and microwave antennas."

CHUBB, LEWIS W., AIEE Fellow 1921. IRE Fellow 1940.

CHUBBUCK, H. EUGENE, AIEE Fellow 1913.

CHUBBUCK, LEONARD BURROWS, AIEE Fellow 1943.

CHUTE, GEORGE M., AIEE Fellow 1946.

CIMORELLI, JOSEPH T., IRE Fellow 1962 "For contributions to the development of electron tubes."

CISSMA, VOLNEY J., AIEE Fellow 1957 "For his contributions to the engineering of a large power transmission system."

CLAPP, JAMES K., IRE Fellow 1933.

CLARK, ALVA BENSON, AIEE Fellow 1930. IRE Fellow 1956 "For early development and leadership in the field of telephone transmission systems."

CLARK, FARLEY G., AIEE Fellow 1913.

CLARK, FRANK MARSHALL, AIEE Fellow 1960 "For contributions to development of electrical insulation through the application of chemical principles."

CLARK, GEORGE H., IRE Fellow 1915.

CLARK, HENRY WOODMAN, AIEE Fellow 1948.

CLARK, J.W., IRE Fellow 1961 "For contributions to research on the effects of radiation."

CLARK, JAMES C., AIEE Fellow 1957 "For his contributions in the field of electrical engineering education."

CLARK, WALTER G., AIEE Fellow 1924.

CLARK, WILLIAM RUSSELL, AIEE Fellow 1961 "For contributions by invention, development and design to recording, measuring and control instruments and systems."

CLARKE, CHARLES L., AIEE Fellow 1912.

CLARKE, EDITH, AIEE Fellow 1948.

CLARKE, HENRY ANABLE, AIEE Fellow 1959 "For contributions to carrier current relaying and ultra high speed power circuit reclosing techniques."

CLARKE, HIRAM OPIE, JR., AIEE Fellow 1941.

CLARKSON, ALBERT J., AIEE Fellow 1934.

CLAVIER, ANDRE GABRIEL, AIEE Fellow 1953 "For pioneer work in research, development and engineering in communication in the microwave field."

CLAYTOR, WILLIAM GRAHAM, AIEE Fellow 1946.

CLEARY, LEO H., AIEE Fellow 1942.

CLEM, JOSEPH EARL, AIEE Fellow 1938.

CLEMENT, EDWARD E., AIEE Fellow 1913.

CLEMENT, LOUIS M., IRE Fellow 1926.

CLIFFORD, H.E., AIEE Fellow 1913.

CLOKE, PAUL, AIEE Fellow 1938.

CLOKEY, ALLISON A., AIEE Fellow 1928.

COATES, WILLIAM A., AIEE Fellow 1926.

COBINE, JAMES DILLON, IRE Fellow 1957 "For contributions to engineering education and gaseous electronics

research." AIEE Fellow 1962 "For contributions to the understanding of Gaseous Conduction and Arc Discharges."

COCHRANE, HARRY HAMILTON, AIEE Fellow 1961 "For contributions to economical power generation and transmission for an extended system."

CODDING, HENRY W., AIEE Fellow 1944.

COE, REGINALD THORNTON, AIEE Fellow 1948.

COFFIN, JOHN RUSKIN, AIEE Fellow 1959 "For his contributions to engineering management and economic evaluation and design of large power projects."

COHEN, LOUIS, IRE Fellow 1915. AIEE Fellow 1919.

COHN, SEYMOUR B., IRE Fellow 1959 "For contributions to the theory and design of microwave components."

COIT, NORMAN H., AIEE Fellow 1940.

COLDWELL, ORIN B., AIEE Fellow 1912.

COLE, FRED H., AIEE Fellow 1948.

COLEMAN, HARRY CHARLES, AIEE Fellow 1948.

COLEMAN, JAMES O'REILLY, AIEE Fellow 1950.

COLEMAN, JOHN B., IRE Fellow 1948 "For his contributions through development, design and technical direction of work in the field of radio transmitters." AIEE Fellow 1956 "For contributions to the development of radio transmitters and standards in the radio engineering field."

COLEMAN, O.K., AIEE Fellow 1944.

COLES, EDMUND P., AIEE Fellow 1935.

COLLBOHM, MAX H., AIEE Fellow 1913.

COLLIER, WILLIAM R., AIEE Fellow 1920.

COLLINS, HAROLD WILSON, AIEE Fellow 1948.

COLLISON, PERCE B., IRE Fellow 1962 "For unusually helpful activities as a Charter Member of the Institute, and many early efforts toward the production of its PROCEEDINGS."

COLPITS, EDWIN HENRY, AIEE Fellow 1912. IRE Fellow 1926.

COLTON, ROGER BALDWIN, AIEE Fellow 1946. IRE Fellow 1954 "For his pioneering contributions to the development and application of radar in the U.S. Army."

COMPTON, KARL T., AIEE Fellow 1931.

COMSTOCK, G.C., IRE Fellow 1957 "For contributions to the development and application of navigational radar."

COMSTOCK, L.K., AIEE Fellow 1912.

CONBLIN, L.H., AIEE Fellow 1912.

CONDIT, B.C., AIEE Fellow 1913.

CONDON, EARL STILSON, AIEE Fellow 1961 "For contributions to power insulator improvement and protective and distribution equipment for aircraft."

CONDON, EDWARD U., IRE Fellow 1953 "In recognition of his outstanding contributions and distinguished leadership in the development of microwave radar and in the atomic energy program."

CONN, CHARLES F., AIEE Fellow 1914.

CONNELL, DOUGLAS J., IEEE Fellow 1966 "For outstanding engineering proficiency, leadership and administrative attainments in the utility industry."

CONRAD, ALBERT G., AIEE Fellow 1949.

CONRAD, EDWARD E., IEEE Fellow 1981 "For technical leadership in the understanding and characterization of transient nuclear radiation effects on insulators."

CONRAD, FRANK, IRE Fellow 1927. AIEE Fellow 1937.

CONRAD, NICHOLAS J., AIEE Fellow 1921.

CONWELL, ROLLIN N., AIEE Fellow 1931.

COOK, ARTHUR L., AIEE Fellow 1951.

COOK, EARLE F., IRE Fellow 1962 "For leadership in military communications and electronics."

COOK, ELLSWORTH D., IRE Fellow 1942.

COOK, JAMES A., AIEE Fellow 1930.

COOK, LEE E., AIEE Fellow 1945.

COOLIDGE, WILLIAM D., AIEE Fellow 1955 "For notable research and scientific achievement in electrical engineering particularly the drawing of tungsten and in the development of the X-Ray."

COONEY, WILLIAM HENRY, AIEE Fellow 1949.

COOPER, ARCHIBALD JOHN, AIEE Fellow 1961 "For contributions toward the practical application of electrical machinery."

COOPER, ASHTON BURTON, AIEE Fellow 1933.

COOVER, MERVIN S., AIEE Fellow 1942.

COPELAND, CLEM A., AIEE Fellow 1940.

COPLEY, ALMON W., AIEE Fellow 1926.

CORCORAN, GEORGE F., AIEE Fellow 1950. IRE Fellow 1956 "For contributions to electrical engineering education and to the associated literature."

CORDRAY, RICHARD E., AIEE Fellow 1949.

COREY, FRED B., AIEE Fellow 1912.

CORNEY, CHESTER ALBERT, AIEE Fellow 1944.

CORWITH, HOWARD P., IRE Fellow 1952 "For his leadership in the development of radio-telegraph, landline and submarine cable communications."

CORWITH, HOWARD POST, AIEE Fellow 1951.

CORY, CLARENCE L., AIEE Fellow 1913.

CORY, MERTON M., AIEE Fellow 1940.

COSENTINE, LOUIS GEORGE, AIEE Fellow 1962 "For contributions to electrified production in industrial plants."

COSTA, HELIO, IRE Fellow 1958 "For his contributions to aeronautical radio."

COSTER, MAURICE, AIEE Fellow 1912.

COTTER, WILLIAM F., AIEE Fellow 1947.

COULBOURN, EDWARD R., AIEE Fellow 1956 "For his achievements in the design of an extensive electric production, transmission and distribution system."

COWAN, FRANK A., AIEE Fellow 1945. IRE Fellow 1955 "For his contributions to long-distance communication, particularly in the development of television network facilities."

COWLES, ALFRED H., AIEE Fellow 1912.

COX, FRANK P., AIEE Fellow 1913.

COX, H.H., AIEE Fellow 1953 "In recognition of pioneering work in the field of large high-voltage systems."

COX, JEROME R., JR., IEEE Fellow 1977 "For contributions to the application of computers to clinical medicine."

COX, JOSEPH H., AIEE Fellow 1957 "For contributions to lightning protection through study of lightning phenomena, and to the development of mercury arc rectifiers."

COX, VERNON L., AIEE Fellow 1950.

CRAFT, EDWARD B., IRE Fellow 1920. AIEE Fellow 1926.

CRAFT, FRANCIS M., AIEE Fellow 1933.

CRAFT, L. MORGAN, IRE Fellow 1962 "For leadership in the electronics industry."

CRAIG, CLEO FRANK, AIEE Fellow 1950.

CRAIG, PALMER H., AIEE Fellow 1938. IRE Fellow 1962 "For contributions to engineering education."

CRAIGHEAD, JAMES R., AIEE Fellow 1917.

CRAMER, C.H., AIEE Fellow 1953 "For outstanding contributions to the ocean cable transmission arts and specifically for directing the development and design of electronic undersea repeaters."

CRARY, SELDEN B., AIEE Fellow 1945.

CRAVATH, JAMES R., AIEE Fellow 1920.

CRAVEN, T.A.M., IRE Fellow 1929.

CRAWFORD, DAVID F., AIEE Fellow 1912.

CRAWFORD, MAGNUS T., AIEE Fellow 1922.

CRAWFORD, WILLIAM ELON, AIEE Fellow 1936.

CREAMER, EDGAR M., JR., IEEE Fellow 1969 "For contributions to time division multiplexing, color television, and solid-state circuit development."

CREAMER, WALTER JOSEPH, AIEE Fellow 1960 "For contributions to electrical communication and to electrical education and literature."

CREEDON, JOHN E., IEEE Fellow 1980 "For leadership in and contributions to high-energy pulse power engineering."

CREIGHTON, ELMER E.F., AIEE Fellow 1925.

CREIM, BENJAMIN W., AIEE Fellow 1931.

CREVER, FREDERICK EMANUEL, AIEE Fellow 1951.

CRILEY, WALTER, AIEE Fellow 1960 "For contributions to personal professional development of engineering students."

CRITTENDEN, EUGENE C., AIEE Fellow 1944.

CROMWELL, LEWIS W., AIEE Fellow 1916.

CROMWELL, PAUL C., IEEE Fellow 1965 "For contributions to electrical engineering education and research."

CRONIN, LATHAN DENNIS, AIEE Fellow 1961 "For contributions to distribution and high voltage power cable systems."

CROSBY, MURRAY GRIMSHAW, IRE Fellow 1943 "For his contributions to the development of high-frequency radio aids in air navigation and meteorology." AIEE Fellow 1951.

CROTHERS, HAROLD M., AIEE Fellow 1950.

CROWELL, GEORGE F., AIEE Fellow 1942.

CROWLEY, HENRY J., AIEE Fellow 1919.

CRUICKSHANK, JOHN POWNALL, AIEE Fellow 1962 "For contributions to the design, operation and construction of a large transmission system."

CRUMLEY, HOWARD LEE, AIEE Fellow 1948.

CRUMP, LINDELL L., AIEE Fellow 1942.

CRUMPTON, WILLIAM JAIRUS, AIEE Fellow 1945.

CULSHAW, WILLIAM, IEEE Fellow 1965 "For contributions related to microwave optics, interferometers, and gas lasers."

CULVER, CHARLES AARON, AIEE Fellow 1922. IRE Fellow 1954 "For his contributions to the art and literature of electronic and allied fields."

CULVER, LAURENCE ROSENCRANS, AIEE Fellow 1935.

CUMMING, LAURENCE G., IEEE Fellow 1966 "For contributions to the engineering profession through leadership in organizing and stimulating the growth of Groups and Technical Committees of the IEEE."

CURRIE, HARRY ALLAN, AIEE Fellow 1929.

CURRIN, HUGH PERKINS, AIEE Fellow 1945.

CURRY, WALTER ANDREW, AIEE Fellow 1937.

CURTIS, HARVEY L., AIEE Fellow 1926.

CURTIS, LESLIE FORREST, AIEE Fellow 1929.

CURTIS, THOMAS EUGENE, AIEE Fellow 1962 "For contributions to the development and standardization of power fuses, distribution cutouts, and power switching equipment."

CUSHING, HARRY C., JR., AIEE Fellow 1912.

CUSHING, HARVEY MORSE, AIEE Fellow 1935.

CUTTING, FULTON, IRE Fellow 1921. AIEE Fellow 1955 "For services to electronic engineering in the fields of research, management and education."

D

DAHL, OTTO GUSTAV C., AIEE Fellow 1933.

DALY, CHARLES J., AIEE Fellow 1946.

DAMBLY, HAROLD ALTHOUSE, AIEE Fellow 1942.

DAMON, JOHN CHURCHILL, AIEE Fellow 1950.

DANA, ALAN STANDISH, AIEE Fellow 1948.

DANA, HOMER J., AIEE Fellow 1962 "For contributions to the field of engineering research and expermentation."

DANIELS, RAYMOND S., AIEE Fellow 1953 "For engineering leadership and technical competence in analyzing long range requirements of large electric power systems and in the design and execution of plant expansion projects."

DANN, WALTER M., AIEE Fellow 1926.

DARLAND, ALVIN F., AIEE Fellow 1948.

DARLING, ALAN G., AIEE Fellow 1955 "For his contributions toward improved efficiency and operation of large industrial power generating systems."

DARLINGTON, FREDRICK, AIEE Fellow 1913.

DARNELL, P.S., IRE Fellow 1961 "For contributions to the field of component engineering."

DARROW, LEO HARVEY, AIEE Fellow 1950.

DART, HARRY F., AIEE Fellow 1958 "For contributions in the research and development of electron tubes." IRE Fellow 1959 "For contributions to the electronic profession."

DASHER, BENJAMIN J., IRE Fellow 1959 "For pioneering contributions in the field of network synthesis and engineering education."

DATES, HENRY B., AIEE Fellow 1932.

DAVENPORT, JOHN ADRIAN, AIEE Fellow 1951.

DAVID, BRUCE WILLIAM, AIEE Fellow 1940.

DAVIDSON, WARD F., AIEE Fellow 1926.

DAVIES, CHARLES E., AIEE Fellow 1921.

DAVIS, ALBERT G., AIEE Fellow 1912.

DAVIS, ERNEST W., AIEE Fellow 1934.

DAVIS, GEORGE S., IRE Fellow 1916.

DAVIS, MINOR M., AIEE Fellow 1912.

DAVIS, ROWLAND FENNER, AIEE Fellow 1961 "For contributions in guiding the development of a wide range of telephone products for use on customers' premises."

DAVIS, T.M., IRE Fellow 1956 "For contributions in the field of military radio communication."

DAWES, CHESTER LAURENS, AIEE Fellow 1935.

DAWSON, J.W., IRE Fellow 1955 "For his contributions to the advancement of scientific and engineering knowledge."

DAWSON, WILLIAM FRANCIS, AIEE Fellow 1929.

DEAKIN, GERALD, AIEE Fellow 1944.

DEAN, HAROLD C., AIEE Fellow 1930.

DEAN, SAMUEL MILLS, AIEE Fellow 1946.

DEAN, WILLIAM J., AIEE Fellow 1912.

DEANS, WILLIAM, AIEE Fellow 1930.

DEARBORN, RICHARD HAROLD, AIEE Fellow 1930.

DEARDORFF, R.W., IRE Fellow 1951 "For his technical contributions to communications and his work on behalf of the radio engineering profession."

DE ARTIGAS, JOSE ANTONIO, AIEE Fellow 1922.

DE BEECH, ALBERT V., AIEE Fellow 1939.

DEBELLIS, ALDO MICHAEL, AIEE Fellow 1948.

DEBLIEUX, EARL VERNE, AIEE Fellow 1949.

DECOLA, RINALDO, IRE Fellow 1957 "For contributions to the fields of military electronics and to television receivers."

DEE, THOMAS CHRISTOPHER, AIEE Fellow 1945.

DEFANDORF, FRANCIS MARION, AIEE Fellow 1942.

DE FOREST, CORNELIUS W., AIEE Fellow 1936.

DE FOREST, LEE, AIEE Fellow 1918. IRE Fellow 1918.

DE GROOT, CORNELIUS-JOHANNES, IRE Fellow 1916.

DE KIEP, JAMES, AIEE Fellow 1954 "For his contributions to the development of electric generators and motors."

DE KLERK, JOHN, IEEE Fellow 1974 "For contributions to the development of ultrasonic technology."

DE LANTY, BENJAMIN F., AIEE Fellow 1944.

DEL BUSTO, ALFONSO FERNANDEZ, AIEE Fellow 1962 "For contributions to the engineering, construction and operation of electric utility systems in Mexico."

DEL CAMP, SCIPIO M., IRE Fellow 1959 "For contributions in the development and standardization of electronic components."

DEL'HOMME, LAURENCE K., AIEE Fellow 1956 "For contributions to the development of an electric power system in an area having a rapid growth."

DELLINGER, J.H., IRE Fellow 1923.

DEL MAR, WILLIAM A., AIEE Fellow 1920.

DEMAREST, CHARLES SIDNEY, AIEE Fellow 1933.

DEMERIT, MERRILL, AIEE Fellow 1950.

DEMUTH, ORIN ANTON, AIEE Fellow 1946.

DENNIS, ROBERT E., AIEE Fellow 1930.

DENTON, ALPHEUS P., AIEE Fellow 1927.

DENTON, FRANCIS M., AIEE Fellow 1928.

DEROSA, LOUIS A., IRE Fellow 1952 "For his contributions in the fields of electronic direction-finding techniques, acoustics and aids to air navigation."

DERR, WILLARD ALLAN, AIEE Fellow 1962 "For contributions to remote control and supervisory systems."

DERRICK, CHARLES L., AIEE Fellow 1962 "For contributions to the economic planning, engineering and operation of a utility system."

DESEVERSKY, ALEXANDER P., IEEE Fellow 1970 "For pioneering efforts in the development of aircraft electrical controls and flight instruments, and for the application of electron physics to ionic propulsion and air pollution reduciton."

DEVORE, LLOYD T., IRE Fellow 1952 "For his contributions in the field of electronic systems engineering, and in the direction of research and development."

DEWARS, ALLEN GUTHRIE, AIEE Fellow 1944.

DEWEESE, FRED C., AIEE Fellow 1950.

DEWEY, CLYDE GILMAN, AIEE Fellow 1962 "For improvement in protective relay equipment and mechanical rectifier switching."

DEWITT, DAVID, IEEE Fellow 1968 "For his contributions to the development and manufacture of semiconductor devices."

DEXTER, JOHN F., IEEE Fellow 1965 "For contributions to the development and evaluation of high temperature electrical insulation material."

D'HUMY, FERNAND EMILE, AIEE Fellow 1930. IRE Fellow 1949 "In recognition of his long service in the communicatitons field and for pioneering in the application of radio relays to telegraph message service."

DIAMOND, HARRY, AIEE Fellow 1947.

DIBBLE, BARRY, AIEE Fellow 1949.

DIBBLEE, JOHN, AIEE Fellow 1944.

DICK, W.A., AIEE Fellow 1913.

DICKENSON, ROBERT CHARLES, AIEE Fellow 1948.

DICKINSON, W.N., AIEE Fellow 1912.

DICKINSON, WILLIAM NOBLE, AIEE Fellow 1935.

DIEDERICH, PETER, AIEE Fellow 1942.

DIEHL, CARL W., AIEE Fellow 1959 "For contributions to the design of equipment and procedures for testing distribution and power transformers."

DIEHL, GEORGE SANDY, AIEE Fellow 1949.

DIEHL, WILLIAM F., IRE Fellow 1936.

DILLARD, EDWARD W., AIEE Fellow 1946.

DILLON, JOHN F., IRE Fellow 1924.

DINGER, HAROLD EUGENE, IRE Fellow 1958 "For his contributions to the reduction of radio interference." AIEE Fellow 1961 "For his contribuitons in the fields of radio propagation and interference."

DINGLEY, EDWARD N., JR., IRE Fellow 1956 "For contributions in the fields of electronic guidance and detection systems." AIEE Fellow 1957 "For his many contributions in the field of electrical engineering to the defense of the United States."

DION, A.A., AIEE Fellow 1913.

DISHAL, MILTON, IRE Fellow 1957 "For research and invention in microwave waveguide techniques."

DISQUE, ROBERT C., AIEE Fellow 1950.

DISSMEYER, EDWARD F., AIEE Fellow 1947.

DITORO, MICHAEL J., IRE Fellow 1958 "For his contributions to electroacoustics."

DIXON, AMOS F., AIEE Fellow 1926.

DOANE, SAMUEL E., AIEE Fellow 1912.

DOBLE, FRANK CURRIER, AIEE Fellow 1947.

DOBSON, WILLIAM PERCY, AIEE Fellow 1943.

DODD, ROY LYON, AIEE Fellow 1941.

DODGE, CHESTER CARLTON, AIEE Fellow 1951.

DOGGETT, LEONARD ALLISON, AIEE Fellow 1936.

DOHERTY, HENRY L., AIEE Fellow 1913.

DOHERTY, ROBERT ERNEST, AIEE Fellow 1939.

DOLKART, LEO, AIEE Fellow 1958 "For contributions to design of diversified industrial distribution and lighting systems."

DOMMERQUE, F.J., AIEE Fellow 1913.

DON CARLOS, HENRY C., AIEE Fellow 1918.

DONNELL, PHILIP STONE, AIEE Fellow 1937.

DOOLITTLE, CLARENCE E., AIEE Fellow 1913.

DOOLITTLE, F.M., IRE Fellow 1954 "For his pioneering work in radio communication and radio broadcasting."

DORFMAN, L.O., AIEE Fellow 1949.

DORSEY, HERBERT GROVE, AIEE Fellow 1934. IRE Fellow 1934.

DOUGLAS, JOHN FREDERICK HOWARD, D, AIEE Fellow 1938.

DOUGLASS, FORREST S., AIEE Fellow 1945.

DOUGLASS, GEORGE WILLIAM, AIEE Fellow 1949.

DOVJIKOV, ALEXANDER, AIEE Fellow 1949.

DOW, ALEX, AIEE Fellow 1913.

DOW, JAMES CHASE, AIEE Fellow 1938.

DOW, JENNINGS B., IRE Fellow 1942 "For engineering leadership in the development and procurement of radio equipment for the United States Navy."

DOWNING, PAUL M., AIEE Fellow 1942.

DOWNS, GEORGE WALLINGFORD, AIEE Fellow 1959 "For contributions to refinement of acoustical and data evaluation apparatus and systems."

DOWS, CHESTER L., AIEE Fellow 1932.

DOYLE, EDGAR D., AIEE Fellow 1927.

DOYLE, HAROLD KINYON, AIEE Fellow 1945.

DRABELLE, JOHN MITCHELL, AIEE Fellow 1950.

DRAKE, CHESTER W., AIEE Fellow 1945.

DRAKE, F.H., IRE Fellow 1957 "For contributions to the field of airborne communications equipment."

DRAKE, HERBERT W., AIEE Fellow 1925.

DREHER, CARL, IRE Fellow 1928. AIEE Fellow 1933.

DRUM, A.L., AIEE Fellow 1913.

DUBILIER, WILLIAM, AIEE Fellow 1925. IRE Fellow 1929.

DUBRIDGE, LEE, IRE Fellow 1942 "For engineering and administrative leadership in the development and application of new radio techniques."

DUDDEL, WILLIAM, IRE Fellow 1915.

DUDLEY, ADOLPHUS M., AIEE Fellow 1913.

DUDLEY, H.W., IEEE Fellow 1964 "For contributions to the fields of speech theory, speech signal processing and speech synthesis."

DUER, JOHN VAN BUREN, AIEE Fellow 1929.

DUKE, CLIFFORD A., AIEE Fellow 1957 "For his contributions to the organization of a testing laboratory and staff for a large power system."

DUMONT, ALLEN B., IRE Fellow 1931, AIEE Fellow 1943.

DUNCAN, LOUIS, AIEE Fellow 1913.

DUNKLE, WILLIAM FREEMONT, AIEE Fellow 1958 "For contributions to electrical system testing techniques particularly in the field of dielectrics."

DUNLAP, W.K., AIEE Fellow 1912.

DUNN, GANO, AIEE Fellow 1912. IRE Fellow 1915.

DUNN, W.L., IRE Fellow 1954 "For his contribution in the development of radio receivers."

DUTSCHMANN, KARL THEODORE, AIEE Fellow 1960 "For the prompt and efficient integration of many newly developed communication systems into a large and complex telephone plant."

DUVALL, W. CLINTON, AIEE Fellow 1938.

DWIGHT, HERBERT BRISTOL, AIEE Fellow 1926.

DYCHE, HOWARD EDWARD, AIEE Fellow 1942.

E

EALES, HERBERT W., AIEE Fellow 1925.

EARLE, RALPH HILLIARD, AIEE Fellow 1959 "For contributions to development of equipment for electric distribution."

EARLY, EDWARD DONALD, AIEE Fellow 1962 "For his inventions and contributions to the theory of economic dispatch of electric power systems."

EASTHAM, MELVILLE, IRE Fellow 1925. AIEE Fellow 1946.

EASTMAN, AUSTIN V., IRE Fellow 1941. AIEE Fellow 1946.

EASTOM, FRANK A., AIEE Fellow 1951.

EASTWOOD, ARTHOR C., AIEE Fellow 1913.

EATON, MILTON, AIEE Fellow 1958 "For contributions to the design and automatic control of electric steam generators."

EATON, WILLIAM W., IEEE Fellow 1969 "For contributions to the coupling of U.S. Government information to the civilian technology."

EAVES, A.J., IEEE Fellow 1964 "For pioneering developments of telegraph transmission systems, radio transmitters and sound amplifying systems."

ECKERSLEY, P.P., IRE Fellow 1925.

ECKERSLEY, T.L., IRE Fellow 1930 "(1944 citation) For his outstanding contributions to the theory and practice of radio-wave-propagation research. Both his approach to the problem from the stand-point of practical communications and his computation of radiated fields are achievements of lasting value acclaimed by the whole radio world and form a monument of which he may be justly proud."

ECKERT, J.P., JR., IRE Fellow 1956 "For contributions to electronic digital computation."

EDDY, WILTON NATHANIEL, AIEE Fellow 1940.

EDGAR, CHARLES L., AIEE Fellow 1912.

EDISON, OSKAR EDWIN, AIEE Fellow 1945.

EDLEFSEN, N.E., IRE Fellow 1954 "For his contribution to radar, and his leadership in the development of military electronic equipment."

EDMONDSON, FRANCIS CHARLES, AIEE Fellow 1940.

EDSON, JAMES O., IEEE Fellow 1969 "For contributions to multiplex communications."

EDSTROM, J. SIGFRID, AIEE Fellow 1914.

EDWARDS, ANDREW WOOD, AIEE Fellow 1962 "For contributions in the field of distribution apparatus and for engineering applications of digital computers."

EDWARDS, CHARLES P., IRE Fellow 1915.

EDWARDS, EVAN JAMES, AIEE Fellow 1939.

EDWARDS, GEORGE D., AIEE Fellow 1951.

EDWARDS, J. PAULDING, AIEE Fellow 1913.

EDWARDS, PAUL GRIFFITH, AIEE Fellow 1959 "For his contributions to the development of carrier, cable and microwave communication systems." IRE Fellow 1960, "For his contributions in the field of telephone communication systems."

EGLI, JOHN J., IEEE Fellow 1967 "For contributions to wave propagation, electromagnetic compatibility and advanced radio communications."

EGLIN, WILLIAM C.L., AIEE Fellow 1912.

EHRET, CORNELIUS D., IRE Fellow 1916.

EICHBERG, FRIEDRICH, AIEE Fellow 1928.

EINTHOVEN, WILLEM F., IRE Fellow 1938.

ELDRED, BYRON E., AIEE Fellow 1930.

ELDRED, W. NOEL, IEEE Fellow 1970 "For contributions in designing radar equipment, transmitters, and vacuum tubes; for leadership in the marketing of electronic instrumentation."

ELDREDGE, K.R., IEEE Fellow 1964 "For contributions to pattern recognition and magnetic character reading systems."

ELDREDGE, MARK, AIEE Fellow 1933.

ELLEFSON, B.S., IRE Fellow 1948 "For his contributions to cathode-ray tube development, proximity-fuze tube design, and wartime electronic research."

ELLESTAD, IRWIN M., AIEE Fellow 1960 "For contributions to the solution of fundamental engineering problems of inductive coordination and telephone transmission."

ELLETT, ALEXANDER, IRE Fellow 1962 "For contributions to military electronics and television."

ELLICOT, EDWARD B., AIEE Fellow 1916.

ELLIOTT, HAROLD F., IRE Fellow 1951 "For contributions of important mechanical advances in both the civil and the military electronic arts."

ELLIOTT, ROBERT DRANE, AIEE Fellow 1959 "For contributions to the design and contruction of large electric utility systems."

ELLIS, ELVARADO LEROY, AIEE Fellow 1919.

ELLIS, FREDERICK W., AIEE Fellow 1920.

ELLWOOD, WALTER B., IEEE Fellow 1966 "For contributions to the field of magnetics."

ELMEN, GUSTAF W., AIEE Fellow 1927.

ELSDON-DEW, WILLIAM, AIEE Fellow 1914.

ELWELL, CYRIL F., IRE Fellow 1918.

EMANUELI, LUIGI P., AIEE Fellow 1956 "For the outstanding development of oil-filled cable which made practical underground power transmission al extra high voltages."

EMBREE, CLAYTON J., AIEE Fellow 1918.

EMERSON, CHERRY L., AIEE Fellow 1933.

ENFIELD, WILLIAM LESTER, AIEE Fellow 1932.

ENGLUND, CARL R., IRE Fellow 1928.

ENGSTROM, HOWARD T., IRE Fellow 1959 "For contribution in the development and utilization of high speed computers."

ENOS, HOWARD A., AIEE Fellow 1954 "For developments in power distribution systems, particularly low-cost rural electrification."

ENSIGN, ORVILLE H., AIEE Fellow 1913.

EPPLEY, MARION, AIEE Fellow 1952 "For contributions in the field of high-accuracy electrical measurements."

EPSTEIN, DAVID W., IRE Fellow 1952 "In recognition of his contributions in the field of cathode-ray tube development and application."

ERIKSON, PER ENGELBERT, AIEE Fellow 1928.

ERNST, CARL CLARK, AIEE Fellow 1961 "For contributions to the development of a rapidly expanding and changing electric power system."

ERSKINE-MURRAY, JAMES, IRE Fellow 1915.

ESHBACH, OVID W., AIEE Fellow 1937.

ESHY, WILLIAM, AIEE Fellow 1913.

ESPLEY, D.C., IRE Fellow 1954 "For his creative contributions to microwave and television techniques in England."

ESTERLINE, JOHN WALTER, AIEE Fellow 1941.

EVANS, CLARENCE TURNER, AIEE Fellow 1947.

EVANS, HERBERT S., AIEE Fellow 1938.

EVANS, LLEWELLYN, AIEE Fellow 1941.

EVANS, PORTER H., IRE Fellow 1931.

EVANS, ROBERT D., AIEE Fellow 1940.

EVANS, WALTER C., IRE Fellow 1945 "In recognition of his past contributions to radio and his present active participation in the affairs of the Institute."

EVERIT, EDWARD H., AIEE Fellow 1912.

EWING, DRESSEL D., AIEE Fellow 1929.

F

FAHY, FRANK P., AIEE Fellow 1929.

FAIR, IRWIN E., IEEE Fellow 1967 "For distinguished scientific contributions and technical leadership in piezoelectric resonators and ultrasonic delay devices."

FAIRFIELD, RONALD MCLEOD, AIEE Fellow 1957 "For contributions to the development, improvement and manufacture of high voltage electric power cables."

FAIRMAN, FRANCIS E., JR., AIEE Fellow 1951.

FAIRMAN, JAMES F., AIEE Fellow 1935.

FALLON, GLOVER PATTERSON, AIEE Fellow 1961 "For contributions to the engineering of a major electric utility system and to the development of microwave and radio communication."

FARMER, F.M., AIEE Fellow 1913.

FARMER, G. EVERETT, AIEE Fellow 1960 "For contributions to the development of a comprehensive communication system for a large power system."

FARNHAM, DALE, AIEE Fellow 1953 "For contributions to high voltage underground and submarine practice in transmission and distribution systems."

FARNSWORTH, PHILIP, IRE Fellow 1915.

FARNSWORTH, PHILO T., IRE Fellow 1939.

FARON, FRANK A., AIEE Fellow 1951.

FARRAND, DUDLEY, AIEE Fellow 1912.

FARRELL, J.J., IRE Fellow 1948 "For his contributions to the design of radio and radar transmitters, and for his leadership in the establishing of industry standards."

FARRINGTON, JOHN F., IRE Fellow 1931.

FARRY, O.T., AIEE Fellow 1953 "For original work in the development and design of power transformers, unit substations and their accessories."

FASSBENDER, HEINRICH, IRE Fellow 1930.

FAUCETT, IRVING THOMPSON, AIEE Fellow 1928.

FAUIOLI, G., AIEE Fellow 1912.

FAULKNER, HARRY, IRE Fellow 1953 "For his engineering achievements in the field of world-wide radio communication, and his contributions to international agreement on telecommunication practices."

FAUS, HAROLD T., AIEE Fellow 1952 "For outstanding creative ability which has resulted in many distinctive contributions, particularly in the development and application of magnetic materials, together with advances in the design and performance of instruments, meters and associated systems."

FAWCETT, CHARLES D., AIEE Fellow 1951.

FECHHEIMER, CARL J., AIEE Fellow 1914.

FEDER, JOSEPH BROWN, AIEE Fellow 1947.

FEIKER, FREDERICK MORRIS, AIEE Fellow 1942.

FEISSEL, H. G., IEEE Fellow 1970 "For contributions to original design in the field of digital computers."

FELDMAN, CARL B., IRE Fellow 1943 "For his investigations of the characteristics of radio waves and his development in antennas and receiving systems."

FERGUSEN, LOUIS A., AIEE Fellow 1912.

FERGUSON, J.D., AIEE Fellow 1953 "For executive leadership and engineering contributions to the development of electric communication in the Republic of Ireland."

FERGUSON, JOHN G., AIEE Fellow 1957 "In recognition of his contributions in the fields of primary electrical measurements and electronic measuring techniques."

FERGUSON, O.J., AIEE Fellow 1913.

FERNOW, BERNARD E., AIEE Fellow 1920.

FERRI, LOUIS F., AIEE Fellow 1951.

FERRIE, GENERAL, IRE Fellow 1917.

FERRILL, RICHARD MILLEDGE, AIEE Fellow 1950.

FERRIS, LIVINGSTON POLK, AIEE Fellow 1934.

FETT, GILBERT H., IRE Fellow 1958 "For his contributions to the knowledge of circuit analysis."

FIELD, A.B., AIEE Fellow 1913.

FIELD, CROSBY, AIEE Fellow 1922.

FIELD, KENNETH SELLERS, AIEE Fellow 1962 "For contributions to the construction, maintenance, and operation of utility electric distribution and sub-transmission systems."

FIELD, ROBERT F., IRE Fellow 1947 "Who, as an engineer and physicist, improved methods and standards in alternating-current measurements."

FIELD, STEPHEN D., AIEE Fellow 1912.

FIELDS, ERNEST S., AIEE Fellow 1929.

FIFER, WILLIAM HARTAGE, AIEE Fellow 1948.

FILIPOWSKY, RICHARD, IRE Fellow 1961 "For contributions in the application of information theory to communication systems."

FINCH, J.L., IRE Fellow 1954 "For his contributions to radio communications, and particularly those associated with VLF antenna design."

FINCH, TUDOR R., IEEE Fellow 1967 "For contributions in solid state circuits and for leadership in related professional activities."

FINDLAY, JOHN H., IEEE Fellow 1966 "For contributions to development of electron tubes and devices for atomic weapons."

FINE, HARRY, IEEE Fellow 1968 "For contributions to radio propagation and television and space communications."

FINLAY, WALTER STEVENSON, JR., AIEE Fellow 1921.

FINZI, LEO ALDO, AIEE Fellow 1959 "For contributions to electrical engineering education and the theory of magnetic amplifiers."

FISCHER, LOUIS ENGELMANN, AIEE Fellow 1929.

FISH, F.A., AIEE Fellow 1913.

FISH, WALTER C., AIEE Fellow 1913.

FISHER, HENRY W., AIEE Fellow 1912.

FISHMAN, SOLOMON, AIEE Fellow 1959 "For contributions to mathematical techniques in solving problems and teaching electrical engineering."

FISK, ERNEST T., IRE Fellow 1926.

FISK, J.B., IRE Fellow 1955 "For his contributions to the development of the magnetron and his leadership in basic electronic research."

FISK, WALTER B., AIEE Fellow 1955 "For contributions to the design, installation and operation of inside and outside plants in a large metropolitan electric system."

FISKE, WARREN H., AIEE Fellow 1912.

FISKEN, JOHN B., AIEE Fellow 1913.

FITZGERALD, ALAN STEWART, AIEE Fellow 1962 "For contributions to power systems relaying and to development of nonlinear magnetic circuit applications."

FITZGERALD, ARTHUR E., AIEE Fellow 1956 "For his contributions to the teaching of electrical engineering and the the analysis of electric power system stability and operation."

FLANDERS, MILTON M., AIEE Fellow 1942.

FLANIGEN, JOHN MONTEITH, AIEE Fellow 1949.

FLATH, EARL H., AIEE Fellow 1943.

FLEAGER, CLARENCE E., AIEE Fellow 1926.

FLEMING, ARTHUR P.M., AIEE Fellow 1934.

FLETCHER, HARVEY, AIEE Fellow 1930.

FLOOD, HENRY, JR., AIEE Fellow 1930.

FLOY, HENRY, AIEE Fellow 1912.

FLOYD, GEORGE D., AIEE Fellow 1956 "For his planning and direction of the technical growth of a large electric utility system."

FOLTZ, LEROY STEWART, AIEE Fellow 1946.

FORBES, E.D., IRE Fellow 1914.

FORBES, H.C., IRE Fellow 1948 "For his contributions as an engineer and executive in the field of home and automobile broadcast receivers and military radio equipment."

FORD, ARTHUR H., AIEE Fellow 1912.

FORD, GAYLON T., IEEE Fellow 1965 "For contributions to the development of grid-controlled electron tubes."

FORMAN, ALEXANDER HARDIE, AIEE Fellow 1943.

FORSTER, DONALD C., IEEE Fellow 1971 "For contributions to low-noise electron guns, lasers and high power millimeter-wavelength sources."

FORSTER, ERIC O., IEEE Fellow 1979 "For contributions in the understanding of high-voltage conduction and breakdown phenomena and for leadership in related professional activities."

FORTENBAUGH, S.B., AIEE Fellow 1913.

FORTESCUE, CHARLES L., AIEE Fellow 1921.

FOSTER, HORATIO, AIEE Fellow 1912.

FOSTER, R.M., IRE Fellow 1954 "For basic mathematical contributions to modern network theory."

FOSTER, WILLIAM J., AIEE Fellow 1916.

FOULKROD, RAYMOND, AIEE Fellow 1949.

FOULTON, FRED, AIEE Fellow 1951.

FOUNTAIN, LAWRENCE L., AIEE Fellow 1951.

FOWLER, CHARLES VERN, AIEE Fellow 1948.

FOX, EDWIN GORDON, AIEE Fellow 1944.

FOX, JOHN CONDRY, AIEE Fellow 1945.

FRAENZ, KURT O., IEEE Fellow 1967 "For his contributions to the theory of noise in linear and nonlinear circuits and to antenna theory."

FRAMPTON, ARTHUR H., AIEE Fellow 1945.

FRANK, JOHN J., AIEE Fellow 1919.

FRANKLIN, CHARLES WHITE, AIEE Fellow 1944.

FRANKLIN, MILTON W., AIEE Fellow 1912.

FRANKLIN, W.S., AIEE Fellow 1913.

FRASER, DANIEL M., AIEE Fellow 1920.

FRASER, J.W., AIEE Fellow 1913.

FREDERICK, HALSEY A., AIEE Fellow 1928.

FREED, JOSEPH D.R., IRE Fellow 1928.

FREEDMAN, W.H., AIEE Fellow 1913.

FREEHAFER, CARL R., AIEE Fellow 1954 "For inspiring leadership contributing to effective solution of difficult problems in communication engineering."

FRIEDLANDER, ERICH S., IEEE Fellow 1973 "For services in the field of transmission, and for the design and development of magnetically saturating control and stabilizing devices."

FRIEND, A.W., IRE Fellow 1954 "For his contributions to tropospheric echo research, and to the development of magnetic materials and circuits."

FRIES, JOENS ELIAS, AIEE Fellow 1915.

FRIIS, HARELD TRAP, IRE Fellow 1929. AIEE Fellow 1941.

FRITZ, LORENZ J., AIEE Fellow 1949.

FRUTH, HAL F., IEEE Fellow 1966 "For contributions in the application of electro-chemistry to electronic engineering and active participation of professional affairs."

FRY, WILLIAM J., IEEE Fellow 1968 "For contributions to biophysics and neurological research by means of ultrasonics technologies."

FUGILL, PERCIVAL A., AIEE Fellow 1955 "For contributions to overall planning and design of a large electric utility system of metalclad equipment, control of noise, and application of computing devices."

FUNK, NEVIN E., AIEE Fellow 1934.

FURST, WALTER A., AIEE Fellow 1951.

FYNN, VALERIE, AIEE Fellow 1912.

G

GABY, FREDERICK A., AIEE Fellow 1918.

GAFFORD, BURNS N., AIEE Fellow 1956 "For his contributions as a teacher and administrator in a University and for his promotion of a better understanding of utility power problems."

GAHAM, EARL A., AIEE Fellow 1920.

GAILLARD, LAWRENCE LEE, AIEE Fellow 1913.

GAKLE, WENDOLYN F., AIEE Fellow 1955 "For pioneering developments in transformer design; particularly the use of grain-oriented steel in cores."

GALBRAITH, RALPH A., IEEE Fellow 1965 "For contributions to engineering education."

GALINDO, VICTOR, IEEE Fellow 1977 "For contributions to the theory and analysis of phased array and reflector antennas."

GALUSHA, DON L., AIEE Fellow 1919.

GAMBLE, LESTER RAYMOND, AIEE Fellow 1939.

GANNETT, DANFORTH K., IRE Fellow 1959 "For contributions to the transmission of television and sound broadcasting signals."

GANZ, ALBERT, AIEE Fellow 1912.

GARDETT, HELMUTH C., AIEE Fellow 1941.

GARDNER, MURRAY F., AIEE Fellow 1951.

GARFIELD, A.S., AIEE Fellow 1913.

GARIN, ALEXIS NICHOLAS, AIEE Fellow 1949.

GARMAN, CHARLES PENROSE, AIEE Fellow 1946.

GARMAN, RAYMOND L., IRE Fellow 1958 "For his leadership in research and development of electronic devices."

GASTON, JOHN RAOUL, AIEE Fellow 1951.

GATEWOOD, ARTHUR R., AIEE Fellow 1954 "For his contribution to the development and application of national and international standards relating to the installation of electrical equipment on shipboard."

GATY, LEWIS R., AIEE Fellow 1950.

GAULT, JAMES SHERMAN, AIEE Fellow 1946.

GAY, FRAIZER WALKER, AIEE Fellow 1932.

GAY, GEORGE F., AIEE Fellow 1921.

GAYLORD, JAMES MASON, AIEE Fellow 1935.

GEAR, HARRY B., AIEE Fellow 1920.

GEAR, ROBERT S., AIEE Fellow 1955 "For contributions to the development of power generation and distribution in a large area."

GEBHARD, LOUIS A., IRE Fellow 1949 "In recognition of his outstanding contributions to radio communication, radar, radio navigation, and radio countermeasures, and of his organization and direction of naval research and development in the fields."

GEIGER, JAMES MARION, AIEE Fellow 1950.

GEORGE, EVERETT E., AIEE Fellow 1936.

GEORGE, ROSCOE H., IEEE Fellow 1966 "For pioneering work in cathode-ray tubes and television systems."

GERELL, GORDON W., AIEE Fellow 1954 "For contributions to the improvement of electric service through advances in the application of protective relaying, and in preventive maintenances of oil insulated apparatus."

GERRY, M.H., JR., AIEE Fellow 1913.

GERSZONOWICS, SEGISMUNDO, AIEE Fellow 1950.

GEYGER, WILLIAM A., AIEE Fellow 1960 "For contributions in the field of electrical measurements and magnetic amplifiers."

GHEITH, FATHY, IEEE Fellow 1966 "For contributions to telecommunications and broadcasting in Egypt, Syria, Ethiopia, and the Middle East."

GHEN, MELVILLE W., AIEE Fellow 1948.

GHERARDI, BANCROFT, AIEE Fellow 1912.

GHERKY, WILLIAM, AIEE Fellow 1913.

GIBBS, GEORGE, AIEE Fellow 1912.

GIBBS, H.P., AIEE Fellow 1913.

GIBBS, JESSE BERTHOLD, AIEE Fellow 1962 "For contributions to design of instrument transformers."

GIBSON, JOHN J., AIEE Fellow 1918.

GIDLUND, HILDING FERDINAND, AIEE Fellow 1960 "For contributions to development and operation of a large electric distribution system."

GIERING, PERCIVAL L., AIEE Fellow 1951.

GIFFORD, R.P., IEEE Fellow 1964 "For leadership in communications techniques and practice."

GILBERT, JOHN JOSEPH, AIEE Fellow 1951.

GILES, MERRITT A., AIEE Fellow 1962 "For contributions to construction and operating methods of an electric power system."

GILL, A.J., IRE Fellow 1940.

GILL, F., AIEE Fellow 1913.

GILL, LESTER W., AIEE Fellow 1921.

GILLE, WILLIS HOWARD, AIEE Fellow 1958 "For contributions to the field of electrical control mechanisms, particularly for flight control."

GILLESPIE, FONTAINE MAURY, AIEE Fellow 1938.

GILLETT, GLENN D., IRE Fellow 1934.

GILMAN, FRANCIS L., AIEE Fellow 1913.

GILSON, WESLEY J., AIEE Fellow 1950.

GILT, CARL M., AIEE Fellow 1935.

GIROUX, CARL HUNTING, AIEE Fellow 1944.

GLASCOW, R.S., IRE Fellow 1955 "For his contributions to the field of engineering education."

GLASGOW, ROY S., AIEE Fellow 1948.

GLASSCO, JOHN GIRDLESTONE, AIEE Fellow 1928.

GLENTZER, KENNETH VERLEY, AIEE Fellow 1950.

GLINSKI, GEORGE S., IRE Fellow 1960 "For contributions to the field of computation."

GOAR, ROY MITCHELL, AIEE Fellow 1962 "For executive guidance as well as personal technical contributions to the design of electric generating stations."

GODLEY, P.F., IRE Fellow 1961 "For pioneering in the short wave art."

GODSHO, A.P., AIEE Fellow 1953 "For distinguished service in contributing significantly to the high standard of performance of the telephone and radio facilities serving an important area of the nation."

GOETZENBERGER, RALPH LEON, AIEE Fellow 1948.

GOLDMARK, PETER C., IRE Fellow 1942 "For his contribution to the development of practical color television." AIEE Fellow 1954 "For contributions to television and to phonographic recording."

GOLDSBOROUGH, SHIRLEY LEROY, AIEE Fellow 1958 "For contributions to the art of high speed relaying and telemetering of power systems."

GOLDSBOROUGH, W.E., AIEE Fellow 1912.

GOLDSCHMIDT, RUDOLPH, IRE Fellow 1915.

GOLDSMITH, ALFRED N., IRE Fellow 1915. AIEE Fellow 1920.

GOLDUP, T.E., IRE Fellow 1955 "For his pioneering achievements in the design and development of thermionic tubes and his contributions to the technical and administrative counsels of the British radio industry."

GONSETH, JULES EMMABLE, AIEE Fellow 1945.

GOODALE, JOSIAH ELMER, AIEE Fellow 1934.

GOODMAN, W.G.T., AIEE Fellow 1913.

GOODWIN, HAROLD, JR., AIEE Fellow 1923.

GOODWIN, VICTOR EARL, AIEE Fellow 1948.

GOODWIN, W.N., JR., AIEE Fellow 1913. IRE Fellow 1957 "For contributions in the field of electrical measuring instruments."

GORDON, WILLIAM GORDON, AIEE Fellow 1916.

GORHAM, J.E., IRE Fellow 1954 "For his leadership and contributions in the field of high-power electron tubes for military applications."

GOSS, FLOYD L., IEEE Fellow 1969 "For contributions to regional power system design and operation."

GOSSLER, P.G., AIEE Fellow 1913.

GOUBAU, GEORG, IRE Fellow 1957 "For contributions to the field of microwave transmission."

GOULD, EDWUARD F., AIEE Fellow 1913.

GOULD, GORDON T., JR., IEEE Fellow 1969 "For contributions and leadership in the development of command, control, and communications systems."

GRABBE, EUGENE M., IEEE Fellow 1966 "For contributions in the field of automation and the control application of digital computers."

GRACE, SERGIUS P., AIEE Fellow 1921.

GRAF, A.W., IRE Fellow 1955 "For his contributions to the radio engineering profession."

GRAFF, JOSEPH WILFORD, AIEE Fellow 1959 "For contributions to power system planning."

GRAFTIO, HENRY, AIEE Fellow 1913.

GRAHAM, EARL ADDISON, AIEE Fellow 1920.

GRAHAM, VIRGIL M., IRE Fellow 1935.

GRAHAM, WILLIAM P., AIEE Fellow 1923.

GRANFIELD, THOMAS H., AIEE Fellow 1955 "For skill in handling the engineering problems of an important communications enterprise, and for leadership in the development and great enlargement of its engineering organization."

GRAY, A.R., IRE Fellow 1961 "For contributions to reliability."

GRAY, CLYDE D., AIEE Fellow 1920.

GRAY, ROY WILLIAM, AIEE Fellow 1928.

GRAY, WILLARD F., IEEE Fellow 1969 "For leadership in engineering education and research, and for contributions to university administration."

GRAYBILL, KENNETH WAYNE, AIEE Fellow 1945.

GREAVES, V. FORD, IRE Fellow 1928.

GREBE, ALFRED H., IRE Fellow 1927.

GREEN, ALFRED L., IRE Fellow 1938.

GREEN, CHARLES F., AIEE Fellow 1961 "For contributions to precise control in the field of aviation, marine and space aircraft."

GREEN, CHARLES WILLIAM, AIEE Fellow 1940.

GREEN, ESTILL I., AIEE Fellow 1946. IRE Fellow 1955 "For his contributions in the development of communication systems and apparatus components."

GREEN, STANLEY, AIEE Fellow 1946.

GREENIDGE, CHARLES A., AIEE Fellow 1913.

GREER, LANIER, AIEE Fellow 1953 "For outstanding contributions to the design of direct-current motors, as a result of which their utility in railway and industrial applications has been significantly advanced."

GREIBACH, E.H., IRE Fellow 1961 "For contributions to high sensitivity measuring instruments."

GREIG, JAMES, IEEE Fellow 1968 "For his contributions in the educational field, and research on the behavior of magnetic materials, computers and automatic control processes."

GRIEMSMANN, JOHN W.E., IRE Fellow 1959 "For contributions to microwave research."

GRIER, LOUIS N., AIEE Fellow 1954 "For his contributions to the design and development of electric power supply systems in the aluminum industry and for improvements in the application of electric power to the fabrication of aluminum."

GRIFFITH, HARVEY CONRAD, AIEE Fellow 1945.

GRIGSBY, LOGAN C., AIEE Fellow 1945.

GRIMSLEY, ANDREW HOWARD, AIEE Fellow 1949.

GRISCOM, SAMUEL B., AIEE Fellow 1948.

GRISSINGER, ELWOOD, AIEE Fellow 1914.

GRISSINGER, GEORGE GORDON, AIEE Fellow 1948.

GROENEVELD MEIJER, NICOLAAS E., AIEE Fellow 1933.

GRONDAHL, LARS OLAI, AIEE Fellow 1947.

GROSS, GERALD, IRE Fellow 1957 "For contributions to international regulation of telecommunication."

GROSS, ISAAC WATTS, AIEE Fellow 1945.

GROSSMAN, ALEXANDER J., IRE Fellow 1962 "For contributions to circuit design."

GROVES, WILLIAM M., JR., AIEE Fellow 1945.

GRUEHR, ANATOLE R., AIEE Fellow 1957 "For his application of economic theory to engineering design."

GRUEN, W.J., IRE Fellow 1957 "For contributions to the improvement of television receivers."

GUANELLA, GUSTAV, IEEE Fellow 1965 "For contributions to the electronics art and for leadership in research."

GUBIN, SAMUEL, IEEE Fellow 1970 "For contributions to communications technology for ground-based and space applications."

GUENARD, P.R., IRE Fellow 1955 "For his scientific and technical contributions in the field of microwave tubes."

GUILLEMIN, ERNST ADOLPH, IRE Fellow 1949 "For outstanding work in the field of electric circuit analysis and synthesis, and for his inspired leadership as a teacher." AIEE Fellow 1951.

GUNBY, FRANK MCCLELLAN, AIEE Fellow 1948.

GUNDLACH, WILLIAM E., AIEE Fellow 1936.

GUNN, ROSS, IRE Fellow 1949 "For his long service and many technical contributions in the radio and electronics fields."

GUPTA, BIRENDRA CHANDRA, AIEE Fellow 1923.

GUSTAFSON, GILBERT E., IRE Fellow 1940.

GUTMANN, LUDWIG, AIEE Fellow 1913.

GUY, RAYMOND F., IRE Fellow 1939. AIEE Fellow 1959 "For contributions to the Technical development of radio and television network broadcasting."

GUYER, EDWIN MICHAEL, AIEE Fellow 1962 "For contributions in the application of electricity to glass working."

H

HAANTJES, JOHANNES, IEEE Fellow 1965 "For contributions to research and development, and the standardization of television systems in Europe."

HACKBUSH, RALPH A., IRE Fellow 1937.

HADAWAY, W.S., JR., AIEE Fellow 1913.

HAGENGUTH, JULIUS H., AIEE Fellow 1951.

HAGGERTY, PATRICK E., IRE Fellow 1958 "For leadership in the advancement of semi-conductor devices."

HAGLUND, HAKON H., AIEE Fellow 1954 "For contributions to the mechanization of ocean telegraph cable operation."

HAINES, THOMAS HENRY, AIEE Fellow 1944.

HALBERG, MAYNARD N., AIEE Fellow 1951.

HALDANE, THOMAS GRAEME NELSON, AIEE Fellow 1950.

HALL, CHESTER I., AIEE Fellow 1920.

HALL, DAVID D.C., AIEE Fellow 1918.

HALL, HARRY Y., AIEE Fellow 1927.

HALL, N.I., IRE Fellow 1956 "For contributions to the measurement of the velocity of radio waves and the design of radar and guidance systems."

HALL, WALTER A., AIEE Fellow 1912.

HALL, WAYNE CURTIS, AIEE Fellow 1958 "For notable contributions to improvement in economy of electric power generation."

HALL, WESLEY B., AIEE Fellow 1947.

HALLBORG, HENRY EMMANUEL, IRE Fellow 1927. AIEE Fellow 1929.

HALMAN, THEODORE RITSON, AIEE Fellow 1960 "For contributions to protective relaying for electrical system performance."

HALPENNY, ROBERT HARIAN, AIEE Fellow 1942.

HALPERT, PERCY, IEEE Fellow 1967 "For contributions to the theory and design of automatic controls and instrumentation for aircraft."

HAMANN, ADOLPH MARTIN, AIEE Fellow 1934.

HAMDI, ABDULLAH FEYZI, AIEE Fellow 1946.

HAMILTON, GEORGE A., AIEE Fellow 1913.

HAMILTON, GEORGE WELLINGTON, AIEE Fellow 1929.

HAMILTON, HAROLD CYRLE, AIEE Fellow 1945.

HAMILTON, JAMES HUGH, AIEE Fellow 1951.

HAMILTON, ROBERT FRANK, AIEE Fellow 1946.

HAMMER, EDWIN W., AIEE Fellow 1912.

HAMMER, OWEN S.C., AIEE Fellow 1962 "For contributions to planning, development and engineering of a large transmission system."

HAMMOND, EDMUND B., JR., IEEE Fellow 1973 "For contributions to the field of gun and missile armament systems and aircraft navigation instruments."

HAMMOND, JOHN H., JR., IRE Fellow 1959 "For pioneering work in radio control and communication."

HAMMOND, W.P., AIEE Fellow 1950.

HANCHETT, GEORGE T., AIEE Fellow 1913.

HANCOCK, MYRON SCOTT, AIEE Fellow 1950.

HANSELL, CLARENCE WESTON, IRE Fellow 1945 "For his pioneer work in the development and application of equipment for the ever higher frequencies employed for radio communication." AIEE Fellow 1951.

HANSEN, EDMUND H., IRE Fellow 1959 "For contributions to the development of motion picture sound recording and reproduction."

HANSEN, KLAUS LOBECK, AIEE Fellow 1934.

HANSEN, WILLIAM W., IRE Fellow 1947 "In recognition of his many contributions to the theory of tubes, networks, circuits, and antennas."

HANSON, O.B., IRE Fellow 1941.

HANSSEN, I.E., AIEE Fellow 1913.

HANSSON, EDWIN, AIEE Fellow 1945.

HAPGOOD, KENNETH ELLSWORTH, AIEE Fellow 1944.

HARDAWAY, WARREN DUNHAM, AIEE Fellow 1951.

HARDIN, LEWIS H., AIEE Fellow 1954 "For achievement in the electrical design of large hydro electric generating stations and associated transmission lines."

HARDING, ARTHUR L., AIEE Fellow 1931.

HARDING, CHARLES FRANCIS, AIEE Fellow 1914.

HARDING, WILLIAM KNOWLTON, AIEE Fellow 1961 "For contributions to the design and construction of transmission lines and substations."

HARDY, HOWARD C., IRE Fellow 1960 "For contributions to electroacoustics."

HARMAN, WILLIS W., IEEE Fellow 1965 "For contributions to engineering education in the fields of electron dynamics, linear systems and communication theory."

HARNETT, DANIEL EDWARD, IRE Fellow 1942 "In recognition of his constructive leadership and direction of radio engineering projects." AIEE Fellow 1955 "For engineering leadership in peacetime and military radio design and for contributions to standardization in monochrome and color television."

HARPER, H.R., AIEE Fellow 1913.

HARPER, J.L., AIEE Fellow 1913.

HARRELL, FREDERICK EDMUND, AIEE Fellow 1940.

HARRER, JOSEPH M., IEEE Fellow 1967 "For contributions to the development of nuclear reactor control technology and experiments with boiling water reactors leading to power plant applications."

HARRIES, GEORGE H., AIEE Fellow 1922.

HARRINGTON, CHARLES A., AIEE Fellow 1939.

HARRINGTON, HOWARD L., AIEE Fellow 1950.

HARRIS, D.B., IRE Fellow 1956 "For contributions to telephonic communication practices and to the organization of electronic research and development."

HARRIS, FORD W., AIEE Fellow 1913.

HARRIS, W. A., IRE Fellow 1955 "For his contributions to the development of frequency converter tubes and to the understanding of fluctuation phenomena in electronic tubes."

HARRISON, A.E., IRE Fellow 1955 "For his contributions as a teacher, author, and engineer, especially in the field of klystrons."

HARRISON, ARTHUR MCCREDIE, AIEE Fellow 1958 "For contributions in the field of large rotating electrical machinery."

HARRISON, JAMES, AIEE Fellow 1925.

HARRISON, WARD, AIEE Fellow 1936.

HARRISON, WILLIAM H., AIEE Fellow 1931.

HART, HARRY U., AIEE Fellow 1913.

HART, LESTER C., AIEE Fellow 1949.

HART, PERCY E., AIEE Fellow 1925.

HARTE, CHARLES RUFUS, AIEE Fellow 1946.

HARTIG, HENRY EDWARD, AIEE Fellow 1948. IRE Fellow 1952 "For his achievements as a teacher, his research in the field of acoustics, and his contributions to the underwater sound program during World War II."

HARTLEY, RALPH VINTON LYON, IRE Fellow 1928. AIEE Fellow 1951.

HARVEY, HAROLD GIFFORD, AIEE Fellow 1933.

HARVEY, HENRY FRANKLIN, JR., AIEE Fellow 1949.

HARWOOD, PAISLEY B., AIEE Fellow 1942.

HASTINGS, CHARLES E., IRE Fellow 1961 "For contributions to radio surveying and navigation." AIEE Fellow 1962 "For development of radar equipment for rapid measurement of distances with great accuracy."

HAUSMANN, ERICK, AIEE Fellow 1918.

HAVILL, O.A., AIEE Fellow 1916.

HAWLEY, JOHN T., AIEE Fellow 1955 "For contributions to the application of electricity to the sanitary system of a large area."

HAWLEY, KENT ALLEN, AIEE Fellow 1935.

HAYES, HAMMOND V., AIEE Fellow 1913.

HAYES, LESLIE W., IRE Fellow 1938.

HAYES, STEPHEN Q., AIEE Fellow 1912.

HAYLER, GEORGE ERNEST, JR., AIEE Fellow 1931.

HAYWARD, R.J., AIEE Fellow 1912.

HAZELTINE, HAROLD L., AIEE Fellow 1940.

HAZELTINE, LOUIS A., IRE Fellow 1921. AIEE Fellow 1921.

HAZEN, HAROLD LOCKE, AIEE Fellow 1943. IRE Fellow 1962 "For contribution in the field of servomechanism theory and in education."

HEADRICK, LEWIS B., IRE Fellow 1953 "For contributions and leadership in the development of cathode-ray tubes and television camera tubes."

HEATH, H.E., AIEE Fellow 1913.

HEATHER, H.J.S., AIEE Fellow 1913.

HECTOR, L.G., IRE Fellow 1961 "For contributions in the field of electronic devices."

HEFELE, JOHN R., IEEE Fellow 1965 "For contributions to research and engineering on visual communica-

tions."

HEFFELMAN, MALCOLM C., AIEE Fellow 1962 "For contributions to the development of a integrated power system in sparse territory."

HEFFNER, HUBERT, IRE Fellow 1962 "For contributions to the theory of backward wave oscillators and parametric amplifiers."

HEHRE, FREDERICK W., AIEE Fellow 1929.

HEIDENREICH, ALLAN H., AIEE Fellow 1927.

HEIDENREICH, GEORGE EDWARD, SR., AIEE Fellow 1950.

HEINTZ, RALPH M., IRE Fellow 1930. AIEE Fellow 1945.

HEINZE, CARL A., AIEE Fellow 1926.

HEISING, RAYMOND A., IRE Fellow 1923. AIEE Fellow 1939.

HEITMANN, EDWARD, AIEE Fellow 1913.

HEITZLER, ALBERT H., AIEE Fellow 1938.

HEJDA, CHARLES JOSEPH, AIEE Fellow 1944.

HELLMUND, R.E., AIEE Fellow 1913.

HELPBRINGER, J., AIEE Fellow 1926.

HENDERSON, R.H., AIEE Fellow 1913.

HENDERSON, SAMUEL FARRELL, AIEE Fellow 1960 "For contributions to the improvement of induction motors."

HENDRICKS, CHESTER I., AIEE Fellow 1955 "For contributions to the design of electric distribution systems."

HENLINE, HENRY HARRISON, AIEE Fellow 1943.

HENNINGSEN, EARLE S., AIEE Fellow 1939.

HENSHAW, A.W., AIEE Fellow 1913.

HENTZ, ROBERT A., AIEE Fellow 1929.

HERBST, PHILIP J., IRE Fellow 1953 "For his contributions to numerous branches of radio engineering, particularly television and air navigation."

HERDT, L.A., AIEE Fellow 1912.

HERGENROTHER, R.C., IRE Fellow 1956 "For contributions in electron optics and storage tube development."

HERING, CARL, AIEE Fellow 1912.

HERLITZ, IVAR, AIEE Fellow 1956 "For distinguished contributions in the field of high voltage engineering."

HERRICK, CHARLES H., AIEE Fellow 1912.

HERRON, JAMES HERVEY, AIEE Fellow 1941.

HERTOG, GERHARD, IRE Fellow 1955 "For his contributions to radioactive instrumentation for geological survey and medical application."

HERTZ, STANTON S., AIEE Fellow 1939.

HERTZOG, EDWARD F., IRE Fellow 1959 "For contributions to military electronics."

HERZ, ALFRED, AIEE Fellow 1921.

HERZOG, EUGENE, AIEE Fellow 1948.

HESS, WENDELL F., AIEE Fellow 1948.

HESS, WILLIAM T., AIEE Fellow 1956 "For his contributions in the development of electric power production, transmission and distribution systems."

HESSEL, JOHN, IRE Fellow 1953 "For his contributions to military radio communications, particularly in the fields of man-carried combat radios and radio relay equipment."

HESSLER, VICTOR PETER, AIEE Fellow 1943.

HESTER, EDGAR ALLEN, AIEE Fellow 1947.

HEWLETT, E.M., AIEE Fellow 1917.

HEYD, J.W., IRE Fellow 1961 "For contributions to the application of atomic energy."

HIBBARD, ANGUS S., AIEE Fellow 1912.

HIBBARD, HARRY L., AIEE Fellow 1921.

HIBBARD, TRUMAN, AIEE Fellow 1926.

HIBBEN, SAMUEL GALLOWAY, AIEE Fellow 1961 "For contributions to advancement of the theory and practice of illumination engineering."

HICKERNELL, LATIMER FARRINGTON, AIEE Fellow 1934.

HICKMAN, R.W., IRE Fellow 1961 "For contributions to electronics education."

HICKOK, ROBERT D., AIEE Fellow 1932.

HIGBIE, HENRY HAROLD, AIEE Fellow 1943.

HIGGINS, NATHAN BERT, AIEE Fellow 1948.

HIGHFIELD, JOHN SOMERVILLE, AIEE Fellow 1942.

HILL, ALLAN W., AIEE Fellow 1948.

HILL, ARTHUR P., AIEE Fellow 1957 "For contributions to the development o wire and radio communications, sound motion pictures and accoustical technology."

HILL, CHARLES FRANKLIN, AIEE Fellow 1939.

HILL, E. ROWLAND, AIEE Fellow 1912.

HILL, FREDERICK WILLIAM LANDIS, NDIS, AIEE Fellow 1934.

HILL, G. LESLIE, IEEE Fellow 1965 "For contributions to power circuit protection and the testing of electrical insulation."

HILL, GUY, IRE Fellow 1915.

HILL, HALBERT P., AIEE Fellow 1922.

HILL, LELAND H., AIEE Fellow 1938.

HINES, CLAUDE MANNING, AIEE Fellow 1962 "For achievements in the development anad design of control equipment in the field of transportation."

HINNERS, FRANK A., IRE Fellow 1926.

HINSON, NOEL BERTRAM, AIEE Fellow 1935.

HIRSCH, C.J., IRE Fellow 1951 "For his contributions in the fields of radio broacast receivers, electronic aids to navigation, and electronic computers."

HIRSHFELD, CLARENCE FLOYD, AIEE Fellow 1936.

HOBART, H.M., AIEE Fellow 1912.

HOBART, K.E., AIEE Fellow 1926.

HOBBINS, WILLIAM D., AIEE Fellow 1924.

HOBBS, MAURICE HILL, AIEE Fellow 1948.

HOBSON, JESSE EDWARD, AIEE Fellow 1948. IRE Fellow 1956 "For leadership in organization for research."

HOCKADAY, OLIN S., AIEE Fellow 1944.

HODNETTE, JOHN K., AIEE Fellow 1942.

HODSON, HERBERT ORVILLE, AIEE Fellow 1962 "For contributions to the engineering, construction, and operations of a rapidly growing electric utility system."

HODTUM, JOSEPH B., AIEE Fellow 1949.

HOEFLE, ALOIS, AIEE Fellow 1957 "For his contributions to the restoration of electric power in Europe while on occupational duty with the Army of the United States and for the application of higher distribution voltages to the Toledo Edison System."

HOEPPNER, HENRY LOUIS, AIEE Fellow 1943.

HOGAN, JOHN V.L., JR., IRE Fellow 1915, AIEE Fellow 1954 "For his pioneer accomplishments in research, development, and engineering in the field of radio communications."

HOGG, THOMAS HENRY, AIEE Fellow 1938.

HOK, GUNNER, IRE Fellow 1957 "For contributions to electronic science and education."

HOLBROOK, HENRY STANLEY, AIEE Fellow 1931.

HOLCOMBE, ERNEST S., AIEE Fellow 1927.

HOLDEN, E.B., JR., AIEE Fellow 1913.

HOLLAND, LEWIS NELSON, IRE Fellow 1959 "For contributions to engineering education." AIEE Fellow 1960 "For contributions to electrical engineering curricula especially the incorporation of UHF pedagody."

HOLLAND, MELVIN G., IEEE Fellow 1976 "For contributions to the field of surface and bulk acoustic wave devices and technology."

HOLLAND, WALTER E., IRE Fellow 1936.

HOLLAND, WAYMAN A., AIEE Fellow 1949.

HOLLIDAY, THEODORE BERGEN, AIEE Fellow 1946.

HOLLISTER, FRANCIS HIEL, AIEE Fellow 1943.

HOLLMANN, HANS E., IRE Fellow 1953 "For his fundamental analyses and inventions in the fields of electron tubes and high-frequency circuits."

HOLMAN, HAROLD RUSSELL, AIEE Fellow 1949.

HOLMES, HOWARD A., AIEE Fellow 1939.

HOLMES, RALPH S., IRE Fellow 1952 "In recognition of his early contributions to the development of television and his work in the field of television standards."

HOLSLAG, CLAUDE JOSEPH, AIEE Fellow 1941.

HOLTZ, FREDERICK CARL, AIEE Fellow 1947.

HOMAN, JOHN GREENLEAF, AIEE Fellow 1940.

HONAMAN, RICHARD K., AIEE Fellow 1936.

HOOKE, ROBERT GAY, AIEE Fellow 1951.

HOOPER, S.C., IRE Fellow 1928.

HOOPER, WILLIAM L., AIEE Fellow 1912.

HOOVEN, MORRIS D., AIEE Fellow 1944.

HOPE, HARRY M., AIEE Fellow 1912.

HOPKINS, NEVIL MONROE, AIEE Fellow 1913.

HOPKINS, RAYMOND A., AIEE Fellow 1943.

HORELICK, SAMUEL, AIEE Fellow 1948.

HORLE, LAWRENCE C.F., IRE Fellow 1925. AIEE Fellow 1935.

HORN, CHARLES W., IRE Fellow 1930.

HORN, HEINZ, IRE Fellow 1959 "For contributions in the development of ultra-high-frequency and submarine cables."

HORNER, S. GEORGE L., IRE Fellow 1957 "For contributions to radio communications in Canadian Northern and Arctic regions."

HORNFECK, ANTHONY J., AIEE Fellow 1955 "For contributions to telemetering and to instrumentation for process and generating station control."

HORNOR, H.A., AIEE Fellow 1913.

HORTON, ARTHUR W., JR., IRE Fellow 1959 "For contributions to long distance telephony, electronic switching and military electronics."

HORTON, J. WARREN, AIEE Fellow 1928. IRE Fellow 1929.

HOSMER, SIDNEY, AIEE Fellow 1912.

HOUCK, HAROLD EDGAR, AIEE Fellow 1959 "For contributions to the electrical aspects of industrial, chemical and nuclear projects."

HOUGH, EUGENE LAWRENCE, AIEE Fellow 1948.

HOUSEKEEPER, WILLIAM G., AIEE Fellow 1925.

HOUSLEY, J. ELMER, AIEE Fellow 1943.

HOVEY, LINDSAY AM., IEEE Fellow 1970 "For contributions in hydro-speed governors."

HOWARD, DAVID GOODALE, AIEE Fellow 1961 "For his contribution to the education in electrical engineering of the midshipmen of the U.S. Naval Academy."

HOWE, G.W.O., IRE Fellow 1955 "For his pioneering work in radio and his outstanding contributions to engineering education."

HOWE, K.L., AIEE Fellow 1936.

HOWE, WILFRED H., IEEE Fellow 1965 "For pioneering development of industrial process instrumentation, especially in the paper, rubber and plastics field."

HOWELL, JOHN W., AIEE Fellow 1912.

HOWELLS, PAUL W., IRE Fellow 1960 "For contributions to color television."

HOWES, ROBERT, AIEE Fellow 1913.

HOWLAND, LEWIS A., AIEE Fellow 1913.

HOXIE, GEORGE L., AIEE Fellow 1913.

HOYT, GERALD A., AIEE Fellow 1961 "For contributions to naval ordnance torpedoes and missiles."

HUBBARD, HORACE STIMPSON, AIEE Fellow 1948.

HUBLEY, GEORGE W., AIEE Fellow 1926.

HUDSON, RALPH G., AIEE Fellow 1941.

HUFFMAN, DAVID A., IRE Fellow 1962 "For contributions to switching theory and information theory."

HUGHES, CALVIN T., AIEE Fellow 1946.

HUGHES, MARTIN C., AIEE Fellow 1943.

HULL, A.W., IRE Fellow 1944 "In recognition of his many contributions to the development and design of electron tubes both for radio and industrial applications."

HULL, ARTHUR HENRY, AIEE Fellow 1943.

HULL, BLAKE D., AIEE Fellow 1939.

HULL, JOHN IRVING, AIEE Fellow 1943.

HULL, JOSEPH F., IRE Fellow 1962 "For contributions to the advancement of the art of crossed field microwave devices."

HULL, LEWIS M., IRE Fellow 1928. AIEE Fellow 1954 "For his many contributions to radio communication and to the utilization of vacuum tubes therein."

HUMPHREY, C.W., AIEE Fellow 1913.

HUMPHREY, HENRY H., AIEE Fellow 1913.

HUMPHREYS, ALEX C., AIEE Fellow 1912.

HUMPHREYS, CLIFFORD W., AIEE Fellow 1913.

HUND, AUGUST, IRE Fellow 1927.

HUNT, FRED L., AIEE Fellow 1917.

HUNT, FRED V., IRE Fellow 1947 "For his work as a scientist, teacher, and administrator and his contributions to acoustics."

HUNT, HENRY JAMES, AIEE Fellow 1962 "For contributions to improvements in the design of hydroelectric plants."

HUNT, JOHN HERMAN, AIEE Fellow 1943.

HUNTER, EUGENE MURRAY, AIEE Fellow 1948.

HUNTER, PHILIP VASSAR, AIEE Fellow 1933.

HUNTER, THEODORE A., IRE Fellow 1953 "For his contributions to the design of stable tunable oscillators, and for his able promotion of the activities of the Institute."

HUNTING, F.S., AIEE Fellow 1913.

HUNTINGTON, D.L., AIEE Fellow 1913.

HUNTLEY, HAROLD R., AIEE Fellow 1951. IRE Fellow 1960 "For contributions to communications engineering."

HURD, OWEN WILSON, AIEE Fellow 1958 "For contributions to design, operation and management of large electric power systems."

HUTCHINSON, CARY T., AIEE Fellow 1912.

HUTCHINSON, CHASE, AIEE Fellow 1951.

HUTTON, C.W., AIEE Fellow 1913.

HYDE, HARRY, AIEE Fellow 1962 "For contributions to the technical and management phases of a large municipal electrical system."

HYNES, FRANCIS B., AIEE Fellow 1931.

I

IDE, CHARLES EDWARD, AIEE Fellow 1951.

IHLDER, JOHN D., AIEE Fellow 1912.

IMLAY, L.E., AIEE Fellow 1913.

INCH, SYDNEY RICHARD, AIEE Fellow 1938.

INGLEDOW, THOMAS, AIEE Fellow 1951.

INGLES, HARRY C., AIEE Fellow 1948.

INGVALDSEN, BERNT I., IEEE Fellow 1968 "For his prominent roles in developing and expanding the electrical industry, and in serving his government."

INSULL, SAMUEL, AIEE Fellow 1912.

ISHIKAWA, KIYOSHI, IEEE Fellow 1965 "For contributions to electrical insulating techniques and materials, and to the development of the atomic energy industry."

ISRAEL, DORMAN, IRE Fellow 1942 "In recognition of his leadership in the engineering of broadcast receivers."

IVES, HERBERT E., AIEE Fellow 1929.

IVES, JAMES E., IRE Fellow 1915.

IWAMI, KAZUO, IEEE Fellow 1982 "For leadership in applying electronic technology to consumer products."

J

JACK, HUGH, AIEE Fellow 1949.

JACKSON, DUGALD C., AIEE Fellow 1912.

JACKSON, DUGALD C., JR., AIEE Fellow 1930.

JACKSON, JOHN PRICE, AIEE Fellow 1912.

JACKSON, WILLIAM B., AIEE Fellow 1913.

JACKSON, WILLIAM E., IRE Fellow 1951 "For his technical contributions to electronic aids to air navigation, and control, and for his administrative contribution in the Civil Aeromautics Administration."

JACKSON, WILLIS, IRE Fellow 1950 "For his services as an educator and for his many contributions to the literature in both the radio and electrical fields."

JACOBS, ERNEST H., AIEE Fellow 1917.

JAFFE, HANS, IRE Fellow 1962 "For contributions in the field of pieoelectric-crystals and piezoelectric-ceramic materials."

JAFFE, J.L., IEEE Fellow 1964 "For contributions to the development of microwave electronic equipment."

JAMES, HENRY D., AIEE Fellow 1912.

JAMMIESON, BERTRAND, AIEE Fellow 1922.

JANES, R.B., IRE Fellow 1955 "For his contributions to the development of improved camera tubes."

JANSKY, C.M., JR., IRE Fellow 1928.

JANSKY, CYRIL M., AIEE Fellow 1932.

JANSKY, KARL G., IRE Fellow 1947 "For his researches in the realm of cosmic and circuit noise affecting radio communication."

JAQUES, CLOYCE A., AIEE Fellow 1951.

JARVIS, KENNETH WARNER, IRE Fellow 1934. AIEE Fellow 1950.

JASIK, HENRY, IRE Fellow 1958 "For his contributions to the theory and design of VHF and microwave antennas."

JEFFERIES, ERNEST SMITH, AIEE Fellow 1924.

JEFFERS, C.L., IRE Fellow 1961 "For contributions to domestic and foreign broadcasting and to antenna development."

JEFFREY, FRASER, AIEE Fellow 1924.

JEFFRIES, ZAY, AIEE Fellow 1942.

JENNY, DIETRICH A., IRE Fellow 1962 "For research to the field of semiconductors."

JENSEN, J.C., IRE Fellow 1956 "For research in the field of static charges on aircraft and their relation to radio communication."

JENSON, AXEL G., IRE Fellow 1942 "For his contributions to the development of the short-wave transatlantic telephone, for his contributions to the development of broad-band, multichannel telephony, and for his further contributions to the art of television."

JETT, EWELL K., IRE Fellow 1939.

JEWETT, FRANK B., AIEE Fellow 1912. IRE Fellow 1920.

JOHANNESEN, SVEND E., AIEE Fellow 1913.

JOHN, KURT W., AIEE Fellow 1962 "For contributions to electrification in the rubber industry."

JOHNSON, CARL E., AIEE Fellow 1925.

JOHNSON, ERNEST A., AIEE Fellow 1958 "For his contributions to conversion of manual telephone systems to dial operations."

JOHNSON, ERNEST E., AIEE Fellow 1947.

JOHNSON, EWART FRANK, AIEE Fellow 1948.

JOHNSON, EWELL C., JR., IEEE Fellow 1967 "For contributions to the field of control systems technology, numerically controlled machine tool systems and automation."

JOHNSON, F.B., AIEE Fellow 1953 "For contribution to the improvement of commutation in rotary converters and to the development of remote control for urban substations."

JOHNSON, FRANCIS ELLIS, AIEE Fellow 1931.

JOHNSON, GEORGE EDWARD, AIEE Fellow 1959 "For contributions to power systems for coordinated operations under public management."

JOHNSON, HARWICK, IRE Fellow 1960 "For contributions to the development of electron devices."

JOHNSON, J. ALLEN, AIEE Fellow 1927.

JOHNSON, J. HUGO, AIEE Fellow 1945.

JOHNSON, J.B., IRE Fellow 1951 "For his discovery of the fundamental limiting noise in electric circuits and his early development in cathode ray tubes."

JOHNSON, ROYCE E., AIEE Fellow 1945.

JOHNSON, TOMLINSON F., JR., AIEE Fellow 1933.

JOHNSON, WILLIAM RODGERS, AIEE Fellow 1959 "For contributions to the development of major electric generating and transmission systems."

JOLLIFFE, CHARLES BYRON, IRE Fellow 1930. AIEE Fellow 1949.

JOLLIFFE, JOHN P., AIEE Fellow 1949.

JOLLYMAN, JOSIAH PICKARD, AIEE Fellow 1930.

JONES, ARTHUR LUCAS, AIEE Fellow 1938.

JONES, BASSETT, AIEE Fellow 1930.

JONES, BENSON M., AIEE Fellow 1942.

JONES, CHARLES RAMEY, AIEE Fellow 1948.

JONES, GEORGE H., AIEE Fellow 1912.

JONES, J. COLEMAN, AIEE Fellow 1948.

JONES, JOHN WANDESFORD, AIEE Fellow 1961 "For contributions to the basic development of a major electric generating and transmission system and to standards for rotating machines."

JONES, JOSEPH S., AIEE Fellow 1925.

JONES, LESTER L., IRE Fellow 1925.

JONES, RAY DEVERE, AIEE Fellow 1960 "For contributions to extended range and life of small motors, generators and dynamotors for militaty application."

JONES, REGINALD LAMONT, AIEE Fellow 1931.

JONES, ROBERT A., AIEE Fellow 1951.

JONES, THOMAS F., JR., IRE Fellow 1962 "For contributions to electrical engineering education."

JOSLIN, WILLIAM M., AIEE Fellow 1957 "For his development and inspiration of young electrical engineers and his engineering achievements in the development of a large electric utility system."

JUCARONE, NICHOLAS T., AIEE Fellow 1950.

JUHNKE, PAUL B., AIEE Fellow 1936.

JUMP, GEORGE H., AIEE Fellow 1946.

JUNKERSTELD, PETER, AIEE Fellow 1912.

K

KAEBURN, LESLIE K., IEEE Fellow 1965 "For his important work in medical electronics."

KAHLER, WILLIAM V., AIEE Fellow 1946.

KALB, WARREN C., AIEE Fellow 1940.

KALLMAN, HEINZ E., IRE Fellow 1959 "For contributions to transient response analysis of networks and to instrumentation."

KALMUS, HENRY P., IRE Fellow 1956 "For contributions in the field of electromechanical devices and electronic measurement instruments."

KAMMERMAN, JOHN OSCAR, AIEE Fellow 1950.

KAPRIELIAN, ZOHRAB A., IEEE Fellow 1971 "For contributions in the fields of electromagnetic theory and artificial dielectrics."

KARAPETOFF, VLADIMER, AIEE Fellow 1912.

KARLIN, JOHN E., IEEE Fellow 1965 "For contributions to the understanding of man-machine problems and their relation to telephony."

KAROLUS, AUGUST, IEEE Fellow 1970 "For work in television and facsimile-transission, in electro-optical methods for measuring distances, and in development of time and frequency standards."

KARPLUS, EDUARD, IRE Fellow 1938.

KARR, J. HAROLD, AIEE Fellow 1960 "For contributions to design of electric motors and automotive directional signals."

KARTAK, FRANZ AUGUST, AIEE Fellow 1933.

KASPAR, FRANK P., AIEE Fellow 1960 "For contributions to design and operating components of a large electric utility system."

KATLE, EDWIN B., AIEE Fellow 1912.

KATZIN, MARTIN, IRE Fellow 1955 "For his contributions to the knowledge of microwave propagation."

KAUFMAN, JACK, IEEE Fellow 1967 "For pioneering contributions and leadership in electron tube design, production and application particularly to communications."

KAUFMAN, RICHARD HAROLD, AIEE Fellow 1948.

KEATH, HOWARD BASCOMB, AIEE Fellow 1939.

KEENAN, GEORGE M., AIEE Fellow 1929.

KEHOE, ARTHUR H., AIEE Fellow 1925.

KEILLER, THOMAS MITCHELL, AIEE Fellow 1962 "For contributions in distribution planning and engineering to the developmemt of a rapidly growing utility."

KEINATH, GEORGE, AIEE Fellow 1961 "For contributions to instrumentation for measurement and in recording of physical quantities by electrical means."

KEITH, ALEXANDER E., AIEE Fellow 1913.

KEITH, FAY ELLSWORTH, AIEE Fellow 1956 "For his contributions to the engineering management field of the electrical construction industry."

KELLEHER, KENNETH S., IEEE Fellow 1969 "For contributions to microwave optical scanners and to the military electronics industry."

KELLER, ERNEST A., IRE Fellow 1960 "For contributions to sound recording and telephone switching systems."

KELLEY, WILL G., AIEE Fellow 1926.

KELLY, JOHN F., AIEE Fellow 1912.

KELLY, MERVIN J., AIEE Fellow 1931. IRE Fellow 1938.

KELLY, WILLIAM, AIEE Fellow 1925.

KELMAN, JOSEPH NELSON, AIEE Fellow 1931.

KELSCH, R.S., AIEE Fellow 1912.

KENDALL, BURTON W., AIEE Fellow 1929.

KENDRICK, WILLIAM HENRY, AIEE Fellow 1950.

KENNEDY, LUKE FRANCIS, AIEE Fellow 1949.

KENNEDY, M.E., IRE Fellow 1956 "For contributions in the flood control and civil defense communication systems."

KENNEDY, TERRENCE O., AIEE Fellow 1939.

KENNELLY, ARTHUR E., AIEE Fellow 1913. IRE Fellow 1928.

KENNEY, NORWOOD D., IEEE Fellow 1969 "For outstanding contributions in the wire and cable field, particularly in the pioneering work connected with he development and application of solid dielectrics for high voltages."

KENRICK, GLEASON W., IRE Fellow 1933.

KENYON, ALONZO F., IEEE Fellow 1967 "For contributions to the development and application of automated electrical drive systems for steel rolling mills."

KEOHLER, GLENN, IRE Fellow 1956 "For contributions to engineering education and educational broadcasting."

KEPLER, LEONARD, AIEE Fellow 1923.

KERR, DONALD E., IEEE Fellow 1966 "For leadership and research in microwave propagation and plasma physics."

KERR, HENRY H., AIEE Fellow 1944.

KERR, WILLIAM E., AIEE Fellow 1948.

KETTERING, CHARLES FRANKLIN, AIEE Fellow 1914.

KIDDER, HARRY ALVIN, AIEE Fellow 1929.

KIEBERT, MARTIN V., JR., IRE Fellow 1959 "For contributions to telemetry and automatic data reduction."

KIERSTEAD, FRIEND HANS, AIEE Fellow 1949.

KIFER, EDWIN HENRY, AIEE Fellow 1945.

KILBY, HUBERT ST. CLAIR, AIEE Fellow 1948.

KILGOUR, C.E., IRE Fellow 1951 "For his contributions in the application of graphical methods to the analysis of detector and output tube performance."

KILGOUR, HAMILTON, AIEE Fellow 1913.

KILLIAN, T.J., IRE Fellow 1951 "For his enlightened guidance of basic scientific research."

KIM, W.H., IEEE Fellow 1964 "For contributions to network theory, particularly network topology."

KIMBERLY, EMERSON E., AIEE Fellow 1954 "For contributions in the field of electrical engineering education."

KIMBRELL, MARVIN REA, AIEE Fellow 1947.

KING, ARCHIE P., IRE Fellow 1962 "For contributions to waveguide theory and application."

KING, ARTHUR C., AIEE Fellow 1920.

KING, MORELAND, AIEE Fellow 1925.

KINMAN, T.H., IRE Fellow 1957 "For contributions in high-frequency research and semiconductor devices."

KINN, THEODORE P., IRE Fellow 1959 "For contributions in the field of industrial heating."

KINNARD, ISAAC F., AIEE Fellow 1943.

KINSLEY, CARL, AIEE Fellow 1935.

KINTER, SAMUEL M., IRE Fellow 1915.

KINZLY, NELSON T., AIEE Fellow 1951.

KIRKE, HAROLD L., IRE Fellow 1945 "For his services to broadcasting in the British Isles and in particular for his leadership in the research activities of the British Broadcasting Corporation."

KIRKNER, HARRY L., AIEE Fellow 1913.

KIRKWOOD, L.R., IEEE Fellow 1964 "For contributions to color television receiver design."

KIRSTEN, K. FRIEDRICH J., AIEE Fellow 1930.

KISTLER, ROY E., AIEE Fellow 1946.

KITTREDGE, LINUS E., AIEE Fellow 1947.

KITTRIDGE, CARLYLE, AIEE Fellow 1922.

KLAUBER, LAURENCE M., AIEE Fellow 1923.

KLINCK, J. HENRY, AIEE Fellow 1913.

KLOEFFLER, ROYCE G., AIEE Fellow 1932. IRE Fellow 1953 "In recognition of his achievements as an educator and author in the fields of electrical and electronic engineering."

KLOSS, MAX, AIEE Fellow 1926.

KLUMPP, J.B., AIEE Fellow 1913.

KNAPP, PETER RICHARD, AIEE Fellow 1938.

KNICKERBOCKER, WALTER G., AIEE Fellow 1946.

KNIGHT, CHARLES D., AIEE Fellow 1913.

KNIGHT, FRED D., AIEE Fellow 1950.

KNIGHT, GEORGE LAURENCE, AIEE Fellow 1917.

KNIGHT, NORMAN, AIEE Fellow 1944.

KNIPMEYER, CLARENCE CARL, AIEE Fellow 1951.

KNOLL, MAX, IRE Fellow 1962 "For work on cathode ray tubes, in particular the electron microscope and image storage tubes."

KNOWLES, DEWEY D., AIEE Fellow 1957 "For his contributions in research, design and application in the field of gas-filled tubes and for his leadership in the electronic industry."

KNOWLTON, ARCHER E., AIEE Fellow 1930.

KNOX, CHARLES E., AIEE Fellow 1913.

KNUDSEN, KNUD JOHANNES, AIEE Fellow 1944.

KNUTZ, WILLIAM HENRY, AIEE Fellow 1948.

KOCH, MARSHALL MCKINLEY, AIEE Fellow 1957 "For administrative achievements in directing the engineering, construction and operation activities of a rapidly expanding electric utility system."

KOCHENBERGER, RALPH J., IEEE Fellow 1967 "For contributions to engineering education and to the advancement of nonlinear control theory and its application."

KOCK, WINSTON E., IRE Fellow 1952 "In recognition of his contributions in the field of electromagnetic-wave lenses and antennas."

KOENIG, HERMAN CHARLES, AIEE Fellow 1951.

KOENIG, HERMAN E., IEEE Fellow 1973 "For leadership in the use of system concepts in electrical engineering curricula, and n pioneering their use in the socioeconomic and biological disciplines."

KOGA, ISSAC, IRE Fellow 1957 "For contributions in quartz crystal techniques and engineering education."

KOHLHAAS, HERMAN THEODORE, AIEE Fellow 1948.

KOINER, C.W., AIEE Fellow 1912.

KOLSTER, FREDERICK A., IRE Fellow 1916.

KOMIVES, LASZLO I., AIEE Fellow 1954 "For contributions to the development and application of high-voltage transmission cable systems."

KOMPFNER, RUDOLPH, IRE Fellow 1950 "For his research in electron tube theory and particularly for his original contributions to the concepts of the traveling-wave amplifier."

KONGSTED, LUDVIG P., AIEE Fellow 1929.

KOONTZ, JOHN ANDREWS, AIEE Fellow 1949.

KORNER, A. JULIUS, AIEE Fellow 1939.

KOSITZKY, GUSTAV ADOLPH, AIEE Fellow 1929.

KOUWENHOVEN, WILLIAM B., AIEE Fellow 1934.

KOZANOWSKI, HENRY N., IRE Fellow 1962 "For early contributions to electronic circuits."

KRAFT, CHRIS H., AIEE Fellow 1931.

KRAMER, ERNST, IEEE Fellow 1965 "For contributions to aircraft radio navigation."

KRASIK, SIDNEY, IRE Fellow 1962 "For contributions toward the development of microwave devices and nuclear power plants."

KRAUSS, HERBERT L., IRE Fellow 1962 "For contributions to electrical engineering education."

KREER, JOHN C., IRE Fellow 1962 "For contributions to standardization."

KROGER, F.H., IRE Fellow 1928.

KRON, GABRIEL, AIEE Fellow 1951.

KRUG, FREDERICK, AIEE Fellow 1936.

KRUM, HOWARD L., AIEE Fellow 1927.

KRUPY, ALEXANDER JOSEPH, AIEE Fellow 1946.

KUEHNI, HANS PAUL, AIEE Fellow 1960 "For contributions to measuring and recording techniques for electrical and physical quantities."

KUHL, FRANK P., AIEE Fellow 1962 "For contributions to electric utility station design and lighting."

KUHN, CLARENCE WILBUR, AIEE Fellow 1939.

KUHN, R.J., IEEE Fellow 1964 "For contributions in the fields of corrosion and cathodic protection."

KULIKOWSKI, EDWIN FRANK, AIEE Fellow 1959 "For contributions fo the application of electronics to military uses, particularly to automatic processing of combat data."

KULJIAN, HARRY ASDOUR, AIEE Fellow 1962 "For contributions to the design and construction of thermal electric generating stations."

KUO, BENJAMIN CHUNG-I, IEEE Fellow 1974 "For contributions to control system theory and practice."

KURTICHANOF, LEONARD E., AIEE Fellow 1934.

KURTZ, EDWIN, AIEE Fellow 1929.

KURTZ, WILLIAM OVERTON, AIEE Fellow 1940.

KUSTERS, NORBERT L., IEEE Fellow 1965 "For leadership in research and development and for contributions in the field of electrical measurements."

KYNOCH, JAMES, AIEE Fellow 1918.

L

LABIN, EMILE, IEEE Fellow 1971 "For contributions in television, radar, pulse communication, radio location, and UHF vacuum tubes."

LABORDE, MAURICE E., AIEE Fellow 1956 "For his contributions in the development and application of extra high voltage cables to underground power transmission."

LACHICOTTE, FRANCIS WILLIAMS, AIEE Fellow 1950.

LACKIE, WILLIAM WALKER, AIEE Fellow 1916.

LACOMBE, CHARLES F., AIEE Fellow 1913.

LACY, THOMAS NORMAN, AIEE Fellow 1944.

LAFFOON, CARTHRAE MERRETTE, AIEE Fellow 1945.

LAKE, EDWARD N., AIEE Fellow 1912.

LAKIN, C. EARL, AIEE Fellow 1947.

LAMAR, ROBERT W., AIEE Fellow 1936.

LAMB, FRANCIS XAVIER, AIEE Fellow 1958 "For contributions to electrical measurements by invention and by design, especially for exacting and rugged militaty service."

LAMMERS, E.S., JR., AIEE Fellow 1953 "For many inventions in the field of steel rolling mills and for the achievement of electronics to industrial processes."

LAMOTTE, WILLIAM REED, AIEE Fellow 1951.

LAMSON, HORATIO W., IRE Fellow 1933.

LANCASTER, W.C., AIEE Fellow 1913.

LANDIS, GEORGE G., AIEE Fellow 1940.

LANDIS, GEORGE H., AIEE Fellow 1949.

LANDON, V.D., IRE Fellow 1951 "For contributions to communication system, filter and noise theory."

LANE, FRANCIS HOWARD, AIEE Fellow 1937.

LANE, RICHARD KELVIN, AIEE Fellow 1943.

LANG, WILMARTH Y., IEEE Fellow 1967 "For broadranging contributions to the development of telegraph, data transmission, and teletypewriter systems."

LANGDELL, JOSEPH CHESTER, AIEE Fellow 1947.

LANGE, ERNEST O., AIEE Fellow 1950.

LANGLEY, GORDON R., AIEE Fellow 1923.

LANGLOIS-BERTHELOT, RICHARD, IEEE Fellow 1969 "For contributions to research, production, education, and international activities in electrical engineering."

LANGMUIR, IRVING, IRE Fellow 1922.

LANGSDORF, ALEXANDER S., AIEE Fellow 1912.

LANGSTAFF, HAROLD, 1943 Fellow A.P..

LANGTON, JOHN, AIEE Fellow 1912.

LANIER, ALEXANDER CARTWRIGHT, AIEE Fellow 1939.

LANK, WILLIAM JEFFERIS, AIEE Fellow 1947.

LANPHIER, ROBERT C., AIEE Fellow 1913.

LANSIGN, VAN RENSSELAER, AIEE Fellow 1913.

LAPOE, ALBERT EDWARD, AIEE Fellow 1948.

LAPORT, EDMUND A., IRE Fellow 1953 "For his contributions to transmitter development, international communication systems and to antenna engineering."

LARDNER, HENRY A., AIEE Fellow 1913.

LARNER, RAY A., AIEE Fellow 1955 "For his contribution in the field of relaying and metering as applied to a far flung power system."

LARSEN, CHRISTIAN JULIUS, AIEE Fellow 1929.

LARSH, EVERETT PAUL, AIEE Fellow 1958 "For inventions and administration in the field of electrical apparatus."

LARSON, LEE RUSSEL, AIEE Fellow 1962 "For contributions in the development of electrical and electronic systems for aircraft and missile system."

LATEY, HARRY N., AIEE Fellow 1913.

LATIMER, CHESTER W., IRE Fellow 1929.

LAURIE, RUSSELL M., AIEE Fellow 1955 "For development of interconnected pool operation of large electric power systems on an international scale."

LAW, RUSSELL R., IRE Fellow 1953 "For his contributions to vacuum tube and television development."

LAWRENCE, RALPH R., AIEE Fellow 1928.

LAWS, FRANK ARTHUR, AIEE Fellow 1928.

LAWSON, CHARLES S., AIEE Fellow 1921.

LAWTON, ARTHUR H., AIEE Fellow 1919.

LAWTON, FREDERIC L., AIEE Fellow 1952 "For outstanding achievements in the development and operation of large power plants, especially hydroelectric plants in Canada and other countries, thereby contributing materially to the solution of the problem of electric power supply posed by the enormous growth of the aluminum industry."

LAYMAN, W.A., AIEE Fellow 1912.

LEAR, WILLIAM P., SR., IEEE Fellow 1967 "For pioneering efforts and continuing contributions to the field of electronics in aircraft communications and automatic flight control systems."

LEBENBAUM, PAUL, AIEE Fellow 1942.

LECLAIR, TITUS GEORGE, AIEE Fellow 1940.

LEE, ALBERT G., IRE Fellow 1929.

LEE, CLAUDIUS, AIEE Fellow 1946.

LEE, EDGAR M., IEEE Fellow 1968 "For contributions to design, specification and safety of electronic equipment and components, and co-ordination of industry and government in electronic development, production and inspection."

LEE, EVERETT S., AIEE Fellow 1930.

LEE, HAROLD R., AIEE Fellow 1950.

LEE, LOUIS R., AIEE Fellow 1926.

LEE, WILLIAM S., AIEE Fellow 1913.

LEE, Y.W., IRE Fellow 1961 "For contributions to communication theory and engineering education."

LEEDS, L.M., IRE Fellow 1957 "For contributions to the design of television equipment for studio and industrial applications."

LEEDS, MORRIS E., AIEE Fellow 1926.

LEEDY, H. A., IRE Fellow 1961 "For contributions to electronic research management."

LEERBURGER, FRANKLIN J., AIEE Fellow 1951.

A CENTURY OF HONORS

LEFEVER, ORLAND LESTER, AIEE Fellow 1944.

LEHDE, PENDLETON E., IRE Fellow 1927.

LEITCH, JOHN DAVIE, AIEE Fellow 1948.

LELAND, GEORGE HAROLD, AIEE Fellow 1944.

LEMP, HERMANN, JR., AIEE Fellow 1913.

LENSNER, HERBERT W., IEEE Fellow 1968 "For outstanding contributions in the field of protective realying for electric power systems, utilizing power-line carrier and microwave pilot channels."

LENZ, CHARLES OTTO, AIEE Fellow 1916.

LENZER, EMIL, IRE Fellow 1961 "For leadership in military communication-electronics."

LEONARD, C. STEPHEN, IEEE Fellow 1965 "For contributions to the fields of measurement, instrumentation and controls."

LEONARD, H. WARD, AIEE Fellow 1913.

LEONARD, MERRILL GRANT, AIEE Fellow 1962 "For contributions to components of distribution transformers, particularly of the protected type."

LEONARZ, E., AIEE Fellow 1913.

LE VEE, CLARENCE HUGHS, AIEE Fellow 1950.

LEVI, GUSTAV, AIEE Fellow 1950.

LEVINGER, DAVID, AIEE Fellow 1950.

LEVY, DAVID H., AIEE Fellow 1947.

LEWIS, CHARLES HENRY, AIEE Fellow 1949.

LEWIS, GEORGE, IRE Fellow 1951 "In recognition of important pioneering contributions including the objective coordination of many management and engineering activities."

LEWIS, WALTER W., AIEE Fellow 1938.

LEWIS, WARREN B., AIEE Fellow 1913.

LEYTON, E.M., IEEE Fellow 1964 "For contributions to the development of television transmitters and video tape systems."

LIAO, T.W., IEEE Fellow 1964 "For contributions to the fields of insulating structures, corona and radio influence, gas insulated electric apparatus and the mass accelerator."

LIBBEY, J.H., AIEE Fellow 1912.

LIBBY, T.M., IRE Fellow 1955 "For his technical contributions and long service in the field of communications."

LICHTENBERG, CHESTER, AIEE Fellow 1949.

LIDDLE, URNER, IRE Fellow 1955 "For his contributions to the establishment, promotion and integration of government sponsored nuclear research in academic institutions."

LIEB, JOHN W., AIEE Fellow 1913.

LIGHT, PHILIP H., AIEE Fellow 1959 "For contributions to electric power transmission."

LIGHTRIPE, JAMES A., AIEE Fellow 1913.

LILJENROTH, FRANS G., AIEE Fellow 1921.

LINCOLN, EDWIN S., AIEE Fellow 1930.

LINCOLN, JAMES F., AIEE Fellow 1939.

LINCOLN, JOHN CROMWELL, AIEE Fellow 1932.

LINCOLN, PAUL M., AIEE Fellow 1912.

LINDE, LEONARD JOSEPH, AIEE Fellow 1951.

LINDELL, SIGURD I., IEEE Fellow 1965 "For creativeness and guidance in the development, design and utilization of high voltage fuses and interrupter gear."

LINDENBLAD, N.E., IRE Fellow 1948 "For his many contributions to the theory and design of transmitting antennas and related equipment."

LINDER, E.G., IRE Fellow 1955 "For his contributions to microwave electronics."

LINDSAY, RICHARD W., AIEE Fellow 1944.

LINDSAY, WILLIAM W., JR., IRE Fellow 1939.

LINKS, GEORGE F., AIEE Fellow 1954 "For contributions to coordinated fault protection of distribution systems."

LINNEY, RALPH WALL, AIEE Fellow 1951.

LINVILL, WILLIAM K., IEEE Fellow 1968 "For contributions to sampled-data and computer control systems, and to systems analysis techniques."

LIPPINCOTT, DONALD K., IRE Fellow 1959 "For outstanding service in the field of patent law."

LISBERGER, SYLVAN JOSEPH, AIEE Fellow 1939.

LITTLE, D.S., IEEE Fellow 1964 "For contributions to electronic communication and to reliability and safety in aviation."

LITTLETON, JESSE TALBOT, JR., AIEE Fellow 1928.

LITTON, CHARLES V., IRE Fellow 1947 "For his contributions to theory and practice in the field of high vacuum techniques, including processes and precision devices for the production of electron tubes."

LIVERSIDGE, HORACE P., AIEE Fellow 1943.

LIWSCHITZ, MICHAEL, AIEE Fellow 1951.

LLEWELLYN, FREDERICK B., IRE Fellow 1938.

LLOYD, MORTON G., AIEE Fellow 1912.

LLOYD, TOM COX, AIEE Fellow 1958 "For contributions to the art of electrical motor design."

LOBO, GUSTAVE, AIEE Fellow 1912.

LOCKWOOD, THOMAS D., AIEE Fellow 1912.

LOEW, EDGAR ALLAN, AIEE Fellow 1936.

LOGAN, FRANK G., AIEE Fellow 1956 "In recognition of pioneering work in the field of self saturating reactors."

LOGAN, HENRY, AIEE Fellow 1943.

LOGAN, KIRK HAROLD, AIEE Fellow 1943.

LOIZEAUX, ALFRED S., AIEE Fellow 1929.

LOMAX, CLARENCE E., AIEE Fellow 1946.

LONGACRE, ANDREW, IRE Fellow 1962 "For originality and resourcefulness in developing military electronic equipment."

LOOMIS, A.L., IRE Fellow 1944 "In recognition of his work in the application of electronic techniques to medical research and for contributions to microwave development."

LOOMIS, ROBERT OLWINE, AIEE Fellow 1958 "For contributions to the planning and development of a large electric power system."

LORD, CHARLES E., AIEE Fellow 1912.

LORY, MARION RICHARDS, AIEE Fellow 1960 "For contributions to design of large synchronous motors and electric couplings."

LOTT, HARRY ABLE, AIEE Fellow 1958 "For contributions to the planning and operation of a major power system and the operation of major power system interconnections."

LOUD, FRANCIS MARTIN, AIEE Fellow 1950.

LOUGHLIN, WILLIAM D., IRE Fellow 1935.

LOUGHRIDGE, CLYDE HIGHBEE, AIEE Fellow 1945.

LOUIS, HENRY CHARLES, AIEE Fellow 1948.

LOVEJOY, J.R., AIEE Fellow 1913.

LOVELADY, MARCUS HENRY, AIEE Fellow 1960 "For contributions to the Southwest Power Pool Interconnected System and to engineering and management of an expanding power system."

LOVELL, ALFRED HENRY, AIEE Fellow 1939.

LOVELL, CLEMENS MALON, AIEE Fellow 1944.

LOVELL, WILLIAM VAIL, AIEE Fellow 1932.

LOWE, HERBERT LAURENCE, AIEE Fellow 1947.

LOWENBERG, MAURICE JOSEPH, AIEE Fellow 1946.

LOWRY, HITER H., AIEE Fellow 1929.

LOZIER, ROBERT TEN EYCK, AIEE Fellow 1914.

LUBKIN, SAMUEL, IRE Fellow 1961 "For applications of digital computers to airborne systems."

LUCE, DONALD C., AIEE Fellow 1949.

LUCK, DAVID G.C., IRE Fellow 1951 "For his early development of the omni-directional radio range."

LUCK, FRED T.W., AIEE Fellow 1912.

LUCKIESH, MATTHEW, AIEE Fellow 1940.

LUKOFF, HERMAN, IEEE Fellow 1970 "For pioneering in the development of digital computers and digital input and output devices."

LUNDY, AYRES D., AIEE Fellow 1920.

LUTZ, SAMUEL G., IRE Fellow 1962 "For improvements in long distance communications techniques and methods of miniaturizing electronic circuitry."

LYFORD, OLIVER S., JR., AIEE Fellow 1913.

LYMAN, JAMES, AIEE Fellow 1913.

LYMAN, WALTER JORDAN, AIEE Fellow 1943.

LYNDON, LAMAR, AIEE Fellow 1917.

LYNN, CLARENCE, AIEE Fellow 1950.

LYNN, SCOTT, AIEE Fellow 1929.

LYON, JESSE D., AIEE Fellow 1917.

LYON, WALDO VINTON, AIEE Fellow 1933.

LYONS, HAROLD, IRE Fellow 1958 "For his contributions to the development of atomic frequency standards."

LYONS, WALTER, IEEE Fellow 1965 "For contributions to telecommunications systems, particularly in diversity reception and spectrum conservation."

LYTLE, C. MYRON, AIEE Fellow 1951.

M

MACCALLA, C.S., AIEE Fellow 1913.

MACCUTCHEON, A., AIEE Fellow 1926.

MACDONALD, ANGUS A., IRE Fellow 1962 "For contributions to spectrum conservation in VHF and UHF mobile radio."

MACDONALD, WILLIAM A., IRE Fellow 1953 "For his engineering contributions and his sustained leadership in electronic development."

MACGAHAN, PAUL, AIEE Fellow 1942.

MACGREGOR, JOHN ROY, AIEE Fellow 1939.

MACKAVANAGH, THOMAS J., AIEE Fellow 1934.

MAC KEOWN, SAMUEL S., IRE Fellow 1940.

MACLACHLAN, WILLS, AIEE Fellow 1921.

MACLEOD, DONALD RIDGWAY, AIEE Fellow 1948.

MACLEOD, HECTOR JOHN, AIEE Fellow 1945.

MACNAUGHTON, ARCHIBALD K., AIEE Fellow 1937.

MAGALHAES, FRANK V., AIEE Fellow 1919.

MAGINNISS, FRANK J., AIEE Fellow 1960 "For contributions to the application of digital computers in solution of engineering problems."

MAGNUSKI, HENRY, IEEE Fellow 1965 "For contributions to very high frequency and microwave radio communications."

MAGNUSSON, C. EDWARD, AIEE Fellow 1913.

MAHAN, JAMES S., AIEE Fellow 1927.

MAHONEY, HARRY PAUL, AIEE Fellow 1947.

MAHONEY, JOSEPH N., AIEE Fellow 1918.

MAHOOD, EDWIN TERRELL, AIEE Fellow 1936.

MAILEY, ROY D., AIEE Fellow 1930.

MAILLARD, ALBERT LUDOVIC, AIEE Fellow 1948.

MAILLOUX, C.O., AIEE Fellow 1912.

MALLETT, JOHN P., AIEE Fellow 1913.

MALOFF, IOURY G., IRE Fellow 1936.

MALONEY, CHARLES WILLIAM, AIEE Fellow 1961 "For contributions to the design of electric generating stations and to a major system frequency conversion."

MALTER, LOUIS, IRE Fellow 1952 "In recognition of his many contributions in the fields of vacuum and gas-filled tubes."

MALTI, MICHEL GEORGE, AIEE Fellow 1959 "For contributions to electrical engineering education through his teaching and publications."

MANIER, ROBERT FRANCIS, AIEE Fellow 1960 "For contributions to the thermal performance of motors particularly, in enclosed forms and small ratings."

MANSON, GEORGE K., AIEE Fellow 1923.

MANSON, RAY HERBERT, IRE Fellow 1930. AIEE Fellow 1944.

MANZ, OTTO W., JR., IEEE Fellow 1966 "For contributions to research, design, fabrication and testing of power and communication cables."

MAPES, LELAND RUSSELL, AIEE Fellow 1937.

MARBURG, L.C., AIEE Fellow 1913.

MARCONI, GUGLIERMO, IRE Fellow 1916.

MARKS, LOUIS B., AIEE Fellow 1913.

MARKS, RALPH L., IEEE Fellow 1966 "For contributions to communications techniques and systems and upgrading of circuit standards for global services."

MARKS, WILLIAM S., JR., IRE Fellow 1952 "For his contributions and leadership in the field of military communications, and his pioneering work in radio-relay systems."

MARRIOTT, ROBERT H., IRE Fellow 1915. AIEE Fellow 1926.

MARRISON, WARREN A., IRE Fellow 1958 "For the achievement of high precision in the measurement of frequency and time by the use of piezoelectric materials."

MARRS, R.E., IEEE Fellow 1964 "For contributions to electrification and automation of the steel industry."

MARSH, HARRY H., JR., AIEE Fellow 1948.

MARSHALL, ALFRED C., AIEE Fellow 1929.

MARSHALL, DONALD E., AIEE Fellow 1957 "For basic contributions to gas type electronic tubes and ignitron rectifiers."

MARSHALL, L.C., IEEE Fellow 1964 "For contributions in the fields of linear accelerators and high power microwave sources."

MARTI, OTHMAR KARL, AIEE Fellow 1939.

MARTIN, CLARENCE ARTHUR, AIEE Fellow 1959 "For management of electrical construction of the steel, oil and chemical industries."

MARTIN, EDWARD R., AIEE Fellow 1947.

MARTIN, TALBOT GRANT, AIEE Fellow 1932.

MARTIN, WILLIAM HENNICK, AIEE Fellow 1930.

MARTON, LADISLAUS S., IEEE Fellow 1969 "For pioneer work in electron physics (electron microscopy, electron optics, electron interferences and scattering) and for distinguished editorship."

MARVIN, HARRY B., IRE Fellow 1950 "For his contributions to the measurements art and for pioneering work in FM, television and allied fields."

MARYATT, ELMER F., AIEE Fellow 1947.

MASLIN, ALBERT JOSEPH, AIEE Fellow 1959 "For contributions to development of power transformers particularly for use with mercury rectifiers."

MASON, C. RUSSELL, AIEE Fellow 1958 "For contributions to development and application of power system protective apparatus."

MASON, SAMUEL J., IRE Fellow 1959 "For contributions in the fields of active networks and engineering education."

MASON, WAYNE, IEEE Fellow 1965 "For contributions to international agreements on maritime radio communications and safety."

MASSINGILL, JENNINGS A., IEEE Fellow 1978 "For contributions to the design of large steam turbine generators."

MATHES, ROBERT CARL, AIEE Fellow 1941.

MATHEWS, CLAUDE L., AIEE Fellow 1927.

MATHEWSON, HAROLD GRANT, AIEE Fellow 1949.

MATHEY, E.D., AIEE Fellow 1919.

MATHIEU, GASTON A., IRE Fellow 1934.

MATTHEWS, CLAUDE LEVERING, AIEE Fellow 1927.

MATTHEWS, HOWARD DAVID, AIEE Fellow 1948.

MATTHEWS, WHITNEY, AIEE Fellow 1961 "For contributions to the engineering application of Solid State phenomena and to telemetry systems."

MATTHIES, WILLIAM H., AIEE Fellow 1929.

MAUCHLY, J.W., IRE Fellow 1957 "For pioneering contributions to the field of electronic computers."

MAVER, WILLIAM, JR., AIEE Fellow 1912.

MAVOR, H.A., AIEE Fellow 1913.

MAXIAN, JOHN, AIEE Fellow 1957 "For pioneering work in the application of electrical engineering principles to the design and development of electrical systems in aircraft."

MAXWELL, ALEXANDER, AIEE Fellow 1937.

MAYER, EMIL E., IRE Fellow 1915. AIEE Fellow 1935.

MAYOTT, CLARENCE WILBUR, AIEE Fellow 1950.

MAZEN, NATALIS, AIEE Fellow 1922.

MCADAM, JOHN C., IEEE Fellow 1969 "For original and continued engineering accomplishment in the field of heat dissipation, leading to greater reliabiltiy of electronic equipment."

MCALLISTER, A.S., AIEE Fellow 1912.

MCAULEY, P.H., IEEE Fellow 1964 "For contribution to the development, evaluation and standardization of high-voltage insulation."

MCBERTY, FRANK ROBERT, AIEE Fellow 1918.

MCCANN, WILLIAM R., AIEE Fellow 1924.

MCCARTHY, J.B., AIEE Fellow 1927.

MCCARTY, WILLARD ROBERT, AIEE Fellow 1959 "For contributions to techniques of high voltage and corona tests of power transformers."

MCCLELLAN, WILLIAM, AIEE Fellow 1912.

MCCONAHEY, WILLIAM M., AIEE Fellow 1922.

MCCORD, A. RAY, IEEE Fellow 1983 "For contributions to the development of radar and infrared reconnaissance systems and leadership in the electronics industry."

MCCORMACK, J. ELIOT, AIEE Fellow 1944.

MCCORMICK, BRADLEY T., AIEE Fellow 1913.

MCCOY, DAVID O., IRE Fellow 1959 "For research in the field of radio astronomy."

MCCOY, ROBERT LESLIE, AIEE Fellow 1951.

MCCOY, WALTER E., AIEE Fellow 1913.

MCCREARY, HAROLD JAMES, AIEE Fellow 1946.

MCCROSKY, JAMES WARREN, AIEE Fellow 1915.

MCCUE, J.J. GERALD, IEEE Fellow 1971 "For contributions to the field of ultrasonics and for editorial leadership."

MCCULLOCH, W.S., IEEE Fellow 1964 "For researches into the information handling processes of the nervous system."

MCCURDY, RALPH GORDON, AIEE Fellow 1934.

MCDAVITT, MARCELLUS B., AIEE Fellow 1951.

MCDONALD, ARTHUR S., IRE Fellow 1941.

MCDONALD, JOHN A., AIEE Fellow 1951.

MCDOWELL, C.S., AIEE Fellow 1918.

MCEACHRON, KARL B., AIEE Fellow 1937.

MC ELROY, PAUL K., IRE Fellow 1931.

MCELROY, PAUL K., IRE Fellow 1931.

MCFARLAND, THOMAS CLAIR, AIEE Fellow 1945.

MCFARLIN, JOHN ROBERT, AIEE Fellow 1951.

MCGOVERN, WILLIAM R., AIEE Fellow 1917.

MCGRATH, M. H., AIEE Fellow 1959 "For contributions to progressive development of high voltage power cables, particularly the pipe-enclosed gas pressure type."

MCHENRY, MORRIS JAMES, AIEE Fellow 1947.

MCILWAIN, KNOX, IRE Fellow 1948 "For his contribution to the technical literature of radio and his activity in the field of radio aids to navigation."

MCIVER, GEORGE W., AIEE Fellow 1926.

MCKAY, C.R., AIEE Fellow 1913.

MCKEARIN, JAMES PATRICK, AIEE Fellow 1938.

MCKEE, EDD RUTHVEN, AIEE Fellow 1962 "For contrubutions to engineering education."

MCLACHLAN, WILLARD J., AIEE Fellow 1951.

MCLAUGHLIN, HAROLD ANDREW, AIEE Fellow 1962 "For contributions to the development of principles and equipment for power system protection."

MCLEAN, DAVID A., IRE Fellow 1962 "For contributions to the fields of capacitors and dielectric material."

MCLEAN, WILLIAM B., IEEE Fellow 1973 "For contributions and inspiring leadership in missile and ocean technology."

MCLEER, CHARLES B., AIEE Fellow 1920.

MCLELLAN, WILLIAM, AIEE Fellow 1913.

MCLENEGAN, D.W., AIEE Fellow 1953 "For important developments and contributions in air conditioning, aircraft head exchanges, transformer cooling and heat pumps."

MCLEOD, WILLARD W., JR., IEEE Fellow 1972 "For contributions to cross-field technology in high-frequency vacuum tubes for increased coherency."

MCMANUS, CLIFFORD BRASWELL, AIEE Fellow 1958 "For contributions to reinforcement and expansion of electric utility systems serving a large area."

MCMEEN, S.G., AIEE Fellow 1912.

MC MICHAEL, LESLIE, IRE Fellow 1925.

MCMILLAN, FRED O., AIEE Fellow 1932.

MCNALL, JOHN W., IEEE Fellow 1970 "For contributions in microwave technology, and for direction of research and development in light sources."

MCNEELY, JOHN K., AIEE Fellow 1931.

MC NICOL, DONALD, IRE Fellow 1924.

MCNICOL, DONALD, AIEE Fellow 1918. IRE Fellow 1924.

MCPHAIL, HARVEY F., AIEE Fellow 1942.

MCQUARRIE, JAMES L., AIEE Fellow 1926.

MCRAE, JAMES W., IRE Fellow 1947 "For outstanding work in the planning of research and development programs in radar and countermeasures and for his researches in radio transmitting methods."

MEACHAM, L.A., IRE Fellow 1948 "For his contributions in the fields of radar range measurement and pulse code modulation."

MEAD, DANIEL W., AIEE Fellow 1913.

MEESE, WILLIAM G., IEEE Fellow 1978 "For research management in the electric power industry."

MEISSNER, ALEXANDER, IRE Fellow 1915.

MEKELBURG, EARL F., AIEE Fellow 1957 "For contributions to the design and application of industrial control."

MELLETT, JOHN EDWARD, AIEE Fellow 1949.

MELSON, SYDNEY WILLIAM, AIEE Fellow 1926.

MEREDITH, WYNN, AIEE Fellow 1913.

MERGENTHALER, ADOLF H., AIEE Fellow 1959 "For contributions to planning of economic power system generating capacity."

MERRELL, EDWIN JOSEPH, AIEE Fellow 1958 "For contributions to the art of high voltage cables."

MERRILL, BARRETT M., AIEE Fellow 1928.

MERRILL, FRANK WINTHROP, AIEE Fellow 1950.

MERRILL, JOSEPH F., AIEE Fellow 1913.

MERSHON, RALPH D., AIEE Fellow 1912.

MERTZ, PIERRE, IRE Fellow 1950 "In recognition of his important contributions to the fundamental concepts of television transmission and reception."

MERWIN, LOUIS TUNIS, AIEE Fellow 1933.

MERWIN, ROBERT B., IEEE Fellow 1975 "For development of ferrite core memories and computer hardware and software programs."

MERZ, CHARLES H., AIEE Fellow 1913.

MESERVE, WILBUR E., IEEE Fellow 1965 "For contributions to engineering education."

MESZAR, JOHN, AIEE Fellow 1959 "For contributions in the field of automatic telephone interconnecting technology."

MEYER, EDWARD B., AIEE Fellow 1927.

MEYER, FRANK J., AIEE Fellow 1936.

MEYER, H. F., IEEE Fellow 1964 "For contributions to the development of advanced communication systems for the Armed Forces."

MEYERHOFF, LOUIS, AIEE Fellow 1958 "For contributions to the development of high voltage cables."

MICHALOWICZ, JOSEPH CASIMIR, AIEE Fellow 1959 "For contributions to the education of electrical engineers."

MICHELL, HUMPHREY G., AIEE Fellow 1949.

MIER, C. WALKER, AIEE Fellow 1939.

MIKINA, STANLEY JOSEPH, AIEE Fellow 1948.

MILBURN, LOYAL R., AIEE Fellow 1939.

MILDNER, RAYMOND C., IEEE Fellow 1973 "For contributions in the research, development, and manufacture of high-voltage, line communicationm and radio frequency cables."

MILES, PAUL D., IEEE Fellow 1966 "For contributions to the efficient utilization of the radio spectrum and management of frequency allocations."

MILLAN, WALTER H., AIEE Fellow 1927.

MILLARD, THOMAS O., AIEE Fellow 1951.

MILLER, ALTEN S., AIEE Fellow 1912.

MILLER, ARTHUR, IEEE Fellow 1971 "For contributions to data amplification and recording in medical and industrial application."

MILLER, B.F., IRE Fellow 1956 "For improvements in sound recording techniques, and for the application of electronics in isotope separation."

MILLER, CHARLES J., JR., AIEE Fellow 1956 "For his contributions to the knowledge of factors influencing high-voltage phenomena."

MILLER, CHARLEY WALTER, AIEE Fellow 1962 "For contributions in the field of large power transformer design, and in high voltage insulation."

MILLER, DECATUR S., AIEE Fellow 1913.

MILLER, J. EDGAR, AIEE Fellow 1913.

MILLER, JAMES SHANNON, JR., AIEE Fellow 1947.

MILLER, JOHN HAROLD, AIEE Fellow 1942. IRE Fellow 1950 "For his long activity and many contributions in the field of electrical metering and measuring technique."

MILLER, JOHN LEONARD, AIEE Fellow 1947.

MILLER, JOHN M., IRE Fellow 1920.

MILLER, KEMPSTER B., AIEE Fellow 1927.

MILLER, LLOYD E., AIEE Fellow 1945.

MILLER, WILLIAM COOK, AIEE Fellow 1943.

MILLER, WILLIAM J., AIEE Fellow 1944.

MILLIKEN, HUMPHREYS, AIEE Fellow 1939.

MILLIS, WALTER T., IEEE Fellow 1966 "For contributions to the electron tube art and standardization."

MILLS, GEORGE ARTHUS, AIEE Fellow 1945.

MILLS, JOHN, AIEE Fellow 1923. IRE Fellow 1930.

MILLS, WILLIAM F., IRE Fellow 1925.

MILNE, GORDON R., AIEE Fellow 1949.

MILNER, CALRISLE K., AIEE Fellow 1957 "For his contributions to executive and technological leadership of a communication enterprize."

MILNOR, JOSEPH WILLARD, AIEE Fellow 1930.

MILTON, TALIAFERRO, AIEE Fellow 1920.

MIMNO, HARRY R., IRE Fellow 1958 "For his leadership in ionospheric research."

MINER, DOUGLAS FULLER, AIEE Fellow 1940.

MINNIUM, BYRON B., IRE Fellow 1959 "For the development of electronic components and for service to the radio engineering profession."

MINOR, EDWARD E., JR., AIEE Fellow 1948.

MINSKY, MARVIN L., IEEE Fellow 1968 "For his research and educational leadership in the field of artificial intelligence and heuristic programming."

MINTON, JOHN P., IRE Fellow 1923.

MIRANDA, FREDERICK J., IEEE Fellow 1974 "For contributions to the design and development of extra-high-voltage cables."

MIRICK, CARLOS BROWN, AIEE Fellow 1947.

MITCHELL, JAMES, AIEE Fellow 1916.

MITCHELL, WILLIAM E., AIEE Fellow 1922.

MITSCHRICH, MELVILLE F., AIEE Fellow 1945.

MITSUDA, RYOTARO, AIEE Fellow 1920.

MITSURA, SAYEKI, IRE Fellow 1916.

MITTAG, ALBERT H., AIEE Fellow 1956 "For his contributions in the field of power rectification."

MITTELMANN, EUGENE, IRE Fellow 1957 "For pioneering in the field of industrial electronics."

MIYAZAKI, GENTARO, IEEE Fellow 1978 "For contributions to color television receiver development and leadership in international communications."

MOLE, HARVEY E., AIEE Fellow 1913.

MOLINA, EDWARD C., AIEE Fellow 1951.

MOLLER, ROLF, IRE Fellow 1956 "For contributions to the development of television in Germany."

MOLLERUS, FRED J., SR., AIEE Fellow 1962 "For contributions to the design and operation of electrical systems for atomic energy plants."

MOLNER, J.P., IRE Fellow 1960 "For contributions to gaseous and solid-state electron devices."

MOLONEY, THOMAS O., AIEE Fellow 1945.

MONK, NEWTON, IEEE Fellow 1964 "For contributions to mobile radio telephone systems for railroads."

MONROE, WILLIAM S., AIEE Fellow 1925.

MONTGOMERY, A.W., IRE Fellow 1954 "For his leadership in radio and telecommunication research in England, and his services in the international liaison in these fields."

MONTGOMERY, TERRYL B., AIEE Fellow 1949.

MONTIETH, ALEXANDER CRAWFORD, AIEE Fellow 1945.

MONTSINGER, VINCENT M., AIEE Fellow 1929.

MOODY, V.D., AIEE Fellow 1912.

MOODY, WALTER S., AIEE Fellow 1912.

MOON, PARRY H., AIEE Fellow 1934.

MOORE, D. MCFARLAN, AIEE Fellow 1912.

MOORE, EDWARD ROYAL, AIEE Fellow 1958 "For his contributions in the planning and development of a large electric power system."

MOORE, EDWARD T., AIEE Fellow 1921.

MOORE, ERNEST ELLSWORTH, AIEE Fellow 1931.

MOORE, J. BURTON, IRE Fellow 1960 "For improvements in communications coding techniques."

MOORE, LEE M., AIEE Fellow 1945.

MOORE, LEONARD J., AIEE Fellow 1942.

MOORE, R.C., IRE Fellow 1955 "For his contributions to television circuitry."

MOORE, ROBERT B., IEEE Fellow 1975 "For leadership and technical contributions to the advancement of systems engineering in serving the needs of industry."

MOORE, WILLIAM E., AIEE Fellow 1913.

MORACK, MARVIN M., AIEE Fellow 1953 "For contributions in the development of mercury arc rectifiers, and their application to electronic frequency conversion and variable speed power drives."

MORECOCK, EARL M., AIEE Fellow 1948.

MORECROFT, JOHN H., AIEE Fellow 1919. IRE Fellow 1922.

MOREHEAD, JOHN M., AIEE Fellow 1919.

MOREHOUSE, LYMAN F., AIEE Fellow 1920.

MORELAND, EDWARD L., AIEE Fellow 1921.

MORELAND, HENRY D., AIEE Fellow 1948.

MORENO, THEODORE, IRE Fellow 1960 "For contributions in the field of microwave electronics."

MORGAN, SAMUEL P., IRE Fellow 1962 "For contributions to electromagnetic theory."

MORGAN, THEODORE HARDING, AIEE Fellow 1939.

MORGAN, WENDAL ALTON, AIEE Fellow 1951.

MORITA, TETSU, IEEE Fellow 1966 "For contributions in the fields of antennas and the ineteraction of electromagnetic energy with plasmas."

MORITZ, CHARLES H., AIEE Fellow 1913.

MORKENSEN, NIELS L., AIEE Fellow 1920.

MORLOCK, W.J., IRE Fellow 1957 "For pioneering work in sound systems, and for contributions to engineering management."

MORRILL, WESTON, IEEE Fellow 1967 "For his outstanding contributions in the development of magnetic core materials that have drastically reduced the cost of producing and transmitting electrical power."

MORRIS, SAMUEL B., AIEE Fellow 1957 "For his achievements in engineering education and in the development and operation of a large metropolitan water and power system."

MORRISON, JOHN WILLIAM, AIEE Fellow 1943.

MORRISON, LAURENCE W., IEEE Fellow 1970 "For contributions and leadership in the development of large electronic systems."

MORROW, L.W.W., AIEE Fellow 1925.

MORSE, GEORGE HART, AIEE Fellow 1922.

MORSS, EVERETT, AIEE Fellow 1913.

MORTON, JACK A., IRE Fellow 1953 "For his leadership and contributions to the engineering physics and development of transistors and of wide-band microwave electron tubes."

MORTON, ROBERT B., AIEE Fellow 1920.

MORTON, WALTER B., AIEE Fellow 1942.

MOSES, GRAHAM LEE, AIEE Fellow 1950.

MOSES, PERCIVAL R., AIEE Fellow 1912.

MOSMAN, CHARLES T., AIEE Fellow 1913.

MOSSAY, PAUL A., AIEE Fellow 1928.

MOTTER, W.N., AIEE Fellow 1912.

MOULTRON, JAMES SEYMOUR, AIEE Fellow 1949.

MOULTROP, IRVING E., AIEE Fellow 1929.

MOUNTAIN, JOHN T., AIEE Fellow 1920.

MOUROMTSEFF, ILIA E., IRE Fellow 1947 "In recognition of his contributions to vacuum tube development, particularly transmitting tubes."

MOWBRAY, WILLIAM J., AIEE Fellow 1931.

MOXLEY, STEPHEN D., JR., IEEE Fellow 1975 "For contributions to electronic geophysical exploration, air traffic control technology, and engineering management."

MOYERS, WALTER RAYMOND, JR., AIEE Fellow 1962 "For contributions to the development and management of a large electric utility system."

MUDGE, A.L., AIEE Fellow 1912.

MUELLER, GEORGE VERNON, AIEE Fellow 1950.

MUIR, ROY CUMMINGS, AIEE Fellow 1936.

MULLEN, JAMES J., AIEE Fellow 1944.

MULLER, CORNELIS A., IEEE Fellow 1973 "For contribution to the knowledge of the the structure of galexies."

MULLER, HENRY NICHOLAS, JR., AIEE Fellow 1949.

MULLER, HENRY NIKOLA, AIEE Fellow 1947.

MULLER, JOHN H., IEEE Fellow 1968 "For contributions in the extending radiocommunication to global telegraph and telephone systems."

MURPHY, FRANCIS H., AIEE Fellow 1930.

MURPHY, GEORGE R., AIEE Fellow 1920.

MURPHY, JOHN, AIEE Fellow 1913.

MURPHY, O.J., IEEE Fellow 1964 "For contributions to the art of communication system signaling and switching control."

MURRAY, ALBERT F., IRE Fellow 1938.

MURRAY, THOMAS E., AIEE Fellow 1912.

MURRAY, THOMAS EDWARD, JR., AIEE Fellow 1933.

MURRAY, WILLIAM ARTHUR, AIEE Fellow 1944.

MURRAY, WILLIAM S., AIEE Fellow 1912.

MUSGROVE, ALBERT MILLARD, AIEE Fellow 1945.

MYERS, BASIL R., IEEE Fellow 1969 "For outstanding leadership in engineering education and administration."

MYERS, OSCAR, IEEE Fellow 1966 "For contributions to the development of common-control telephone switching systems and direct distance dialing."

N

NAEF, OTTO, AIEE Fellow 1961 "For contributions in the fields of arc extinction and insulation coordination which contributed to successful design of high voltage power systems, especially for extra high voltage circuit breakers."

NAETER, ALBRECHT, AIEE Fellow 1953 "For achievement in coordinating effective technical teaching with educational administration and dirction of collegiate research."

NANCE, HORACE H., AIEE Fellow 1947.

NARBUT, PAUL, AIEE Fellow 1962 "For contributions in the thermal field of transformers, and the development of transformer insulation and cooling systems."

NARESKY, JOSEPH J., IEEE Fellow 1976 "For contributions to the quantification of electronic equipment reliability."

NASH, GEORGE HOWARD, AIEE Fellow 1920.

NASH, JOHN P., IEEE Fellow 1969 "For contributions to the design and programming of computers, and leadership in industrial applications of information processing."

NASON, HORACE EDWARD, AIEE Fellow 1951.

NATHAN, EUGENE J., AIEE Fellow 1940. .

NEALL, N.J., AIEE Fellow 1912.

NEHER, JOHN HUTCHINS, AIEE Fellow 1958 "For contributions to protective relay schemes and research on the thermal factors in underground power cables."

NEIDIG, RALPH E., AIEE Fellow . 19 "For contributions to planned development and operating reliability of an electric power system."

NEILER, SAMUEL G., AIEE Fellow 1913.

NELSON, ARTHUR R., AIEE Fellow 1955 "In recognition of his engineering and administrative achievement in the overall guidance of the continued development and expansion of one of the nations largest transmission and distribution systems."

NELSON, EDWARD L., IRE Fellow 1938.

NELSON, JOHN MAGNUS, AIEE Fellow 1962 "For contributions to the engineering and management of a large electric public utility."

NELSON, RICHARD B., IRE Fellow 1962 "For contributions to high power wideband klystrons."

NESBIT, WILLIAM, AIEE Fellow 1929.

NESTEL, WERNER M., IEEE Fellow 1965 "For achievements in radio and electronics, especially FM radio broadcasting and television networks."

NETHERCUT, DONALD WARREN, AIEE Fellow 1945.

NEWBOLD, R.M., AIEE Fellow 1912.

NEWBURY, FRANK D., AIEE Fellow 1921.

NEWMAN, JOHN MILTON, AIEE Fellow 1949.

NEWTON, JOHN M., SR., AIEE Fellow 1937.

NICHOLS, LOUIS C., AIEE Fellow 1913.

NICHOLS, W.A., IRE Fellow 1955 "For his contributions to the construction of the national radio system in Canada."

NICHOLSON, S.L., AIEE Fellow 1913.

NICOLL, F.H., IRE Fellow 1961 "For contributions in electron optics and electroluminescence."

NIKLAS, WILFRID F., IEEE Fellow 1971 "For technical contributions in image intensification devices, and leadership in the application of these devices for medical diagnostic purposes."

NIMS, ALBERT ARMSTRONG, AIEE Fellow 1953 "For outstanding contributions to the electrical engineering profession in the theory and design of arc welding equipments, armature windings of electric machines, and in electric network analysis. Also, outstanding contributions to engineering education."

NIMS, FREDERICK, AIEE Fellow 1912.

NIPPES, IRA S., AIEE Fellow 1949.

NISSEN, JACOB PREBENSEN, AIEE Fellow 1922.

NIWA, YASUJIRO, IRE Fellow 1954 "For his leadership in radio engineering in Japan, and his contributions to vocational education."

NOAKES, FRANK, AIEE Fellow 1962 "For contributions to electrical engineering education and research."

NOBLE, DANIEL EARL, IRE Fellow 1947 "In recognition of his contributions to the design and application of very-high frequency voice communication systems for police and other emergency services."

NOBLE, HARRELL V., IRE Fellow 1958 "For leadership in the development of electron tubes and components."

NOBLE, PAUL O., AIEE Fellow 1949.

NOE, JAMES BRYAN, AIEE Fellow 1930.

NOEGGERATH, J.E., AIEE Fellow 1913.

NOEST, JOHN G., AIEE Fellow 1951.

NOODLEMAN, SAMUEL, IEEE Fellow 1977 "For contributions to the design of electric motors."

NORDELL, ROBERT, AIEE Fellow 1961 "For contribution to the design and manufacturing of high voltage cables and for the development of welded aluminum sheaths for cables."

NORDLUND, RAYMOND J., IEEE Fellow 1970 "For research in aerospace digital computers, microelectronic technology, and advanced navigation and guidance equipment; and for pioneering effort in system analysis."

NORELL, ELMER G., AIEE Fellow 1958 "For his contributions to the advancement of large scale electrical power generation."

NORINDER, HAROLD, AIEE Fellow 1956 "For his preeminent contributions to lightning protection of electrical circuits and to high voltage power transmission."

NORMAN, HENRY M.P., IRE Fellow 1915.

NORRIS, ERIC DOUGLAS TOBIAS, AIEE Fellow 1930.

NORRIS, FERRIS WALDO, AIEE Fellow 1945.

NORRIS, HENRY N., AIEE Fellow 1912.

NORTHMORE, EMANUEL RICHARD, AIEE Fellow 1928.

NORTHRUP, EDWIN F., AIEE Fellow 1913.

NORTON, EDWARD L., IRE Fellow 1962 "For fundamental contributions to circuit theory."

NORTON, FRANK R., IRE Fellow 1960 "For contributions to radar television."

NORTON, WILLIAM J., AIEE Fellow 1922.

NOTT, GORDON EUGENE, AIEE Fellow 1945.

NOTTINGHAM, W.B., IRE Fellow 1956 "For basic studies, teaching, and leadership in the field of physical electronics."

NOTTORF, WILLIAM E.A., AIEE Fellow 1930.

NUNAN, JAMES KNEELAND, AIEE Fellow 1962 "For contributions to the direction of scientific military and industrial engineering projects."

NYE, HENRY V., AIEE Fellow 1940.

NYQUIST, HARRY, AIEE Fellow 1951. IRE Fellow 1952 "In recognition of his fundamental contributions to physical and mathematical sciences in the field of communications."

NYSTROM, C.W., AIEE Fellow 1945.

O

O'BANION, ALBERT LEE, AIEE Fellow 1950.

OBOUKHOFF, NICOLAS MICHAILOVICH, AIEE Fellow 1938.

OCH, HENRY G., IEEE Fellow 1968 "For his contributions to analog and digital computer design and utilization in military systems."

O'CONNOR, ROGER RUSSELL, AIEE Fellow 1950.

ODARENKO, TODOS M., IRE Fellow 1959 "For contributions to the development of radio transmission systems, techniques and components."

OEHLER, ALFRED G., AIEE Fellow 1926.

OESTERREICH, EDMUND W., AIEE Fellow 1942.

OETINGER, HERBERT WILLIAM, AIEE Fellow 1952.

OGDEN, HAROLD STEPHEN, AIEE Fellow 1962 "For contributions to electric locomotive and multiple unit control systems."

OHTSUKI, TAKASHI, IEEE Fellow 1967 "For his contributions to the theory and techniques of electric power transmission and his services to engineering education."

OLDACRE, MARMION S., AIEE Fellow 1947.

OLIVE, G.W., IRE Fellow 1951 "For major contributions to the development of radio broadcasting in Canada, and in particular, for the engineering and organization of Canadian Overseas broadcast service during the war."

OLLENDORF, FRANZ, IRE Fellow 1960 "For contributions to electromagnetic theory and engineering education."

OLSEN, MAXWELL L., AIEE Fellow 1951.

OLSON, H.F., IRE Fellow 1949 "For his outstanding developments and publications in the fields of acoustics and underwater sound."

OLSON, ROBERT W., IRE Fellow 1959 "For leadership in geophysical research."

O'NEILL, GEORGE, IRE Fellow 1949 "For his work in electron tube theory and design."

OPLINGER, KIRK AUGUSTUS, AIEE Fellow 1948.

ORSETTICH, ROBERT, AIEE Fellow 1926.

ORTUSI, JEAN A., IEEE Fellow 1970 "For pioneering work in wave-guide theory, for original studies of circuit analysis and synthesis, and for contributions to the education of electronic engineers."

OSBORN, JOSEPH AUGUSTUS, AIEE Fellow 1912.

OSBORNE, LOYALL A., AIEE Fellow 1912.

OSENBAUGH, CHESTER LAMOTT, AIEE Fellow 1958 "For design and administrative achievements in directing the engineering, construction and operation of an electric utility system."

OSGOOD, FARLEY, AIEE Fellow 1912.

OSTERGREN, CLARENCE N., IEEE Fellow 1964 "For contributions to engineering economics in public utilities."

OSTLINE, JOHN ELLIS, AIEE Fellow 1946.

OSTRANDER, JOHN K., AIEE Fellow 1954 "In recognition of extensive contributions made to large industrial and public utility power projects."

OSWALD, ARTHUR A., IRE Fellow 1928. AIEE Fellow 1930.

OTTEN, HARRY CHARLES, AIEE Fellow 1962 "For contributions in the field of electric public utiltiy system operation and power interchange."

OUDIN, JEAN M.A.H., IEEE Fellow 1971 "For contribution to the development of high-voltage cables and submarine cables for power and communications."

OWENS, JAMES W., AIEE Fellow 1927.

OXLEY, ALLEN B., IRE Fellow 1953 "For his contributions and sustained leadership in radio engineering and manufacturing in Canada."

P

PACENT, LOUIS GERARD, IRE Fellow 1927. AIEE Fellow 1930.

PACKARD, L.E., IRE Fellow 1957 "For contributions in the field of electronic instrumentation."

PAGE, ARCHIBALD, AIEE Fellow 1929.

PAGE, GLENN I., AIEE Fellow 1945.

PAINE, F.B.H., AIEE Fellow 1912.

PAINE, SIDNEY B., AIEE Fellow 1912.

PALANDRI, GIUSEPPE L., IEEE Fellow 1966 "For contributions to the development of power cable systems."

PALMER, GEORGE W., AIEE Fellow 1912.

PALMER, GRANVILLE ERNST, AIEE Fellow 1936.

PALMER, RALPH L., IEEE Fellow 1967 "For technical and managerial leadership in the development and design of digital computers."

PALMER, RAY, AIEE Fellow 1915.

PALUEV, KONSTANTIN K., AIEE Fellow 1948.

PANNEL, ERNEST V., AIEE Fellow 1925.

PANNILL, CHARLES JACKSON, IRE Fellow 1929.

PANTER, THOMAS ALFRED, AIEE Fellow 1926.

PARKER, HENRY W., IRE Fellow 1952 "In recognition of his contributions in the field of electron tube design."

PARKER, JOHN C., AIEE Fellow 1912.

PARKER, JOHN CLIFTON, AIEE Fellow 1951.

PARKER, RALZEMOND DRAKE, AIEE Fellow 1946. IRE Fellow 1959 "For pioneer developments of teleprinter and data transmission systems."

PARKHURST, CHARLES W., AIEE Fellow 1913.

PARKS, CHARLES WELLMAN, AIEE Fellow 1923.

PARRACK, VASCO R., AIEE Fellow 1949.

PARSHALL, HORACE F., AIEE Fellow 1913.

PATCHELL, WILLIAM H., AIEE Fellow 1913.

PATTERSON, GEORGE W., AIEE Fellow 1913. IRE Fellow 1959 "For leadership in the development of machine logic and switching theory."

PATTERSON, JAMES GEORGE, AIEE Fellow 1944.

PATTERSON, LEROY R., AIEE Fellow 1957 "For his contributions to the planning and development of an electric power system and related interconnections."

PATTERSON, LUCIUS LAMAR, AIEE Fellow 1951.

PATTISON, DONALD R., AIEE Fellow 1948.

PATTISON, FRANK, AIEE Fellow 1913.

PAULSEN, ALFRED G., AIEE Fellow 1928.

PAULUS, CLARENCE FRANCIS, AIEE Fellow 1960 "For contributions to the technology and design of large thermal electric generating stations."

PAXSON, HOUSTON RANDOLPH, AIEE Fellow 1951.

PAXTON, EARL B., AIEE Fellow 1954 "For contributions to the formulation and organization of standards in the electrical field and promotion of their acceptance and use by industry, and their coordination with other national and international codes."

PAXTON, ROBERT, AIEE Fellow 1948.

PAYNE, BASIL THEODORE, AIEE Fellow 1962 "For contributions in electrical substation design and standardization."

PEARCE, CULLEN THOMAS, AIEE Fellow 1949.

PEARCE, JAMES G., IEEE Fellow 1973 "For contributions to the development of electronic telephone switching systems."

PEARCE, S.L., AIEE Fellow 1912.

PEARSON, EDGAR FORD, AIEE Fellow 1937.

PEASE, EUGENE I., AIEE Fellow 1951.

PEASLEE, W.D.A., AIEE Fellow 1927.

PECK, BERT H., AIEE Fellow 1923.

PECK, EMERSON P., AIEE Fellow 1923.

PECK, HENRY W., AIEE Fellow 1913.

PECK, J.S., AIEE Fellow 1912.

PEDERSEN, PEDER OLUF, IRE Fellow 1915. AIEE Fellow 1920.

PEDERSEN, SVEND A.C., IEEE Fellow 1970 "For contributions in the development of measurement methods applicable to radio receivers, and for leadership in international standardization."

PEEBLES, JOHN B., AIEE Fellow 1933.

PEEK, FRANK W., AIEE Fellow 1925.

PEIRCE, ALBERT EDWIN, AIEE Fellow 1917.

PEN DELL, C.W., AIEE Fellow 1912.

PENDER, HAROLD, AIEE Fellow 1914.

PENMAN, ROY FRANKLIN, AIEE Fellow 1944.

PENN, MARION, AIEE Fellow 1935.

PENNELL, WALTER O., AIEE Fellow 1922.

PENNIMAN, ABBOTT L., JR., AIEE Fellow 1943.

PERKINS, CHARLES ALBERT, AIEE Fellow 1940.

PERRY, IRVING DAVIDSON, AIEE Fellow 1959 "For contributions to the use of microwaves for utility system communication and protection."

PERRY, LESLIE LAURENCE, AIEE Fellow 1912.

PERRY, WILLIAM W., AIEE Fellow 1946.

PERSONS, JAMES TURNER, AIEE Fellow 1951.

PESHELL, WILLIAM, AIEE Fellow 1913.

PESSION, GIUSSEPPE, IRE Fellow 1929.

PETERS, JOHN FINDLEY, AIEE Fellow 1923.

PETERSON, ARTHUR C., IRE Fellow 1962 "For contributions to short-wave telephony."

PETERSON, EUGENE, AIEE Fellow 1951. IRE Fellow 1959 "For analysis and application of nonlinear devices in communication."

PETERSON, JOHN BEVILL, AIEE Fellow 1948.

PETERSON, R.S., AIEE Fellow 1953 "For outstanding ingenuity and important contributions in the field of underground transmission and distrubution of large blocks of power at high voltages in the form of improved types of insulated cables and cable accessories."

PETERSON, WILLIAM SIMON, AIEE Fellow 1946.

PETURA, FRANK J., AIEE Fellow 1931.

PETZINGER, AMBROSE J., AIEE Fellow 1957 "For original developments in the field of electrical measurements."

PHILLIPS, LEO A., AIEE Fellow 1913.

PIATT, WILLIAM MCKINNEY, AIEE Fellow 1939.

PICKARD, GREENLEAF W., AIEE Fellow 1912. IRE Fellow 1916.

PICKING, JAY WILFRED, AIEE Fellow 1960 "For contributions to development of automation by means of electronically controlled systems."

PIEPLOW, HANSWERNER K., IEEE Fellow 1965 "For pioneering work in electoacoustics, photographic recording, and tape recording."

PIERCE, ARTHUR G., AIEE Fellow 1920.

PIERCE, D. LAWRENCE, IEEE Fellow 1966 "For his inventions and leadership in developing and standardizing industrial control equipments."

PIERCE, DANA, AIEE Fellow 1928.

PIERCE, GEORGE W., IRE Fellow 1915.

PIERCE, HOMER J., AIEE Fellow 1951.

PIKE, O.W., IRE Fellow 1951 "For his pioneering contributions and his leadership in the development and design of industrial and radio tubes."

PIKLER, ARMIN H., AIEE Fellow 1913.

PIKLER, HENRY, AIEE Fellow 1921.

PILLIOD, JAMES J., AIEE Fellow 1934.

PINTO, ANTHONY, AIEE Fellow 1951.

PIPER, WILLIAM JOHN, AIEE Fellow 1961 "For contributions to the field of power metering."

PIPES, LOUIS A., IRE Fellow 1958 "For his contributions to the fields fo electronics and engineering mathematics."

PLUMB, HENRY HAMMOND, AIEE Fellow 1951.

PLUMB, HYLON THERON, AIEE Fellow 1941.

PODOLSKY, LEON, AIEE Fellow 1956, IRE Fellow 1957 "For his leadership and contributions in the field of electronics and electrical component engineering."

POLHEMUS, JAMES H., AIEE Fellow 1946.

POLLARD, ALFRED LEON, AIEE Fellow 1959 "For contributions to development and operating management of a large electric utility."

POLLARD, N.L., AIEE Fellow 1913.

POLLOCK, SAMUEL HURLEY, AIEE Fellow 1962 "For contributions to the application and control of capacitors of Power Systems."

PONIATOFF, ALEXANDER M., IEEE Fellow 1965 "For outstanding leadership in the magnetic recording industry."

PONTE, M. J. H., IRE Fellow 1954 "For his contributions to high-power electron tubes, and his unswerving leadership in electronic research in France."

POOLE, ARTHUR BARNARD, AIEE Fellow 1959 "For contributions to creative design and manufacture of refined timing mechanisms by electrical means."

POOLE, C.O., AIEE Fellow 1913.

POOLE, CECIL P., AIEE Fellow 1913.

POOLE, ROBERT E., AIEE Fellow 1953 "In recognition of able administration of scientific and engineering projects, and contributions to the development of radio telephony and television, radar, and atomic weapons."

PORSKIEVIES, ANTHONY J., AIEE Fellow 1925.

PORTER, FREDERICK MEREDITH, AIEE Fellow 1960 "For contributions to hydrogen cooling of synchronous condensers and generators, excitation systems and high voltage transmission."

PORTER, GEORGE A., AIEE Fellow 1961 "For contributions to engineering, organization and professional development in a large electric utility."

PORTER, JOSEPH F., AIEE Fellow 1933.

PORTER, LAWRENCE COPELAND, AIEE Fellow 1936.

PORTER, ROLAND GUYER, AIEE Fellow 1944.

PORTS, D.C., IRE Fellow 1961 "For contributions to research in electromagnetic propagation."

PORTS, EARL GEORGE, AIEE Fellow 1946.

POST, GEORGE GILBERT, AIEE Fellow 1933.

POTE, ALFRED J., IRE Fellow 1959 "For contributions to high power accelerators, radio frequency heating and communication systems."

POTTER, CHARLES P., AIEE Fellow 1929.

POTTER, JOHN T., IEEE Fellow 1971 "For contributions to the technology of electronic data-processing equipment, and for leadership in the design and production of such equipment."

POTTER, RALPH K., IRE Fellow 1941.

POUARD, MICHEL J., IEEE Fellow 1978 "For contributions to the development and testing of high-voltage switchgear and to high-voltage and high-current testing."

POULSEN, VALDEMAR, IRE Fellow 1915.

POWEL, CHARLES ALFRED, AIEE Fellow 1941.

POWELL, ALVIN L., AIEE Fellow 1926.

POWELL, WILLIAM H., AIEE Fellow 1912.

PRATT, HARADEN, IRE Fellow 1929. AIEE Fellow 1937.

PRATT, WILLIAM H., AIEE Fellow 1913.

PRICE, L.M., IRE Fellow 1949 "For his contributions to the development, production and application of electron tubes in Canada."

PRICE, LESLIE D., AIEE Fellow 1951.

PRICE, WILLIAM S., IEEE Fellow 1970 "For contributions to the electrical, mechanical, and environmental aspects of EHV transmission."

PRIEST, CONAN A., IRE Fellow 1947 "For his contributions as an engineer, executive and organizer in the field of radio transmitter development and design."

PRIEST, DONALD H., IEEE Fellow 1965 "For contributions in extending the limits on power and frequency in communications and radar."

PRIHAR, ZVI, IEEE Fellow 1964 "For contributions to communications systems, and for applications of operations analysis to communications networks."

PUCHSTEIN, ALBERT F., AIEE Fellow 1951.

PUFFER, WILLIAM L., AIEE Fellow 1913.

PUNCHARD, J.C.R., IRE Fellow 1953 "For his contributions to the design and development of communication and radar equipment, particularly for the Canadian Armed Services."

PUPIN, MICHAEL L., IRE Fellow 1915.

PUTMAN, HENRY ST. CLAIR, AIEE Fellow 1912.

PYE, HARVEY NORRIS, AIEE Fellow 1950.

Q

QUARLES, DONALD AUBREY, AIEE Fellow 1941. IRE Fellow 1954 "For his distinguished service in the administration of major technical programs in the fields of communications, military electronics and atomic ordnance."

QUINAN, GEORGE E., AIEE Fellow 1918.

R

RADDIN, ELLERY H., AIEE Fellow 1956 "For his accomplishments in applying engineering principles to the development of new light sources."

RADER, JOHN FRANK, AIEE Fellow 1959 "For contribuions to design, construction and operation of an electric utility system."

RADER, RAY, AIEE Fellow 1950.

RADFORD, WILLIAM HENRY, IRE Fellow 1954 "In recognition of his contributions through teaching and research in the field of radio communications." AIEE Fellow 1959 "For his contributions to national defense through education and communications."

RANDOLPH, L.S., AIEE Fellow 1913.

RANGER, RICHARD HOWLAND, IRE Fellow 1928. AIEE Fellow 1951.

RAO, VEPA V.L., IEEE Fellow 1965 "For contributions to engineering education in India."

RAPPAPORT, PAUL, IEEE Fellow 1970 "For contributions to energy-conversion devices, and for leadership in the professional organization of this field."

RATHBUN, HAROLD VERNON, AIEE Fellow 1961 "For contributions to the design and development of transmission and distribution systems in Central Kansas."

RAUTH, ADOLPH WILLIAM, AIEE Fellow 1948.

RAWCLIFFE, GORDON H., IEEE Fellow 1970 "For contributions to electrical-machines development, particularly in the field of pole-amplitude modulation."

RAYMOND, FRANCOIS H., IRE Fellow 1962 "For contributions to the analysis and development of electronic computers and automation systems."

REA, WILTON TREAT, AIEE Fellow 1962 "For contributions to telegraph, teletype and data transmission systems."

READ, ERNEST KENNETH, AIEE Fellow 1948.

READY, LESTER S., AIEE Fellow 1936.

REAGAN, MAURICE E., AIEE Fellow 1941.

REBER, HENRY LINTON, AIEE Fellow 1913.

REBER, SAMUEL, AIEE Fellow 1912. IRE Fellow 1923.

REDDING, CHARLES SUMMERFIELD, AIEE Fellow 1948.

REED, HARRISON P., AIEE Fellow 1921.

REED, HENRY R., AIEE Fellow 1947. IRE Fellow 1960 "For contributions to engineering education."

REED, M.B., IRE Fellow 1961 "For contributions to the understanding of electrical networks."

REED, W. EDGAR, AIEE Fellow 1913.

REES, JOHN BARTOW, AIEE Fellow 1932.

REEVE, JOHN, IEEE Fellow 1981 "For contributions to high-voltage direct current transmission systems."

REIBER, ALBERT H., AIEE Fellow 1940.

REID, CHARLES ROY, AIEE Fellow 1944.

REID, HARRY, AIEE Fellow 1931.

REID, JOHN D., IRE Fellow 1949 "For his developments in radio frequency circuits."

REIGER, SIEGFRIED H., IEEE Fellow 1969 "For contributions to data transmission and to the establishment of the first commercial communications satellite system."

REIMERS, THEODORE D., AIEE Fellow 1956 "For his contributions to design of electric production, transmission and distribution systems for a highly concentrated metropolitan load area."

REINARTZ, JOHN L., IRE Fellow 1958 "For his early work on radio wave propagation."

REINHARD, LOUIS F., AIEE Fellow 1920.

REIST, HENRY G., AIEE Fellow 1913.

REISZ, EUGENE, IRE Fellow 1954 "In recognition of his pioneering contribution to the development of grid-controlled electronic tubes."

RENO, C.S., AIEE Fellow 1913.

RENSHAW, DAVID E., AIEE Fellow 1946.

REOCH, ALEXANDER E., IRE Fellow 1916.

REPPY, W. HOWARD, AIEE Fellow 1912.

REUKEMA, L. E., IRE Fellow 1951 "For research in electromagnetic radiation and a distinguished record in the teaching of electronics."

REYNOLDS, HERBERT B., AIEE Fellow 1947.

RHEA, T.R., IEEE Fellow 1964 "For contributions to the electrification of industry, and to the teaching and training of engineers."

RHODES, FREDERICK L., AIEE Fellow 1913.

RHODES, GEORGE I., AIEE Fellow 1918.

RHODES, SAM R., AIEE Fellow 1941.

RHODES, SAMUEL G., AIEE Fellow 1913.

RIAD, MAHNOUD M., IEEE Fellow 1972 "For contributions to the development of telecommunications networks and leadership in engineering education."

RICE, CALVIN W., AIEE Fellow 1912.

RICE, CHESTER W., IRE Fellow 1928. AIEE Fellow 1939.

RICE, EDWIN WILBUR, JR., AIEE Fellow 1913.

RICE, RALPH H., AIEE Fellow 1920.

RICHARDS, WILLIAM E., AIEE Fellow 1926.

RICHARDSON, DONALD ELMER, AIEE Fellow 1950.

RICHARDSON, W.G., IRE Fellow 1955 "For his contributions to the art of broadcasting, both sound and television, in Canada."

RICHEY, ALBERT S., AIEE Fellow 1912.

RICHMOND, HAROLD BOURS, IRE Fellow 1929. AIEE Fellow 1940.

RICKER, CHARLES W., AIEE Fellow 1919.

RICKER, CLAIRE WILLIAM, AIEE Fellow 1945.

RICKETTS, FORREST E., AIEE Fellow 1950.

RIDDILE, JOHN SCOTT, AIEE Fellow 1921.

RIDDLE, FRANK H., AIEE Fellow 1947.

RIDENOUR, L.N., IRE Fellow 1955 "For his stimulating leadership in the field of electronic engineering."

RIED, HARRY, AIEE Fellow 1931.

RIESZ, ROBERT R., IEEE Fellow 1966 "For basic contributions in the areas of speech, hearing and the human factor in communication systems."

RIETZ, EARL B., IEEE Fellow 1969 "For contributions to the formulation of circuit breaker testing standards and procedures."

RIGGS, ARTHUR FORREST, AIEE Fellow 1945.

RIGGS, W.M., AIEE Fellow 1913.

RIVES, T.C., IRE Fellow 1954 "For his leadership in military electronic research and development."

RIXSE, JOHN H., JR., IEEE Fellow 1975 "For contributions to rural electrification in the United States and in developing nations."

ROBB, W.L., AIEE Fellow 1913.

ROBERTS, EDWARD A., IEEE Fellow 1983 "For contributions to frequency control and advancements in magnetic recording technology."

ROBERTS, EDWARD ALEXANDER, AIEE Fellow 1951.

ROBERTSON, BURTIS L., AIEE Fellow 1943.

ROBERTSON, ELGIN BARNETT, AIEE Fellow 1945.

ROBERTSON, S.D., IRE Fellow 1961 "For contributions to engineering in the microwave field."

ROBINSON, CHARLES A., AIEE Fellow 1922.

ROBINSON, DURGHT P., AIEE Fellow 1913.

ROBINSON, EDGAR W., AIEE Fellow 1956 "For his engineering and administrative achievements in the design and operation of a large steam and hydroelectric power utility."

ROBINSON, HENRY B., AIEE Fellow 1950.

ROBINSON, J.K., AIEE Fellow 1912.

ROBINSON, LEWIS T., AIEE Fellow 1912.

ROBINSON, LLOYD N., AIEE Fellow 1925.

ROBINSON, PRESTON, IEEE Fellow 1965 "For contributions to the development of electrochemical capacitors."

ROBINSON, WILLIAM, AIEE Fellow 1913.

ROBISON, ARCH R., AIEE Fellow 1941.

ROBISON, SAMUEL S., IRE Fellow 1915.

ROBLEY, ROY R., AIEE Fellow 1945.

ROCKWELL, RONALD J., IRE Fellow 1945 "For active work in the affairs of the Instute and in the engineering of high-power international broadcast transmitters." AIEE Fellow 1956 "For his contributions in the field of high power radio broadcasting."

ROCKWOOD, ALAN C., IRE Fellow 1959 "For contributions to standardization of electron devices."

RODE, NORMAN FREDERICK, AIEE Fellow 1943.

RODMAN, WALTER SHELDON, AIEE Fellow 1928.

ROE, FREEMONT LEE, AIEE Fellow 1958 "For contributions to electric power system development in pioneer territory."

ROE, RALPH COATS, AIEE Fellow 1933.

ROEHL, C.E., AIEE Fellow 1912.

ROETKEN, ALFRED A., IRE Fellow 1953 "For his contributions to the development of transcontinental microwave radio realy system and the development of single-sideband radio telephone receivers."

ROGERS, CLARENCE ELLAMS, AIEE Fellow 1936.

ROGOFF, JULIAN, AIEE Fellow 1961 "For contributions to improvement of electrical connections and to design and development of circuit protective devices."

ROLLER, F.W., AIEE Fellow 1912.

ROLLINS, NORMAN A., AIEE Fellow 1940.

ROLLOW, JAMES GRADY, AIEE Fellow 1942.

ROMANOWITZ, H. ALEX, AIEE Fellow 1962 "For contributions to engineering education and to raising the standards of technical institutes."

ROMIG, HARRY G., IEEE Fellow 1965 "For contributions to quality control."

ROMNES, HAAKON I., AIEE Fellow 1957 "For his leadership in and contributions to communication engineering." IRE Fellow 1961 "For contributions in the field of communications engineering and management."

RONCI, V.L., IRE Fellow 1961 "For contributions to the development of electron devices."

ROOS, OSCAR C., IRE Fellow 1917.

ROPER, DENNEY W., AIEE Fellow 1914.

RORDEN, HAROLD L., AIEE Fellow 1957 "For his contributions to control of lightning and surges of power transmission circuits."

ROSA, EDWARD B., AIEE Fellow 1913.

ROSE, ARTHUR FISHER, AIEE Fellow 1951.

ROSE, G.M., JR., IRE Fellow 1954 "For his contributions to vacuum tube research, design, and manufacture, and his transistor developments."

ROSENBLITH, WALTER A., IEEE Fellow 1973 "For contributions to the quantification of sensory and neural functions, and for professional and educational activities linking engineering and living systems."

ROSENBUSCH, GILBERT, AIEE Fellow 1912.

ROSENTHAL, LEON W., AIEE Fellow 1919.

ROSS, IRVINE EMERSON, JR., AIEE Fellow 1960 "For contributions to development of commutations at high altitudes."

ROSS, J.D., AIEE Fellow 1912.

ROSS, JAMES HARVEY, AIEE Fellow 1936.

ROSS, MALCOM D., AIEE Fellow 1948.

ROSS, NORMAN N., AIEE Fellow 1912.

ROSS, PHILIP N., IEEE Fellow 1972 "For his contributions to nuclear propulsion systems and nuclear power plants."

ROSS, R.A., AIEE Fellow 1912.

ROSS, RALPH H., AIEE Fellow 1943.

ROSSELOT, GERALD A., IEEE Fellow 1965 "For contributions to engineering management research and education."

ROSSMAN, ALLEN M., AIEE Fellow 1926.

ROTERS, HERBERT C., IEEE Fellow 1972 "For his contributions to teaching and practical applications of magnetic materials and devices and particularly for his development of the hysterisis motor."

ROTHE, HORST, IRE Fellow 1959 "For contributions in the fields of electron tubes and the theory of noise."

ROTTY, OSCAR J., AIEE Fellow 1945.

ROULSTON, KENNETH I., IEEE Fellow 1968 "For contributions to the improvement of scintillation counter techniques both in nuclear physics and geo-physics."

ROWE, BERTRAND P., AIEE Fellow 1913.

ROWE, HARTLEY, AIEE Fellow 1948.

ROWELL, RALPH M., IEEE Fellow 1968 "For outstanding engineering contributions in the field of electrical measurements and technical leadership in the development, design and application of electrical recording and indicating instruments."

ROY, JAMES A.S., AIEE Fellow 1947.

RUCKER, B. PARKS, AIEE Fellow 1913.

RUDENBERG, REINHOLD, AIEE Fellow 1950.

RUDGE, WILLIAM JEROME, JR., AIEE Fellow 1951.

RUFFNER, CHARLES S., AIEE Fellow 1912.

RUMSEY, AUSTIN L., AIEE Fellow 1943.

RUNYON, FREDERICK O., AIEE Fellow 1923.

RUPIN, MICHAEL, AIEE Fellow 1915.

RUSH, WALTER A., IRE Fellow 1924.

RUSHMORE, D.B., AIEE Fellow 1913.

RUSSELL, JOHN BENJAMIN, JR., AIEE Fellow 1951. IRE Fellow 1958 "For his management of electronic development and research organizations."

RUTH, EDWARD S., AIEE Fellow 1945.

RYAN, HARRIS J., AIEE Fellow 1923.

RYAN, PHILIP, AIEE Fellow 1959 "For contributions to creation and production of electrical controls."

RYBNER, JORGEN, IRE Fellow 1953 "For outstanding contributions to research and teaching in the field of telecommunications."

RYERSON, WILLIAM S., AIEE Fellow 1912.

S

SAATHOFF, GEORGE WEBSTER, AIEE Fellow 1930.

SABBAH, HASSAN C.A., AIEE Fellow 1933.

SACHS, JOSEPH, AIEE Fellow 1912.

SAH, ADAM PEN-TUNG, AIEE Fellow 1945.

SAKRISON, DAVID J., IEEE Fellow 1974 "For contributions in research and teaching in communication theory and its applications."

SALINGER, H.W.G., IRE Fellow 1951 "For his original contributions to electron optic and filter theory."

SALMON, VINCENT, IRE Fellow 1962 "For contributions to acoustics."

SALSBURY, RAYMOND J., AIEE Fellow 1955 "For contributions to protection, standardization, testing and development of power system equipment."

SALTON, HARRY D., AIEE Fellow 1948.

SAMUELS, MAURICE, AIEE Fellow 1924.

SANDERSON, CLARENCE H., AIEE Fellow 1918.

SANFORD, FRANK E., AIEE Fellow 1946.

SARAGA, WOLJA, IRE Fellow 1962 "For contributions to network theory and its application in communications."

SARGENT, F.C., AIEE Fellow 1912.

SARNOFF, DAVID, IRE Fellow 1917. AIEE Fellow 1951.

SAUNDERS, CECIL CLIFTON, AIEE Fellow 1959 "For contributions to the electrical aspects of industrial, chemical and nuclear projects."

SAUNDERS, LLEWELLYN, AIEE Fellow 1952.

SAVANT, DOMENICO P., AIEE Fellow 1935.

SAWTELLE, E.M., AIEE Fellow 1912.

SAWYER, MARK A., AIEE Fellow 1948.

SAWYER, W.H., AIEE Fellow 1913.

SAYLES, EDGAR VAN SYCKEL, AIEE Fellow 1938.

SCATTERGOOD, EZRA FREDERICK, AIEE Fellow 1913.

SCHADE, OTTO H., IRE Fellow 1951 "For his many contributions in the development of circuits and electronic devices used in television."

SCHAELCHLIN, WALTER, AIEE Fellow 1956 "For his control engineering achievements in the field of industry and marine propulsion."

SCHAIRER, O.S., IRE Fellow 1957 "For distinguished service in fostering and administering electronic research."

SCHAMBERBER, SANFORD O., AIEE Fellow 1949.

SCHEFFER, SEBASTIAN L., AIEE Fellow 1956 "For his contributions to control of transportation of bulk materials by rail and belt."

SCHEFFLER, FREDERICK A., AIEE Fellow 1912.

SCHELDORF, MARVEL W., IRE Fellow 1958 "For his contributions to the fields of radio frequency transmission lines and radiating systems."

SCHENK, M.H., IRE Fellow 1957 "For pioneering achievements in naval aviation electronics."

SCHICK, D. FREDERICK, AIEE Fellow 1913.

SCHIFF, MARTIN, AIEE Fellow 1939.

SCHIFREEN, CLEMENT SOLOMON, AIEE Fellow 1958 "For contributions toward extended life and load ratings of underground power cables."

SCHILDHAUER, EDWARD, AIEE Fellow 1913.

SCHILLER, A.R., AIEE Fellow 1953 "For distinction as an executive of a large electric utility and particularly for engineering, design, construction, and successful operation of a fast growing system, including the pioneering of the first completley intergrated binary-cycle (mercury-steam) plant."

SCHILLING, E.W., AIEE Fellow 1944.

SCHLEICHER, GEORGE BERTHOLD, AIEE Fellow 1961 "For contributions to the development of transformer-loss compensation techniques for effecting economies in high tension metering."

SCHLESINGER, KURT, IRE Fellow 1954 "For many contributions to cathode-ray tubes and television circuitry."

SCHMIDT, MARVIN L., AIEE Fellow 1949.

SCHMIDT, WALTER CARL, AIEE Fellow 1962 "For contributions to the development of wood pole high voltage transmission."

SCHMIT, DOMINIC F., IRE Fellow 1951 "For outstanding leadership and direction in many phases of radio engineering."

SCHMITT, GILBERT EUGENE, AIEE Fellow 1949.

SCHNEEBERGER, GROVER B., AIEE Fellow 1945.

SCHNURE, FRED OSCAR, AIEE Fellow 1949.

SCHOEN, A.M., AIEE Fellow 1913.

SCHOEPF, T.H., AIEE Fellow 1913.

SCHOLZ, CARL E., AIEE Fellow 1955 "For his design of high-power radio transmission systems and for contributions to international radiotelegraph operations." IRE Fellow 1956 "For contributions to international radio communication."

SCHOOLFIELD, HARRISON HERBERT, AIEE Fellow 1940.

SCHOUTEN, JAN F., IEEE Fellow 1969 "For contributions to research in visual and auditory perception and to telecommunications."

SCHRADER, H.J., IRE Fellow 1954 "For his application of imformation theory to navigation and television systems."

SCHREGARDUS, DIRK, AIEE Fellow 1940.

SCHREIBER, E.H., IRE Fellow 1957 "For contributions to radio and television broadcasting."

SCHRIBNER, CHARLES E., AIEE Fellow 1913.

SCHROEDER, GIULIO, AIEE Fellow 1918.

SCHUCHARDT, R.F., AIEE Fellow 1912.

SCHUCK, O.H., IRE Fellow 1957 "For contributions in the field of instrumentation and control."

SCHULKE, HERBERT A., JR., IEEE Fellow 1973 "For technical leadership and major contributions to the planning, research, development, and performance of worldwide communications systems."

SCHULTZ, NORMAN RUDOLPH, AIEE Fellow 1962 "For contributions to the analysis and planning of electric utility systems."

SCHULZ, THOMAS NORBERG, AIEE Fellow 1920.

SCHURIG, OTTO R., AIEE Fellow 1955 "For contributions to increased knowledge and understanding in several areas of electrical engineering; particularly of the electromagnetic forces and temperature rises of buses and conductors carrying heavy currents."

SCHWAGER, AUGUST CHARLES, AIEE Fellow 1944.

SCHWAN, CLARENCE FRED, AIEE Fellow 1960 "For contributions to design of rotation machinery, especially for high-frequency inducton heating."

SCHWARZ, B.A., IRE Fellow 1955 "For his contributions to the development and production of automobile radio."

SCHWEITZER, EDMUND O., AIEE Fellow 1920.

SCLATER, I.H., AIEE Fellow 1927.

SCOTT, A.H., IEEE Fellow 1964 "For contributions to precision electrical measurements, particularly in the field of dielectrics."

SCOTT, CARL FORSE, AIEE Fellow 1947.

SCOTT, CHARLES F., AIEE Fellow 1925.

SCOTT, HERMON H., IRE Fellow 1952 "For his contributions to acoustic measurement and the reduction of noise in audio reproduction."

SCOTT, J.C.W., IRE Fellow 1957 "For contributions in the field of ionospheric propagation."

SCOTT, ROBERT CLYDE, AIEE Fellow 1939.

SCOTT, ROBERT J., AIEE Fellow 1913.

SCOTT-TAGGART, JOHN, IRE Fellow 1935. AIEE Fellow 1938.

SCUDDER, FREDERICK J., AIEE Fellow 1947.

SEALEY, WILLIAM C., AIEE Fellow 1948.

SEAMANS, ROBERT C., JR., IEEE Fellow 1970 "For leadership in the application of electronics to guidance and control problems, and for direction of space programs."

SEARBY, NORMAN H., IEEE Fellow 1968 "For his contributions to research and development in the appllocation of electronics to guided missile projects."

SEARING, EMERY DEFOREST, AIEE Fellow 1928.

SEARING, HUDSON R., AIEE Fellow 1930.

SEBAST, FREDERICK M., AIEE Fellow 1951.

SEDDON, J. CARL, IRE Fellow 1958 "For his theoretical studies and rocket-borne experiments leading to a better understanding of the ionosphere."

SEEGER, EDWIN WILBUR, AIEE Fellow 1936.

SEELEY, H.T., IEEE Fellow 1964 "For contributions to the design and application of relays for power system protection."

SEELEY, STUART W., IRE Fellow 1943 "For the development of practical apparatus for the radio industry, particularly in the fields of television and high-frequency broadcast reception."

SEELEY, WALTER J., AIEE Fellow 1945.

SEELYE, HOWARD P., AIEE Fellow 1943.

SEIBT, GEORGE, IRE Fellow 1915.

SEINYARD, WILLIAM O., IRE Fellow 1945 "In recognition of his work in promoting electronics and the affairs of the Institute, particularly in his district."

SELBY, MYRON C., IEEE Fellow 1968 "For outstanding scientific achievement in high-frequency radioelectronics measurement techniques and standards, and for promoting, initiating and implenting national and international standardization activities and cooperation."

SELLERS, LESTER RALPH, AIEE Fellow 1958 "For contributions to coordination of design of large hydroelectric and steam plants."

SESSIONS, E.O., AIEE Fellow 1913.

SEVER, GEORGE F., AIEE Fellow 1912.

SHAAD, GEORGE C., AIEE Fellow 1913.

SHACKELFORD, BENJAMIN E., IRE Fellow 1938, AIEE Fellow 1949.

SHAFFER, LEE JAY, AIEE Fellow 1960 "For contributions to the development of personnel for communications engineering."

SHANKLIN, GEORGE BRYAN, AIEE Fellow 1946.

SHARP, CLAYTON H., AIEE Fellow 1912.

SHARP, HUBERT, AIEE Fellow 1951.

SHAW, G.R., IRE Fellow 1954 "For his technical contributions to the manufacture of radio tubes and their standardization."

SHAW, HENRY S., IRE Fellow 1935.

SHEALS, VINCENT ALLEN, AIEE Fellow 1950.

SHELBY, ROBERT EVART, IRE Fellow 1948 "For his many contributions to sound and television broadcasting." AIEE Fellow 1951.

SHELDON, S.R., AIEE Fellow 1913.

SHELDON, SAMUEL, AIEE Fellow 1913.

SHEPARD, FRANCIS H., AIEE Fellow 1924.

SHEPARD, ROBERT BLANCHARD, AIEE Fellow 1936.

SHEPHERD, JUDSON O'DONALD, AIEE Fellow 1950.

SHEPHERD, ROBERT V., AIEE Fellow 1957 "For his contributions to the design of large electric motors and generators."

SHOEMAKER, GUY T., AIEE Fellow 1939.

SHOEMAKER, RICHARD WOOLSEY, AIEE Fellow 1941.

SHOVER, BARTON R., AIEE Fellow 1912.

SHREEVE, HERBERT EDWARD, AIEE Fellow 1930.

SHULER, WILLIAM, AIEE Fellow 1944.

SHUTE, EMMETT R., AIEE Fellow 1945.

SHUTT, CHARLES COURTER, AIEE Fellow 1950.

SIBLEY, EDGAR DOW, AIEE Fellow 1936.

SIBLEY, ROBERT, AIEE Fellow 1913.

SIEGEL, K.M., IEEE Fellow 1964 "For contributions in the field of scattering of electromagnetic waves by complex targets."

SIEGFRIED, VICTOR, AIEE Fellow 1957 "For his contributions to the field of dielectrics and cable insulation."

SIEGMUND, HUMPHREYS OLIVER, AIEE Fellow 1945.

SILING, P.F., IRE Fellow 1951 "For contributions to the establishment and administration of national and international radio frequency allocations."

SILLS, HUBERT R., AIEE Fellow 1955 "For his contributions to the design of large hydroelectric generators."

SILSBEE, FRANCIS B., AIEE Fellow 1942.

SILVER, ARTHUR E., AIEE Fellow 1926.

SILVER, SAMUEL, IRE Fellow 1954 "For research in the field of electromagnetic radiation."

SIMMONS, B.F., AIEE Fellow 1913.

SIMMONS, DONALD MACLAREN, AIEE Fellow 1928.

SIMON, ARTHUR, AIEE Fellow 1913.

SIMON, EMIL J., IRE Fellow 1915.

SIMPSON, FREDERICK G., IRE Fellow 1915.

SIMPSON, LEROY C., IRE Fellow 1959 "For contributions to broadcasting and telecommunication."

SIMS, WILLIAM F., AIEE Fellow 1933.

SINCLAIR, CARROLL T., AIEE Fellow 1936.

SINCLAIR, H.A., AIEE Fellow 1912.

SINDEBAND, M.L., AIEE Fellow 1926.

SINGER, FRED J., AIEE Fellow 1951.

SINK, ROBERT L., IRE Fellow 1960 "For contributions to the field of digital instrumentaion systems."

SINNETT, C.M., IRE Fellow 1955 "For his contributions in the field of electronic circuitry."

SISSON, CHARLES E., AIEE Fellow 1944.

SITZ, EARL LEROY, AIEE Fellow 1961 "For contributions to the teaching of Electrical Engineering."

SKEETS, WILFRED FRANCIS, AIEE Fellow 1943.

SKIFF, WARNER M., AIEE Fellow 1939.

SKIFTER, H.R., IRE Fellow 1951 "For his technical contributions to radio broascasting and leadership of electronics research and development."

SKINKER, MURRAY FONTAINE, AIEE Fellow 1934.

SKINNER, CHARLES E., AIEE Fellow 1912.

SKIRROW, JOHN F., AIEE Fellow 1925.

SLEE, JOHN A., IRE Fellow 1929.

SLEEMAN, HECTOR, AIEE Fellow 1936.

SLEEPER, HARVEY PRESCOTT, AIEE Fellow 1949.

SLEPIAN, JOSEPH, AIEE Fellow 1927. IRE Fellow 1945 "For his contributions as scientist, engineer and inventor in the field of electronics."

SLICHTER, WALTER I., AIEE Fellow 1912.

SLIGANT, STANLEY, AIEE Fellow 1927.

SLOAN, MATTHEW SCOTT, AIEE Fellow 1930.

SMALE, J.A., IRE Fellow 1958 "For his pioneering work in telecommunications."

SMALL, ALVAH, AIEE Fellow 1943.

SMALLEY, DAVID D., AIEE Fellow 1947.

SMITH, ARTHUR BESSEY, AIEE Fellow 1922.

SMITH, BOB H., IEEE Fellow 1969 "For pioneering contributions to large radio-frequency-type particle accelerators, and for developing unique methods of resonator analysis in the accelerator field."

SMITH, CHARLES G., IEEE Fellow 1969 "For fundamental contributions to the development and understanding of plasma phenomena and devices."

SMITH, ERNEST F., AIEE Fellow 1921.

SMITH, HAROLD B., AIEE Fellow 1913.

SMITH, HAROLD LEONARD, AIEE Fellow 1942.

SMITH, HOMER KEPHART, AIEE Fellow 1944.

SMITH, HOWARD H., IEEE Fellow 1967 "For pioneering work in voice frequency signalling system transmission standards."

SMITH, J.E., IRE Fellow 1955 "For his contributions to the art of radio communications."

SMITH, JOHN HERBERT, AIEE Fellow 1962 "For contributions to electrical system design and project administration."

SMITH, JULIAN C., AIEE Fellow 1912.

SMITH, KILBURN M., AIEE Fellow 1953 "For contributions to the field of power system engineering, particularly in the poineering of high-voltage transmission of bulk power in densely loaded areas."

SMITH, LOUIS GOLDEN, AIEE Fellow 1934.

SMITH, MALCOLM W., AIEE Fellow 1956 "For his contributions to the economical construction of transmission and generation facilities of a large electric power system."

SMITH, PHILIP T., IRE Fellow 1960 "For contributions to the development of high-power transmitting tubes."

SMITH, ROBERT S., JR., AIEE Fellow 1957 "For contributions to the development of protective devices for electric power distribution systems."

SMITH, SAMUEL BAILEY, AIEE Fellow 1961 "For contributions to the development of interconnected high voltage power systems for the electric utility industry."

SMITH, VICTOR G., IEEE Fellow 1966 "For his work in applied mathematics in the solution of systems engineering problems."

SMITH, WALTER CHARLES, AIEE Fellow 1940.

SMITH, WALTER CURTIS, AIEE Fellow 1949.

SMITH, WILLIAM LINCOLN, AIEE Fellow 1940.

SMITH, WILLIE RALPH, AIEE Fellow 1930.

SMITH-ROSE, R.L., IRE Fellow 1944 "In recognition of his pioneer work in the field of direction finding and radio propagation, allied to his leadership of an outstanding radio research group."

SMOUROFF, ALEXANDER, AIEE Fellow 1924.

SMULLIN, L.D., IRE Fellow 1957 "For contributions in the field of microwave tubes."

SMYTH, J.B., IRE Fellow 1956 "For contributions and leadership in electromagnetic propagation research."

SNAVELY, A. BOWMAN, AIEE Fellow 1951.

SNELL, JOHN F.C., AIEE Fellow 1912.

SNIDER, GEORGE E., AIEE Fellow 1939.

SNOOK, H. CLYDE, AIEE Fellow 1920.

SNOW, WILLIAM B., IRE Fellow 1962 "For contributions to sound systems."

SNYDER, CLAYTON EDWARD, AIEE Fellow 1950.

SNYDER, FRANKLIN LLOYD, AIEE Fellow 1951.

SOGGE, RICHARD C., AIEE Fellow 1953 "For achievement in national and international electrical standardization."

SOMERVILLE, GARTH GERAINT, AIEE Fellow 1960 "For contributions to improvement of magnetic cores for power transformers."

SORENSEN, ROYAL W., AIEE Fellow 1919.

SOTHMAN, P.W., AIEE Fellow 1917.

SOUTHWORTH, GEORGE C., IRE Fellow 1942.

SPAGNOLETTI, PHILIP H., IEEE Fellow 1968 "For contributions to the design of short-wave broadcasting equipment and single side-band telecommunications equipment."

SPALDING, SAMUEL ALBERT, AIEE Fellow 1929.

SPANGENBERG, KARL, IRE Fellow 1949 "For his many technical contributions, particularly his analytical work on vacuum tubes."

SPARKES, HARRY P., AIEE Fellow 1942.

SPAULDING, GEORGE WHITTIER, AIEE Fellow 1950.

SPEIGHT, HERBERT, AIEE Fellow 1951.

SPENCER, CLARENCE G., AIEE Fellow 1924.

SPENCER, P.L., IRE Fellow 1955 "For his contributions to the design and development of electron tubes."

SPENCER, PAUL, AIEE Fellow 1912.

SPENCER, ROY C., IRE Fellow 1960 "For contributions to the theory of microwave antennas."

SPOONER, THOMAS, AIEE Fellow 1929.

SPORN, PHILIP, AIEE Fellow 1930.

SPRACKLEN, EMERY EVERLIEGH, AIEE Fellow 1941.

SPRAGUE, FRANK, AIEE Fellow 1912.

SPRINGER, F.W., AIEE Fellow 1913.

SPRONG, SEVERN D., AIEE Fellow 1912.

SPURCK, ROBERT MICHAEL, AIEE Fellow 1937.

SQUIER, GEORGE O., IRE Fellow 1916. AIEE Fellow 1919.

STAEGE, STEPHEN A., AIEE Fellow 1919.

STAFFORD, RICHARD THOMAS, AIEE Fellow 1947.

STAHL, EDWARD C.M., AIEE Fellow 1951.

STAHL, NICHOLAS, AIEE Fellow 1913.

STANFORD, ALAN GRIFFITH, AIEE Fellow 1950.

STANLEY, ROY MORGAN, AIEE Fellow 1923.

STANLEY, WILLIAM, AIEE Fellow 1913.

STANSBURY, CARROLL, AIEE Fellow 1938.

STANSEL, NUMAN REID, AIEE Fellow 1945.

STARBIRD, LEVI C., AIEE Fellow 1944.

STARK, E.E., AIEE Fellow 1913.

STARR, JAMES HAMMOND, AIEE Fellow 1946.

ST. CLAIR, HARRY PRENTICE, AIEE Fellow 1944.

STEBBINS, THEODORE, AIEE Fellow 1913.

STEEDE, JACK HOSFORD, AIEE Fellow 1962 "For contributions to planning, engineering and operation of an electric utility."

STEEL, WILLIAM ARTHUR, IRE Fellow 1950 "For his contributions in Canada in advancing development of military radio broadcasting and international communication."

STEEN, HALFDAN, AIEE Fellow 1915.

STEIN, ADAM, JR., IRE Fellow 1917.

STEIN, I. MELVILLE, AIEE Fellow 1939.

STEINBERG, MAX JACOB, AIEE Fellow 1948.

STEINMAYER, ALWIN G., AIEE Fellow 1957 "For contributions to equipment for power distribution systems particularly in the field of cutouts and fuses."

STEINMETZ, CHARLES P., AIEE Fellow 1912.

STEMMONS, BEVERLY L., AIEE Fellow 1951.

STERLING, GEORGE E., IRE Fellow 1949 "In recognition of his long public service in the radio communication field and, in particular, for the organization and operation of radio wartime intelligence activities, which were of significant importance."

STEVENS, ARCHIE M., IRE Fellow 1931.

STEVENS, FRANK J., AIEE Fellow 1944.

STEVENS, J. FRANKLIN, AIEE Fellow 1913.

STEVENS, THEODORE, AIEE Fellow 1938.

STEVENS, WILLIAM C., AIEE Fellow 1920.

STEVENSON, ALEXANDER R., JR., AIEE Fellow 1937.

STEVENSON, JAMES ROSS, AIEE Fellow 1946.

STEWART, WELBY, IEEE Fellow 1969 "For contributions in the fields of audio, magnetic recording, and communication."

STICKNEY, GEORGE H., AIEE Fellow 1924.

STIGANT, STANLEY AUSTEN, AIEE Fellow 1938.

STILL, ALFRED, AIEE Fellow 1914.

STILLWELL, LEWIS B., AIEE Fellow 1912.

STINER, H. WRAY, AIEE Fellow 1943.

STITZER, ARTHUR B., AIEE Fellow 1913.

STIVENDER, EDWARD H., AIEE Fellow 1949.

STOCKER, CLOSMAN P., AIEE Fellow 1955 "For his developments of static ringing machines and dial tone generators."

STODDART, RICHARD R., IRE Fellow 1958 "For his development of instrumentation for the measurement of radio interference."

STOEKLE, ERWIN R., AIEE Fellow 1933.

STOELTING, HERMAN O., IEEE Fellow 1976 "For contributions to the electric power industry in the field of surge arrester design and standardization."

STONE, CHARLES W., AIEE Fellow 1912.

STONE, EDMUND CUSHING, AIEE Fellow 1930.

STONE, ELLERY W., IRE Fellow 1924.

STONE, HENRY A., JR., IEEE Fellow 1968 "For contributions to the development of advanced componenets for communications systems."

STONE, JOHN STONE, IRE Fellow 1915.

STOPPELMANN, FRED HENRY, AIEE Fellow 1951.

STORER, N.W., AIEE Fellow 1913.

STORER, SIMON B., AIEE Fellow 1913.

STOUT, MELVILLE B., AIEE Fellow 1950.

STRANDBERG, HERBERT V., AIEE Fellow 1956 "For his accomplishments in the development of a transmission and distribution system to supply electricity to a metropolitan area."

STRASBOURGER, JULIUS C., AIEE Fellow 1960 "For contributions to mechanization of electric utility system construction and maintenance."

STRATLON, SAMUEL W., AIEE Fellow 1924.

STRELZOFF, JOSEPH A., AIEE Fellow 1961 "For contributions to the engineering education and the teaching of Electro-magnetic Theory."

STRENG, LEWIS S., AIEE Fellow 1926.

STRIEBY, MAURICE EDWARD, AIEE Fellow 1941. IRE Fellow 1961.

STRINGFIELD, THEODORE FRANK, AIEE Fellow 1958 "For his contributions to high frequency powering of communication systems."

STROBERG, ERIC ARVID, AIEE Fellow 1951.

STRONG, FREDERICK G., AIEE Fellow 1913.

STROVER, JAMES R., AIEE Fellow 1953 "For contributions to the electrical development of an electric utility particularly in transmission and distribution."

STUART, DONALD M., IRE Fellow 1962 "For contributions to electronic aids to air navigation."

STUART, HARVE R., AIEE Fellow 1913.

STUEVE, WINFRED H., AIEE Fellow 1954 "For contributions to the use of electric power in the petroleum industry; particularly to the movement of oil and gas through pipe lines."

STUMPF, MALCOLM W., IEEE Fellow 1968 "For contributions to original design and direction of pioneering development of publicly acceptable overhead urban transmission lines."

SUHR, FRED W., IEEE Fellow 1966 "For outstanding contributions to the design of fractional horsepower induction machines."

SUMMERHAYES, HENRY ROSWELL, AIEE Fellow 1939.

SUMNER, WILLIAM ANDROSS, AIEE Fellow 1959 "For contributions to design of self-protective distribution transformers and to use of grain-oriented sheet steel transformer cores."

SUNDIUS, HERBERT WILLIAM, AIEE Fellow 1947.

SUNDT, EDWARD V., IEEE Fellow 1968 "For contributions to the understanding of circuit interruption phenomena."

SUNNY, BERNARD E., AIEE Fellow 1913.

SUNSTEIN, DAVID E., IRE Fellow 1958 "For his contributions to the field of airborne radar."

SUOPE, GERARD, AIEE Fellow 1922.

SUTHERLAND, GEORGE, AIEE Fellow 1927.

SVOBODA, ANTONIN, IEEE Fellow 1968 "For his contributions in logical design, mechanical design and his fundamental work on residual class number systems."

SWANSTROM, FRANK N., AIEE Fellow 1943.

SWEETNAM, A.H., AIEE Fellow 1948.

SWENSON, B.V., AIEE Fellow 1913.

SWENSON, GEORGE W., AIEE Fellow 1936.

SWINTON, A.A. CAMPBELL, AIEE Fellow 1913. IRE Fellow 1914.

SWISS, JACK, IEEE Fellow 1970 "For leadership and contributions in insulating materials and processes."

SWOPE, GERARD, AIEE Fellow 1922.

SYKES, WILFRED, AIEE Fellow 1914.

T

TAIT, FRANK M., AIEE Fellow 1912.

TALBOT, EMMETT D., AIEE Fellow 1936.

TAPPAN, FRANK G., AIEE Fellow 1936.

TAPSCOTT, RALPH HENRY, AIEE Fellow 1929.

TAPY, RALPH WILVER, AIEE Fellow 1948.

TARBOUX, JOSEPH G., AIEE Fellow 1943.

TATE, THOMAS ROUSE, AIEE Fellow 1935.

TATUM, FINLEY W., AIEE Fellow 1956 "For his contributions in the field of engineering education."

TATUM, LEWIS L., AIEE Fellow 1913.

TAYLOR, ALBERT HOYT, IRE Fellow 1920. AIEE Fellow 1946.

TAYLOR, ALBERT LEROY, AIEE Fellow 1941.

TAYLOR, ARTHUR Y., IEEE Fellow 1970 "For leadership in the planning and management of nuclear power."

TAYLOR, CHARLES H., IRE Fellow 1915.

TAYLOR, EDWARD, AIEE Fellow 1922.

TAYLOR, HAMILTON DANA, AIEE Fellow 1949.

TAYLOR, JOHN BELLAMY, AIEE Fellow 1914.

TAYLOR, JOHN J.W., AIEE Fellow 1913.

TAYLOR, NORMAN H., IEEE Fellow 1970 "For pioneer work in computer logic, circuitry and techniques, electronic system reliabilty, and in the application of computers to real-time control."

TAYLOR, THOMAS A., AIEE Fellow 1955 "For leadership in solution of inductive coordination problems and in the field of communication engineering."

TAYLOR, WILLIAM T., AIEE Fellow 1912.

TEAGUE, WILLIAM LUDY, AIEE Fellow 1950.

TEBO, GORDON BUCHANAN, AIEE Fellow 1961 "For contributions to research and to developments relating to the safe use of electric power."

TELLEGEN, B.D.H., IRE Fellow 1955 "For his contributions and teachings in the field of vacuum tubes and communication networks."

TEMPLE, W.E.S., AIEE Fellow 1913.

TEMPLIN, JOHN R., AIEE Fellow 1927.

TENNEY, HARRY W., AIEE Fellow 1943.

TERMAN, FREDERICK E., IRE Fellow 1937, AIEE Fellow 1945.

TERRELL, CHARLES FOSTER, AIEE Fellow 1946.

TERRELL, W.D., IRE Fellow 1929.

TERRY, CLARK ARTHUR, AIEE Fellow 1938.

TERWILLIGER, CHARLES VAN ORDEN, AIEE Fellow 1944.

TESLA, NICOLA, AIEE Fellow 1917.

THACKER, MANEKLAL S., AIEE Fellow 1949.

THALER, J.A., AIEE Fellow 1913.

THAMES, WILLIAM M., IRE Fellow 1962 "For contributions to military electronics."

THOMAS, DONALD E., IEEE Fellow 1967 "For contributions to communication devices and systems."

THOMAS, EARL ROGER, AIEE Fellow 1949.

THOMAS, GEORGE N., AIEE Fellow 1926.

THOMAS, HENRY P., IRE Fellow 1951 "In recognition of his pioneering contributions in the development of high-frequency and microwave equipment and communication systems."

THOMAS, JAMES WINTHROP, AIEE Fellow 1920.

THOMAS, PERCY H., AIEE Fellow 1912.

THOMAS, RALPH L., AIEE Fellow 1944.

THOMAS, TALMAGE DEWITT, AIEE Fellow 1961 "For contributions in guiding the design and construction of a large electric utility system."

THOMAS, WILLIAM A., AIEE Fellow 1913.

THOMPSON, BROWDER J., IRE Fellow 1938.

THOMPSON, GORDON, AIEE Fellow 1946.

THOMPSON, HENRY HAVELOCK, AIEE Fellow 1947.

THOMPSON, JOSEPH S., AIEE Fellow 1938.

THOMPSON, WILBUR H., AIEE Fellow 1913.

THOMS, ALEXANDER P., AIEE Fellow 1926.

THOMSON, ELIHU, AIEE Fellow 1913.

THORESEN, JETMUND J., AIEE Fellow 1921.

THORNTON, FRANK, JR., AIEE Fellow 1921.

THURNAUER, E., AIEE Fellow 1913.

THURSTON, ERNEST B., AIEE Fellow 1923.

TIMBIE, WILLIAM H., AIEE Fellow 1924.

TIMMERMAN, A.H., AIEE Fellow 1912.

TISCHENDORFER, F., AIEE Fellow 1913.

TITUS, OLCOTT WOOD, AIEE Fellow 1944.

TODD, JAMES M., AIEE Fellow 1945.

TOEPFER, A.H., IEEE Fellow 1964 "For his contributions to atomic reactors."

TOLMIE, J.R., IRE Fellow 1955 "For his early contributions to radio."

TOLSON, W.A., IRE Fellow 1956 "For invention and development in the field of television receivers and military target tracking."

TOMBERG, VICTOR T., IEEE Fellow 1970 "For pioneering work in medical electronics, particularly radiation hazards."

TOMPKINS, FREDERICK NEALE, AIEE Fellow 1944.

TORCHIO, PHILIP, AIEE Fellow 1912.

TORGERSEN, HAROLD, AIEE Fellow 1959 "For contributions to engineering education."

TOTH, EMERICK, IRE Fellow 1958 "For his contributions to the field of military communications."

TOWN, GEORGE R., IRE Fellow 1950 "For his contributions in radio receiver engineering and research." AIEE Fellow 1954 "For contributions as an engineer, teacher, research administrator, and author in the fields of radio and television."

TOWNLEY, CALVERT, AIEE Fellow 1912.

TOWNSEND, STANLEY C., AIEE Fellow 1957 "For his contributions to the production and transmission of electric power."

TRACY, ATLEC H., AIEE Fellow 1923.

TRACY, GORDON FREDERICK, AIEE Fellow 1961 "For contributions to electrical engineering education."

TRADUP, ALBERT, AIEE Fellow 1951.

TRAFFORD, E.W., AIEE Fellow 1913.

TRAVER, OLIVER CLAGGETT, AIEE Fellow 1943.

TREANOR, EDWARD DONELLY, AIEE Fellow 1948.

TREAT, ROBERT, AIEE Fellow 1939.

TREVOR, BERTRAM A., IRE Fellow 1953 "For contributions in the field of pulse multiplex communication systems and long-range receiving equipment."

TRICE, THOMAS WHEELER, AIEE Fellow 1961 "For contributions to the management of engineering and operations of a major electric utility system."

TRIPLETT, HUGH A., AIEE Fellow 1948.

TROLESE, LOUIS G., IRE Fellow 1960 "For contributions to the understanding of radio wave propagation."

TRONE, DIMITRI, AIEE Fellow 1953 "For pioneering activities in electric power generation on the electrification of transportation system in foreign fields particularly in Russia and Brazil."

TROTTER, HERBERT, JR., IEEE Fellow 1971 "For leadership in communication technology."

TROUARD, SIDNEY E., AIEE Fellow 1962 "For contributions to corrosion protection."

TRUCKSESS, DAVID E., AIEE Fellow 1958 "For his contributions to power supplies for high frequency communication systems."

TRUEBLOOD, HOWARD M., AIEE Fellow 1940.

TUCKER, CARLTON EVERETT, AIEE Fellow 1939.

TULLER, W.G., IRE Fellow 1955 "For contributions to the advance of theoretical analysis of information theory and its practical application."

TUNIS, WILLIAM C., IRE Fellow 1951 "For his contributions to fire control radar and his guidance of postwar electronic development for military purposes."

TUPPER, B.R., IRE Fellow 1954 "In recognition of his application of radio techniques to the extension of toll-telephone services in Canada."

TURNER, HUBERT M., IRE Fellow 1937. AIEE Fellow 1944.

TURNER, WILLIAM S., AIEE Fellow 1913.

TUVE, MERLE ANTHONY, IRE Fellow 1945 "For pioneering in the experimental probing of the ionosphere and giving to the world the first publication of the experimental proof of the existence of the ionosphere; and for having initiated at an early date the pulse method of probing by reflection which is the basis of modern radar."

TYKOCINER, J.T., IEEE Fellow 1964 "For his pioneering contributions to radio science."

TYVAND, JAMES ADDISON, AIEE Fellow 1945.

U

UCHIDA, HIDENARI, IEEE Fellow 1974 "For contributions to the theory and practice of VHF and UHF antennas."

UEHLING, FRITZ FREDERICK, AIEE Fellow 1950.

UHL, WILLIAM F., AIEE Fellow 1955 "For his contributions to the development and design of hydroelectric generating plants."

UHR, IRWIN A., AIEE Fellow 1950.

ULLRICH, EDWARD H., IRE Fellow 1939.

ULREY, DAYTON, IRE Fellow 1950 "For pioneering research and for administrative and technical contributions to the development of special purpose and power tubes."

UMANSKY, LEONID A., AIEE Fellow 1945.

UNDERHILL, CHARLES REGINALD, AIEE Fellow 1914.

UNDERHILL, GEORGE HAROLD, AIEE Fellow 1961 "For contributions to the development of common neutral primary distribution aerial cable systems."

V

VAITSES, GREGORY S., AIEE Fellow 1949.

VAN ANTWERP, GEORGE STEWART, AIEE Fellow 1951.

VANCE, ARTHUR W., IEEE Fellow 1973 "For pioneering contributions to television, electronic instruments, regulated power supplies, weapons systems, and simulators."

VAN DER BIJL, HENDRIK J., IRE Fellow 1928.

VAN DER POL, BALTH, JR., IRE Fellow 1929.

VANDERSLUIS, WARREN M., AIEE Fellow 1929.

VAN DEUSEN, GEORGE L., IRE Fellow 1958 "For his direction of research and development of military electronics equipment."

VAN DUUREN, H.C. A., IRE Fellow 1961 "For contributions toward the reliability of radio communication."

VAN DYCK, ARTHUR F., IRE Fellow 1925.

VAN DYKE, K.S., IRE Fellow 1944 "In recognition of his work in research on characteristics of Piezo-Electric Crystals and their application to frequency control."

VANN NESS, BARTOW, JR., AIEE Fellow 1947.

VAN SOEST, JOHANNES L., IRE Fellow 1962 "For contributions to defense research in the Netherlands."

VAN VALKENBURG, HERMON LEACH, AIEE Fellow 1939.

VARIAN, RUSSELL H., IRE Fellow 1952 "For his contributions in the field of applied physics and, particularly, in the field of velocity-modulated tubes."

VAUGHAN, DANIEL C., AIEE Fellow 1948.

VAUGHAN, H. EARLE, IRE Fellow 1962 "For contributions to the development of high-speed switching systems."

VAUGHAN, JOHN F., AIEE Fellow 1912.

VAUGHN, FRANCIS A., AIEE Fellow 1913.

VERNIER, CHARLES, AIEE Fellow 1927.

VICKERS, HERBERT, AIEE Fellow 1934.

VINAL, GEORGE W., AIEE Fellow 1942.

VINCENT, HAROLD BLANCHARD, AIEE Fellow 1930.

VIVIAN, JOSEPH H., AIEE Fellow 1948.

VOGEL, WILHELM, AIEE Fellow 1950.

VOKAC, CHARLES W., IEEE Fellow 1967 "For his contributions to the electric-arc furnace, education and standardization."

VOLKMER, THEODORE FREDERICK, AIEE Fellow 1962 "For his contribution to the design and standardization of transformers."

VON KROGH, KARL KONORA, AIEE Fellow 1917.

VREELAND, FREDERICK K., AIEE Fellow 1913. IRE Fellow 1926.

W

WADA, MASAMOBU, IEEE Fellow 1980 "For contributions to the development of display devices, and for leadership in engineering education."

WADDELL, CHARLES E., AIEE Fellow 1912.

WADDINGTON, CHARLES VERNE, AIEE Fellow 1949.

WAGNER, CHARLES F., AIEE Fellow 1940.

WAGNER, DONALD M., AIEE Fellow 1927.

WAGNER, EDWARD A., AIEE Fellow 1927.

WAGNER, HOMER H., AIEE Fellow 1943.

WAGNER, MILTON HENRY, AIEE Fellow 1930.

WAGNER, WALTER CALVIN, AIEE Fellow 1941.

WAGSTAFF, ROBERT R., AIEE Fellow 1958 "For his contributions to design and execution of large power projects."

WAITE, LESLIE OSGOOD, AIEE Fellow 1950.

WALDSCHMITT, JOSEPH A., IEEE Fellow 1968 "For technical and administrative contributions to advanced world-wide communications systems."

WALL, HARRY RUTHERFORD, AIEE Fellow 1960 "For contributions to development and operation of a large utility system."

WALLACE, JAMES M., IEEE Fellow 1966 "For contributions in electrical distribution equipment and technical direction in this field."

WALLACE, JOHN N., AIEE Fellow 1920.

WALLAU, H.L., AIEE Fellow 1913.

WALLER, ALFRED E., AIEE Fellow 1921.

WALLIS, JOHN, AIEE Fellow 1956 "For his accomplishments in power transmission and distribution design and in system planning."

WALLS, JOHN A., AIEE Fellow 1913.

WALTON, PERCY JAMES, AIEE Fellow 1944.

WARD, O.M., AIEE Fellow 1941.

WARDER, WALTER, AIEE Fellow 1913.

WARE, JOHN S., AIEE Fellow 1944.

WARE, PERRY H., IEEE Fellow 1969 "For work in the field of solid dielectrics, and for ability to communicate such knowledge to the industry."

WARNECKE, R.R., IRE Fellow 1950 "For his engineering and research contributions to vacuum tube theory and design in France."

WARNER, JOHN C., IRE Fellow 1933.

WARNER, KENNETH B., IRE Fellow 1936.

WARNER, ROBERT W., AIEE Fellow 1939.

WARNER, RUSSELL GILLETTE, AIEE Fellow 1945.

WARREN, HENRY ELLIS, AIEE Fellow 1948.

WARREN, HOWARD S., AIEE Fellow 1913.

WARREN, MEAD, JR., IEEE Fellow 1967 "For contributions to power generation and transmission, in the advancement of communication, telemetry, and preventive maintenance."

WARRINGTON, A. VAN COURTLANDT, AIEE Fellow 1949.

WASHINGTON, BOWDEN, IRE Fellow 1923.

WASSERMAN, CHARLES, AIEE Fellow 1962 "For contributions to the design and application of power system controls."

WATANABE, MICHAEL S., IEEE Fellow 1973 "For contributions to probabilistic information processing and the foundations of cybernetics."

WATERMAN, FRANK N., AIEE Fellow 1912.

WATERS, GRANVILLE, AIEE Fellow 1921.

WATERS, JAMES S., AIEE Fellow 1949.

WATERS, W.L., AIEE Fellow 1913.

WATERSON, KARL W., AIEE Fellow 1922.

WATKINS, SAMUEL S., AIEE Fellow 1946.

WATSON, ARTHUR EUGENE, AIEE Fellow 1937.

WATSON, THOMAS AUGUSTUS, AIEE Fellow 1915.

WATSON-WATT, ROBERT, IRE Fellow 1947 "For his early contributions to radio and for his pioneering work in radar."

WATTERSON, HAROLD E., IRE Fellow 1923.

WATTS, FRANK WILMER, AIEE Fellow 1945.

WATTS, GEORGE W., AIEE Fellow 1912.

WAUGH, LESTER R., AIEE Fellow 1933.

WAY, SYLVESTER BEDELL, AIEE Fellow 1938.

WAYNICK, A.H., IRE Fellow 1957 "For contributions in radio transmission, ionosphere research, and engineering education."

WEAGANT, ROY A., IRE Fellow 1915.

WEAVER, WILLIAM D., AIEE Fellow 1913.

WEBB, W.L., IRE Fellow 1954 "For early contributions and engineering leadership in radio direction finding, radar, and radio communication."

WEBER, CLIFFARD A.M., AIEE Fellow 1922.

WEBSTER, ARTHUR GORDON, AIEE Fellow 1919.

WEEKS, EDWIN R., AIEE Fellow 1913.

WEICHSEL, HANS, AIEE Fellow 1921.

WEIL, JOSEPH, AIEE Fellow 1936. IRE Fellow 1957 "For contributions to engineering education."

WEIL, ROBERT T., JR., AIEE Fellow 1961 "For contributions to engineering education."

WEINBACH, MENDEL P., AIEE Fellow 1939.

WEIR, I.R., IRE Fellow 1949 "For his pioneering work in the development and application of transmitting equipment for higher frequencies and higher power."

WEISS, FERNAND CARL, AIEE Fellow 1951.

WEISS, H.G., IEEE Fellow 1964 "For contributions to the development of high-power radar."

WEISS, HERBERT, IEEE Fellow 1977 "For contributions to the development and applications of semiconducting galvanomagnetic devices."

WELLER, GEORGE LOUIS, AIEE Fellow 1931.

WELLER, HARRY W., AIEE Fellow 1912.

WELLS, G.A., AIEE Fellow 1912.

WELLS, HARRY W., IRE Fellow 1953 "In recognition of his contributions to ionospheric research and the organization of a world-wide network of ionospheric stations."

WELLS, WALTER F., AIEE Fellow 1912.

WELLWOOD, ARTHUR RUSSELL, AIEE Fellow 1936.

WELSCH, JAMES P., IEEE Fellow 1968 "For his contributions to thermal measurement and cooling of electronic equipment."

WENDELL, EDWARD N., IRE Fellow 1947 "For his contributions to the development and production of radio systems for navigating and landing airplanes by instrument."

WENNER, FRANK, AIEE Fellow 1926.

WERLY, BERLYN MCINTYRE, AIEE Fellow 1950.

WERNER, CALVIN J., AIEE Fellow 1950.

WEST, CHARLES P., AIEE Fellow 1946.

WEST, GIFFORD BETHEL, AIEE Fellow 1946.

WEST, HARRY R., AIEE Fellow 1951.

WETMORE, WILLIAM F., AIEE Fellow 1955 "For contributions to power system engineering of a large utility and to the development of standards fo circuit breakers and rotating machinery."

WHEELER, KIMBALL L., AIEE Fellow 1959 "For contributions to the development of a large power system and the rehabilitation of power systems in Korea."

WHEELER, LYNDE P., IRE Fellow 1928.

WHEELER, SCHUYLER SKOATS, JR., AIEE Fellow 1912.

WHELCHEL, CORNELIUS C., AIEE Fellow 1960 "For contributions to design of large steam turbines and power plants and evolution of practical nuclear generation."

WHIPPLE, CLYDE COLBURN, AIEE Fellow 1951.

WHITAKER, U.A., AIEE Fellow 1962 "For contributions to development and mass production of wire connection devices."

WHITAKER, WILLIAM GORDON H., AIEE Fellow 1944.

WHITE, WILLIAM COMINGS, IRE Fellow 1940. AIEE Fellow 1947.

WHITE, WILLIAM CROMBIE, AIEE Fellow 1958 "For his contributions to the education of engineers."

WHITEHEAD, JOHN B., AIEE Fellow 1912.

WHITEHURST, ROLAND, AIEE Fellow 1943.

WHITING, DONALD FAIRFAX, AIEE Fellow 1930.

WHITLOW, GEORGE S., AIEE Fellow 1954 "For contributions in the development of efficient transmission systems and the applications of the heat pump."

WHITMAN, EZRA B., AIEE Fellow 1951.

WHITMORE, HAROLD B., IEEE Fellow 1969 "For pioneering contributions and creative leadership toward improved functioning of the American patent system."

WHITNEY, RICH D., AIEE Fellow 1925.

WICKENDEN, WILLIAM E., AIEE Fellow 1939.

WIENER, FRANCIS M., IRE Fellow 1962 "For research in electroacoustics."

WILBRAHAM, ROSSITER W., AIEE Fellow 1945.

WILCOX, NORMON T., AIEE Fellow 1912.

WILCOX, THOMAS W., IEEE Fellow 1970 "For contributions to high-voltage circuit breakers."

WILDER, CLIFTON WHITE, AIEE Fellow 1914.

WILDER, WILFORD D., IEEE Fellow 1966 "For contributions to the development of equipment and procedures for power control in large electric supply systems."

WILDER, WILLARD S., AIEE Fellow 1960 "For contributions to direction of design of power plants, transmission and distribution."

WILDMAN, LEONARD D., IRE Fellow 1915.

WILKENSON, JOHN REED, AIEE Fellow 1943.

WILKENSON, KENNETH L., AIEE Fellow 1930.

WILKERSON, JEFFERSON R., IEEE Fellow 1967 "For his many contributions to the techniques of electronic intelligence and electronic warfare."

WILKINS, ARNOLD F., IRE Fellow 1958 "For his contributions ot research and short wave direction finding."

WILKINS, ROY, AIEE Fellow 1929.

WILKINSON, KENNETH L., AIEE Fellow 1930.

WILLEY, FRANK W., AIEE Fellow 1924.

WILLIAMS, ALBERT J., JR., IRE Fellow 1955, AIEE Fellow 1961 "For his contributions to the field of self-balancing recorders of electrical quantities, and for his invention, research, and development to practice of null-balance system of precise electrical measurement."

WILLIAMS, ALLISON RIDLEY, AIEE Fellow 1935.

WILLIAMS, ARTHUR, AIEE Fellow 1913.

WILLIAMS, EARL C., AIEE Fellow 1951.

WILLIAMS, EVERARD MOTT, IRE Fellow 1954 "For his contributions to the development of military electronic equipment." AIEE Fellow 1958 "For contributions to the theory of communication and the development of electric spark machine tools."

WILLIAMS, F.C., IRE Fellow 1957 "For contributions to the theory of noise in vacuum tubes, and in the fields of radar and digital computers."

WILLIAMS, SAMUEL B., AIEE Fellow 1926, IRE Fellow 1957 "For contributions to telephone switching systems and to computers."

WILLIAMSON, R.B., AIEE Fellow 1912.

WILLICUTT, FREDERICK WEISS, AIEE Fellow 1948.

WILLIS, CLODIUN HARRIS, AIEE Fellow 1942.

WILLIS, EDWARD J., AIEE Fellow 1913.

WILLIS, FRED W., AIEE Fellow 1919.

WILLS, W.P., IEEE Fellow 1964 "For contributions to the field of instrumentation particularly recorders and controllers."

WILSON, FREMONT, AIEE Fellow 1912.

WILSON, H.S., AIEE Fellow 1913.

WILSON, J.R., IRE Fellow 1954 "For his stimulating leadership in research on electron tubes and devices, and in their development and manufacture."

WILSON, ROBERT L., AIEE Fellow 1913.

WILSON, ROBERT M., AIEE Fellow 1913.

WILSON, WILLIAM, IRE Fellow 1928.

WILTSE, STANLEY B., AIEE Fellow 1949.

WINCHESTER, ALBERT E., AIEE Fellow 1912.

WINFIELD, FREDERICK C., AIEE Fellow 1951.

WINKLER, GEORGE HOWE, AIEE Fellow 1930.

WINKLER, GERNOT M.R., IEEE Fellow 1970 "For contributions to precision frequency control and time measurement."

WINNE, HARRY A., AIEE Fellow 1945.

WINOGRAD, HAROLD, AIEE Fellow 1946.

WINTRINGHAM, W.T., IRE Fellow 1951 "For his studies of colotimetry and color television, and for his work on television interference problems."

WISE, ROGER M., IRE Fellow 1937.

WISEMAN, ROBERT J., AIEE Fellow 1927.

WISHARD, WILLIAM W., AIEE Fellow 1953 "For outstanding contributions and originality in the application of high-voltage subtransmission systems to high density load areas, and for sound engineering judgement and able leadership of an engineering organization of a large utility system."

WISNER, RAYMOND R., AIEE Fellow 1956 "For his application of electrical engineering to the design of power plants and the problem of industry."

WITHINGTON, SIDNEY, AIEE Fellow 1924.

WITTING, EDWARD G., IRE Fellow 1962 "For services and contributions to military electronics."

WOHLGEMUTH, MELVILLE J., AIEE Fellow 1937.

WOLCOTT, C.F., JR., IRE Fellow 1957 "For contributions as an engineer and executive, and for active participation in Instute affairs."

WOLF, EDGAR FAHRNEY, AIEE Fellow 1962 "For contributions to the reliability and safety of electric utility equipment."

WOLFF, IRVING, IRE Fellow 1942 "For basic research in centimeter wave radio and for application of it to the development of navigation instruments."

WOOD, ELVIN MORLEY, AIEE Fellow 1943.

WOOD, GEORGE R., AIEE Fellow 1912.

WOOD, HENRY BLAKE, AIEE Fellow 1938.

WOOD, JAMES J., AIEE Fellow 1918.

WOOD, JOSEPH D., AIEE Fellow 1956 "For contributions to the development and application of switch-gear and its components."

WOODBRIDGE, J. LESTER, AIEE Fellow 1921.

WOODFORD, LAURANCE G., AIEE Fellow 1942.

WOODMANSEE, FAY, AIEE Fellow 1913.

WOODROW, HARRY R., AIEE Fellow 1923.

WOODS, FRANCIS LAWRENCE, AIEE Fellow 1960 "For contributions to the field of network and distribution transformer engineering."

WOODS, JOHN C., AIEE Fellow 1953 "For contributions in the field of electrical control of large turbogenerators and improvements in substation design contributing to their reliability and economy."

WOODYARD, JOHN R., IRE Fellow 1962 "For contributions to electronic engineering education."

WOOLDRIDGE, WILLIAM JOHN, AIEE Fellow 1938.

WOOLSTON, LOUIS F.B., AIEE Fellow 1929.

WOOLVERTON, ROBERT B., IRE Fellow 1915.

WOONTON, G.A., IRE Fellow 1951 "For long service as a teacher of radio engineering and for varied contributions through research in the field of electronics."

WORK, WILLIAM ROTH, AIEE Fellow 1940.

WORTH, B.G., AIEE Fellow 1912.

WRAY, J.G., AIEE Fellow 1913.

WRIGHT, ERNEST M., AIEE Fellow 1943.

WRIGHT, R.I., AIEE Fellow 1913.

WRIGHT, RALPH H., AIEE Fellow 1946.

WRIGLEY, GEORGE, AIEE Fellow 1953 "For instituting pioneer and revolutionary engineering changes in the textile industry resuting in the early establishment of the trend to individual motor drive."

WULFING, HARRY E., AIEE Fellow 1944.

WULFSBERG, ARTHUR H., IEEE Fellow 1969 "For contributions in the development of modern aerospace communication and radio navigation."

WUSTENEY, HERBERT H., IEEE Fellow 1970 "For work in development and design of telegraph and data transmission equipment."

WYATT, FRANCIS DALE, AIEE Fellow 1949.

WYATT, KENNETH SAPWELL, AIEE Fellow 1948.

WYCKOFF, R.D., IRE Fellow 1955 "For his contributions to geophysical instrumentation, and the development of guided missiles."

WYLIE, LAURENCE, AIEE Fellow 1958 "For pioneering contributions to the design, construction and operation of an electrified railroad and its equipment."

WYMAN, FRANK T., AIEE Fellow 1921.
WYMAN, MAYNARD BROWN, AIEE Fellow 1942.
WYMAN, WALTER SCOTT, AIEE Fellow 1919.
WYNKOOP, H.S., AIEE Fellow 1912.
WYNNE, FRANCIS E., AIEE Fellow 1920.

Y

YAGI, HIDETSUGU, IRE Fellow 1924.
YARDLEY, J.L.M., AIEE Fellow 1917.
YARNALL, JOSEPH H., AIEE Fellow 1946.
YASUKAWA, DAIGORO, IEEE Fellow 1966 "For his contributions in the electrical industry in both engineering and manufacturing."
YERK, RALPH G., AIEE Fellow 1961 "For contributions to design of wood transmission, distribution and substation structures."
YERKES, EARL P., AIEE Fellow 1940.
YOKURA, MORINOSUKE, AIEE Fellow 1921.
YORKE, GEORGE MARSHALL, AIEE Fellow 1929.
YOUNG, C. GRIFFITH, AIEE Fellow 1913.
YOUNG, HERBERT W., AIEE Fellow 1925.

YOUNG, LEO C., IRE Fellow 1943 "In recognition of his pioneer work on the causes of the vagaries of short wave propagation and for his invaluable research laboratory." AIEE Fellow 1948.
YOUNG, OWEN D., IRE Fellow 1928.

Z

ZAHL, HAROLD A., IRE Fellow 1950 "For his early contributions to radar and for his guidance of the Army Signal Corps research program in the transition from war to peace."
ZENNECK, JONATHAN, IRE Fellow 1914.
ZIMMERMAN, CARL P., IEEE Fellow 1967 "For his outstanding contributions to the development of high voltage power systems and equipment."
ZOTTU, P.D., IRE Fellow 1944 "For his contributions in the field of high frequency heating, particularly in the application of dielectric heating in industry."
ZWERLING, STANLEY, IEEE Fellow 1972 "For leadership in electronics standardization and reliability engineering."
ZWORYKIN, VALDIMIR KOSMA, IRE Fellow 1938. AIEE Fellow 1945.

Part III
Present Fellows

A

AARON, M. ROBERT, Fellow 1968. Born: August 21, 1922, Philadelphia, Pa. Degrees: B.S.E.E., 1949, M.S.E.E., 1951, University of Pennsylvania.

Fellow Award "For contributions to the analysis of PCM systems"; Alexander Graham Bell Medal(corecipient), 1978. Other Awards: National Academy of Engineering, 1979.

AARONS, JULES, Fellow 1973. Born: October 3, 1921. Degrees: B.S., 1942, City College of the City University of New York; M.A., 1951, Boston University; Ph.D., 1954, University of Paris.

Fellow Award "For contributions to the understanding of propagation effects on satellite communications, and for advances in solar radio astronomy"; Harry Diamond Award, IEEE, 1982. Other Awards: Exceptional Civilian Service Award, U.S. Air Force, 1969; Townsend Harris Award, City College of the City University of New York, 1970.

ABBOTT, CHARLES T., AIEE Fellow 1962. Born: December 16, 1905, Hershey, Nebr. Degrees: B.S., 1930, Massachusetts Institute of Technology.

Fellow Award "For contributions to electric utility engineering, operation, and management."

ABBOTT, HENRY H., Fellow 1967. Born: May 27, 1901, Columbus, Ohio. Degrees: B.E.E., 1923, E.E., 1930, Ohio State University.

Fellow Award "For contributions to private branch exchange switching systems." Other Awards: Sigma Xi; 37 U.S. patents.

ABBOTT, THOMAS A., AIEE Fellow 1950. Born: March 9, 1901, Kansas City, Mo. Degrees: B.S.E.E., 1923, M.S., 1925, Washington University; E.E., 1930, Yale University.

Other Awards: Fellow, Instrument Society of America, 1965; Special Service Award, Western Society of Engineers, 1966.

ABETTI, PIER A., Fellow 1982. Born: February 2, 1921. Degrees: Dr.(E.E.and Indust.Eng.), 1945, University of Pisa, Italy; M.S.E.E., 1948, Ph.D.(E.E.), 1952, Illnois Institute of Technology.

Fellow Award "For the development of electromagnetic models of transformers, the use of digital computers for the design of power transformers, and leadership in extra-high-voltage power transmission"; First Prize, AIEE Northeast District, 1953. Other Awards: Coffin Award, General Electric, 1952; Eta Kappa Nu Recognition Award, 1953; Recognition Award, Italian Historial Society of America, 1954; International Prize Montefiore, Liege, Belgium, 1955.

ABRAHAM, GEORGE, Fellow 1964. Born: July 15, 1918, New York, N.Y. Degrees: Sc.B., 1940, Brown University; S.M., 1942, Harvard University; Ph.D., 1972, University of Maryland.

Fellow Award "For research on solid-state phenomena and for contributions to graduate engineering education"; IRE Distinguished Service Award, 1963; USAB Citation of Honor, IEEE, 1979; IEEE Distinguished Service Award, 1980; Harry Diamond Memorial Award, IEEE, 1981, "For development of multistable semiconductor devices and integrated circuits and their application to military systems"; IEEE Centennial Medal, 1984. Other Awards: Sigma Xi, 1947; Sigma Pi Sigma, 1948; Scientific Research Society of America, 1949; Sigma Tau, 1955; Group Achievement Award, Navy Fleet Ballistic Missile Program, 1961; Fellow, American Association for the Advancement of Science, 1964; Fellow, Washington Academy of Sciences, 1964; Tau Beta Pi, 1967; Eta Kappa Nu, 1967; Navy Edison Award, 1971; Navy Research Publications Award, 1975; Fellow, New York Academy of Sciences, 1976; Washington Society of Engineers Award, 1981; District of Columbia Science Citation, 1982.

ABRAMS, RICHARD L., Fellow 1982. Born: April 20, 1941, Cleveland, OH. Degrees: B.(Eng. Physics), 1964, Ph.D.(Applied Physics), 1968, Cornell University.

Fellow Award "For contributions to optical communications and nonlinear optics and for technical leadership in the development of CO_2 waveguide lasers." Other Awards: L.A. Hyland Patent Award, Hughes Aircraft Company, 1981; Fellow, Optical Society of America, 1982; Tau Beta Pi; Phi Kappa Phi.

ABRAMSON, NORMAN, Fellow 1973. Born: April 1, 1932, Boston, Mass. Degrees: A.B., 1953, Harvard University; M.A., 1955, University of California at Los Angeles; Ph.D., 1958, Stanford University.

Fellow Award "For contributions to engineering education, coding theory, pattern recognition, and computer networks"; Region 6 Achievement Award, IEEE, 1972, "For contributions to information theory and coding systems."

ADACHI, SABURO, Fellow 1984. Born: September 2, 1930, Yamagata Prefecture, Japan. Degrees: B.E., 1953, M.E., 1955, Ph.D., 1958, Tohoku University.

Fellow Award "For contributions to the theory and practice of antennas in plasma." Other Awards: Hattori-Hoko Prize for Research Achievement, 1971; Research Promotion Award for Scientific Measurements, 1978; Paper Award, IECE of Japan, 1981.

ADAIR, GEORGE P., IRE Fellow 1947. Born: December 8, 1903, Rancho, Tex. Degrees: B.S., 1926, Texas A and M University.

Fellow Award "For his technical direction of matters relating to allocation of radio frequencies." Other Awards: Honorary Life Member, International Municipal Signal Association, 1944; Tau Beta Pi, 1949.

ADAM, STEPHEN F., Fellow 1981. Born: February 28, 1929, Budapest, Hungary. Degrees: B.S. (Mech. Eng.), 1952, M.S. (Mech. Eng.), 1952, M.S.(E.E.), 1955, Ph.D.(E.E.), 1955, Polytechnic Institute of Budapest.

Fellow Award "For contributions to and technical leadership in the theory and application of automatic network analyzers"; Member, IEEE Microwave Theory and Techniques Society ADCOM, 1973-1983; Member, IEEE, Microwave Theory and Techniques Society Transactions Editorial Board; President, IEEE Microwave Theory and Techniques Society, 1980; Member, MTT-S Intnl. Symp. Technical Program Committee, 1974-1983; Member, Executive Committee, CPEM, IEEE Instrumentation and Measurement Society, 1974-1980; Member, Technical Activities Board, 1979-1980; Member, ED and ASSP Societies; Member, IEEE Fields Awards Committee since 1979; Chairman, Morris E. Leeds Awards Subcommittee, 1983-1984; Chairman, 1984 International Microwave Symposium; 1983-1984 Distinguished Microwave Lecturer, IEEE Microwave Theory and Teconiques Society; Member, IEEE-TAB Transnational Relations Committee, 1981-82; Chairman, IEM Society Awards Committee, 1984-. Other Awards: Member, Joint Industry Research Committee for the Standardization of Miniature Precision Coaxial Connectors, sponsored by U.S. Dept. of Commerce, 1968-1972; Member, Association of Old Crows, 1978-1982; Past Secretariat, International Electrotechnical Commission IEC-TC66-WG5.

ADAMS, ROBERT J., IRE Fellow 1961. Born: May 24, 1915, Solon, Iowa. Degrees: B.A., 1936, State University of Iowa; Ph.D., 1941, University of Wisconsin.

Fellow Award "For contributions to the design of radar antennas, and for direction of research and development of naval radar systems." Other Awards: Meritorious Civilian Service Award, U.S. Dept. of Navy, 1945 and 1976.

ADCOCK, WILLIS A., Fellow 1965. Born: November 25, 1922, St. Johns, Quebec, Canada. Degrees: B.S., 1943, Hobart College; Ph.D., 1948, Brown University;

M.L.A., Southern Methodist University, 1975.

Fellow Award "For contributions to the advancement of silicon material and device technology." Other Awards: Member, National Academy of Engineering, 1974; Principal Fellow, Texas Instruments Incorporated, 1978; Fellow, American Assn. for the Advancement of Science, 1981.

ADEN, ARTHUR L., Fellow 1966. Born: February 1, 1924, Cullom, Ill. Degrees: M.A., 1948, M.E.S., 1949, Ph.D., 1950, Harvard University.

Fellow Award "For contributions in the field of electromagnetic scattering and for leadership in research and development of devices and instrumentation for military and aerospace applications"; Engineering Achievement Award, Arizona Section, 1963, "For outstanding contributions to electronics." Other Awards: Gordon McKay Scholar, Harvard University, 1947-1948; Pre-doctoral Fellow, National Research Council, 1948-1950.

ADEY, W. ROSS, Fellow 1970. Born: January 31, 1922, Adelaide, Australia. Degrees: M.B., 1943, B.S., 1943, M.D., 1949, School of Medicine, University of Adelaide.

Fellow Award "For electronic techniques and principles in understanding brain organization and neural mechanisms." Other Awards: Nuffield Foundation Dominion Fellow in Medicine, University of Oxford, 1950; Fellow, Royal Society of London and Nuffield Foundation, 1956; Judson Herrick Fellow, American Association of Anatomists, 1963; Fellow, American Academy of Arts and Sciences, 1965.

ADLER, BENJAMIN, IRE Fellow 1962. Born: November 10, 1903, New York, N.Y. Degrees: E.E., 1926, D.Eng.(Hon.), 1967, Polytechnic Institute of Brooklyn.

Fellow Award "For contributions toward the effective utilization of the UHF spectrum." Other Awards: Distinguished Alumnus Award, Polytechnic Institute of Brooklyn, 1965; Fellow, American Association for the Advancement of Science, 1966; Registered Professional Engineer, New York State since 1951.

ADLER, RICHARD B., IRE Fellow 1960. Born: May 9, 1922, New York, N.Y. Degrees: S.B., 1943, Sc.D., 1949, Massachusetts Institute of Technology.

Fellow Award "For contribution to engineering education and to research in electronics." Other Awards: Premium Award, Royal Aeronautical Society, 1955; Fellow, American Academy of Arts and Sciences, 1964.

ADLER, ROBERT, Fellow 1951. Born: December 4, 1913, Vienna, Austria. Degrees: Ph.D., 1937, University of Vienna.

Fellow Award "For his developments of transmission and detection devices for frequency-modulated signals and of electromechanical filter systems"; Outstanding Achievement Award, Consumer Electronics Group,

1970; Outstanding Technical Paper Award, Chicago Spring Conference, 1974; Edison Medal, IEEE, 1980; Achievement Award, IEEE Sonics and Ultrasonics Group, 1981. Other Awards: Fellow, American Association for the Advancement of Science, 1951; Inventor of the Year Award, George Washington University's Patent, Trademark and Copyright Research Institute, 1967; Member, National Academy of Engineering, 1967.

AEIN, JOSEPH M., Fellow 1984. Born: January 26, 1936, Washington, D.C. Degrees: S.B./S.M.(E.E.), 1958, M.I.T.; Ph.D.(E.E.), 1962, Purdue.

Fellow Award "For contributions to the theory of multiple access, interference analyses, and capacity allocation as applied to communication satellites."

AGARWAL, PAUL D., Fellow 1975. Born: January 21, 1924, Ambala, India. Degrees: B.Sc., E.E., M.E.(HONS), 1944, Benares Hindu University; M.S.E.E., 1949, Illinois Institute of Technology; Ph.D., 1958, Polytechnic Institute of Brooklyn.

Fellow Award "For contributions to the design of rotating machines, controls, and applications to vehicular electric drive"; District Prize Paper Award, AIEE, 1960, 1962. Other Awards: Prize Paper, Sigma Xi, 1959.

AGGARWAL, JAGDISHKUMAR K., Fellow 1976. Born: November 19, 1936, Amritsar, India. Degrees: B.Sc., 1956, Bombay University; B.Eng., 1960, Liverpool University; M.S., 1961, Ph.D., 1964, University of Illinois.

Fellow Award "For contributions to time delay systems and digital filters." Other Awards: Second Annual Pattern Recognition Society Best Paper Award, 1975; John J. McKetta Energy Professor of Electrical Engineering, University of Texas, 1981-present.

AGUSTA, BENJAMIN, Fellow 1976. Born: July 1, 1931, Brooklyn, N.Y. Degrees: B.S.E.E., 1952, M.S.E.E., 1954, Massachusetts Institute of Technology; Ph.D.(E.E.)(Solid State Physics minor), 1964, Syracuse University.

Fellow Award "For contributions to monolithic semiconductor memory devices, processes, and applications." Other Awards: IBM Outstanding Contribution Award for Monolithic Memory Development Work, 1967; IBM Outstanding Contribution Award for Monolithic Memory Development and P.E. Work, 1969; Special IBM Outstanding Invention Award, 1969, 1970; Sixth Plateau Patent Invention Award, 1972

AICHER, LOUIS C., Fellow 1971. Born: December 25, 1912, Aberdeen, Idaho. Degrees: B.S.E.E., 1935, M.S.E.E., 1936, Kansas State University.

Fellow Award "For contributions in power transformer design relative to impulse testing techniques and analysis, corona measurement, and fault detection."

AIGRAIN, PIERRE R., IRE Fellow 1961. Born: September 28, 1924, Poitiers, France. Degrees: D.Sc., 1948, Carnegie Institute of Technology; D.Sc., 1950, University of Paris.

Fellow Award "For contributions to the theory and application of solid-state devices." Other Awards: Tate Award, American Institute, 1981.

AILLERET, PIERRE M., AIEE Fellow 1958. Born: March 10, 1900, Vienne en Arthies, Seine et Oise, France. Degrees: 1921, Ecole Polytechnique; 1924, Ecole Nationale des Ponts et Chaussees; 1925, Ecole Superieure d'Electricite.

Fellow Award "For contributions to planning, execution and interconnection of a nation-wide electric system." Other Awards: Societe Francaise des Electriciens; Societe des Ingenieurs Civils de France; Faraday Medal, British Institution of Electrical Engineers; Ingenieur General Honoraire des Ponts et Chaussees; President d'honneur de l'Union technique de l'Electricite; President d'honneur de l'Association Francaise de Normalisation; Past President, International Electrotechnical Commission, 1967-1970; Foreign associate member, U.S. National Academy of Engineering.

AKERS, SHELDON B., Fellow 1975. Born: October 22, 1926, Washington, D.C. Degrees: B.S.E.E., 1948, M.A., 1952, University of Maryland.

Fellow Award "For contributions to switching circuit theory and design automation."

AKIYAMA, MORIO, Fellow 1978. Born: September 16, 1907. Degrees: Bachelor of Engineering, 1932, Tokyo University; Ph.D. of Engineering, 1945, Tohoku University.

Fellow Award "For contributions to the analysis of variable parametric networks and applications of the method to power engineering and electrodynamics."

ALBERIGI-QUARANTA, ALESSANDRO Q., Fellow 1978. Born: June 15, 1927, Reggio E., Italy. Degrees: Laurea in Fisica, 1948, University of Roma.

Fellow Award "For contributions and technical leadership in the fields of nuclear electronics and charge transport in semiconductors."

ALBERT, ARTHUR L., AIEE Fellow 1941, IRE Fellow 1950. Born: March 13, 1899, Forest Grove, Oreg. Degrees: B.S. 1923, M.S., 1926, E.E.,1938, Oregon State University.

IRE Fellow Award "For his contribution to electronics as a teacher and writer."

ALBRECHT, PAUL F., Fellow 1980. Born: January 31, 1930. Degrees: B.S.(E.E.), B.S.(Bus. Admin.), 1952, University of Colorado.

Fellow Award "For outstanding contributions to the development, application, and teaching of reliability methods for analysis of large-scale electric power systems."

ALEXANDER, DONALD C., Fellow 1980. Born: September 1, 1921. Degrees: B.S.(E.E.) (high distinction), 1943, Worcester Polytechnic Institute.

Fellow Award "For outstanding technical leadership in the development of electrical insulation for cables."

ALEXANDER, THOMAS W., JR., AIEE Fellow 1944. Born: September 6, 1892, Pittsburgh, Pa. Degrees: Carnegie Institute of Technology (now Carnegie-Mellon University).

ALEXEFF, IGOR, Fellow 1981. Born: January 5, 1931, Pittsburgh, PA. Degrees: B.A., 1952, Harvard University; M.S., 1955, Ph.D., 1959, University of Wisconsin.

Fellow Award "For innovation in plasma engineering and controlled thermonuclear fusion research." Other Awards: Fellow, American Physical Society; Listed in "Who's Who in the South and Southwest," "Who's Who in America," "Who's Who in the World"; Weinberg Award, Oak Ridge National Laboratories; Member, MENSA; Chairman, NPSS PSAC EXCOM; Vice President, NPSS; Secretary-Treasurer, APS plasma physics division; Vice President, Southern Appalachian Science Fair; President, Tennessee Inventor's Association; P.E. License, Tenn.; 6 patents; Visiting Professor, Japan, India, South Africa, Brazil.

ALFORD, ANDREW, IRE Fellow 1942. Born: August 5, 1904, Samara, Russia. Degrees: A.B., 1924, University of California; Doctor of Applied Science(Hon.), 1975, Ohio University.

Fellow Award "For contributions to the theory of radiation and the design of short-wave antennas"; Pioneer Award, 1965, "In recognition of inventions of antennas which have found wide use in radio navigation systems, including VOR and instrument landing systems. The term 'Alford Loop' has become a byword in the art. He has also made other notable technical contributions to radio navigation systems and to antennas for a variety of purposes." Other Awards: Presidential Certificate of Merit, National Defense Research Committee, 1948; Scott Helt Award, 1978; National Invention Hall of Fame, 1983.

ALFVEN, HANNES O. G., IRE Fellow 1960. Born: May 30, 1908, Norrkoping, Sweden. Degrees: Dr.Phil., 1934, University of Uppsala, Sweden.

Fellow Award "For contributions to the understanding of the properties of plasma." Other Awards: Nobel Prize in Physics, 1970; Franklin Medal; Lomonosov Medal.

ALLEN, GORDON Y. R., Fellow 1976. Born: February 5, 1920. Degrees: E.E., 1945, Royal Air Force College, London.

Fellow Award "For contributions to safety and reliability of wire communications facilities in high-voltage environments"; IEEE Member, Power Systems Communications Committee and Chairman of its Wire Line Subcommittee; Member, Power System Relaying Committee and its Pilot Wire Relaying Performance WG, its Relaying Electrical Environment Subcommittee, its Relaying Channels Subcommittee and its Audio Tone WG; Member, Power System Instrumentation Committee WG Revising Standard No. 81; Member, Transmission Systems Committee of the Communications Society and Chairman of its Inductive Coordination, Interference and Protection Subcommittee; Member, Standards Board Coordinating Committee on Telecommunication; Member, Standards Board Coordinating Committee on Dispersed Storage and Generation; Past Member, IEEE Standards Board. Other Awards: Member, CIGRE and Canadian Delegate, CIGRE Study Group No. 36, Interference, Delegates to C.C.I.T.T.; Member, CCITT Study Group V on Telephone Protection, Member, Directives Editing Board; Canadian Delegate, TC 77 on EMI/EMC of Networks; Member, CSA Telecommunications Committee and CSA Committee on Electrical Equipment Connected to the Telecommunications Network, Member, Network Protection Subcommittee and the Inductive Coordination Committee; Member, Inductive Coordination Committee.

ALLEN, JAMES L., Fellow 1977. Born: September 25, 1936, Graceville, Fla. Degrees: B.E.E., 1959, M.S.E.E., 1961, Ph.D., 1966, Georgia Institute of Technology.

Fellow Award "For contributions to the characterization, analysis, and application of ferrimagnetic materials in microwave devices." Other Awards: Sigma Xi; Tau Beta Pi; Florida West Coast Engineer of the Year, Florida Engineering Society, 1978; Sigma Xi Outstanding Research Scholar (Forida West Coast), 1980.

ALLEN, JOHN L., Fellow 1972. Born: June 13, 1931, Estherville, Iowa. Degrees: B.S., 1958, Pennsylvania State University; S.M., 1962, Ph.D., 1968, Massachusetts Institute of Technology.

Fellow Award "For contributions to array antenna theory and radar systems development"; MTT National Lecturer on Microwave Theory and Techniques, 1973. Other Awards: Distinguished Civilian Service Award, U.S. Navy, 1975.

ALLEN, JONATHAN, Fellow 1978. Born: June 4, 1934. Degrees: A.B., 1956, Dartmouth College; M.S., 1957, Thayer School of Engineering; Ph.D., 1968, Massachusetts Institute of Technology.

Fellow Award "For contributions to the design of computer architecture for signal processing and to the synthesis of speech from text."

ALLEN, JOSEPH W., AIEE Fellow 1953. Born: March 12, 1893, Carbondale, Pa. Degrees: B.S.E.E., 1915, Bucknell University.

Fellow Award "For pioneering and outstanding achievement in the development of electric systems and equipment for aircraft."

ALMASI, GEORGE S., Fellow 1982. Degrees: B.S.E.E., Syracuse University; M.S.E.E., Ph.D.(E.E.), Massachusetts Institute of Technology.

Fellow Award "For contributions to the development of magnetic bubble device technology." Other Awards: Three IBM Outstanding Innovation and Invention Awards.

ALMON, CHARLES P., JR., AIEE Fellow 1951. Born: August 2, 1905, Russellville, Ala. Degrees: B.S.E.E., 1928, University of Alabama.

ALSBERG, DIETRICH A., Fellow 1979. Born: June 5, 1917, Kassel, Germany. Degrees: B.S.E.E., 1938, Stuttgart Technical University.

Fellow Award "For the development of low-loss millimeter waveguides and techniques of their electrical measurement."

ALTHOUSE, JAMES M., IRE Fellow 1962. Born: February 9, 1900, Jewell Co., Kans. Degrees: A.B., 1923, A.M., 1927 (Physics), University of Missouri.

Fellow Award "For contribution to airborne electronics." Other Awards: Meritorious Civilian Award, U.S. Air Force Aeronautical Systems Division, 1948.

ALTSCHULER, HELMUT M., Fellow 1967. Born: February 13, 1922, Mannheim, Germany. Degrees: B.E.E., 1947, M.E.E., 1949, Ph.D., 1963, Polytechnic Institute of Brooklyn.

Fellow Award "For contributions to microwave network theory and measurements."

ALTSHULER, EDWARD E., Fellow 1984. Born: January 10, 1931, Boston, MA. Degrees: B.Sc., 1953, Northeastern Univ.; M.Sc., 1954, Tufts Univ.; Ph.D., 1960, Harvard Univ.

Fellow Award "For contributions to the understanding of tropospheric effects on millimeter wave propagation." Other Awards: U.S. Air Force Scientific Achievement Awards, 1973, 1979, 1980, 1981.

AMDAHL, GENE M., Fellow 1970. Born: November 16, 1922, Flandreau, S.Dak. Degrees: B.S., 1948, South Dakota State University; Ph.D. 1952, University of Wisconsin; D.Eng.(Hon.), 1974, South Dakota State University; D.Sc.(Hon.), 1979, University of Wisconsin-Madison; D.Sc.(Hon.), 1980, Luther College.

Fellow Award "For contributions to the design of large-scale digital computers"; W. Wallace McDowell 1976, "For outstanding contributions to technical design and computer architecture." Other Awards: Member, National Academy of Engineering, 1967; IBM Fellow, 1965; Distinguished Alumni Award, South Dakota State University, 1970; DPMA Computer Sciences Man-of-the-Year Award, 1976; Distinguished Service Citation, University of Wisconsin, 1976; Michelson-Morley Award, Case Western Reserve University, 1977; Distinguished Fellow, British Computing Society, 1980; Entrepeneur of the Year, University of Southern California, 1981; Conspicuous Achievement Award, Association for Systems Management, 1982.

AMELIO, GILBERT F., Fellow 1978. Born: New York, N.Y. Degrees: B.S., 1965, M.S., 1967, Ph.D., 1968 (all in Physics), Georgia Institute of Technology.

Fellow Award "For pioneering technical and managerial contributions in the field of charge-coupled devices."

AMER, SALAH, Fellow 1966. Born: July 24, 1912. Degrees: B.Sc., 1936, Cairo University.

Fellow Award "For contributions to the establishment and development of radio and television broadcasting, the electronics industry and electronic engineeering education in the United Arab Republic."

AMEY, WILLIAM G., Fellow 1967. Born: February 24, 1918, Baltimore, Md. Degrees: B.E., 1938, D.E., 1947, The Johns Hopkins University.

Fellow Award "For valuable contributions to the field of basic electrical measurements, and as a director of research in this field." Other Awards: Fellow, Instrument Society of America, 1979.

AMITAY, NOACH, Fellow 1983. Born: April 30, 1930, Tel-Aviv, Israel. Degrees: B.Sc.(cum laude), 1952, Dipl. Ing., 1953, E.E., Technion, Israel Inst. of Technology; M.Sc., 1957, Ph.D., 1960, Carnegie Inst. of Technology.

Fellow Award "For contributions to the design and application of satellite based phased array antennas." Other Awards: Member, International Scientific Radio Union.

AMUNDSON, ROALD HARRY, Fellow 1979. Born: October 18, 1913. Degrees: B.S.M.E., 1935, University of Wisconsin, Madison.

Fellow Award "For contributions to high-voltage fuse design and standardization."

ANCKER-JOHNSON, BETSY, Fellow 1975. Born: April 29, 1927, St. Louis, Mo. Degrees: B.A., 1949, Wellesley College; Ph.D., 1953, Tuebingen University, Germany; Honorary Doctor of Law Degree, Bates College; Honorary Doctor of Science Degree, Polytechnic Institute of New York; Honorary Doctor of Science Degree, University of Southern California.

Fellow Award "For contributions to the understanding of plasmas in solids and to the development of governmental science policy." Other Awards: Fellow, American Physical Society; Recipient, Three National Science Foundation Grants; Member, National Academy of Engineering; Fellowship, American Association of University Women; Horton Hollowell Fellowship; Phi Beta Kappa; Sigma Xi; Performance Excellence Award, The Boeing Company.

ANDERSON, A. EUGENE, IRE Fellow 1959. Born: April 22, 1916, Lima, Ohio. Degrees: B.S., M.S., 1939, Ohio State University.

Fellow Award "For contributions to transistor switching circuits." Other Awards: Fellow, American Association for the Advancement of Science; Distinguished Engineer Award, Pennsylvania Society of Professional Engineers, 1965; Distinguished Alumnus Award, Ohio State University, 1968.

ANDERSON, ALBERT S., AIEE Fellow 1950. Born: June 6, 1902, Gloucester, Mass. Degrees: B.S., 1924, M.S., 1925, Massachusetts Institute of Technology.

ANDERSON, ARTHUR G., Fellow 1980. Born: November 22, 1926. Degrees: B.S.(Physics), 1949, University of San Francisco; M.S.(Math.), 1951, Northwestern University; Ph.D.(Physics), 1958, New York University.

Fellow Award "For outstanding leadership of computer research and development and personal contributions to the growth of computer technology."

ANDERSON, ARVID E., AIEE Fellow 1940. Born: May 26, 1897, Falmouth, Mass. Degrees: B.S.E.E., 1920, E.E., 1943, D.Eng.(Hon.), 1966, Worcester Polytechnic Institute.

Other Awards: Charles A. Coffin Award, General Electric, 1942; Fellow, AAAS; Tau Beta Pi; Sigma Xi.

ANDERSON, BRIAN D.O., Fellow 1975. Born: January 15, 1941, Sydney, Australia. Degrees: B.S., 1961, B.E., 1963, University of Sydney; Ph.D., 1966, Stanford University.

Fellow Award "For contributions to quadratic optimal control and stability theory, and for leadership in electrical engineering education." Other Awards: Fellow, Australian Academy of Science, 1974; Fellow, Australian Academy of Technological Sciences, 1980.

ANDERSON, CLIFFORD N., IRE Fellow 1934. Born: September 22, 1895, Scandinavia, Wis. Degrees: Ph.B., 1919, M.S., 1920, University of Wisconsin.

ANDERSON, DEAN B., Fellow 1972. Born: October 22, 1921, Billings, Mont. Degrees: B.S.E.E., 1948, Montana State University.

Fellow Award "For contributions to optical waveguides and optical parametric amplification"; IEEE Centennial Medal, 1984.

ANDERSON, EARL I., IRE Fellow 1954. Born: April 19, 1907, Chicago, Ill.

Fellow Award "For his contributions to the improvement and simplification of radio and television circuitry"; Award, Professional Group on Broadcast and Television Receivers, 1963, "In recognition of his outstanding contribution to the progress of our Group and as an expression of our high esteem." Other Awards: Certificate of Appreciation, War Dept., 1947.

ANDERSON, JOHN G., Fellow 1965. Born: August 21, 1922, Dante, Va. Degrees: B.S., 1943, Virginia Polytechnic Institute.

Fellow Award "For contributions to extra-high voltage transmission line engineering and the development of high voltage instrumentation."

ANDERSON, JOHN M., Fellow 1975. Born: October 9, 1924, Kansas City, Mo. Degrees: B.S.E.E., 1947,

M.S.E.E., 1948, Ph.D., 1955, University of Illinois.

Fellow Award "For contributions to the understanding of electromagnetic wave and plasma interaction." Other Awards: Fellow, American Physical Society.

ANDERSON, LAWRENCE K., Fellow 1974. Born: October 2, 1935, Toronto, Canada. Degrees: B.Eng., 1957, McGill University; M.Sc., 1959, Ph.D., 1962, Stanford University.

Fellow Award "For contributions in the field of holographic optical memories."

ANDERSON, PAUL M., Fellow 1981. Born: January 22, 1926, Des Moines, IA. Degrees: B.S., 1949, M.S., 1958, Ph.D., 1961, Iowa State University.

Fellow Award "For contributions to power system engineering education and research in system planning and operation." Other Awards: Eta Kappa Nu; Phi Kappa Phi; Pi Mu Epsilon; Sigma Xi; Faculty Citation, Iowa State University, 1979; Listed in "Who's Who in the West," 1980; Arizona State University Endowed Professorship, 1980; Professional Achievement Citation in Engineering, Iowa State University, 1981.

ANDERSON, RICHARD L., Fellow 1971. Born: February 4, 1927, Minneapolis, Minn. Degrees: B.S., 1950, M.S., 1952, University of Minnesota; Ph.D., 1960, Syracuse University; D.Sc.(Hon.), 1969, University of Sao Paulo.

Fellow Award "For contributions to semiconductors and to engineering education." Other Awards: First Brazilian Prize in Microelectronics, 1980.

ANDERSON, ROY E., Fellow 1974. Born: October 30, 1918, Batavia, Ill. Degrees: B.A., 1943, Augustana College; M.S.E.E., 1952, Union College.

Fellow Award 1970; "For contributions to the early development of communications and navigation technology for satellite applications." Other Awards: Coolidge Fellowship Award, 1970; Fellow, AAAS, 1977.

ANDERSON, STANLEY WILLIAM, Fellow 1979. Born: August 2, 1921. Degrees: B.S.E.E., 1949, M.S.E.E., 1950, Ph.D., 1957, Illinois Institute of Technology.

Fellow Award "For adapting operations research techniques to electric utility power and operation problems."

ANDREWS, FREDERICK T., JR., Fellow 1973. Born: October 6, 1926, Palmerton, Pa. Degrees: B.S.E.E., 1948, Pennsylvania State University.

Fellow Award "For contributions to digital transmission systems, and to transmission objectives and standards"; Edwin Howard Armstrong Achievement Award, 1980; IEEE Communications Society Vice President Technical Affairs, 1982-1983; IEEE Communications Society Vice President, 1984.

ANDREWS, GUILLERMO J., IRE Fellow 1962. Born: March 28, 1911, Buenos Aires, Argentina.

Fellow Award "For leadership in electronic engineering in Argentina and service to the Institute of Radio

Engineers."

ANGELAKOS, DIOGENES J., Fellow 1971. Born: July 3, 1919, Chicago, Ill. Degrees: B.S., 1942, University of Notre Dame; M.S., 1946, Ph.D., 1950, Harvard University.

Fellow Award "For contributions to antenna research, to engineering education, and for administration in university research." Other Awards: John Simon Guggenheim Fellow, 1957.

ANGELINI, ARNALDO M., Fellow 1968. Born: February 2, 1909, Force(Ascoli Piceno), Italy. Degrees: Diplome d'ingenieur electromecanicien, 1930, Institut Polytechnique de Liege, Belgium; "Libera docenza"-(E.E.), 1936, University of Rome.

Fellow Award "For contributions to the development of electric power systems, particularly nuclear power, and to engineering education." Other Awards: "Pugno Vanoni Award," 1944, "Emanuele Jona Award," 1966, "Claudio Castellani Award," 1970, Italian Association of Electrical Engineers; Member, Accademia Nazionale dei Lincei, Italy, 1967; Fellow, American Nuclear Society, 1967; Professor honoris causa, Polytechnic Institute of New York, 1975; Foreign Associate, National Academy of Engineering, 1976.

ANGELL, JAMES B., Fellow 1966. Born: December 25, 1924, Staten Island, N.Y. Degrees: S.B., 1946, S.M., 1946, Sc.D., 1952, Massachusetts Institute of Technology.

Fellow Award "For technical, professional, and educational contributions in solid-state electronics." Other Awards: Outstanding Service Award, Department of Electrical Engineering, Stanford University, 1983.

ANGELLO, STEPHEN J., IRE Fellow 1959. Born: March 2, 1918, Haddonfield, N.J. Degrees. B.S.E.E., 1939, M.S.E.E., 1940, Ph.D., 1942, University of Pennsylvania.

Fellow Award "For contributions in the field of solid-state rectifiers." Other Awards: A. Atwater Kent Award, University of Pennsylvania, 1939.

ANGWIN, BRUCE S., Fellow 1966. Born: November 24,1919, Oakland, Calif. Degrees: B.S.E.E., 1941, University of California.

Fellow Award "For dynamic leadership in the growth of the electronics profession"; Recognition Award, IRE, Los Angeles Section, 1956; WESCON Recognition Award, 1962; Recognition Award, Region 6, 1967. Other Awards: Contributions to TV Industry, Society of Television Engineers, 1957; Recognition for Development of Specialized Audio Support Equipment, U.S. Navy,1971.

ANNESTRAND, STIG A., Fellow 1978. Born: September 18, 1933, Husby, Sweden. Degrees: M.S.(E.E.), 1958, Royal Institute of Technology, Stockholm, Sweden.

Fellow Award "For leadership and contributions to

the art of high voltage ac and dc power transmission"; First Prize Paper, Portland Section, IEEE, 1970; Recognition Award, National Working Group, 1973. Other Awards: Order of the Big Thunder, Sweden, 1967.

ANTHONY, CLARENCE W., Fellow 1967. Born: January 3,1908, Marlow, Okla. Degrees: B.S., 1930, University of Oklahoma.

Fellow Award "For contributions to the transfer of large blocks of power between large geographic areas."

ANTON, NICHOLAS G., IRE Fellow 1956. Born: December 14, 1906, Trieste, Italy. Degrees: Dipl., 1926, Technical Institute Leonardo Da Vinci, Trieste, Italy; 1926-1928, Columbia University.

Fellow Award "For contributions to the design and production of power, counter, and voltage regulator tubes." Other Awards: Fellow, New York Academy of Sciences; Fellow, American Physical Society; Fellow, American Association for the Advancement of Science; Presidential Certificate of Appreciation, Office of Civil and Defense Mobilization; American Society of Mechanical Engineers; Society for Nondestructive Testing; Mathematical Society; Fellow, New York Academy of Medicine; Listed in: "Who's Who in Engineering," 1980-1983, "Who's Who in America," 1980-1983, "Who's Who in the World," 1980-1983; Wisdom Hall of Fame, 1981-1982.

ANTONIOU, ANDREAS, Fellow 1982. Born: March 3, 1938, Cyprus. Degrees: B.Sc.(Eng) (with honors), 1963, Ph.D., 1966, University of London, U.K.

Fellow Award "For contributions to active and digital filters, and to electrical engineering education." Other Awards: Ambrose Flemming Award, Institution of Electrical Engineers, U.K., 1969; Fellow, Institution of Electrical Engineers, U.K., 1979.

AOKI, MASANAO, Fellow 1976. Born: May 14, 1931, Hiroshima, Japan. Degrees: B.S., 1953, M.S., 1955, University of Tokyo; Ph.D., 1960, University of California, Los Angeles; Dr.Sc., 1966, Tokyo Institute of Technology.

Fellow Award "For contributions to the theory and application of control systems." Other Awards: Fellow, Econometric Society, 1979.

APPLEBAUM, SIDNEY P., Fellow 1966. Born: August 31, 1923, New York, N.Y. Degrees: B.E.E., 1944, City College of the City University of New York; M.E.E., 1956, Syracuse University.

Fellow Award "For invention and analytical contributions in the field of signal processing."

APPLEMAN, W. ROSS, AIEE Fellow 1951. Born: April 25, 1906, Manila, Philippine Islands. Degrees: B.S., 1928, E.E., 1936, University of Illinois.

Awards: Transaction Paper Prize, AIEE, Dayton Section, 1937, for "The Cause and Elimination of Noise in Small Motors"; Chairman, Dayton Ohio Section, AIEE, 1948; National Rotating Machinery Committee;

DC and Single Phase Sub-Committee, IEEE. Other Awards: Listed in: "Who's Who in Engineering"; "Who's Who in Commerce and Industry," "Who's Who in the Midwest"; "Who's Who in Technology Today."

APSTEIN, MAURICE, IRE Fellow 1959. Born: May 5, 1910, Bridgeport, Conn. Degrees: B.S., 1932, City College of the City University of New York; M.E.A., George Washington University; Ph.D., American University.

Fellow Award "For contributions to the improvement of electronic ordnance devices"; Harry Diamond Memorial Award, 1969. Other Awards: Gold Medal, U.S. Dept. of Commerce, 1951; Dept. of Army R & D Award, 1967; Gold Medal For Exceptional Civilian Service, Dept. of Army, 1972; Oak Leaf,1974; National Academy of Engineering, 1977.

ARAM, NATHAN W., Fellow 1966. Born: October 5, 1916, Moline, Ill. Degrees: B.S.E.E., 1939, Purdue University.

Fellow Award "For contribution to the development of FM, FM multiplex, and television systems." Other Awards: Distinguished Alumnus Award, Purdue University, 1964.

ARAMS, FRANK R., Fellow 1967. Born: October 18, 1925, Danzig. Degrees: B.S.E., 1947, University of Michigan; M.S., 1948, Harvard University; M.S., 1953, Stevens Institute of Technology; Ph.D.E.E., 1961, Polytechnic Institute of New York.

Fellow Award "For contributions to microwave quantum devices."

ARCHAMBAULT, JEAN-JACQUES, Fellow 1982. Born: March 21, 1919, Montreal, Canada. Degrees: B.Asc.E., Ecole Polytechnique, Montreal.

Fellow Award "For contributions in EHV transmission and in establishment and promulgation of standards for EHV equipment and systems"; William M. Habirshaw Award, IEEE; IEEE Centennial Medal, 1984. Other Awards: Archambault Medal, Association Can-Francaise pour l'Advancement des Sciences.

ARDITI, M., IRE Fellow 1962. Born: March 1, 1913, Paris, France. Degrees: 1935, Ecole Superieure Physique; 1936, Ecole Superieure Electricite; M.A., 1939, Ph.D., 1958, Sorbonne, Paris.

Fellow Award "For contributions to development of gas cell atomic clocks." Other Awards: American Physical Society.

ARIMOTO, SUGURU, Fellow 1983. Born: August 3, 1936, Hiroshima City, Japan. Degrees: B.Sc., 1959, Kyoto University; Dr.E., 1967, University of Tokyo.

Fellow Award "For contributions to information theory and linear systems theory"; IEEE Information Theory Group Paper Award, 1974. Other Awards: SICE Paper Award, 1968, 1976.

ARMSTRONG, EMERSON A., AIEE Fellow 1958. Born: January 20, 1890, Toledo, Ohio. Degrees: B.S., 1911,

E.E., 1922, Michigan State University.

Fellow Award "For contributions to diversified electrical applications in industrial processes."

ARNAUD, JOHN P., IRE Fellow 1953. Born: September 1, 1905, London, England. Degrees: Ind.Eng.(Power Engineering), 1929, School of Engineering, University of Buenos Aires, Argentina.

Fellow Award "For his outstanding work in telecommunication development, and educational activity in South America." Other Awards: Honor Diploma Scholarship (post-graduate courses in East Pittsburgh), Westinghouse International, 1930-1931; Scholarship, Columbia University, 1931-1932.

ARONOW, SAUL, Fellow 1979. Born: October 4, 1917, New York, N.Y. Degrees: B.E.E., 1939, Cooper Union; M.S., 1948, Ph.D., 1953, Harvard University.

Fellow Award "For contributions to nuclear medical instrumentation and for innovative leadership in establishing clinical engineering in hospitals." Other Awards: Fellow, National Science Foundation, 1950; Fulbright Fellow, 1969; Gano Dunn Medal, Cooper Union, 1979.

ASBURY, CARL E., AIEE Fellow 1961. Born: June 7, 1912, Tamaroa, Ill. Degrees: B.S.E.E., 1938, University of Illinois.

Fellow Award "For contributions to the field of power engineering and to national and international standards."

ASELTINE, JOHN A., Fellow 1966. Born: April 12, 1925, Calif. Degrees: B.A., 1947, Universiy of California; M.S., 1949, Ph.D., 1952, University of California at Los Angeles.

Fellow Award "For leadership in systems analysis and contributions to national defense."

ASH, ERIC A., Fellow 1968. Born: January 31, 1928, Berlin, Germany. Degrees: B.Sc., 1948, Ph.D., 1952, D.Sc., 1974, London University.

Fellow Award "For significant contributions to microwave tubes, solid-state microwave devices and electron optics." Other Awards: FRS, 1977; F.Eng., 1978.

ASHBY, ROBERT M., IRE Fellow 1955. Born: May 9, 1912, American Fork, Utah. Degrees: A.B., 1934, M.A., 1939, Brigham Young University; Ph.D., 1942, University of Wisconsin.

Fellow Award "For his contributions to radar detection theory and integration of fire control-flight control systems for aircraft."

ASHKIN, ARTHUR, Fellow 1976. Born: September 2, 1922, Brooklyn, N.Y. Degrees: A.B.(Physics), 1947, Columbia College; Ph.D.(Physics), 1952, Cornell University.

Fellow Award "For contributions to the theory and application of nonlinear interactions at microwave and optical frequencies." Other Awards: Fellow, American Physical Society, 1967; Fellow, Optical Society of America, 1983.

ASHMORE, J., AIEE Fellow 1953. Born: July 26, 1897, Manchester, Lancashire, England. Degrees: 1916, Royal Technical College, Salford, England.

Fellow Award "For distinction as an executive of a large organization specializing in the construction, modernization, and repair of electric systems, switchgear testing facilities and lighting installations." Other Awards: Member, Institution of Electrical Engineers, U.K.; Associate, Institution of Mechanical Engineers.

ASTIN, ALLEN V., IRE Fellow 1954. Born: June 12, 1904, Salt Lake City, Utah. Degrees: A.B., 1925, University of Utah; M.Sc., 1926, Ph.D., 1928, New York University; D.Sc.(Hon.), 1953, Lehigh University; D.Sc.(Hon.), 1958, George Washington University; D.Sc.(Hon.), 1960, New York University.

Fellow Award "In recognition of his distinguished leadership and administration of science and engineering"; Harry Diamond Memorial Prize Award, IEEE, 1970, "For outstanding contributions and inspiring technical leadership in instrument and measurement technology." Other Awards: His Majesty's Medal for Service, U.K., 1947; Presidential Certificate of Merit, 1948; Gold Medal, U.S. Dept. of Commerce, 1952; American Philosophical Society, 1958, Vice President, 1977-1980; National Academy of Sciences, 1960, Home Secretary, 1971-1975; National Civil Service Award, 1960; Rockefeller Public Service Award, 1963; Scott Gold Medal, American Ordnance Association, 1964; Executive's Award, American Society for Testing Materials, 1965; Award, Scientific Apparatus Makers Association, 1973; Standards Medal, American National Standards Institute, 1969; Astin-Polk International Standards Medal, American National Standards Institute, 1974; Officer, French Legion of Honor, 1977; American Society for Testing and Materials; Standards Engineering Society; American Society of Heating, Refrigeration, and Air Conditioning Engineers, Inc.; American Institute of Aeronautics and Astronautics; Instrument Society of America; American Dental Association.

ASTRAHAN, MORTON M., Fellow 1969. Born: December 5, 1924, Chicago, Ill. Degrees: B.S.E.E., 1945, Northwestern Technological Institute; M.S.E.E., 1946, California Institute of Technology; Ph.D.E.E., 1949, Northwestern Technological Institute.

Fellow Award "For leadership, organization and management of professional activities in the computer field, and for contributions to computer system design."

ASTROM, KARL JOHAN, Fellow 1979. Born: 1934. Degrees: M.Sc. Eng.(Physics), 1957, Ph.D.(Mathematics and Control), 1960, Royal Institute of Technology, Stockholm, Sweden.

Fellow Award "For comprehensive contributions to stochastic control theory." Other Awards: Callender Silver Medal, Institute of Measurement and Control,

London, 1980.

ATAL, BISHNU S., Fellow 1982. Born: May 10, 1933, Kanpur, India. Degrees: B.Sc.(honors), 1952, University of Lucknow, India; Diploma, 1955, Indian Institute of Science, Bangalore; Ph.D.(E.E.), 1968, Polytechnic Institute of Brooklyn.

Fellow Award "For contributions to the theory of linear prediction and its applications to speech processing"; Achievement Award, IEEE Acoustics, Speech, and Signal Processing Society, 1975; Senior Award, IEEE Acoustic, Speech, and Signal Processing Society, 1979. Other Awards: Fellow, Acoustical Society of America.

ATCHISON, F. STANLEY, Fellow 1967. Born: January 3, 1918, Stoddard Co., Mo. Degrees: B.A., 1938, Southeast Missouri State University; M.S., 1940, Ph.D., 1942, State University of Iowa.

Fellow Award "For contributions to computer components and missile guidance."

ATHANS, MICHAEL, Fellow 1973. Born: May 3, 1937, Drama, Greece. Degrees: B.S.E.E., 1958, M.S.E.E., 1959, Ph.D., 1961, University of California at Berkeley.

Fellow Award "For contributions to the theory and application of optimal control and estimation." Other Awards: Donald P. Eckman Award, American Automatic Control Council, 1964; Frederick E. Terman Award, American Society for Engineering Education, 1969; Fellow, American Association for the Advancement of Science, 1977; Education Award, American Automatic Control Council, 1980.

ATTINGER, ERNST O., Fellow 1978. Born: December 27, 1922, Zurich, Switzerland. Degrees: M.D., University of Zurich, 1947; M.S.(Biomedical Instrumentation), Drexel University, 1961; Ph.D.(Biomedical Engineering), University of Pennsylvania, 1965.

Fellow Award "For pioneering applications of electrical engineering methods to medicine." Other Awards: Board of Visitors, University of Virginia Research Prize, 1976.

ATWELL, CLARENCE A., AIEE Fellow 1950. Born: January 26, 1891, Fairfield, Nebr. Degrees: B.Sc., 1914, E.E., 1930, University of Nebraska.

Awards: First Paper Prize, AIEE, General Application Division, 1955. Other Awards: Sigma Xi, 1914.

AUERBACH, ISAAC L., IRE Fellow 1958. Born: October 9, 1921, Philadelphia, Pa. Degrees: B.S., 1943, Drexel University; M.S., 1947, Harvard University.

Fellow Award "For his contributions to the development and application of computer techniques." Other Awards: Sigma Xi, 1942; Eta Kappa Nu; Instrument Society of America; Grand Medal, Paris, 1959; Fellow, American Association for the Advancement of Science, 1969; Alumni Citation, Drexel University, 1961; Fellow, British Computer Society, 1970; National Academy of Engineering, 1974; Honorary Member, Information Processing Society of Japan, 1974; Distinguished Fellow, British Computer Society, 1975; Founder, First President and Honorary Life Member, International Federation for Information Processing; Department of Commerce Technical Advisory Board, 1978-81.

AUGUSTADT, HERBERT W., Fellow 1968. Born: November 17, 1906. Degrees: M.A., 1941, Columbia University; B.S.E.E., 1941, University of North Dakota.

Fellow Award "For contributions to naval air-defense systems and leadership in development of command and control equipment."

AULD, BERTRAM A., Fellow 1973. Born: 1922, Wei-Hwei-Fu, China. Degrees: B.S.E.E., 1946, University of British Columbia; Ph.D., 1952, Stanford University.

Fellow Award "For contributions to the theory of microwave ferrite devices and microwave acoustics"; IRE, 1959, Microwave Prize for Paper, "The Synthesis of Symmetrical Waveguide Circulators."

AVILA, CHARLES F., AIEE Fellow 1961. Born: September 17, 1906, Taunton, Mass. Degrees: B.S.(Elec. Eng. and Bus. Admin.), 1929, Harvard University; LL.D., 1963, University of Massachusetts.

Fellow Award "For his contributions to the design and improvement of underground electric distribution systems"; IEEE Edison Medal, 1968, "For his early contribution to underground transmission, for his continuing guidance in the field of electrical research and for his positive leadership in the development of the electrical utility industry." Other Awards: Member Emeritus, National Academy of Engineering, 1968; New England Award, Engineering Societies of New England, Inc., 1983.

AVINS, JACK, IRE Fellow 1957. Born: March 18, 1911, New York, N.Y. Degrees: A.B., 1932, Columbia University; M.E.E., 1949, Polytechnic Institute of Brooklyn.

Fellow Award "For contributions to the development of television receivers"; Outstanding Contributions Award, IEEE Consumer Electronics, 1976, "For his outstanding contribution and loyalty to the Group, particularly in the editing and promulgating of standards." Other Awards: RCA David Sarnoff Achievement Award, 1971; 50 U.S. patents.

AVIZIENIS, ALGIRDAS A., Fellow 1973. Born: July 8, 1932, Kaunas, Lithuania. Degrees: B.S., 1954, M.S., 1955, Ph.D., 1960, University of Illinois at Urbana.

Fellow Award "For fundamental contributions in the field of fault-tolerant computing"; Honor Roll, IEEE Computer Group, 1968. Other Awards: Tau Beta Pi; Eta Kappa Nu; Pi Mu Epsilon; Sigma Xi; RCA Fellow, Digital Computer Laboratory, University of Illinois; Gold Medal, Pan-American Congress of Mechanical and Electronic Engineering and Related Branches, for the paper, "Redundant Number Systems for Electronic Digital Computers," 1965; Apollo Achievement Award, U.S. National Aeronautics and Space Administration, 1969.

AXELBAND, ELLIOT I., Fellow 1980. Born: June 6, 1937, New York, NY. Degrees: B.S.E.E., 1958, Cooper Union; M.S.E.E., 1960, University of Southern California; Ph.D., 1966, University of California at Los Angeles.

Fellow Award "For fundamental contributions to the theory of distributed parameter systems, and for outstanding leadership of spacecraft and missile programs."

AXELBY, GEORGE S., IRE Fellow 1961. Born: March 7, 1922, Thomaston, Conn. Degrees: B.S., 1950, University of Connecticut; M.S. 1951, Yale University.

Fellow Award "For achievements in electronics and automatic control systems; and for promoting progress of the IRE Professional Groups at the local and national levels"; Special Service Award, Automatic Control Group, 1968; Distinguished Member, IEEE Control Systems Society, 1983. Other Awards: Fellow, American Association for the Advancement of Science.

AYER, WILLIAM E., Fellow 1969. Born: November 12, 1921, Phoenix, Ariz. Degrees: A.B., 1943, A.M., 1944, E.E., 1948, Ph.D., 1951, Stanford University.

Fellow Award "For creative research, inspirational leadership, and sound engineering in the development and production of sophisticated electronic countermeasure systems."

B

BACHMAN, C. H., IRE Fellow 1955. Born: December 8, 1908, Cedar Falls, Iowa. Degrees: B.S., 1932, M.S., 1933, Ph.D., 1935, Iowa State University.

Fellow Award "For his contributions in the field of electron physics." Other Awards: Fulbright Lectureship, University of Calcutta, 1959; Fellow, American Physical Society; Fellow, AAAS; Past President, American Vacuum Society.

BACHMAN, WILLIAM S., IRE Fellow 1956. Born: October 29, IRE 1908, Williamsport, Pa. Degrees: E.E., 1932, Cornell University.

Fellow Award "For contributions to the recording and reproducing of sound." Other Awards: Emile Berliner Award, Audio Engineering Society; Fellow, Audio Engineering Society.

BACHYNSKI, MORREL P., Fellow 1977. Born: July 19, 1930, Bienfait, Saskatchewan, Canada. Degrees: B.Eng., 1952, M.Sc., 1953, University of Saskatchewan; Ph.D., 1955, McGill University.

Fellow Award "For contributions to the fields of electromagnetic waves and plasmas." Other Awards: David Sarnoff Outstanding Achievement Award, RCA, 1963; Fellow, Canadian Aeronautics and Space Institute, 1966; Fellow, Royal Society of Canada, 1967; Fellow,

American Physical Society, 1968; Prix Scientifique du Quebec, 1973; Canada Enterprise Award, 1977; Queens 25th Anniversary Silver Medal, 1977.

BACKUS, CHARLES E., Fellow 1981. Born: September 17, 1937, Wadestown, WV. Degrees: B.S.M.E., 1959, Ohio University; M.S., 1961, Ph.D., 1965, University of Arizona.

Fellow Award "For contributions to photovoltaics and education in advanced energy conversion." Other Awards: Faculty Achievement Award, Arizona State University, 1976.

BAERWALD, HANS G., IRE Fellow 1962. Born: April 5, 1904, Breslau, Germany. Degrees: Dipl.Ing., 1928, Dr., 1930, Breslau Institute of Technology.

Fellow Award "For contributions to network theory and to the theory of piezoelectric devices." Other Awards: Fellow, Acoustical Society of America, 1956.

BAHDER, GEORGE, Fellow 1979. Born: January 17, 1925, Warsaw, Poland. Degrees: B.S., Engineering School, Warsaw, 1946; M.S., 1949, Ph.D., 1957, Warsaw Politechnique.

Fellow Award "For contributions to the understanding of electrical and electrochemical voltage breakdown of extruded and laminar dielectric cables." Other Awards: Member, CIGRE, 1962.

BAILEY, F. MEADE, Fellow 1972. Born: January 10, 1916, Yachow, West China. Degrees: B.S., 1937, University of Rochester; M.S., 1939, Ph.D., 1940, Iowa State College.

Fellow Award "For contributions to engineering technology in electronic and electrical control systems." Other Awards: Steinmetz Award, General Electric Company, 1973.

BAILEY, R. COOPER, AIEE Fellow 1937. Born: May 11, 1898, Emporia, Va. Degrees: B.S., 1921, Virginia Polytechnic Institute.

BAILEY, STUART L., IRE Fellow 1943. Born: October 7, 1905, Minneapolis, Minn. Degrees: B.S., 1927, M.S., 1928, University of Minnesota.

Fellow Award "For pioneering accomplishment in the application of radio engineering principles to the solution of technical problems in broadcasting." Other Awards: Outstanding Achievement Award, University of Minnesota, 1956.

BAILEY, WILLIAM F., IRE Fellow 1954. Born: May 4, 1911, Buffalo, N.Y. Degrees: M.E., 1933, M.S., 1941, Stevens Institute of Technology.

Fellow Award "For his contributions to the theory, practice, and standardization of television."

BAIRD, JACK A., Fellow 1969. Born: May 27, 1921, Omaha, Tex. Degrees: B.S.E.E., 1943, Texas A. & M. University; M.S.E.E., 1950, Stevens Institute of Technology; Ph.D. E.E., 1952, Texas A. & M. University.

Fellow Award "For fundamental contributions to the philosophy of systems engineering management in the communication field."

BAKER, CHARLES H., Fellow 1967. Born: June 2, 1910, Chicago, Ill. Degrees: 1936, Northwestern University.

Fellow Award "For contributions to high-voltage fuses and interrupters."

BAKKEN, EARL ELMER, Fellow 1974. Born: January 10, 1924, Minneapolis, Minn. Degrees: B.E.E., 1948, University of Minnesota.

Fellow Award "For leadership in the application of electrical engineering to clinical medicine." Other Awards: Member, Instrument Society of America; Member, Association for the Advancement of Medical Instrumentation; Outstanding Achievement Award, University of Minnesota, 1981; Minnesota Business Hall of Fame Award, 1978; Associate Member, NASPE, 1982.

BALABANIAN, NORMAN, Fellow 1965. Born: August 13, 1922, New London, Conn. Degrees: Sci. Dipl., Aleppo College, Syria; B.E.E., 1949, M.E.E., 1951, Ph.D., 1954, Syracuse University.

Fellow Award "For contributions to engineering education and circuit theory"; IEEE Centennial Medal, 1984. Other Awards: Annual Science Award, Armenian Students Association of America, 1963; Sword of Damocles Award, Syracuse University, 1967; Annual Peace Award, Syracuse Peace Council, 1967; Annual Civil Liberties Award, Central New York Chapter, ACLU, 1968; Fellow, American Association for the Advancement of Science, 1972.

BALAKRISHNAN, A. V., Fellow 1966.

Fellow Award "For contributions to the theory of communication."

BALDWIN, CLARENCE J., Fellow 1967. Born: August 8, 1929, San Antonio, Tex. Degrees: B.S., 1951, M.S., 1952, University of Texas; E.E., 1957, Massachusetts Institute of Technology; PMD, 1969, Harvard Business School.

Fellow Award "For contributions to the planning of power generation, transmission and distribution systems"; Fortescue Fellow, AIEE, 1951-1952. Other Awards: Lamme Scholar, Westinghouse Electric 1956-1957; Outstanding Young Electrical Engineer Award, Eta Kappa Nu, 1961; Distinguished Engineering Graduate, University of Texas, 1967.

BALDWIN, GARY L., Fellow 1982. Born: October 12, 1943, El Centro, CA. Degrees: B.S.E.E., 1966, M.S.E.E., 1967, Ph.D., 1970, University of California at Berkeley.

Fellow Award "For developments of integrated circuits providing high performance and precision in digital communications."

BALDWIN, M. STANLEY, Fellow 1981. Born: December 20, 1921, St. Louis, MO. Degrees: B.S.E.E., 1949, Washington University at St. Louis.

Fellow Award "For the development of analysis techniques for power system performance during system disturbances"; Power Generation Committee, Station Design Subcommittee, IEEE Auxiliaries Systems Working Group; Prize Paper Award, IEEE Power Engineering Society, 1979; Prize Paper Award, IEEE Power Generation Committee. Other Awards: Registered Professional Engineer, Pennsylvania, Ohio.

BALIGA, B. JAYANT, Fellow 1983. Born: April 28, 1948, Madras, India. Degrees: B.Tech., 1969, Indian Institute of Technology (Madras); M.S., 1971, Ph.D., 1974, Rensselaer Polytechnic Institute.

Fellow Award "For contributions to the development of power semiconductor devices"; Region I Award, IEEE, 1982. Other Awards: Phillips India Award, 1969, I.I.T. Special Merit Prize, 1969, IBM Fellowship, 1972, Allen B. Dumont Prize, 1974, Sigma Xi, 1972; Coolidge Fellow Award, 1983; Dushman Award, 1983; IR 100 Award, 1983.

BALLATO, ARTHUR, Fellow 1981. Born: October 15, 1936, Astoria, NY. Degrees: S.B., 1958, Massachusetts Institute of Technology; M.S., 1962, Rutgers University; Ph.D., 1972, Polytechnic Institute of New York.

Fellow Award "For contributions to the theory of piezoelectric crystals and frequency control."

BALLENGER, WILLIAM M., AIEE Fellow 1950. Born: November 15, 1902, Greer, S.C. Degrees: B.S.E.E., 1923, Clemson University.

BANCKER, ELBERT H., AIEE Fellow 1948. Born: February 28, 1896, Brooklyn, N.Y. Degrees: A.B., 1916, Williams College; B.S., 1918, Massachusetts Institute of Technology; B.S., 1921, Harvard University.

BANDLER, JOHN W., Fellow 1978. Born: November 9, 1941, Jerusalem, Palestine. Degrees: B.Sc.(Eng.), 1963, Ph.D., 1967, D.Sc.(Eng.), 1976, University of London.

Fellow Award "For contributions to computer-oriented microwave and circuit practices."

BANGERT, JOHN T., Fellow 1964. Born: January 8, 1919, Chicago, Ill. Degrees: B.S.E.E., 1942, University of Michigan; M.S.E.E., 1947, Stevens Institute of Technology.

Fellow Award "For contributions to the advancement of network design through the use of computers."

BANY, HERMAN, AIEE Fellow 1949. Born: May 12, 1896, Tripoli, Iowa. Degrees: B.S.E.E., 1918, Iowa State University.

Other Awards: Professional Achievement Citation in Engineering, Iowa State University, 1968; Anson Marston Medal, Iowa State University, 1973.

BARAS, JOHN S., Fellow 1984. Born: March 13, 1948, Piraeus, Greece. Degrees: B.S.(E.E.), 1970, National Technical University of Athens, Greece; M.S., 1971, Ph.D., 1973, Harvard University (Applied Mathematics).

Fellow Award "For contributions to distributed parameter systems theory, quantum and nonlinear estimation, and control of queueing systems"; Outstanding Paper Award, IEEE Control Systems Society, 1980. Other Awards: Naval Research Laboratory Research Publication Award, 1978.

BARBE, DAVID FRANKLIN, Fellow 1978. Born: May 26, 1939. Degrees: B.S.E.E.(with high honors), 1962, M.S.E.E., 1964, West Virginia University; Ph.D.E.E., 1969, The Johns Hopkins University.

Fellow Award "For contributions to the theory, understanding, and development of charge-coupled devices." Other Awards: Naval Research Laboratory Research Publication Award, 1975; Dept. of Defense Citation, 1979; Senior Executive Service, 1979.

BARBER, MARK ROY, Fellow 1975. Born: July 23, 1931, Wellington, New Zealand. Degrees: B.Sc., 1954, B.E.(Hon.), 1955, University of New Zealand; Ph.D., 1959, University of Cambridge, England.

Fellow Award "For contributions and leadership in microwave, medical and digital electronics."

BAR-DAVID, ISRAEL, Fellow 1980. Born: April 4, 1930. Degrees: B.Sc., 1954, E.E., 1955, M.Sc., 1959, Technion, Israel; Sc.D., 1965, Massachusetts Institute of Technology.

Fellow Award "For contributions to detection theory as applied to optical communications."

BARGELLINI, PIER L., Fellow 1976. Born: February 7, 1914, Florence, Italy. Degrees: Pre-engineering diploma, 1932, University of Florence; Doctorate(E.E.), 1937, Polytechnic of Turin; M.S.E.E., 1949, Cornell University.

Fellow Award "For contributions to satellite communications." Other Awards: Fellow, American Institute of International Education, N.Y.C., 1948; City of Columbus, Ohio; Prize, Institute of International Communications, Genoa, Italy, 1975.

BARKAN, PHILIP, Fellow 1972. Born: March 29, 1925, Boston, Mass. Degrees: B.S., 1946, Tufts University; M.S., 1948, University of Michigan; Ph.D., 1953, Pennsylvania State University.

Fellow Award "For his contribution to circuit breakers, particularly in the areas of dynamics of mechanical systems and protection." Other Awards: Charles P. Steinmetz Award, General Electric Co., 1973; Member, National Academy of Engineering, 1980.

BARKER, RICHARD C., Fellow 1979. Born: March 27, 1926, Bridgeport, Conn. Degrees: B.E., 1950, M.E., 1951, Ph.D., 1955, Yale University.

Fellow Award "For contributions to research, teaching, and international cooperation in the field of magnetics." Other Awards: Alexander von Humboldt Senior American Scientist Award, 1975.

BARKLE, JOHN E., Fellow 1970. Born: March 16, 1918, Edgewood, Pa. Degrees: B.S.E.E., 1939, Carnegie-Mellon University.

Fellow Award "For contributions to high-voltage systems."

BARLOW, HAROLD M., IRE Fellow 1956. Born: November 15, 1899, London, England. Degrees: B.Sc., 1921, Ph.D., 1924, University of London; D.Sc., 1971, Heriot Watt University, Edinburgh; D.Eng., 1973, Sheffield University, England.

Fellow Award "For contributions to engineering education, telecommunication, and high-frequency techniques." Other Awards: Kelvin Premium, 1930; Fleming Premium, Institution of Electrical Engineers, U.K., 1952; Extra Premiums, Institution of Electrical Engineers, U.K., 1954, 1957; Fellow, Royal Society, London, 1958; Foreign Member, Polish Academy of Sciences, 1966; Faraday Medal, Institution of Electrical Engineers, U.K., 1967; Dellinger Gold Medal of the International Union of Radio Science, Ottawa, 1969; Honorary Fellow, I.E.R.E., London, 1971; Honorary Member, Institute of Electronics and Communication Engineers of Japan, 1973; Sir Harold Hartley Medal, Institute of Measurement and Control, 1973; Founder Member, Fellowship of Engineering, Council of Engineering Institutions, London, S.W.1; Foreign Associate, National Academy of Engineering (U.S.), 1979.

BARNES, EARL C., AIEE Fellow 1961. Born: May 20, 1918, Ravenna, Ohio. Degrees: B.S.E.E., 1940, Ohio University; M.S.E.E., 1949, Case Institute of Technology.

Fellow Award "For contributions to the design and engineering development of electric motors."

BARNES, FRANK S., Fellow 1970. Born: July 31, 1932, Pasadena, Calif. Degrees: B.S.E.E., 1954, Princeton University; M.S.E.E., 1955, E.E., 1956, Ph.D., 1958, Stanford University.

Fellow Award "For leadership in engineering education, and for outstanding contributions to laser technology and the design of xenon-krypton flashlamps." Other Awards: Curtis McGraw Research Award, American Society for Engineering Education, 1965; Fellow, American Association for the Advancement of Science, 1976; Robert L. Stearns Award, University of Colorado, 1980.

BARNES, HOWARD C., Fellow 1965. Born: May 28, 1912, Terre Haute, Ind. Degrees: B.S.E.E., 1934, Doctor of Engineering(Hon.), 1977, Rose-Hulman Institute of Technology.

Fellow Award "For contributions to the protection and control of power systems"; William M. Habirshaw Award (co-recipient), 1979; President, IEEE Power Engineering Society, 1971-72. Other Awards: National Academy of Engineering, 1974; Meritorious Service Award, 1975; Power-Life Award, 1977; Fellow, American Association for the Advancement of Science, 1980; Attwood Associate, CIGRE, 1982.

BARNES, JAMES A., Fellow 1978. Born: December 7, 1933. Degrees: B.S., 1956, Ph.D., 1966, University of Colorado; M.S., 1958, Stanford University; M.B.A., 1979, University of Denver.

Fellow Award "For contributions to and leadership in measurements of time and frequency and the promulgation of standards." Other Awards: Gold Medal, U.S. Dept. of Commerce.

BARNES, THOMAS D., AIEE Fellow 1962. Born: March 2, 1904, Ainsworth, Iowa. Degrees: Certificate, 1926, Bliss Electrical School.

Fellow Award "For contributions to the electric metering art"; Chief U.S. delegate to the International Electrotechnical Commission, representing IEEE, NEMA, EEI, Budapest, Hungary, 1970. Other Awards: Gold Medal Award, Bliss Electrical School, 1926.

BARNETT, HOWARD G., AIEE Fellow 1951. Born: January 20, 1908, Iantha, Mo. Degrees: B.Sc., 1931, M.Sc., 1934, Oregon State University.

BARON, SHELDON, Fellow 1982. Born: May 13, 1934, Brooklyn, New York. Degrees: B.S., 1955, Brooklyn College; M.A., 1961, William and Mary College; Ph.D.(Applied Math.), 1966, Harvard University.

Fellow Award "For leadership in and contributions to the application of control theory to manual control and the modeling of human operator response"; Administration Committee, IEEE Systems, Man, and Cybernetics Society, 1971-1973; Secretary-Treasurer, IEEE Control Systems Society, 1982-present; Program Committee, IEEE Conf. on Decision Control, 1972; Program Chairman, IEEE Conf. on Systems, Man, and Cybernetics, 1973. Member, American Institute of Aeronautics and Astronautics; President, Harvard Society of Engineers and Scientists, 1976-1978; Scientific Advisory Group, US Army Missile Command, 1975-1977; Editorial Advisory Board, Journal of Cybernetics and Information Sciences, 1976-1982; Member, National Academy of Sciences, National Research Council Working Group on Simulation, 1982-1983; Chairman, National Academy of Sciences; National Research Council, Working Group on Human Performance Modeling, 1983-present.

BARROW, BRUCE B., Fellow 1970. Born: April 12, 1929, Mahoning Township, Pa. Degrees: B.S.E.E., 1950, M.S.E.E., 1950, Carnegie-Mellon University; E.E., 1956, Massachusetts Institute of Technology; Dr. of Technical Science, 1962, Delft Technological University; Advanced Management Program, 1970, Harvard Business School.

Fellow Award "For contributions to the field of standardization and to communication theory and practice." Other Awards: Fellow, American Association for the Advancement of Science, 1973.

BAR-SHALOM, YAAKOV, Fellow 1984. Born: May 11, 1941, Timisoara, Romania. Degrees: B.S., 1963, M.S., 1967, Technion, Haifa, Israel; Ph.D., 1970, Princeton University.

Fellow Award "For contributions to the theory of stochastic systems and of multitarget tracking."

BARSTOW, JOHN M., IRE Fellow 1959. Born: July 7, 1901, Topeka, Kans. Degrees: B.S. 1923, Washburn University; M.S., 1924, University of Kansas.

Fellow Award "For contributions to the transmission of mono-chrome and color television." Other Awards: Sigma Xi, 1924; Certificate of Appreciation, Radio-Electronics-Television Manufacturers Association, 1954.

BARTHOLD, LIONEL O., Fellow 1972. Born: March 20, 1926, Great Barrington, Mass. Degrees: B.S., 1950, Northwestern University.

Fellow Award "For contribution to EHV and UHV technology"; First Prize Paper, National Power Division, AIEE, 1956, (with co-authors I.B. Johnson and A.J. Schultz), for "Switching Surges and Arrester Performance on High-Voltage Stations"; First Prize Paper, National Power Division, AIEE, 1961 (with co-author G.K. Carter), for "Digital Traveling Wave Solutions, Part I: Single-Phase Equivalent"; Region I, Award, IEEE, 1978; President, IEEE Power Engineering Society, 1981, 1982 & 1983. Other Awards: Member, CIGRE; Chairman, CIGRE Study Committee 41, 1976-1982; National Academy of Engineering; Order of the Yugoslav Flag with Wreath, 1977.

BARTNIKAS, RAY, Fellow 1977. Born: January 25, 1936. Degrees: B.A.Sc.(E.E.), 1958, University of Toronto; M.Eng.(E.E.), 1962, Ph.D.(E.E.), 1964, McGill University, Montreal.

Fellow Award "For contributions to the field of dielectric and corona loss mechanisms in electrical insulating systems." Other Awards: American Society for Testing and Materials, Committee on Electrical Insulation.

BARTOLINI, ROBERT A., Fellow 1984. Born: April 4, 1942, Waterbury, CT. Degrees: B.S.E.E., 1964, Villanova University; M.S.E.E., 1966, Case Western Reserve University; Ph.D., 1972, University of Pennsylvania.

Fellow Award "For contributions to the development of optical recording media and systems." Other Awards: RCA Laboratories Achievement Award, 1970, 1975, 1979; Society of Information Display (SID) Best Paper Award, 1979.

BARTON, DAVID K., Fellow 1971. Born: September 21, 1927, Greenwich, Conn. Degrees: A.B., 1949, Harvard College.

Fellow Award "For contributions to precision tracking radar and radar systems engineering"; M. Barry Carlton Award, IRE-PGMIL, for Paper "The Future of

Pulse Radar for Missile and Space Range Instrumentation." Other Awards: David Sarnoff Outstanding Achievement Award in Engineering, RCA, 1958.

BARTON, LOY E., IRE Fellow 1956. Born: November 7, 1897, Fayetteville, Ark. Degrees: B.E.E., 1921, E.E., 1925, University of Arkansas.

Fellow Award "For contributions to radio engineering, including inventions in Class-B amplification." Other Awards: Distinguished Alumni Award, University of Arkansas, 1947; Engineering Hall of Fame, University of Arkansas, 1963.

BARTON, T. F., AIEE Fellow 1930. Born: December 25, 1885, Orangeburg, S.C. Degrees: Clemson University.

BARTON, THOMAS H., Fellow 1978. Born: May 28, 1926, Sheffield, England. Degrees: B.Eng.(Elec.), 1947, Ph.D., 1949, D.Eng., 1968, University of Sheffield.

Fellow Award "For contributions to the field of rotating machine theory and electronic drive dynamics." Other Awards: Fellow, Institution of Electrical Engineers, London; Sigma Xi.

BARUCH, JORDAN J., Fellow 1967. Born: August 21, 1923, Brooklyn, N.Y. Degrees: B.S., 1947, M.S., 1948, Sc.D., 1950, Massachusetts Institute of Technology; LL.D.(Hon.), 1979, Franklin Pierce Law School.

Fellow Award "For contributions to the application of computers to the information-handling needs of the medical community." Other Awards: Celotex Fellow in Acoustics, 1948; Armstron Cork Fellow in Acoustics, 1949; Fellow, Acoustical Society of America, 1950; Outstanding Young Electrical Engineer Award, Eta Kappa Nu, 1956; Fellow, New York Academy of Science, 1961; Fellow, American Academy of Arts and Sciences, 1963; Member, National Academy of Engineering, 1974; Assistant Secretary of Commerce for Science and Technology, 1977-1981.

BARZILAI, GIORGIO, Fellow 1978. Born: June 23, 1911. Degrees: Laurea in Ingegneria Industriale, 1935, University of Rome; M.Sc.(E.E.), 1947, University of Birmingham.

Fellow Award "For contributions to the field of electromagnetic theory and to engineering education."

BASAR, TAMER, Fellow 1983. Born: January 19, 1946, Istanbul, Turkey. Degrees: B.S.E.E., 1969, Robert College, Istanbul, Turkey; M.S., 1970, M.Phil., 1971, Ph.D., 1972, Yale University.

Fellow Award "For contributions to multiperson decisionmaking and deterministic and stochastic dynamic game theory." Other Awards: TUBITAK (Scientific and Technical Research Council of Turkey) Young Scientist Award in Applied Mathematics, 1976; Sedat Simavi Foundation Award in Mathematical Sciences (Turkey), 1979.

BASS, MICHAEL, Fellow 1981. Born: October 24, 1939, Bronx, NY. Degrees: B.S., 1960, Carnegie Mel-

Ion; M.S., 1962, Ph.D., 1964, University of Michigan.

Fellow Award "For contributions to quantum electronics related to laser material interactions and optical properties of matter." Other Awards: Distinguished Researcher Award, U.S.C. School of Engineering, 1979; Fellow, Optical Society of America, 1978.

BAST, RAY R., Fellow 1981. Born: March 10, 1925, Kutztown, PA. Degrees: B.S.E.E., 1950, Lehigh University.

Fellow Award "For the development of methods for extending the thermal operating capabilities of electrical power apparatus."

BATCHELDER, LAURENCE, IRE Fellow 1957. Born: October 26, 1906, Cambridge, Mass. Degrees: A.B., 1928, M.S., 1929, Harvard University.

Fellow Award "For contributions to the design and development of sonar equipment." Other Awards: Fellow, Acoustical Society of America, 1949; Distinguished Service Citation, Acoustical Society of America, 1972; Fellow, American Association for the Advancement of Science.

BATCHELOR, JOHN W., Fellow 1965. Born: November 17, 1914, Winchester, Ind. Degrees: A.B., 1935, Butler University; B.S.E.E., 1937, Purdue University.

Fellow Award "For contributions to the design and development of very large turbine-driven generators"; Nikola Tesla Award, 1979. Other Awards: National Academy of Engineering, 1980.

BATE, GEOFFREY, Fellow 1979. Born: March 30, 1929, Sheffield, England. Degrees: B.Sc.(Hons. Physics), 1949, Ph.D.(Physics), 1952, University of Sheffield.

Fellow Award "For significant contributions to the understanding of magnetic recording materials and devices"; Distinguished Lecturer, IEEE Magnetics Society, 1980; IEEE Centennial Medal, 1984. Other Awards: American Physical Society; Sigma Xi; 15 Patents; IBM Patent Award (5th level), 1976.

BATEMAN, ROSS, IRE Fellow 1959. Born: July 22, 1912, Toledo, Oreg. Degrees: B.S., 1934, Oregon State University.

Fellow Award "For contributions in the field of ionospheric scatter propagation."

BATES, RICHARD H.T., Fellow 1980. Born: July 8, 1929, Sheffield, England. Degrees: B.Sc., 1952, D.Sc., 1972, University of London.

Fellow Award "For creative contributions to electromagnetic imaging and its applications." Other Awards: Fellow, Royal Society of New Zealand, 1976; E. R. Cooper Memorial Prize, Royal Society of New Zealand, 1980; Mechaelis Memorial Prize, University of Otago, 1980; Visiting Member, Institution of Biomedical Engineers (Australia), 1982; President, Australasian College of Physical Scientists in Medicine, 1981-83.

BATHE, C. E., AIEE Fellow 1951. Born: September 28, 1904, Lenapah, Indian Terr., Okla. Degrees: B.S., 1925, A.B., 1933.

Other Awards: Sigma Tau, 1924; Phi Beta Kappa, 1933; Tau Beta Pi, 1956; Eta Kappa Nu, 1957.

BAUDRY, RENE, Fellow 1964. Born: May 23, 1899, France. Degrees: E., 1919, Ecole Nationale D'Arts et Metiers, Aix-en-Provence, France.

Fellow Award "For contributions to the mechanical development and design of large rotating machines"; First Paper Prize, AIEE, Power Division, 1951, for "Improved Cooling of Turbine Generator Windings"; Lamme Medal, 1966, "For his significant contributions to the design of large generators." Other Awards: Certificate of Commendation, U.S. Dept. of Navy; Fellow, American Society of Mechanical Engineers; Rene A. Baudry Patent Award, Westinghouse Electric Corp., 1982; Rene Baudry Award, est. by the Westinghouse Electric Steam Generator Division, 1982.

BAUER, FREDERICK, Fellow 1980. Born: August 30, 1920, Detroit, MI. Degrees: B.S.E.E., 1941, M.S., 1949, Wayne State University.

Fellow Award "For accomplishments in unification of worldwide vehicular radio frequency interference standards, and innovations in the technology of electromagnetic compatibility"; Director, Michigan Section, AIEE, 1953-1955; Director, Southeast Michigan Section, IEEE, 1965-1967; Prize Paper Award, IEEE Electromagnetic Compatibility Group, 1973. Other Awards: Tau Beta Pi, 1941; Registered Electrical and Mechanical Engineer, Michigan, 1945; Award of Merit, American Association for State and Local History, 1962; President, Engineering Society of Detroit; Distinguished Alumni Award, Wayne State University, 1970; Award for Outstanding Contributions, Society of Automotive Engineers, 1971, 1984; Distinguished Service Award, Automobile Manufacturers Association, 1971; Fellow, Engineering Society of Detroit, 1974; Ford Community Service Award, 1975; Technical Advisor to the U.S. National Committee, International Electrotehnical Commission, 1964 to present; Chairman, Electromagnetic Radiation Subcommittee, Society of Automotive Engineers; U.S. Representative to International Special Committee on Radio Interference Subcommittee D; Member, Committee C-63 American National Standards Institute; Advisor, Canadian Standards Association.

BAUGHMAN, G. W., AIEE Fellow 1957. Born: February 11, 1900, Gilboa, Ohio. Degrees: B.E.E., 1920, E.E., 1924, Ohio State University.

Fellow Award "For his contributions to the design and application of railway control systems"; Elmer A. Sperry Award, 1971. Other Awards: Eta Kappa Nu, 1919; Tau Beta Pi, 1920; Sigma Xi, 1935; Fellow, Institute Railway Signal Engineers, Great Britain; Lamme Medal, Ohio State University, 1978; Who's Who in Engineering; Who's Who in the World; Who's Who in America; Anerican Men of Science.

BAUM, CARL E., Fellow 1984. Born: February 6, 1940, Binghamton, N.Y. Degrees: B.S.(Eng.), 1962, M.S.(E.E.), 1963, Ph.D.(E.E.), 1969, Caltech.

Fellow Award "For pioneering the singularity expansion method and electromagnetic topology in electromagnetic theory, and for development of EMP simulation and electromagnetic sensors"; Distinguished Lecturer, IEEE Antennas and Propagation Society, 1977-78. Other Awards: Director and President, SUMMA Foundation; President, Electromagnetics Society; Membership, Commissions Band E, U.S. National Committee (USNC) of International Union of Radio Science (URSI); Tau Beta Pi, 1959; Honeywell Award, Best Undergraduate in Engineering, 1962; Sigma Xi, 1962; Air Force Research and Development Award, 1970.

BAUM, WILLARD U., Fellow 1968. Born: September 21, 1906, Perkasie, Pa. Degrees: B.S.E.E., 1930, Drexel Institute.

Fellow Award "For contributions to the planning and development of large utility systems." Other Awards: Engineer of the Year, Pennsylvania Society of Professional Engineers, 1965.

BAXANDALL, FRANK M., AIEE Fellow 1962. Born: July 23, 1900, Fall River, Wis. Degrees: Ph.B., 1923, Ripon College; B.S., 1924, University of Wisconsin.

Fellow Award "For contributions to the use of electricity in the reduction of magnesium and chlorine." Other Awards: Life Member, Michigan Society of Professional Engineers, 1967; Life Member, National Society of Professional Engineers, 1970.

BAYLY, BEN DE F., IRE Fellow 1947. Born: June 20, 1903, London, Ontario, Canada. Degrees: B.A.Sc., 1930, University of Toronto.

Fellow Award "For his contributions as a teacher in the field of radio communications and his service in the coordination of communications security." Other Awards: Member of the Order of the British Empire, 1946.

BEAM, ROBERT E., IRE Fellow 1956. Born: July 11, 1914, Cambridge, Ohio. Degrees: B.E.E., 1937, Ohio State University; Ph.D., 1940, Iowa State University.

Fellow Award "For contributions to education and research in the fields of microwave theory and techniques."

BEAM, WALTER R., Fellow 1968. Born: August 27, 1928, Richmond, Va. Degrees: B.S.E.E., 1947, M.S.E.E., 1950, Ph.D., 1953, University of Maryland.

Fellow Award "For his leadership and numerous technical contributions in the fields of microwave tube design, thin film technology, and engineering education." Other Awards: Eta Kappa Nu Award, 1957. Outstanding Achievement Award, RCA Laboratories, 1956; Exceptional Civilian Service Awards, U.S. Air

Force, 1979, 1981.

BEARD, ARTHUR D., Fellow 1974. Born: February 19, 1926, New York, N.Y. Degrees: B.E.E., 1947, M.E.E., 1949, Rensselaer Polytechnic Institute.

Fellow Award "For contributions to the development or magnetic storage devices and solid-state techniques for data processing." Other Awards: Eta Kappa Nu, 1946; Tau Beta Pi, 1946; Sigma Xi, 1949.

BEARD, CHARLES I., Fellow 1968. Born: November 30, 1916, Ambridge, Pa. Degrees: B.S.E.E., 1938, Carnegie Institute of Technology; Ph.D.(Physics), 1948, Massachusetts Institute of Technology.

Fellow Award "For contributions to the understanding of microwave scattering from random media"; John T. Bolljahn Memorial Award, IEEE, 1962; Certificate of Achievement, IEEE, 1967. Other Awards: Sigma Xi; Tau Beta Pi; Phi Kappa Phi.

BEARDSLEY, KENNETH D., AIEE Fellow 1961. Born: July 18, 1909, Boston, Mass. Degrees: B.S., 1929, M.S., 1930, Massachusetts Institute of Technology.

Fellow Award "For contribution to the development of cores and magnetic materials for distribution transformer."

BEATTIE, WILLIAM C., Fellow 1967. Born: December 26, 1904, London, England.

Fellow Award "For his outstanding contributions to metropolitan power engineering." Other Awards: Fellow, American Society of Mechanical Engineers, 1965.

BECK, ALFRED C., IRE Fellow 1958. Born: July 26, 1905, Granville, N.Y. Degrees: E.E., 1927, Rensselaer Polytechnic Institute.

Fellow Award "For his contributions to microwave antennas and transmission lines." Other Awards: Sigma Xi; Tau Beta Pi.

BECK, ARNOLD H. W., IRE Fellow 1959. Born: August 7, 1916, Hethersett, Norfolk, England. Degrees: B.Sc., 1937, University College, London; M.A., 1959, University of Cambridge.

Fellow Award "For contributions to the development of the thermionic valve." Other Awards: Fellow, Corpus Christi College, Cambridge, 1961; Fellow, University College, London, 1979.

BECKEN, EUGENE D., Fellow 1971. Born: April 29, 1911, Thief River Falls, Minn. Degrees: B.S.E.E., 1932, University of North Dakota; M.S.E.E., 1933, University of Minnesota; M.S., 1952, Massachusetts Institute of Technology.

Fellow Award "For contributions to management of engineering and to the use of computers in international communications." Other Awards: Sloan Fellow, Massachusetts Institute of Technology, 1951; DeForest Audion Gold Medal, Veteran Wireless Operators Association, 1975; Sioux Award for Distinguished Service and Outstanding Achievements, University of North Dakota, 1975; Fellow, Radio Club of America, 1976;

Lobdell Distinguished Service Award, Massachusetts Institute of Technology, 1979; Sigma Tau, 1932; Sigma Xi, 1982.

BECKETT, JOHN C., AIEE Fellow 1962. Born: August 6, 1917, Goldfield, Nev. Degrees: A.B., 1938, E.E., 1941, Stanford University.

Fellow Award "For contributions to the application of electrical techniques to medicine and biology, to advances in electric heating and electric measurement, and for public service;" First Paper Prize, AIEE, Section, 1958. Other Awards: Phi Beta Kappa, 1938; Tau Beta Pi, 1938; Outstanding Electrical Engineer Award, ENGINEERS' WEEK, 1956; ANSI Board of Directors, 1981-83.

BECKMANN, PETR, Fellow 1973. Born: November 13, 1924, Prague, Czechoslovakia. Degrees: M.Sc., 1949, Ph.D., 1955, Prague Technical University; Dr.Sc., 1961, Czechoslovak Academy of Sciences.

Fellow Award "For contributions to the theory of electromagnetic scattering, and to engineering education."

BECKWITH, ROBERT W., Fellow 1974. Born: July 25, 1919, Kent, Ohio. Degrees: B.S.E.E., 1941, Case Western Reserve; M.E.E., 1951, Syracuse University.

Fellow Award "For contributions in application of electronic technology to the development of frequency shift keying techniques for data transmission and remote circuit breaker tripping." Other Awards: Sigma Xi; Theta Tau, 1939; Eta Kappa Nu, 1940; Tau Beta Pi, 1940.

BECKWITH, STERLING, AIEE Fellow 1949. Born: 1905, Carthage, Mo. Degrees: A.B., 1927, Stanford University; M.S., 1931, University of Pittsburgh; Ph.D., 1933, California Institute of Technology.

Awards: Lamme Medal, AIEE, 1958, "In recognition of his contributions to the arts and science of design and application of rotating electric machines." Other Awards: Sigma Xi, 1932; Tau Beta Pi, 1946; Eta Kappa Nu, 1947.

BEDFORD, ALDA V., IRE Fellow 1950. Born: January 6, 1904, Winters, Tex. Degrees: B.S., 1925, University of Texas; M.S., 1929, Union College.

Fellow Award "For his many contributions to sound recording and to present-day television"; Vladimir K. Zworykin Award, IRE, 1954, "For his contributions to the principle of mixed highs and its application to color television." Other Awards: Modern Pioneer Award, National Association of Manufacturers, 1940; Research Award, Radio Corporation of America, 1949; Incentive Awards, Radio Corporation of America, 1950, 1951; David Sarnoff Gold Medal, Society of Motion Picture and Television Engineers, 1967.

BEDFORD, B. D., AIEE Fellow 1951. Born: September 22, 1906, Winters, Tex. Degrees: B.S., 1927, University of Texas.

BEDROSIAN, EDWARD, Fellow 1977. Born: May 22, 1922, Chicago, Ill. Degrees: B.S.E.E., 1949, M.S., 1950, PhD., 1953, Northwestern University.

Fellow Award "For contributions to the fields of nonlinear circuit analysis and communication systems." Other Awards: Fairbanks-Morse Award, 1947, 1948; Airborne Instruments Award, 1949-1950.

BEDROSIAN, SAMUEL D., Fellow 1975. Born: March 24, 1921. Degrees: A.B., 1942, State University of New York (Albany); M.E.E., 1951, Polytechnic Institute of New York; Ph.D., 1961, University of Pennsylvania.

Fellow Award "For contributions in graph theory applications to networks and systems." Other Awards: U.S. Army Commendation, Bikini Atoll, 1946; Eta Kappa Nu; Sigma Xi, 1963; ASA Kabakjian Award, 1976; NAVELEX Research Chair, Naval Postgraduate School, 1980-1981.

BEEHLER, JAMES E., Fellow 1974. Born: March 11, 1923, Marshall County, Ind. Degrees: B.S.E.E., 1944, M.S.E.E., 1948, Purdue University.

Fellow Award "For leadership in improving the design and performance of high-speed circuit breaker technology"; Switchgear Committee Award, IEEE, 1980; IEEE Centennial Medal, 1984.

BEEMAN, DONALD, AIEE Fellow 1949. Born: June 27, 1907, Morton, Ill. Degrees: B.A., 1929, Stanford University.

BEERS, G. L., IRE Fellow 1947. Born: May 13, 1899, Indiana, Pa. Degrees: B.S., 1921, Sc.D.(Hon.), 1947, Gettysburg College.

Fellow Award "For his numerous advances in circuits and systems leading to improved radio broadcasting, particularly in the receiver field." Other Awards: Modern Pioneer Award, National Association of Manufacturers, 1940; Fellow, Society of Motion Picture and Television Engineers; Distinguished Alumni Award, Gettysburg College; Licensed Professional Engineer, New Jersey; 80 U.S. Patents; Member, National Television Systems Center.

BEGOVICH, NICHOLAS A., Fellow 1965. Born: November 29, 1921, Oakland, Calif. Degrees: B.S., 1943, M.S., 1944, Ph.D., 1948, California Institute of Technology.

Fellow Award "For contributions to controlled phased array radars."

BEGUN, SEMI J., IRE Fellow 1952. Born: December 2, 1905, Free City of Danzig. Degrees: M.S., 1929, Ph.D., 1933, Institute of Technology, Berlin, Germany.

Fellow Award "In recognition of his contributions to the field of magnetic recording." Other Awards: Presidential Certificate of Merit, 1948; Emile Berliner Award, Audio Engineering Society, 1956; John H. Potts Award, Audio Engineering Society, 1960; Fellow, Acoustical Society of America.

BEHNKE, WALLACE BLANCHARD, Fellow 1978. Born: February 5, 1926. Degrees: B.S., 1945, B.S.E.E., 1947, Northwestern University.

Fellow Award "For contributions in developing economical nuclear power and the fast breeder reactors." Other Awards: National Academy of Engineering, 1980.

BEHREND, WILLIAM L., Fellow 1975. Born: January 11, 1923, Wisconsin Rapids, Wis. Degrees: B.S.E.E., 1946, M.S.E.E., 1947, University of Wisconsin.

Fellow Award "For contributions to television transmitter systems and measurement methods"; Scott Helt Award, 1971. Other Awards: Sigma Xi, 1950; RCA Corporation Awards, 1950, 1956, 1959, 1963, 1975.

BEILER, ALBERT H., AIEE Fellow 1962. Born: August 1, 1898, Vienna, Austria. Degrees: B.S., 1925, E.E, 1930, Cooper Union.

Fellow Award "For contributions to the application of power station electric auxiliaries and to the education and training of operating personnel."

BEKEY, GEORGE ALBERT, Fellow 1972. Born: June 19, 1928, Bratislava, Czechoslovakia. Degrees: B.S.E.E., 1950, University of California at Berkeley; M.S.E.E., 1952, Ph.D., 1962, University of California at Los Angeles.

Fellow Award "For contributions in the areas of hybrid computation, man-machine systems, and biomedical engineering." Other Awards: Tau Beta Pi; Eta Kappa Nu; Pi Kappa Delta; Sigma Xi; National Lecturer, Sigma Xi, 1976-1977; Charter Member, Biomedical Engineering Society.

BELANGER, PIERRE R., Fellow 1984. Born: August 18, 1937, Montreal, Quebec. Degrees: B.Eng.(Eng. Physics), 1959, McGill University; M.S.E.E., 1961, M.I.T.; Ph.D., 1964, M.I.T.

Fellow Award "For contributions to the application of modern control theory to industrial processes."

BELEVITCH, VITOLD, IRE Fellow 1961. Born: March 2, 1921, Helsinki, Finland. Degrees: E.E., M.E., 1942, D.A.Sc., 1945, University of Louvain, Belgium; Doctorate(Hon.), 1975, Technical University of Munich, Germany; Doctorate(Hon.), 1978, Polytechnic Lausanne, Switzerland.

Fellow Award "For contributions to the general theory of lumped networks"; IEEE Centennial Medal, 1984.

BELL, CHESTER GORDON, Fellow 1974. Born: August 19, 1934, Kirksville, Mo. Degrees: B.S.E.E., 1956, M.S.E.E., 1957, Massachusetts Institute of Technology.

Fellow Award "For contributions to the design of time-sharing computer systems, and for education in the understanding of computer structures." Other Awards: Mellon Institute Award, Carnegie-Mellon University, 1973; McDowell Award, 1975; National

Academy of Engineering, 1977; Eckert-Mauchly Award, 1982.

BELL, JOHN F., Fellow 1966. Born: January 8, 1914, Cass Co., Ind. Degrees: B.S.E.E., Purdue University.

Fellow Award "For pioneering in the development of UHF television tuners."

BELL, JOHN W., IRE Fellow 1953.

Fellow Award "For outstanding contributions to the design and development of military radar and cathode-ray direction-finding equipment in Canada."

BELL, RONALD L., Fellow 1972. Born: September 5, 1924, Alnwick, England. Degrees: Ph.D. (E.E.), 1949, University of Durham, England.

Fellow Award "For contributions to infrared photoemission." Other Awards: Member, American Physical Society.

BELLANGER, MAURICE G., Fellow 1984. Born: June 21, 1941, France. Degrees: Engineer Enst., 1965; Doctorat D'Etat, 1981, Universite Paris-Sud.

Fellow Award "For contributions to the theory of digital filtering, and the applications to communication systems"; Leonard Abraham Paper Award, IEEE Communications Society, 1978. Other Awards: Grand Prix Ferrie de L'Electronique, Paris, 1982.

BELLASCHI, PETER L., AIEE Fellow 1940. Born: February 13, 1903, Piedmont, Italy. Degrees: B.S., 1926, M.S., 1928, Massachusetts Institute of Technology; D.Sc., 1940, Washington and Jefferson College.

Awards: Paper Prize, Honorary Mention, AIEE, 1932, for "Impulse Generators for Transformer Testing"; William M. Habirshaw Award, IEEE, 1982, "For contributions in the field of transmission and distribution of electric power and to the development of extra-high voltage apparatus and systems"; Certificate of Appreciation, Administrative Committee, IEEE Power Engineering Society for 50 Years of Service to the Transformers Committee, 1982. Other Awards: Eta Kappa Nu Award, 1936; Award of Merit, Westinghouse Electric, 1937.

BELLER, CLARENCE J., AIEE Fellow 1949. Born: February 17, 1902, Cleveland, Ohio. Degrees: B.S., 1924, Case Institute of Technology; A.M.P., 1949, Harvard Business School.

BELLO, PHILLIP A., Fellow 1970. Born: October 22, 1929, Lynn, Mass. Degrees: B.S.E.E., 1953, Northeastern University; S.M., 1955, Sc.D. 1959, Massachusetts Institute of Technology.

Fellow Award "For contributions to statistical communication theory, particularly in the analysis of time-varying channels."

BELOCK, HARRY D., Fellow 1968. Born: April 10, 1908, New York, N.Y. Degrees: B.S., 1939, Sc.D., 1967, Pratt Institute.

Fellow Award "For pioneering in weapons systems, analog computers and controls for instrumentation and simulation, and leadership in engineering management." Other Awards: Tau Beta Pi; Fellow, Brooklyn Engineers' Club; Distinguished Public Service Award, U.S. Navy.

BELOHOUBEK, ERWIN F., Fellow 1975. Born: May 7, 1929, Vienna, Austria. Degrees: Diplom Ing., 1953, Ph.D. 1955, Technical University, Vienna.

Fellow Award "For contributions to microwave solid-state amplifiers and to microwave tubes"; Microwave Application Award, IEEE Microwave Theory and Techniques Society, 1980. Other Awards: Outstanding Performance Award, RCA, 1963; Laboratories Achievement Award, RCA, 1967; Best Paper Award, DOD/AOC Technical Symposium, 1980.

BENDER, WELCOME W., Fellow 1965. Born: November 30, 1915, Elizabeth, N.J. Degrees: B.S., 1938, M.S., 1939, Massachusetts Institute of Technology.

Fellow Award "For contributions to the organization and direction of industrial research." Other Awards: Associate Fellow, American Institute of Aeronautics and Astronautics, 1962.

BENEDICT, R. RALPH, AIEE Fellow 1956. Born: October 1, 1904, Medford, Wis. Degrees: B.S., 1925, M.S., 1926, University of Wisconsin.

Fellow Award "For his teaching and contributions to the engineering literature in the field of industrial electronics." Other Awards: Guest Professor and Group Leader, Bengal Engrg. College on A.I.D. program, 1954-1956; 1961-1963; Benjamin Smith Reynolds Teaching Award, 1969.

BENEKING, HEINZ, Fellow 1984. Born: March 28, 1924, Frankfurt/Main, Germany. Degrees: Dipl.-Phys., 1951, Dr.rer.nat., 1952, University of Hamburg, Germany.

Fellow Award "For innovation in the field of compound semiconductor technology and devices, especially for work on heterostructure bipolar transistors." Other Awards: Gauss Medal, Braunschweigische Wissenschaftliche Gesellschaft, 1984.

BENHAM, T. A., IRE Fellow 1962. Born: December 30, 1914, Hartford, Conn. Degrees: B.S., 1938, M.S., 1945, Haverford College.

Fellow Award "For contributions to electronics for the blind." Other Awards: Presidential Citation, U.S. Dept. of Labor, 1951; Handicapped Pennsylvanian of the Year, 1969; Professor Emeritus, Haverford College.

BENNETT, C. LEONARD, Fellow 1982. Born: October 5, 1939, Lowell, MA. Degrees: B.S.E.E., 1961, Lowell Technological Institute; M.S., 1964, North Carolina State University; Ph.D., 1968, Purdue University.

Fellow Award "For contributions to time domain analysis of electromagnetic radiators."

BENNETT, RALPH D., AIEE Fellow 1935, IRE Fellow 1952. Born: June 30, 1900, Williamson, N.Y. Degrees:

B.S., 1921, M.S., 1923, Union College; Ph.D., 1925, University of Chicago; Sc.D.(Hon.), 1945, Union College.

IRE Fellow Award "For his contributions to the administration of research in Government service and to the measurement art as physicist, engineer, and educator." Awards: Chairman, Washington Section, AIEE, 1948-1949; Patron, Washington Section, IEEE, 1967. Other Awards: Charles A. Coffin Fellow, University of Chicago, 1923-1925; Sigma Xi; National Research Fellow, Princeton 1926-7, California Institute of Technology, 1927-8; Fellow, American Physical Society, 1924; Officer of the British Empire, 1946; Distinguished Civilian Service Award, U.S. Dept. of the Navy, 1950; Meritorious Public Service Award, City of San Francisco, 1976; Distinguished Public Service Award, U.S. Dept. of the Navy, 1977.

BENNETT, ROBERT R., Fellow 1965. Born: May 7, 1926, Spokane, Wash. Degrees: B.S., 1945, M.S., 1947, Ph.D., 1949, California Institute of Technology.

Fellow Award "For contributions to the design of missile guidance and control systems, and for his technical leadership."

BENNETT, WILLIAM RALPH, JR., Fellow 1974. Born: January 30, 1930, Jersey City, N.J. Degrees: A.B., 1951, Princeton University; Ph.D., 1959, Columbia University; M.A.(Hon.), 1965, Yale University; D.Sc.(Hon.), 1975, University of New Haven.

Fellow Award "For contributions to the realization and understanding of gas lasers"; Morris N. Liebmann Award, IEEE, 1965, "For basic contributions to the theoretical understanding and experimental realization of oscillators at optical frequencies." Other Awards: Fellow, American Physical Society; Fellow, Optical Society of America; Sloan Foundation Fellow, 1963; Guggenheim Foundation Fellow, 1967; Sigma Xi; Charles Baldwin Sawyer Chair in Engineering and Applied Science, Yale University, 1973; Voted one of "Ten Best Teachers at Yale University," Yale Student Course Critique, 1974, 1975, 1976; Western Electric Fund Award, American Society of Engineering Educators for Outstanding Teaching, 1977; Outstanding Patent Award, Research and Development Council of New Jersey, 1977; Master of Silliman College, Yale University, 1981.

BENNON, SAUL, Fellow 1966. Born: August 9, 1914, Philadelphia, Pa. Degrees: B.S.E.E., 1936, M.S.E.E., 1937, University of Pennsylvania.

Fellow Award "For contributions to the design of transformers and calculations of performance, especially in the fields of impulse voltage distribution." Other Awards: Order of Merit, Westinghouse Electric, 1945.

BENSON, FRANK A., Fellow 1976. Born: November 21, 1921, Grange-over-Sands, Cumbria, United Kingdom. Degrees: B.Eng., 1942, M.Eng., 1945, University of Liverpool; Ph.D., 1952, D.Eng., 1957, University of Sheffield.

Fellow Award "For advances in knowledge of guided waves and ionized gases."

BERANEK, LEO L., IRE Fellow 1952. Born: September 15, 1914, Solon, Iowa. Degrees: B.A., 1936, Cornell College; ; M.Sc., 1937, D.Sc., 1940, Harvard University; D.Sc.(Hon.), 1946, Cornell College; Dr.Engrg.(Hon.), 1971, Worcester Polytechnic Institute; D. Commercial Sc.(Hon.), 1979, Suffolk University; LL.D., Emerson College, 1982.

Fellow Award "For his contributions in research, teaching and administration in the fields of acoustics and speech communication." Other Awards: Silver Commemorative Medal, Groupement des Acousticiens de Langue Francaise, 1966; Bienniel Award, Acoustical Society of America, 1944; Presidential Certificate of Merit, 1948; Wallace Clement Sabine Award, Acoustical Society of America, 1961; Gold Medal Award, Audio Engineering Society, 1971; Gold Medal Award, Acoustical Society of America, 1975; Abe Lincoln Award honoring Broadcasters, Radio and Television Commission, 1976; Member, American Academy of Arts and Sciences; National Academy of Engineering; Distinguished Community Service Award, Greater Boston Chamber of Commerce, 1980; Distinguished Service Award, Massachusetts Broadcasters Association, 1980.

BERBERICH, LEO J., AIEE Fellow 1945. Born: November 19, 1906, Petersburg, Va. Degrees: B.E., 1928, D.E., 1931, The Johns Hopkins University.

Awards: Chairman, Pittsburgh Section, IEEE. Other Awards: Tau Beta Pi; Sigma Xi; Special Outstanding Patent Awards, Westinghouse Electric, 1941, 1954.

BERESKIN, ALEXANDER B., IRE Fellow 1958. Born: November 15, 1912, San Francisco, Calif. Degrees: B.S., 1935, M.S., 1941, University of Cincinnati.

Fellow Award "For contributions to electronic circuitry"; Achievement Award, IRE, Professional Group on Audio, 1959, "For outstanding contributions to audio technology documented by papers in IRE publications"; IEEE Centennial Award, 1984. Other Awards: Distinguished Engineer, The Technical and Scientific Societies Council of Cincinnati, Ohio, 1976; American Society for Engineering Education Delos Award, 1981; Western Electric Fund Award, American Society for Engineering Education, 1983; Award of Merit, Cincinnati Chapter Eta Kappa Nu, 1983.

BERG, DANIEL, Fellow 1974. Born: June 1, 1929, New York, N.Y. Degrees: B.S., 1950, City College of The City University of New York; M.S., 1951, Ph.D. 1953, Yale University.

Fellow Award "For contributions to the science and the art of electrical insulation." Other Awards: Belden

Medal, 1949; Fellow, AAAS, 1975; National Academy of Engineering, 1976; Wilbur Cross Medal, Yale University, 1983; Fellow, American Institute of Chemists, 1979.

BERGEN, ARTHUR R., Fellow 1984. Born: April 7, 1923, New York, N.Y. Degrees: B.E.E., 1944, City College of New York; M.S.E.E., 1953, Sc.D., 1958, Columbia University.

Fellow Award "For contributions to system control theory and its practical application, and for teaching."

BERGER, TOBY, Fellow 1978. Born: New York City, N.Y. Degrees: Bachelor of Engineering (Electrical), 1962, Yale University; Master of Science, 1964, Ph.D.(Applied Mathematics), 1966, Harvard University.

Fellow Award "For contributions to information theory and engineering education." Other Awards: Guggenheim Fellowship, 1975-76; Fellow, Japan Society for Promotion of Science, 1980-1981; Fellow, Ministry of Education, Peoples Republic of China, 1981; Frederick E. Terman Award, American Society for Engineering Education, 1982.

BERGER, U. SYLVESTER, Fellow 1966. Born: April 7, 1915, Dayton, Ohio. Degrees: B.Sc.(E.E.), 1937, Ohio State University.

Fellow Award "For contributions in the field of VHF, UHF and microwave radio system analysis, development and design."

BERGH, ARPAD A., Fellow 1981. Born: April 26, 1930, Hungary. Degrees: M.S.(Chem.), 1952, University of Szeged, Hungary; Ph.D.(Phys.Chem.), 1959, University of Pennsylvania.

Fellow Award "For leadership in the advancement of Optoelectronics and the technology of light-emitting diodes."

BERGLUND, C. NEIL, Fellow 1983. Born: July 21, 1938, Fort William, Ontario, Canada. Degrees: B.Sc., 1960, Queen's University; M.S.E.E., 1961, M.I.T.; Ph.D.(Elec. Eng.), 1964, Stanford University.

Fellow Award "For contributions to metal-oxide-semiconductor interface physics and devices." Other Awards: nine scholarships; Gold Medal, Association of Professional Engineers of Ontario.

BERKOWITZ, MILTON, Fellow 1978. Born: January 10, 1930, New York, N.Y. Degrees: B.A.(Physics and Math.), 1951, New York University; M.S.(Eng.), 1965, University of Pennsylvania.

Fellow Award "For leadership and contributions to space communication and aerospace electronic systems development."

BERKOWITZ, RAYMOND S., Fellow 1981, Born: February 21, 1923, Philadelphia, PA. Degrees: B.S.(E.E.), 1943, M.S.(E.E.), 1948, Ph.D., 1951, University of Pennsylvania.

Fellow Award "For contributions to advanced data

processing techniques in modern radar and engineering education." Other Awards: Lady Davis Fellow, Technion, Haifa, Israel, 1976-1977.

BERLEKAMP, ELWYN R., Fellow 1972. Born: September 6, 1940, Dover, Ohio. Degrees: B.S., 1962, M.S., 1964, Ph.D., 1964, Massachusetts Institute of Technology.

Fellow Award "For contributions to information theory, particularly in the field of algebraic coding"; Best Research Paper Award, for book, "Algebraic Coding Theory," IEEE, Information Theory Group, 1968. Other Awards: Outstanding Young Electrical Engineer Award, Eta Kappa Nu, 1971; Member, National Academy of Engineering, 1977.

BERLINCOURT, DON A., Fellow 1972. Born: January 10, 1924, Fremont, Ohio. Degrees: B.S.E.E., 1949, Case Institute of Technology; M.S., 1950, Harvard University.

Fellow Award "For development and characterization of new piezoelectric materials that have had major impact on transducer and wave filter technology." Other Awards: Fellow, Acoustical Society of America.

BERMAN, BARUCH, Fellow 1979. Born: November 10, 1925, Tel Aviv, Israel. Degrees: B.S.E.E., 1947, (additional Ing. diploma in industrial technology), 1948, Israel Institute of Technology; M.S.E.E., 1957, Ph.D. (Eng. Sc.), 1960, Columbia University.

Fellow Award "For contributions to solid-state power converter technology"; Achievement Award, Region Six, IEEE, 1979; Engineer of the Year Award, South Bay Harbor Section, IEEE, 1979-1980; Excellence Citation for Accomplishments as 1981-1982 Region 6 Awards Chairman. Other Awards: Fellow, Institute for the Advancement of Engineering; Meritorious Service to Engineering Award, National Society of Professional Engineers; Chairman Award, California Society of Professional Engineers; Registered Professional Engineer, Control Systems, California.

BERNSTEIN, ARTHUR J., Fellow 1981. Born: May 28, 1937, New York, NY. Degrees: A.B., 1957, B.S., 1958, M.S.E.E., 1959, Ph.D., 1962, Columbia University.

Fellow Award "For contributions to the theory and development of software for large-scale dispersed interactive operating systems"; Best Paper Award, IEEE Transactions on Computers, 1975; Distinguished Visitors Program, IEEE Computer Society, 1979-1981.

BERNSTEIN, RALPH, Fellow 1983. Born: February 20, 1933, Zweibrucken, Germany. Degrees: B.S.E.E., 1956, University of Connecticut; M.S.E.E., 1960, Syracuse University.

Fellow Award "For contributions to digital image processing technique of earth observation sensor data and applications to operational systems." Other Awards: Medal for Exceptional Scientific Achievement, NASA, 1974; Outstanding Contribution Award, IBM,

1974; Orbiting Astronomical Observatory Group Achievement Award, NASA, 1972.

BERRY, WILLIAM L., AIEE Fellow 1960. Born: December 7, 1909, New York, N.Y. Degrees: B.S., 1932, M.S., 1933, California Institute of Technology.

Fellow Award "For contributions to the development of electric systems for aircraft and support equipment." Other Awards: Chairman, AIEE Committees; Author of eight Technical Papers, Electrical and Radio Systems.

BERTRAM, JOHN E., Fellow 1984.

Fellow Award "For technical and managerial leadership in the development and manufacturing of advanced data processors."

BERTRAM, SIDNEY, Fellow 1967. Born: July 7, 1913, Winnipeg, Manitoba, Canada. Degrees: A.A., 1932, Los Angeles City College; B.S.(with honor), 1938, California Institute of Technology; M.S., 1941, Ph.D., 1951, Ohio State University.

Fellow Award "For contributions to sonar and to stereomapping and map compilation systems." Other Awards: Certificate of Commendation, U.S. Dept. of Navy, 1947; Fairchild Award, American Society of Photogrammetry, 1969.

BERTSEKAS, DIMITRI P., Fellow 1984. Born: July 9, 1942, Athens, Greece. Degrees: B.Sc.(E.E.M.E.), 1965, National Technical University of Athens, Greece; M.S.E.E., 1969, George Washington Univ.; Ph.D., 1971, M.I.T.

Fellow Award "For contributions to optimization, data communication networks, and distributed control."

BESSON, PIERRE A. J., Fellow 1970. Born: December 5, 1901, Privas, (Ardeche), France. Degrees: Ecole Polytechnique, Paris; Ecole Nationale des Ponts et Chaussees; Ecole Superieure D'Electricite; Licence d'Electrotechnique Generale, Faculte des Sciences de Paris.

Fellow Award "For research, management, and teaching in the electrical power and electronics fields." Other Awards: Officer, French Legion of Honor; Silver Gilt Medal, French Military Transmissions.

BETTERSWORTH, THOMAS A., Fellow 1969. Born: April 26, 1911, Hollister, Calif. Degrees: B.S.E.E., 1934, University of California.

Fellow Award "For pioneering leadership in the field of electric power distribution."

BEUTLER, FREDERICK J., Fellow 1980. Born: October 3, 1926, Berlin, Germany. Degrees: S.B., 1949, S.M., 1951, Massachusetts Institute of Technology; Ph.D., 1957, California Institute of Technology.

Fellow Award "For contributions to stochastic process theory and its engineering applications." Other Awards: Tau Beta Pi (Eminent Engineer), 1981.

BEVERAGE, HAROLD H., IRE Fellow 1928, AIEE Fellow 1948. Born: October 14, 1893, North Haven, Me. Degrees: B.S., 1915, D.Eng., 1938, University of Maine.

Awards: Morris N. Liebmann Award, IRE, 1923, "For development of directive receiving antennas"; Medal of Honor, IRE, 1945; Lamme Medal, AIEE, 1957, "For his pioneering and outstanding engineering achievements in the conception and application of principles basic to progress in national and world-wide radio communication"; Achievement Award, IRE, Professional Group on Communications Systems, 1958. Other Awards: Armstrong Medal, Radio Club of America, 1938; Modern Pioneer Award, National Association of Manufacturers, 1940; Certificate of Appreciation, Signal Corp., 1944; Presidential Certificate of Merit, 1948; Fellow, American Association for the Advancement of Science, 1954; Eminent Member, Eta Kappa Nu, 1955; Honorary Member, Tau Beta Pi, 1959; Marconi Gold Medal, Veteran Wireless Operators Association, 1974; Pioneer Award, Radio Club of America, 1976; Alumni Career Award, University of Maine, 1976.

BEWLEY, LOYAL V., AIEE Fellow 1947. Born: December 19, 1898, Republic, Wash. Degrees: B.S.E.E., 1923, University of Washington; M.S.E.E., 1928, Union University.

Awards: National Best Paper Prize in Theory and Research, 1932; Lamme Medal, 1964, "For meritorious achievement in the theoretical analysis of high voltage surges resulting in an advancement of insulation design and improvement of protection for machines, transformers, station apparatus and transmission lines." Other Awards: Charles A. Coffin Award, General Electric Co., 1934; Hillman Award, Lehigh University, 1960.

BEYER, ROBERT T., Fellow 1969. Born: January 27, 1920. Degrees: A.B.(Math.), 1943, Hofstra University; Ph.D.(Physics), 1945, Cornell University.

Fellow Award "For contributions to ultrasonic propagation in fluids"; Ad Com, IEEE Sonics and Ultrasonics Group, 1964-67. Other Awards: Distinguished Service Citation, Acoustical Society of America, 1978; Gold Medal, Acoustical Society of America, 1984; Member, Sigma Xi; Fellow, American Physical Society, Acoustical Society of America.

BHAUMIK, MANI L., Born: January 5, 1932, Calcutta, India. Degrees: B.S.(Physics), 1951, M.S.(Physics), 1953, University of Calcutta; Ph.D.(Applied Physics), 1958, Indian Institute of Technology.

Fellow Award "For contributions to the research and development of high-energy lasers and new laser systems."

BIANCHI DI CASTELBIANCO, FRANCO, Fellow 1966. Born: December 3, 1905, Florence, Italy. Degrees: Dr., 1927, Pisa Engineering School.

Fellow Award "For contributions to the technology of electric transmission towers and for achievements in the field of electric power transmission."

BIARD, JAMES R., Fellow 1968. Born: May 20, 1931, Paris, Texas. Degrees: B.S.E.E., M.S.E.E., Ph.D., Texas A&M University.

Fellow Award "For outstanding contributions in the field of optoelectronics."

BIAS, FRANK J., Fellow 1966. Born: October 1, 1919, Des Moines, Iowa. Degrees: B.S., 1941, Iowa State College.

Fellow Award "For contributions to the development, design and standardization of high power broadcasting transmitters."

BIBBER, HAROLD W., AIEE Fellow 1946. Born: March 12, 1899, Gloucester, Mass. Degrees: B.S., 1920, Massachusetts Institute of Technology.

BIBERMAN, LUCIEN M., Fellow 1975. Born: May 31, 1919. Degrees: B.S., 1940, Rennselaer Polytechnic Institute.

Fellow Award "For contributions to infrared technology for remote sensing, viewing, and guiding." Other Awards: Fellow, Optical Sociey of America; Fellow, Society for Information Display; Fellow, Society of Photo-Optical Instrumentation Engineers.

BICKART, THEODORE A., Fellow 1977. Born: August 25, 1935, New York, N.Y. Degrees: B.E.S., 1957, M.S.E., 1958, D.Eng., 1960, The Johns Hopkins University.

Fellow Award "For contributions to theory and education in circuits and systems."

BIDARD, RENE A., Fellow 1971. Born: June 14, 1909, Paris, France. Degrees: 1932, Ecole Centrale des Arts et Manufactures; 1934, Ecole Superieure d'Electricite.

Fellow Award "For contributions to the applications of thermodynamics and to the development and design of large turbogenerators." Other Awards: Former Member of the Board, Societe Francaise d'Energie Nucleaire; Honorary Professor, Ecole Nationale Superieure d'Aeronautique, 1973; Former Member of the Board, Ecole Centrale des Arts et Manufactures Paris, and Ecole Superieure d'Electricite, Paris; Former Member of the Board, Societe des Ingenieurs et Scientifiques de France; and former President of its Section; Electrical and Electronical Industries (Section Electricity).

BIEDENBACH, JOSEPH M., Fellow 1977. Born: January 29, 1927. Degrees: M.S.(Education), 1950, 1951, University of Illinois; M.S.(Physics), 1956, University of Michigan; Ph.D., 1964, Michigan State University; M.B.A., 1973, Southern Illinois University.

Fellow Award "For contributions to the continuing education of engineers"; Meritorious Service Award, IEEE Education Society, 1979; Fellow, American Society for Engineering Education, 1982; Outstanding Service Award, IEEE Region 3, 1982. Other Awards: Distinguished Young Man of the Year, Chamber of Commerce, 1960; Silver Beaver Award, Boy Scouts of America, 1961; Distinguished Service Award, Continuing-Engineering Studies Division, American Society for Engineering Education, 1976; Distinguished Service Award, American Society for Engineering Education, 1977; Outstanding Service Award, ASEE-Southeastern Section, 1982.

BIGELOW, ROBERT O., Fellow 1973. Born: March 28, 1926, Boston, Mass. Degrees: B.S., 1950, M.S., 1950, Massachusetts Institute of Technology.

Fellow Award "For development of digital computer applications to the planning, design, reliability, and protection of bulk power systems." Other Awards: Young Engineer of the Year, Massachusetts Society of Professional Engineers, 1960.

BILLINTON, ROY, Fellow 1978. Born: September 14, 1935, Leeds, England. Degrees: B.Sc., 1960, M.Sc., 1963, University of Manitoba; Ph.D., 1967, D.Sc., 1975, University of Saskatchewan.

Fellow Award "For contributions to development and education in power system reliability evaluation." Other Awards: Sir George Nelson Award, Engineering Institute of Canada, 1965, 1967; Ross Medal, Engineering Institute of Canada, 1972; Fellow, Royal Society of Canada, 1980; Fellow, Engineering Institute of Canada, 1981.

BIORCI, GIUSEPPE, Fellow 1976. Born: August 7, 1930, Brescia, Italy. Degrees: D.E.E., 1953, University of Genoa, Italy.

Fellow Award "For contributions to network theory"; Volta Fellow, AIEE, academic year 1954-55. Other Awards: Bonavera Award, Academy of Science of Torino, Italy, 1958; Vallauri Award, Salone Internazionale della Tecnica, 1960; Panzarasa Award, Associazione Electrotecnica Italiana, 1960.

BIRCH, LELAND W., AIEE Fellow 1952. Born: June 27, 1895, Columbus, Ohio. Degrees: B.E.E., 1917, E.E., 1929, Ohio State University.

Fellow Award "For outstanding contributions and ingenuity in the development of overhead distribution and contact wire systems as used by electrified railroads, open-pit mining operations, street railways and trolley coach systems"; First Paper Prize, AIEE, General Applications Group, 1948, for "Are the Overhead Distribution Costs Retarding Railroad Electrification?"

BIRCHARD, WAYNE E., AIEE Fellow 1951. Born: January 2, 1911, Council Bluffs, Iowa. Degrees: B.S.E.E., 1931, M.S.E.E., 1932, Iowa State University.

BIRDSALL, CHARLES K., IRE Fellow 1962. Born: November 19,1925, New York, N.Y. Degrees: B.S.E.(E.E.), B.S.E.(Engr.Math.), 1946, M.S.E.(E.E.), 1948, University of Michigan; Ph.D., 1951, Stanford University.

Fellow Award "For contributions to research on trav-

eling-wave tubes." Other Awards: Miller Professor, University of California at Berkeley, 1963-1964; National Science Foundation Grantee, U.S.-Japan Cooperative Science Program, 1966; Fellow, American Physical Society; Founding Chairman, Energy and Resources Graduate Group, University of California at Berkeley, 1972-1974; Fellow, American Association for the Advancement of Science; Senior Research Fellow, University of Reading, England, 1976; Visiting Chevron Energy Professor, California Institute of Technology, 1982; Research Associate, Institute of Plasma Physics, Nagoya Univerity, Japan, 1981-1982; 24 patents; 60 articles.

BIRINGER, PAUL P., Fellow 1970. Born: October 1, 1924, Marosvasarhely, Hungary. Degrees: Dipl.Eng., 1947, Technical University, Budapest; M.A.Sc., 1951, Royal Institute of Technology, Stockholm; Ph.D., 1956, University of Toronto.

Fellow Award "For educational leadership, and for technical contributions in the analysis of nonlinear magnetic devices"; Canadian District Prize, AIEE, (With Hausen and Slemon), 1958 Prize Paper Award, IEEE Industry Applications Society, 1979. Other Awards: Pleyel Award for Research, 1951; Senior Research Fellowship, Canadian National Research Council, 1966; Sons of Martha Medal, Association of Professional Engineers, 1968; Certificate of Award, American Society of Mechanical Engineers, 1972; Senior Research Fellowship, Japan Society for Promotion of Science, 1979.

BIRMINGHAM, HENRY P., Fellow 1966. Born: March 17, 1920, Newburgh, N.Y. Degrees: Ed.B., 1942, Rhode Island College.

Fellow Award "For contributions to knowledge and techniques related to the design of man-machine systems." Other Awards: Naval Ordnance Development Award; Meritorious Civilian Service Award.

BIRNBAUM, GEORGE, Fellow 1970. Born: July 16, 1919, New York, N.Y. Degrees: M.S., 1949, Ph.D., 1956, George Washington University.

Fellow Award "For contributions to the understanding of the interaction of electromagnetic radiation with the gaseous constituents of planetary atmospheres." Other Awards: Fellow, American Physical Society.

BISCHOFF, ALFRED F. H., Fellow 1969. Born: November 2, 1912, Richmond Hill, N.Y. Degrees: B.S.E.E., 1935, Union College.

Fellow Award "For outstanding technical contributions to the Loran System, nuclear weapons and Apollo reliabilty, as well as for his inspirational leadership in encouraging personnel technical growth and promoting communication and education." Other Awards: U.S. Navy Citation, 1942; Boy Scout Silver Beaver Award, 1953; Associate Fellow, AIAA, 1967.

BITZER, DONALD L., Born: January 1, 1934, East St. Louis, IL. Degrees: B.S., 1955, M.S., 1956, Ph.D., 1960, University of Illnois.

Fellow Award "For contributions to computer-assisted instruction systems and to the introduction of devices such as plasma display panels used in these systems." Other Awards: Industrial Research 100 Award, 1966; Vladmir K. Zworykin Award, National Academy of Engineering, 1973; National Academy of Engineering, 1974; Bobby C. Connelly Memorial Award, Miami Valley Computer Association, 1973; DPMA Computer Science Man of the Year Award, 1975; Chester F. Carlson Award, American Society for Engineering Education, 1981; Lincoln Academy, 1982; SID Recognition Award, 1979; Fellow, AAAS, 1983.

BIXBY, WILLIAM H., AIEE Fellow 1960. Born: December 28, 1906, Indianapolis, Ind. Degrees: B.S.E., 1930, B.S.E.E., 1930, M.S., 1931, Ph.D., 1933, University of Michigan; M.M.E., 1935, Chrysler Institute of Engineering.

Fellow Award "For contribution to automatic voltage regulated power supply." Other Awards: Fellow, American Association for the Advancement of Science, 1954.

BLAAUW, GERRIT A., Fellow 1972. Born: July 17, 1924, The Hague, Netherlands. Degrees: B.S.E.E., 1948, Lafayette College; Ph.D., 1952, Harvard University.

Fellow Award "For contributions to computer architecture, particularly for demonstrations of the distinction among the architecture, implementation, and realization of computing systems." Other Awards: De Groot Award, 1979; Member, Royal Netherlands Academy of Science, 1982.

BLACHMAN, NELSON M., Fellow 1966. Born: October 27, 1923, Cleveland, Ohio. Degrees: B.S.,1943, Case School of Applied Science; A.M., 1947, Ph.D., 1947, Harvard University.

Fellow Award "For contributions to statistical communication theory and to information theory"; Prize Paper Award for Communication Theory, IEEE Communications Society, 1976. Other Awards: Ordnance Development Award, U.S. Navy, 1945; Fulbright Lecturer, Madrid, 1964-1965; Fellow, American Association for the Advancement of Science, 1970; Fellow, Institution of Electrical Engineers, 1974; Member, New York Academy of Sciences, 1979; Commissions C and E, U.S. National Committe of Union Radioscientifique Internationale, 1958; Sigma Xi, 1943.

BLACK, HAROLD S., AIEE Fellow 1941, IRE Fellow 1948. Born: April 14, 1898, Leominister, Mass. Degrees: B.S., 1921, D.Eng., 1955, Worcester Polytechnic Institute.

IRE Fellow Award "For his work on negative feedback amplifier and for his application of pulse techniques to radio-communication systems"; First Paper Prize, AIEE, 1934, for "Stabilized Feedback Amplifiers"; AIEE, 1941, "For his work on negative feedback amplifiers"; Lamme Medal, AIEE, 1957, "For his many outstanding contributions to telecommunications and allied electronic arts, especially the invention of the negative feedback amplifier and the successful development and application of the negative feedback amplification principle." Other Awards: Modern Pioneer Award, National Association of Manufacturers, 1940; John Price Wetherill Medal, Franklin Institute, 1941; Certificate of Appreciation, War Dept., 1946; Research Corporation Scientific Award, 1952; John H. Potts Memorial Award, Audio Engineering Society, 1959; Fellow, AAAS, 1954; Tau Beta Pi, 1920; Sigma Xi, 1921; Medal, Engineers Club of Philadelphia, 1961; Wisdom Award of Honor, Wisdom Society, 1969; Wisdom Hall of Fame, 1970; American Institute of Aeronautics and Astronautics, 1970; Award, Franklin Institute, 1971; Award, New York Academy of Sciences, 1971; Award, New Jersey Academy of Science, 1971; International Platform Association, 1972; Award, Creativity Recognition, 1972; Award, Intercontinental Who's Who in Community Service, 1973; Award, Intercontinental Biographical Association, 1973; Award, American Bicentennial Research Institute, 1974; Award, American Bicentennial Research Institute, 1974; National Board of Sponsors of the Institute for American Strategy, 1974; Award, International Institute of Community Service, 1975; Award, World-Wide Academy of Scholars, 1975; Founder, Center for International Security Studies, 1977; Charter Member, Coalition for Peace Through Strength, 1978; American Biographical Institute, 1979; National Inventors Hall of Fame, 1981; Robert H. Goddard Award, Worcester Polytechnic Institute, 1981.

BLACKBURN, J. LEWIS, Fellow 1970. Born: October 2, 1913, Kansas City, Mo. Degrees: B.S.E.E., 1935, University of Illinois.

Fellow Award "For contributions in the application of protective relaying to large electric power systems"; Distinguished Service Award, IEEE Power Engineering Society, Power System Relaying Committee, 1978; Outstanding Teaching Award, Educational Activities Board, IEEE, 1980; IEEE Centennial Medal, 1984. Other Awards: Westinghouse Order of Merit, 1971; Westinghouse Auditorium-Class Room, Coral Springs, Fla., named "Lewis Blackburn Room", 1980.

BLACKMAN, RALPH B., Fellow 1964. Born: August 29, 1904, Mangaldan, Pangasinan, Philippines. Degrees: B.Sc., 1926, California Institute of Technology.

Fellow Award "For contributions to circuit theory and data processing."

BLACKSMITH, PHILIPP, Fellow 1979. Born: October 12, 1921, Liebling, Romania. Degrees: B.S.(Physics), 1949, Carnegie Institute of Technology; M.S.E.E., 1963, Air Force Institute of Technology.

Fellow Award "For leadership in developing technology for advanced antennas and electromagnetic sensor systems." Other Awards: Air Force Meritorious Civilian Service Award, 1963; Presidential Citation, 1964.

BLACKWELL, WILLIAM A., Fellow 1982. Born: May 17, 1920, Fort Worth, TX. Degrees: B.S.E.E., 1949, Texas Technological College; M.S., 1952, University of Illinois; Ph.D., 1958, Michigan State University.

Fellow Award "For contributions to electrical engineering education"; IEEE Centennial Medal, 1984. Other Awards: Distinguished Service Award, Electrical Engineering Alumni Association, University of Illinois, 1979; Executive of the Year Award, New River Valley Chapter, National Secretaries Association, 1979; Outstanding Service Award, Southeastern Center for Electrical Education, Inc., 1980.

BLAHUT, RICHARD E., Fellow 1981. Born: June 9, 1937, Orange, NJ. Degrees: B.S.(E.E.), 1960, Massachusetts Institute of Technology; M.S.(Physics), 1964, Stevens Institute of Technology; Ph.D.(E.E.), 1972, Cornell University.

Fellow Award "For the development of passive surveillance systems and for contributions to information theory and error control codes"; Outstanding Paper Award, IEEE Information Theory Group, 1974; Board of Governors, IEEE Information Theory Group, 1978-1984; President, IEEE Information Theory Group, 1982. Other Awards: IBM Fellow, 1980; Courtesy Professor of Electrical Engineering, Cornell University.

BLAKE, DAVID K., AIEE Fellow 1951. Born: September 19, 1896, Monroe, La.

Other Awards: Charles A. Coffin Foundation Award, 1931; Sigma Tau, Lambda Chapter, University of Kansas, 1950.

BLAKE, GORDON A., IRE Fellow 1961. Born: July 22, 1910, Charles City, Iowa. Degrees: B.S., 1931, United States Military Academy; Air War College, 1948.

Fellow Award "For leadership in military electronics and communications." Other Awards: Honorary Life Member, Armed Forces Communications Electronics Association, 1953; Meritorious Service Award, Armed Forces Communications Electronics Association, 1962; Silver Star, Distinguished Flying Cross; Air Medal with Oak Leaf Cluster; Distinguished Service Medals with Oak Leaf Cluster, 1959, 1965; Joint Service Commendation Medal, 1965; Distinguished Intelligence Medal, 1965.

BLAKE, LAMONT V., Fellow 1979. Born: November 7, 1913, Somerville, Mass. Degrees: B.S., 1935, Massachusetts State College; M.S., 1950, University of Maryland.

Fellow Award "For contributions to the theory and practice of radar range-performance analysis." Other Awards: Applied Science Award, Naval Research Laboratory Chapter, Research Society of America, 1963.

BLAKELY, ROBERT T., Fellow 1971. Born: February 12, 1912, Brooklyn, N.Y. Degrees: E.E., 1935, Rensselaer Polytechnic Institute; M.E.E., 1946, Brooklyn Polytechnic Institute.

Fellow Award "For contributions in the computer field and in military computer and information systems." Other Awards: Palmer C. Ricketts Award, Rensselaer Polytechnic Institute, 1935; Navy Commendation Ribbon, Navy Unit Citation, Two Commendation Letters, U.S. Navy; Sigma Xi; 79 U.S., Foreign Patents; Tau Beta Pi Eminent Engineer, Citadel Military College, 1982; Licensed Prof. Engineer.

BLAKESLEE, THEODORE M., AIEE Fellow 1948. Born: March 21, 1902, Kalamazoo, Mich. Degrees: A.B., 1924, Western State Teachers College; M.S., 1927, E.E., 1938, University of Southern California.

BLANKENBURG, R. CARTER, Fellow 1966. Born: February 7, 1905, San Diego, Calif. Degrees: B.S.E.E., 1927, California Institute of Technology.

Fellow Award "For improvement in designs and application of aluminum in underground power systems."

BLANKENSHIP, GILMER L., Fellow 1984. Born: September 11, 1945, Berkeley, WV. Degrees: S.B., 1967, S.M., 1969, Ph.D., 1971, M.I.T.

Fellow Award "For contributions in stochastic control and stability theory and in education."

BLECHER, FRANKLIN H., Fellow 1964. Born: February 24, 1929, Brooklyn, N. Y. Degrees: B.E.E., 1949, M.E.E., 1950, D.E.E., 1955, Polytechnic Institute of Brooklyn.

Fellow Award "For contributions to the design of solid-state circuits and their application to communication systems"; Browder J. Thompson Award, IRE, 1959. Other Awards: National Academy of Engineering, 1979; Distinguished Alumnus Citation, Polytechnic Institute of New York, 1980.

BLEWETT, JOHN P., Fellow 1964. Born: April 12, 1910, Toronto, Ontario, Canada. Degrees: B.A., 1932, M.A., 1933, University of Toronto; Ph.D., 1936, Princeton University.

Fellow Award "For contributions in the field of high-energy particle accelerators"; Merit Award, IEEE Nuclear and Plasma Sciences Society. Other Awards: Fellow, American Physical Society; Fellow, New York Academy of Sciences; Fellow, American Association for the Advancement of Science.

BLOCH, ERICH, Fellow 1980. Born: January 9, 1925, Sulzburg, Germany. Degree: B.S.E.E., 1952, University of Buffalo, New York.

Fellow Award "For technical and managerial contributions to computer component technology and production"; Outstanding Award, Mid-Hudson Valley Section, IEEE. Other Awards: Member, National Academy of Engineering, 1980; Honorary Member, SME, 1983.

BLOEMBERGEN, NICOLAAS, Fellow 1964. Born: March 11, 1920, Dordrecht, The Netherlands. Degrees: Phil.Cand., 1941, Phil.Dr., 1943, University of Utrecht; Dr.Phil., 1948, University of Leiden.

Fellow Award "For fundamental contributions to masers and lasers"; Morris N. Liebmann Award, IRE, 1959, "For important fundamental contributions to the maser"; IEEE Medal of Honor, 1983. Other Awards: Oliver E. Buckley Prize, American Physical Society, 1958; National Academy of Sciences, 1960; Stuart Ballantine Medal, Franklin Institute, 1961; National Medal of Science, presented by the President of the U.S.A., 1974; Lorentz Medal, Royal Dutch Academy of Sciences, 1978; Frederick Ives Medal, Optical Society of America, 1979; Nobel Prize for Physics, Royal Swedish Academy of Sciences, 1981.

BLOOM, LOUIS R., Fellow 1964. Born: May 9, 1914, Chicago, Ill. Degrees: B.S., 1938, M.S., 1941, University of Illinois.

Fellow Award "For contributions to the development and design of microwave and millimeter-wave tubes." Other Awards: Sigma Xi, 1941; Fellow, American Association for the Advancement of Science, 1964.

BLOOMQUIST, WALTER C., AIEE Fellow 1951. Born: Chisholm, Minn. Degrees: B.S.E.E., 1932, M.S.E.E., 1934, B.B.A., 1935, E.E., 1950, University of Minnesota.

Awards: IEEE Pulp and Paper Award, 1974; IEEE Industrial and Commercial Power Systems Achievement Award, 1974. Other Awards: Fellow, Technical Association of the Pulp and Paper Industry; Engineering Division Award, Technical Association of the Pulp and Paper Industry, 1970.

BLOOR, W. SPENCER, Fellow 1975. Born: October 16, 1918, Trenton, N.J. Degrees: B.S.E.E., 1940, D.Eng.(-Hon.), 1981, Lafayette College.

Fellow Award "For leadership in the synthesis and application of complex multilevel multivariable electronic control systems"; Achievement Award, IEEE Industrial Electronics and Control Instrumentation Society, 1980; IEEE Centennial Medal, 1984. Other Awards: Fellow, Instrument Society of America, 1971; Fellow, American Association for the Advancement of Science, 1975; National Academy of Engineering, 1979; Engineer of the Year in the Delaware Valley, 1980.

BLUM, FRED A., Fellow 1982. Born: November 30, 1939, Austin, TX. Degrees: B.S.(Physics), 1962, University of Texas at Austin; M.S.(Physics), 1964, Ph.D.(Physics), 1968, California Institute Technology.

Fellow Award "For leadership in and contributions to the development of high-speed electronic and optoelectronic devices using III-V compounds."

BLUM, MANUEL, Fellow 1983. Born: April 26, 1938, Caracas, Venezuela. Degrees: B.S., 1959, M.S., 1960, Ph.D., 1963, M.I.T.

Fellow Award "For fundamental contributions to the abstract theory of computational complexity." Other Awards: Sloan Foundation Fellowship, 1972-73; U.C. Berkeley Distinguished Teaching Award, 1977; Fellow, AAAS, 1982.

BOAST, WARREN B., AIEE Fellow 1957. Born: December 13, 1909, Topeka, Kans. Degrees: B.S., 1933, M.S., 1934, University of Kansas; Ph.D., 1936, Iowa State University.

Fellow Award "For his contributions to electrical engineering as an educator, administrator and author"; Meritorious Service Award, IEEE Education Society, 1978. Other Awards: American Men of Science, 1960; Anson Marston Distinguished Professor, 1964; Faculty Citation, 1971; Fellow, Illuminating Engineering Society, 1976; Listed in "Who's Who in the World," 1976; Marston Medal, Iowa State University, 1980.

BOBECK, ANDREW H., Fellow 1971. Born: October 1, 1926, Tower Hill, Pa. Degrees: B.S., 1948, M.S., 1949, Doctor Engrg.(Hon.), 1972, Purdue University.

Fellow Award "For contributions to fundamental magnetic memory and logic devices"; W.R.G. Baker Award (co-recipient), IEEE, 1971; Morris N. Liebmann Award (co-recipient), IEEE, 1975, "For the concept and development of single-walled magnetic domains (magnetic bubbles), and for recogniton of their importance to memory technology"; Distinguished Speakers Program, IEEE Magnetics Society, 1979. Other Awards: Distinguished Engineering Alumnus, Purdue University, 1968; Stuart Ballantine Medal, Franklin Institute, 1973; National Academy of Engineering, 1975; Valdemar Poulsen Gold Medal, Danish Academy of Technical Sciences, 1976; Annual Technology Achievement Award, Electronics Magazine, 1979; Achievement Award, Industrial Research Institute, 1980; American Institute of Physics Prize, 1980.

BOEHNE, EUGENE W., AIEE Fellow 1943. Born: June 2, 1905, Laramie, Wyo. Degrees: B.S., 1926, Texas A and M University; M.S., 1928, Massachusetts Institute of Technology; E.E., 1940, D.Eng., 1948, Texas A and M University.

Awards: National Paper Prize, AIEE, 1944; National Paper Prize, IEEE, Power Division, 1965; Four Paper Prizes, Philadelphia Section; William M. Habirshaw Award, IEEE, 1973, "For his studies and evaluation of the performance of EHV transmission facilities under varying conditions and for his contributions to studies and analysis of system behavior." Other Awards: Charles A. Coffin Awards, General Electric, 1937, 1940; Outstanding Young Electrical Engineer, Honorable Mention, Eta Kappa Nu, 1936; Director of Cooperative Course in Electrical Engineering, Massachusetts Institute of Technology, 1947-1960; Eta Kappa Nu; Sigma Xi; Tau Beta Pi; CIGRE; 25 Year Citation, Science and Arts Committee, Franklin Institute, 1967.

BOERNER, MANFRED, Fellow 1982. Born: March 16, 1929, Saxonia, Germany. Degrees: Diplom-Physiker, 1954, Freie Universitat, West Berlin; Dr.rer.nat., 1959, Technische Universitat Muenchen.

Fellow Award "For contributions to and technical leadership in optical communication technology and for developments of mechanical frequency filters." Other Awards: NTG Prize, German Communications Society, 1962.

BOERNER, WOLFGANG-MARTIN, Fellow 1984. Born: July 26, 1937, Finschhafen, Papua New Guinea. Degrees: ABITUR 1958, August von Platen Gymnasium, ANSBACH, FRG; Dipl.-Ing., 1963, Technische Universitat Munchen; Ph.D., 1963, Moore School of Electrical Engineering, University of Pennsylvania.

Fellow Award "For advancement in inverse methods in sensing systems and in high-resolution broad band Doppler radar polarimetry." Other Awards: Bad Honeff Scholarship, 1958-64; Fulbright Exchange Fellow, 1963-67; U. of Penn. Dissertation Year Scholarship, 1966-67; A. von Humboldt Scientist, 1975-78; Winnipeg Rh Institute Award for Outstanding Contributions to Scholarship and Research in the Natural Sciences, 1976; The Government of Netherlands Visiting Professorship, 1978-79; HUMBOLDT FELLOW, 1978; U.S. Congressional Advisory Board: Peace Through Strength, 1982; Directorship for NATO-ARW on Inverse Methods in Electromagnetic Imaging, 1983; NATO Senior Scientist Award, 1983.

BOESCH, FRANCIS T., Fellow 1979. Born: September 28, 1936, New York, N.Y. Degrees: B.S.E.E., 1957, M.S.E.E., 1960, Ph.D., 1963, Polytechnic Institute of Brooklyn.

Fellow Award "For contributions to the application of network theory to invulnerable communication nets."

BOICE, WILLIAM K., AIEE Fellow 1951. Born: July 17, 1913, Lansing, Mich. Degrees: B.S.E.E., 1935, University of Michigan.

Other Awards: Member, NEMA.

BOLIE, VICTOR W., Fellow 1972. Born: July 23, 1924, Silverton, Oreg. Degrees: B.S., 1949, M.S., 1950, Ph.D., 1952, Iowa State University; B.A., 1957, Coe College; M.A., 1959, Stanford University.

Fellow Award "For contributions to bio-medical engineering education." Other Awards: Member, American Society for Engineering Education; Member, National Society of Professional Engineers; Member, American Physiological Society; Eta Kappa Nu; Sigma Xi; Phi Kappa Phi; Pi Mu Epsilon; Phi Eta Sigma; United States Air

Force Gold Ring Academic Achievement Award, 1944; National Science Foundation Senior Postdoctoral Fellow Award, 1958-59; Mark L. Morris Animal Foundation Research Director Award, 1961; Albrecht Naeter Professor Chair, Oklahoma State University, 1966-1971; Certificate of Appreciation, American Society for Engineering Education, 1970; 35 U.S. Patents, 1957-1984; Member, National Society of Professional Engineers; American Physiological Society; Eta Kappa Nu; Sigma Xi, Phi Kappa Phi; Pi Mu Epsilon.

BOLINDER, E. FOLKE, Fellow 1975. Born: August 11, 1922. Degrees: Civ.Ing., 1946, Tekn.Lic., 1954, Tekn.Dr., 1959, Royal Institute of Technology, Stockholm.

Fellow Award "For contribution to microwave transmission line theory." Other Awards: Fellow, Sweden-American Foundation, 1951; Member, New York Academy of Sciences, 1961; Member, Scientific Research Society of America, 1958-64; Member, Engineering Society of Gothenburg, 1967.

BOLL, HARRY J., Fellow 1981. Born: March 5, 1930, St. Bonifacius, MN. Degrees: B.S.E.E., 1956, M.S.(E.E.), 1958, Ph.D.(E.E.), 1962, University of Minnesota.

Fellow Award "For the development of novel semiconductor devices and leadership in large-scale integrated circuit design."

BOLLMEIER, EMIL WAYNE, Fellow 1971. Born: January 16, 1925, Hurst, Ill. Degrees: B.S.Ch.E., 1947, University of Nebraska.

Fellow Award "For research and development of basic mechanical connecting techniques and insulating materials for communications and electrical distribution." Other Awards: Sigma Tau, 1946; Sigma Xi, 1956; Carlton Society, 1966; Bollmeier's Fellow Award.

BOND, DONALD S., IRE Fellow 1954. Born: June 28, 1909, Libertyville, Ill. Degrees: S.B., 1929, S.M., 1931, University of Chicago.

Fellow Award "For his contributions to the development of communication and navigation apparatus and systems." Other Awards: Fellow, American Association for the Advancement of Science, 1940.

BOND, FREDERICK E., Fellow 1974. Born: January 10, 1920, Philadelphia, Pa. Degrees: B.S.E.E., 1941, Drexel Institute of Technology; M.S.E.E., 1950, Rutgers University; D.E.E., 1956, Polytechnic Institute of New York.

Fellow Award "For contributions to the development of satellite systems for military communications." Other Awards: Sigma Xi, 1950.

BONN, THEODORE H., Fellow 1965. Born: May 27, 1923, Philadelphia, Pa. Degrees: B.S.E.E., 1943, M.S.E.E., 1947, University of Pennsylvania.

Fellow Award "For contributions to the field of magnetic devices and computer memories." Other Awards: Scientific Research Society of America (RESA); Tau Beta Pi; Eta Kappa Nu; Pi Mu Epsilon; Sigma Tau.

BONNETT, L. B., AIEE Fellow 1940. Born: November 11, 1889, Geneva, N.Y. Degrees: E.E., 1910, Syracuse University.

Other Awards: Tau Beta Pi.

BOOKER, HENRY G., IRE Fellow 1953. Born: December 14, 1910, Barking, Essex, England. Degrees: B.A., 1933, Ph.D., 1936, University of Cambridge.

Fellow Award "For his theoretical research in electromagnetism and radio wave propagation." Other Awards: Member, National Academy of Sciences, 1960; Honorary President, International Union of Radio Science, 1978.

BOONE, E. MILTON, IRE Fellow 1956. Born: February 17, 1903, Millersburg, Ky. Degrees: B.A., 1926, M.S. 1932, University of Colorado; M.S., 1937, University of Michigan.

Fellow Award "For contributions as an educator and research investigator in the field of electronics." Other Awards: Eta Kappa Nu; Phi Beta Kappa; Tau Beta Pi; Sigma Xi; Distinguished Teaching Award, Electrical Engineering Dept., Ohio State University, 1964, 1969, 1972; University Alumni Award, 1970.

BOOTH, TAYLOR L., Fellow 1975. Born: September 22, 1933, Middletown, Conn. Degrees: B.S.E., 1955, M.S., 1956, Ph.D., 1962, University of Connecticut.

Fellow Award "For leadership in the development of curriculum, laboratories, and textbooks for computer science programs"; Fortesque Fellowship, AIEE, 1955; Editor, IEEE Transactions on Computers, 1978-1982; Board of Governors, IEEE Computer Society, 1980; Secretary, IEEE Computer Society, 1981; First Vice-President, IEEE Computer Society, 1982-1983; Vice Pres., Educational Activities, Computer Society. Other Awards: Frederick Emmons Terman Award, American Society for Engineering Education, 1972.

BOOTHROYD, ALBERT R., Fellow 1969. Born: June 21, 1925, Hudderfield, England. Degrees: B.Sc., 1946, Ph.D., 1951, Imperial College, London.

Fellow Award "For special work on solid-state techniques and their applications."

BOOTON, RICHARD C., JR., Fellow 1969. Born: July 26, 1926, Dallas, Tex. Degrees: B.S., M.S., 1948, Texas A&M University; Sc.D., 1952, Massachusetts Institute of Technology.

Fellow Award "For the development of techniques for the analysis and optimal design of time-varying and nonlinear guidance and control systems"; Browder J. Thompson Memorial Prize Award, IRE, 1953.

BORDOGNA, JOSEPH, Fellow 1976. Born: March 22, 1933, Scranton, Pa. Degrees: B.S.E.E., 1955, University of Pennsylvania; S.M., 1960, Massachusetts Institute of Technology; Ph.D., 1964, University of Pennsyl-

vania.

Fellow Award "For innovations in engineering and technical education"; Achievement Award, IEEE Education Society, 1983. Other Awards: George Westinghouse Award, American Society for Engineering Education, 1974; Alfred Fitler Moore Professorship, University of Pennsylvania, 1979; Engineer of the Year, Greater Philadelphia Area, 1984.

BORGIOTTI, GIORGIO V., Fellow 1979. Born: November 23, 1932, Rome, Italy. Degrees: Dr. Eng.(Electrical Engineering), 1957, University of Rome, Rome, Italy.

Fellow Award "For the development of Fourier transform modal methods for analysis of phased-array antennas"; Special Recognition Paper Award, IEEE Group on Antennas and Propagation, 1968.

BORN, WILLIAM T., IRE Fellow 1957. Born: August 3, 1903, Chicago, Ill. Degrees: B.S., 1925, University of Chicago.

Fellow Award "For applications of electronic techniques to geophysical exploration." Other Awards: Honorary Member, Society of Exploration Geophysicists, 1959; Honorary Member, Tulsa Geophysical Society, 1965; Professional Achievement Award, University of Chicago Alumni Association, 1975.

BORST, DAVID W., Fellow 1978. Born: August 5, 1918, Jacksonville, Fla. Degrees: Sc.B.E.E., 1940, Brown University.

Fellow Award "For contributions in the application of power semiconductor devices"; Achievement Award, IEEE Industry Applications Society, 1978; Papers and Publication Chairman, AIEE Conference on Rectifiers in Industry, 1957; Program Chairman, AIEE Third Conference on Rectifiers in Industry, 1962; Secretary, Power Semiconductor Committee, IAS, 1966-68; Chairman, Power Semiconductor Committee, IAS, 1968-70; Publication Chairman, International Semiconductor Power Converter Conference, 1972; Chairman, Transactions Papers Working Group, Power Semiconductor Committee, IAS, 1974-1976; Secretary, International Semiconductor Power Converter Conference, 1977; Chairman, Industrial Power Conversion Systems Dept., IAS, 1977-1980; Honorary General Chairman, International Semiconductor Power Converter Conference, 1982; Member-at-Large, Industry Applications Society, 1981-83. Other Awards: Member, JEDEC Committee JC-22 (formerly JS-14) 1963 to date; Chairman, JEDEC Committee JC-22, 1974 to date; Chairman, Welding Control Task Group of JC-22, 1968 to date; Member, NEMA Joint Sections Committee on Spacings for Power Semiconductors, 1973 to date; USA Delegate, International Electrotechnical Commission, Technical Committee 47 on Semiconductor Devices and Integrated Circuits; Leningrad, 1969, Ottawa, Canada, 1978, Orlando, Florida, 1980, Montreux, Switzerland, 1981.

BOSE, AMAR G., Fellow 1972. Born: November 2, 1929, Philadelphia, Pa. Degrees: S.B., 1952, S.M., 1952, Sc.D., 1956, Massachusetts Institute of Technology.

Fellow Award "For contributions to loudspeaker design, two-state amplifier-modulators, and nonlinear systems." Other Awards: Baker Memorial Award, Massachusetts Institute of Technology, 1963-64; Western Electric Fund Award, N.E. Section, American Society for Engineering Education, 1965; Member, Audio Engineering Society; Sigma Xi; Tau Beta Pi; Eta Kappa Nu.

BOSE, JOHN H., Fellow 1969. Born: March 26, 1912, New York, N.Y. Degrees: B.S., 1934, E.E., 1935, Columbia School of Engineering.

Fellow Award "For contributions to the advancement of frequency-modulated communication systems." Other Awards: Armstrong Medal, Radio Club of America, 1959.

BOSE, NIRMAL KUMAR, Fellow 1981. Born: August 19, 1940, India. Degrees: B.Tech.(with honors), 1961, Indian Institute of Technology; M.S., 1963, Cornell University; Ph.D., 1967, Syracuse University.

Fellow Award "For contributions to multidimensional systems theory and circuits and systems education"; Associate Editor, IEEE Transactions on Circuits and Systems, 1979; Chairman, Technical Committee on Education, IEEE Circuits and Systems Society, since 1980; Editor of "Multidimensional Systems: Theory and Applications," IEEE Press, N.Y., 1979. Other Awards: Recipient, C.N.R.S.-I.R.I.A. Grant, Toulouse, France, 1975; Recipient, D.A.A.D. Grant, Ruhr Universitaet, Bochum, F.R. Germany, 1978.

BOSSART, PAUL N., IRE Fellow 1956. Born: May 10, 1896, Buffalo, N.Y. Degrees: B.S., 1916, Carnegie Institute of Technology; E.E., 1920, Columbia University; Ph.D., 1933, University of Pittsburgh.

Fellow Award "For contributions to railway safety and operating efficiency through electronic communications." Other Awards: James Scholar, Columbia University, 1916.

BOSSHART, WILLIAM R., Fellow 1983. Born: April 16, 1926, Astoria, Oregon. Degrees: B.Sc.(E.E.), 1949, Oregon State University.

Fellow Award "For leadership in the operation of large power system interconnections."

BOSTWICK, MYRON A., AIEE Fellow 1958. Born: February 15, 1900, Spokane, Wash. Degrees: B.S., 1926, Washington State University.

Fellow Award "For contributions to relay protection of electric power systems"; Chairman, IEEE Pacific Coast Relays Sub-Committee IEEE, 1954-1961; Chairman, Portland Section, IEEE, 1964. Other Awards: Registered Professional Engineer, Oregon; Tau Beta Pi; Sigma Tau; Phi Kappa Phi.

BOTTS, JOHN C., Fellow 1978. Born: June 21, 1923, Akron, Ohio. Degrees: B.S.(Physics and Mathematics), Kent State University.

Fellow Award "For contributions to the development, testing, application, and standardization of insulation systems for rotating apparatus."

BOUGHTON, W. V., AIEE Fellow 1948. Born: March 6, 1900, Buffalo, N.Y. Degrees: 1919, Tennessee Military Institute; 1919-1921, Univ. of So. Calif.

BOUGHTWOOD, JOHN E., Fellow 1965. Born: Medford, Mass. Degrees: B.E.E., 1930, Northeastern University.

Fellow Award "For contributions to frequency division multiplex and FM data and telegraph transmission." Other Awards: F.E. d'Humy Award, Western Union, 1957.

BOULET, LIONEL, Fellow 1979. Born: July 29, 1919, Quebec City, P.Q, Canada. Degrees: B.A.(cum laude), 1938, B.Sc.(summa cum laude), 1944, Laval University; M.Sc., 1947, University of Illinois; Honoris Causa Doctorate, 1968, Sir George William University, 1971, University of Ottawa, 1972, Laval University, 1977, McGill University, 1977, Universite' du Quebec, 1979.

Fellow Award "For leadership in the establishment and management of the Research Institute of Hydro-Quebec"; Award, Montreal Section, IEEE, 1943; Prize, IEEE, 1944; General McNaughton Award, Region 7, 1983. Other Awards: Prize of the Corporation of Professional Engineers of Quebec, 1943; Prize, Engineering Institute of Canada, 1943; Sigma Xi, 1947; Scientific Prize, Province of Quebec, 1969; Fellow, Engineering Institute of Canada; Fellow, Institut Canadien des Ingenieurs, 1973; Officer, Order du Canada, 1975; Fellow, The Royal Society of Canada, 1976; Canadian Research Manager Association Award, 1983; University of Illinois Electrical Engineering Alumni Award, 1983.

BOURICIUS, WILLARD G., Fellow 1977. Born: September 23, 1920, Omaha, Nebr. Degrees: B.A., 1942, Hastings College; M.A., 1945, University of Wisconsin; Ph.D., 1949, Yale University.

Fellow Award "For contributions to design and diagnosis procedures leading to increased reliability of computer systems."

BOURNE, HENRY C., JR., Fellow 1979. Born: December 31, 1921, Tarboro, N.C. Degrees: S.B., 1947, S.M., 1948, Sc.D., 1952, Massachusetts Institute of Technology.

Fellow Award "For contributions to the theory and application of magnetic thin films and magnetic amplifiers and to electrical engineering education"; Prize Paper, AIEE, 8th District. Other Awards: NSF Science Faculty Fellowship, 1960-61; Honorary Research Associate, University College, London; Sigma Xi; Tau Beta Pi; Eta Kappa Nu; ODK; Beta Gamma Sigma; Fellow,

American Association for the Advancement of Science.

BOUYOUCOS, JOHN VINTON, Fellow 1978. Born: November 9, 1926. Degrees: A.B., 1949, S.M., 1951, Ph.D., 1955, Harvard University; Smaller Company Management Program Certificate, 1976, Harvard Business School.

Fellow Award "For contributions to the field of hydrodynamic energy conversion devices." Other Awards: Fellow, Acoustical Society of America, 1961; Associate Member, Institute of Noise Control Engineers, 1972; Rochester Patent Law Association Inventors' Award, 1973; Member, Society for Exploration Geophysicists, 1976; Member, Audio Engineering Society, 1978.

BOWDEN, BERTRAM VIVIAN, Fellow 1970. Born: January 18, 1910, Derbyshire, England. Degrees: M.A., 1931, Ph.D., 1934, Cambridge University.

Fellow Award "For contributions in the field of radar and digital computers"; Pioneer Award, Aerospace and Electronics Systems Group, IEEE, 1973. Other Awards: Fellow, Institute of Electrical Engineers, U.K.

BOWERS, KLAUS D., Fellow 1975. Born: December 27, 1929, Stettin, Germany. Degrees: B.A., M.A., Ph.D., 1947-1953, University of Oxford, England.

Fellow Award "For contributions to and management of solid-state optical, magnetic, and semiconductor research and development."

BOWHILL, SIDNEY A., Fellow 1965. Born: August 6, 1927, Dover, Kent, England. Degrees: B.A., 1948, M.A., 1950, Ph.D., 1954, University of Cambridge.

Fellow Award "For contributions to aeronomy and the physics of the ionosphere." Other Awards: National Academy of Engineering, 1971; Fellow, American Geophysical Union; Fellow, American Astronomical Society; Fellow, American Association for the Advancement of Science.

BOWIE, ROBERT M., IRE Fellow 1948. Born: August 24, 1906, Table Rock, Nebr. Degrees: B.S., 1929, M.S. 1931, Ph.D., 1933, Iowa State University.

Fellow Award "For his contributions in the field of micro-wave techniques, spectroscopic methods and standards, and for his development of means to avoid the effect of ion bombardment on cathode-ray tube screens."

BOWLES, EDWARD L., AIEE Fellow 1933, IRE Fellow 1947. Born: December 9, 1897, Westphalia, Mo. Degrees: B.S., 1920, Washington University; M.S., 1922, Massachusetts Institute of Technology; D.Sc.(Hon.), 1945, Norwich University.

IRE Fellow Award "For his activities in making possible the maximum practical use of advanced radio equipment in military operations and for his work in the educational field." Other Awards: Distinguished Service Medal, 1945; Honorary Commander of the Order of the British Empire; Presidential Medal of Merit, 1948; Distinguished Alumni Citation, Washington Uni-

versity, 1955.

BOWLES, KENNETH L., Fellow 1966. Born: February 20, 1929, Bronxville, N.Y. Degrees: B.E.P., 1951, M.E.E., 1953, Ph.D., 1955, Cornell University.

Fellow Award "For contributions to ionospheric physics, and use of incoherent scatter for the study of the ionosphere and exosphere."

BOYD, GARY D., Fellow 1976. Born: September 14, 1932, Los Angeles, Calif. Degrees: B.S.E.E., 1954, M.S.E.E., 1955, Ph.D.(E.E.)(Physics minor), 1959, California Institute of Technology.

Fellow Award "For contributions to theory and practice in nonlinear optical phenomena"; IEEE Awards Board.

BOYD, JOSEPH A., Fellow 1977. Born: March 25, 1921, Oscar, Ky. Degrees: B.S.E.E., 1946, M.S.E.E., 1949, University of Kentucky; Ph.D., 1954, University of Michigan.

Fellow Award "For engineering and managerial contributions to communications and information handling systems."

BOYER, JOHN L., Fellow 1970. Born: June 29, 1919, Houston, Tex. Degrees: B.S.E.E., 1942, Rice University.

Fellow Award "For achievement in expanding power-converter technology and the development of high-power semiconductor devices."

BOYKIN, JOHN R., Fellow 1972. Born: February 24, 1913, Bowling Green, Ky. Degrees: B.S.E.E., 1936, North Carolina State College.

Fellow Award "For contributions to communications and navigation systems featuring all solid-state, high power, and optimal spectrum use, and to early FM broadcast technology."

BOYLE, WILLARD S., Fellow 1971. Born: August 9, 1924, Amherst, Nova Scotia, Canada. Degrees: B.S., 1947, M.S., 1948, Ph.D., 1950, McGill University.

Fellow Award "For scientific contributions, inventions, and technical leadership in semiconductor electronics"; Morris N. Liebmann Award (with G.E. Smith), IEEE, 1974, "For the invention of the charge-coupled device (CCD) and leadership in the field of metal oxide semiconductor (MOS) device physics." Other Awards: Fellow, N.R.C., 1948; Stuart Ballantine Award, Franklin Institute, 1973; Fellow, American Physical Society, 1974; Member, National Academy of Engineering, 1974.

BRACEWELL, RONALD N., IRE Fellow 1961. Born: July 22, 1921, Sydney, Australia. Degrees: B.Sc., 1940, B.E., 1942, M.E., 1948, Sydney University; Ph.D. 1950, University of Cambridge.

Fellow Award "For contributions to radio astronomy." Other Awards: Duddell Premium, Institution of Electrical Engineers, London, 1951, 1952; Pollock Memorial Lecturer, University of Sydney, Australia,

1978. First Lewis M. Terman Professor and Fellow in Electrical Engineering, 1974-1979.

BRACKEN, JOHN F., Fellow 1971. Born: December 9, 1907, Chicago, Ill. Degrees: B.S., 1929, University of Illinois.

Fellow Award "For contributions in power distribution systems engineering."

BRADBURD, ERVIN M., Fellow 1968. Born: May 29, 1920, Philadelphia, Pennsylvania. Degrees: B.S.E.E., 1941, M.S.E.E., 1943, Columbia University.

Fellow Award "For contributions to communication transmission systems and techniques.

BRADLEY, WILLIAM E., IRE Fellow 1953. Born: January 7, 1913, Lansdowne, Pa. Degrees: B.S., 1937, University of Pennsylvania.

Fellow Award "In recognition of his technical and original analytical methods in the fields of frequency modulation detection"; Outstanding Engineer of the Year Award, IRE, Philadelphia Section, 1962, "For his services to the United States Government and for his many contributions to electronic communications." Other Awards; Certificate of Commendation, U.S. Dept. of Navy, 1947; Honorary Citation, University of Pennsylvania, 1955.

BRADY, BRYCE, AIEE Fellow 1951. Born: September 8, 1903, Blackburn, Okla. Degrees: B.S.E.E., 1927, University of Oklahoma.

Other Awards: Sigma Tau, 1926; Tau Beta Pi, 1927; Eta Kappa Nu, 1952; Life Member, National Society of Professional Engineers, 1969; Life Member, Oklahoma Society of Professional Engineers, 1969; Member, Engineering Club of Oklahoma City.

BRADY, FRANK B., Fellow 1969. Born: June 1, 1914, Pomeroy, Ohio. Degrees: 1933-38, University of Cincinnati.

Fellow Award "For contributions to the development and of use of aircraft approach and landing"; Guest Editor, IRE Transactions on Aeronautical and Navigational Electronics Special Issue on Instrument Approach and Landing, June 1959; Guest Editor, IEEE Proceedings Special Issue on Global Navigation, October 1983. Other Awards: War Department Medal of Freedom, 1947; Air Transport Association Citation, Radio Technical Commission for Aeronautics, 1969.

BRAINARD, D. EDWARD, AIEE Fellow 1951. Born: January 22, 1901, Syracuse, N.Y. Degrees: E.E., 1923, Cornell University.

BRAINERD, JOHN G., AIEE Fellow 1947, IRE Fellow 1951. Born: August 7, 1904. Degrees: B.S., 1925, D.Sc., 1934, University of Pennsylvania.

IRE Fellow Award "For contributions to the technical literature, to the teaching profession, and to the art of electronic computation"; Philadelphia Section Award, 1968, "For his contributions, stimulation and leadership in behalf of the Philadelphia Section IEEE and wel-

fare of the engineering fraternity in the Delaware Valley"; Founders Medal, IEEE, 1975, "For leadership of great value to the profession." Other Awards: Ordnance Department Award, 1946; Honeywell Engineering and Science Medal, 1969; Organized and was in charge of ENIAC Project; University Professor of Engineering and Professor of History of Technology (Honorary Title), University of Pennsylvania, 1969.

BRANCATO, EMANUEL L., AIEE Fellow 1957. Born: November 3, 1914, New York, N.Y. Degrees: B.A., 1936, B.S., 1937, M.S., 1938, Columbia University.

Fellow Award "For basic research contributions in the field of electric insulation." Other Awards: Fellow, Washington Academy of Sciences, 1966; Navy Meritorious Civilian Service Award, 1978; Citation, Congressional Record (House), City of New Carrollton, Maryland, 1979.

BRAND, FRANK A., Fellow 1967. Born: June 26, 1924, Brooklyn, N.Y. Degrees: B.S., 1950, M.S., 1958, Polytechnic Institute of Brooklyn; Ph.D., 1970, University of California at Los Angeles.

Fellow Award "For contributions in the field of microwave semiconductor and quantum electronic devices." Other Awards: Electronics Command Outstanding Performance Appraisal, 1960, 1961, 1967, 1969, 1970; Electronics Command Technical Leadership Award, 1964; Eta Kappa Nu, 1971.

BRANDON, MERWIN M., AIEE Fellow 1944. Born: November 14, 1898, Pinckneyville, Miss. Degrees: B.S., 1919, E.E., 1936, Mississippi State University.

Other Awards: American Society for Testing Materials; Standards Engineers Citation, 1963; Honorary Member, International Association of Electrical Inspectors, 1963; A.S.A. Citation, 1963; Honorary Member, National Fire Protection Association, 1964; N.F.P.A. Distinguished Service Award.

BRAVERMAN, NATHANIEL, Fellow 1964. Born: July 4, 1915, New York, N.Y. Degrees: B.E.E., 1937, City College of the City University of New York; M.S.E.E., 1939, Columbia University.

Fellow Award "For contributions to planning, development and application of air navigation systems and techniques"; Annual Paper Prize, IRE, Dayton Section, 1957. Other Awards: Citation, Radio Technical Commission for Aeronautics, 1948; Special Service Award, FAA, 1967.

BRAYTON, ROBERT K., Fellow 1981. Born: October 23, 1933. Degrees: B.S.E.E., 1956, Iowa State University; Ph.D.(Math.), 1961, Massachusetts Institute of Technology.

Fellow Award "For pioneering work in the theory of nonlinear networks, stability theory, and sparse matrix techniques"; Best Paper Award, IEEE Circuits and Systems Society, 1971. Other Awards: Fellow, American Association for the Advancement of Science; IBM Outstanding Invention Award, 1970; IBM Outstanding Innovation Award, 1982.

BREITWIESER, CHARLES J., IRE Fellow 1958. Born: September 23, 1910, Colorado. Degrees: B.S., 1932, University of North Dakota; M.S., 1934, California Institute of Technology; D.Sc., 1949, University of North Dakota.

Fellow Award "For his contribution to the field of missile guidance"; First Paper Prize, AIEE, 1945, for "Constant speed drives for aircraft alternators." Other Awards: Award, Mayor's Committee on Municipal Finance (City of San Diego), 1971; Citation, Joint Atomic Forces in the South Pacific, 1963.

BRENNAN, LAWRENCE E., Fellow 1983. Born: January 29, 1927, Oak Park, Illinois. Degrees: B.Sc.(E.E.), 1948, Ph.D.(E.E.), 1951, Univ. of Illinois.

Fellow Award "For contributions to the theory and concepts of adaptive arrays."

BRENNER, EGON, Fellow 1977. Born: July 1, 1925, Vienna, Austria. Degrees: B.E.E., 1944, City College of the City University of New York; M.E.E., 1949, D.E.E., 1955, Polytechnic Institute of Brooklyn.

Fellow Award "For contributions to engineering education."

BRERETON, DONALD S., Fellow 1971. Born: June 29, 1925, Terre Haute, Ind. Degrees: B.S.E.E., 1946, University of Colorado.

Fellow Award "For contributions to industrial power systems and to the communication of engineering information"; Prize Paper Award, Industry Applications Society, 1968; Achievement Award, Industrial and Commercial Power Systems Committee, 1971; Outstanding Achievement Award, Industry Applications Society, 1971; IEEE Standards Medallion, 1977; Engineering Award, Power Systems Sector, General Electric, 1981; IEEE Centennial Medal, 1984.

BRESLER, AARON D., Fellow 1975. Born: June 20, 1924, Bronx, N.Y. Degrees: B.E.E., 1944, City College of New York; M.E.E., 1951, D.E.E., 1959, Polytechnic Institute of Brooklyn.

Fellow Award "For leadership in antenna engineering, and for contributions to the theory of propagation in anisotropic waveguides."

BREUER, GLENN D., Fellow 1983. Born: August 17, 1926, Berkeley, California. Degrees: B.Sc.(E.E.), 1948, University of California, Berkeley; M.E.E., 1954, Rensselaer Polytechnic Institute.

Fellow Award "For applications of systems techniques to HVDC systems."

BREWER, GEORGE R., Fellow 1965. Born: September 10, 1922, New Albany, Ind. Degrees: B.E.E., 1943, University of Louisville; M.E.E., 1948, M.S., 1949, Ph.D., 1951, University of Michigan.

Fellow Award "For contributions to electron dynamics and ion propulsion."

BRICK, DONALD B., Fellow 1982. Born: October 1, 1927, Brooklyn, NY. Degrees: A.B.(cum laude), 1950, M.S., 1951, Ph.D., 1954, Harvard University.

Fellow Award "For contributions to the theory and application of pattern recognition and for technical leadership in command, control, and communications systems"; Chairman, IEEE Systems, Man, and Cybernetics Society, 1969-1970. Other Awards: Sigma Xi, 1954; President, 1980-1981, Chairman, Lexington-Concord Chapter, Armed Forces Communications and Electronics Association, 1982.

BRIDGES, JACK E., Fellow 1974. Born: January 6, 1925, Denver, Colo. Degrees: B.S.E.E. 1945, M.S.E.E., 1947, University of Colorado.

Fellow Award "For development of techniques and standards to control radio frequency interference, and for contributions to receiver design to suppress common-channel interference and noise"; Browder J. Thompson Memorial Prize Award, 1956, for paper, September, 1954, Proceedings of the IRE, "Detection of Television Signals in Thermal Noise"; Certificate of Achievement, Group on Electromagnetic Compatibility, 1972; Certificate of Appreciation, Group on Electromagnetic Compatibility, 1976; Prize Paper Award, IEEE Power Engineering Society, 1980. Other Awards: Tau Beta Pi; Eta Kappa Nu; Sigma Xi.

BRIDGES, JAMES M., IRE Fellow 1959. Born: August 29, 1906, Ellsworth, Me. Degrees: B.S.E.E., 1928, D.Eng.(Hon.), 1963, University of Maine.

Fellow Award "For contributions to precision tracking radar and component reliability."

BRIDGES, WILLIAM B., Fellow 1970. Born: November 29, 1934. Degrees: B.S., 1956, M.S., 1957, Ph.D., 1962, University of California at Berkeley.

Fellow Award "For contributions to the theory and development of gas lasers, and for studies of space-change instabilities in diodes." Other Awards: Sherman Fairchild Distinguished Scholar, California Institute of Technology, 1974-75; Fellow, Optical Society of America, 1976; Member, National Academy of Engineering, 1977; Eta Kappa Nu; Tau Beta Pi; Sigma Xi; Phi Beta Kappa, National Academy of Sciences, 1981.

BRIDGLAND, CHARLES J., Fellow 1966. Born: December 18, 1908, Barrie, Ontario, Canada. Degrees: B.A.Sc., 1933, M.A.Sc., 1934, University of Toronto.

Fellow Award "For contributions to the development of a trans-continental communications system in Canada."

BRIGGS, MAYNARD R., IRE Fellow 1954. Born: July 19, 1904, St. Paul, Minn. Degrees: B.E.E., 1929, University of Minnesota.

Fellow Award "For his contributions to the design of radio transmitters, and his leadership in the establishment of industry standards"; Director, Region 2, IRE, 1962. Other Awards: Certificate of Appreciation, Office of Scientific Research and Development, 1945; Naval Ordinance Development Award, 1946; Certificate of Commendation, 1947; Bureau of Ships, Navy Department; Chairman, Transmitter Section, Radio-Electronics, Television MFGRS. ASSN., 1946-1953.

BRIGHAM, E. ORAN, Fellow 1979. Born: September 13, 1940, Stamford, Tex. Degrees: B.S.E.E., 1963, M.S.E.E., 1964, Ph.D., 1967, University of Texas at Austin; M.S.A., 1971, George Washington University.

Fellow Award "For leadership in and contributions to the development of automated electronic reconnaissance systems." Other Awards: Tau Beta Phi; Eta Kappa Nu; Listed in "Who's Who in the South and Southwest"; Who's Who in America"; Who's Who in Engineering"; "Jane's Who's Who in Aviation and Aerospace"; "Who's Who in California"; "International Who's Who in Engineering."

BRISKMAN, ROBERT D., Fellow 1975. Born: October 15, 1932, New York, N.Y. Degrees: B.S.E., 1954, Princeton University; M.S.E., 1961, University of Maryland.

Fellow Award "For contributions to the development of communications satellite systems"; USAB Citation of Honor, 1979; EASCON Founder Award, 1980. Other Awards: Commendation Medal, U.S. Army, 1958; Apollo Achievement Award, NASA, 1969.

BROBECK, WILLIAM M., Fellow 1976. Born: July 5, 1908, Berkeley, Calif. Degrees: A.B., 1930, Stanford University; M.Sc., 1933, Massachusetts Institute of Technology; D.Sc.(Hon.), 1971, University of California at Berkeley.

Fellow Award "For contributions to the particle accelerator field."

BROCKETT, ROGER W., Fellow 1974. Born: October 22, 1938, Wadsworth, Ohio. Degrees: B.S., 1960, M.S., 1962, Ph.D., 1964, Case Institute of Technology.

Fellow Award "For developments in the theory of control on manifolds, and for contributions to frequency response methods in stability, to linear theory, and to engineering education." Other Awards: D.P. Eckman Award, American Automatic Control Council, 1967; John Simon Guggenheim Fellowship, 1975.

BRODERSEN, ROBERT WILLIAM, Fellow 1982. Born: November 1, 1945. Degrees: B.S.E.E., 1966, California State Polytechnic Institute; M.S.E.E., 1966, Ph.D., 1972, Massachusetts Institute of Technology.

Fellow Award "For contributions to the development of integrated circuits for signal processing."

BRODZINSKY, ALBERT, Fellow 1978. Born: July 7, 1920, Buffalo, N.Y. Degrees: B.E.E., 1942, Cornell University; M.S., 1951, University of Maryland.

Fellow Award "For technical contributions and leadership in government electronics research."

BRONWELL, ARTHUR B., IRE Fellow 1956. Born: August 18, 1909, Chicago, Ill. Degrees: B.S., 1933, M.S.,

1936, Illinois Institute of Technology; M.B.A., 1947, Northwestern University; LL.D.(Hon.), 1955, Northeastern University; D.Sc.(Hon.), 1958, Wayne State University.

Fellow Award: "For contributions to radio science as teacher and author." Other Awards: Distinguished Alumni Award, Illinois institute of Technology, 1956; Engineer of Distinction, Engineers Joint Council; Cited in Library Journal List of 100 Best Books Published for "Science and Technology in The World of the Future," 1970; Dean of Engineering Emeritus, University of Connecticut, 1978; Member, Council on Foreign Relations (New York); Formerly Exec. Secy., American Soc. for Engineering Education; Who's Who in the World; Who's Who in America; International Writer's Directory.

BROOKNER, ELI, Fellow 1972. Born: April 2, 1931, New York, N.Y. Degrees: B.E.E., 1953, College of the City of New York; M.S.E.E., 1955, D.Sc., 1962, Columbia University.

Fellow Award "For contributions to radar signal processing and wave propagation in random media." Other Awards: Annual $1000 Journal Premium Award, The Franklin Institute, 1966; Associate Fellow, American Institute of Areonautics and Astronautics; Eta Kappa Nu; Tau Beta Pi; "American Men and Women of Science"; "Who's Who in the East"; "Leaders in Electronics"; "Who's Who in Technology Today"; Commission C, International Scientific Radio Union; Who's Who in Engineering; Jane's "Who's Who in Aviation and Aerospace: U.S. Edition."

BROOKS, FREDERICK E., JR., IRE Fellow 1958. Born: October 14, 1916. Mineola, N.Y. Degrees: B.S.E.E., 1940, University of Kansas; D.Eng., 1944, Yale University.

Fellow Award "For his contributions to antenna systems research and development."

BROOKS, FREDERICK P., JR., Fellow 1968. Born: April 19, 1931, Durham, N.C. Degrees: B.A., 1953, Duke University; M.S., 1955, Ph.D., 1956, Harvard University.

Fellow Award "For leadership in the design of computers which significantly extends their capability in scientific, commercial and military applications"; McDowell Award for Outstanding Contributions to the Computer Art, IEEE Computer Group, 1970; Computer Pioneer Award, IEEE Computer Society, 1982. Other Awards: Computer Sciences Man-of-the-Year Award, Data Processing Management Association, 1970: Guggenheim Fellowship, 1975; Member, National Academy of Engineering; Fellow, American Academy of Arts and Sciences; Member, U.S. Defense Science Board; Kenan Professor and Chairman, Computer Science Dept., University of North Carolina, Chapel Hill.

BROOKS, H. W., AIEE Fellow 1928. Born: July 19, 1890, Jackson, Tenn. Degrees: 1911, Cornell University.

BROOKS, JOSIAH A., AIEE Fellow 1934. Born: January 26, 1901, Lincoln, Nebr. Degrees: B.S., 1922, University of Nebraska.

Other Awards: Sigma Xi; Sigma Tau.

BROTHERS, JAMES T., IRE Fellow 1959. Born: April 30, 1908, New York, N.Y.

Fellow Award "For contributions and leadership in the field of electronic components." Other Awards: Fellow, American Society for Testing and Materials; Radio Fall Meeting Plaque, E.I.A., 1964.

BROWER, R. FRANK, AIEE Fellow 1949. Born: May 24, 1899, Concord, N.C. Degrees: A.B., 1920, Duke University.

BROWN, DAVID R., Fellow 1969. Born: October 31, 1923, Los Angeles, Calif. Degrees: B.S.E.E., 1944, University of Washington; S.M.E.E., 1947, Massachusetts Institute of Technology.

Fellow Award "For leadership in the development of digital-computer components and the design and operation of large-scale information processing systems." Other Awards: Certificate for Exceptional Service, Naval Ordnance Development, 1946.

BROWN, DELMAR L., AIEE Fellow 1958. Born: March 28, 1909, Silverton, Oreg. Degrees: B.S.M.E., 1931, Oregon State University.

Fellow Award "For contributions to testing and performance of power system components."

BROWN, GEORGE H., IRE Fellow 1941, AIEE Fellow 1948. Born: October 14, 1908, North Milwaukee, Wis. Degrees: B.S., 1930, M.S., 1931, Ph.D., 1933, E.E., 1942, University of Wisconsin; Doctor Engrg.(Hon.), 1968, University of Rhode Island.

IRE Fellow Award "For studies and publications in the field of radio antennas"; Edison Medal, 1967, "For a meritorious career distinguished by significant engineering contributions to antenna development, electromagnetic propagation, the broadcast industry, the art of radio frequency heating, and color television." Other Awards: Modern Pioneer Award, National Association of Manufacturers, 1940; Certificate of Appreciation, War Dept., 1945; Fellow, American Association for the Advancement of Science, 1958; Distinguished Service Citation, University of Wisconsin, 1962; National Academy of Engineering, 1965; Citation, Fourth International Television Symposium, 1965; Eminent Member, Eta Kappa Nu, 1967; Deforest Audion Award, V.W.O.A., 1968; Fellow, Royal Television Society, 1970; Shoenberg Memorial Lecture, Royal Institution, 1972; Marconi Centenary Lecture, AAAS, 1974; Silver Beaver and Silver Antelope, Boy Scouts of America; David Sarnoff Award for Outstanding Achievement in Radio and Television, University of Arizona, 1980.

BROWN, GORDON S., AIEE Fellow 1950, IRE Fellow 1955. Born: August 30, 1907, Drummoyne, New South Wales, Australia. Degrees: S.B., 1931, S.M., 1934, Sc.D., 1938, Massachusetts Institute of Technology; D.Eng.(Hon.), 1958, Purdue University; D.Sc.(Hon.), 1964, Dartmouth College; D.Eng. Technics (Hon.), 1965, Technical University, Denmark; D.Eng.(Hon.), 1967, Southern Methodist University; D.Eng.(Hon.), 1968, Stevens Institute of Technology.

IRE Fellow Award "For his contributions to automatic control systems and to engineering education"; Education Medal, AIEE, 1959. Other Awards: Naval Ordnance Development Award, 1945; Presidential Certificate of Merit, 1948; Fellow, American Academy of Arts and Sciences, 1950; Westinghouse Award, American Society for Engineering Education, 1952; Lamme Medal, American Society for Engineering Education, 1959; Eminent Member, Eta Kappa Nu, 1963; National Academy of Engineering, 1965; Joseph Marie Jacquard Memorial Award, Numerical Control Society, 1970; Ingold C. Jackson Professor of Electrical Engineering, 1969; Institute Professor, Massachusetts Institute of Technology, 1973; Robert Fletcher Award, Thayer School and Dartmouth College, 1976; Rufus Oldenburger Medal, Automatic Control Division, American Society of Mechanical Engineers, 1977.

BROWN, HOMER E., Fellow 1976. Born: April 14, 1909, Humboldt, Minn. Degrees: B.S.E.E., 1930, University of Minnesota.

Fellow Award "For contributions to the application of computers in the electric utility industry." Other Awards: Outstanding Achievement Award, University of Minnesota, 1974; Fellow, Institute of Electrical Engineers, London, 1975.

BROWN, J. E., IRE Fellow 1949. Born: September 11, 1902, Greenport, N.Y.

Fellow Award "For his contributions in the field of broadcast receiver design"; Award, IEEE, 1969, "For contributions to consumer electronics." Other Awards: Award, National Association of Broadcasters, 1968.

BROWN, JACK H. U., Fellow 1977. Born: November 16, 1918, Nixon, Tex. Degrees: B.S., 1931, Southwest Texas State University; Ph.D., 1948, Rutgers University.

Fellow Award "For contributions to biomedical engineering theory and practice." Other Awards: National Academy of Engineering, 1945; Gerard Swope Research Award, 1947-1948; Fulbright Fellow, 1967; Research Award, Sigma Xi, 1968; Special Team Award, NASA, 1978.

BROWN, JOHN, Fellow 1973. Born: July 17, 1923, Auchterderran, Fife, Scotland. Degrees: M.A., 1944, Edinburgh; Ph.D., 1954, D.Sc., 1960, London.

Fellow Award "For achievements in research and teaching of electrical and electronic engineering, especially in the fields of microwaves and electronic devices." Other Awards: Radio and Telecommunications

Section Premiums, Institution of Electrical Engineers, 1954, 1957; Blumlein Browne-Willans Premium, 1966; CBE, 1982.

BROWN, JOHN L., JR., Fellow 1970. Born: March 6, 1925, Ellenville, N.Y. Degrees: B.S., 1948, Ohio University; Ph.D., 1953, Brown University.

Fellow Award "For contributions in the areas of statistical communication theory and underwater acoustics."

BROWN, PHILIP G., Fellow 1984. Born: November 16, 1919, Hood River, OR. Degrees: B.Sc.(E.E.), 1947, Oregon State Univ.

Fellow Award "For contributions to the control, protection, design, and performance analysis of gas, hydro, and steam turbine generators"; Prize Paper Award, IEEE Power Engineering Society, 1977.

BROWN, ROSS D., AIEE Fellow 1954. Born: December 15, 1897, Morgantown, W.Va. Degrees: B.S.E.E., 1922, West Virginia University.

Fellow Award "For his initiative in direction and engineering development of a statewide electric power system."

BROWN, WILLIAM C., IRE Fellow 1959. Born: May 1916, Lewis, Iowa. Degrees: B.S., 1937, Iowa State University; M.S., 1941, Massachusetts Institute of Technology.

Fellow Award "For contributions in the field of microwave tubes." Other Awards: Naval Ordnance Development Award, 1945; Presidential Certificate of Commendation, 1947; Citation for Meritorious Civilian Service, Department of Defense, 1968; Professional Achievement Citation, Iowa State Univ., 1972; Fellow, International Microwave Power Institute, 1979.

BROWN, WILLIAM F., JR., Fellow 1968. Born: September 21, 1904, Lyon Mountain, N.Y. Degrees: B.A., 1925, Cornell University; Ph.D., 1937, Columbia University.

Fellow Award "For contributions to the field of magnetism, especially micromagnetics, and to engineering education"; IEEE Centennial Medal, 1984. Other Awards: Fellow, American Physical Society, 1938; Fellow, American Association for the Advancement of Science, 1940; Fellow, New York Academy of Sciences, 1959; Meritorious Civilian Service Award, U.S. Navy, 1945; Fulbright Scholar, 1962; A. Cressy Morrison Award, New York Academy of Sciences, 1967; Honorary Life Member, Magnetics Society, 1974.

BROWN, WILLIAM M., Fellow 1971. Born: February 14, 1932, Wheeling, W.Va. Degrees: B.S.E.E., 1952, West Virginia University; M.S.E.E., 1955, D.Eng., 1957, The Johns Hopkins University.

Fellow Award "For contributions to linear systems analysis, random processes, and fine resolution radar."

BROWNE, THOMAS E., JR., AIEE Fellow 1958. Born: June 30, 1908, Murfreesboro, N.C. Degrees: B.S., 1928, North Carolina State University; M.S., 1933, University of Pittsburgh; Ph.D., 1936, California Institute of Technology.

Fellow Award "For research contributions to rapid arc quenching for high voltage circuit interrupters"; Dedication of Special Issue on Arc Plasmas, IEEE Transactions on Plasma Science, 1980.

BROWNLEE, THEODORE, AIEE Fellow 1957. Born: May 17, 1903, Cambridge, N.Y. Degrees: E.E., 1924, Rensselaer Polytechnic Institute.

Fellow Award "For contributions to the development of equipment and techniques used in high voltage impulse testing."

BROWNLEE, WILLIAM R., AIEE Fellow 1951. Born: December 1, 1905, Cleveland, Tenn. Degrees: B.S., 1927, E.E., 1954, University of Arizona.

Awards: Outstanding Engineer, IEEE Region III, 1970. Other Awards: Engineer of the Year, Birmingham, Alabama, 1968; Tau Beta Pi; Phi Kappa Phi.

BRUCK, GEORGE, Fellow 1965. Born: October 20, 1904, Budapest, Hungary. Degrees: Federal Degree, 1927, Technische Hochschule Vienna, Austria.

Fellow Award "For contributions to military electronic systems and components and to electro-optical devices."

BRUDNER, HARVEY J., Fellow 1978. Born: May 29, 1931, Brooklyn, N.Y. Degrees: B.S.(in Engineering-Physics, cum laude), 1952, M.S., 1954, Ph.D., 1959, New York University.

Fellow Award "For leadership in the development and application of computers and electronic, audiovisual systems in education and training"; Committee on Solar Photo-Voltaic Standards, IEEE, 1980-1981. Other Awards: Tau Beta Pi, 1950; Sigma Pi Sigma, 1951; Founder's Day Achievement Award, New York University, 1960; Guest Scientist, Rockefeller Institute, 1959-1960; Sigma Xi, 1962; Listed in: "Who's Who in the East," 1979-1982; "Who's Who in America," 1976-1984; "Who's Who in the World," 1980-1982; "Who's Who in Finance and Industry," 1977-1982; "Who's Who in Engineering," 1979-1984; "Who's Who in Training and Development," 1976-1982; "American Men and Women of Science," 1970-1982; Steering Committee, National Science Foundation, 1977-1978; Contributor, House Committee on Science and Technology, 1977; Computer Advisory Committee, 1965-1982, Middlesex County College, New Jersey; Long Range Planning Committee, Computer Advisory Committee, Highland Park Public Schools, 1976-1984; Editorial Advisory Board, Technological Horizons in Education Journal, 1974-1984; Columnist, Educational Technology Magazine, 1981-1984; Datamation, 1980-1984.

BRUENE, WARREN B., IRE Fellow 1961. Born: November 1, 1916, Beaman, Iowa. Degrees: B.S.E.E., 1938,

Iowa State University.

Fellow Award "For advancing single-sideband radio communications." Other Awards: Engineer of the Year Award, Preston Trail Chapter of Texas Society of Professional Engineers (TSPE), 1975; Engineer of the Year Award, Rockwell International, 1982.

BRUNCKE, HARRY P., Fellow 1970. Born: January 2, 1905, St. Paul, Minn. Degrees: B.E.E., 1930, University of Minnesota.

Fellow Award "For leadership in power-system protection and maintenance"; IEEE Centennial Medal, 1984. Other Awards: Golden Jubilee Pioneer, North Central Electrical League, 1981.

BRUNING, JOHN H., Fellow 1981. Born: October 9, 1942, Newark, NJ. Degrees: B.S.(E.E.), 1964, Pennsylvania State University; M.S.(E.E.), 1967, Ph.D.(E.E.), 1969, University of Illinois.

Fellow Award "For contributions to optical measurement technology leading to advances in semiconductor microlithography."

BRUNZELL, GEORGE M., Fellow 1964. Born: March 16, 1909, Reynolds, Idaho. Degrees: B.S.E.E., 1936, University of Idaho; E.E.(Hon.), 1960, International Correspondence School; D.Sc.(Hon.), 1970, University of Idaho.

Fellow Award "For outstanding engineering proficiency, leadership, and administrative attainments"; Engineer of the Year Award, AIEE, Spokane Chapter, 1961. Other Awards: Achievement Award, Delta Tau Delta, 1966; Honorary Life Member, American Gas Association; Honorary Chief, Flat Head Indians, 1968; Honorary Life Member, Spokane Indian Tribes; "Who's Who in America," 1968-1971; Member, National Society of Professional Engineers; Listed in "International Registry of Who's Who" (Geneva, Switzerland).

BRUTON, LEONARD THOMAS, Fellow 1981. Born: September 9, 1942, London, England. Degrees: B.S.(E.E.), 1964, University of London; M.Eng.(E.E.), 1967, Carleton University, Ottawa, Canada; Ph.D.(E.E.), 1970, University of Newcastle Upon Tyne, England.

Fellow Award "For contributions to the theory and design of active circuits." Other Awards: Western Electric Fund Award, Pacific Northwest, American Society for Engineering Education, 1978.

BRYANS, H.B., AIEE Fellow 1918.

BRYANT, JOHN H., Fellow 1967. Degrees: B.S.E.E., 1942, Texas A and M University; M.S., 1947, Ph.D., 1949, University of Illinois.

Fellow Award "For contributions to the engineering of microwave tubes and circuits." Other Awards: Distinguished Alumnus Award, University of Illnois, College of Engineering, 1980.

BUCHHOLZ, WERNER, Fellow 1965. Born: October 24, 1922, Detmold, Germany. Degrees: B.A.Sc., 1945, M.A.Sc., 1947, University of Toronto; Ph.D., 1950, California Institute of Technology.

Fellow Award "For broad contributions to the field of electronic computers."

BUCHSBAUM, SOLOMON J., Fellow 1972. Born: December 4, 1929, Poland. Degrees: B.Sc., 1952, M.Sc., 1953, McGill University; Ph.D., 1957, Massachusetts Institute of Technology.

Fellow Award "For contributions to gaseous and solid-state plasma physics and laser applications." Other Awards: Fellow, American Association for the Advancement of Science; Fellow, American Academy of Arts and Sciences; Fellow, American Physical Society; Member, National Academy of Engineering; Member, National Academy of Sciences; Secretary of Defense Medal for Outstanding Public Service, 1977; Secretary of Energy Award for Exceptional Public Service, 1981.

BUCY, J. FRED, Fellow 1974. Born: July 29, 1928, Tahoka, Tex. Degrees: B.A.(Physics), 1951, Texas Tech University; M.A.(Physics), 1953, University of Texas.

Fellow Award "For leadership in the development of semiconductor electronics and early development of solid-state systems for oil explorations." Other Awards: Distinguished Engineers Award, Texas Tech University, 1972; Member, National Academy of Engineering, 1974.

BUCY, RICHARD S., Fellow 1975. Born: July 20, 1935, Washington, D.C. Degrees: S.B., 1957, Massachusetts Institute of Technology; Ph.D., 1963, University of California at Berkeley.

Fellow Award "For developments in linear and nonlinear filtering theory." Other Awards: Professeur Associee, Government of France, 1972; Senior U.S. Scientist Fellowship, Alexander von Humboldt Foundation, 1975.

BUIE, JAMES L., Fellow 1973. Born: February 21, 1920, Los Angeles, Calif. Degrees: A.A., 1940, Los Angeles City College; B.S.E.E., 1950, University of Southern California.

Fellow Award "For contributions to high-power, high-frequency transistor technology and innovations in the field of integrated circuits." Other Awards: Eta Kappa Nu, 1949; Tau Beta Pi, 1949; Distinguished Flying Cross, U.S. Navy; Air Medal, U.S. Navy.

BULLINGTON, KENNETH, IRE Fellow 1955. Born: January 11, 1913, Guthrie, Okla. Degrees: B.S., 1936, University of New Mexico; S.M., 1937, Massachusetts Institute of Technology.

Fellow Award "For his contributions to the field of radio propagation"; Morris N. Liebmann Award, IRE, 1956, "For his contributions to the knowledge of tropospheric transmission beyond the horizon and to the application of the principles of such transmission to practical communications systems"; Achievement

Award, IRE Professional Group on Communications, 1959. Other Awards: Stuart Ballantine Medal, Franklin Institute, 1956.

BURGESS, JOHN S., Fellow 1966. Born: May 1, 1918, Milwaukee, Wis. Degrees: B.S., 1940, St. Lawrence University; M.S., 1942, University of Notre Dame; Ph.D., 1949, Ohio State University.

Fellow Award "For contributions and leadership in electronics research and development and engineeering education." Other Awards: duPont Fellow, Ohio State University, 1948; St. Lawrence University Alumni Citation, 1968; Exceptional Civilian Service Award, U.S. Dept. of Air Force, 1965, 1971, 1980.

BURKE, CHARLES T., IRE Fellow 1937. Born: August 4, 1902, Watertown, Mass. Degrees: S.B., 1924, S.M., 1924, Massachusetts Institute of Technology.

BURNS, GERALD, Fellow 1983. Born: Oct. 5, 1932, New York, New York. Degrees: B.S., 1954, RPI; M.S., 1957, Ph.D., 1962, Columbia University.

Fellow Award "For spectroscopic studies in injection lasers, solid-state laser materials, ferroelectrics, and superionic conductors." Other Awards: Fellow, American Physical Society; Ed., Solid State Commun.

BURNS, LESLIE L., JR., Fellow 1967. Born: June 12, 1923, Houston, Tex. Degrees: B.S., 1944, Texas A and M University; M.S., Ph.D., 1952, Princeton University.

Fellow Award "For contributions to superconductive computer memories and to color television." Other Awards: Sigma Xi; David Sarnoff Outstanding Achievement Award.

BURRELL, RANDOLPH W., AIEE Fellow 1959. Born: April 27, 1905, Millwood, N.Y. Degrees: E.E., 1925, D.Eng., 1928, Rensselaer Polytechnic Institute.

Fellow Award "For his contributions in the field of power cable engineering and soil thermal resistivity research." Other Awards: Sigma Xi, 1928; Tau Beta Pi, 1960.

BURRUS, CHARLES A., JR., Fellow 1974. Born: July 16, 1927, Shelby, N.C. Degrees: B.S.(cum laude), 1950, Davidson College; M.S., 1951, Emory University; Ph.D., 1955, Duke University.

Fellow Award "For contributions to semiconductor technology for wide-band transmission systems and radio astronomy." Other Awards: Fellow, American Physical Society, 1975; Fellow, AAAS, 1976; Fellow, Optical Society of America, 1979; David Richardson Medal, Optical Society of America, 1982; Distinguished Technical Staff Award, Bell Laboratories, 1982.

BURRUS, CHARLES S., Fellow 1981. Born: October 9, 1934, Abilene, TX. Degrees: B.A., 1957, B.S.E.E., 1958, M.S., 1960, Rice University, Ph.D., Stanford University, 1965.

Fellow Award "For contributions to signal processing and electrical engineering education"; Senior Paper Award, IEEE Acoustics, Speech, and Signal Processing

Society, 1975. Other Awards: Alexander Von Humboldt Foundation Senior Award, 1975; Prize for Excellence in Teaching, Rice University; Senior Fulbright Fellow, 1979; Humboldt Reinvitation Award, 1982.

BUSH, RICHARD M., AIEE Fellow 1946. Born: September 28, 1907, Jalapa, Vera Cruz, Mexico. Degrees: M.S.E.E., Private Instructor.

BUSSGANG, JULIAN J., Fellow 1973. Born: March 26, 1925, Lwow, Poland. Degrees: B.Sc.(Eng.), 1949, University of London; S.M.E.E., 1951, Massachusetts Institute of Technology; Ph.D.(App.Physics), 1955, Harvard University.

Fellow Award "For contributions to sequential detection theory, radar techniques, and statistical communication theory."

BUTLER, CHALMERS M., Fellow 1983. Born: July 31, 1935, Hartsville, South Carolina. Degrees: B.Sc., 1957, M.Sc., 1959, Clemson University; Ph.D., 1962, University of Wisconsin.

Fellow Award "For contributions to aperture theory and to numerical techniques for solving electromagnetic boundary value problems." Other Awards: Outstanding Engineering Teacher Award, University of Mississippi, 1977-78; Western Electric Fund Award, ASEE, 1974; Best EMP Paper Award, 1975-78; Sigma Xi; Tau Beta Pi; Eta Kappa Nu.

BUTMAN, STANLEY A., Fellow 1984. Born: July 10, 1937, Radzin, Poland. Degrees: B.Eng., 1960, McGill University; M.S.E.E., 1962, M.I.T.; Ph.D., 1967, Caltech.

Fellow Award "For leadership and contributions in the field of communications theory and its application to space exploration." Other Awards: NASA Exceptional Service Medal, 1977; NASA Group Achievement Awards: Seasat Synthetic Aperture Radar Team, 1979, Voyager Mission Operations System Design and Development, 1981, Radio Frequency Surveillance System, 1983.

BUTTON, KENNETH J., Fellow 1982. Born: October 11, 1922, Rochester, NY. Degrees: B.S.(with distinction), 1950, M.S., 1952, University of Rochester.

Fellow Award "For pioneering work on microwave ferrites and ferrite devices"; Distinguished Service Award, IEEE Microwave Theory and Techniques Society, 1980; Certificate of Recognition, IEEE Microwave Theory and Techniques Society, 1981.

BYERLY, RICHARD T., Fellow 1984. Born: November 22, 1922, Laverne, OK. Degrees: B.S., Texas A&M University, College Station, Tex.; M.S., University of Pittsburgh, Pittsburgh, Pa.

Fellow Award "For contributions to power systems analysis in synchronous stability and fault calculations."

BYLOFF, ROBERT CONRAD, Fellow 1978. Born: April 6, 1928. Degrees: B.S.(E.E.), 1952, University of California at Los Angeles; Certificate, 1967, University of Wisconsin (Finance Institute).

Fellow Award "For contributions and technical leadership in applied magnetics and in the development of ultra-high-speed rotating machinery."

C

CACHAT, JOHN F., Fellow 1975. Born: September 1, 1916, Cleveland, Ohio. Degrees: B.S.E.E., 1939, M.B.A., 1952, Case Western Reserve University.

Fellow Award "For development of new technology for induction heating of seam-welded high-pressure pipelines."

CACHERIS, JOHN C., Fellow 1964. Born: May 18, 1916. Degrees: B.S.E.E., 1946, Carnegie Institute of Technology; M.S., 1953, University of Maryland.

Fellow Award "For contributions to advancement of microwave technology, particularly in the application of microwave ferrites"; Annual Award, IRE, Phoenix Section, 1961, "For outstanding contributions to electronics." Other Awards: Certificate of Achievement, U.S. Dept. of the Army, 1956.

CAHN, CHARLES R., Fellow 1971. Born: October 7, 1929, Syracuse, N.Y. Degrees: B.E.E., 1949, M.E.E., 1951, Ph.D., 1955, Syracuse University.

Fellow Award "For contributions to the theory of wideband digital communication systems"; Transactions Contribution Award, IRE Professional Group on Communications Systems, 1959; Prize Paper Award, IEEE Communications Society, 1978. Other Awards: Sigma Xi, 1955.

CAIN, BERNARD M., AIEE Fellow 1958. Born: April 20, 1908, Hastings, Mich. Degrees: B.S.E.E., 1929, M.S., 1930, University of Michigan.

Fellow Award "For contributions to design of small motors and improved cooling and performance of large generators." Other Awards: Tau Beta Pi, 1928; Phi Kappa Phi, 1929; Sigma Xi, 1930; Fellow, American Society of Mechanical Engineers, 1964.

CALAHAN, DONALD A., Fellow 1973. Born: February 23, 1935, Cincinnati, Ohio. Degrees: B.S.E.E., 1957, Notre Dame University; M.S.E.E., 1958, Ph.D., 1960, University of Illinois.

Fellow Award "For leadership in the use of digital computers as an aid in circuit design."

CALECA, VINCENT, Fellow 1979. Born: May 27, 1919, New York, N.Y. Degrees: B.E.E.(with distinction), 1949, Cornell University.

Fellow Award "For contributions to the design, protection, and operation of high-voltage ac and dc power transmission systems." Other Awards: Past International Secretary, CIGRE Study Comm. of Transmission Systems; Eta Kappa Nu; Who's Who in Engineering, 1980.

CALVERLEY, THOMAS E., Fellow 1984.

Fellow Award "For technical leadership in the development and implementation of high voltage dc transmission systems."

CAMPANELLA, SAMUEL JOSEPH, Fellow 1978. Degrees: B.S.E.E., Ph.D., Catholic University of America; M.S.E.E., University of Maryland.

Fellow Award "For contributions to signal processing and satellite communications." Other Awards: Fellow, American Association for the Advancement of Science; Sigma Xi; Phi Eta Sigma; American Institute of Aeronautics and Astronautics.

CAMPBELL, FRANCIS J., Fellow 1983. Born: July 29, 1924, Toledo, Ohio. Degrees: B.S.(Chem. Eng.), 1948, University of Toledo.

Fellow Award "For leadership in research and standardization in the field of ionizing radiation effects on electrical insulation." Other Awards: Fellow, American Institute of Chemists, 1980; Research Publication Award, Naval Research Laboratory, 1982.

CAMPBELL, HAROLD E., AIEE Fellow 1961. Born: January 15, 1913, Lynchburg, Va. Degrees: B.S.E.E., 1935, Virginia Polytechnic Institute.

Fellow Award "For contributions in the field of electric power distribution engineering."

CAMRAS, MARVIN, IRE Fellow 1952, AIEE Fellow 1959. Born: January 1, 1916, Chicago, Ill. Degrees: B.S., 1940, Armour Institute of Technology; M.S., 1942, LL.D., 1968, Illinois Institute of Technology.

IRE Fellow Award "In recognition of his contributions to the field of magnetic recording"; AIEE Fellow Award "For contributions to the science and art of magnetic recording"; Achievement Award, IRE, Professional Group on Audio, 1958, "For contributions to audio technology as documented in publications of the IRE"; Consumer Electronics Award, Broadcast and Television Receivers Group, 1964, "For outstanding contribution to consumer electronics"; IEEE Professional Group on Broadcast & Television Receivers, Papers Award, 1964, "For outstanding paper published in the TRANSACTIONS." Other Awards: JOURNAL SMPTE Paper Award, Society of Motion Picture and Television Engineers, 1947; Alumni Distinguished Service Award, Illinois Institute of Technology, 1948; HKN Young Engineers Award, 1948; Fellow, Acoustical Society of America, 1951; John Scott Medal, Franklin Institute, 1955; Citation, Indiana Technical College, 1958; Fellow, American Association for the Advancement of Science, 1961; Product Award, INDUSTRAL RESEARCH MAGAZINE, 1966; John S. Potts (Gold Medal) Memorial Award, Audio Engineering Society, 1969; Honorary Member, AES, 1970; Fellow, SMPTE, 1971; Merit Award, Chicago Technical Societies Council, 1973; Member, National Academy of Engineering, 1976; IIT Alumni Medal, 1978; Washington Award,

Western Society of Engineers, 1979; Inventor of the Year Award, Patent Law Association of Chicago, 1979; IIT Hall of Fame, 1981; Pioneers of Electronics, Foothills Electronics Museum, 1981; Academy of the Chicago Association of Technical Societies, 1981.

CANAVACIOL, FRANK E., AIEE Fellow 1949. Born: January 16, 1896, New York, N.Y. Degrees: E.E., 1918, Polytechnic Institute of Brooklyn; Dr. Eng.(Hon), 1980, Polytechnic Institute of New York.

Other Awards: Professional Engineer, State of N.Y., 1922; E.E. Examiner Municipal Civil Service, 1924-1936, City of NY; Examiner, P.E. Exam, State of N.Y., 1927-1962.

CANDY, JAMES C., Fellow 1976. Born: September 27, 1929, Crickhowell, South Wales, United Kingdom. Degrees: B.Sc., 1951, Ph.D., 1955, University of North Wales.

Fellow Award "For contributions to high-speed digital coders and video picture processing."

CANFIELD, WRIGHT, AIEE Fellow 1962. Born: April 11, 1907, Yale, Okla. Degrees: B.S., 1929, Oklahoma State University.

Fellow Award "For contribution to a major electrical utility system through management of design, construction, and operation." Other Awards: Engineer of the Year Award, Tulsa Chapter, Oklahoma Society of Professional Engineers, 1972.

CANNON, ROBERT S., AIEE Fellow 1957. Born: February 15, 1905, Foreman, Ark. Degrees: B.S., 1926, University of Illinois.

Fellow Award "For important contributions in the application of advanced electrical systems to all phases of refined petroleum product pipeline operations,"

CANTWELL, JOHN L., AIEE Fellow 1957. Born: September 3, 1906, Wilmington, N.C. Degrees: B.S.E.E., 1927, University of North Carolina; M.S.E.E., 1929, Massachusetts Institute of Technology.

Fellow Award "For his contributions to functional testing and the evaluation of transformers and transformer insulating materials"; Best Paper Award, Pittsfield Section, AIEE, 1936. Other Awards: Managerial Award, General Electric Company, 1963.

CAPON, JACK, Fellow 1975. Born: April 28, 1932, New York, N.Y. Degrees: B.E.E., 1953, City College of The City University of New York; M.S.E.E., 1955, Massachusetts Institute of Technology; Ph.D., ·1959, Columbia University.

Fellow Award "For contributions in statistical techniques to electronics and geophysics in high-resolution spectral analysis and nonparametric detection." Other Awards: Eta Kappa Nu, 1952; Tau Beta Pi, 1953; Sigma Xi, 1955; Elected Member, Sigma Xi, 1976; Fellow, American Association for the Advancement of Science, 1976.

CARASSA, FRANCESCO, Fellow 1979. Born: March 7, 1922, Naples, Italy. Degrees: Doctor in Electrotechnical Engineering, 1946, Politecnico di Torino.

Fellow Award "For development and implementation of wide-band radio links and for contributions to space communication." Other Awards: Premio Internazionale delle Comunicazioni, 1971; Premio Nazionale G. Marconi, 1974; Premio migliore pubblicazione AEI, 1978; Medaglia d'oro per i bene meriti della cultura, della scuola e dell' arte, 1972; 9th Marconi International Fellowship, 1983.

CAREIL, FRANCOIS M., AIEE Fellow 1958. Born: February 17, 1894, Pantin, Seine, France. Degrees: Ing-.Dipl., 1919, Ecole Polytechnique; Ing. Civil des Mines, 1921, Ecole des Mines de Paris; Dr.(Hon.), 1953, Ecole Polytechnique de L'Universite de Lusanne.

Fellow Award "For contributions to stable, secured operations of an extra high voltage continental interconnected system"; Paper Prize, AIEE, 1960, for "Two examples of industrial research in France relating to the transmission of electrical energy."

CARLIN, HERBERT J., IRE Fellow 1956. Born: May 1, 1917, New York, N.Y. Degrees: B.S., 1938, M.S., 1940, Columbia University; D.E.E., 1947, Polytechnic Institute of Brooklyn.

Fellow Award "For advances in microwave network synthesis." Other Awards: Outstanding Achievement Award, Air Force Systems Command, 1965; Senior Postdoctoral Fellow, National Science Foundation, 1964-1965; NATO Fellowship, University of Genoa (summer), 1975; Visiting Scientist, Centre National D'Etudes des Telecommunications, 1979-1980; Visiting Professor, MIT, 1972-73; Visiting Professor, Tianjiu University (China), summer 1982; Preston Lewis Professor of Engineering, Cornell University.

CARLIN, P. J., AIEE Fellow 1948. Born: February 5, 1894, Pittsburgh, Pa. Degrees: 1922, Pratt Institute.

CARLSON, WENDELL L., IRE Fellow 1949. Born: January 1, 1897, Jamestown, N.Y. Degrees: Bliss Electric School, 1918.

Fellow Award "In recognition of his contributions over many years to the development of radio receivers and their components." Other Awards: Charles A. Coffin Award, General Electric, 1925; Modern Pioneer Award.

CARNES, WILLIAM T., Fellow 1972. Born: February 17, 1915, Kansas City, Mo. Degrees: B.S.E.E., 1937, University of Kansas.

Fellow Award "For contributions to the technology, international standardization, and reliability of air transport avionics."

CARPENTIER, MICHEL H., Fellow 1978. Born: January 16, 1931, Estaires, France. Degrees: Ecole Polytechnique, Paris, 1952; E.N.S.A.E., Paris, 1955; Ecole Superieure D'Electricite, Paris, 1956.

Fellow Award "For pioneering work in the fields of radars and information processing." Other Awards: Grand Prix de L'Electronique General Ferrie, 1969; Chevalier de L'Ordre National du Merite, 1978.

CARR, PAUL H., Fellow 1979. Born: May 12, 1935, Boston, Mass. Degrees: B.S., 1957, M.S., 1961, Massachusetts Institute of Technology; Ph.D., 1966, Brandeis University.

Fellow Award "For contributions to microwave acoustics and their use as signal-processing components." Other Awards: Marcus D. O'Day Memorial Award, 1967; Guenter Loeser Memorial Award, 1973; Air Force Systems Command Outstanding Technical Achievement of the Quarter Award, 1976.

CARRARA, GIANGUIDO, Fellow 1972. Born: February 23, 1930, Brescia, Italy. Degrees: D.E.E., 1953, Polytechnical Institute of Milano.

Fellow Award "For contribution to the analysis of the behavior of dielectric systems and advancements to the science of insulation coordination in high-voltage networks, particularly under transient disturbances."

CARREL, ROBERT L., Fellow 1973. Born: February 8, 1933, Fort Wayne, Ind. Degrees: B.S.E.E., 1955, Purdue University; M.S.E.E., 1958, Ph.D., 1961, University of Illinois.

Fellow Award "For contributions to antenna research and development"; Member, Board of Directors, MIDCON Convention Director, MIDCON Conference, 1982. Other Awards: Eta Kappa Nu; Tau Beta Pi; Sigma Xi; Member, Armed Forces Communications and Electronics Association; Member, Association of Old Crows; Member, Board of Diretors, Southern Methodist University Foundation for Science and Engineering; Founder and Executive Vice President, Electrospace Systems, Incorporated.

CARROLL, THOMAS J., IRE Fellow 1962. Born: April 26, 1912, Pittsburgh, Pa. Degrees: A.B., 1932, University of Pittsburgh; Ph.D., 1936, Yale University.

Fellow Award "For contributions to the understanding of tropospheric propagation."

CARSON, VICTOR S., IRE Fellow 1955. Born: June 29, 1908, Pocatello, Idaho. Degrees: B.S., 1938, Oregon State University; Eng., 1940, Ph.D., 1946, Stanford University.

Fellow Award "For his contributions to the development and analysis of long-range aeronautical electronic navigation systems."

CARTER, GEOFFREY W., Fellow 1969. Born: May 21, 1909, Rugby, England. Degrees: B.A., 1931, M.A., 1937, University of Cambridge, England.

Fellow Award "For contributions to the application of electromagnetic theory and mathematics to electrical engineering design and analysis, and for particular success in electrical engineering education." Other Awards: Fellow, IEE, 1947.

CARTER, GORDON K., AIEE Fellow 1961. Born: May 13, 1912, Richmond, Va. Degrees: B.S.E., 1934, E.E., 1935, University of Virginia.

Fellow Award "For contributions to the development of advanced engineering computational methods in diversified fields"; First Paper Prize, AIEE, Power Division, 1960; First Paper Prize, AIEE, District 5, 1961. Other Awards: Mac Wade Award, 1952.

CARTER, WILLIAM C., Fellow 1975. Born: January 16, 1917, Waterville, Me. Degrees: B.A., 1938, Colby College; Ph.D., 1947, Harvard University.

Fellow Award "For contributions to fault-tolerant computing"; Computer Society Honor Roll, 1974, "For his efforts as Guest Editor of the IEEE Transactions on Computers and for his contributions to the International Symposia on Fault Tolerant Computing"; Computer Society Distinguished Visitor, 1972, 1973; Computer Society Certificate of Appreciation for "Contributions to the Technical Committee on Fault Tolerant Computing," 1982. Other Awards: Rhodes Scholarship, 1938; Phi Beta Kappa, 1938; Sigma Xi, 1941; ACM National Lecturer, 1963-1964, 1968-1969; NASA Certificate of Recognition, 1972, 1973; IBM 6th Level Invention Achievement Award, 1983; Who's Who, 1982, 1983; NASA Advisory Committee on Space and Systems Technology, 1982, 1983.

CARVILLE, T. ELLSWORTH M., AIEE Fellow 1947. Born: October 20, 1903, Leeds, Me. Degrees: B.S., 1924, E.E., 1933, University of Maine.

CASABONA, ANTHONY M., Fellow 1968. Born: June 16, 1920, New York, N.Y. Degrees: B.S.E.E., 1941, School of Technology, City College of The City University of New York.

Fellow Award "For his pioneering technical and managerial contributions to navigation and instrument landing systems." Other Awards: Eta Kappa Nu, 1941; Tau Beta Pi, 1941.

CASASENT, DAVID P., Fellow 1979. Born: December 8, 1942, Washington, D.C. Degrees: B.S.E.E., 1964, M.S.E.E., 1965, Ph.D. (E.E.), 1969, University of Illinois.

Fellow Award "For theoretical and engineering contributions to coherent optical data processing." Barry Carlton Outstanding Paper Award, IEEE Aerospace and Electronic Systems Society, 1976; Special Recognition Award, IEEE Electron Devices Society, Pittsburgh Chapter. Other Awards: Fellow, Society of Photo-Optical Instrumentation Engineers, 1977; Service Citation, Society of Photo-Optical Instrumentation Engineers, 1976, 1978; Outstanding Paper Award, American Institute of Aeronautics and Astronautics, 1979; Fellow, Optical Society of America, 1980; George Westinghouse Chair, Carnegie Mellon University, 1981; Governor, Society of Photo-Optical Engineers, 1983.

CASAZZA, JOHN A., Fellow 1975. Born: January 3, 1924, Brooklyn, N.Y. Degrees: B.E.E., 1945, Cornell University.

Fellow Award "For contributions to new technical and economic analyses in electric power system planning"; IEEE Spectrum Editorial Board, 1975-1979; Chairman, IEEE energy Committee, 1981-1983; Member, IEEE Individual Benefits and Services Committee. Other Awards: Chairman, U.S. Technical Committee, CIGRE, 1974-1981; Chairman, Edison Electric Institute System Planning Committee, 1972-1974; "Who's Who in Engineering," "Who's Who in the East."

CASE, RICHARD P., Fellow 1975. Born: May 13, 1935, Akron, Ohio. Degrees: B.S.E.E., 1956, Case Institute of Technology.

Fellow Award "For leadership in the design and development of large-scale computers, and for contributions to systems architecture."

CASEY, H. CRAIG, JR., Fellow 1984. Born: December 4, 1934, Houston, TX. Degrees: B.S.E.E., 1957, Oklahoma State University; Ph.D., 1964, State University.

Fellow Award "For contributions to III-V compounds in understanding emission based on the basic optical and impurity behavior."

CASTENSCHIOLD, RENE, Fellow 1979. Born: February 7, 1923, Mount Kisco, N.Y. Degrees: B.E.E., 1944, Pratt Institute.

Fellow Award "For contributions to the development of automatic transfer switches and emergency power generator control"; Member, IEEE Standards Board, 1983-1984. Other Awards: Member, Distinguished Alumni Board of Visitors, Pratt Institute, 1979; Listed in: "Who's Who in Engineering," 1980-1982, "Who's Who in Technology Today," 1980-1982, "American Men and Women of Science," 1982, "Who's Who in the East," 1983-1984; 9 Patents; Member, National Society of Professional Engineers; Registered Professional Engineer, New York, New Jersey.

CASWELL, HOLLIS L., Fellow 1984. Born: August 15, 1931, Nashville, TN. Degrees: B.A., M.S., S.M., Ph.D.

Fellow Award "For technical contributions and leadership in computer semiconductor technologies"; IEEE Mid-Hudson Section Technical Contribution Award, 1982.

CASWELL, R. W., AIEE Fellow 1962. Born: September 21, 1902, West Point, Ind. Degrees: B.S., 1924, University of Illinois.

Fellow Award "For contributions to the development of high voltage electrical systems."

CATENACCI, GIORGIO, Fellow 1969. Born: March 25, 1927, Milano, Italy. Degrees: D.Eng., 1950, Politecnico di Milano.

Fellow Award "For contributions to the analysis of the transient behavior of electric power systems."

CAULTON, MARTIN, Fellow 1980. Born: August 28, 1925. Degrees: B.S.(Physics), 1950, M.S.(Physics), 1952, Renssalaer Polytechnic Institute; Ph.D.(Physics), 1954, Renssalaer; Thesis at Brookhaven National Laboratory.

Fellow Award "For technical contributions and leadership in development of microwave integrated circuits and high-power transistors." Other Awards: IR-100 Award, 1969; RCA Laboratories Outstanding Achievement Award, 1967, 1970.

CAVE, JERE S., Fellow 1971. Born: September 5, 1914, Youngstown, Ohio. Degrees: B.S.E.E., 1936, Vanderbilt University.

Fellow Award "For contributions to engineering and management in the economic application of technology to telecommunications networks."

CAWEIN, MADISON, IRE Fellow 1954. Born: March 18, 1904, Louisville, Ky. Degrees: B.S.(Physics), 1924, University of Kentucky.

Fellow Award "For his contributions to television theory and circuitry."

CEDERBAUM, ISRAEL, IRE Fellow 1962. Born: February 4, 1910, Warsaw, Poland. Degrees: M.Sc., 1930, University of Warsaw; E.E., 1934, Polytechnic Institute, Warsaw; Ph.D., 1956, University of London.

Fellow Award "For contributions to applications of matrix algebra to network theory."

CHADWICK, JOHN W., JR., Fellow 1983. Born: June 30, 1923, New Bern, North Carolina. Degrees: B.E.E., 1947, North Carolina State College.

Fellow Award "For contributions to protection and metering of high-voltage power systems utilizing both solid-state and electromechanical components."

CHADWICK, JOSEPH H., JR., Fellow 1967. Born: April 9, 1925, San Pedro, Calif. Degrees: B.S.E.E., 1944, California Institute of Technology; M.S.E.E., 1947, Ph.D., 1951, Stanford University.

Fellow Award "For contributions to advanced marine instrument and control systems." Other Awards: Capt. Joseph H. Linnard Prize, Society of Naval Architects and Marine Engineers, 1956.

CHADWICK, WALLACE L., AIEE Fellow 1956. Born: December 4, 1897, Loring, Kans. Degrees: D.Eng.Sc.(Hon.), 1965, University of Redlands.

Fellow Award "In recognition of his engineering and administrative achievements in directing design and construction in one of the nation's large power systems." Other Awards: 75th Anniversary Award, American Society of Mechanical Engineers, Southern California Section, 1955; Engineer of the Year Award, Los Angeles Engineering Societies, 1962; Philip Sprague Award, Instrument Society of America, 1963; President, American Society of Civil Engineers, 1964-1965; Member, National Academy of Engineering, 1964; Golden Beavers Award, 1969; Rickey Award,

American Society of Civil Engineers, 1974; "Construction's Man of the Year" Award, Engineering News-Record, 1978; Centennial Honorary Membership, American Society of Mechanical Engineers, 1980.

CHAFFEE, J. G., IRE Fellow 1954. Born: March 6, 1901, Hackensack, N.J. Degrees: S.B., 1923, Massachusetts Institute of Technology.

Fellow Award "For his contributions to transmission and relay systems and frequency modulation."

CHAMBERS, CARL C., AIEE Fellow 1951, IRE Fellow 1959. Born: May 8, 1907, Philadelphia, Pa. Degrees: B.S., 1929, Dickinson College; Sc.D., 1934, University of Pennsylvania; D.Sc.(Hon.), 1972, Dickinson College.

IRE Fellow Award "For leadership in electrical engineering education"; Philadelphia Section Award, IEEE, 1973, "For his long standing service to the IEEE on local and national levels, his leadership in the development of international standards, and his service to the Delaware Valley Community through his outstanding and progressive leadership in engineering education." Other Awards: Army-Navy E, 1947; Philadelphia Engineer of the Year Award, 1966; Member, National Academy of Engineering, 1970; D. Robert Yarnall Award, University of Pennsylvania Engineering Alumni Society, 1971; Allan R. Cullimore Award, Newark College of Engineering, 1973.

CHAMBERS, DUDLEY E., AIEE Fellow 1948, IRE Fellow 1948. Born: November 27, 1905, Kansas City, Mo. Degrees: B.A., 1927, E.E., 1928, Stanford University.

CHAMBERS, FRED, AIEE Fellow 1962. Born: February 17, 1912, Carbon Hill, Ala. Degrees: B.S., 1930, Auburn University.

Fellow Award "For contributions to system planning, coordination and engineering administration of a large public utility system." Other Awards: Engineer of the Decade, Engineers of Tennessee, 1960-1969; "Who's Who in America," "Who's Who in the World," "Who's Who in Engineering."

CHAMBERS, JOSEPH A., IRE Fellow 1955. Born: August 21, 1903, Union, S.C. Degrees: B.S., 1924, Clemson University.

Fellow Award "For his contributions to the development of high power broadcast transmitters and military electronic equipment."

CHAN, SHU-PARK, Fellow 1983. Born: October 10, 1929, Canton, China. Degrees: B.Sc.(E.E.), 1955, Virginia Military Institute; M.S.(E.E.), 1957, Ph.D.(E.E.), 1963, University of Illinois, Urbana-Champaign.

Fellow Award "For contributions to the application of graph theory to networks and leadership in engineering education." Other Awards: President's Outstanding Teacher Award 1979, University of Santa Clara; Honorary Professor, 1980, University of Hong Kong; Honorary Professor, 1982, Anhuei University, Hefei, China; California Senate Commendation (Resolution

#527) for contributions in E.E. and in community, 1972; Special Chair, National Taiwan University, 1973; Honorary Consultant, National Science Council, Taiwan, 1981; Tamkang Chair, Tamkang University, Taiwan, 1981; Electrical Engineering Distinguished Alumnus Award, University of Illinois, 1983.

CHANCE, BRITTON, IRE Fellow 1954. Born: July 24, 1913, Wilkes-Barre, Pa. Degrees: B.S., 1935, M.S., 1936, Ph.D., 1940, University of Pennsylvania; Ph.D., 1942, Sc.D. 1952, University of Cambridge; M.D.(Hon.), Karolinska Institutets, Stockholm, Sweden; D.Sc.(Hon.), 1974, Medical College of Ohio at Toledo; M.D.(Hon.), 1976, Semmelweis University, Budapest; D.Sc.(Hon.), 1977, Hahneman Medical College, Philadelphia.

Fellow Award "For his contributions to radar development, and to the application of electronics to biophysics", Morlock Award, IRE, Professional Group on Biomedical Electronics, 1961, Other Awards: Paul Lewis Award, 1950; Presidential Certificate of Merit, 1950; Harvey Lecturer, 1954; Phillips Lecturer, 1956, 1965; Pepper Lecturer, 1957; Genootschaps Medaille, Dutch Academy of Science, 1965; Keilin Lecturer, British Biochemical Society, 1966; Harrison Howe Award, American Biochemistry Congress, 1981; Member, Royal Academy of Arts and Scienes, Upsala; Leopoldina Academy, Halle; Bavarian Academy of Sciences; American Philosophical Society; Explorer's Club. Chemical Society, 1966; Franklin Medal, Franklin Institute, 1966; Overseas Fellow, Churchill College, Cambridge University, 1966; Pennsylvania Award for Excellence (Life Sciences), 1968; Philadelphia Section Award & Nichols Award, New York Section, American Chemical Society, 1969, 1970; Heineken Medal, Netherlands Academy of Science and Letters, 1970; Redfearn Lecturer, 1970; Gairdner Award, 1972; Festschrift Symposium, Stockholm, 1973; Semmelweis Medal, Budapest, 1974; Foreign Associate, Argentine National Academy of Science, 1975; Foreign Member, Max-Planck-Gesellschaft zur Forderung der Wissenschaften, Germany, 1975; Presidential Lecturer, University of Pennsylvania, 1975; 35th Richtmeter Memorial Lecture of Association of Asphalt Paving Technologists, 1976; 2nd Julius Jackson Memorial Lecture, Wayne State University, Sigma Xi, 1976; ISCO Award, 1976; DaCosta Oration, Philadelphia County Medical College, 1976; Philip Morris Lecturer, 1978; Honorary Vice President, International Union of Pure and Applied Biophysics, 1979; Rudolf-Lemberg Memorial Lecture Series, Australia, 1979; Invited Plenary Lecturer, Society of General Physiologists, International Congress of Physiological Sciences, Budapest, 1980; Invited Lecturer, First European Bioenergetics Congress, Urbino, 1980; Hastings Lecture, Scripps Clinic and Research Foundation, 1980; Invited Lecturer Gordon Conference on Energy Coupling Mechanisms, 1981; Invited Plenary Lecturer, VII International Biophysics Congress, II Pan American Biochemistry Congress, 1981; Foreign Member, Royal Society of London, 1981.

CHANG, HERBERT Y., Fellow 1976. Born: November 25, 1937, Shanghai, China. Degrees: B.S.E.E., 1960, M.S.E.E., 1962, Ph.D.(E.E.), 1964, University of Illinois.

Fellow Award "For contributions in fault-tolerant computing and design automation."

CHANG, SHELDON S. L., IRE Fellow 1962. Born: January 20, 1920, Peking, China. Degrees: B.S., 1942, National Southwest Associated University, China; M.S., 1944, National Tsing Hua University, China; Ph.D., 1947, Purdue University.

Fellow Award "For contributions to information theory and to control systems theory."

CHANG, WILLIAM S.C., Fellow 1978. Born: April 4, 1931, China. Degrees: B.S.E.E., 1952, M.S.E.E., 1953, University of Michigan; Ph.D., 1957, Brown University.

Fellow Award "For contributions to optoelectronics and intergrated optics." Other Awards: Distinguished Professional Achievement Award, University of Michigan, 1978; Fellow, Optical Society of America, 1982.

CHAPIN, EDWARD W., IRE Fellow 1962. Born: July 23, 1908, Freeburg, Ill. Degrees: B.S., 1930, University of Illinois.

Fellow Award "For contributions to the measurement and reduction of radio frequency interference and to the formulation of standards, rules and regulations in this field."

CHARAP, STANLEY H., Fellow 1984. Born: April 21, 1932, Brooklyn, N.Y. Degrees: B.S.(Physics), 1953, Brooklyn College; Ph.D., 1959, Rutgers University.

Fellow Award "For contributions to the advancement of the theory and application of magnetism."

CHARLTON, OAKLEE E., AIEE Fellow 1956. Born: October 19, 1897, Christiansburg, Va. Degrees: B.S.E.E., Massachusetts Institute of Technology.

Fellow Award "For his engineering achievements in the planning of a large interconnected power system."

CHARTIER, VERNON L., Fellow 1980. Born: February 14, 1939, Fort Morgan, CO. Degrees: B.S.E.E., 1963, B.S.(Bus.), 1963, University of Colorado.

Fellow Award "For contributions to the understanding of corona phenomena associated with high-voltage power transmission lines."

CHARYK, JOSEPH V., Fellow 1980. Born: September 9, 1920, Canmore, Alberta, Canada. Degrees: B.Sc., 1942, University of Alberta, Canada; M.S., 1943, Ph.D., 1946, California Institute of Technology; L.L.D.(Hon.), 1946, University of Alberta; D.Eng.(Hon.), 1974, University of Bologna.

Fellow Award "For leadership in the development

and application of communications satellite systems." Other Awards: Fellow, American Institute of Aeronautics and Astronautics; Member, National Academy of Engineering, Member, International Academy of Astronautics; Distinguished Service Medal, Department of Defense, 1963; Lloyd V. Berkner Space Utilization Award, American Astronautical Society, 1967; Guglielmo Marconi International Award, 1974; Theodore Von Karman Award, American Institute of Aeronautics and Astronautics, 1977; Goddard Astronautics Award, American Institute of Aeronautics and Astronautics, 1978.

CHASE, ROBERT L., Fellow 1973. Born: March 19, 1926, Brooklyn, N.Y. Degrees: B.S.E.E., 1945, Columbia University; M.E.E., 1947, Cornell University; Ph.D., 1973, Uppsala University, Sweden.

Fellow Award "For contributions to nuclear instrumentation and measurement techniques."

CHATTERJEE, SYAMA D., Fellow 1984. Born: June 29, 1909, Calcutta, India. Degrees: M.Sc., 1932, D.Sc., 1946, Calcutta University.

Fellow Award "For contributions to the understanding of spontaneous fission of uranium, and for electronics education." Other Awards: Sir John Woodburn Gold Medal, Indian Association for the Cultivation of Science, 1948; Sir Charles Elliot Medal & Prize, Royal Asiatic Society, 1951; Prof M.N. Saha Memorial Gold Medal, Asiatic Society, 1981; Fellow, Indian National Science Academy, 1955; Fellow and Ex-President, Indian Physical Society, 1935-; Post-Doctoral Research Fellow, National Research Council, Ottawa, Canada, 1949-51; Royalsociety Bursary, Davy-Faraday Laboratory U.K., 1958-59; Visiting Professor Technical University, Munich, Germany, 1964-65; Professor Emeritus, Jadavpur University, Calcutta, 1978.

CHEEK, ROBERT C., AIEE Fellow 1962. Born: November 14, 1917, Charleston, S.C. Degrees: B.S.E.E., 1941, Georgia Institute of Technology; M.S., 1943, University of Pittsburgh.

Fellow Award "For contributions to power system communication and applications of electronics to industry." Other Awards: Outstanding Young Electrical Engineer Award, Eta Kappa Nu, 1949: Order of Merit, Westinghouse Electric, 1963; Computer Sciences Man of the Year, Data Processing Management Association, 1972.

CHELOTTI, EDWARD F., Fellow 1983. Born: May 23, 1924, Chicago, Ill. Degrees: B.Sc.E.E., 1945, Northwestern University.

Fellow Award "For contributions to the integration of computer technology into power plant design and for leadership in the development of industry standards"; IEEE Standards Medallion Award, 1978; IEEE Power Generation Committee Outstanding Service Award, 1982.

CHEN, DI, Fellow 1975. Born: March 15, 1929, Chekiang, China. Degrees: B.S., 1953, National Taiwan University; M.S., 1956, University of Minnesota; Ph.D., 1959, Stanford University.

Fellow Award "For contributions to magnetooptic materials and their application in data storage." Other Awards: Honeywell H.W. Sweatt Engineers and Scientists Award, 1971; Honeywell Fellow, 1979.

CHEN, FRANCIS F., Fellow 1980. Born: November 18, 1929, Canton, China. Degrees: A.B., 1950, M.A., 1951, Ph.D., 1954, Harvard University.

Fellow Award "For significant contributions to plasma diagnostics, and to the understanding of plasma instabilities and anomalous transport phenomena."

CHEN, KAN, Fellow 1973. Born: August 28, 1928, Hong Kong. Degrees: B.E.E., 1950, Cornell University; S.M., 1951, Sc.D., 1954, Massachusetts Institute of Technology.

Fellow Award "For contributions in automatic control and system sciences, and the application of systems engineering to societal problems"; First Prize for Technical Paper, AIEE Industrial Division, 1956; Best Paper Award, IEEE Transactions on Education, 1978. Other Awards: Paul G. Goebel Professor of Advanced Technology, University of Michigan, 1971-73.

CHEN, KAO, Fellow 1983. Born: March 21, 1919, Shanghai, China. Degrees: B.S.E.E., 1942, Jiao Tong University, China; M.S.E.E., 1948, Harvard University; Completed advanced graduate studies at Polytechnic Institute of Brooklyn, 1952.

Fellow Award "For leadership and contributions in the design and development of industrial power distribution systems and energy-saving illumination systems"; Production and Application of Light Committee Prize Paper Awards, IEEE, 1979,1980,1981,1982,-1983; Industry Applications Society Best Papers Award, 1981, 1983; IEEE Centennial Medal, 1984. Other Awards: Federation of British Industries Scholar, 1945-47; Westinghouse Engineering Achievement Award, 1971; "Who's Who in America", 1984-1985; "International Who's Who in Engineering", 1984; "Personalities of America", 1984; Registered Professional Engineer, New York, New Jersey.

CHEN, KUN-MU, Fellow 1976. Born: February 3, 1933, Taipei, Taiwan. Degrees: B.S.E.E., 1955, National Taiwan University; M.S.(Applied Physics), 1958, Ph.D.(Applied Physics), 1960, Harvard University.

Fellow Award "For contributions to electromagnetics and plasmas." Other Awards: Distinguished Faculty Award, Michigan State University, 1976; Fellow, American Association for the Advancement of Science, 1977; Excellent Achievement Award, Taiwanese-American Foundation, 1983; Member, Sigma Xi; Member, Phi Kappa Phi; Member, Tau Beta Pi; Member, Commission A.B.C., International Union of Radio Science.

CHEN, MO-SHING, Fellow 1978. Born: August 20, 1931, China. Degrees: B.S., 1954, National Taiwan University; M.S., 1958, Ph.D., 1962 (all in Electrical Engineering), University of Texas at Austin.

Fellow Award "For contributions to education and research in power system engineering." Other Awards: Exxon Award, 1974; Outstanding Research Award, University of Texas at Arlington, 1975; First Power Engineering Educator Award, Edison Electric Institute, 1976; Western Electric Fund Award, American Society for Engineering Education, 1977; Distinguished Honorary Alumnus Award, University of Texas at Arlington, 1979; U.S.A. Achievement Award, Chinese Institute of Engineers, 1979.

CHEN, TIEN-CHI, Fellow 1977. Born: November 12, 1928, Hong Kong. Degrees: Sc.B.(Chemistry), 1950, Brown University; M.A.(Chemistry), 1952, Ph.D.(Physics), 1957, Duke University.

Fellow Award "For contributions to computer organization and multicomputer systems"; IEEE Computer Society Honor Roll, 1978. Other Awards: Outstanding Contribution Award, IBM, 1964, 1971; Invention Achievement Award, IBM, 1973, 1974, 1975, 1977; Phi Lambda Upsilon; Sigma Pi Sigma; Sigma Xi; Phi Beta Kappa.

CHEN, WAI-KAI, Fellow 1977. Born: December 23, 1936, Nanking, China. Degrees: B.S.E.E., 1960, M.S.E.E., 1961, Ohio University; Ph.D., 1964, University of Illinois at Urbana-Champaign.

Fellow Award "For contributions to graph and network theory." Other Awards: University Fellow, University of Illinois, 1961; C. T. Loo Fellow, China Institute of America, 1962; Lester R. Ford Award, Mathematical Association of America, 1967; Research Institute Fellow Award, Ohio University, 1972; Outstanding Educators of America Award, 1973; Baker Fund Award, Ohio University, 1974; Excellence in Teaching Award, College of Engineering, Ohio University, 1975; Fellow Award, American Association for the Advancement of Science, 1978; Distinguished Professor Award, Ohio University, 1978; Baker Research Award, Ohio University, 1978.

CHEN, WAYNE H., Fellow 1969. Born: December 13, 1922, Soochow, China. Degrees: B.S.E.E., 1944, National Chiao Tung University; M.S.E.E., 1949, Ph.D., 1952, University of Washington.

Fellow Award "For contributions in the fields of network theory and switching circuits, and for vigorous leadership in engineering education." Other Awards: Sigma Xi, 1954; Florida Blue Key Outstanding Faculty Award, 1960; Outstanding Publication Award, Chia Hsin Cement Co. Cultural Fund, 1964; Sigma Tau, 1965; Eta Kappa Nu, 1967; Tau Beta Pi, 1966; Teacher-Scholar Award, University of Florida, 1971; Epsilon

Lambda Chi, 1974; Phi Tau Phi, 1976; Omicron Delta Kappa, 1977; Phi Kappa Phi, 1981.

CHENG, DAVID K., IRE Fellow 1960. Born: January 10, 1918, Kiangsu, China. Degrees: B.S.E.E., 1938, National Chiao Tung University, China; S.M., 1944, Sc.D., 1946, Harvard University.

Fellow Award "For contributions to engineering education and antenna theory"; Best Paper Prize, Syracuse Section, 1961, 1964, 1965, 1966, 1969, 1970, 1980; European Distinguished Lecturer, IEEE Antennas and Propagation Society, 1975-1976. Other Awards: Charles Storrow Scholar, Harvard University, 1944-1945; Gordon McKay Scholar, Harvard University, 1945-1946; Best Original Paper Award, National Electronics Conference, 1960; Fellow, John Simon Guggenheim Foundation, 1960-1961; Research Award, Sigma Xi, Syracuse Chapter, 1962; Annual Achievement Award, Chinese Institute of Engineers, 1962; Fellow, Institution of Electrical Engineers, U.K., 75; Distinguished Service Award, Phi Tau Phi Scholastic Honor Society of America, 1975; Eta Kappa Nu; Sigma Xi, Phi Tau Phi; Distinguished Engineer, Li Institution of Science and Technology; Centennial Professor, Syracuse University, 1970; Chancellor's Citation for Exceptional Academic Achievement, Syracuse University, 1980; U.S. Representative, Institution of Electrical Engineers, U.K., 1980-83.

CHEO, PETER KONG-LIANG, Fellow 1984. Born: February 2, 1930, Nanking, China. Degrees: B.Sc., 1951, Aurora College; M.Sc., 1953, Virginia Polytech State University; Ph.D., 1963, Ohio State University.

Fellow Award "For developments in high-resolution tunable infrared and far-infrared sources for remote sensing of the atmosphere and for inspection of high-voltage cable insulation." Other Awards: NSF Faculty Fellow Award, 1960; Am. Optical Society Fellow Award, 1976; United Technologies Award, 1982.

CHESTER, ARTHUR N., Fellow 1981. Born: August 5, 1940, Seattle, WA. Degrees: B.S.(Physics), 1961, University of Texas at Austin; Ph.D.(Theoretical Physics), 1965, California Institute of Technology.

Fellow Award "For contributions and technical leadership in gas laser technology."

CHESTNUT, HAROLD, AIEE Fellow 1962. Born: November 25, 1917, Albany, N.Y. Degrees: B.S.E.E., 1939, M.S.E.E., 1940, Massachusetts Institute of Technology; D.Eng.(Hon.), 1966, Case Institute of Technology; D.Eng.(Hon.), 1972, Villanova.

Fellow Award "For contributions to the theory and design of control systems"; First Paper Prize, AIEE, Industry Group, 1948. Other Awards: Fellow, AAAS, 1972; Fellow, ISA, 1974; Member, National Academy of Engineering, 1974; Case Centennial Scholar, 1980; Honda Prize, 1981.

CHIEN, YI-TZUU, Fellow 1980. Born: August 21, 1938. Degrees: B.S.E.E., 1960, National Taiwan University; M.S.E.E., 1964, Ph.D., 1967, Purdue University.

Fellow Award "For contributions to computer science education and research in statistical pattern recognition." Other Awards: Program Director of Intelligent Systems, Computer Science Section, National Science Foundation, 1982-83.

CHILDERS, DONALD GENE, Fellow 1976. Born: February 11, 1935, The Dalles, Oreg. Degrees: B.S., 1958, M.S., 1959, Ph.D., 1964, University of Southern California.

Fellow Award "For contributions to electroencephalography"; William J. Morlock Award, IEEE Group on Biomedical Engineering. Other Awards: Distinguished Faculty Award, University of Florida, 1970; Western Electric Award, American Society for Engineering Education, Southeastern Section, 1971-1972; Award for Research, American Society for Engineering Education, Southeastern Section, 1972-1973; Award for Outstanding Service, University of Florida College of Engineering, 1974-1975; George Westinghouse Award, American Society for Engineering Education, 1975.

CHILES, JOHN H., JR., AIEE Fellow 1948. Born: June 16, 1903, Brentwood, Tenn. Degrees: B.S., 1925, Virginia Polytechnic Institute.

Other Awards: Order of Merit, Westinghouse Electric, 1944; Outstanding Service Commendation, U.S. Dept. of Navy, Bureau of Ships, 1948; Tau Beta Pi; Vice President, Westinghouse Electric Corporation.

CHINN, HOWARD A., IRE Fellow 1945, AIEE Fellow 1958. Born: January 5, 1906, New York, N.Y. Degrees: S.B., 1927, S.M., 1929, Massachusetts Institute of Technology.

IRE Fellow Award "For his contributions to improved broadcasting"; AIEE Fellow Award "For his contributions to the development of measuring and monitoring equipment in the audio and video broadcasting fields." Other Awards: Presidential Certificate of Merit, 1949; John H. Potts Award, Audio Engineering Society, 1950; Fellow, Audio Engineering Society; Fellow, Acoustical Society of America; Fellow, Society of Motion Picture and Television Engineers; Engineering Achievement Award, National Association of Broadcasters.

CHIPMAN, ROBERT A., Fellow 1970. Born: April 28, 1912, Winnipeg, Manitoba, Canada. Degrees: B.Sc., 1932, University of Manitoba; M.Eng., 1933, McGill University; Ph.D., 1939, Cambridge university.

Fellow Award "For contributions to high-frequency and microwave measurements." Other Awards: Western Electric ASEE Teaching Award, 1967.

CHIRLIAN, PAUL M., Fellow 1979. Born: April 29, 1930, New York, N.Y. Degrees: B.E.E., 1950, M.E.E., 1952, Eng. ScD., 1956, New York University.

Fellow Award "For contributions to engineering education through the writing of innovative engineering textbooks." Other Awards: Eta Kappa Nu, 1952; Sigma Xi, 1956; Tau Beta Pi, 1960.

CHISHOLM, DONALD A., Fellow 1968. Born: May 7, 1927, Toronto, Ontario, Canada. Degrees: B.A.Sc., 1949, M.A., 1950, Ph.D., 1952, D.Sc., 1980, University of Toronto; D.Eng., 1979, University of Western Ontario.

Fellow Award "For his technical contribution in the field of microwave electron tubes, and his outstanding leadership in electron and optical device development." Other Awards: Member, Science Council of Canada; Chairman, Board of Governors of Ontario Research Foundation.

CHITTICK, KENNETH A., IRE Fellow 1953. Born: November 6, 1903, Old Bridge, N.J. Degrees: B.S., 1925, Rutgers University.

Fellow Award "For his contributions to radio and television engineering, and for his meritorious work on many important institute and industry technical committees." Other Awards: National Geographical Society; Fellow, Radio Club of America.

CHO, ALFRED Y., Fellow 1981. Born: July 10, 1937, Beijing, China. Degrees: B.S.E.E., 1960, M.S.E.E., 1961, Ph.D., 1968, University of Illinois.

Fellow Award "For pioneering work in the development of molecular beam epitaxy and its applications to microwave and optoelectronic devices"; Morris N. Liebmann Memorial Award, 1982. Other Awards: Electronics Division Award, Electrochemical Society, 1977; American Physical Society International Prize for New Materials, 1982; Distinguished Technical Staff Award, Bell Laboratories, 1982.

CHO, ZANG-HEE, Fellow 1983. Born: July 15, 1936, Seoul, Korea. Degrees: B.Sc.(E.E.), 1960, M.Sc.(E.E.), 1962, Seoul National University; Ph.D., 1966, Uppsala University.

Fellow Award "For contributions to computerized tomography and its application to positron emission, X-ray transmission, and nuclear magnetic resonance tomography." Other Awards: Distinguished Scientist Award, International Workshop on Physics and Engineering in Medical Imaging, 1982, Asilomar, California.

CHODOROW, MARVIN, IRE Fellow 1954. Born: July 16, 1913, Buffalo, N.Y. Degrees: B.A., 1934, University of Buffalo; Ph.D., 1939, Massachusetts Institute of Technology; L.L.D.(Hon.), 1972, University of Glasgow.

Fellow Award "For his contributions to the theory and design of klystron tubes"; W.R.G. Baker Award, IRE, 1962, for "A high-efficiency klystron with distributed interaction"; Lamme Medal, IEEE, 1982. Other Awards: Member, National Academy of Engineering,

1967; Member, National Academy of Sciences, 1971; American Academy of Arts and Sciences, 1972.

CHOW, C. K., Fellow 1971. Born: December 28, 1928, Hong Kong. Degrees: M.E.E., 1950, Ph.D., 1953, Cornell University.

Fellow Award "For contributions to pattern recognition."

CHOW, WOO F., Fellow 1969. Born: June 7, 1923, Shanghai, China. Degrees: B.S.E.E., 1945, Ta Tung University; M.S.E.E., 1949, Ph.D., 1952, University of Minnesota.

Fellow Award "For outstanding contributions to solid-state electronics." Other Awards: Outstanding Paper Award, I.S.S.C.C., 1960.

CHRISTALDI, PETER S., IRE Fellow 1952. Born: November 26, 1914, Philadelphia, Pa. Degrees: E.E., 1935, Ph.D., 1938, Rensselaer Polytechnic Institute.

Fellow Award "In recognition of his contributions in the field of cathode-ray devices." Other Awards: Fellow, Radio Club of America.

CHRISTIANSEN, DONALD D., Fellow 1981. Born: June 23, 1927, Plainfield, NJ. Degrees: B.E.E., 1950, Cornell University.

Fellow Award "For contributions to professional communication in electrical and electronics technology"; IEEE Executive Committee Commendation, 1979; IEEE Directors' Commendation, 1980; IEEE Directors' Commendation, 1982. Other Awards: Commendation for Outstanding Accomplishment, Council of Engineering and Scientific Society Executives, 1977; Recipient, Citation and Medal for the Advancement of Culture, Flanders Academy of Arts, Science, and Literature, 1980.

CHU, J. CHUAN, Fellow 1967. Born: July 14, 1919, Tientsin, China. Degrees: B.S., 1942, University of Minnesota; M.S., 1945, University of Pennsylvania.

Fellow Award "For contributions to electronic computation and data processing"; Computer Pioneer Award, IEEE, 1982. Other Awards: CIE Achievement Award, 1973.

CHU, TA-SHING, Fellow 1978. Born: July 18, 1934, Shanghai, China. Degrees: B.S.(E.E.), 1955, National Taiwan University; M.S., 1957, Ph.D., 1960, Ohio State University.

Fellow Award "For contributions to dual-polarization radio transmission, and to propagation of radio and light waves in precipitation." Other Awards: Distinguished Technical Staff Award, Bell Laboratories, 1983; Sigma Xi; Pi Mu Epsilon.

CHU, WESLEY W., Fellow 1978. Degrees: B.S., 1960, M.S., 1961, University of Michigan; Ph.D.(in Electrical Engineering), 1966, Stanford University.

Fellow Award "For contributions to multiplexing techniques of and file allocation in computers"; Distinguished Visitor, IEEE, 1973-74; Associate Editor, IEEE

Transactions on Computers, 1977-1982; Assoc. Editor, Journal of Computer Network; Assoc. Editor, Journal of Electrical Engineering and Computer Science; Meritorious Service Award, IEEE, 1982. Other Awards: National Lecturer, Association for Computing Machinery, 1972-73; Tau Beta Pi; Eta Kappa Nu; Phi Tau Phi; Sigma Xi.

CHUA, LEON O., Fellow 1974. Born: June 28, 1936, Tarlac, Philippines, of Chinese nationality. Degrees: B.S.E.E., 1959, Mapua Institute of Technology; S.M.E.E., 1962, Massachusetts Institute of Technology; Ph.D., 1964, University of Illinois.

Fellow Award "For contribution to nonlinear network theory"; Browder J. Thompson Memorial Prize Award (with R.A. Rohrer), 1967, for paper "On the Dynamic Equations of a Class of Nonlinear Networks"; W.R.G. Baker Prize Award, 1973, for paper "Memristor-The Missing Circuit Element"; IEEE Society on Circuits and Systems, Best Paper Prize, 1973. Other Awards: Frederick Emmons Terman Award, 1974; Outstanding Paper Award, Asilomar Conference on Circuits, Systems, and Computers, 1974; Miller Research Professor, 1976; Senior Visiting Fellow, Cambridge University, 1982; Alexander Humboldt Senior U.S. Scientist Award, 1983; Honorary doctorate (Doctor Honoris Causa), Ecole Polytechnique, Lausanne, Switzerland, 1983; Senior Visiting Fellowship, Japan Society for Promotion of Science, 1984.

CHYNOWETH, ALAN G., Fellow 1980. Born: November 18, 1927, Harrow, England. Degrees: B.S.(Physics), 1948, Ph.D.(Physics), 1950, King's College, London U., England.

Fellow Award "For development of innovative experimental techniques in semiconductor material and device physics"; W.R.G. Baker Prize Award, IEEE, 1967.

CISLER, WALKER L., AIEE Fellow 1947. Born: October 8, 1897, Marietta, Ohio. Degrees: M.E., 1922, Cornell University.

Awards: Edison Medal, IEEE, 1965, "For his achievements in the power industry, including the development of nuclear power; for his services to his country and to international understanding, including the application of electric power to economic growth among all nations; and for his broad contributions to the profession of engineering"; John Fritz Medal, IEEE, ASME, ASCE, AICE, AIMMPE, 1966, "For vision and accomplishment in the field of energy conversion and distribution; for skillful leadership in corporate and professional management; for civic awareness and personal contribution within his home community; for distinguished public service on an international scale." Other Awards: Fellow, American Institute of Management, 1952; George Westinghouse Gold Medal, American Society of Mechanical Engineers, 1954; Henry Gantt Gold Medal, American Society of Mechanical Engineers, American Management Association, 1955; Washington Award, Western Society of Engineers, 1957; George W. Goethals Award, Society of American Military Engineers, 1958; Award of Merit, American Institute of Consulting Engineers, 1960; Hoover Medal, American Society of Mechanical Engineers, 1962; William Metcalf Award, Engineers Society of Western Pennsylvania, 1963; Award, National Society of Professional Engineers, 1963; Honorary Member, American Society of Mechanical Engineers, 1966; Distinguished Service Member, Engineering Society of Detroit, 1969; Eminent Member, Eta Kappa Nu, 1972; Fellow, Society of American Military Engineers, 1973; Fellow, Engineering Society of Detroit, 1974; Horace H. Rackham Humanitarian Award, Engineering Society of Detroit, 1976; Centennial Award, American Society of Mechanical Engineers, 1980; Distinguished Service Award, Society of American Military Engineers, 1980.

CLADE, JACQUES J., Fellow 1983. Born: April 13, 1933, Paris, France. Degrees: Ecole Polytechnique, Ecole Superieure d'Electricite, France.

Fellow Award "For leadership and contributions in transmission line design, dc links, and power system operation and planning"; IEEE Centennial Medal, 1984. Other Awards: Medaille Blondel, 1971.

CLAPP, RICHARD G., IRE Fellow 1962. Born: March 3, 1911, Hanover, N.H. Degrees: B.S., 1931, M.S., 1932, University of Pennsylvania.

Fellow Award "For contribution to color television display systems."

CLARE, CARL P., AIEE Fellow 1952. Born: May 25, 1903. Degrees: B.S.E.E., 1927, D.Sc., 1962, University of Idaho; Dr.Eng., South Dakota School of Mines and Technology, 1970.

CLARK, CLAYTON, Fellow 1970. Born: March 9, 1912, Hyde Park, Utah. Degrees: B.S., 1933, Utah State University; E.E., 1946, Ph.D., 1957, Stanford University.

Fellow Award "For contributions to ionospheric research and engineering education"; Community Service Award, Utah Section, IEEE; IEEE Centennial Medal, 1984. Other Awards: Distinguished Service Award, Utah State University Alumni Association, 1980.

CLARK, E. GARY, Fellow 1970. Born: December 23, 1927. Degrees: B.S., 1948, College of William and Mary.

Fellow Award "For contributions in the fields of television receiver design, computer circuits, and advances in electronic data processing." Other Awards: Member, Association for Computing Machinery; Member, Research Society of America.

CLARK, JOHN F., Fellow 1976. Born: December 12, 1920, Reading, Pa. Degrees: B.S.E.E., 1942, Lehigh University; M.S.(Math.), 1946, George Washington University; E.E., 1947, Lehigh University; Ph.D.(Physics), 1956, University of Maryland.

Fellow Award "For technical leadership of space programs." Other Awards: Medal for Outstanding Leadership, NASA, 1966; Medal for Distinguished Service, NASA, 1969; Fellow, American Astronautical Society, 1970; Collier Trophy (corecipient), 1974; Fellow, American Institute of Aeronautics and Astronautics, 1976.

CLARK, LEMORE W., AIEE Fellow 1949. Born: April 9, 1901, Evansville, Wis. Degrees: B.S., 1923, University of Wisconsin.

CLARK, PETER O., Fellow 1980. Born: June 5, 1938, Ottawa, Canada. Degrees: B. Eng., 1960, McGill University; M.Sc., 1961, Ph.D., 1964, California Institute of Technology.

Fellow Award "For contributions and technical leadership in the development of high-energy laser technology"; IEEE Centennial Medal, 1984. Other Awards: Secretary of Defense Meritorious Service Medal; OSD Certificate for Outstanding Performance.

CLARK, RALPH L., IRE Fellow 1958. Born: June 2, 1908, East Jordan, Mich. Degrees: B.S.E.E., 1930, Michigan State University.

Fellow Award "For outstanding work in electronic countermeasures." Other Awards: Certificate of Commendation, U.S. Dept. of Navy.

CLARK, ROBERT N., Fellow 1983. Born: April 17, 1925, Ann Arbor, Michigan. Degrees: B.S.E.(EE), 1950, M.S.E.(EE), 1951, Michigan; Ph.D., 1969, Stanford.

Fellow Award "For contributions to engineering education and the practical application of control theory." Other Awards: Science Faculty Fellowship, National Science Foundation, 1966-68.

CLARK, TREVOR H., IRE Fellow 1960. Born: July 16, 1909, Haviland, Kans. Degrees: A.B., 1930, Friends University; M.S., 1933, University of Michigan.

Fellow Award "For contributions to military electronics." Other Awards: Certificate of Merit, Office of Scientific Research and Development, 1945; Certificate of Commendation, U.S. Dept. of Navy, 1947; Associate Fellow, American Institute of Aeronautics and Astronautics; Silver Certificate, Acoustical Society of America, 1966; Governor's Citation, State of Maryland, 1969.

CLARKE, DAVID D., AIEE Fellow 1933. Born: June 7, 1887, Grand Rapids, Mich. Degrees: 1910, University of Michigan.

CLARRICOATS, PETER J. B., Fellow 1968. Born: April 6, 1932, London, England. Degrees: B.Sc., 1953, Ph.D., 1958, D.Sc., 1968, London University.

Fellow Award "For contributions in the field of guided electromagnetic wave propagation." Other Awards: Electronic Division Premium, 1961-62, Coopers Hill Memorial Prize, 1964, Marconi Premium, Institution of Electrical Engineers, 1974; Fellowship of Engineering,

1983.

CLAVIN, ALVIN, Fellow 1968. Born: June 17, 1924, Los Angeles, Calif. Degrees: B.S.E.E., 1947, M.S.E.E., 1954, University of California at Los Angeles.

Fellow Award "For contributions to the development of microwave and antenna technique and the field of technical management"; President, IEEE Microwave Theory and Techniques Society, 1972; Distinguished Service Certificate, IEEE Microwave Theory and Techniques Society, 1977; Chairman, Steering Committee, IEEE Microwave Theory and Techniques Society Symposium, 1981; Certificate of Recognition, IEEE MTT-S, 1982; Distinguished Service Award, IEEE MTT-S, 1984; IEEE Centennial Medal, 1984. Other Awards: Fellow, Institute for Advancement of Engineering, 1974.

CLAYTON, J. PAUL, AIEE Fellow 1936. Born: October 3, 1888, Sterling, Ill. Degrees: B.E., 1909, Tulane University; M.E., 1911, University of Illinois; D.E., 1942, Tulane University.

Other Awards: American Society of Mechanical Engineers; Sigma Xi.

CLAYTON, JAMES M., JR., Fellow 1967. Born: July 18, 1920, Columbus, Ga. Degrees: B.S., 1944, University of New York; B.S., 1947, Auburn University.

Fellow Award "For his contributions to the design of lightning protection equipment for electric power systems, and the development of guides for application to transmission lines and substations"; IEEE Centennial Medal, 1984. Other Awards: Service Award, National Fire Protection Association, 1966.

CLAYTON, ROBERT J., Fellow 1976. Born: October 30, 1915. Degrees: B.A.(Natural Sciences Tripos), 1937, Cambridge University; D.Sc.(Hon.), University of Aston, Birmingham, U.K.; D.Sc.(Hon.), University of Salford, U.K.; D.Sc.(Hon.), City University, London, U.K.

Fellow Award "For contributions to airborne microwave system design." Other Awards: Radio Section Premium, John Hopkinson Premium, Institution of Electrical Engineers, U.K.

CLEETON, CLAUD E., IRE Fellow 1955. Born: December 11, 1907, Missouri. Degrees: B.S., 1928, Northeast Missouri State Teachers College; M.S., 1930, University of Missouri; Ph.D., 1935, University of Michigan.

Fellow Award "For his contributions to microwave spectroscopy and electronic identification systems." Other Awards: Presidential Certificate of Merit, 1946; Distinguished Civilian Service Award, U.S. Navy, 1969.

CLEWELL, DAYTON H., IRE Fellow 1954. Born: December 15, 1912, Berwick, Pa. Degrees: B.S., 1933, Ph.D., 1936, Massachusetts Institute of Technology.

Fellow Award "For his research in the application of instrumentation to petroleum exploration and production." Other Awards: Environmental Conservation Distinguished Service Award, AIMME, 1974; National Academy of Engineering, 1976.

CLINGERMAN, B.H., AIEE Fellow 1944.

COALES, JOHN F., Fellow 1968. Born: September 14, 1907, Birmingham, England. Degrees: B.A., 1929, M.A., 1934, Cambridge University; D.Sc.(Hon.), 1970, City University; D.Tech.(Hon.), 1977, Loughborough; D.Eng.(Hon.), 1978, Sheffield.

Fellow Award "For his outstanding contributions in the field of electronics and control engineering, and particularly for major contributions to the development of automation and computer interests in the United Kingdom." Other Awards: OBE, 1945; Fellow, Royal Society, 1970; Harold Hartley Medal, 1971; Honorary Member, Institute of Measurement and Control, 1971; Honorary Fellow, Hatfield Polytechnic, 1971; CBE, 1974; Founder Member, U.K. Fellowship of Engineering, 1976; Emeritus Fellow, Clare Hall, Cambridge, England; Emeritus Professor of Engineering, University of Cambridge; Foreign Member Serbian Academy of Sciences, 1981; Giorgio Quazza Medal, International Federation of Automatic Control, 1981; Honda Prize, 1982.

COATES, CLARENCE L., JR., Fellow 1974. Born: November 5, 1923, Hastings, Nebr. Degrees: B.S.E.E., 1944, M.S., 1948, University of Kansas: Ph.D., 1953, University of Illinois.

Fellow Award "For contributions to switching theory and leadership in engineering research and education." Other Awards: Fellow, American Association for the Advancement of Science, 1980.

COATES, ROBERT J., Fellow 1971. Born: May 8, 1922, Lansing, Mich. Degrees: B.S.E.E., 1943, Michigan State University; M.S.E.E., 1948, University of Maryland; Ph.D.(Physics), 1957, The Johns Hopkins University.

Fellow Award "For contributions in millimeter-wave radio astronomy, and developments in aerospace technology." Other Awards: Tau Beta Pi, 1943; Phi Kappa Phi, 1943; Sigma Xi, 1957; Outstanding Performance Award, U.S. Naval Research Laboratory, 1959; Exceptional Performance Award, Goddard Space Flight Center, 1971.

COCKRELL, WILLIAM D., AIEE Fellow 1955. Born: September 2, 1906, Tallahassee, Fla. Degrees: B.S.E.E., 1928, E.E., 1951, University of Florida.

Fellow Award "For contributions to the advancement of electric techniques, resulting in practical industrial and control systems." Other Awards: Charles A. Coffin Award, General Electric, 1942; Outstanding Community Service Award, Elfun Society (General Electric), 1978; 50 patents.

COGBILL, BELL A., AIEE Fellow 1954. Born: March 18, 1909, Marianna, Ark. Degrees: B.S.E.E., 1931,

M.S., 1936, University of Tennessee.

Fellow Award "For contributions to a better understanding of the factors encountered in designing large transformers, and to the advanced technical education of young engineers."

COGGESHALL, IVAN S., IRE Fellow 1942, AIEE Fellow 1948. Born: September 30, 1896, Newport, R.I. Degrees: D.Eng.(Hon.), 1951, Worcester Polytechnic Institute.

IRE Fellow Award "For his contributions to the welfare of the Institute and the engineering profession"; IEEE Communication Systems Group's Achievement Award for "Leadership in the integration of wire and radio media," 1963; The Haraden Pratt Award, 1978.

COHEN, ARNOLD A., Fellow 1964. Born: August 1, 1914, Duluth, Minn. Degrees: B.E.E., 1935, M.S., 1938, Ph.D., 1947, University of Minnesota.

Fellow Award "For pioneering achievement on computers and storage devices and sustained service to the profession in this field." Other Awards: Valuable Invention Citation, American Patent Law Association and Minnesota Patent Law Association, 1962.

COHEN, MELVIN I., Fellow 1984. Born: June 25, 1936, New York, N.Y. Degrees: S.B.(M.E.), 1957, S.M.(M.E.), 1958, M.I.T.; Ph.D.(mechanics), 1964, Rensselaer Polytechnic Institute.

Fellow Award "For leadership in research, development, and implementation of lightguides for optical communications and lasers for use as a tool in materials processing."

COHEN, ROBERT M., Fellow 1967. Born: February 2, 1921. Degrees: B.S.M.E., 1949, Stevens Institute of Technology.

Fellow Award "For significant contributions to the application of electron tubes and semiconductor devices in radio and television receivers." Other Awards: Award of Merit, Radio Corporation of America, 1957.

COHN, MARVIN, Fellow 1974. Born: September 25, 1928. Degrees: B.S.E.E., 1950, M.S.E.E., 1951, Illinois Institute of Technology; D.Eng., 1960, The Johns Hopkins University.

Fellow Award "For contributions to millimeter wave technology, low-noise microwave integrated circuitry, electromagnetic surface wave excitation, and transmission." Other Awards: Eta Kappa Nu; Tau Beta Pi; Sigma Xi.

COHN, NATHAN, AIEE Fellow 1954. Born: January 2, 1907, Hartford, Conn. Degrees: S.B., 1927, Massachusetts Institute of Technology; D.Eng(Hon.), 1976, Rensselaer Polytechnic Institute.

Fellow Award "For his contribution to the art of load and frequency control of large power systems"; Lamme Medal, IEEE, 1968, "For meritorious achievement in the field of automatic control of power generation and frequency for interconnected electric power systems";

Edison Medal, IEEE 1982. Other Awards: John Price Wetherill Medal, Franklin Institute, 1968; Albert F. Sperry Medal, ISA, 1968; Pennsylvania Engineer of the Year, 1969; National Academy of Engineering, 1969; Honorary Member, Instrument Society of America, 1976; Association Award, Scientific Apparatus Makers, 1977.

COHN, SEYMOUR B., IRE Fellow 1959. Born: October 21, 1920, Stamford, Conn. Degrees: B.E., 1942, Yale University; M.S., 1946, Ph.D., 1948, Harvard University.

Fellow Award "For contributions to the theory and design of microwave components"; Microwave Prize, IEEE Microwave Theory and Techniques Society, 1964, "For many significant contributions to the field of endeavor of the IEEE GMTT in numerous published papers"; Lamme Medal, IEEE, 1974, "For outstanding contributions to the theory and practice of microwave component design"; Career Award, IEEE Microwave Theory and Techniques Society, 1980. Other Awards: Annual Award, Yale Engineering Association, 1954.

COLBORN, HARRY W., Fellow 1983. Born: May 27, 1921, Pittsburgh, Pa. Degrees: B.S.E.E., 1951, Carnegie-Mellon University; Grad., 1958, Oak Ridge School of Reactor Technology.

Fellow Award "For contributions to the planning of reliable interconnected high-voltage transmission systems."

COLBY, PAUL S., AIEE Fellow 1958. Born: October 24, 1905, Junction City, Kans. Degrees: B.S.E.E., 1929, Kansas State University.

Fellow Award "For contributions to planning and execution of power transmission and distribution"; Chairman, Wichita Section, IEEE, 1940; Chairman, East Carolina Subsection, IEEE; Chairman, North Carolina Section, IEEE, 1963. Other Awards: Phi Kappa Phi, Kansas State University, 1929; Faculty Scholarship Gold Medal, Kansas State University, 1929.

COLCLASER, R. GERALD, JR., Fellow 1976. Born: September 21, 1933, Wilkinsburg, Pa. Degrees: B.S.E.E., 1956, University of Cincinnati; M.S.E.E., 1961, D.Sc., 1968, University of Pittsburgh.

Fellow Award "For contributions to high-power circuit breaker technology."

COLDREN, LARRY A., Fellow 1982. Born: January 1, 1946, Mifflintown, PA. Degrees: B.A. and B.S.(E.E.and Physics), 1968, Bucknell University; M.S.E.E., 1969, Ph.D.(E.E.), 1972, Stanford University.

Fellow Award "For contributions to surface acoustic wave resonator filters, long delay lines, and monolithic acoustoelectric signal processing devices"; Best Paper Award, IEEE Sonics and Ultrasonics Society, 1979. Other Awards: 15 Patents; Phi Beta Kappa; Tau Beta Pi; Pi Mu Epsilon; Sigma Pi Sigma.

COLE, RALPH I., IRE Fellow 1958. Born: August 17, 1905, St. Louis, Mo. Degrees: B.S.E.E., 1927, Washington University; M.S. (Physics), 1936, Rutgers University.

Fellow Award "For his contributions to military engineering management"; First Honorary Life Member, IEEE Engineering Management Society, 1979; IEEE Centennial Medal, 1984. Other Awards: Legion of Merit, U.S. Air Force, 1945.

COLEMAN, PAUL D., IRE Fellow 1962. Born: June 4, 1918, Stoystown, Pa. Degrees: A.B., 1940, D.Sc.(Hon.), 1978, Susquehanna University; M.S., 1942, Pennsylvania State University; Ph.D., 1951, Massachusetts Institute of Technology.

Fellow Award "For contributions to millimeter and submillimeter waves." Other Awards: Meritorious Civilian Award, A.A.F., 1946; Fellow, Optical Society of America; American Physical Society; Alumni Achievement Medal, Susquehanna University, 1980; Sigma Xi; Pi Mu Epsilon; Eta Kappa Nu.

COLLIN, ROBERT E., Fellow 1972. Born: October 24, 1928, Canada. Degrees: B.Sc., 1951, University of Saskatchewan; Ph.D., 1954, University of London.

Fellow Award "For contribution to microwave engineering and to education." Other Awards: Junior Technical Achievement Award, Cleveland Technical Societies Council, 1964.

COLLINS, ARTHUR A., IRE Fellow 1953. Born: September 9, 1909, Kingfisher, Okla. Degrees: Sc.D.(Hon.), 1954, Coe College; D.Eng.(Hon.), 1968, Brooklyn Polytechnic Institute; D.Eng.(Hon.), 1973, Southern Methodist University; Sc.D.(Hon.), 1974, Mount Mercy College.

Fellow Award "For his contributions to the art of transmitter design and precision remote tuning systems, and for his farsighted engineering direction"; Armstrong Achievement Award, IEEE, 1979. Other Awards: Distinguished Service Award, U.S. Dept. of Navy, 1962; Distinguished Service Award, Iowa Broadcasters Association, 1966; Distinguished Service Award, University of Iowa, 1967; National Academy of Engineering, 1968; Outstanding Achievement Award, American Electronics Industries, 1977; Armstrong Medal, Radio Club of America, 1977; Quadrato della Radio Association, Guglielmo Marconi Foundation, 1978; David Sarnoff Award, Armed Forces Communications and Electronics Association, 1979; Medal of Honor; Electronic Industries Association, 1980.

COLLINS, JEFFREY H., Fellow 1980. Born: April 22, 1930, Luton, England. Degrees: B.Sc., 1951, M.Sc., 1954, University of London.

Fellow Award "For contributions to the field of magnetic and surface acoustic wave delay line technology, and to engineering education." Other Awards: Bulgin Premium, Institute of Electronics and Radio Engineers,

UK, 1974, 1977; Marconi Premium, Institute of Electronics and Radio Engineers, UK, 1976; Hewlett-Packard Europhysics Prize, 1979.

COLLINS, ROBERT J., Fellow 1969. Born: July 23, 1923, Philadelphia, Pa. Degrees: B.A., 1947, M.S., 1948, University of Michigan; Ph.D., 1953, Purdue University.

Fellow Award "For contributions to solid-state optics; in particular, optics and lasers."

COLTMAN, JOHN W., IRE Fellow 1955. Born: July 19, 1915, Cleveland, Ohio. Degrees: B.S., 1937, Case Institute of Technology; M.S., 1939, Ph.D., 1941, University of Illinois.

Fellow Award "For his contributions to the fields of microwave techniques, x-ray applications and nuclear studies." Other Awards: Fellow, American Physical Society, 1954; Edward Longstreth Medal, Franklin Institute, 1960; Order of Merit, Westinghouse, 1968; Roentgen Medal, Remscheid, Germany, 1970; National Academy of Engineering, 1976; Gold Medal, Radiological Society of North America, 1982.

COMPTON, RALPH T., JR., Fellow 1984. Born: July 26, 1935, St. Louis, MO. Degrees: S.B., 1958, M.I.T.; M.Sc., 1961, Ph.D., 1964, Ohio State Univ. (all in E.E.).

Fellow Award "For contributions to adaptive array theory and use of adaptive arrays in communication systems"; M. Barry Carlton Award, IEEE Aerospace and Electronic Systems Society, 1983. Other Awards: Battelle Memorial Institute Graduate Fellow, 1961-62; National Science Foundation Postdoctoral Fellow, 1967-1968; Ohio State University Electroscience Laboratory Outstanding Paper Awards, 1978, 1980, 1982; Senior Research Award, Engineering College, Ohio State University, 1983.

COMSTOCK, RICHARD L., Fellow 1975. Born: August 9, 1932, Butte, Mont. Degrees: B.S.E.E., 1956, M.S.E.E., 1957, University of California; Ph.D., 1962, Stanford University.

Fellow Award "For developments in magnetic storage and magnetic microwave devices."

CONCORDIA, CHARLES, AIEE Fellow 1947. Born: June 20, 1908, Schenectady, N.Y. Degrees: D.Sc., 1971, Union College.

Awards: Lamme Medal, AIEE, 1961, "For meritorious achievements in the design of electrical machinery; more specifically, for analyses of synchronous machine characteristics leading to improved designs and for exceptional contributions to the application and control of machines used in electric power systems." Other Awards: Charles A. Coffin Award, General Electric, 1942; Fellow, American Society of Mechanical Engineers, 1957; Engineer of the Year Award, Schenectady Professional Engineers Society, 1963; Steinmetz Award, General Electric, 1973; Fellow, American Association for the Advancement of Science; Member,

National Society of Professional Engineers; National Academy of Engineering, 1978; Professional Engineer, New York and Florida; Founding Member, Association for Computing Machinery; Member, Conference Internationale des Grands Reseaux Electriques.

CONE, DONALD I., AIEE Fellow 1942. Born: July 30, 1891, Eureka, Calif. Degrees: B.S.E.E., 1913, University of California at Berkeley.

CONRAD, ALBERT G., AIEE Fellow 1949. Born: May 19, 1902, Norwalk, Ohio. Degrees: B.E.E., 1925, M.S., 1927, Ohio State University; E.E., 1931, Yale University.

Other Awards: Fellow, American Association for the Advancement of Science.

CONRAD, EDWARD E., Fellow 1981. Born: June 11, 1927. Degrees: B.A., 1950, M.S., 1955, Ph.D., 1970, University of Maryland.

Fellow Award "For technical leadership in the understanding and characterization of trasient nuclear radiation effects on insulators."

CONVERTI, VINCENZO, Fellow 1982. Born: November 27, 1925, Roseto, Italy. Degrees: B.S.E.E., 1952, M.S.E.E., 1956, University of Arizona.

Fellow Award "For contributions to the design, planning, and operational functions of integrated power systems by the creation and use of new concepts in computer applications."

CONWELL, ESTHER M., Fellow 1980. Born: May 23, 1922. Degrees: B.A., 1942, City University of New York; M.S., 1945, University of Rochester; Ph.D., 1948, University of Chicago.

Fellow Award "For contributions to semiconductor theory, particularly transport in both low and high electric fields." Other Awards: National Academy of Engineering, 1980.

COOK, CHARLES E., Fellow 1972. Born: October 27, 1926, New York, N.Y. Degrees: S.B., 1949, Harvard College; M.E.E., 1954, Polytechnic Institute of Brooklyn.

Fellow Award "For contributions to signal processing theory and radar design." Other Awards: Sigma Xi, 1960.

COOK, GERALD, Fellow 1984. Born: October 31, 1937, Hazard, KY. Degrees: B.S., 1961, Virginia Polytechnic Institute and State University; S.M., 1962, E.E., 1963, Sc.D., 1965, M.I.T.

Fellow Award "For contributions to the practical application of modern control theory"; IEEE Centennial Medal, 1984. Other Awards: Dept. of Army Certificate of Achievement, 1981; Outstanding Research Award, Southeastern Section, ASEE, 1971; Technical Achievement Award, USAF Office of Aerospace Research, 1968; Junior Officer of the Year, F.J. Seiler Research Laboratory, USAF Office of Aerospace Research, 1968.

COOKE, WILLIAM T., Fellow 1965. Born: November 3, 1903, Oberlin, Ohio. Degrees: B.A., 1925, M.A., 1926, University of Michigan.

Fellow Award "For pioneering in the development and use of microwave klystrons." Other Awards: Epsilon Chi, 1930; Sigma Xi, 1931.

COOKSON, ALAN H., Fellow 1983. Born: July 3, 1939, London, England. Degrees: B.Sc.(Eng.), Electrical Engineering(First Class Honors) 1961, Queen Mary College, London, Ph.D., 1965, Queen Mary College, London.

Fellow Award "For contributions to the understanding of high-voltage compressed gas breakdown and the development of new higher voltage compressed gas transmission line systems."

COOKSON, ALBERT E., Fellow 1969. Born: October 30, 1921, Needham, Mass. Degrees: B.S.E.E., 1943, Northeastern University; M.S.E.E., 1951, Massachusetts Institute of Technology; Sc.D.(Hon.), 1974, Gordon College.

Fellow Award "For contributions in high-speed data transmission and processing for military command-control, and for application of communication satellites to provide worldwide military communications."

COOLEY, AUSTIN G., IRE Fellow 1955. Born: February 9, 1900, Seattle, Wash. Degree: E.E., 1973, University of Alaska.

Fellow Award "For his contributions to facsimile transmission methods." Other Awards: Certificate of Appreciation, War Dept., 1946; Marconi Medal, Veteran Wireless Operators Association, 1956; Scroll, Air Transport Association, 1965; 1982 Nevada Inventor of the Year Appreciation Award, Nevada Innovation & Technology Council.

COOLEY, JAMES W., Fellow 1981. Born: September 18, 1926. Degrees: B.A.(Arts), 1949, Manhattan College; M.A.(Math.), 1951, Ph.D.(Applied Math.), 1961, Columbia University.

Fellow Award "For development of the fast Fourier transform."

COOPER, ANDREW R., Fellow 1976. Born: October 1, 1902, Rotherham, Yorkshire, England. Degrees: Assoc.Eng., 1922, Sheffield University; M.Eng.(honoris causa), 1950, Liverpool University.

Fellow Award "For contributions to high-voltage network transmission"; Meritorious Service Award, IEEE Power Engineering Society, 1972. Other Awards: John Hopkinson Award, 1948, Willans Medal, 1950, Faraday Lecturer, 1952-1954, Institution of Electrical Engineers; Thornton Medal, Association of Mining and Mechanical Engineers, 1954; Past President, CIGRE, 1966-1972; Bernard Price Memorial Lectures, South Africa, 1970, S.F. Inst.E., S.F. Inst. of Elect. Eng., F. Eng.; C.B.E.

COOPER, FRANKLIN S., Fellow 1969. Born: April 29, 1908, Robinson, Ill. Degrees: B.S., 1931, University of Illinois; Ph.D., 1936, Massachusetts Institute of Technology; D.Sc.(Hon.), 1976, Yale University.

Fellow Award "For studies in the correlation of speech units from the physical, physiological, psychological and phonetic aspects"; Pioneer in Speech Communication (IEEE), 1972, "For his discoveries and leadership in fundamental and applied research on speech analysis-synthesis and speech perception." Other Awards: Sigma Xi; Fellow, Acoustical Society of America; Honors of the Association, ASHA; President's Certificate of Merit, 1948; Fellow, Calhoun College, 1971; Warren Medal, Society of Experimental Psychology, 1975; Silver Medal in Speech Communication, Acoustical Society of America, 1975; Fletcher-Stevens Award, Brigham Young University, 1977.

COOPER, GEORGE R., Fellow 1969. Born: November 29, 1921, Connersville, Ind. Degrees: B.S.E.E., 1943, M.S.E.E., 1945, Ph.D., 1949, Purdue University.

Fellow Award "For contributions to electrical engineering education, and research in communication theory."

COOPER, H. WARREN, III, Fellow 1970. Born: July 19, 1920, Chicago, Ill. Degrees: B.S.E.E., 1947, New Mexico State University; M.S.E.E., 1948, Stanford University.

Fellow Award "For contributions to antenna development, microwave integrated circuit development, and the application of microwave techniques to aircraft landing systems."

COOPER, MARTIN, Fellow 1976. Born: December 26, 1928, Chicago, Ill. Degrees: B.S.E.E., 1950, M.S.E.E., 1957, Illinois Institute of Technology.

Fellow Award "For contributions to radiotelephony"; IEEE Committee on Information and Computer Policy; Past Chairman IEEE Vehicular Technology Society, Chicago Section; President, IEEE Vehicular Technology Society; IEEE Centennial Medal, 1984. Other Awards: Fellow, Radio Club of America; Advisory Board, U. of Ill., Chicago and Champaign; President's Council, Illinois Institute of Technology.

COPELAND, JOHN A., Fellow 1983. Born: February 6, 1941, Atlanta, Georgia. Degrees: B.S., 1962, M.S., 1963, Ph.D., 1965, Georgia Institute of Technology.

Fellow Award "For contributions to the development of optically coupled semiconductor logic circuits"; Morris N. Liebmann Award, IEEE, 1970; Best Paper Award, International Solid State Circuits Conf., 1967.

COPPOLA, ANTHONY, Fellow 1982. Born: July 14, 1935, Brooklyn, NY. Degrees: B.A.(Physics), 1956, M.S.(Eng.Admin.), 1966, Syracuse University.

Fellow Award "For contributions to improving the reliability and maintainability of electronic systems and components." Other Awards: P. K. McElroy Award for

Best Paper, Reliability and Maintainability Symposium, 1979; Air Force Scientific Achievement Awards, 1978, 1981.

CORBATO, FERNANDO J., Fellow 1975. Born: July 1, 1926, Oakland, Calif. Degrees: B.S., 1950, California Institute of Technology; Ph.D., 1956, Massachusetts Institute of Technology.

Fellow Award "For contributions to the development of multiple-access computer systems"; W. W. McDowell Award, IEEE Computer Group, 1966, "For his pioneering work in organizing and spearheading the early development of the first practical large-scale time-sharing mputer system, and for his tireless efforts in providing direction for the entire time-sharing concept"; Harry Goode Memorial Award, American Federation of Information Processing Societies, 1980; The Computer Pioneer Award, IEEE Computer Society, 1982.

CORFIELD, R. J., AIEE Fellow 1951. Born: October 24, 1898, Mercur, Utah. Degrees: 1915, Hawley School of Engineering; M.E., 1916, Bliss Electrical School.

CORNELL, LLOYD P., JR., Fellow 1970. Born: May 27, 1920, San Francisco, Calif. Degrees: B.S.E.E., 1942, University of California at Berkeley.

Fellow Award "For leadership in the engineering of a large utility communication network."

CORRINGTON, MURLAN S., IRE Fellow 1954. Born: May 26, 1913, Bristol, S.D. Degrees: B.S., 1934, South Dakota School of Mines and Technology; M.Sc., 1936, Ohio State University.

Fellow Award "For mathematical analysis of frequency modulation, and of transients in networks and loudspeakers"; Award, Professional Group on Audio, 1958, 1968; IEEE Philadelphia Section Award, 1974, "For contributions to mathematical analysis of circuits and systems and for service to the IEEE"; IEEE Solid State Circuits Council, 1975, "For many years of distinguished service to the International Solid-State Circuits Conference." Other Awards: Fellow, Acoustical Society of America, 1957.

CORRY, ANDREW F., Fellow 1974. Born: October 28, 1922, Lynn, Mass. Degrees: B.S., 1947, Massachusetts Institute of Technology; AMP, 1966, Harvard University.

Fellow Award "For contributions to underground transmission technology"; William M. Habirshaw Award, IEEE, 1983; IEEE Centennial Medal, 1984. Other Awards: National Academy of Engineering, 1978.

CORTELLI, JOHN A., AIEE Fellow 1962. Born: September 6, 1905, Italy. Degrees: B.S., 1928, Case Institute of Technology.

Fellow Award "For contributions to development and manufacture of industrial control devices."

CORY, WILLIAM E., Fellow 1971. Born: April 5, 1927, Dallas, Tex. Degrees: B.S.E.E., 1950, Texas A&M University; M.S., 1959, University of California at Los Angeles.

Fellow Award "For contributions in the fields of electromagnetic compatibility and systems analysis"; Certificate of Acknowledgement, IEEE Electromagnetic Compatibility Society, 1975; Certificate of Appreciation, IEEE Electromagnetic Compatibility Society, 1976; Lawrence G. Cummings Award, IEEE Electromagnetic Compatibility Society, 1983. Other Awards: Special Service Award, U.S. Air Force Security Service, 1957.

COSBY, LYNWOOD A., Fellow 1974. Born: June 11, 1928, Richmond, Va. Degrees: B.S., 1949, University of Richmond; M.S., 1951, Virginia Polytechnic Institute.

Fellow Award "For leadership in electronic systems concepts and technology." Other Awards: Sigma Xi Thesis Award, 1951; Distinguished Civilian Service Award, U.S. Navy, 1958; ASNE Gold Medal Award, 1968; Dept. of Defense Distinguished Civilian Service Award, 1974.

COSENTINO, ADOLFO T., IRE Fellow 1940. Born: November 18, 1898, Buenos Aires, Argentina. Degrees: National School, Argentina.

COSTAIN, CECIL C., Fellow 1981. Born: June 16, 1922, Ponoka, Alberta, Canada. Degrees: B.A., 1941, B.A.(Honors), 1946, M.A., 1947, University of Saskatchewan; Ph.D., 1951, University of Cambridge.

Fellow Award "For leadership in the development of primary frequency standards and two-way time transfer techniques via geostationary satellites." Other Awards: Fellow, Royal Society of Canada, 1974; President, Canadian Association of Physicists, 1980; D.S.C. 1945.

COSTAS, JOHN P., Fellow 1965. Born: September 16, 1923, Wabash, Ind. Degrees: B.S.E.E., 1944, M.S.E.E., 1947, Purdue University; Sc.D., 1951, Massachusetts Institute of Technology.

Fellow Award "For contributions to communications theory and techniques."

COSTRELL, LOUIS, Fellow 1968. Born: June 26,-1915, Bangor, Me. Degrees: B.S., 1939, University of Maine; M.S., 1949, University of Maryland.

Fellow Award "For outstanding contributions to the development and standardization of nuclear instruments"; Award, IEEE, 1972, "For outstanding achievement through standards"; IEEE Harry Diamond Memorial Award, 1975, "For outstanding achievements in nuclear radiation measurement techniques and related instrumentation standardization"; Merit Award, IEEE Nuclear and Plasma Sciences Society, 1975, "In recognition of outstanding contributions in the field of nuclear science." Other Awards: Meritorious Service Silver Medal Award, U.S. Dept. of Commerce, 1955; Special Service Award, U.S. Dept. of Commerce, 1963; Distinguished Achievement Gold Medal Award, U.S. Dept. of Commerce, 1968; Edward Bennett Rosa Award, National Bureau of Standards, 1979.

COTELLESSA, ROBERT F., Fellow 1977. Born: June 7, 1923, Passaic, N.J. Degrees: M.E., 1944, M.S.(Math. and Physical Sc.), 1949, Stevens Institute of Technology; Ph.D., 1962, Columbia University.

Fellow Award "For contributions to engineering leadership and education"; Citation of Honor, U.S. Activities Board, IEEE, 1979; IEEE Centennial Medal, 1984. Other Awards: Fellow, American Association for the Advancement of Science, 1980.

COTSWORTH, A., III, Fellow 1967. Born: February 9, 1919, Oak Park, Ill. Degrees: B.S.E.E., 1941, Cornell University.

Fellow Award "For technical contributions to the development of radio and television receivers and ultra-high-frequency television."

COTTONY, HERMAN V., IRE Fellow 1962. Born: March 27, 1909, Nizhni-Novgorod, Russia. Degrees: B.S., 1932, Cooper Union; M.S., 1933, Columbia University; E.E., 1946, Cooper Union.

Fellow Award "For contributions to antenna research and measurement." Other Awards: Meritorious Service Award, U.S. Dept. of Commerce, 1961; Outstanding Publication Awards, U.S. Department of Commerce, National Bureau of Standards, 1959, 1965, 1972.

COULTER, WALLACE H., Fellow 1984. Born: Little Rock, AR. Degrees: Sc.D.(Hon.), 1979, Westminister College, Mo.; D.Eng.(Hon.), 1979, University of Miami, Fl.; Sc.D.(Hon.), 1979, Clarkson College, N.Y.

Fellow Award "For developments in automated instrumentation for clinical hematology and contributions to the technology of cytological instrumentation"; Morris E. Leeds Award, IEEE, 1980. Other Awards: John Scott Award, City of Philadelphia, 1960; Sigma Xi.

COVER, THOMAS M., Fellow 1974. Born: August 7, 1938, Pasadena, Calif. Degrees: B.S., 1960, Massachusetts Institute of Technology; M.S., 1961, Ph.D., 1964, Stanford University.

Fellow Award "For contributions to pattern recognition, learning theory, and information theory"; IEEE Information Theory Group, Outstanding Paper Award, 1972-1973; President, IEEE Information Theory Group, 1972; Board of Governors, IEEE Information Theory Group. Other Awards: Fellow, Institute of Mathematical Statistics, 1981.

COX, C. RUSSELL, Fellow 1969. Born: March 18, 1916, Chicago, Ill. Degrees: B.S., 1937, M.Sc., 1939, M.B.A., 1950, University of Chicago.

Fellow Award "For contributions in antenna and transmission-line design, and leadership in electronic industry standardization."

COX, DONALD C., Fellow 1979. Born: November 22, 1937, Lincoln, Nebr. Degrees: B.S., 1959, M.S., 1960, University of Nebraska at Lincoln; Ph.D., 1968, Stanford University; D.Sc.(Hon.), 1963, University of Nebraska at Lincoln.

Fellow Award "For contributions to the understanding of radio propagation effects in mobile telephone and satellite communications systems." Other Awards: Prize Guglielmo Marconi, 1983; Sigma Xi; Sigma Tau; Eta Kappa Nu; Pi Mu Epsilon.

COX, FRANCIS A., AIEE Fellow 1951. Born: May 8, 1907, Braidwood, Ill. Degrees: B.S.E.E., 1928, University of Illinois.

COX, HENRY, Fellow 1983. Born: March 7, 1935, Philadelphia, PA. Degrees: B.S., 1956, Holy Cross College; Sc.D., 1963, Massachusetts Institute of Technology.

Fellow Award "For technical leadership in underwater research and development." Other Awards: Gold Medal, ASNE, 1979; Fellow, ASA, 1975.

COZZENS, BRADLEY, AIEE Fellow 1948. Born: November 20, 1903, San Jose, Calif. Degrees: B.S., 1925, University of the Pacific; M.S.E.E., 1927, Stanford University.

Other Awards: Sigma Xi, 1927; James D. Donovan Personal Achievement Award, American Public Power Association (APPA), 1969.

CRAGON, HARVEY G., Fellow 1983. Born: April 21, 1929, Ruston, Louisiana. Degrees: B.S.E.E., 1950, Louisiana Polytechnic Institute.

Fellow Award "For the development of computer architectures suitable for state-of-the-art integrated circuit fabrication."

CRAIG, PALMER M., IRE Fellow 1954. Born: January 29, 1904, Cherry Hill, Md. Degrees: B.S.E.E., 1927, University of Delaware.

Fellow Award "In recognition of his leadership in the design of radio and television receivers." Other Awards: Naval Ordnance Award, 1945; Certificate of Commendation, U.S. Dept. of Navy, 1947.

CRAIN, CULLEN M., IRE Fellow 1958. Born: September 10, 1920, Goodnight, Tex. Degrees: B.S.E.E., 1942, M.S.E.E., 1947, Ph.D., 1952, University of Texas.

Fellow Award "For his contributions to microwave atmospheric refractometry." Other Awards: National Academy of Engineering, 1980.

CRANE, CARL C., AIEE Fellow 1950. Born: September 10, 1901, Hillsdale, Mich. Degrees: B.S.E.E., 1922, University of Michigan.

Membership Chairman, Washington DC Section, AIEE, 1943-44; Section Chairman, Madison Wisconsin Section, AIEE, 1948-49; Chairman, Great Lakes District Meeting, Madison, Wisconsin, 1951. Other Awards: Fellow, American Society of Civil Engineers, 1950; Eta Kappa Nu, 1958; Registered Professional Engineer, 1932-82.

CRANE, HEWITT D., IRE Fellow 1962. Born: April 27, 1927, Jersey City, N.J. Degrees: B.S., 1947, Columbia University; Ph.D., 1960, Stanford University.

Fellow Award "For pioneering research in the field of computer techniques, particularly in the areas of all-magnetic digital logic and neuristor logic", IRE Award, 1962, "For pioneering research in the field of computer techniques." Other Awards: IR-100 Award, INDUSTRIAL RESEARCH magazine, 1974, 1976; Award for Scientific Achievement, NASA, 1970.

CRANE, ROBERT K., Fellow 1980. Born: December 9, 1935, Worcester, MA. Degrees: B.S.(E.E.), 1957, M.S.(E.E.), 1959, Ph.D.(E.E.), 1970, Worcester Polytechnic Institute.

Fellow Award "For contributions to satellite communications."

CRAPUCHETTES, PAUL W., IRE Fellow 1960. Born: February 12, 1917, San Francisco, Calif. Degrees: B.S., 1940, University of California at Berkeley; Graduate, Advanced Course in Engineering (3 yrs.) General Electric, Schenectady, N.Y.

Fellow Award "For contributions to microwave tube development." Other Awards: Tau Beta Pi; Honorary Member, Cal Alpha, University of California at Berkeley, 1965.

CRAWFORD, ARTHUR B., IRE Fellow 1952. Born: February 26, 1907, Graysville, Ohio. Degrees: B.S.E.E., 1928, Ohio State University.

Fellow Award "In recognition of his contributions in the field of high-frequency and microwave propagation."

CRAWFORD, FREDERICK W., Fellow 1972. Born: July 28, 1931. Degrees: B.Sc., 1952, M.Sc., 1958, University of London; Ph.D., 1955, Dip.Ed., 1956, D.Eng., 1965, University of Liverpool; D.Sc., 1975, University of London.

Fellow Award "For contributions to the understanding of wave and instability phenomena in plasmas." Other Awards: Fellow, Institute of Physics, 1964; Fellow, Institution of Electrical Engineers, 1965; Fellow, American Physical Society, 1965; Fellow, AAAS, 1971; Fellow, Institute of Mathematics and Its Applications, 1978.

CREED, FRANK C., Fellow 1974. Born: April 3, 1921. Degrees: B.Sc., 1945, Queen's University; Ph.D., 1952, University of London.

Fellow Award: "For contributions to the theory and practice of high-voltage impulse measurement and to the development of standards"; IEEE Power Group Award, 1971; Morris E. Leeds Award, IEEE, 1981.

CRELLIN, EARLE A., AIEE Fellow 1928. Born: July 8, 1889, Pleasanton, Calif. Degrees: A.B., 1911, Engineer, 1927, Stanford University.

Other Awards: Sigma Xi, 1939.

CRICHLOW, WILLIAM Q., Fellow 1967. Born: August 1, 1917, Murfreesboro, Tenn. Degrees: B.E., 1939, Vanderbilt University.

Fellow Award "For contributions in the field of measurement and prediction of atmospheric radio noise, and their applications to the evaluation of telecommunications systems."

CRIPPEN, REID P., AIEE Fellow 1938. Born: June 10, 1897, Tarpon Springs, Fla. Degrees: B.S., 1921(summa cum laude), M.S., 1924, University of California at Berkeley. Other Awards: Tau Beta Pi; Eta Kappa Nu; Sigma Xi.

CRISTAL, EDWARD G., Fellow 1980. Born: January 27, 1935, St. Louis, MO. Degrees: A.B., 1957, B.S., 1957, M.S., 1958, Washington University St. Louis; Ph.D., 1961, University of Wisconsin.

Fellow Award "For significant contributions to the theory, analysis, and design of microwave filters, directional couplers, and equalizers"; Microwave Applications Award, IEEE Microwave Theory and Techniques Society, 1973.

CRITZAS, DEMOSTHENES J., Fellow 1963. Born: July 30, 1892, Smyrna, Turkey. Degrees: B.S.E.E., 1918, Cooper Union Institute of Science and Art; E.E., 1922, M.S., 1936, Polytechnic Institute of Brooklyn.

CROCHIERE, RONALD E., Fellow 1982. Born: September 28, 1945, Wausau, WI. Degrees: B.S., 1967, Milwaukee School of Engineering; M.S., 1968, Ph.D., 1974, Massachusetts Institute of Technology.

Fellow Award "For contributions to the field of digital signal processing and its application to digital encoding of speech"; Paper Award, IEEE Acoustics, Speech, and Signal Processing, Society, 1976; Past Secretary-Treasurer and Vice President, IEEE Acoustics, Speech, and Signal Processing Society; Current President, IEEE Acoustics, Speech, and Signal Processing Society, 1983.

CRONVICH, JAMES A., AIEE Fellow 1959. Born: October 26, 1914, New Orleans, La. Degrees: B.E., 1935, M.S., 1937, Tulane University; S.M., 1938, Massachusetts Institute of Technology.

Fellow Award "For contributions to medical practice through electronic applications and to teaching in this area of relationship."

CROSS, L. ERIC, Fellow 1984. Born: August 14, 1923, Leeds, England. Degrees: B.Sc. (Hons.Physics), Leeds University; Ph.D., Leeds University.

Fellow Award "For leadership in the field of ferroelectric materials, and for the development of electrostrictive devices and composite piezoelectric transducers." Other Awards: Leeds University Scholar, ICI Fellowship; American Ceramic Society Electronics Division, Award Outstanding Paper, 1957-1967; The Pennsylvania State University, University Scholars Medal, 1981.

CROWDES, GEORGE J., AIEE Fellow 1946. Born: June 28, 1899, Roxbury, Mass.

Other Awards: Award, War Dept., Signal Corps, 1946; Certificate of Commendation, U.S. Dept. of Navy, 1947.

CROWELL, CLARENCE R., Fellow 1974. Born: July 29, 1928, Sweetsburg, Quebec, Canada. Degrees: B.A., 1949, M.Sc., 1951, Ph.D., 1955, McGill University.

Fellow Award "For contributions to charge transport and charge storage in semiconductors and metal-semiconductor systems." Other Awards: Alexander Von Humboldt U.S. Senior Scientist Award, 1974.

CROWELL, MERTON H., Fellow 1973. Born: June 5, 1932. Degrees: B.S.E.E., 1956, Pennsylvania State University; M.S., 1960, New York University; Ph.D., 1966, Polytechnic Institute of Brooklyn.

Fellow Award "For leadership in the development of silicon diode arrays and contributions to camera tubes, storage tubes, and intensifiers."

CROWLEY, THOMAS H., Fellow 1971. Born: June 7, 1924, Bowling Green, Ohio. Degrees: B.E.E., 1948, M.A., 1950, Ph.D., 1954, Ohio State University.

Fellow Award "For contributions to computing techniques, switching theory, machine-aided design, and the development of large software systems." Other Awards: College of Engineering Distinguished Alumnus, Ohio State University, 1970; Outstanding Civilian Service Medal, U.S. Army, 1975.

CRUZ, JOSE B., Fellow 1968. Born: September 17, 1932, Bacolod City, Philippines. Degrees: B.S.E.E., 1953, University of the Philippines; S.M., 1956, Massachusetts Institute of Technology; Ph.D., 1959, University of Illinois.

Fellow Award "For significant contributions in circuit theory and the sensitivity analysis of control systems." Other Awards: Curtis W. McGraw Research Award, American Society for Engineering Education, 1972; Member, National Academy of Engineering, 1980; Founding Member, Phil-American Academy of Science and Engineering, 1980; Halliburton Engineering Education Leadership Award, 1981; Most Outstanding Filipino in the Midwest, USA, in Technology, 1981; Most Outstanding Engineer and Scientist, Philippine Engineers and Scientists Organization, 1982.

CRYMBLE, A. CARTER, AIEE Fellow 1946. Born: January 15, 1897, Bristol, Va. Degrees: B.S.E.E., 1919, Georgia Institute of Technology.

Other Awards: Engineer of the Year Award, Upper East Tennessee Society of Professional Engineers, 1963; Outstanding Achievement Award, Tennessee Society of Professional Engineers, 1967; Member,

American Society of Mechanical Engineers; Public Service Award, American Society of Mechanical Engineers, Region IV, 1972; Distinguished Service Award, Southern Zone NCEE, 1976; Award for Dedicated Service, Tennessee Society of Professional Engineers, 1976; Life Member, National Society of Professional Engineers; Life Member, Tennessee Society of Professional Engineers; Order of the Engineer, 1971; Tennessee State Board of Architectural and Engineering Examiners, 1963-1975.

CULLEN, ALEXANDER L., Fellow 1967. Born: April 30, 1920, Lincoln, England. Degrees: B.Sc., 1940, Ph.D., 1951, D.Sc., 1965, University of London.

Fellow Award "For his contributions to the fields of microwave power measurements and the propagation of electromagnetic waves in plasmas." Other Awards: Kelvin Premium, Ambrose Fleming Premium, Duddell Premium, Oliver Lodge Premium, Faraday Medal, Institution of Electrical Engineers, U.K.; Fellow, Institute of Physics, Institution of Electrical Engineers; Fellow, Royal Society of London, 1977; Fellow, Fellowship of Engineering, 1977.

CULLUM, A. E., JR., IRE Fellow 1948. Born: September 26, 1909, Abilene, Tex. Degrees: B.S.(Comm.), 1931, Massachusetts Institute of Technology.

Fellow Award "For his contributions to the wartime radio-counter-measures program." Awards: Presidential Certificate of Merit, 1948; National Academy of Engineering, 1970; Engineering Hall of Achievement, 1982.

CUMMINGS, PAUL G., Fellow 1983. Born: October 28, 1926, Wessington, South Dakota. Degrees: B.S.E.E., 1950, South Dakota State University.

Fellow Award "For technical leadership and contributions in the establishment of industry standards for rotating machinery." Other Awards: Rotating Machinery Committee Outstanding Service Award, 1982; IAS Prize Paper Award, 1978; PCIC Prize Paper Award, 1977, 1978; Honorable Mention, 1981, 1982.

CURRIE, MALCOLM R., IRE Fellow 1962. Born: March 13, 1927, Spokane, Wash. Degrees: A.B., 1949, M.S., 1951, Ph.D., 1954, University of California at Berkeley.

Fellow Award "For outstanding contributions to microwave tube research and to education." Other Awards: Phi Beta Kappa; Nation's Outstanding Young Electrical Engineer, 1958, Eta Kappa Nu; National Academy of Engineering, 1971; Fellow, IEE; Fellow, AIAA; Defense Distinguished Public Service Medal; NASA Distinguished Service Medal; National Intelligence Distinguished Service Medal; French Legion of Honor; Distinguished Engineering Alumnus of University of California, Berkeley Award of 1982.

CURRY, A. MALCOLM, Fellow 1977, Born: October 9, 1915, Dandridge, Tenn. Degrees: B.S.E.E., 1937, Tri-

State College; B.S.E.E., 1940, University of Tennessee.

Fellow Award "For contributions to the design of electrical systems for aluminum rolling mills."

CURRY, THOMAS FORTSON, Fellow 1978. Born: November 22, 1926, Thomasville, Ga. Degrees: B.E.E., 1949, Georgia Tech; M.S.E.E., 1954, Pennsylvania State University; Ph.D.E.E., 1959, Carnegie Institute of Technology.

Fellow Award "For the development of remotely controlled electronic reconnaissance systems"; IEEE Centennial Medal, 1984. Other Awards: Sigma Xi; Eta Kappa Nu; Tau Beta Pi; Board of Directors, Association of Old Crows, 1976-; Best Paper Presentation Award, DoD/AOC Symposium, 1977; Bell Telephone Laboratories Fellowship in Electrical Communication.

CURTIS, ORLIE L., JR., Fellow 1976. Born: February 27, 1934, Hutchinson, Kans. Degrees: B.A.(Physics), 1954, Union College, Lincoln, Nebraska; M.S.(Physics), 1956, Purdue University; Ph.D.(Physics), 1961, University of Tennessee; J.D., 1977, University of Southern California.

Fellow Award "For advancing understanding of high-speed digital signal processing and the effects of radiation on semiconductors"; Outstanding Paper Award, IEEE Radiation Effects Conference, 1972, 1974. Other Awards: Fellow, American Physical Society; Listed in: "Outstanding Young Men of America," 1967, "Who's Who in the West," 1976, Order of the COIF, 1977; "Who Who in America," 1982.

CURTISS, ARTHUR N., IRE Fellow 1960. Born: March 27, 1906, Buffalo, N.Y. Degrees: B.S., 1927, University of Pittsburgh; Dipl.(Executive Management), 1954, University of California at Los Angeles.

Fellow Award "For contributions to radio and radar technology." Other Awards: Award of Merit, Radio Corporation of American, 1954; Silver Beaver Award, Boy Scouts of America, 1968; Gerard B. Lambert Award, 1970; Paul Harris Fellow of Rotary, International, 1974; Fellow, Rider College, 1971.

CUTKOSKY, ROBERT D., Fellow 1974. Born: October 24, 1933. Degrees: B.S., 1955, Massachusetts Institute of Technology.

Fellow Award "For contributions to precise electrical measurements through absolute ampere and ohm determinations, and for contributions to the development of capacitance standards of high stability"; Morris E. Leeds Award, 1979. Other Awards: Outstanding Young Scientist from Washington, D.C. Area, Washington Council of Engineering and Architectural Societies, 1962; Achievement Award, Washington Academy of Sciences, 1967; Department of Commerce Gold Medal, 1981.

CUTLER, C. CHAPIN, IRE Fellow 1955. Born: December 16, 1914, Springfield, Mass. Degrees: B.Sc., Doctorate(Hon.), Worcester Polytechnic Institute.

Fellow Award "For his research on microwave antennas and tubes"; Edison Medal, IEEE, 1981. Other Awards: National Academy of Sciences,1976; National Academy of Engineering, 1970; Robert H. Goddard Distinguished Alumni Award, Worcester Polytechnic Institute, 1982.

CUTLER, LEONARD S., Fellow 1974. Born: January 10, 1928, Los Angeles, Calif. Degrees: B.S., 1958, M.S., 1960, Ph.D., 1966, Stanford University.

Fellow Award "For contributions to the design of atomic frequency standards and to the theory and measurement of frequency stability."

CUTRONA, LOUIS J., IRE Fellow 1962. Born: March 11, 1915, Buffalo, N.Y. Degrees: B.A., 1936, Cornell University; M.A., 1938, Ph.D., 1940, University of Illinios.

Fellow Award "For contributions to airborne radar and optical data-processing." Other Awards: Fellow, Optical Society of America, 1973.

D

DACEY, GEORGE C., Fellow 1964. Born: January 23, 1921, Chicago, Ill. Degrees: B.S.E.E., 1942, University of Illinois; Ph.D., 1951, California Institute of Technology.

Fellow Award "For contributions in the field of solid-state devices and in research management"; Outstanding Paper Prize 1961, Solid-State Conference. Other Awards: Fellow, American Physical Society, 1962; Member, National Academy of Engineering, 1973.

DAHL, HELMER H., Fellow 1968. Born: June 17, 1908, Sarpsborg, Norway. Degrees: B.S.E.E., 1931, Technical University of Norway.

Fellow Award "For scientific contributions in the fields of acoustics and microwave techniques." Other Awards: Commander of the Royal Order of St. Olav, 1952; Prize of Per Kures Fund, 1968.

DAHLKE, WALTER E., Fellow 1967. Born: August 24, 1910, Berlin, West Germany. Degrees: Dr.Phil., 1936, University of Berlin, Dr.Habil., 1939, University of Jena, East Germany; Prof.(Hon.), 1961, Technische Hochschule Karlsruhe, Germany.

Fellow Award "For contributions to the theory of noisy networks and the improvements in the understanding of microwave devices and networks."

DAKIN, THOMAS W., Fellow 1968. Born: May 15, 1915, Minneapolis, Minn. Degrees: A.B., 1935, University of Minnesota; M.S., 1938, Michigan State University; Ph.D., 1941, Harvard University.

Fellow Award "For achievements and leadership in basic research on the thermal evaluation of insulation, corona detection and measurement, and the influence of corona discharges on electrical insulation"; Distinguished Technical Achievement Award, IEEE Electrical Insulation Society, 1978. Other Awards: Order of Merit

Award, Westinghouse Electric Corporation, 1978; Special Meritorious Patent Award, Westinghouse Electric Corporation, 1979; National Academy of Engineering, 1981.

DALLAS, JAMES P., AIEE Fellow 1959. Born: December 2, 1904, Charlestown, Wash.

Fellow Award "For contributions to the development of electrical equipment for operation in the rigorous environment encountered by modern aircraft."

DALLOS, PETER, Fellow 1981. Born: November 26, 1934, Budapest, Hungary. Degrees: B.S.E.E., 1958, Illinois Institute of Technology; M.S.E.E., 1959, Ph.D.E.E.(Biomed. Eng.), 1962, Northwestern University.

Fellow Award "For contributions to the understanding of the peripheral auditory system." Other Awards: Beltone Award, Beltone Institute for Hearing Research, 1977; Guggenheim Fellowship, 1977-1978.

DALMAN, G. CONRAD, Fellow 1965. Born: April 7, 1917, Winnipeg, Manitoba, Canada. Degrees: B.E.E., 1940, City College of the City University of New York; M.E.E., 1947, D.E.E., 1949, Polytechnic Institute of Brooklyn.

Fellow Award "For contributions in the field of microwave oscillators and power amplifiers." Other Awards: Certificate of Distinction, Polytechnic Institute of Brooklyn, 1957; Fellow, American Association for the Advancement of Science, 1978; School of Electrical Engineering, Cornell University Award for "Excellence in Teaching", 1977.

DALZIEL, CHARLES F., AIEE Fellow 1957. Born: June 6, 1904, San Francisco, Calif. Degrees: B.S., 1927, M.S., 1934, E.E., 1935, University of California at Berkeley.

Fellow Award "For effective teaching of electrical engineering and original research on the effects of electrical currents in the human body"; Second Prize Paper, AIEE, General Applications Division, 1950, for "Effect of Frequency on Perception Currents"; First Paper Prize, AIEE, San Francisco Section, 1960, for "Underfrequency Protection of Power Systems for Power System Relief-Load Shedding-System Splitting"; First Prize Paper from Industry and General Applications Group, 1969, for "Re-evaluation of Lethal Electric Currents"; Achievement Award, Industry and General Applications Group, 1970. Other Awards: Eta Kappa Nu; Tau Beta Pi; Sigma Xi; Certificate, Office of Scientific Research and Development, National Defense Research Committee; U.S. Dept. of Navy Citation, War Dept., 1947; Certificate of Commendation, Washington, D.C., 1947, 1954; Certificate, Office of Scientific Research and Development, 1947; Fulbright Visiting Professor, Istituto Electtrotecnica Nazionale Galileo Ferraris, Turin, Italy, 1951-1952; Citation, California Disaster Office, 1964; Honorary Member, American Society of Safety Engineers, 1979; First Prize Paper, National Safety Council, 1980.

DAMON, RICHARD W., Fellow 1968. Born: May 14, 1923, Concord, Mass. Degrees: B.S., 1944, M.A., 1947, Ph.D., 1952, Harvard University.

Fellow Award "For contributions to the field of microwave solid-state devices"; National Microwave Lecturer, IEEE Microwave Theory and Techniques Group, 1969. Other Awards: Fellow, American Physical Society, 1969; Sigma Xi; Visiting Lecturer in Applied Physics, Harvard University, 1962; U.S. Commission I, International Union of Radio Sciences, 1970-; Certificate of Appreciation, Advisory Group on Electron Devices, Department of Defense, 1980.

DANDENO, PAUL L., Fellow 1975. Born: April 17, 1921, Toronto, Ontario, Canada. Degrees: B.A.Sc., 1943, University of Toronto; M.S., 1950, Illinois Institute of Technology.

Fellow Award "For contribution to the development and application of digital computer programs for the steady-state and transient analysis of large interconnected electric power systems."

DANIELS, FRED B., Fellow 1965. Born: June 30, 1901, Grand Rapids, Mich. Degrees: B.A., 1933, M.A., 1934, University of Nebraska; Ph.D., 1938, University of Texas.

Fellow Award "For pioneer work on radar signals reflected from the moon and contributions to radio communications." Other Awards: Meritorious Civilian Service Award, U.S. Dept. of Army, 1958; Exceptional Civilian Service Award, U.S. Dept. of Army, 1965.

DANIELS, REXFORD, Fellow 1980. Born: June 16, 1898. Degrees: Mech. Eng., Ph.B., 1920, Yale Sheffield Scientific School.

Fellow Award "For innovative concepts and leadership in the beneficial uses of nonionizing electromagnetic energy and its potential dangers."

DANKO, STANLEY F., Fellow 1966. Born: January 6, 1916, New York, N.Y. Degrees: B.S., 1937, Cooper Union.

Fellow Award "For conception and development of solder-dipped printed wiring for military and commercial use, and for his pioneering work in microminiaturization." Other Awards: Secretary of the Army Award, 1957; Meritorious Civilian Service Award, U.S. Army, 1957.

DANNER, RONALD F., AIEE Fellow 1945. Born: November 2, 1897, Astoria, Ill. Degrees: B.S., 1920, Oklahoma University.

Other Awards: Eta Kappa Nu, 1951; Life Member, National Society of Professional Engineers, 1964; Life Member, Oklahoma Society of Professional Engineers, 1964.

DARLINGTON, SIDNEY, IRE Fellow 1957. Born: July 18, 1906, Pittsburgh, Pa. Degrees: B.S., 1928, Harvard University, B.S., 1929, Massachusetts Institute of Technology; Ph.D., 1940, Columbia University.

Fellow Award "For contributions to network theory and to guidance and control systems"; IEEE Edison Medal, 1975, "For basic contributions to network theory and for important inventions in radar systems and electronic circuits"; IEEE Medal of Honor, 1981. Other Awards: National Academy of Engineering, 1975; National Academy of Sciences, 1978.

DARVENIZA, MAT, Fellow 1979. Born: November 3, 1932, Innisfail, Australia. Degrees: B.E., 1953, University of Queensland; Ph.D., 1959, University of London; D.Eng., 1980, University of Queensland.

Fellow Award "For contributions to the engineering analysis of lightning effects on electric power transmission systems." Other Awards: Institution Award, Institution of Engineers, Australia, 1962, 1969; Electrical Engineering Prize, Institution of Engineers, Australia, 1966, 1968, 1975; Fellow, Institution of Engineers, Australia, 1968; Professor (Personal Chair), University of Queensland, 1980; Fellow, Australian Academy of Technological Sciences, 1982.

DAVENPORT, HOWARD H., AIEE Fellow 1960. Born: July 22, 1910, Houston, Tex. Degrees: B.S., 1932, University of Texas.

Fellow Award "For contributions to military and civil telephone communications." Other Awards: Eta Kappa Nu.

DAVENPORT, LEE L., Fellow 1978. Born: December 31, 1915, Schenectady, N.Y. Degrees: B.S.(Physics), 1937, Union College; M.S.(Physics), 1940, Ph.D.(Physics), 1946, University of Pittsburgh.

Fellow Award "For leadership in industrial research and advanced development." Other Awards: National Academy of Engineering; Presidential Citation, 1946; La Croix Medal, 1963.

DAVENPORT, WILBUR B., JR., IRE Fellow 1958. Born: July 27, 1920, Philadelphia, Pa. Degrees: B.E.E., 1941, Auburn University; S.M., 1943, Sc.D., 1950, Massachusetts Institute of Technology.

Fellow Award "For his contributions to statistical methods in communications systems"; Pioneer Award, IEEE, 1981. Other Awards: Fellow, American Association for the Advancement of Science, 1958; Certificate of Commendation, U.S. Dept. of Navy, 1960; Member, National Academy of Engineering, 1975; Fellow, American Academy of Arts and Sciences, 1977.

DAVEY, JAMES R., Fellow 1966. Born: February 19, 1912, Jackson, Mich. Degrees: B.S.E.E., 1936, University of Michigan.

Fellow Award "For contributions to telegraph and data transmission systems."

DAVID, EDWARD E., JR., IRE Fellow 1962. Born: January 25, 1925, Wilmington, N.C. Degrees: B.E.E., 1945, Georgia Institute of Technology; S.M., Sc.D., 1950,

Massachusetts Institute of Technology; Doctorates(Hon.): 1971, Stevens Institute of Technology; 1971, Polytechnic Institute of Brooklyn; 1971, University of Michigan; 1972, Carnegie-Mellon University; 1973, Lehigh University; 1973, University of Illinois at Chicago Circle; 1978, Rose-Hulman Institute of Technology; University of Florida, 1982; Rensselaer Polytechnic Institute, 1982.

Fellow Award "For contributions to the understanding and exploitation of speech and hearing in communication." Other Awards: Eta Kappa Nu, 1954; Fellow, Audio Engineering Society; G.W. McCarthy-ANAK Award, 1958; Fellow, Acoustical Society of America, 1959; Outstanding Young Man of the Year, New Jersey Junior Chamber of Commerce Award, 1959; Fellow, American Academy of Arts and Sciences, 1966; National Academy of Engineering, 1966; President's Award of Merit, ASME, 1971; Harold Pender Award, Moore School, University of Pennsylvania, 1972; North Carolina Award, 1972; Member, National Academy of Sciences; Fellow, Acoustical Society of America; Society of Research Administrators Award, 1980; New Jersey Science/Technology Medal, 1982; Industrial Research Institute Medal, 1983.

DAVID, PIERRE B. F., Fellow 1968. Born: October 18, 1897, Limoges, Hte Vienne, France. Degrees: 1919, Ecole Polytechnique, Paris; 1929, Docteur Sciences Physiques.

Fellow Award "For his contributions to radio and radar techniques, and his outstanding performance as teacher and author of textbooks on telecommunications." Other Awards: Prix Jean S. Bares, 1932; Medaille d'honneur des Transmission, 1976; Member, Academie de Marine, 1976.

DAVIDS, HUGH H., Fellow 1970. Born: 1906, England. Degrees: B.S.E.E., 1928, M.S.E.E., 1932, University of Pennsylvania.

Fellow Award "For contributions to international standardization of measurement methods and specifications for radio equipment."

DAVIDSON, EDWARD S., Fellow 1984. Born: December 27, 1939, Boston, MA. Degrees: B.A.(Mathematics), 1961, Harvard University; M.S.(Communication Sciences), 1962, University of Michigan; Ph.D.(Elec. Eng.), 1978, University of Illinois.

Fellow Award "For contributions to the use of pipeline structures in computer architecture"; Computer Society Honor Roll. Other Awards: Listing in American Men of Science; Who's Who in Technology.

DAVIES, GOMER L., Fellow 1967. Born: November 24, 1905. Degrees: B.S., 1929, Case Institute of Technology.

Fellow Award "For contributions to and leadership in advancing the technology of aircraft navigation and electronic equipment." Other Awards: Instrument Society of America; Sigma Xi, 1929.

DAVIES, KENNETH, Fellow 1984. Born: January 28, 1928, Merthyr Tydfil, Great Britain. Degrees: B.Sc.(Physics), 1949, Ph.D.(Physics), 1953, University of Wales.

Fellow Award "For contributions to understanding of the ionospheric propagation of radio waves and to experimental studies of the structure and physics of the ionosphere."

DAVIES, LOUIS W., Fellow 1981. Born: August 27, 1923, Sydney, Australia. Degrees: B.Sc., 1948, University of Sydney, Australia; D. Phil., 1951, Oxford University, England.

Fellow Award "For leadership of industrial and university solid-state electronics research"; Fellow Committee, 1983. Other Awards: Fellow, Australian Academy of Technological Sciences, 1975; Fellow, Australian Academy of Science, 1976; Officer In The General Division, Order of Australia, 1978.

DAVIS, CARL G., Fellow 1983. Born: November 30, 1937, St. Louis, Mo. Degrees: B.A.E., 1961, Georgia Institute of Technology; M.S.A.E., 1966, M.S.M.H., Ph.D., 1972, University of Alabama.

Fellow Award "For contributions to software development technology and distributed data processing." Other Awards: U.S. Army R&D Achievement Award, 1981; Outstanding Performance Awards, 1977, 1980, 1981; AIAA Prize Paper, SE Region, Grad Div., 1967.

DAVIS, CLAUD M., Fellow 1983. Born: Aug. 23, 1924, Water Valley, Mississippi. Degrees: B.S.E.E., 1950, Oklahoma State; S.M. (Applied Math), 1961, Harvard University.

Fellow Award "For contributions to the architecture and development of a large fault-tolerant computer system for air traffic control."

DAVIS, EDWARD M., Fellow 1972. Born: November 8, 1933, Pittsburgh, Pa. Degrees: B.S., 1955, Carnegie Institute of Technology; M.S., 1956, California Institute of Technology; Ph.D., 1958, Stanford University.

Fellow Award "For his contributions in the development of computer components and circuits and pioneering work in integrated circuits." Other Awards: Outstanding Electrical Engineer Award, Eta Kappa Nu, 1965; Outstanding Paper Presentation, International Solid State Circuits Conference, 1965; Outstanding Invention Achievement Award, IBM, 1965, 1969; Outstanding Contribution Award, IBM, 1965; Third Level Invention Achievement Award, IBM, 1970; Carnegie Mellon Outstanding Contribution to Engineering and Management Alumni Association Award, 1978; Eta Kappa Nu; Tau Beta Pi; Sigma Xi.

DAVIS, HARRY I., IRE Fellow 1955. Born: December 2, 1909, New York, N.Y. Degrees; B.S., 1931, E.E. 1933, City College of the City University of New York;

M.E.E., 1948, Sc.D., 1973, Polytechnic Institute of Brooklyn.

Fellow Award "For his contributions to the development of electronic aerial navigation systems"; Harry Diamond Award, 1968, "For his outstanding contributions in the conception and development of innovations in military electronics." Other Awards: Sigma Xi; Fellow, American Optical Society; Fellow, American Association for the Advancement of Science; Man of the Year Award, Air Force Association, 1970; Distinguished Alumnus Award, Polytechnic Institute of Brooklyn, 1972, Citation of Honor, Air Force Association, 1969; George W. Goddard Award; Contributor, Encyclopaedia Brittanica.

DAVIS, JOHN F., IRE Fellow 1962. Born: October 2, 1917, Montreal, Quebec, Canada. Degrees: B.Eng., 1942, M.Eng., 1949, M.D., C.M., 1950, McGill University.

Fellow Award "For contributions to measurement techniques in medical electronics." Other Awards: Life Member, Order of Engineers of Quebec, 1980; Emeritus Member, Biomedical Engineering Society, 1982.

DAVIS, JOHN W., AIEE Fellow 1962. Born: December 12, 1898, Owenton, Ky. Degrees: B.S., 1921, Georgetown College; D.E.(Hon.), 1960, Clemson University.

Fellow Award "For contributions to communication engineering in the development of young engineers." Other Awards: Outstanding Service Award, National Society of Professional Engineers, Georgia, 1966.

DAVIS, L. BERKLEY, Fellow 1965. Born: October 27, 1911, Lewisport, Ky. Degrees; 1934, D.Sc.(Hon.), 1973, University of Kentucky; Doctor of Humanities(-Hon.), 1966, Brescia College.

Fellow Award "For leadership in the electronic component industry." Other Awards: Medal of Honor, Electronic Industries Association, 1963.

DAVIS, LUTHER, JR., Fellow 1973. Born: July 12, 1922, Mineola, N.Y. Degrees: B.S., 1942, Ph.D., 1949, Massachusetts Institute of Technology.

Fellow Award "For contributions to the technology of semiconductor, ferrite, and ferroelectric devices,"

DAVISON, EDWARD J., Fellow 1978. Born: September 12, 1938, Toronto, Ont., Canada. Degrees: A.R.C.T., 1958, Royal Conservatory of Music, Toronto; B.A.Sc., 1960, M.A., 1961, University of Toronto; Ph.D., 1964, Sc.D., 1977, University of Cambridge, England.

Fellow Award "For contributions to control system theory"; Honorary Mention Paper Award, IEEE Control Systems Society, 1973-1974; Outstanding Paper Award, IEEE, Control Systems Society, 1979-1980; President, IEEE Control Systems Society, 1983; IEEE Centennial Medal, 1984. Other Awards: Killiam Research Fellowship, 1979-1980, 1981-1983; E.W.R. Steacie Memorial Fellowship, 1974-77; Athlone Fellowship, 1961-63; Fellow, Royal Society of Canada,

1977; Consulting Engineer, Association of Professional Engineers of the Province of Ontario, 1979-present (P.Eng.).

DAVISSON, LEE D., Fellow 1976. Born: June 16, 1936, Evanston, Ill. Degrees: B.S.E., 1958, Princeton University; M.S.E., 1961, Ph.D., 1964, University of California at Los Angeles.

Fellow Award "For contributions to data compression in communications"; Prize Paper Award, IEEE Information Theory Group, 1976 (coauthor R. M. Gray). Other Awards: Honorable Mention, Outstanding Young Electrical Engineer in the U.S., Eta Kappa Nu, 1968.

D AZZO, JOHN J., Fellow 1983. Born: November 30, 1919, New York, N.Y. Degrees: B.E.E., 1941, City College of New York; M.S.E.E., 1950, Ohio State University, Ph.D., 1978, University of Salford, England.

Fellow Award "For contributions to engineering education and research in control systems." Other Awards: Outstanding Engineer, Affiliate Societies of Dayton, Ohio, 1961.

DEAL, BRUCE E., Fellow 1984. Born: September 20, 1927, Lincoln, NE. Degrees: A.B., 1950, Nebr. Wesleyan Univ.; M.S., 1953, Ph.D., 1955, Iowa State Univ.

Fellow Award "For advances in the understanding of the silicon-silicon dioxide interface that led to the development of stable MOS device technology"; I.E.E.E. Electronics Insulation Conf. Paper Award, 1973. Other Awards: Fellow, A.A.A.S., 1966; Electrochemical Society Electronics Division Award, 1974; Franklin Institute Certificate of Merit, 1975; Electrochemical Society T. D. Callinan Award, 1982.

DEAN, CHARLES E., AIEE Fellow 1955, IRE Fellow 1960. Born: May 23, 1898, S.C. Degrees: A.B., 1921, Harvard University; M.A., 1923, Columbia University; Ph.D., 1927, The Johns Hopkins University.

AIEE Fellow Award "For contributions to literature in the field of communications, particularly television"; IRE Fellow Award "For contributions in the field of radio and television receivers." Other Awards: Certificate of Commendation, U.S. Dept. of Navy, Bureau of Ships, 1947.

DEAN, WILFRID, JR., Fellow 1977. Born: October 11, 1920, Princeton, N.J. Degrees: B.S.E.E., 1942, Carnegie Institute of Technology.

Fellow Award "For leadership in engineering and management of the nation's radio frequency spectrum." Other Awards: Distinguished Civilian Service Award, Dept. of the Navy, 1965; Outstanding Performance Award, Executive Office of the President, 1970; Distinguished Service Award, Executive Office of the President, 1975.

DEARDORFF, HAROLD E., AIEE Fellow 1951. Born: January 21, 1899, Dayton, Ohio, Degrees: E.E., 1922, University of Cincinnati.

Other Awards: Eta Kappa Nu; Tau Beta Pi.

DE BETTENCOURT, JOSEPH T., IRE Fellow 1959. Born: June 9, 1912, Washington, D.C. Degrees: B.S.E.E., 1932, M.S., 1934, Catholic University of America; Sc.M., 1937, Sc.D., 1949, Harvard University.

Fellow Award "For contributions to radio wave propagation theory and systems"; Best Paper Prize, 1963. Other Awards: Commendation Ribbon for Meritorious Service, Secretary of Navy, 1946; Fellow, American Association for the Advancement of Science, 1959; Annual Alumni Award, Catholic University of America, 1966.

DE BOER, DEWEY J., AIEE Fellow 1948. Born: December 22, 1897, S.Dak. Degrees: B.S.E.E., 1922, South Dakota State University.

Other Awards: Honorary Life Member, Nebraska Engineering Society, 1959; Honorary Life Member, American Public Power Association, 1961; Certificate of Appreciation, U.S. Junior Chamber of Commerce, 1963; Certificate of Public Service, Federal Power Commission, 1964; Award of Merit, Alumni Association, South Dakota State University, 1970; Distinguished Engineer Award, College of Engineering, South Dakota State University, 1978; Honorary Life Member, Beta Chapter of South Dakota, Tau Beta Pi, 1978.

DE CLARIS, NICHOLAS, Fellow 1970. Born: January 1, 1931, Greece. Degrees: B.S., 1952, Texas A & M; M.S., 1954, Sc.D., 1959, Massachusetts Institute of Technology.

Fellow Award "For contributions to network synthesis, system theory, and electrical engineering education." Other Awards: Jesse H. Jones Award for Achievement, Texas, 1952; Outstanding Educators of America, 1972.

DEES, JULIAN W., Fellow 1984. Born: February 20, 1933, Henderson, NC. Degrees: M.S. (Electrical Engineering), 1955, University of Cincinnati; B.S.(Administrative Engineering), 1954, B.S.(Radio Eng.), 1953, Tri-State College.

Fellow Award "For advancing infrared and millimeter wave instrumentation through the application of the metal-oxide-metal detector"; Chairman Emeritus, Professional Program Committee SOUTHCON/84; Chairman, Professional Program Committee SOUTHCON/83; Vice Chairman, Professional Program Committee SOUTHCON/84 and SOUTHCON/82; Member, Technical Program Committee and Session Chairman, 1982 IEEE MTT-S International Microwave Symposium, Dallas, TX. Other Awards: Central Florida Engineer-of-the-Year Award, 1970; Author-of-the-Year Award, Martin Marietta Corporation, 1967; Sigma Xi; Registered Professional Engineer, State of Georgia; Session Chairman, Solar Power Satellite Symposium, Toulouse, France, 1980; Executive Secretary, International Conference on Submillimeter Waves, 1974, 1976; Member, Program Committee, International Microwave

Symposium, 1974.

DE FIGUEIREDO, RUI J. P., Fellow 1976. Born: April 19, 1929, Degrees: S.B., 1950, S.M., 1952, Massachusetts Institute of Technology; Ph.D., 1959, Harvard University.

Fellow Award "For contributions to nonlinear system theory and to the application of spline functions to signal processing theory"; First Prize, IEEE Student Branch Best Paper Competition, MIT, 1950. Other Awards: Citation of Merit for Outstanding Service, Portuguese Government; Member, AAAS; Member, Society for Industrial and Applied Mathematics; Eta Kappa Nu; Research Fellowship Award, Mathematics Research Center, University of Wisconsin, 1972-1973; Research Grants: National Science Foundation, Air Force Office of Scientific Research, NASA, Office of Naval Research, Advanced Research Projects Agency; Listed in: "Who's Who in the Southwest," "American Men of Science."

DE LANGE, OWEN E., Fellow 1966. Born: August 12, 1906, Koosharem, Utah. Degrees: B.S., 1930, University of Utah; M.A., 1937, Columbia University.

Fellow Award "For basic contributions to broadband FM systems, high speed PCM systems and radar."

DELLA-GIOVANNA, CIRO, Fellow 1973. Born: 1916, Italy. Degrees: Dr. Industrial Eng., 1938, University of Genoa.

Fellow Award "For leadership in the development of telecommunications systems in Italy." Other Awards: Member, Italian Electrotechnical Association; Member, Federation of Telecommunication Engineers of the European Community; Honorary Academic Member and Scientific Adviser, SFIPI-UNUPCE-AITRI; Member, "Quadrato Della Radio."

DELLA TORRE, EDWARD, Fellow 1979. Born: March 31, 1934, Milan, Italy. Degrees: B.E.E., 1954, Brooklyn Polytechnic Institute; M.Sc.(Electrical Engineering), 1956, Princeton University; M.Sc.(Physics), 1961, Rutgers University; D.Eng.Sc., 1964, Columbia University.

Fellow Award "For contributions to the theory of magnetic devices and to the teaching of electromagnetic field theory."

DELLIS, PAUL L., AIEE Fellow 1951. Born: September 19, 1895, Couvin, Belgium. Degrees: E.E., 1921, University of Liege, Belgium.

Other Awards: American Society of Mechanical Engineers, 1945.

DE LOACH, BERNARD C., JR., Fellow 1972. Born: February 19, 1930, Birmingham, Ala. Degrees: B.S., 1951, M.S., 1952, Auburn University; Ph.D., 1956, Ohio State University.

Fellow Award "For contributions to solid-state microwave power sources"; David Sarnoff Award, 1975, "For contribution to and leadership in the development of the impact avalanche and transit time (IMPATT) device." Other Awards: Ohio State University Fellow, 1955-1956; Pi Mu Epsilon; Sigma Xi; Sigma Pi Sigma; Stuart Ballantine Medal, The Franklin Institute, 1975.

DELORAINE, E. M., IRE Fellow 1941, AIEE Fellow 1948. Born: May 16, 1898, Clichy, Seine, France. Degrees: 1921, Ecole Superieure de Physique et Chimie, Paris; Dr.Ing., 1945, University of Paris.

Other Awards: International Communications Award, 1966; Commandeur du Merite National, France; Officier du Merite Postal, France; Officier de la Legion d'Honneur, France; Commandeur du Merite National, Italy.

DE LUCCIA, E. ROBERT, AIEE Fellow 1962. Born: September 20, 1904, Brighton, Mass. Degrees: B.S., 1927, Massachusetts Institute of Technology.

Fellow Award "For contributions in the fields of electric generation transmission." Other Awards: American Society of Civil Engineers; Society of American Military Engineers; American Nuclear Society; American Geophysical Union; "Oregon Engineer of the Year."

DE MARIA, ANTHONY J., Fellow 1978. Born: October 30, 1931, Santa Cruce, Italy. Degrees: B.S.E.E., 1956, Ph.D.E.E., 1965, University of Connecticut; M.S., 1960, Renssalear Polytechnic Institute.

Fellow Award "For pioneering contribution to acoustic-optics and picosecond pulse laser development"; Morris N. Liebman Award, IEEE, 1980; Distinguished Service Award, IEEE Quant. Electronics and Applications Society, 1982; IEEE Centennial Medal, 1984. Other Awards: President, Optical Society of America, 1982; Fellow, SPIE; 4 United Technologies Corp. Outstanding Technical Achievement Awards. Other Awards: Fellow, Optical Society of America; National Academy of Engineering; Connecticut Academy of Science and Engineering; Distinguished Alumni Award, Distinguished Engineering Alumni Award, University of Connecticut, 1978; Davies Metal's Award, Rensselaer Polytechnic Institute, 1980; Sherman Fairchield Distinguished Scholar, California Institute of Technology, 1982-1983.

DE MELLO, F. PAUL, Fellow 1974. Born: July 20, 1927, Goa, Port., India. Degrees B.S.E.E., M.S.E.E., 1948, Massachusetts Institute of Technology.

Fellow Award "For development of new methods of simulation of power plant and power system dynamic phenomena and their application to practical problems in system design and operation"; AIEE Prize Paper Award, Schenectady Section, 1959. Other Awards: Ralph Cordiner Award, General Electric, 1963.

DEMPSTER, BURGESS, IRE Fellow 1957. Born: November 12, 1907, Berkeley, Calif. Degrees: B.S.E.E., 1929, University of California at Berkeley.

Fellow Award "For pioneering in loudspeaker production, and contributions to engineering management." Other Awards: Western Electronic Manufacturers Association.

DENBROCK, FRANK A., Fellow 1981. Born: July 6, 1925, Coldwater, MI. Degrees: B.S.E.E., 1948, Indiana Institute of Technology.

Fellow Award "For contributions to transmission line and substation technology and the development of safety standards"; IEEE Transmission and Distribution Distinguished Service Award, 1978; MSPE Engineer of the Year Award, Jackson Chapter, 1979. Affiliations; American Society of Civil Engineers; National Society of Professional Engineers, State Vice President, Michigan Society of Professional Engineers, 1980-84; President-Elect, M.S.P.E., 1984; U.S. State Dept. HVdc Technical Exchange Team to Russia, 1975; IEEE PES Technical Delegation of Power Engineers to People's Republic of China, 1978; U.S. Dept of Energy/Italy, Joint UHV Program Executive Committee, 1980-85; China Association for Science and Technology/National Rural Electric Cooperation Association (NRECA) Team to improve agricultural production in China by more efficient use of electricity, 1982; USA Representative on CIGRE S/C 23-Substations since 1976; Chairman, IEEE Substations Committee; Chairman, T&D Committee Standards Subcommittee; Secretary, PES Technical Council; Chairman S/C5 ANSI-C2 Committee on the National Electrical Safety Code.

DENKHAUS, WALTER F., AIEE Fellow 1962. Born: June 23, 1905, Colwyn, Pa. Degrees: B.S., 1928, E.E., 1933, Swarthmore College.

Fellow Award "For contributions to the long-range planning and expansion of communications systems."

DENNARD, ROBERT H., Fellow 1980. Born: September 5, 1932, Terrell, TX. Degrees: B.S., 1954, M.S., 1956, Southern Methodist University; Ph.D., 1958, Carnegie-Mellon University.

Fellow Award "For advances in the state of the art of MOSFET devices and circuits"; Cledo Brunetti Award, IEEE, 1982. Other Awards: IBM Fellow, 1979.

DENNING, PETER J., Fellow 1982. Born: January 6, 1942, New York, NY. Degrees: B.E.E., 1964, Manhattan College; M.S.E.E., 1965, Ph.D.(E.E.), 1968, Massachusetts Institute of Technology.

Fellow Award "For contributions to the understanding of virtual memory systems and to the development of the working set concept." Other Awards: Best Paper, Association for Computing Machinery, 1968; Princeton Engineering Faculty Award, 1971; Best Paper, American Federation Information Processing Society, 1972; Recognition of Service Award, Association for Computing Machinery, 1974.

DENNIS, JACK B., Fellow 1973. Born: 1931, New Jersey. Degrees: B.S., 1954, M.S., 1954, D.Sc., 1958, Massachusetts Institute of Technology.

Fellow Award "For contributions to the design of memory systems research, and educational contributions in the computer field." Other Awards: Member, Association for Computing Machinery; Tau Beta Pi; Eta Kappa Nu; Sigma Xi.

DENNIS, JAMES L., IRE Fellow 1959. Born: July 29, 1915, Quincy, Ill. Degrees; B.S., 1937, University of Illinois.

Fellow Award "For contributions and leadership in the field of aeronautical navigation." Other Awards: Exceptional Civilian Service Award, U.S. Dept. of Air Force, 1950.

DENNY, WILLIAM M., AIEE Fellow 1951. Born: November 18, 1899, Crockett, Tex. Degrees: B.S.E.E., 1921, Texas A and M University.

DENTON, RICHARD T., Fellow 1975. Born: July 13, 1932, York, Pa. Degrees: B.S., 1953, M.S., 1954, Pennsylvania State University; Ph.D., 1961, University of Michigan.

Fellow Award "For demonstration of the data transmission capacity of an optical transmission system, and for contributions to ferromagnetic parametric amplifiers." Other Awards: Fellow, National Science Foundation, 1953; Fellow, National Elelctronics Conference, 1957.

DEPP, WALLACE A., Fellow 1970. Born: December 22, 1914, Ky. Degrees: B.S.E.E., 1936, M.S.E.E., 1937, University of Illinois.

Fellow Award "For contributions in the design of gaseous discharge devices and the development of electronic digital systems." Other Awards: Tau Beta Pi Fellowship, 1936-1937; Eta Kappa Nu, Outstanding Young Electrical Engineer, Honorable Mention, 1945; Distinguished Alumnus Awards, Electrical Engineering Alumni Association, University of Illinois, 1974.

DERTINGER, ELLSWORTH F., Fellow 1965. Born: September 10, 1920, Sparrows Point, Md. Degrees: B.S.E.E. 1948, University of Alabama.

Fellow Award "For leadership in the early implementation and teaching of reliability engineering." Other Awards: Fellow, American Society for Quality Control, 1967.

DERTOUZOS, MICHAEL L., Fellow 1976. Born: November 5, 1936, Athens, Greece. Degrees: B.S.E.E., 1957, M.S.E.E., 1959, University of Arkansas; Ph.D.(E.E.), 1964, Massachusetts Institute of Technology.

Fellow Award "For contributions to education in computer science and electrical engineering"; Browder J. Thompson Memorial Prize Award, IEEE, 1968. Other Awards: Fulbright Scholar, 1954-1956; Ford Postdoctoral Fellow, Massachusetts Institute of Technology, 1964-1966; Terman Education Award, American Society for Engineering Education, 1975; Corresponding Member, Academy of Athens, 1981; Arkansas Academy of Electrical Engineering, 1981.

DESCHAMPS, GEORGES A., IRE Fellow 1960. Born: October 18, 1911, Vendome, France. Degrees: License Math., 1932, License Phys., 1933, Agregation Math., 1934, Ecole Normale Superieure, University of Paris.

Fellow Award "For contributions to analysis of microwave components." Other Awards: National Academy of Engineering, 1978.

DE SHONG, JAMES A., JR., Fellow 1967. Born: May 28, 1917, Citypoint, Va. Degrees: Graduate, 1938, General Motors Institute; B.S.E.E., 1940, M.S.E., 1941, Purdue University.

Fellow Award "For contributions to the art of nuclear reactor control and to the development of nuclear and high energy physics instrumentation"; Award, IRE, 1945, "In grateful appreciation of the financial support given to radio and electronics in the creation of a headquarters building for The Institute of Radio Engineers, Inc." Other Awards: Fellow, A.A.A.S., 1968.

DESOER, CHARLES A., Fellow 1964. Born: January 11, 1926, Brussels, Belgium. Degrees: Radio-Engineer, University of Liege, Belgium; Sc.D., Massachusetts Institute of Technology; D.Sc.(honoris causa), 1976, University of Liege.

Fellow Award "For contributions to control theory, circuit theory, and engineering education"; Best Paper Prize, Joint Automatic Control Conference, 1962; IEEE Education Medal, 1975; Outstanding Paper Award (Hon. Mention), IEEE Transactions on Automatic Control, 1979. Other Awards: Miller Research Professorship, University of California at Berkeley, 1967-1968; Medal, University of Liege, 1970; Guggenheim Fellowship, 1970-1971; Distinguished Teaching Award, University of California at Berkeley, 1971; Montefiore Prize, 1975; National Academy of Engineering, 1977; Fellow, AAAS, 1977; Education Award, American Automatic Control Council, 1983.

DETURK, ELMER F., AIEE Fellow 1954. Born: October 4, 1899, West Reading, Pa. Degrees: E.E., Lehigh University.

Fellow Award "For contributions to design of substations, transmission, and distribution facilities for a power system expanding rapidly to meet execeptional load growth."

DEUTSCH, SID, Fellow 1975. Born: September 19, 1918, New York, N.Y. Degrees: B.E.E., 1941, Cooper Union; M.E.E., 1947, D.E.E., 1955, Polytechnic Institute of Brooklyn.

Fellow Award "For development of models of the nervous system and prosthetic devices." Other Awards: Fellow, Society for Information Display, 1967.

DEVEY, GILBERT B., Fellow 1976. Born: January 5, 1921, Swissvale, Pa. Degrees: B.S., 1946, Massachusetts Institute of Technology.

Fellow Award "For technical leadership in biomedical engineering." Other Awards: Annual Award, Association for the Advancement of Medical Instrumentation Foundation, 1975.

DEWAN, SHASHI B., Fellow 1982. Born: April 16, 1941, Jampur, India. Degrees: B.Sc.(E.E.)(with honors), 1960, Chandigarh, India; M.(E.E.), 1961, University of Roorkee, India; M.Sc.(E.E.), 1964, Ph.D.(E.E.), 1966, Toronto, Canada.

Fellow Award "For research and education in power electronics"; Bill Newell Power Electronics Award, IEEE, 1979. Other Awards: Gold Medal, University of Roorkee, India, 1962; Ford Fellowship, 1963-1965; Killam Fellowship, 1980-1982.

DEWILDE, PATRICK M., Fellow 1982. Born: January 17th, 1943, Korbeek-Lo, Belgium. Degrees: E.E., 1966, University of Leuven; Lic. Math., 1968, Brussels; Ph.D.(E.E.), 1970, Stanford University.

Fellow Award "For contributions to network theory, especially the synthesis of scattering matrices." Other Awards: KVIV, 1967; ONR Postdoctoral Fellow, 1971; Eta Kappa Nu, 1980.

DEWITT, JOHN H., JR., IRE Fellow 1951. Born: February 20, 1906, Nashville, Tenn. Degrees: Duncan College Preparatory School, 1924.

Fellow Award "For his achievements in the field of radio broadcast engineering and for his demonstration of radar reflections from the moon." Other Awards: Engineering Achievement Award, National Association of Broadcasters, 1964; Distinguished Alumnus, Vanderbilt University Engineering School, 1974.

D'HEEDENE, ALBERT R., IRE Fellow 1962. Born: June 18, 1904, Brussels, Belgium. Degrees: B.S., 1924, New York University.

Fellow Award "For contributions to the development of networks for communication systems."

DIAMOND, FRED I., Fellow 1975. Born: December 13, 1925, Brooklyn, N.Y. Degrees: S.B., 1950, Massachusetts Institute of Technology; M.E.E., 1953, Ph.D., 1966, Syracuse University.

Fellow Award "For contributions to the development of signal processing technology for radar and communications." Other Awards: Air Force Decoration for Exceptional Civilian Service.

DIBLE, HARVEY J., Fellow 1940. Born: October 5, 1894, Verna, Pa. Degrees: B.S., 1916, University of Pittsburgh.

DIBNER, BERN, AIEE Fellow 1942. Born: August 18, 1897. Degrees: E.E., 1921, D.Eng., 1959, Polytechnic Institute of New York; D.S.T., 1976, Technion; D.Sc., 1977, Brandeis University; D.Sc., 1981, University of Bridgeport.

Other Awards: Life Member, America Association for the Advancement of Science; Fellow, American Academy of Arts and Sciences; Leonardo da Vinci Medal, Society of the History of Technology, 1974; Smithsonian

Gold Medal, 1976; Sarton Medal, History of Science Society, 1976; Eli Whitney Award, Connecticut Society of Patent Attorneys, 1973; Sir Thomas More Medal, University of San Francisco, 1983.

DI CENZO, COLIN D., Fellow 1977. Born: July 26, 1923, Hamilton, Canada. Degrees: B.Sc.(E.E.), 1952, University of New Brunswick; D.I.C.(E.E.), 1953, Imperial College of Science and Technology, London; M.Sc.(E.E.), 1957, University of New Brunswick.

Fellow Award "For contributions to the development and design of detection and control systems." Other Awards: Canadian Decoration and Clasps, 1953, 1963, 1974; Centennial Medal, 1967; Decorated Member, Order of Canada, 1972; Award, Public Servants Invention Act, Canada, 1973; Fellow, Engineering Institute of Canada, 1975; Engineering Medal, Association of Professional Engineers of Ontario, 1976; Julian C. Smith Medal, Engineering Institute of Canada, 1977; Queen's Silver Jubilee Medal, 1977; Governor-General's Commemorative Medal, 1978; Professor Emeritus, McMaster University, 1980; Merit Award, Association of Professional Engineers of Newfoundland, 1982.

DICKENS, LAWRENCE E., Fellow 1977. Born: December 8, 1932, North Kingstown, R.I. Degrees: B.S.E.E., 1960, M.S.E., 1962, Dr.Eng., 1964, The Johns Hopkins University.

Fellow Award "For contributions to the field of low-noise microwave mixers and parametric amplifiers."

DICKEY, FRANK R., JR., Fellow 1966. Born: April 10, 1918, San Antonio, Tex. Degrees: B.S., 1939, University of Texas; M.A., 1946, Ph.D., 1951, Harvard University.

Fellow Award "For advances in the radar field particularly in antennas for airborne high resolution radars, and the correlation air navigator."

DICKIESON, ALTON C., AIEE Fellow 1959. Born: August 16, 1905, New York, N.Y.

Fellow Award "For contributions to long distance multi-channel communication and to underwater guidance systems"; IEEE Communication Technology Group, 1971 Achievement Award, "For outstanding contributions to terrestrial and satellite telephone system transmission and to air defense communication and detection system development"; Mervin J. Kelly Award, 1975, "For meritorious achievement in the field of telecommunications." Other Awards: Naval Ordnance Development Award, 1945; Gen. H.H. Arnold Award, Air Force Association, 1962; Gen. Hoyt S. Vandenberg Award, Air Society, 1963; Associate Fellow, AIAA; Member, National Academy of Engineering, 1970; Distinguished Public Service Award, U.S. Navy, 1973.

DICKINSON, WILLARD H., Fellow 1970. Born: May 28, 1907, Emery, S.Dak. Degrees: B.S.E.E., 1929, University of South Dakota; M.E.E., 1945, Polytechnic Institute of New York.

Fellow Award "For contributions to industrial power system design."

DICKSON, JAMES F., III, Fellow 1979. Born: May 4, 1924, Boston, Mass. Degrees: A.B., 1944, Dartmouth College; M.D., 1947, Harvard Medical School; M.S., 1961, Drexel University.

Fellow Award "For leadership in the application of engineering science to biomedical research and development and the delivery of health care."

DI DOMENICO, MAURO, JR., Fellow 1977. Born: January 12, 1937, Bronx, N.Y. Degrees: B.S., 1958, M.S., 1959, Ph.D., 1963, Stanford University.

Fellow Award "For contributions to optoelectronic devices for optical fiber transmission systems." Other Awards: Fellow, American Physical Society, 1976; Member: Tau Beta Pi, Sigma Xi.

DIEMOND, CLIFFORD C., AIEE Fellow 1960. Born: August 9, 1914, Pasco, Wash. Degrees: B.S., 1935, University of Washington.

Fellow Award "For contributions to power system protection." Other Awards: Meritorious Service Award, U.S. Dept. of Interior, 1965; Distinguished Service Award, U.S. Dept of the Interior, 1974.

DIETRICH, RICHARD E., Fellow 1981. Born: April 1, 1922, Junction City, KS. Degrees: B.S.E.E., 1948, Oregon State University.

Fellow Award "For contributions to power system protection and the development of shielding and grounding standards for EHV control circuits." Other Awards: Meritorious Service Award, Dept. of Interior, 1976.

DIETZOLD, ROBERT L., IRE Fellow 1955. Born: August 3, 1904, Akron, Ohio. Degrees: S.B., 1925, Massachusetts Institute of Technology; Ph.B., 1927, Yale University.

Fellow Award "For his application of mathematics to network design and military problems."

DI FRANCO, JULIUS V., Fellow 1975. Born: June 16, 1925, New York, N.Y. Degrees: B.S., 1950, M.S., 1956, Columbia University.

Fellow Award "For contributions to phased array radar theory and to multidimensional detection of radar targets." Other Awards: Tau Beta Pi.

DIKE, SHELDON H., IRE Fellow 1961. Born: Ocotober 23, 1916, Atlantic City, N.J. Degrees; B.S.E.E., 1941, University of New Mexico; Ph.D., 1951, The Johns Hopkins University.

Fellow Award "For contributions in the analysis and development of electronic systems." Other Awards: Naval Ordnance Development Award, 1945.

DILL, FREDERICK H., Fellow 1978. Born: March 1, 1932. Degrees: B.S.(Physics), 1954, M.S., 1956, Ph.D.E.E., 1958, Carnegie Institute of Technology.

Fellow Award "For contribution to semiconductor device and process research"; President, IEEE Electron Devices Society, 1982, 1983.

DILL, HANS G., Fellow 1973. Born: May 27, 1927, Switzerland. Degrees: M.S.E.E., 1953, Ph.D., 1967, Swiss Federal Institute of Technology.

Fellow Award "For contributions to semiconductor devices, especially insulated gate field-effect transistors, and to the development of ion implantation processes." Other Awards: Lawrence A. Hyland Patent Award, Hughes Aircraft, 1970.

DILLARD, JOSEPH K., AIEE Fellow 1962. Born: May 10, 1917, South Carolina. Degrees: B.S.E.E., 1948, Georgia Institute of Technology; M.S.E.E., 1950, Massachusetts Institute of Technology.

Fellow Award "For contributions to electric power system engineering in conversion and tranmission research"; President, IEEE, 1976. Other Awards: Sigma Xi, 1950; Order of Merit, Westinghouse Electric, 1973; National Academy of Engineering, 1975.

DILLOW, N. EUGENE, AIEE Fellow 1961. Born: August 2, 1918, Dongola, Ill. Degrees: B.S., 1939, University of Illinois.

Fellow Award "For contribution toward the optimization of transformer design through the use of computers and standardized components."

DINGMAN, JAMES E., AIEE Fellow 1954. Born: August 19, 1901, Baltimore, Md. Degrees: B.S. 1921, University of Maryland; D.Eng.(Hon.), 1960, University of Maryland.

Fellow Award "For contributions in both technical work and in leadership of engineering groups in the field of communication."

DIRECTOR, STEPHEN W., Fellow 1978. Born: June 28, 1943, Brooklyn, N.Y. Degrees: B.S., 1965, State University of New York at Stony Brook; M.S., 1967, Ph.D., 1968, University of California, Berkeley.

Fellow Award "For pioneering work in computer-aided circuit design and for contributions to engineering education." Best Paper Award, IEEE Professional Group on Circuit Theory, 1970; W.R.G. Baker Prize Award, 1979; IEEE Centennial Medal, 1984. Other Awards: Distinguished Service Award, College of Engineering, University of Florida, 1974; Frederick Emmons Terman Award, American Society of Engineering Education, 1976; Whitaker Chair, Dept. of Electrical Engineering, Carnegie-Mellon University, 1980.

DISSON, STANLEY B., Fellow 1969. Born: August 30, 1928, Philadelphia, Pa. Degrees: B.S.E.E., 1950, University of Pennsylvania.

Fellow Award "For contributions to digital magnetic techniques, and for leadership in the development of data processing systems." Other Awards: Tau Beta Pi, Eta Kappa Nu, Sigma Tau, 1949; Engineers Club of Philadelphia Award, 1950.

DIXON, N. REX, Fellow 1983. Born: February 24, 1932, Ecorse, MI. Degrees: B.A., 1958, Western Michigan University, Kalamazoo; M.A., 1960, Indiana University, Bloomington; Ph.D., 1966, Stanford University, Palo Alto.

Fellow Award "For contributions to speech synthesis and speech recognition"; Editor, IEEE Press Book on Automatic Speech and Speaker Recognition; Chairman, Standing Committee on Conference Standards, IEEE-ASSPS AdCom; Member, Administrative Committee, IEEE-ASSPS, 1976-79; Technical Program Chairman, 1977 IEEE International Conference on Acoustics, Speech and Signal Processing, Hartford, CT; Chairman, Technical Committee on Speech Processing, IEEE-ASSPS, 1977-80; Associate Editor and Member of Editorial Board, IEEE Transactions on Acoustics, Speech, and Signal Processing, 1977-81; General Chairman, 1979 IEEE-ASSPS Workshop on Automatic Speech Recognition, Carnegie-Mellon Institute, Pittsburgh, PA; Meritorious Service Award, IEEE Acoustics, Speech, and Signal Processing Society, 1979; Vice-President, IEEE Acoustics, Speech, and Signal Processing Society, 1979-81; President, IEEE Acoustics, Speech, and Signal Processing Society, 1982-83; IEEE Centennial Medal, 1984; General Chairman, 1985 IEEE International Conference on Acoustics, Speech, and Signal Processing, Tampa, FL. Other Awards: Visiting Professor, Engineering Research, Brown University, Providence, RI; Outstanding Innovation Award, IBM Corporation, 1976; Member, National Research Council Committee on Computerized Speech Recognition, 1983-84.

DIXON, RICHARD W., Fellow 1977. Born: September 25, 1936, Hubbard, Oreg. Degrees: A.B.(summa cum laude), 1958, Harvard College; M.A., 1960, Ph.D., 1965, Harvard University.

Fellow Award "For contributions to the theory and realization of acoustooptical modulators."

DOBA, STEPHEN, JR., IRE Fellow 1960. Born: May 27, 1907, New York, N.Y.

Fellow Award "For contributions in the field of television signal transmission."

DODDS, WELLESLEY J., IRE Fellow 1961. Born: October 18, 1915, Faulkton, S.D. Degrees: B.S., 1938, South Dakota State University; M.S., 1941, University of Kansas.

Fellow Award "For technical contributions and leadership in the development of microwave tubes." Other Awards: Slosson Scholar, Univ. of Kansas, 1940-1941; RCA Research Award, 1953; RCA Inventor Award, 1974; Sigma Xi; Sigma Pi Sigma; Pi Mu Epsilon; Member, American Society for Quality Control.

DODINGTON, SVEN H.M., Fellow 1966. Born: May 22, 1912, Vancouver, British Columbia, Canada. Degrees: A.B., 1934, Stanford University.

Fellow Award "For contributions and leadership in the development of radio navigational aids"; Pioneer Award, IEEE Aerospace and Electronic Systems Society, 1980. Other Awards: Outstanding Patent Award, N.J. Council of Research and Development, 1967; Volare Award, Airline Avionics Institute, 1967; Inventor of the Year Award, N.Y. Patent Law Association, 1982.

DOEHLER, OSKAR F., Fellow 1973. Born: 1913, Schwarzenbek, Germany. Degrees: Ph.D., 1938, University of Hamburg, Germany.

Fellow Award "For contributions and technical leadership in the field of high-power microwave tubes, and especially crossed-field devices." Other Awards: Silver Medal, Syndicat General de la Construction Technique, France, 1968; Certificate of Achievement, Association of Old Crows, AOC National Convention, Anaheim, California, 1970.

DOELZ, MELVIN L., IRE Fellow 1961. Born: December 30, 1918, Minneapolis, Minn. Degrees: B.E.E., 1941, University of Minnesota; M.S., 1948, State University of Iowa.

Fellow Award "For contributions to mechanical radio-frequency filter development and predicted-wave digital data communication."

DOEPPNER, THOMAS W., Fellow 1979. Born: May 22, 1920, Berlin, Germany. Degrees: B.S.E.E., Kansas State University; M.S.E.E., University of California at Berkeley.

Fellow Award "For leadership in advancing electromagnetic compatibility in the design, development, and operation of military telecommunication systems"; Patron Award, Washington, DC Section, IEEE, 1983; IEEE Centennial Medal, 1984. Other Awards: Bronze Star Medal, 1966; Medal of Honor, First Class, Republic of Vietnam, 1966; Legion of Merit, with Oak Leaf Cluster, 1969, 1973; Listed in "Who's Who in Engineering, 1980; "Men of Achievement," 1981.

DOERING, HERBERT K., Fellow 1974. Born: February 10, 1911, Vienna, Austria. Degrees: Dipl.-Ing., 1934, Dr.techn., 1936, TH Wien; Habilitation, 1948, TH Stuttgart.

Fellow Award "For contributions to microwave electronics and to education." Other Awards: Member, Rheinisch-Westfaelische Akademie der Wissenschaften, 1969; Member, Oesterreichische Akademie der Wissenschaften, 1971; VDE Ehrenring, 1972; W. Exner Medaille, 1975.

DOHERTY, WILLIAM H., IRE Fellow 1944. Born: August 21, 1907, Cambridge, Mass. Degrees: S.B., 1927, S.M., 1928, Harvard University; Sc.D.(Hon.), 1950, Catholic University of America.

Fellow Award "For his contributions to the design of radio transmitters"; Morris N. Liebmann Award, IRE, 1937, "For his improvement in the efficiency of radio-frequency power amplifiers." Other Awards: Naval Ordnance Development Award, 1945; Fellow Award, American Association for the Advancement of Science, 1978; Tau Beta Pi; Sigma Xi.

DOME, ROBERT B., IRE Fellow 1948. Born: October 12, 1905, Tell City, Ind. Degrees: B.S.(in E.E.), 1926, Purdue University; M.S.(in E.E.), 1929, Union University.

Fellow Award "For his many contributions to the profession and for his accomplishments in the training of young engineers"; Morris Liebmann Award, IRE, 1951, "For many technical contributions to the profession, but notably his contributions to the inter-carrier sound system of television reception, wide-band phase-shift networks and various simplifying innovations in FM receiver circuits." Other Awards: Sigma Xi, 1926; Charles A. Coffin Award, General Electric, 1936, 1947.

DOMMEL, HERMANN W., Fellow 1979. Born: December 13, 1933, Hennenbach, West Germany. Degrees: Dipl. Ing., 1959, Dr. Ing., 1962, Technical University, Munich, Germany.

Fellow Award: "For contributions to the development and application of digital computer programs for the analysis of complex electric systems network problems."

DONALDSON, MERLE R., Fellow 1970. Born: April 7, 1920, Silverdale, Kans. Degrees: B.E.E., 1946, M.S.E.E., 1947, Ph.D., 1959, Georgia Institute of Technology.

Fellow Award "For contributions to cyclotron instrumentation, and RF high-voltage measurements." Other Awards: Florida West Coast Engineer-of-the-Year Award, 1966.

DONALDSON, W. LYLE, Fellow 1969. Born: May 1, 1915. Degrees: B.S.E.E., 1938, Texas Tech University.

Fellow Award "For contributions to nondestructive testing and evaluation of materials." Other Awards: Distinguished Engineer, Texas Tech University, 1969; Fellow, American Society of Nondestructive Testing, 1975.

DOPAZO, JORGE F., Fellow 1977. Born: April 2, 1921, Havana, Cuba. Degrees: M.S.E.E., 1950, University of Havana.

Fellow Award "For contributions to computer methods for analysis, real-time monitoring, and economic operation of power systems"; Prize Paper Award, Power Engineering Society, 1981. Other Awards: Member, Sigma Xi.

DORATO, PETER, Fellow 1977. Born: December 17, 1932, New York, N.Y. Degrees: B.E.E., 1956, City College of the City University of New York; M.S.E.E., 1957, Columbia University; D.E.E., 1961, Polytechnic Institute of Brooklyn.

Fellow Award "For contributions to sensitivity analysis and design in automatic control systems."

DORF, RICHARD C., Fellow 1973. Born: December 27, 1933, Bronx, N.Y. Degrees: B.S.E.E., 1955, Clarkson College of Technology; M.S., 1957, University of Colorado; Ph.D., 1961, U.S. Naval Postgraduate School.

Fellow Award "For leadership in engineering education, and for contributions to control theory." Other Awards: American Council on Education Fellow in Academic Administration, 1968; Clarkson College Alumni Award, 1979.

DORGELO, EDUARD G., Fellow 1969. Born: June 7, 1908, Rotterdam, Netherlands. Degrees: B.S., 1926, M.S., 1930, Teacher's College.

Fellow Award "For contributions to the design and technology of high-frequency and microwave power sources."

DORNE, ARTHUR, IRE Fellow 1962. Born: April 6, 1917, Philadelphia, Pa. Degrees: B.A., 1938, B.S., 1939, University of Pennsylvania.

Fellow Award "For contributions to the design of antennas, particularly for aircraft."

DORROS, IRWIN, Fellow 1979. Born: October 3, 1929, Brooklyn, N.Y. Degrees: S.B., 1956, S.M., 1956, Massachusetts Institute of Technology; Eng.Sc.D., 1962, Columbia University.

Fellow Award "For the management of engineering projects associated with integrated nationwide telecommunications." Other Awards: Karl Taylor Compton Prize, Massachusetts Institute of Technology, 1955.

DORTORT, ISADORE K., AIEE Fellow 1962. Born: March 8, 1904, Basel, Switzerland. Degrees: B.S., 1927, University of Pennsylvania.

Fellow Award "For achievements in power conversion equipment"; AIEE Certificate, Philadelphia Section, 1962. Other Awards: 45 U.S. patents and foreign patents.

DOUGAL, ARWIN A., Fellow 1974. Born: November 22, 1926, Dunlap, Iowa. Degrees: B.S., 1952, Iowa State University; M.S., 1955, Ph.D., 1957, University of Illinois.

Fellow Award "For contributions to the theory and teaching of plasma science and physical electronics"; Special Award by Southwestern IEEE Conference and Exhibition, 1967, "For outstanding service as program chairman"; IEEE Centennial Medal, 1984. Other Awards: Teaching Excellence Award, University of Texas Students' Assn., 1962, 1963; Fellow, American Physical Society, 1967; Outstanding Graduate Adviser Award, University of Texas Graduate Engineering Council, 1971; Professional Achievement Citation in Engineering, College of Engineering, Iowa State University, 1975; Distinguished Adviser Award, Student Engineering Council, University of Texas, 1977; Teaching Achievement Award, Student Engineering Council, Uni-

versity of Texas, 1977.

DOUGHERTY, JOHN J., Fellow 1974. Born: August 15, 1924, Philadelphia, Pa. Degrees: B.S.E.E., 1950, Villanova University.

Fellow Award "For contributions to dc transmission technology and the application of low-cost, high-voltage cables."

DOW, WILLIAM G., AIEE Fellow 1949, IRE Fellow 1950. Born: September 30, 1895, Faribault, Minn. Degrees: B.S.E., 1916, E.E., 1917, University of Minnesota; M.S.E., 1929, University of Michigan; Dr.Sc. (Hon.), 1980, University of Colorado.

Fellow Award "For outstanding contributions to the teaching and understanding of electronics through the organization of educational material and the stimulation of students and others of critical thought"; Education Medal, 1963, "For his outstanding contributions to teaching, to educational theory, and to research, and his inspiring leadership in faculty development." Other Awards: Outstanding Achievement Award, University of Minnesota, 1961; Distinguished Faculty Achievement Award, University of Michigan, 1962; Fellow, Engineering Society of Detroit, 1975; Fellow, American Association for the Advancement of Science, 1980.

DOWIS, WILLIAM J., AIEE Fellow 1956. Born: August 24, 1908, Idaho Springs, Colo. Degrees: B.S., 1930, University of Colorado.

Fellow Award "For his contributions in the application of electrical engineering to the field of nuclear reactors."

DOWNING, WILLIAM C., JR., AIEE Fellow 1959. Born: December 6, 1902, Terre Haute, Ind. Degrees: B.S., 1924, Yale University.

Fellow Award "For contributions to development of the thermal converter, particulary for demand metering, totalization and load control."

DOXEY, WILLIE L., Fellow 1964. Born: February 21, 1912, Montgomery, La. Degrees: B.A., 1934, Northwestern State University, La.; M.A., 1939, Louisiana State University.

Fellow Award "For leadership in research and development of electronic materials and devices"; National Chairman, IRE Prof. Group Military Electronics, 1962. Other Awards: Meritorious Civilian Service Award, U.S. Dept. of Army, 1965; President, Ft. Monmouth Chapter, Armed Forces Communications and Electronics Association, 1967; Trustee Emeritus, Monmouth College; Board of Directors, Assn. of U.S. Army; Board of Directors Award, National Aerospace & Electronics Systems, 1975.

DOYLE, WILLIAM D., Fellow 1983. Born: June 5, 1935, Dorchester, Massachusetts. Degrees: B.S., 1957, M.S., 1959, Boston College; Ph.D., 1964, Temple University (all in Physics).

Fellow Award "For contributions to the development

of advanced magnetic memories"; Distinguished Lecturer, IEEE Magnetics Society, 1982-1983.

DRANGEID, KARSTEN E., Fellow 1973. Born: March 24, 1925, Norway. Degrees: Elec.Eng., 1951, Swiss Federal Institute of Technology, Zurich, Switzerland.

Fellow Award "For contributions to the development of semiconductor devices"; Director, Region 8, IEEE, 1983-84.

DRAPER, CHARLES S., IRE Fellow 1955. Born: October 2, 1901, Windsor, Mo. Degrees: B.A., 1922, Stanford University; B.S., 1926, S.M., 1928, Sc.D., 1938, Massachusetts Institute of Technology; Ph.D.(Hon.), 1966, Eidgenossishe Technische Hochschule, Switzerland; Dr.(Hon.), 1970, University of Portland; Dr.(Hon.), 1975, University of Missouri-Rolla.

Fellow Award "For his contributions to the theory and practical application of precise instrumentation and to engineering education"; Lamme Medal, IEEE, 1972; Pioneer Award, IEEE, 1978. Other Awards: Sloan Automotive Fellow, Massachusetts Institute of Technology, 1928; Crane Automotive Fellow, Massachusetts Institute of Technology, 1929; S. A. Reed Award, American Institute of Aeronautics and Astronautics, 1946; Presidential Certificate of Merit; 1946; Naval Ordnance Development Award, 1946; Award, Engineering Societies of New England, 1947; Exceptional Civilian Service Award, 1951; Testimonial, American Society of Mechanical Engineers, 1951; Wilbur Wright Lecturer, Royal Aeronautical Society, U.K., 1955; Distinguished Public Service Award, U.S. Dept. of Navy, 1956; Thurlow Award, Institute of Navigation, 1957; Airpower Trophy, Air Force Association, 1957; Holley Medal, American Society of Mechanical Engineers, 1957; Blandy Medal, American Ordnance Association, 1958; American Honorary Fellow, Institute of Aeronautical Sciences, 1959; Godfrey L. Cabot Award, Aero Club of New England, 1959; William Proctor Prize, Scientific Research Society of America, 1959; Exceptional Service Award, U.S. Dept. of Air Force, 1960; Howard N. Potts Award, Franklin Institute, 1960; Distinguished Public Service Award, U.S. Dept. of Navy, 1961; Golden Plate Award, Academy of Achievement, 1961; Distinguished Service Award, University of Missouri, 1962; Award, National Society of Professional Engineers, 1962; Space Flight Award, American Astronautical Society, 1962; Louis W. Hill Space Transportation Award, American Institute of Aeronautics and Astronautics, 1963; Commanders' Award, U.S. Dept. of Air Force, 1964; Montgomery Award, National Society of Aerospace Professionals, 1965; National Medal of Science, Presidential Award, 1965; Wright Brothers Lecturer, American Institute of Aeronautics and Astronautics, 1966; Vincent Bendix Award, American Society for Engineering Education, 1966; Daniel Guggenheim Award, 1967; Exceptional Civilian Service Award, U.S.A.F., 1968; NASA Public

Service Award, 1969; Founders Medal, National Academy of Engineering, 1970; Thomas D. White National Defense Award, U.S. Air Force Academy, 1970; Distinguished Civilian Service Medal, Dept. of Defense, 1970; Charles F. Kettering Award, and Elmer E. Sperry Award, PTC Research Institute, George Washington University, 1970; Distinguished Service Citation, American Ordnance Association, 1971; W. Randolph Lovelace, II Award, American Astronautical Society, 1971; Rufus Oldenburger Award, American Society of Mechanical Engineers, 1971; Award of Merit, American Institute of Consulting Engineers, 1972; Kelvin Gold Medal, Institution of Civil Engineers, London, 1974; Certificate of Merit Award, Polaris-Poseidon Fleet Ballistic Missile Program, 1975; Citation, University of Missouri at Rolla, 1975; International Space Hall of Fame, 1976; Establishment of Dr. Draper Scholarship Fund, MIT, 1976; Diploma and Medal Award, Scientific Council of the Tsiolkovsky State Museum of Cosmanautic History, U.S.S.R., 1977; Elder Statesman Award of Aviation, National Aeronautic Association, 1977; Allan D. Emil Memorial Award, International Astronautical Federation, Prague, 1977; Dr. Robert H. Goddard Trophy, National Space Club, 1978; Draper Chair; Charles Stark Draper Professorship of Aeronautics and Astronautics, MIT, 1978; Bronze Beaver, MIT, 1978; Foreign Associate Member, French Academy of Sciences, 1979; Meritorious Service to Aviation Award, Board of Directors, National Business Aircraft Association, Inc., 1979; Invit. to visit and lecture in China, Int'l Dept., Scientific and Technol. Comm., People's Republic of China, 1980; Eagle Award, American Astronautical Society, 1980; National Inventors Hall of Fame, 1981; New England Inventor of the Year, 1981; Citation, Academy of Applied Science, 1981; Langley Medal, Board of Regents, Smithsonian Institute, 1981.

DREESE, ERWIN E., AIEE Fellow 1962. Born: September 10, 1895, Millbrook, Mich. Degrees: B.S., 1920, M.S., 1922, E.E., 1927, University of Michigan.

Fellow Award "For contributions to electrical engineering education." Other Awards: American Society for Engineering Education; Sigma Xi; Tau Beta Pi; Eta Kappa Nu; Distinguished Service Award and Medal, Ohio State University, 1974.

DRENICK, RUDOLF F., Fellow 1967. Born: August 20, 1914, Vienna, Austria. Degrees: Ph.D., 1939, University of Vienna.

Fellow Award "For his contributions to the development of adaptive servomechanism systems." Other Awards: Faculty Fellow, National Science Foundation, 1964-1965.

DRESSELHAUS, MILDRED S., Fellow 1979. Born: November 11, 1930, Brooklyn, N.Y. Degrees: A.B., 1951, Hunter College; A.M., 1953, Radcliffe College; Ph.D., 1958, University of Chicago; D.Eng.(Hon.), 1976, Worcester Polytechnic Institute; D.Sc. (Hon.),

1980, Smith College; D.Sc.(Hon.), 1982, Hunter College.

Fellow Award "For contributions to the understanding of electronic properties of semiconductors, semimetals, and metals, to electrical engineering education, and to the enhancement of women's opportunities in engineering education." Other Awards: Fellow, American Physical Society; Hall of Fame, Hunter College, 1972; Alumnae Medal, Radcliffe College, 1973; National Academy of Engineering, 1974; American Academy of Arts and Sciences, 1974; Corresponding Member, Brazilian Academy of Science, 1976; Achievement Award, Society of Women Engineers, 1977.

DROGIN, EDWIN M., Fellow 1984. Born: October 26, 1932, Brooklyn, N.Y. Degrees: B.E.E., 1954, The Cooper Union School of Engineering; M.S.E.E., 1958, Columbia University.

Fellow Award "For advances in digital signal processing architecture leading to improvements in the agility and versatility of electronic warfare systems"; Best Paper Award, NAECON, 1982. Other Awards: Plaques and Certificates of Appreciation, for talks on B-1B Defensive Avionics System, Association of Old Crows chapters.

DUANE, JAMES T., Fellow 1971. Born: August 24, 1928, Clinton, Mass. Degrees: S.B.E.E., S.M.E.E., 1954, Massachusetts Institute of Technology; A.M.P., Harvard Business School, 1970.

Fellow Award "For contributions to the design of electrical machinery, to the control of electric motors, and for leadership in design, analysis, and management." Other Awards: Outstanding Young Electrical Engineer, Eta Kappa Nu, 1962.

DU BOSE, MCNEELY, AIEE Fellow 1943.

DUDA, RICHARD O., Fellow 1980. Born: April 27, 1936, Evanston, IL. B.S.(Eng.), 1958, M.S.(Eng.), University of California at Los Angeles; Ph.D.(Elec. Eng.), 1962, Massachusetts Institute of Technology.

Fellow Award "For contributions to the theory and applications of pattern recognition."

DUESTERHOEFT, WILLIAM C., Fellow 1980. Born: December 10, 1921. Degrees: B.S.E.E., 1943, M.S.E.E., 1949, University of Texas; Ph.D., 1953, California Institute of Technology.

Fellow Award "For enduring contributions to power engineering education"; Second Prize Paper Award, Southwest Region, AIEE, 1959. Other Awards: Convair General Dynamics Award for Excellence in Engineering Teaching, 1958; Student Association Teaching Excellence Award, University of Texas, 1961; Student Engineering Council Teaching Award, 1977.

DUFF, WILLIAM G., Fellow 1981. Born: December 16, 1936, Alexandria, VA. Degrees: B.E.E., 1959, George Washington University; M.S.E.E., 1969, Syracuse University; D.Sc.E.E., 1979, Clayton University.

Fellow Award "For the development of design and analysis technology for achieving electromagnetic compatibility at the system level"; President, IEEE Electromagnetic Compatibility Society; Past Secretary, Treasurer, Vice Chairman, Chairman, Washington, D.C. Chapter, IEEE Electromagnetic Compatibility Society; Best Paper Award, AIEE, 1961; Certificate of Appreciation, Electromagnetic Compatibility Symposium, IEEE, 1976; Certificate of Appreciation, IEEE Electromagnetic Compatibility Group Newsletter. Other Awards: Listed in "Who's Who in the South and Southeast," "Notable Americans"; Engineer Alumni Service Award, George Washington University, 1980.

DUFFENDACK, ORA S., IRE Fellow 1956. Born: May 7, 1890, Napoleon, Mo. Degrees: B.S., 1917, University of Chicago; A.M., 1921, Ph.D., 1922, Princeton University.

Fellow Award "For contributions to spectroscopy and gaseous electronics, and for leadership in electronic research."

DUHAMEL, RAYMOND H., IRE Fellow 1962. Born: October 11, 1922, Tuscola, Ill. Degrees: B.S., 1947, M.S., 1948, Ph.D., 1951, University of Illinois.

Fellow Award "For contributions to frequency-independent antennas"; Best Paper Prize, IRE, Professional Group on Antennas and Propagation, 1958.

DUINKER, SIMON, Fellow 1983. Born: October 30, 1924, Batavia. Degrees: Electrical Engineering Degree, 1950; Doctor's Degree Technical Sciences, 1957.

Fellow Award "For contributions to nonlinear circuit theory, transaxial tomography, and industrial leadership."

DUMMER, G. W. A., Fellow 1968. Born: February 25, 1909, Hull, England. Degrees: Manchester College of Science and Technology.

Fellow Award "For contributions to microelectronics and component reliability"; Cledo Brunetti Award, 1979. Other Awards: Webber Premium Paper Award, Institution of Electrical Engineers, 1952; Medal of Freedom with Bronze Palm, U.S., 1946; Member of the Order of the British Empire, 1946; Wakefield Gold Medal, Royal Aeronautical Society, 1964.

DUNCAN, CHARLES C., Fellow 1971. Born: February 10, 1907, Overland, Mo. Degrees: B.S.(E.E.), 1927, Washington University.

Fellow Award "For leadership in the development and improvement of communications." Other Awards: Distinguished Service Cross, Episcopal Diocese of Long Island, 1964; Alumni Citation, Washington University, 1969.

DUNCAN, JAMES W., Fellow 1981. Born: September 15, 1926. Degrees: B.S.(E.E.), 1950, University of Colorado; M.S.(E.E.), 1955, Ph.D.(E.E.), 1958, University of Illinois.

Fellow Award "For contributions to and technical leadership in the development of communications satellite antennas."

DUNCAN, THOMAS C., AIEE Fellow 1962. Born: July 16, 1905, Pittsburgh, Pa. Degrees: E.E., 1927, Cornell University.

Fellow Award "For contributions in the planning and design of electric power systems,"

DUNKI-JACOBS, JOHN R., Fellow 1982. Born: Surabaya, Indonesia. Degrees: B.S.E.E., 1948, Technical College of Amsterdam, The Netherlands.

Fellow Award "For contributions to the design and implementation of industrial power systems"; Prize Paper Award, IEEE PCIC and IAS, 1972; Prize Paper Award, IEEE PCIC, 1976; Best Paper Award, IEEE IAS, 1978; Prize Paper Award, IEEE PCIC, 1981. Other Awards: Management Award, General Electric, 1973, 1978.

DUNLAP, G. WESLEY, AIEE Fellow 1948. Born: April 13, 1911, Gardnerville, Nev. Degrees: B.A., 1931, E.E., 1933, Ph.D., 1936, Stanford University.

Other Awards: Harris J. Ryan High Voltage Research Fellow, Stanford Univ., 1932-35; Alfred Noble Prize, 1942; Edwin Sibley Webster Prof.-Visiting at MIT, 1955-1956.

DUNLAP, W. CRAWFORD, IRE Fellow 1961. Born: July 21, 1918, Denver, Colo. Degrees: B.S., 1938, University of New Mexico; Ph.D., 1943, University of California at Berkeley.

Fellow Award "For contributions to semiconductor research and transistor production techniques." Other Awards: Fellow, American Physical Society, 1957.

DUNN, ANDREW F., Fellow 1971. Born: January 17, 1922, Sydney, Nova Scotia, Canada. Degrees: B.Sc., 1942, M.Sc., 1947, Dalhousie University; Ph.D., 1950, University of Toronto.

Fellow Award "For contributions to the development of electrical standards."

DUNN, DONALD A., Fellow 1975. Born: December 31, 1925, Los Angeles, Calif. Degrees: B.S., 1946, California Institute of Technology; M.S., 1947, L.L.B., 1951, Ph.D., 1956, Stanford University.

Fellow Award "For contributions to the development of telecommunications policy."

DUNN, FLOYD, Fellow 1980. Born: April 14, 1924, Kansas City, MO. Degrees: B.S.E.E., 1949, M.S.E.E. (Acoustics), 1951, Ph.D.E.E. (Biophysics), 1956, University of Illinois at Urbana.

Fellow Award "For contributions to the understanding of the interaction of ultrasonic waves with living tissue." Other Awards: Fellow, Acoustical Society of America, 1963; Special Research Fellowship, National Institutes of Health, 1968; American Cancer Society-Eleanor Roosevelt-International Cancer Fellowship, 1975 and 1982; Fellow, American Institute of Ul-

trasound in Medicine, 1979; Fellow, Institute of Acoustics, United Kingdom, 1980; Fulbright Scholar 1982/3; National Academy of Engineering, 1982; Japan Society for Promotion of Science Fellowship, 1982.

DUNNING, ORVILLE M., IRE Fellow 1955. Born: February 21, 1904, St. Louis, Mo. Degrees: B.S., 1925, Rose Polytechnic Institute.

Fellow Award "For his contributions to the field of sound recording, and his effective organization of engineering effort." Other Awards: Commendation, U.S. Dept. of Navy.

DUPUIS, RENE, AIEE Fellow 1955. Born: May 5, 1898, Pike River, Quebec, Canada. Degrees: B.A., 1919, Laval University, Quebec; Cert. Appl. Mech., 1923, Cert. Appl. Phys., 1924, E.E., 1924, Universite de Nancy, France; D.Sc.(Hon), 1953, Laval University.

Fellow Award "For contribution to the electrical industry as engineer, author, administrator and educator." Other Awards: Brevet des Palmes d'Academie, France, 1948; Professor Emeritus, Laval University, Quebec, 1948; Honorary Director, Hydro-Quebec Institute of Research, Quebec (IREQ), 1976.

DURGIN, HAROLD L., AIEE Fellow 1954. Born: December 7, 1902, Benton, Me. Degrees: B.S., 1924, University of Maine.

Fellow Award "For contributions to power generation and conservation of natural resources."

DURRANI, SAJJAD H., Fellow 1976. Born: August 27, 1928, Jalalpur, Pakistan. Degrees: B.A.(Math), 1946, Government College, Lahore, Pakistan; B.Sc.Eng.(E.E.)(Hons.), 1949, Engineering College, Lahore, Pakistan; M.Sc.Tech.(E.E.), 1953, College of Technology, Manchester, England; Sc.D.(E.E.), 1962, University of New Mexico.

Fellow Award "For contributions to the conceptual design and analysis of space communications systems"; USAB Citation of Honor, 1980; Outstanding Member, Region 2, IEEE, 1982; President, IEEE Aerospace and Electronic Systems Society, 1982, 1983; Director, Division IX (Signals and Applications), 1984, 1985. Other Awards: Associate Fellow, American Institute of Aeronautics and Astronautics; Fellow, Washington Academy of Sciences; Member, Sigma Xi; NASA Special Achievement Awards, 1977, 1978.

DUTTON, JOHN C., Fellow 1982. Born: March 8, 1918, Chicago, IL. Degrees: B.S.E.(with honors), 1939, Swarthmore College; M.S.E.E., 1950, Rensselaer Polytechnic Institute.

Fellow Award "For significant technical leadership in international standards activities relating to transformers, voltages, and safety"; Transformers Committee Award, IEEE, 1978; Industrial Plants Power Systems Committee Prize Award, IEEE, 1980. Other Awards: Sigma Tau, 1939; Engineering Award, General Electric Power Systems Sector, 1979; NEMA, 1982, 1983;

U.S. Technical Adviser to IEC Technical Committees 14 & 14B (Transformers & On-Load-Tap-Changers); Lead U.S. Transformer Delegation (4) to IEC 1983 General Meeting in Tokyo, Japan.

DUTTON, ROBERT W., Fellow 1984. Born: March 24, 1944, Eugene, OR. Degrees: B.S., 1966, M.S., 1967, Ph.D., 1970, University of California, Berkeley.

Fellow Award "For contributions to computer-aided modeling of silicon devices and fabrication processes." Other Awards: Department Citation (UCB), 1966; NATO Fellowship, 1975; Fairchild Fellowship, 1966; NSF Fellowship, 1967-70.

DUYAN, PETER, JR., AIEE Fellow 1962. Born: April 18, 1915, Pittsburgh, Pa. Degrees: B.S.E.E., 1938, University of California.

Fellow Award "For contributions to engineering direction of electrical systems for aircraft." Other Awards: Citation, IAE.

DWON, LARRY, Fellow 1969. Born: May 2, 1913, New York, N.Y. Degrees: E.E., 1935, Cornell University; M.B.A., 1954, New York University Graduate School of Business Administration.

Fellow Award "For contributions to mutual understanding between electric power industry, the professional societies, and the schools of engineering"; Distinguished Service Award, Power Engineering Education Committee of Power Engineering Society, 1977; Region I Citation for Contribution to 1980-1981 Professional Awareness Program; USAB Award for Engineering Professionalism, IEEE, 1982; IEEE Centennial Medal, 1984. Other Awards: Fred Plummer Lecturer, American Welding Society, 1975; Listed in "Engineers of Distinction" and "Who's Who," American Association of Engineering Societies, Distinguished Service Award, 1977, National President, 1958, Eta Kappa Nu; Special Citations, Edison Electric Institute, 1977, 1978; Lectured to over 40 SPAC's since 1979; Numerous papers, chairmanships, and citations in IEEE, ECPD, EBI, HKN and ASEE; Eminent Membership, HKN, 1984; Registered P.E.(1941) in N.Y., (1983) in N.C.

DYER, GIBBS A., AIEE Fellow 1945. Born: December 20, 1893, Dallas, Tex.

Other Awards: Member, Association for Computing Machinery; Member, Society for Industrial and Applied Mathematics; Member, Acoustical Society of America; Tau Beta Pi; Eta Kappa Nu; Sigma Xi.

DYER, IRA, Fellow 1979. Born: June 14, 1925, New York, N.Y. Degrees: S.B., 1949, S.M., 1951, Ph.D., 1954 (all in Physics), Massachusetts Institute of Technology.

Fellow Award "For contributions to the science of acoustics and its applications and for distinguished academic leadership in advancing oceanic engineering and its applications"; Distinguished Technical Contri-

bution Award, IEEE Council on Oceanic Engineering, 1982. Other Awards: Fellow (and Biennial Award), Acoustical Society of America, 1960; National Academy of Engineering, 1976; Meritorious Public Service Award, U.S. Coast Guard, 1979; Fellow, American Association for the Advancement of Science, 1980.

DYER, JOHN N., IRE Fellow 1949. Born: July 14, 1910, Haverhill, Mass. Degrees: B.S., 1931, Massachusetts Institute of Technology.

Fellow Award "For administrative and technical contributions to radio, including polar expedition communications and important wartime radio countermeasures."

DYKE, WALTER P., Fellow 1967. Born: December 9, 1914, Forest Grove, Oreg. Degrees: B.A., 1938, Linfield College; Ph.D., 1946, University of Washington.

Fellow Award "For contributions to the understanding and development of high intensity field emission sources and the application of these sources to pulses x-ray equipment"; Electronic Achivement Award, IRE, Region 7, 1958; 1958; IEEE David Sarnoff Award, 1968, "For pioneering contributions to the development of field emission in x-ray technology." Other Awards: Presidential Certificate of Merit, 1946; Citation, Oregon Academy of Science, 1958; American Physical Society; Oregon Academy of Science; American Association of Physics Teachers; Sigma Xi; Sigma Pi Sigma.

DY LIACCO, TOMAS E., Fellow 1975. Born: November 12, 1920, Naga City, Philippines. Degrees: B.S.E.E., 1940, B.S.M.E., 1941, University of the Philippines; M.S.E.E., 1955, Illinois Institute of Technology; Ph.D., 1968, Case Western Reserve University.

Fellow Award "For contributions in advanced control concepts for the secure and reliable operation of electric power systems"; Prize Paper Award, IEEE Power Group, 1968, for paper "The Adaptive Reliability Control System."

DYSON, JOHN D., Fellow 1974. Born: August 9, 1918, Lemmon, S.Dak. Degrees: B.S., 1940, B.S., 1949, South Dakota State University; M.S., 1950, Ph.D., 1957, University of Illinois.

Fellow Award "For contributions to the development of log-spiral antennas."

E

EACHUS, JOSEPH J., Fellow 1970. Born: November 5, 1911, Anderson, Ind. Degrees: A.B., 1933, Miami University; M.A., 1936, Syracuse University; Ph.D., 1939, University of Illinois.

Fellow Award "For contributions and leadership in the design and application of large-scale information processing systems." Other Awards: Fellow, New York Academy of Sciences, 1966; Fellow, American Association for Advancement of Science, 1972.

EADIE, THOMAS W., Fellow 1968. Degrees: B.Sc., 1923, McGill University; L.L.D.(Hon.), 1959, University of Montreal; L.L.D.(Hon.), 1960, Dalhousie University.

Fellow Award "For outstanding technical, administrative, and executive contributions in the telecommunications industry." Other Awards: Knight of Grace, St. John of Jerusalem, 1956; Human Relations Award, Canadian Council of Christians and Jews, 1961; Julian C. Smith Medal, Engineering Institute of Canada, 1961.

EAGER, GEORGE S., JR., Fellow 1975. Born: September 5, 1915, Baltimore, Md. Degrees: B.E., 1936, D.Eng., 1941, The Johns Hopkins University.

Fellow Award "For the development of an industry-accepted method of corona detection in extruded dielectric power cables"; District First Prize Paper, IEEE, 1958, for paper "Transmission Properties of Polyethylene Insulated Telephone Cables at Voice and Carrier Frequencies."

EARLY, JAMES M., IRE Fellow 1959. Born: July 25, 1922, Syracuse, N.Y. Degrees: B.S., 1943, New York State College of Forestry; M.S., 1948, Ph.D., 1951, Ohio State University.

Fellow Award "For contributions in the development of high frequency transistors"; J.J. Ebers Award, IEEE Electron Devices Society, 1979. Other Awards: TEXNIKOI Outstanding Alumnus Award, Ohio State University, 1967.

EARLY, RUPERT, AIEE Fellow 1943. Born: June 10, 1885, Hillsville, Va. Degrees: B.S., 1907, Virginia Polytechnic Institute.

EARP, CHARLES W., Fellow 1967. Born: July 14, 1905, Cheltenham, Gloucester, England. Degrees: B.A., 1927, University of Cambridge.

Fellow Award "For his contributions to the art of radio direction-finding antenna systems." Other Awards: Kelvin Premium, Institution of Electrical Engineers, U.K., 1948; Officer of the Order of the British Empire, 1966; Heaviside Premium, Institution of Electrical Engineers, U.K., 1967; First J.J. Thomson Medal, Institution of Electrical Engineers, U.K., 1976.

EASTERLING, MAHLON F., Fellow 1973. Born: May 11, 1921, Muskegon Heights, Mich. Degrees: B.S.E.E., 1949, M.S.E.E., 1951, Columbia University.

Fellow Award "For developments in acquirable pseudorandom ranging systems, and coded digital communications for space applications." Other Awards: Exceptional Service Medal, NASA, 1977; Medal for Outstanding Leadership, NASA, 1979.

EASTMAN, LESTER F., Fellow 1969. Born: May 21, 1928, Utica, N.Y. Degrees: B.S.E.E., 1953, M.S., 1955, Ph.D., 1957, Cornell University.

Fellow Award "For contributions to microwave solid-state electronics"; 1983 National Lecturer for IEEE

Electron Device Society; Organizer, IEEE-Cornell biennial conference on microwave solid-state devices, since 1967; Chairman, 1982 International Symposium on Gallium Arsenide and Related Compounds. Other Awards: Director of Joint Services Electronics Program on compound semiconductor materials and devices at Cornell University; Member, U.S. Government Advisory Group on Electron Devices.

EASTON, ELMER C., Fellow 1967. Born: December 23, 1909, Newark, N.J. Degrees: B.S., 1931, M.S., 1933, Lehigh University; Sc.D., 1942, Harvard University; D.Eng.(Hon.), 1965, Lehigh University.

Fellow Award "For his contributions to engineering education."

EASTON, IVAN G., IRE Fellow 1960. Born: November 20, 1916, Sweden. Degrees: B.S.E.E., 1938, Northeastern University; M.S.E.E., 1940, Harvard University.

Fellow Award "For contributions to impedance measuring techniques and equipment"; Award, IEEE Instrumentation and Measurement Society; President, 112E I&M Group, 1953; Chairman, CPEM, 1960, NEREM, 1954, IEC TC66, 1973-1980; Staff Director of Standards, IEEE, 1976-1980; President, U.S. National Committee, International Electrotechnical Commission; Director, American National Standards Institute. Other Awards: Alumni Citation, Northeastern University.

EASTON, ROGER L., Fellow 1977. Born: April 30, 1921. Degrees: A.B., 1943, Middlebury College.

Fellow Award "For contributions to the development of navigation satellites and worldwide precise time transfer." Other Awards: Distinguished Civilian Service Award, 1960; Award, Radio Engineers Professional Group for Reliability and Quality Control, 1960; Publications Award, Naval Research Laboratory, 1970; Applied Science Award, Sigma Xi, 1974.

ECKER, H. ALLEN, Fellow 1977. Born: October 22, 1935, Athens, Ga. Degrees: B.E.E., 1957, M.S.E.E., 1959, Georgia Institute of Technology; Ph.D.(E.E.), 1965, Ohio State University.

Fellow Award "For contributions in the application of radar scattering analyses and measurements to radar discrimination." Other Awards: Armed Forces Communications and Electronics Association Award, 1957; Air Force Association Award, 1957; Air Force Special Act of Service Award, 1963; Air Force Association Engineering Achievement Award, 1963; Air Force Civilian Sustained Superior Performance Award, 1964; Air Force Civilian Outstanding Performance Award, 1965; Georgia Tech Sigma Xi Research Award in Engineering, 1974; Member: Sigma Xi, Phi Kappa Phi, Tau Beta Pi, Eta Kappa Nu.

EDELSON, BURTON I., Fellow 1976 Born: July 31, 1926, New York, N.Y. Degrees: B.S., 1947, United

States Naval Academy; M.S., 1954, Ph.D., 1960, Yale University.

Fellow Award "For contributions to and leadership in the development of satellite communications systems." Other Awards: Howe Research Award, American Society for Metals, 1963; Legion of Merit, U.S. Navy, 1965; Fellow, American Institute of Aeronautics and Astronautics, 1976.

EDEN, MURRAY, Fellow 1973. Born: 1920, New York, N.Y. Degrees: S.B., 1939, City College of The City University of New York; S.M., 1944, Ph.D., 1951, University of Maryland.

Fellow Award "For contributions to the application of technology to the health sciences." Other Awards: Member, Biophysical Society; American Physiological Society; American Society for Engineering Education; Association for Medical Instrumentation; Sigma Xi; Tau Beta Pi.

EDEN, RICHARD C., Fellow 1983. Born: July 10, 1939, Anamosa, Iowa. Degrees: B.S., 1961, Iowa State University; M.S., 1962, California Institute of Technology; Ph.D., 1967, Stanford University.

Fellow Award "For contribution to the development of high speed gallium arsenide integrated circuits and III-V alloy photodetectors." Other Awards: Rockwell International Engineer of the Year Award, 1977.

EDGERTON, HAROLD E., AIEE Fellow 1946, IRE Fellow 1956. Born: April 6, 1903, Fremont, Nebr. Degrees: B.Sc., 1925, University of Nebraska; M.Sc., 1927, D.Sc., 1931, Massachusetts Institute of Technology; D.Eng.(Hon.), 1948, University of Nebraska; LL.D.(Hon.), 1969, Doane College; LL.D.(Hon.), 1969, University of South Carolina; D.Sc.(Hon.), 1979, Washington University in St. Louis.

IRE Fellow Award "For contributions in the application of electronic techniques to high-speed stroboscopic photography." Other Awards: Morris E. Leeds Award, 1965; National Best Paper Prize in Theory and Research, AIEE, 1936; Award, Oceanography Coordinating Committee, IEEE, 1974. Other Awards: Certificate of Appreciation, War Dept.; Citation, University of California; Medal, Royal Photographic Society of London, 1936; Modern Pioneer Award, National Association of Manufacturers, 1940; Howard N. Potts Medal, Franklin Institute, 1941; Medal of Freedom, 1946; Master of Photography, Photographers Association of America, 1949; Joseph A. Sprague Award, National Press Photographic Association, 1949; U.S. Camera Achievement Gold Medal Award, 1951; Photography Magazine Award, 1952; Franklin L. Burr Prize, National Geographic Society, 1953; Progress Medal Award, Photographic Society of America, 1955; 75th Anniversary Citation, Photographers Association of America, 1955; Progress Award, Society of Motion Picture and Television Engineers, 1959; George W. Harris Achievement Award, Photographers Association of America,

1959; New England Engineer of the Year Award, Engineering Societies of New England, 1959; Boston Sea Rovers Award, 1960. Gordon Y. Billard Award, Massachusetts Institute of Technology, 1962; E. I. duPont Gold Medal Award, Society of Motion Picture and Television Engineers, 1962; Man of the Year Award, Industrial Photographers Association of America, 1963; Silver Progress Medal, Royal Photographic Society of Great Britain, 1964; Technical Achievement Award, American Society of Magazine Photographers, 1965; Richardson Medal, Optical Society of America, 1968; Eminent Member Award, Eta Kappa Nu, 1968; John Oliver LaGorce Gold Medal, National Geographic Society, 1968; Alan Gordon Memorial Award, S.P.I.E., 1965; Albert A. Michelson Medal, Franklin Institute, 1969; Distinguished Service Award, University of Nebraska, 1972; NOGI Award, Underwater Society of America, 1971; Holley Medal (with K. Germeshausen), American Society of Mechanical Engineers, 1973; National Medal of Science, 1973; Lockheed Award for Marine Science and Engineering, Marine Science and Technology Society, 1978; Nikon Special Achievement Citation, 1979; Photographic Administrators Award for Technical Achievement, 1980; Professor Emeritus, Massachusetts Institute of Technology, 1968-1981. Fellow: Photography Society of America, Royal Photographic Society of Great Britain, Society of Motion Picture and Television Engineers; Member: Academy of Applied Science, Academy of Underwater Arts and Sciences, American Academy of Arts and Sciences, American Philosophical Society, Eta Kappa Nu, Marine Technology Society, Boston Museum of Science, National Academy of Engineering, National Academy of Sciences, Society of Photographic Engineers, Sigma Tau, Sigma Xi, Woods Hole Oceanographic Institute; Honorary Member, Boston Camera Club; Honorary Member, Boston Sea Rovers; New England Aquarium; Honorary Member, Fairhaven Whalers Club; Photographers Association of New England; Honorary Member, Society of Motion Picture and TV Engineers.

EDSALL, WILLIAM S., AIEE Fellow 1962. Born: May 17, 1890, Bozeman, Mont. Degrees: B.S.E.E., 1911, E.E., 1914, Montana State University; B.C.S., 1916, New York University; D.Eng.(Hon.), 1955, Montana State University.

Fellow Award "For contributions through the invention and development of circuit protective devices."

EDSON, WILLIAM A., IRE Fellow 1957. Born: October 30, 1912, Burchard, Nebr. Degrees: B.S., 1934, M.S., 1935, University of Kansas; D.Sc., 1937, Harvard University.

Fellow Award "For contributions in the fields of education and microwave electronics"; Award, San Francisco Bay Area Council, IEEE, 1980; IEEE Centennial Medal, 1984. Other Awards: Wescon Award, Chairman Executive Committee, 1979.

EDWARDS, MARTIN A., AIEE Fellow 1947, IRE Fellow 1954. Born: March 22, 1905, Chautauqua, Kans. Degrees: B.S.E.E., 1928, B.S.M.E., 1929, M.E., 1934, D.Sc.(Hon.), 1946, Kansas State University.

IRE Fellow Award "For his creative contributions to the development of the amplidyne and other control systems." Other Awards: Charles A. Coffin Awards, General Electric, 1934, 1940, 1949; General Electric Gold Medal to Inventors, "50th Patent"; Invention Fulcrium of Progress.

EDWARDS, ROBERT F., AIEE Fellow 1962. Born: March 21, 1918, Wilkinsburg, Pa. Degrees: B.S., 1940, M.S., 1943, University of Pittsburgh.

Fellow Award "For contributions to the development and design of synchronous machines and means of calculating performance."

EERNISSE, ERROL P., Fellow 1980. Born: February 15, 1940, Rapid City, SD. Degrees: B.S.E.E., 1962, South Dakota State University; M.S.E.E., 1963, Ph.D., 1965, Purdue University; M.(Indust. Admin.), 1974, University of New Mexico.

Fellow Award "For outstanding contributions to the analysis and development of piezoelectric devices." Other Awards: Fellow, American Physical Society, 1981; W.G. Cady Award, Frequency Control Symposium, 1983.

EILERS, CARL G., Fellow 1977. Born: March 21, 1925, Fairbury, Ill. Degrees: B.S.E.E., 1948, Purdue University; M.S.E.E., 1956, Northwestern University.

Fellow Award "For pioneering contributions to the FM stereophonic broadcasting system."

EINSPRUCH, NORMAN G., Fellow 1980. Born: June 27, 1932, New York, NY. Degrees: B.A.(Physics), 1953, Rice University; M.S.(Physics), 1955, University of Colorado; Ph.D.(Applied Math.), 1959, Brown University.

Fellow Award "For research in the transport properties of electronic materials and acoustic effects in solids." Other Awards: Fellow, American Physical Society; Fellow, Acoustical Society of America; Fellow, American Association for the Advancement of Science; Member, American Institute of Industrial Engineers; Member, American Society for Engineering Education; Sigma Xi; Tau Beta Pi; Eta Kappa Nu; Phi Kappa Phi; Tau Sigma Delta; Alpha Pi Mu.

EITEL, WILLIAM W., IRE Fellow 1953. Born: January 16, 1908, San Jose, Calif.

Fellow Award "For his pioneering contributions to power tube design." Other Awards: Distinguished Public Service Award, U.S. Dept. of Navy, 1950; Fellow, Radio Club of America; Sarnoff Award, Radio Club of America; Professional Electrical Engineer, State of California.

EKSTRAND, STURE O., Fellow 1967. Born: January 10, 1910, New York, N.Y. Degrees: B.S.E.E., 1935, Cooper Union.

Fellow Award "For exceptional engineering contributions and technical leadership in the development and standardization of electron tubes and semi-conductor devices."

EL-ABIAD, AHMED H., Fellow 1973. Born: May 24, 1926, Mersa Matruh, Egypt. Degrees: B.Sc.E.E., 1948, Cairo University; M.S.E.E., 1953, Ph.D., 1956, Purdue University.

Fellow Award "For contributions to the analysis of power systems, and to education for the power industry." Other Awards: Fulbright and Smith Mund U.S. Government Scholarship, 1952-1953.

ELFANT, ROBERT F., Fellow 1972. Born: March 12, 1936, New York, N.Y. Degrees: B.S.E.E., 1957, Southern Methodist University; M.E.E., 1959, New York University; Ph.D., 1961, Purdue University.

Fellow Award "For contributions to magnetic memory systems." Other Awards: Outstanding Young Electrical Engineer, Eta Kappa Nu, 1967; Distinguished Engineering Alumni, Purdue University, 1968.

ELGERD, OLLE I., Fellow 1971. Born: March 31, 1925, Mora, Sweden. Degrees: Dipl., 1950, The Royal Institute of Technology, Stockholm, Sweden; Ph.D., 1956, Washington University, St. Louis, MO.

Fellow Award "For contributions in the field of control and electric energy engineering, and leadership in electric power energy education."

ELIAS, PETER, IRE Fellow 1959. Born: November 26, 1923, New Brunswick, N.J. Degrees: S.B., 1944, Massachusetts Institute of Technology; M.A., 1948, M.Eng.Sci., 1949, Ph.D., 1950, Harvard University.

Fellow Award "For contributions to information theory and to engineering education"; Shannon Lecturer, IEEE Information Theory Society, 1977. Other Awards: Lowell Fellow, Harvard University, 1950-1953; Sigma Xi, 1950; Fellow, American Academy of Arts and Sciences, 1961; Eta Kappa Nu, 1964; Member, National Academy of Sciences, 1975; National Academy of Engineering, 1979.

EL-KOSHAIRY, MAHMOUD A. B., Fellow 1977. Born: April 11, 1909, Tanta, Egypt. Degrees: Dipl.(First Class Distinction), 1929, Royal School of Engineering, Cairo; A.C.G.I.(Associate of the City and Guilds Institute), 1932, B.Sc.(First Class Honors), 1933, D.I.C.(Diploma of Imperial College), 1936, Ph.D., 1936, City and Guilds Engineering College, Imperial College, London University.

Fellow Award "For leadership in the development of his Nation's power system, and engineering education." Other Awards: First Order of Decorations of the Republic 1964, Merit 1980, Science and Arts 1980, Medal of Distinguished Service 1972; Fellow, Institution of Electrical Engineers; Fellow, Egyptian Engineers Society; Canadian Solar Energy Society; Chairman,

Egyptian National Committees, CIGRE, I.E.C., W.E.C.

ELLERT, FREDERICK J., Fellow 1973. Born: April 8, 1929, New Britain, Conn. Degrees: B.E.E., 1951, M.E.E., 1952, Ph.D., 1964, Rensselaer Polytechnic Institute.

Fellow Award "For contributions to automatic control and high-voltage direct current transmission." Other Awards: Fellow, Tau Beta Pi, 1951; American Automatic Control Council Prize Paper Award, 1962.

ELLIOTT, ROBERT G., Fellow 1964. Born: October 6, 1907, Craigmont, Idaho. Degrees: B.S., 1928, University of Idaho.

Fellow Award "For contributions to communications services and as an engineering manager."

ELLIOTT, ROBERT S., IRE Fellow 1961. Born: March 9, 1921, New York, N.Y. Degrees: A.B., 1942, B.S., 1943, Columbia University; M.S., 1947, Ph.D., 1952, University of Illinois; M.A., 1971, University of California at Santa Barbara.

Fellow Award "For contributions in the field of electromagnetic waves."

ELLIS, HARRY M., Fellow 1970. Born: January 12, 1923, Vancouver, British Columbia, Canada. Degrees: B.Ap.Sc., 1945, University of British Columbia; M.Sc., 1948, Ph.D., 1951, California Institute of Technology.

Fellow Award "For contributions to electric utility transmission planning."

ELLIS, LYNN W., Fellow 1974. Born: February 27, 1928, San Mateo, Calif. Degrees: B.E.E., 1948, Cornell University; M.Sc., 1954, Stevens Institute of Technology; D.P.S., 1979, Pace University.

Fellow Award "For contribution in introducing solid-state technology in wire and microwave communication transmission and leadership in communications planning"; IEEE Award in International Communication in Honor of Hernand and Sosthenes Behn, 1983. Other Awards: Phi Kappa Phi, 1948; Certificate of Appreciation, U.S. Dept. of Commerce, 1975; Fellow, American Association for the Advancement of Science, 1976.

ELMENDORF, CHARLES H., Fellow 1965. Born: July 1, 1913, Los Angeles, Calif. Degrees: B.S., 1935, M.S., 1936, California Institute of Technology.

Fellow Award "For contributions in the fields of wideband cable systems and transmissions systems development." Other Awards: Member, National Academy of Engineers, 1971.

ELMORE, WALTER A., Fellow 1983. Born: October 2, 1925, Bartlett, Tennessee. Degrees: B.S.(Elec.Eng.), 1949, University of Tennessee.

Fellow Award "For contributions to the advancement of protective relay systems and their applications"; District First Prize Paper Award, AIEE, 1962; Chairman, IEEE Power Systems Relaying Committee, 1983-1984. Other Awards: Tau Beta Pi, Eta Kappa Nu, Phi Kappa Phi, 1949.

ELMS, JAMES C., Fellow 1970. Born: May 16, 1916, East Orange, N.J. Degrees: B.S., 1948, California Institute of Technology; M.A., 1950, University of California at Los Angeles.

Fellow Award "For contributions and leadership in the fields of airborne systems and manned space flight." Other Awards: Assoc. Fellow, AIAA, 1964; NASA Certificate of Appreciation and Special Award, 1964; NASA Exceptional Service Medal, 1969; NASA Outstanding Leadership Medal, 1970; DOT Award for Meritorious Achievement, 1974; National Academy of Engineering, 1974; Explorers Club, 1979.

EL-SAID, MOHAMED A. H., Fellow 1977. Born: May 14, 1916, Aga, Egypt. Degrees: B.Sc.E.E., 1938, M.Sc.E.E., 1943, Ph.D.(E.E.), 1944, Cairo University.

Fellow Award "For leadership in electronics research and engineering in his Country."

ELSPAS, BERNARD, Fellow 1976. Born: July 26, 1925, New York, N.Y. Degrees: B.E.E., City College of the City University of New York; M.E.E., New York University; Ph.D., Stanford University.

Fellow Award "For contributions to information theory and to switching and automata theory."

ELZI, J. A., AIEE Fellow 1948. Born: September 14, 1896, Boulder, Colo. Degrees: B.S., 1918, University of Colorado.

EMBERSON, RICHARD M., Fellow 1967. Born: April 2, 1914, Columbia, Mo. Degrees: A.B., 1931, M.A., 1932, Ph.D., 1936, University of Missouri.

Fellow Award "For contributions to spectrum allocations and to the coordination of engineering and scientific programs"; Director Emeritus, IEEE, 1979.

EMLING, JOHN W., Fellow 1965. Born: June 3, 1904, Erie, Pa. Degrees: B.S.E.E., 1925, University of Pennsylvania.

Fellow Award "For contributions to the principles of the design and rating of communication systems."

ENDICOTT, HAROLD S., AIEE Fellow 1958. Born: December 10, 1901, Fremont, Iowa. Degrees: B.S., 1923, California Institute of Technology.

Fellow Award "For contributions to the techniques and precision of dielectric measurements." Other Awards: Award of Merit, American Society for Testing and Materials, 1968; Fellow, American Society for Testing and Materials.

ENGEL, JOEL S., Fellow 1980. Born: February 4, 1936, New York, NY. Degrees: B.E.E., 1957, City College of New York; M.S.E.E., 1959, Massachusetts Institute of Technology; Ph.D., 1964, Polytechnic Institute of Brooklyn.

Fellow Award "For contributions to the concept and to the implementation of spectrally efficient, cellular mobile telephone systems"; Paper of the Year Award, IEEE Vehicular Technology Group, 1969, 1973.

ENGELBRECHT, RUDOLF, Fellow 1970. Born: April 18, 1928, Atlanta, Ga. Degrees: B.E.E., 1951, M.E.E., 1953, Georgia Institute of Technology; Ph.D., 1979, Oregon State University.

Fellow Award "For contributions to parametric, domain, and other active microwave devices, and for leadership in the integration of microwave circuits."

ENGELER, WILLIAM ERNEST, Fellow 1979. Born: November 13, 1928. Degrees: B.S., 1951, Polytechnic Institute of Brooklyn; M.S., 1958, Ph.D., 1961, Syracuse University.

Fellow Award "For contributions to the understanding of surface-charge transport and its application to the development of new devices."

ENGEN, GLENN F., Fellow 1984. Born: April 26, 1925, Battle Creek, MI. Degrees: B.A., 1947, Andrews University; Ph.D., 1969, University of Colorado.

Fellow Award "For contributions to microwave metrology, particularly the development of the "six-port" measurement technique"; Society Award, IEEE Instrumentation and Measurement Society, 1982. Other Awards: Silver Medal for Meritorious Service, Dept. of Commerce, 1961; Gold Medal for Distinguished Achievement, Dept. of Commerce, 1976; Applied Research Award, National Bureau of Standards, 1981; Woodington Award for Professionalism in Metrology, Measurement Science Conference, 1983.

ENGL, WALTER L., Fellow 1980. Born: April 8, 1926, Regensburg, West Germany. Degrees: Dipl.(Physics), 1949, Dr. rer. nat., 1953, Technische Hochshule Munich; venia legendi, 1961, Technische Hochschule Karlsruhe, Germany.

Fellow Award "For outstanding contributions in integrated circuits design techniques and device modeling." Other Awards: Academy of Science, Northrhine-Westfalia, 1978.

ENGLEMAN, CHRISTIAN L., IRE Fellow 1954. Born: April 3, 1906, Vancouver, Wash. Degrees: B.S., 1930, United States Naval Academy; M.S., 1939, Harvard University.

Fellow Award "For his contributions to the administration of electronic programs of the U.S. Navy."

ENGSTROM, ELMER W., IRE Fellow 1940, AIEE Fellow 1949. Born: August 25, 1901, Minneapolis, Minn. Degrees: B.S., 1923, University of Minnesota; Recipient of 17 honorary doctoral degrees from American colleges and universities and one from a European university.

Awards: Founders Award, 1966, "For his leadership in management and integration of research and development programs and for his foresighted application of the systems engineering concept in bringing television to the public." Other Awards: Silver Plaquette, Royal Swedish Academy of Engineering Sciences, 1949; John Ericsson Medal, American Society of Swedish Engineers, 1956; Industrial Research Institute Medal, 1958; Christopher Columbus International Award in Telecommunications, 1959; Medal for the Advancement of Research, American Society for Metals, 1960; Medal of Honor, Electronic Industries Association, 1962; Award of Merit, Aerospace Electrical Society, 1963; Charles Proteus Steinmetz Medal, National Academy of Engineering, 1965; Commander of the Royal Order of Vasa, King of Sweden, 1965; William Proctor Prize, Scientific Research Society of America, 1966; Progress Medal Award, Society of Motion Picture and Television Engineers, 1955.

ENNIS, BRUCE J., AIEE Fellow 1961. Born: December 9, 1910, Kansas City, Mo. Degrees: B.S., 1933, Massachusetts Institute of Technology.

Fellow Award "For contributions to the design and construction of electric power supply systems." Other Awards: Registered Professional Engineer, Missouri, Maine, New Hampshire, Vermont, Indiana.

ENNS, MARK K., Fellow 1977. Born: October 13, 1931, Hutchinson, Kans. Degrees: B.S.E.E., 1953, Kansas State University; M.S.E.E., 1960, Ph.D., 1967, University of Pittsburgh.

Fellow Award "For contributions to the development of hybrid computers and their application to the operation and control of electric power centers." Other Awards: Westinghouse Lamme Scholarship, 1964; Listed in "Who's Who in America," 1978.

ENNS, WALDO E., AIEE Fellow 1957. Born: November 8, 1901, Trenton, Mo. Degrees: B.S.E.E., 1924, University of California at Berkeley.

Fellow Award "For contributions to the development of a power utility system and the invention of a simplified network analyzer."

ENTZMINGER, JOHN N., JR., Fellow 1981. Born: December 17, 1936, Memphis, TN. Degrees: B.S.(E.E.), 1959, University of South Carolina; M.S.(E.E.), 1968, Syracuse University.

Fellow Award "For technical leadership in the development of weapons systems for the detection, location, and strike of ground targets"; Chairman, Mohawk Valley Section, IEEE, 1979-1980. Other Awards: Tau Beta Pi, 1958; Phi Beta Kappa, 1959; Dept. of Air Force Decoration for Exceptional Civilian Service, 1974; USAF Rome Air Development Center, Harry I. Davis Award for Outstanding Technical Achievement, 1978; Member, Senior Executive Service, 1981.

EPHREMIDES, ANTHONY, Fellow 1984. Born: September 19, 1943, Athens, Greece. Degrees: B.S.E.E., 1967, National Technical University of Athens; M.A., 1969, Ph.D.(E.E.), 1971, Princeton University.

Fellow Award "For contributions to statistical communication theory, modeling and analysis of communication networks, and engineering education." Other Awards: George Corcoran Award, University of Maryland, 1977; Distinguished Program Award, Maryland Association of Higher Education, 1982; CSC, 1979; Prize Paper Awards, NRL, 1981.

EPPERS, WILLIAM C., JR., Fellow 1976. Born: January 17, 1930. Degrees: B.E.E., 1950, D.Eng., 1962, The Johns Hopkins University.

Fellow Award "For technical contributions and leadership in the field of laser technology and applications." Other Awards: Samuel M. Burka Award, 1969; Outstanding Engineer, Dayton Area, 1972; Tau Beta Pi; Eta Kappa Nu; Sigma Xi; Gamma Alpha.

EPPERSON, JOSEPH B., IRE Fellow 1958, AIEE Fellow 1958. Born: May 8, 1910, Charleston, Tenn.

IRE Fellow Award "For his contributions to radio and television broadcasting"; AIEE Fellow Award "For his achievements in the broadcasting field." Other Awards: Tau Beta Pi Award, University of Tennessee, 1957; Engineering Achievement Award, National Association of Broadcasters, 1974.

EPSTEIN, HENRY D., Fellow 1975. Born: April 5, 1927, Frankfurt-main, Germany. Degrees: B.S.E.E., 1948, Brown University; M.S., 1950, Harvard Graduate School of Engineering, Harvard Business School.

Fellow Award "For development of electrical protection devices for consumer safety and pollution control."

ERATH, LOUIS W., IRE Fellow 1962. Born: June 10, 1917, Abbeville, La. Degrees: 1939, Louisiana State University.

Fellow Award "For contributions to seismic and geophysical instruments and to naval ordnance." Other Awards: Meritorious Civilian Service Award, U.S. Dept. of Navy, 1945.

ERDELYI, EDWARD A., AIEE Fellow 1960. Born: July 13, 1908, Hlohovec, Czechoslovakia. Degrees: B.Sc., 1926, Czechoslovakia Technical University, Brno; M.Sc., 1929, German Technical University, Brno; Dipl.-Ing., 1929, Masaryk University, Brno; Ph.D., 1955, University of Michigan; Doctor honoris causa, 1971, Budapest Polytechnic University; Dr.Ing. honoris causa, 1977, Technical University, Braunsechweig.

Fellow Award "For contributions to the theory of noise in electric machinery"; Outstanding Paper Prize, Professional Group on Aerospace, 1963, for "The Non-Linear Potential Equation and its Solution for Highly Saturated Machines"; Paper Prize, Honorable Mention, Professional Group on Aerospace, 1964, for "Non-Linear Vector Potential Equation for Saturated Heteropolar Electrical Machines"; Recognition Award, Rotating Machinery Committee, 1967, "For outstanding contributions to the field of rotating electrical machinery by his work on flux distribution and nonlinear solutions for saturated machines. His tireless efforts in disseminating that knowledge has inspired many in our field." Other Awards: Engineer of the Year

Award, Pennsylvania Engineering Society, 1958; Certificate of Appreciation, American Society of Mechanical Engineers, 1963.

ERICKSON, WILLIAM H., AIEE Fellow 1962. Born: April 4, 1916, McKeesport, Pa. Degrees: B.S., 1938, University of Pittsburgh; M.S., 1946, Carnegie Institute of Technology.

Fellow Award "For contributions to engineering education."

ERICSON, RUDOLPH C., AIEE Fellow 1950. Born: December 6, 1901, Orion, Ill. Degrees: B.S., 1925, E.E., 1946, University of Illinois.

ERNST, EDWARD W., Fellow 1976. Born: August 28, 1924, Great Falls, Mont. Degrees: B.S., 1949, M.S., 1950, Ph.D., 1955, University of Illinois at Urbana-Champaign.

Fellow Award "For contributions to improved methods of laboratory instruction in electrical engineering education"; IEEE Centennial Medal, 1984. Other Awards: Fellow, American Association for the Advancement of Science, 1979; Halliburton Engineering Education Leadership Award, 1983.

ESAKI, LEO, IRE Fellow 1962. Born: March 12, 1925, Osaka, Japan. Degrees: B.S., 1947, Ph.D., 1959, University of Tokyo.

Fellow Award "For contributions to the semiconductor field, and specifically the tunnel diode"; Morris N. Liebmann Award, IRE, 1960, "For important contributions to the theory and technology of solid state devices, particularly as embodied in the tunnel diode." Other Awards: Nishina Memorial Award, 1959; Asahi Press Award, 1960; Tokyo Rayon Foundation Award, 1961; Stuart Ballantine Medal, Franklin Institute, 1961; Award, Japan Academy, 1965; Fellow, American Physical Society, 1967; Fellow, International Business Machines, 1967; Nobel Prize in Physics, 1973; Order of Culture from Japanese Government, 1974; Member, Japan Academy, 1975; Member, National Academy of Science (Foreign Assoc.), 1976; Member, National Academy of Engineering (Foreign Assoc.), 1977; IBM-Japan Board Member, 1976; IBM-Japan Science Institute Governing Board, 1981.

ESHLEMAN, VON R., IRE Fellow 1962. Born: September 17, 1924, Darke Co., Ohio. Degrees: B.S.E.E., 1949, George Washington University; M.S., 1950, Ph.D., 1952, Stanford University.

Fellow Award "For contributions to radar astronomy and communications." Other Awards: Engineer Alumni Achievement Award, 1974, Alumni Achievement Award, 1975, George Washington University; National Academy of Engineering, 1977; NASA Medal for Exceptional Scientific Achievement, 1981.

ESPENSCHIED, LLOYD, IRE Fellow 1924, AIEE Fellow 1930. Born: April 27, 1889, St. Louis, Mo. Degrees: A.E. 1909, D.Sc.(Hon.), 1957, Pratt Institute.

Awards: First Paper Prize, AIEE, 1935, for "Coaxial Cable Wide-Band System"; Medal of Honor, IRE, 1940; Pioneer Award, IEEE, Aerospace and Electronic Group, 1967. Other Awards: First Annual Award, Television Broadcasters Association, 1944; Citation, Broadcast Pioneers, 1954.

ESPERSEN, GEORGE A., IRE Fellow 1956. Born: May 17, 1906, Jersey City, N.J. Degrees: B.S., 1931, New York University.

Fellow Award "For contributions to the fields of thermionic emission and electron tubes." Other Awards: Andrew Carnegie Scholarship, Institute for Theoretical Physics, Copenhagen, Denmark, 1929-1930; Samuel Morse Medal, New York University, 1929.

ESPOSITO, RAFFAELE, Fellow 1975. Born: August 11, 1932, Rome, Italy. Degrees: Dr.Ing., 1956, "Libera Docenza," 1971, University of Rome.

Fellow Award "For contributions to communication theory and system analysis."

ESSELMAN, WALTER H., Fellow 1965. Born: March 19, 1917, Hoboken, N.J. Degrees: B.S., 1938, Newark College of Engineering; M.S., 1944, Stevens Institute of Technology; D.E.E., 1953, Polytechnic Institute of Brooklyn.

Fellow Award "For contributions in the field of automatic elevator controls and reactor power plant instrumentation and design." Other Awards: Edward Weston Award, New Jersey Institute of Technology, 1962; Fellow, American Nuclear Society, 1965; Allan R. Cullimore Award, New Jersey Institute of Technology, 1968.

ESTCOURT, VIVIAN F., Fellow 1983. Born: May 31, 1897, London, England. Degrees: A.B. (Mechanical and Electrical Engineering), 1922, Stanford University.

Fellow Award "For contributions to power plant design and pioneering application of air pollution control equipment and power plant automation"; IEEE Power Generation Committee Award under title of "Honorary Member of the Power Generation Committee," 1982. Other Awards: National Academy of Engineering, 1981; Newcomen Medal (ASME/Franklin Institute), 1966; ASME Honorary Member Award, 1963; ASME Prime Movers Committee Award, 1958; Fellow, ASME, 1954; Registered Professional Electrical and Mechanical Engineer, State of California.

ESTRADA, HERBERT, Fellow 1964. Born: July 25, 1903, Philadelphia, Pa. Degrees: B.S., 1926, M.S., 1927, University of Pennsylvania.

Fellow Award "For contributions to the design and operation of interconnected power generating stations and systems." Other Awards: Fellow, American Society of Mechanical Engineers, 1959.

ESTRIN, GERALD, Fellow 1968. Born: September 9, 1921, New York, N.Y. Degrees: B.S.E.E., 1948,

M.S.E.E., 1949, Ph.D.E.E., 1951, University of Wisconsin.

Fellow Award "For outstanding contributions to university research and education in computers, and to the design and construction of pioneering digital computers." Other Awards: Lipsky Fellowship, 1954-55; Guggenheim Fellowships, 1962-63, 1967; Distinguished Service Citation, University of Wisconsin, 1975; Special Recognition Award, Third Jerusalem Conference on Information Technology, 1978; Chair, Computer Science Department, UCLA, 1979-1982; Governor, Weizmann Institute of Science, Israel, 1971-; Scientific Advisory Council, Gould, Inc., 1981-1982.

ESTRIN, THELMA, Fellow 1977. Born: February 21, 1924, U.S.A. Degrees: B.S.E.E., 1948, M.S.E.E., 1949, Ph.D.(E.E.), 1951, University of Wisconsin.

Fellow Award "For contributions to the design and application of computer systems for neurophysiological and brain research." Other Awards: Distinguished Service Citation, University of Wisconsin, College of Engineering, Madison, 1975; Achievement Award, Society of Women Engineers, 1981.

ESVAL, ORLAND E., AIEE Fellow 1961. Born: December 15, 1904, Renville Co., N.D. Degrees: B.S.E.E., 1929, Iowa State University.

Fellow Award "For contributions in the field of aircraft flight instrumentation"; First Paper Prize, AIEE, 1945, for "Electric Automatic Pilots for Aircraft." Other Awards: Member, International Executive Service Corps.

EVANS, BOB O., Fellow 1971. Born: August 19, 1927. Degrees: B.S.E.E., 1949, Iowa State College.

Fellow Award "For leadership in the development of high-speed digital computers"; Edwin Howard Armstrong Achievement Award, IEEE Communications Society, 1984. Other Awards: NASA Public Service Award, 1969; Meritorious Service Award, Armed Forces Communication and Electronic Association, 1969; Member, National Academy of Engineering, 1970; Professional Achievement Citation in Engineering, Iowa State University, 1971; National Security Agency Citation, 1975.

EVANS, DAVID C., Fellow 1975. Born: February 24, 1924, Salt Lake City, Utah. Degrees: B.S., 1949, Ph.D., 1953, University of Utah.

Fellow Award "For contributions in the field of computer science." Other Awards: Member, National Academy of Engineering; Computer Design Hall of Fame, 1982; Charter Member, Computer Science & Engineering Board, NAE.

EVANS, HAYDEN W., Fellow 1970. Born: November 26, 1913, Springfield, Ohio. Degrees: A.B., 1934, Ohio Wesleyan; B.S., 1936, Michigan University.

Fellow Award "For contributions to radio relay and

satellite communication systems." Other Awards: Member, American Institute of Aeronautics and Astronautics; Tau Beta Pi; Phi Delta Phi; Sigma Xi.

EVANS, JOHN V., Fellow 1968. Born: July 5, 1933, Manchester, England. Degrees: B.Sc., 1954, Ph.D., 1957, Manchester University.

Fellow Award "For numerous contributions to radar astronomy." Other Awards: Appleton Prize in Ionospheric Physics, Royal Society of London, 1975.

EVANS, RALPH A., Fellow 1978. Born: February 2, 1924. Degrees: B.S.(Engineering Physics), 1944, Lehigh University; Ph.D.(Physics), 1954, University of California, Berkeley.

Fellow Award "For contributions to the practice and advancement of reliability theory"; Editor, IEEE Transactions on Reliability. Other Awards: Fellow, American Society for Quality Control, 1972; Editor, ASQC Reliability Review.

EVANS, WILLIAM E., Fellow 1964. Born: August 2, 1921, Louisville, Ky. Degrees: B.E.E., 1943, University of Louisville; M.E.E., 1947, Stanford University.

Fellow Award "For contributions to applications of video techniques and video systems"; Vladimir K. Zworykin Award, IRE, 1963, "For recording of television pictures on paper." Other Awards: Oscar, Academy of Motion Picture Arts and Sciences, 1960.

EVANS, WILLIAM T., AIEE Fellow 1962. Born: July 2, 1912, Corsicana, Tex. Degrees: B.S., 1933, Texas A and M University.

Fellow Award "For contributions to electrical engineering in the field of geophysical exploration." Other Awards: Tau Beta Pi; Armed Forces Communications and Electronics Association; Society of Exploration Geophysicists; Fellow, AAAS.

EVERETT, ROBERT R., Fellow 1969. Born: June 26, 1921. Degrees: B.S.E.E., 1942, Duke University; S.M.E.E., 1943, Massachusetts Institute of Technology.

Fellow Award "For contributions to the development of digital computers and of computer-based real-time control systems for continental air defense."

EVERETT, WOODROW W., JR., Fellow 1976. Born: October 11, 1937, Newton, Miss. Degrees: B.E.E., 1959, George Washington University; M.S., 1965, Ph.D., 1968, Cornell University.

Fellow Award "For contributions to electromagnetic compatibility, and for pioneering advances in university-federal laboratory relations." Other Awards: Commendation Certificate, Air Defense Command, 1962; Air Force Special Service Award, 1964; Air Force Special Act Award, 1965; Outstanding Merit Award, Mohawk Valley Engineer's Executive Council, 1966; Award for Scientific Achievement(with prize), Air Force Systems Command, 1966; State of Tennessee Honorary Citizen Award, 1966; Air Force Special Service

Award, 1967; Air Force Academy Liason Officer Coordinator Award, 1973; Post-Doctoral Program Award, 1967-1975; Outstanding Award, Mohawk Valley Engineer's Executive Council, 1976; Distinguished Service Award, Southeastern Center for Electrical Engineering Education, 1976-77; Listed in: "Who's Who in Industry and Finance," "Who's Who in the East," "Who's Who in the South and Southwest," "Who's Who in America," "Who's Who in the World," "Who's Who in Engineering", "Notable Americans," "Community Leaders of America," "Personalities of the South," "Personalities of America," "The American Scientific Registry," "Dictionary of International Biography," "Men of Achievement," "International Who's Who in Engineering," "Who's Who in Technology Today."

EVERHART, JAMES G., AIEE Fellow 1951. Born: August 29, 1915, Pittsburgh, Pa. Degrees: B.S., 1938, Pennsylvania State University.

Other Awards: Distinguished Service Award, American Management Association, 1973-75.

EVERHART, THOMAS E., Fellow 1969. Born: February 15, 1932, Kansas City, Mo. Degrees: A.B., 1953, Harvard University; M.Sc., 1955, University of California at Los Angeles; Ph.D., 1958, Cambridge University.

Fellow Award "For inspiring teaching, and contributions to scanning electron beam devices"; IEEE Centennial Medal, 1984. Other Awards: Senior Postdoctoral Fellow, National Science Foundation, 1966-67; Miller Research Professorship, 1969-70; John Simon Guggenheim Fellowship, 1974-75; National Academy of Engineering, 1978.

EVERITT, WILLIAM L., AIEE Fellow 1936, IRE Fellow 1938. Born: April 14, 1900, Baltimore, Md. Degrees: E.E., 1922, Cornell University; M.S., 1926, University of Michigan; Ph.D., 1933, Ohio State University; D.Eng.(Hon.), 1959, Bradley University; D.Sc.(Hon.), 1964, Monmouth College; D.Eng.(Hon.), 1964, Tri-State College; D.Ing.(Hon.), 1966, University of the Andes; D.Sc.(Hon.), 1966, Ohio State University; D.Eng.(Hon.), 1967, Michigan Technological University; D.Eng.(Hon.), 1967, University of Michigan; LL.D.(Hon.), 1968, University of Denver; D.Sc.(Hon.), 1969, University of Illinois; D.Eng.(Hon.), 1972, Southern Methodist Universtiy.

Awards: Medal of Honor, IRE, 1954; Medal for Electrical Engineering Education, AIEE, 1957; Mervin J. Kelly Award, IEEE, 1963. Other Awards: Meritorious Civilian Service Award, War Dept., 1946; Lamme Medal, American Society for Engineering Education, 1957; Founding Member, National Academy of Engineers, 1964; Distinguished Prof. Achievement Award, College of Engineering, University of Michigan, 1968; Hall of Fame, American Society for Engineering Education, 1968; National Society of Professional Engineers Award, 1969; Washington Award, Western Society of Engineers, 1971; L.E. Grinter Distinguished Service

Award, Engineers' Council for Professional Development, 1973; Exceptional Service Award, American Society for Engineering Education, 1980.

EWART, DONALD N., Fellow 1978. Born: July 16, 1930. Degrees: B.E.E., 1954, Cornell University; M.S.(E.E.), 1963, Union College.

Fellow Award "For pioneering contributions to the analysis of the control and dynamics of large-scale electric power systems."

EWELL, MANSFIELD MAC, AIEE Fellow 1962. Born: Egypt, Wash. Degrees: B.S., 1924, University of Washington.

Fellow Award "For contributions to the development of electrical apparatus and application to electric utility systems."

EWEN, HAROLD I., Fellow 1970. Born: March 5, 1922, Chicopee, Mass. Degrees: B.A., 1943, Amherst College; M.A., 1948, Ph.D., 1951, Harvard University.

Fellow Award "For research and the application of radio astronomy in connection with radiation from galactic and inter-galactic hydrogen"; Morris E. Leeds Award, 1970, "For contributions to the design of sensitive radiometric systems, and for co-discovery of the 21 cm. spectral line of interstellar hydrogen." Other Awards: Phi Beta Kappa, 1942; Sigma Xi, 1950; Fellow, American Association for the Advancement of Science, 1954; Fellow, American Academy of Arts and Sciences, 1957.

EWING, DOUGLAS H., IRE Fellow 1954. Born: June 24, 1915, Mitchell, Ind. Degrees: A.B., 1935, Butler University; M.S., 1937, Ph.D., 1939, University of Rochester; D.Sc.(Hon.), 1955, Butler University.

Fellow Award "For his contributions to the development of electronic aids to air navigation and traffic control systems."

EYKHOFF, PIETER, Fellow 1979. Born: April 9, 1929, The Hague, The Netherlands. Degrees: B.S.E.E., 1955, M.S.E.E., 1956, Delft University of Technology; Ph.D.(E.E.), 1961, University of California, Berkeley.

Fellow Award "For contributions to identification and parameter estimation in dynamic systems, to control engineering education in The Netherlands, and to international scientific cooperation." Other Awards: Sigma Xi; Visiting Research Fellow, U.S. Academy of Sciences, 1958.

F

FAHNOE, HAROLD H., Fellow 1976. Born: November 9, 1912. Degrees: E.E., 1934, Cornell University.

Fellow Award "For contributions to the development of standards for high-voltage fuses and to their basic design." Other Awards: Member, American National Standards Institute; Member, National Electric Manufacturing Association.

FAIRSTEIN, EDWARD, Fellow 1971. Born: December 14, 1922, Brooklyn, N.Y. Degrees: B.E.E., 1944, City College of The City University of New York.

Fellow Award "For contributions to nuclear instrumentation, particularly in the technology of linear pulse amplification."

FANCHER, H. BRAINARD, Fellow 1970. Born: July 14, 1914, Hartford, Conn. Degrees: B.S.E.E., 1935, Brown University.

Fellow Award "For contributions to television broadcasting and for leadership in the emerging semiconductor industry."

FANG, FRANK F., Fellow 1984. Born: September 11, 1930, Peking, China. Degrees: B.S.(E.E.), 1951, National Taiwan University; M.S., 1954, University of Notre Dame; Ph.D., 1959, University of Illinois, Urbana.

Fellow Award "For discovery and understanding of the two-dimensional properties of silicon inversion layers and for contributions in semiconductor device physics research." Other Awards: Fellow, American Physical Society; John Price Wetherill Medal, Franklin Institute, 1981.

FANO, ROBERT M., IRE Fellow 1954. Born: November 11, 1917, Torino, Italy. Degrees: S.B., 1941, Sc.D., 1947, Massachusetts Institute of Technology.

Fellow Award "For contributions in the fields of information theory and microwave filters"; Shannon Lecturer, IEEE Professional Group on Information Theory, 1976; IEEE Education Medal, 1977, "For leadership in engineering education through teaching and outstanding research in computer science, information theory and electromagnetic theory." Other Awards: "Premio Citta di Columbus, Ohio," Instituto Internazionale delle Comunicazioni, 1969; American Academy of Arts and Sciences, 1958; National Academy of Engineering, 1973; National Academy of Sciences, 1978.

FANTE, RONALD L., Fellow 1979. Born: October 27, 1936, Philadelphia, Pa. Degrees: B.S.E.E., 1958, University of Pennsylvania; M.S.E.E., 1960, Massachusetts Institute of Technology; Ph.D., 1964, Princeton University.

Fellow Award "For contributions to the understanding of electromagnetic wave propagation in turbulent media"; Prize Paper Award, IEEE Antennas and Propagation Society, 1980; Editor, IEEE Transactions on Antennas & Propagation 1983-. Other Awards: Atwater Kent Prize, 1958; Marcus O'Day Prize, 1975; Fellow, Optical Society of America, 1981; Director, Electromagnetics Society, 1981; URSI Commission B, Membership Chairman, 1982-.

FARHAT, NABIL H., Fellow 1981. Born: November 3, 1933. Degrees: B.Sc.(E.E.), 1957, Technion, Israel; M.Sc.(E.E.), 1959, University of Tennessee; Ph.D.,

1963, University of Pennsylvania.

Fellow Award "For work in advancement and understanding of microwave holographs and associated contributions to electrical engineering education."

FARNELL, GERALD W., Fellow 1970. Degrees: B.A.Sc., 1948, University of Toronto; S.M., 1950, Massachusetts Institute of Technology; Ph.D., 1957, McGill University.

Fellow Award "For research in microwave optics and solid-state electronics."

FARNHAM, SHERMAN B., AIEE Fellow 1953. Born: June 23, 1912, New Haven, Conn. Degrees: B.S., 1933, E.E., 1935, Yale University.

Fellow Award "For contributions to the solution of problems in the design and operation of electric power systems, and in the application of equipment thereto."

FARRAND, C. L., IRE Fellow 1931. Born: October 25, 1895, Newark, N.J. Degrees: M.E.(Hon.), 1949, Stevens Institute of Technology.

Other Awards: Fellow, Society of Motion Picture and Television Engineers; American Physical Society; British Kinematograph Society.

FARRAR, CLYDE L., AIEE Fellow 1950. Born: December 30, 1896, Klamath Falls, Oreg. Degrees: B.S.E.E., 1921, E.E., 1926, University of Colorado.

FARRAR, WALTER B., AIEE Fellow 1948. Born: July 13, 1900, Holden, Mo. Degrees: B.S.E.E., 1925, University of Kansas.

FAY, CLIFFORD E., IRE Fellow 1956. Born: December 2, 1903, St. Louis, Mo. Degrees: B.S.E.E., 1925, M.S., 1927, Washington University.

Fellow Award "For contributions to the development of high-power vacuum tubes."

FEE, WALTER F., Fellow 1977. Born: November 6, 1921, Providence, R.I. Degrees: B.S.E.E., 1946, Purdue University.

Fellow Award "For contributions to the management of power systems engineering"; IEEE Centennial Medal, 1984. Other Awards: Distinguished Engineering Alumnus Award, Purdue University, 1979.

FEGLEY, KENNETH A., Fellow 1975. Born: February 14, 1923, Mont Clare, Pa. Degrees: B.S.E.E., 1947, M.S.E.E., 1950, Ph.D., 1955, University of Pennsylvania.

Fellow Award "For contributions in the application of mathematical programming to engineering problems."

FEIN, LOUIS, Fellow 1968. Born: October 1, 1917, Brooklyn, N.Y. Degrees: B.S.(Physics)(Hon.), 1938, Long Island University; M.S.(Physics), 1941, University of Colorado, Ph.D.(Physics), 1947, Brown University.

Fellow Award "For his contributions to the orderly assimilation of computer science into education, tech-

nology and society." Other Awards: Ordinance Development Award, U.S. Navy, 1945; Sigma Xi.

FEINSTEIN, JOSEPH, Fellow 1966. Born: July 8, 1925, New York, N.Y. Degrees: B.E.E., 1944, Cooper Union; M.A., 1947, Columbia University; Ph.D., 1951, New York University.

Fellow Award "For contributions to crossed-field microwave tubes." Other Awards: Member, National Academy of Engineering, 1976.

FELCH, EDWIN P., IRE Fellow 1954. Born: January 10, 1909, Madison, N.J. Degrees: A.B., 1929, Dartmouth College.

Fellow Award "For his contributions in the field of precision measurement and instrumentation of communication circuits"; First Prize Paper, AIEE, 1951. Other Awards: Commanders Award, U.S. Dept. of Air Force, Ballistic Missiles Division, 1961; Outstanding Civilian Service Medal, U.S. Dept. of Army, 1973; The Armstrong Medal, The Radio Club of America, 1983.

FELDMAN, DAVID, Fellow 1972. Born: October 16, 1927, New York. Degrees: B.S.E.E., 1947, M.E.E., 1949, Newark College of Engineering.

Fellow Award "For contributions to electronic components and thin-film technology"; Best Technical Paper, Electronic Components Conference, 1968; Major Contributions Award, IEEE Parts, Hybrids, and Packaging Group, 1977.

FELDTKELLER, ERNST, Fellow 1984. Born: October 19, 1931, Berlin-Charlottenburg, Germany. Degrees: Dipl.-Physiker, 1957, Technische Hochschule Stuttgart; Dr.rer.nat., 1959, Universitat Goettingen; Apl. Professor, 1976, Technische Universitat Munchen.

Fellow Award "For contributions to the understanding of domain structures in permalloy films in the teaching of magnetics." Other Awards: Preis, Deutschen Physikalischen Gesellschaft, 1963; Preis, Nachrichtentechnischen Gesellschaft, 1964.

FELKER, J. H., IRE Fellow 1960. Born: March 14, 1919, Centralia, Ill. Degrees: B.S.E.E., 1941, Washington University.

Fellow Award "For contributions to computer technology." Other Awards: Member, National Academy of Engineering, 1974.

FELLERS, RUFUS G., AIEE Fellow 1961, IRE Fellow 1962. Born: September 26, 1920, Columbia, S.C. Degrees: B.S., 1941, University of South Carolina; Ph.D., 1943, Yale University.

AIEE Fellow Award "For contributions to the advancement of science and education"; IRE Fellow Award "For research at millimeter wavelengths"; Paper Prize, AIEE, District 4, 1960, for "A Circular Polarization Duplexer for Millimeter Waves"; Outstanding Engineer, Region 3, IEEE, 1972. Other Awards: Eta Kappa Nu; Tau Beta Pi; Sigma Xi; Phi Beta Kappa.

FELSEN, LEOPOLD B., IRE Fellow 1962. Born: May 7, 1924, Munich, Germany. Degrees: B.E.E., 1948, M.E.E., 1949, D.E.E., 1952, Polytechnic Institute of Brooklyn; Doctor Tecnices(honoris causa), Technical University of Denmark, 1979.

Fellow Award "For contributions to electromagnetic theory and measurement"; IEEE Paper Award, 1969, for "Transients in Dispersive Media, Part I: Theory," in TRANSACTIONS on Antennas and Propagation; Best Paper Award, IEEE Antennas and Propagation Society, 1974, for "Ray Analysis of Conformal Antenna Arrays" (coauthored with A. Hessel and J. Shapira); Best Paper Award, IEEE Antennas and Propagation Society, 1981; IEEE Centennial Medal, 1984. Other Awards: Premium for paper in PROCEEDINGS, IEE, London, 1964; First Citation for Distinguished Research, Polytechnic Chapter of Sigma Xi, 1973; Guggenheim Fellow, 1973-1974; Balthasar van der Pol Gold Medal, International Board of Officers of URSI, 1975; Member, National Academy of Engineering, 1977; Humboldt Senior Scientist Award, 1982-1983.

FENG, TSE-YUN, Fellow 1980. Born: February 6, 1928, Hangchow, China. Degrees: B.S.E.E., 1950, National Taiwan University; M.S., 1957, Oklahoma State University; Ph.D, 1967, University of Michigan.

Fellow Award "For outstanding contributions to parallel processors and processing"; Best Paper Award, Syracuse Section, IEEE, 1975; Honor Roll Award, IEEE Computer Society, 1978; Special Award, IEEE Computer Society, 1981; IEEE Centennial Medal, 1984. Other Awards: Recognition Plaque, Rome Air Development Center, 1975; Phi Kappa Phi; Tau Beta Pi; Sigma Xi; Eta Kappa Nu; Phi Tau Phi.

FERENCE, MICHAEL, JR., IRE Fellow 1961. Born: November 1911, Whiting, Ind. Degrees: B.S., 1933, M.S., 1934, Ph.D., 1936, University of Chicago; D.Sc.(Hon.), Kenyon College.

Fellow Award "For technical contributions and leadership in military communications and space electronics." Other Awards: Quantrell Prize, University of Chicago, 1943; Exceptional Civilian Service Award, U.S. Dept. of Army, 1953.

FERGUSON, RANDON O., AIEE Fellow 1951.

FERRELL, ENOCH B., IRE Fellow 1953. Born: June 1, 1898, Sedan, Kans. Degrees: B.A., 1920, B.S., 1921, M.A., 1924, University of Oklahoma.

Fellow Award "For his original contributions and research leadership in high-frequency power amplifiers, servomechanisms, telephone switching, and quality control." Other Awards: Fellow, American Society for Quality Control, 1948.

FERRIS, WARREN R., IRE Fellow 1953. Born: May 14, 1904, Terre Haute, Ind. Degrees: B.S.E.E., 1927, Rose Polytechnic Institute; M.S.E.E., 1932, Union College; D.E.E., 1946, Polytechnic Institute of Brooklyn.

Fellow Award "For early contributions in the high-frequency operation of vacuum tubes."

FETTWEIS, ALFRED L.M., Fellow 1975. Born: November 27, 1926, Eupen, Belgium. Degrees: Ingenieur civil electricien, 1951, Docteur en sciences appliquees, 1963, University of Louvain, Belgium.

Fellow Award "For contributions to the theory of switched networks and digital filters"; Darlington Prize Paper Award, IEEE Circuits and Systems Society, 1980; IEEE Centennial Medal, 1984. Other Awards: Prize "Acta Technica Belgica," 1962-1963; Prix de la Fondation George Montefiore, Belgium, 1980.

FEUCHT, DONALD L., Fellow 1978. Born: August 25, 1933, Akron, Ohio. Degrees: B.S.E.E, 1955, Valparaiso University; M.S.E.E., 1956, Ph.D.E.E., 1961, Carnegie Institute of Technology.

Fellow Award "For contributions to semiconductor heterojunctions research." Other Awards: Distinguished Alumni Award, Valparaiso University, 1979.

FFOLLIOTT, CHARLES F., AIEE Fellow 1950. Born: September 18, 1904, New Westminister, British Columbia, Canada. Degrees: E.E., 1927, M.S., 1930, Rensselaer Polytechnic Institute.

Other Awards: Sigma Xi, 1927.

FICK, CLARENCE W., AIEE Fellow 1939. Born: September 30, 1890, St. Joseph, Mo. Degrees: B.S., 1912, University of Illinois.

FICK, CLIFFORD G., IRE Fellow 1951. Born: December 13, 1904, Ida Grove, Iowa. Degrees: B.Sc., 1925, Iowa State University.

Fellow Award "For his leadership and contributions in the development and design of a wide variety of radio and radar equipment."

FIEDLER, GEORGE J., AIEE Fellow 1956, IRE Fellow 1960. Born: March 18, 1904, Bushton, Kans. Degrees: B.S., 1926, Kansas State University; M.S., 1932, University of Kansas; E.E., 1934, Kansas State University.

AIEE Fellow Award "For his contributions to the analysis, synthesis and design of process control systems"; IRE Fellow Award "For contributions to electronic process control systems." Other Awards: Fellow, American Association for the Advancement of Science, 1960; Fellow, Instrument Society of American, 1962.

FIELD, GEORGE S., IRE Fellow 1957. Born: October 23, 1905, England. Degrees: B.Sc., 1929, M.Sc., 1930, D.Sc., 1937, University of Alberta; D.Sc.(Hon.), 1972, Royal Military College, Canada.

Fellow Award "For contributions to ultrasonics and to the defense research program of the Royal Canadian Navy." Other Awards: Fellow, Royal Society of Canada, 1944; Fellow, Acoustical Society of America, 1945; Order of the British Empire, 1946.

FIELD, LESTER M., IRE Fellow 1952. Born: February 9, 1918, Chicago, Ill. Degrees: B.S., 1939, Purdue University; Ph.D., 1944, Stanford University.

Fellow Award "For many technical contributions to the electron tube art, and in particular, the development of new forms of traveling-wave tubes." Other Awards: Eta Kappa Nu; Wilbur Scholar, Tau Beta Pi, 1938; National Academy of Engineering, 1967.

FINCH, WILLIAM G. H., IRE Fellow 1956. Born: June 28, 1897, Birmingham, England. Degrees: Doc. Sc. (Hon.), 1983, Florida Institute of Technology.

Fellow Award "For contributions in the fields of record and radio communications." Other Awards: Registered Patent Attorney, U.S., Canada, 1927; Registered Professional Engineer, New York, 1929; Legion of Merit, U.S. Dept. of Navy, 1946; American Association for the Advancement of Science; Wisdom Award of Honor, 1969; Armstrong Medal, Radio Club of America, 1976; Listed in "Who's Who in America," 1924, "Who's Who in the World," 1979.

FINDLAY, JOHN W., Fellow 1967. Born: October 22, 1915, Kineton, Warwickshire, England. Degrees: B.A., 1937, M.A., 1940, Ph.D., 1950, University of Cambridge.

Fellow Award "For his contributions and leadership in radio astronomy in the development of large antennas and measurements on an absolute scale."

FINE, TERRENCE L., Fellow 1982. Born: March 9, 1939, New York, NY. Degrees: B.E.E., 1958, City College of New York; M.S., 1959, Ph.D., 1963, Harvard University.

Fellow Award "For contributions to the foundations of probabilistic reasoning and its implications for modeling and decision making"; Honorable Mention (Best Paper Category), IEEE Control Systems Society, 1978, 1979.

FINK, DONALD G., IRE Fellow 1947, AIEE Fellow 1951. Born: November 8, 1911, Englewood, N.J. Degrees: B.Sc., 1933, Massachusetts Institute of Technology, M.Sc., 1942, Columbia University.

IRE Fellow Award "In recognition of his espousal of high standards of technical publishing and for his wartime contributions in the field of electronic aids to navigation"; Radio Fall Meeting Plaque, IRE, 1951, "For his many contributions to the development of the television industry"; IEEE Director Emeritus (for life), 1974; IEEE Consumer Electronics Award, 1978; Founders Medal, 1978; Donald G. Fink Prize Paper Award, IEEE, established, 1980. Other Awards: Outstanding Young Electrical Engineer, Eta Kappa Nu, 1940; Premium Award, Institution of Electrical Engineers, U. K., 1944; Medal of Freedom, War Dept., 1946; Presidential Certificate of Merit, 1948; Journal Award, Society of Motion Picture and Television Engineers, 1955; Eminent Member, Eta Kappa Nu, 1965; Fellow, Institution of Electrical Engineers, U.K., 1968; Member, National Academy of Engineering, 1969; Outstanding Civilian

Service Award, U.S. Dept. of Army, 1969; Outstanding Service to Television Citation, International Symposium on TV, Montreux, 1971; Progress Medal, Society of Motion Picture and Television Engineers, 1979.

FINK, LESTER H., Fellow 1973. Born: May 3, 1925, Philadelphia, Pa. Degrees: B.S.E.E., 1950, M.S.E.E., 1961, University of Pennsylvania.

Fellow Award "For contributions to the automatic control of power generation and dispatch, and leadership in underground transmission research." Other Awards: Fellow, Instrument Society of America, 1975; Meritorious Service Award, U.S. Dept. of Energy, 1979; Eta Kappa Nu; Sigma Tau; Tau Beta Pi.

FINK, LYMAN R., IRE Fellow 1953, AIEE Fellow 1962. Born: November 14, 1912, Elk Point, S.D. Degrees: B.S., 1933, M.S., 1934, Ph.D., 1937, University of California.

IRE Fellow Award "For his contributions to the development of armament radar"; AIEE Fellow Award "For contributions to gun control, measuring and computing systems, television, radio, and scientific research." Other Awards: Charles A. Coffin Award, General Electric, 1948.

FINN, HAROLD M., Fellow 1983. Born: June 3, 1922, New York, New York. Degrees: B.E.E., 1950, City College of the City University of New York; M.S.E.E., 1955, University of Maryland; Ph.D., 1965, University of Pennsylvania.

Fellow Award "For the development of innovative radar system detection techniques." Other Awards: Tau Beta Pi; Eta Kappa Nu.

FIRE, PHILIP, Fellow 1980. Born: December 18, 1925, Paterson, NJ. Degrees: B.S. (E.E.), M.S. (E.E.), 1952, Massachusetts Institute of Technology; Engineer, 1959, Stanford University; Ph.D. (E.E.), 1964, Stanford University.

Fellow Award "For fundamental contributions to error-burst-correcting codes."

FIRESTONE, WILLIAM L., Fellow 1965. Born: June 20, 1921, Chicago, Ill. Degrees: B.S.E.E.(with special honors), 1946, University of Colorado; M.S.E.E., 1949, Illinois Institute of Technology; Ph.D., 1952, Northwestern University; Business Administration Certificate, 1955, University of Chicago.

Fellow Award "For contributions to the utilization of the radio frequency spectrum in space communications"; Two Best Paper Prizes, IRE, Professional Group on Vehicular Communications, 1956, 1965. Other Awards: Second Paper Prize, Illinois Institute of Technology, 1949.

FISHER, JOSEPH F., IRE Fellow 1961. Born: February 28, 1911, Philadelphia, Pa. Degrees: Dipl., 1936, Drexel University Evening School (1934-1936); Drexel University Day School (1929-1934).

Fellow Award "For contributions to color television;"

Section Award, IEEE, Philadelphia, 1976, "For his important pioneering contributions in applying television principles to space, medical and instrumentation applications, and his leadership in the IEEE."

FISHER, LAWRENCE E., Fellow 1967. Born: March 31, 1905, Sweetser, Ind. Degrees: B.S., 1927, University of Michigan.

Fellow Award "For contributions to the increased reliability, safety and service continuity of electrical products and systems"; First Chapter Chairman, IEEE Industry and General Applications Group, Connecticut Section. Other Awards: 50 Patents; Listed in "Who's Who in Engineering," "International Blue Book," 1948-1949, "American Men of Science"; Member, National Society of Professional Engineers; Registered Professional Engineer, Connecticut, Michigan.

FITCH, CLYDE J., AIEE Fellow 1959. Born: April 13, 1899, Jersey City, N.J.

Fellow Award "For contributions to electronic machines and to specialized extensions of radio communication." Other Awards: Fellow, International Business Machines Corp., 1963.

FITZGERALD, JOSEPH P., Fellow 1975. Born: September 20, 1923, Sheffield, Pa. Degrees: B.S.E.E., 1950, Pennsylvania State University.

Fellow Award "For contributions to the design of electric power systems and for leadership in the organization and administration of standards"; IEEE Standards Achievement Medal, 1972, "For outstanding achievement through standards." Other Awards: Award of Merit, Cleveland Electronic Conference, 1980.

FITZSIMMONS, LAURENCE G., JR., AIEE Fellow 1962. Born: August 4, 1918, San Francisco, Calif. Degrees: B.S.E.E., 1940, University of California at Berkeley.

Fellow Award "For achievements in the development of communication services." Other Awards: U.S. Naval Reserve Medal, 1950.

FLAMBOURIARIS, JOHN D., AIEE Fellow 1961. Born: November 11, 1902, Zante, Greece. Degrees: M.S., 1925, National Technical University of Athens.

Fellow Award "For contributions to the scientific and economic advancement of Greece through education and the nation's power system."

FLANAGAN, JAMES L., Fellow 1969. Born: August 26, 1925, Greenwood, Miss. Degrees: B.S.E.E., 1948, Mississippi State University; M.S.E.E., 1950, Sc.D., 1955, Massachusetts Institute of Technology.

Fellow Award "For contributions to reduced-bandwidth speech communication systems and to the fundamental understanding of human hearing"; IEEE Audio and Electroacoustics Achievement Award, 1970, "For leadership in electroacoustics ..."; Society Award, IEEE Acoustics, Speech, and Signal Processing Society, 1976, "In recognition of creative technical contributions and leadership in speech communications." Other Awards: Fellow, Rockefeller Foundation, 1952-53; Fellow, Acoustical Society of America, 1958; Air Research and Development Command Patent Awards, 1957, 1958, 1960; Distinguished Service Award in Science, American Speech and Hearing Association, 1977; National Academy of Engineering, 1978; National Academy of Sciences, 1983.

FLASCHEN, STEWARD S., Fellow 1978. Born: May 28, 1926, Berwyn, Ill. Degrees: B.S.(Chemistry), University of Illinois, 1947; M.S.(Chemistry), Miami University of Ohio, 1948; Ph.D.(Geochemistry), Pennsylvania State University, 1952.

Fellow Award "For leadership in consumer, industrial, and component electronics development." Other Awards: Forest Award, American Ceramics Society, 1960; New York Academy of Sciences, 1967; Fellow, American Institute of Chemists, 1969; Electrochemical Society of America; American Chemical Society; Industrial Research Institute.

FLECKENSTEIN, WILLIAM O., Fellow 1971. Born: April 6, 1925, Scranton, Pa. Degrees: B.S.E.E., 1949, Lehigh University.

Fellow Award "For contributions and leadership in the field of data communications." Other Awards: Outstanding Young American Engineer, Honorable Mention, Eta Kappa Nu, 1959.

FLEISCHMANN, HANS H., Fellow 1983. Born: June 2, 1933, Munich, Germany. Degrees: Dipl. Phys., 1959, Technische Hochschule, Munich; Dr.rer.nat., 1962, Technische Hochschule, Munich, Germany.

Fellow Award "For the development of the ion ring compressor and its application to controlled fusion research." Other Awards: Fellow, American Physical Society.

FLEISHER, HAROLD, Fellow 1969. Born: October 12, 1921, Kharkov, Russia. Degrees: B.A., 1942, M.S., 1943, University of Rochester; Ph.D., 1951, Case Institute of Technology.

Fellow Award "For contributions to radar, television, nuclear instrumentation, solid-state computers, and optical electronics." Other Awards: Member, Sigma Xi, 1950; Fellow, IBM, 1974; Certificate of Merit, Office of Scientific Research and Development, 1945.

FLEMING, GEORGE A., AIEE Fellow 1949. Born: August 22, 1893, Williams, Ariz. Degrees: B.S., 1917, University of California.

Other Awards: Meritorious Service Award, U.S. Dept. of Interior, 1961; Citation, Engineering Council of Sacramento Valley, 1966; Sigma Xi; Eta Kappa Nu; Tau Beta Pi.

FLETCHER, JAMES C., Fellow 1969. Born: June 5, 1919, Millburn, N.J. Degrees: A.B., 1940, Columbia University; Ph.D., 1948, California Institute of Technology; D.Sc.(Hon.), 1971, University of Utah;

D.Sc.(Hon.), 1977, Brigham Young University; Doctor of Laws(Hon.), 1978, Lehigh University.

Fellow Award "For contributions to the development of space vehicle systems." Other Awards: First Distinguished Alumni Award, California Institute of Technology, 1966; Achievement Award, American Astronautical Society, 1975; Exceptional Civilian Service Medal, U.S. Air Force, 1977, Department of Energy, 1981; Distinguished Service Medal, NASA, 1978, 1981; John Jay Award, Columbia College, 1979; Fellow, American Astronautics Society, Honorary Fellow, AIAA.

FLETCHER, ROBERT C., Fellow 1973. Born: May 27, 1921, New York, N.Y. Degrees: B.S., 1943, Ph.D., 1949, Massachusetts Institute of Technology.

Fellow Award "For research on microwave beam devices and spin and cyclotron resonances in silicon, and for technical leadership in the development of solid-state devices." Other Awards: Fellow, American Physical Society; Fellow, American Association for the Advancement of Science, 1982.

FLICK, CARL, Fellow 1984. Born: June 22, 1926, Vienna, Austria. Degrees: B.E.E., 1951, M.E.E., 1953, Polytechnic Institute of Brooklyn; Ph.D., University of Pittsburgh, pending Dissertation.

Fellow Award "For innovative developments in the design of superconducting and other advanced generators"; AIEE District Branch Paper Prize, 1951; IEEE Centennial Medal, 1984.

FLORIANI, VIRGILIO, Fellow 1970. Born: June 29, 1906, Cison di Valmarino, Italy. Degrees: Dr.Eng., 1929, Polytechnic of Torino.

Fellow Award "For contribution to the development of radio links and related technology." Other Awards: Member, Associazione Elettrotecnica Italiana.

FLORY, LESLIE E., IRE Fellow 1957. Born: March 17, 1907, Sawyer, Kans. Degrees: B.S.E.E., 1930, University of Kansas.

Fellow Award "For contributions in the fields of industrial television and medical electronics." Other Awards: Overseas Award, Institution of Electrical Engineers, U.K., 1938; Awards, Radio Corporation of America Laboratories, 1949, 1950, 1958; Fellow, Radio Corporation of America Laboratories; Sigma Xi.

FLOWERS, HAROLD L., IRE Fellow 1961. Born: June 25, 1917, Hickory, N.C. Degrees: B.S.E.E., 1938, Duke University; M.S.E., University of Cincinnati.

Fellow Award "For contributions to radar and guidance control systems." Other Awards: Phi Beta Kappa, 1938; Tau Beta Pi, 1939; Sigma Xi, 1940; ADPA Award, 1978; Distinguished Alumnus Awards, University of Cincinnati, 1969, Duke University, 1984.

FLUGUM, ROBERT W., Fellow 1974. Born: June 12, 1924, Madison, Wis. Degrees: B.S.E.E., 1948, University of Wisconsin.

Fellow Award "For contributions to the electric power industry in the field of surge arrester application and standardization"; Prize Paper Award, Power Engineering Society, IEEE, 1973, 1974, 1976. Other Awards: Distinguished Service Citation, University of Wisconsin, 1977; Eta Kappa Nu.

FLUKE, JOHN M., SR., Fellow 1978. Born: December 14, 1911, Tacoma, WA. Degrees: B.S.(E.E.), 1935, University of Washington; M.S.(E.E.), 1936, Massachusetts Institute of Technology.

Fellow Award "For contributions to electronic instrumentation"; Community Service Award, Region 6, 1967; Engineers Council for Professional Development Visiting Committee, 1970; Engineering Education and Accreditation Committee, 1971, 1972. Other Awards: Charles A. Coffin Award, General Electric, 1938; Navy Legion of Merit, 1946; Tau Beta Pi; Honorary Member, Instrument Society of America, 1976; Corporate Leadership Award, 1976; Medal of Achievement; American Electronics Association, 1976.

FLURSCHEIM, CEDRIC H., Fellow 1967. Born: February 4, 1906, Fleet, Hampshire, England. Degrees: B.A., 1927, University of Cambridge; F.Eng., 1977.

Fellow Award "For his contributions to circuit breaker design, and leadership in engineering organization in this field." Other Awards: John Hopkinson Premium, Institution of Electrical Engineers, U.K., 1938.

FLYNN, MICHAEL J., Fellow 1980. Born: May 20, 1934, New York, NY. Degrees: B.S.E.E., 1955, Manhattan College; M.S.E.E., 1960, Syracuse University; Ph.D., 1961, Purdue University.

Fellow Award "For outstanding contributions to the field of computer architecture."

FONDILLER, WILLIAM, AIEE Fellow 1926. Born: December 21, 1885, Russia. Degrees: B.S., 1903, E.E., 1909, M.A., 1913, Columbia University; D.Sc.(Hon.), 1949, Israel Institute of Technology.

Other Awards: Townsend Harris Medal, City College of the City University of New York, 1944; University Medal, Columbia University, 1945; Egleston Medal, Columbia University, 1967; Sigma Xi; Tau Beta Pi; Fellow, American Association for the Advancement of Science; Fellow, New York Academy of Sciences.

FONG, ARTHUR, Fellow 1971. Born: Calif. Degrees: B.S.E.E., 1943, University of California at Berkeley; M.S.E.E., 1968, Stanford University.

Fellow Award "For contributions to microwave measurement techniques and instrument design." Other Awards: Honor Awards, University of California at Berkeley, 1943; Hewlett-Packard Engineering Achievement Award, 1964; Sigma Xi; Eta Kappa Nu; Tau Beta Pi; HP Consultant and Lecturer.

FONTANA, ROBERT E., Fellow 1975. Born: November 26, 1915, Brooklyn, N.Y. Degrees: B.E.E., 1939, New York University; M.S.E.E., 1947, Ph.D., 1949, University of Illinois.

Fellow Award "For leadership in graduate education and research and development in the United States Air Force"; Meritorious Service Award, IEEE Education Society, 1983. Other Awards: Legion of Merit, Research and Development, U.S. Air Force, 1966; Outstanding Teacher Award, Tau Beta Pi, 1968; Legion of Merit, Graduate Education, U.S. Air Force, 1969; Distinguished Alumnus Award, University of Illinois, 1973; Outstanding Engineer and Scientist Award, Dayton Engineers Week, 1980.

FOOTE, JAMES H., AIEE Fellow 1932. Born: November 21, 1891, Jackson, Mich. Degrees: B.S., 1914, Michigan State University; D.Sc., 1958, Wayne State University.

Other Awards: Honorary Member, Michigan Engineering Society, 1954; Centennial Award, Michigan State University, 1955; Howard Coonley Medal, American Standards Association, 1960; Engineer of the Year Award, Michigan Society of Professional Engineers, 1960; Distinguished Alumni Award, Michigan State University, 1961; Honorary Member, American Society for Testing and Materials, 1962; Tau Beta Pi; Eta Kappa Nu; Pi Tau Sigma; Fellow, AAAS.

FORBES, HARLAND C., AIEE Fellow 1944. Born: February 21, 1898, Colebrook, N.H. Degrees: B.S., 1921, University of New Hampshire; S.M., 1923, Massachusetts Institute of Technology; D.E., 1960, University of New Hampshire.

FORGUE, STANLEY V., Fellow 1965. Born: October 6, 1916, Cleveland, Ohio. Degrees: B.S.(Physics), 1939, B.E.E., 1940, M.S.(Physics), 1940, Ohio State University.

Fellow Award "For contributions to the development of television pickup tubes and infrared imaging devices." Other Awards: Five Outstanding Research Awards, RCA Labs, 1942-1972; Distinguished Alumnus Award, Ohio State University, 1970; Registered Professional Engineer, Ohio.

FORMOSO, JOSE F., AIEE Fellow 1962. Born: April 18, 1903, Mexico City, Mexico.

Fellow Award "For contributions to the operation and maintenance of electric utility systems in Mexico."

FORNEY, G. DAVID, JR., Fellow 1973. Born: March 6, 1940, New York, N.Y. Degrees: B.S.E., 1961, Princeton University; M.S., 1963, Sc.D., 1965, Massachusetts Institute of Technology.

Fellow Award "For contributions to the theory and practice of error-correcting codes and data communications"; Information Theory Group, Prize Paper Award, 1970, "For the authorship of an outstanding paper contributing to information and coding theory"; Browder J. Thompson Memorial Prize Award, 1972, "For his paper entitled 'Convolutional Codes I: Algebraic Structure.'" Other Awards: Sigma Xi, 1960; Phi Beta Kappa, 1961; Howard M. Samuel Prize, Princeton

School of Engineering, 1961; NSF Fellowship, 1961-62, 1963-65; National Academy of Engineering, 1983.

FORRESTER, A. THEODORE, Fellow 1967. Born: April 13, 1918, Brooklyn, N.Y. Degrees: A.B., 1938, A.M., 1939, Ph.D., 1942, Cornell University.

Fellow Award "For his contributions to the fields of electromagnetics, photoelectric mixing and ion generation, including implementation in the field of rocket propulsion." Other Awards: American Physical Society Fellowship, 1957; Research Award, American Institute of Aeronautics and Astronautics, 1962.

FORRESTER, JAY W., IRE Fellow 1955. Born: July 14, 1918, Anselmo, Nebr. Degrees: B.Sc.E.E., 1939, University of Nebraska; M.Sc.E.E., 1945, Massachusetts Institute of Technology; D.Eng.(Hon.), 1954, University of Nebraska; D.Sc.(Hon.), 1969, Boston University; D.Eng.(Hon.), 1971, Newark College of Engineering; D.Sc.(Hon.), 1973, Union College; D.Eng.(Hon.), 1974, Notre Dame University; D.Pol.Sc.(Hon.), 1979, Mannheim University, Germany.

Fellow Award "For his contributions to the development and engineering design of high speed digital computers"; IEEE Medal of Honor, 1972, "For exceptional advances in the digital computer through his invention and application of the magnetic-core random-access memory, employing coincident current addressing"; IEEE Systems, Man, and Cybernetics Society, Award for Outstanding Accomplishment, 1972; Charter Recipient, Computer Pioneer Award, IEEE Computer Society, 1982. Other Awards: Book Award, Academy of Management, 1962; National Academy of Engineering, 1967; "Inventor of the Year," George Washington University, 1968; American Academy of Arts and Sciences, 1968; Valdemar Poulsen Gold Medal, Danish Academy of Technical Sciences, 1969; New England Award, Engineering Societies of New England, 1973; Howard N. Potts Medal, Franklin Institute, 1974; Honorary Member, Society of Manufacturing Engineers, 1976; Harry Goode Memorial Award, American Federation of Information Processing Societies, 1977; National Inventor's Hall of Fame, 1979; Common Wealth Award of Distinguished Service, 1979; Fellow, American Assoc. for the Advancement of Science, 1980.

FORSTER, ERIC O., Fellow 1979. Born: October 24, 1918, Austria. Degrees: B.S., M.A., Ph.D. (all in Chemistry), Columbia University.

Fellow Award "For contributions in the understanding of high-voltage conduction and breakdown phenomena and for leadership in related professional activities." Other Awards: U.S. Inventor's Hall of Fame; Honorary Professor of Physics, State University of New Jersey, Rutgers, American Institute of Physics; American Chemical Society.

FORSTER, JULIAN, Fellow 1972. Born: August 31, 1918, New York, N.Y. Degrees: B.S., 1940, City University of New York.

Fellow Award "For contributions to instrumentation and control systems for nuclear power plants"; Leadership and Service Award, IEEE Standards Board and Activities, 1980; Member, Standards Board, Adm. Comm., New Stds. Comm., 1983; Adm. Com., IEEE Nuclear and Plasma Science Society, 1964-1983; Chrmn, Nuclear Power Symposium, 1974-81-83.

FORSTER, WILLIAM H., IRE Fellow 1962. Born: July 11, 1922, Belmar, N.J. Degrees: A.B., 1943, Harvard University; A.M.P., 1959, Harvard Business School.

Fellow Award "For contributions to transistor development." Other Awards: Phi Beta Kappa, 1943; Sigma Xi, 1943; Fellow, Radio Club of America, 1970; Fellow, Physical Society (U.K.), 1970; Busignies Award, Radio Club of America, 1981.

FOSDICK, ELLERY R., AIEE Fellow 1940. Born: February 8, 1900, Lakeside, Wash. Degrees: B.S., 1924, E.E., 1931, Washington State University.

Other Awards: Fellow, American Consulting Engineers Council, 1973.

FOSSUM, JERRY G., Fellow 1983. Born: July 18, 1943, Phoenix, Arizona. Degrees: B.S.E.E., 1966, M.S.E.E., 1969, Ph.D.(E.E.), 1971, University of Arizona.

Fellow Award "For contributions to the theory and technology of silicon solar cells and transistors." Other Awards: Research Unit Award, ASEE(SE Section), 1979; Best Faculty Paper Award, EE Dept., Univ. of Florida, 1979, 1981; Honorary Editorial Advisory Board, Solid-State Electronics, 1979-present; Who's Who in America, 1984.

FOSTER, DUDLEY E., IRE Fellow 1950. Born: December 12, 1900, Newark, N.J. Degrees: E.E., 1922, Cornell University.

Fellow Award "For his contributions and technical direction of work leading to better radio receiver design." Other Awards: Modern Pioneer Award, National Association of Manufacturers, 1940.

FOUST, CLIFFORD M., AIEE Fellow 1947. Born: December 27, 1893. Degrees: B.S., 1921, Carnegie Institute of Technology.

Other Awards: Charles A. Coffin Award, General Electric, 1935; Howard N. Potts Medal, Franklin Institute, 1951; Award of Merit, Carnegie Institute of Technology, 1952.

FOWLER, ALAN B., Fellow 1978. Born: October 15, 1928. Degrees: B.S., 1951, M.S., 1952, Rensselaer Polytechnic Institute; Ph.D., 1958, Harvard University.

Fellow Award "For contributions to the understanding of charge carriers' behavior in inversion layers." Other Awards: John Price Wetherill Medal, Franklin Institute; Alexander von Humboldt Preis; IBM Fellow.

FOWLER, CHARLES A., Fellow 1969. Born: December 17, 1920, Centralia, Ill. Degrees: B.S., 1942, University of Illinois.

Fellow Award "For contributions to the introduction of radar to air traffic control, and the development of advanced radar techniques and their application to complex military problems." Other Awards: Secretary of Defense Meritorious Civilian Service Medal, 1970; Fellow, American Association for the Advancement of Science, 1980; Joint Service Medal, AOC, 1980; Defense Intelligence Agency Exceptional Civilian Service Award, 1982.

FOWLER, GLENN A., IRE Fellow 1960. Born: June 1, 1918, Riverdale, Calif. Degrees: B.S., 1941, University of California.

Fellow Award "For contributions to atomic instrumentation."

FOWLER, WENDELL C., AIEE Fellow 1958. Born: January 31, 1900, Emporia, Kans. Degrees: B.S., 1923, United States Naval Academy.

Fellow Award "For contributions to instrumentation in the field of power."

FOX, A. GARDNER, IRE Fellow 1956. Born: November 22, 1912, Syracuse, N.Y. Degrees: B.S., 1934, M.S., 1935, Massachusetts Institute of Technology.

Fellow Award "For research and invention in microwave wave-guide techniques"; Quantum Electronics Award, 1978; Microwave Career Award, 1979; David Sarnoff Award, 1979. Other Awards: Naval Ordnance Development Award, 1945; Fellow, Optical Society of America.

FRANK, HOWARD, Fellow 1978. Born: June 4, 1941, New York, N.Y. Degrees: B.S.E.E., 1962, University of Miami; M.S., 1964, Ph.D., 1965, Northwestern University.

Fellow Award "For contributions to large-scale network design and analysis"; Leonard G. Abraham Best Paper Award, IEEE Communications Society, 1969. Other Awards: Honorable Mention, The Lanchaster Prize, Operations Research Society of America, 1972.

FRANK, ROBERT L., Fellow 1964. Born: June 8, 1917, Detroit, Mich. Degrees: B.S., 1938, University of Michigan; M.S., 1939, Massachusetts Institute of Technology.

Fellow Award "For contributions to radio navigation and the development of instrumentation for the Loran-C system"; AES Pioneer Award, 1971, "For contributions to low-frequency navigation systems, in particular the conception, design, and implementation of Loran C." Other Awards: Sigma Xi, 1938; Eta Kappa Nu, 1938; Commendation, U.S. Navy Bureau of Ships, 1945; Medal of Merit, Wild Goose Association (Navigation), 1977; Fellow, Engineering Society of Detroit, 1981.

FRANKEL, SIDNEY, IRE Fellow 1957. Born: October 6, 1910, New York, N.Y. Degrees: E.E., 1931, M.S., 1934, Ph.D., 1936, Rensselaer Poyltechnic Institute.

Fellow Award "For contributions to the field of circuit techniques." Other Awards: Sigma Xi, 1931; Eta Kappa Nu, 1963.

FRANKLIN, GENE FARTHING, Fellow 1978. Born: July 25, 1927, Banner Elk, N.C. Degrees: B.S.E.E., 1950, Georgia Institute of Technology; M.S.E.E., 1952, Massachusetts Institute of Technology; D.Eng.Sci.E.E., 1955, Columbia University.

Fellow Award "For leadership in engineering education and for outstanding contributions in control theory and applications."

FRANKLIN, PHILIP J., IRE Fellow 1962. Born: October 25, 1908, Riverside, Calif. Degrees: B.A., 1932, University of California at Los Angeles.

Fellow Award "For pioneering contributions to printed circuit technology." Other Awards: Meritorious Service Award, National Bureau of Standards, 1949; Certificate of Achievement, Harry Diamond Laboratories, 1955; Incentive Award for R & D, DOFL, 1955.

FRANKS, LEWIS E., Fellow 1977. Born: November 8, 1931, San Mateo, Calif. Degrees: B.S.E.E., 1952, Oregon State University; M.S.E.E., 1953, Ph.D., 1957, Stanford University.

Fellow Award "For contributions to the theory and application of periodically variable networks and filters." Other Awards: Hewlett-Packard Fellowship, 1952-1953.

FRAZIER, RICHARD H., AIEE Fellow 1950. Born: May 29, 1900, Bellevue, Pa. Degrees: S.B., 1923, S.M., 1932, Massachusetts Institute of Technology.

Other Awards: Sigma Xi; Eta Kappa Nu; Tau Beta Pi; Member, American Society for Engineering Education; Member, Society for the History of Technology; Member, The Newcomen Society in North America.

FREDENDALL, GORDON L., IRE Fellow 1955. Born: December 20, 1909, Kettle Falls, Wash. Degrees: B.S., 1931, M.S., 1932, Ph.D., 1936, University of Wisconsin.

Fellow Award "For his applications of network analysis and synthesis to television system problems"; Philadelphia Section Award, IEEE, 1980.

FREDERICK, CARL L., Fellow 1971. Born: 1903, Nebr. Degrees: A.B., 1925, Nebraska University; M.A., 1927, George Washington University; D.Sc., 1950, Nebraska Wesleyan University.

Fellow Award "For leadership and contributions in the field of electromagnetic technology and in the development of standards." Other Awards: Naval Ordnance Development Award for Exceptional Service, 1945; Citations, Congressional Record, 1955, 1960; Alumni Achievement Award, Nebraska Wesleyan University, 1960; Who's Who in Engineering, 1982.

FREED, CHARLES, Fellow 1979. Born: March 21, 1926, Budapest, Hungary. Degrees: B.E.E., 1952, New York University; S.M., 1954, E.E., 1958, Massachusetts Institute of Technology.

Fellow Award "For contributions to gas lasers and the pioneering development of ultrastable lasers." Other Awards: Tau Beta Pi; Eta Kappa Nu; Sigma Xi.

FREEDMAN, JEROME, Fellow 1978. Born: August 16, 1916, New York, N.Y. Degrees: B.S.E.E., 1938, College of the City of New York; M.S.E.E., 1951, Polytechnic Institute of Brooklyn.

Fellow Award: "For contributions to the development of radar systems"; Editor's Award, IRE, 1952. Other Awards: Citation from Commanding General, Strategic Air Command, 1972; Commendation from National Delegates, Advisory Group for Aerospace Research and Development; NATO, Avionics Panel, 1968-1979; Associate Fellow, American Institute of Aeronautics and Astronautics, 1978.

FREEMAN, HERBERT, Fellow 1967. Born: December 13, 1925, Frankfurt-am Main, Germany. Degrees: B.S.E.E., 1946, Union College; M.S., 1948, Dr.Eng.Sc., 1956, Columbia University.

Fellow Award "For contributions to the theory of multivariable discrete-time systems and digital computer graphics." Other Awards: Pullman Prize, Union College, 1946; Postdoctoral Fellow, National Science Foundation, 1966; American Association for the Advancement of Science; Guggenheim Fellow, 1972; Silver Core, International Federation for Information Processing, 1974; President, International Association for Pattern Recognition, 1978-80; Treasurer, 1976-78 and 1982-84.

FREI, E. H., Fellow 1967. Born: March 2, 1912, Vienna, Austria. Degrees: Dr.Phil., 1936, University of Vienna.

Fellow Award "For contributions to magnetic transducers and ferromagnetism, medical electronics and education"; Life Member, IEEE Magnetics Society, 1979. Other Awards: Weizmann Prize, Tel-Aviv Municipality, 1956; Member, Computor Pioneer Team, National Computor Conference, 1975; Honorary Fellow, Israel Soc. for Medical & Biological Engineering, 1981; Prof. Emeritus, Weizmann Institute of Science; Editorial Board, Inst. of Cardiovascular Ultrasonography.

FREITAG, HARLOW, Fellow 1981. Born: April 17, 1936, Brooklyn, NY. Degrees: A.B., 1955, New York University; M.S., 1957, Ph.D., 1959, Yale University.

Fellow Award "For contributions to the development of design automation methodologies for large-scale integrated circuitry." Other Awards: IBM Outstanding Innovation Award, 1966.

FRENCH, HEYWARD A., Fellow 1967. Born: August 20, 1921, Brooklyn, N.Y. Degrees: B.S.E.E., 1944, City College of the City University of New York.

Fellow Award "For the concept and development of microwave communication techniques and equipment, and his engineering leadership in the system engineering and implementation of microwave communication systems."

FRENCH, JUDSON CULL, Fellow 1978. Born: September 30, 1922. Degrees: B.S.(cum laude), 1943, American University; M.S., 1949, Harvard University.

Fellow Award "For contributions to the understanding and measurement of semiconductor devices and materials." Other Awards: Distinguished Service Award, National Bureau of Standards, 1964; Silver Medal for Meritorious Service, Dept. of Commerce, 1964; Gold Medal for Exceptional Service, Dept. of Commerce, 1978; Edward Bennett Rosa Award, National Bureau of Standards, 1971; Presidential Rank of Meritorious Executive, Senior Executive Service, 1980.

FREY, HOWARD A., AIEE Fellow 1951. Born: October 30, 1905, Baltimore, Md. Degrees: B.E., 1926, The Johns Hopkins University.

Other Awards: Charles A. Coffin Award, General Electric, 1947.

FRIED, WALTER R., Fellow 1981. Born: April 10, 1923, Vienna, Austria. Degrees: B.S.E.E., 1948, University of Cincinnati; M.S.E.E., 1953, Ohio State University.

Fellow Award "For contributions to and technical leadership in the development of Doppler radar and relative navigation systems."

FRIEDLAENDER, FRITZ J., Fellow 1969. Born: May 7, 1925, Freiburg i. Brg., Germany. Degrees: B.S., 1951, M.S., 1952, Ph.D., 1955, Carnegie Institute of Technology.

Fellow Award "For outstanding contributions to engineering education, to engineering societies, and to the understanding of magnetic devices"; IEEE Centennial Medal, 1984. Other Awards: Senior Award, Special Program, Alexander Von Humboldt Foundation, 1972; Senior Research Fellowship, Japanese Society for the Promotion of Science, 1980.

FRIEDLAND, BERNARD, Fellow 1980. Born: Brooklyn, NY. Degrees: A.B., 1952, B.S., 1953, M.S., 1953, Ph.D., 1957, Columbia University.

Fellow Award "For contributions to the application of modern control theory in navigation, guidance, and control systems." Other Awards: Fellow, American Society of Mechanical Engineers, 1974; Rufus Oldenburger Medal, American Society of Mechanical Engineers, 1982.

FRIEDMAN, ARTHUR D., Fellow 1984. Born: New York, N.Y. Degrees: B.A., B.S., M.S., Ph.D. (all in E.E.), 1961, 1962, 1963, 1965, respectively, Columbia University.

Fellow Award "For contributions to fault-tolerant computing and switching theory and to computer engineering education." Other Awards: Conference Chairman, International Conference on Fault Tolerant Computing, 1977.

FRIEDRICH, ROBERT E., Fellow 1965. Born: July 16, 1918, Pittsburgh, Pa. Degrees: B.S.(Physics and Eng.), 1940, M.S.(Physics), 1953, University of Pittsburgh.

Fellow Award "For developments in the fields of high voltage power circuit breakers"; Certificate of Appreciation, IEEE Power Engineering Society. Other Awards: Special Patent Award, Westinghouse Electric; Professional Eng., Pa; 29 patents.

FRISCH, IVAN T., Fellow 1977. Born: September 21, 1937, Budapest, Hungary. Degrees: B.S.(Physics), 1958, Queens College, New York; B.S.E.E., 1958, M.S.E.E., 1959, Ph.D.(E.E.), 1962, Columbia University.

Fellow Award "For contributions to the development and application of network theory." Other Awards: Peter Trowbridge Fellowship, Columbia University, 1960; Ford Foundation Residency in Engineering Practice at Bell Laboratories, 1965; Guggenheim Fellowship, 1969; Honorable Mention, Lanchester Prize, Operations Research Society of America, 1972; Member, New York Academy of Sciences; Member: Phi Beta Kappa, Tau Beta Pi, Eta Kappa Nu, Sigma Alpha (Queens College Honor Society); Listed in: "American Men of Science," "Who's Who in Computers and Data Processing," "Who's Who in America"; Member, URSI, Commission C.

FRISCHE, CARL A., IRE Fellow 1957. Born: August 13, 1906, Freeport, Kans. Degrees: A.B., 1928, Miami University, Ohio; M.S., 1931, Ph.D., 1932, University of Iowa; D.Sc.(Hon.), 1955, Miami University; D.Sc.(Hon.), 1964, Long Island University.

Fellow Award "For contributions in the field of electronic servocontrols and related devices." Other Awards: Annual Management Award, 1964, Society for the Advancement of Management.

FRITZ, DWAIN E., AIEE Fellow 1962. Born: February 26, 1919, St. Johns, Ohio.

Fellow Award "For contributions to the design and development of aircraft generators and test instrumentation."

FRITZ, H. R., AIEE Fellow 1939. Born: November 12, 1890, St. Louis, Mo. Degrees: E.E., 1914, University of Texas; M.S., 1917, University of Illinois.

FROEHLICH, FRITZ E., Fellow 1975. Born: November 12, 1925, Worms a/Rh., Germany. Degrees: B.S., 1950, M.S., 1952, Ph.D., 1955, Syracuse University.

Fellow Award "For contributions and technical leadership in data communications and in developing new telephone services."

FROHMAN-BENTCHKOWSKY, DOV, Fellow 1982. Born: March 28, 1939, Amsterdam, The Netherlands.

Degrees: B.S.E.E., 1963, Israel Institute of Technology, Haifa; M.S.(E.E. and Comp.Sc.), 1965, Ph.D.(E.E. and Comp.Sc.), 1969, University of California at Berkeley.

Fellow Award "For the development and understanding of programable nonvolatile MOS memories"; Jack A. Morton Award, IEEE, 1982. Other Awards: Member, International Program Committee, International Solid State Circuits Conference.

FROMM, WINFIELD E., Fellow 1965. Born: January 18, 1918, Haddonfield , N.J. Degrees: B.S., 1940, Drexel Institute of Technology; M.E.E., 1948, Polytechnic Institute of Brooklyn; Dr.Humanities(Hon.), 1974, Dowling College; D.Eng.(Hon.), 1974, Polytechnic Institute of New York.

Fellow Award "For contributions to microwave techniques and for technical leadership"; Outstanding Engineering Manager, Long Island Section, IEEE, 1974. Other Awards: Associate Fellow, American Institute of Aeronautics and Astronautics, 1963; Distinguished Citizen Award, Dowling College, 1973; Distinguished Alumnus, Polytechnic Institute of New York; Distinguished Leadership Award, Long Island YMCA, 1977; Distinguished Citizen Award, Suffolk County Boy Scouts of America, 1978; Fellow, Polytechnic Institute of New York; Phi Kappa Phi; Tau Beta Pi; Eta Kappa Nu; Gifford Visiting Executive, Lynchburg College School of Business, 1983.

FROSCH, ROBERT A., Fellow 1969. Born: May 22, 1928, Bronx, N.Y. Degrees: A.B., 1947, M.A., 1949, Ph.D., 1952, Columbia University.

Fellow Award "For outstanding contributions in applying acoustical theory and techniques to practical problems, and in planning and administering research and development programs." Other Awards: Arthur S. Flemming Award, 1966; Navy Distinguished Public Service Award, 1969; National Academy of Engineering, 1971; Meritorious Civilian Service Award, Dept. of Defense, 1972; NASA Distinguished Service Medal, 1981.

FRY, DONALD W., IRE Fellow 1960. Born: November 30, 1910, Weymouth, Dorset, England. Degrees: M.Sc., 1932, University of London.

Fellow Award "For contributions to research in controlled fusion." Other Awards: Fellow, Institution of Electrical Engineers, U.K., 1950; Fellow, Institution of Physics and Physical Society, U.K., 1950; Duddell Medal, Physical Society, U.K., 1950; Fellow, King's College, University of London, 1959.

FRY, FRANCIS JOHN, Fellow 1982. Born: April 2, 1920. Degrees: B.S., 1940, Pennsylvania State University; M.S., 1946, University of Pittsburgh.

Fellow Award "For pioneering contribution to the applications of ultrasonics in biology and medicine."

FRY, THORNTON C., AIEE Fellow 1950. Born: January 7, 1892, Findlay, Ohio. Degrees: A.B., 1912, Findlay

College; M.A., 1913, Ph.D., 1920, University of Wisconsin; D.Sc.(Hon.), 1957, Findlay College.

Other Awards: Presidential Certificate of Merit, 1948; Award for Distinguished Service, Mathematics Association of America, 1982.

FTHENAKIS, EMANUEL J., Fellow 1983. Born: January 30, 1928, Salonica, Greece. Degrees: B.S.M.E., B.S.E.E., 1951, Technical University of Athens; M.S.E.E., 1954, Columbia University.

Fellow Award "For technical leadership and innovative application of technology in the field of space communications." Other Awards: Bicentennial Award, Columbia University, for Outstanding Academic Achievement; EASCON, Man of the Year Award, 1982.

FU, K.S., Fellow 1971. Born: October 2, 1930, Nanking, China. Degrees: B.S., 1953, National Taiwan University; M.A.Sc., 1955, University of Toronto; Ph.D., 1959, University of Illinois.

Fellow Award "For contribution to the pattern recognition and leadership in engineering education"; IEEE Award for Service, 1971; IEEE Computer Society Honor Roll, 1973; Outstanding Paper Award, IEEE Computer Society, 1977; Certificate for Appreciation, IEEE Computer Society, 1977. Other Awards: Guggenheim Foundation Fellow, 1971; Member, National Academy of Engineering, 1976; Academia Sinica, 1978; Senior Research Award, American Society for Engineering Education, 1981; Harry Goode Memorial Award, American Federation of Information Societies, 1982; IEEE Education Medal, 1982.

FUBINI, EUGENE G., IRE Fellow 1954. Born: April 19, 1913, Torino, Italy. Degrees: D.Phys., 1933, University of Rome; D.Eng.(Hon.), 1966, Polytechnic Institute of Brooklyn; D.Sc.(Hon.), 1966, Rensselaer Polytechnic Institute; D.Eng.(Hon.), 1967, Pratt Institute.

Fellow Award "For his many contributions to the analysis of electronic counter measures." Other Awards: National Academy of Engineering; Presidential Certificate of Merit; Two Awards, Defense M Distinguished Public Service; Exceptional Service Medal, Defense Intelligence Agency; Vice Chairman, Defense Science Board.

FUJISAWA, TOSHIO, Fellow 1979. Born: December 27, 1928, Kobe, Japan. Degrees: B.S.E.E., 1952, Ph.D.E.E., 1962, Osaka University.

Fellow Award "For contributions to the theory of filter design and nonlinear circuit analysis"; Guillemin-Cauer Paper Prize Award, IEEE Circuits and Systems Society, 1973.

FUKUI, HATSUAKI, Fellow 1983. Born: December 14, 1927, Yokohama, Japan. Degrees: Diploma, 1949, Miyakojima Technical College, Osaka, Japan; Doctor of Engineering, 1961, Osaka University, Osaka, Japan.

Fellow Award "For contributions to the understanding and design of low noise microwave transistors and transistor amplifiers"; Microwave Prize, IEEE Microwave Theory and Techniques Society, 1980. Other Awards: Inada Prize, Institute of Electrical Communication Engineers of Japan, 1959.

FUKUNAGA, KEINOSUKE, Fellow 1979. Born: July 23, 1930, Himeji-shi, Japan. Degrees: B.S.E.E., 1953, Ph.D., 1962, Kyoto University, Japan; M.S.E.E., 1959, University of Pennsylvania.

Fellow Award "For contributions to statistical pattern recognition." Other Awards: Inada Prize, Institute of Electrical Communication Engineers of Japan, 1962; Ohm Prize, Ohm Book Company, 1964; Prize, Kanto Electrical Association in Japan, 1966.

FULLER, JACKSON F., Fellow 1982. Born: October 9, 1920, Salt Lake City, UT. Degrees: B.S.E.E., 1944, University of Colorado.

Fellow Award "For contributions to utility and industrial power systems and creative approaches to power engineering education"; IEEE Centennial Medal, 1984. Other Awards: Charles A. Coffin Award, 1953; Honorable Mention as Young Engineer, Eta Kappa Nu, 1954; NECA Big Ring, 1975; Edison Electric Institute Power Educator, 1977.

FULLER, JOHN L., AIEE Fellow 1959. Born: November 11, 1913, Cleveland, Ohio. Degrees: B.S.E.E., 1936, Case Institute of Technology.

Fellow Award "For contributions to design, controls and testing of electrical power equipment."

FULLER, LEONARD F., SR., AIEE Fellow 1923, IRE Fellow 1925. Born: August 21, 1890, Portland, Oreg. Degrees: M.E., 1912, Cornell University; Ph.D., 1919, Stanford University.

Awards: Morris N. Liebmann Award, IRE, 1919, "In recognition of the valuable contributions to the radio art embodied in his researches and development work dealing with the high-power Poulsen arc converter"; AIEE National Nominating Committee, 1935; National Committee on Education, 1940; National Committee on Telegraphy and Telephony, 1923; Chairman, San Francisco Section, 1929-30; IRE National Offices: Manager 1928; National Committees: Admissions, 1928-29; Radio Transmitters and Antennas, 1930-31; Sections, 1926-28; Specifications for Steamship Radio Equipment, 1914-15; Standardization, 1916-20, 1930-31; Transmitters and Antennas, 1929-30; Wave Length Regulation, 1916-17, 1919-20; Chairman, San Francisco Section, 1927-28. Other Awards: Sigma Xi, Tau Beta Pi, Eta Kappa Nu; Who's Who in America, 1962-63; Who Was Who in America.

FULLER, RICHARD H., Fellow 1973. Born: March 15, 1928, San Francisco, Calif. Degrees: B.S.E.E., 1952, California Institute of Technology; M.S.E.E., 1954, Massachusetts Institute of Technology; Ph.D., 1963, University of California.

Fellow Award "For contributions to the development

of advanced computer technology."

FULLERTON, DICK P., AIEE Fellow 1962. Born: January 20, 1903, Seattle, Wash. Degrees: A.B., 1924, E.E., 1925, Stanford University.

Fellow Award "For contributions to the advancement of telephone communicaton in a large metropolitan area."

FURFARI, FRANK A., Fellow 1980. Born: February 14, 1915, Morgantown, WV. Degrees: B.S.E.E., 1938, West Virginia University.

Fellow Award "For technical and administrative contributions to the installation, servicing, and maintenance of heavy electrical equipments"; Outstanding Achievement Award, IEEE Industry Applications Society, 1979. Other Awards: Eta Kappa Nu, 1970.

G

GABELMAN, IRVING J., Fellow 1967. Born: November 12, 1918, Brooklyn, N.Y. Degrees: B.A., 1938, Brooklyn College; B.E.E., 1945, City College of the City University of New York; M.E.E., 1948, Polytechnic Institute of Brooklyn; Ph.D., 1961, Syracuse University.

Fellow Award "For contributions to the engineering of real time digital data processing techniques and equipment and their application to military command and control systems." Other Awards: Fellow, American Association for Advancement of Science.

GABRIEL, WILLIAM F., Fellow 1982. Born: October 17, 1925, Sault Ste. Marie, MI. Degrees: B.S.E.E., 1945, M.S.E.E., 1948, Ph.D.(E.E.), 1950, University of Wisconsin.

Fellow Award "For leadership in and contributions to the theory of adaptive antenna arrays and super-resolution processing techniques"; Microwave Prize, IEEE Microwave Theory and Techniques Society, 1966; Distinguished Lecturer, IEEE Antennas and Propagation Society, 1980-1982. Other Awards: Best Paper Award, Naval Research Laboratory, 1977, 1981; Tau Beta Pi; Eta Kappa Nu; Sigma Xi.

GADDY, OSCAR L., Fellow 1974. Born: July 18, 1932, Republic, Mo. Degrees: B.S.E.E., 1957, M.S.E.E., 1959, University of Kansas; Ph.D., 1962, University of Illinois.

Fellow Award "For contributions to the development and understanding of fast response time detectors in both the visible and infrared region of the spectrum."

GAERTNER, WOLFGANG W., Fellow 1974. Born: July 5, 1929. Degrees: Ph.D., 1951, University of Vienna; Dipl.-Ing., 1955, Technische Hochschule Vienna.

Fellow Award "For contribution to the development of microelectronic circuits and systems with low power comsumption, high packing density, and adaptive self-healing features"; Certificate of Merit, PGSET Symposium Committee for Microelectronic, Micropower, Analog and Digital Circuits Function Blocks for Space Application, 1962; Microelectronics Symposium Award for Micropower Microelectronic Systems, 1964. Other Awards: Certificate of Achievement, Dept. of the Army, 1959.

GAFFNEY, FRANCIS J., IRE Fellow 1955. Born: June 27, 1912, England. Degrees: B.S., 1935, Northeastern University.

Fellow Award "For contributions to the field of electrical measurements." Other Awards: Tau Beta Pi, 1937; Sigma Xi, 1947; Commendation, War Dept., U.S. Dept. of the Navy, 1947.

GAHAGAN, JOSEPH E., AIEE Fellow 1956. Born: January 22, 1909, Silver Lake, Pa. Degrees: E.E., 1930, Rensselaer Polytechnic Institute.

Fellow Award "For his contributions to the application of industrial electronics in the field of photographic manufacturing." Other Awards: Sigma Xi; Tau Beta Pi.

GALIL, UZIA, Fellow 1978. Born: April 27, 1925. Degrees: B.Sc.(E.E.), 1947, Dipl. E.E., 1947, Technion-Israel Institute of Technology; M.Sc.E.E., 1953, Purdue University.

Fellow Award "For the pioneering contribution to the establishment of a modern electronic industry in Israel."

GALLAGER, ROBERT G., Fellow 1968. Born: May 29, 1931, Philadelphia, Pa. Degrees: B.S., 1953, University of Pennsylvania; M.S., 1957, Sc.D., 1960, Massachusetts Institute of Technology.

Fellow Award "For contributions to information theory and error correcting codes"; W.R.G. Baker Prize Award, 1966. Other Awards: Member, National Academy of Engineering, 1979.

GALLAGHER, JAMES J., Fellow 1982. Born: November 15, 1922, Albany, NY. B.S., 1948, Siena College; M.S., 1952, Columbia University.

Fellow Award "For original and sustained contributions to molecular spectroscopy and frequency control and for the advancement of millimeter-wave technology."

GALLOTTE, WILLARD A., Fellow 1972. Born: March 16, 1902, Manette (E. Bremerton), Wash. Degrees: B.S.(E.E.), 1924, Worcester Polytechnic Institute.

Fellow Award "For contributions to the transmission of electric energy by overhead lines and submarine cables." Other Awards: Registered Professional Engineer, Washington, 1952.

GALT, JOHN K., Fellow 1973. Born: September 1, 1920, Portland, Oreg. Degrees: A.B., 1941, Reed College; Ph.D., 1947, Massachusetts Institute of Technology.

Fellow Award "For pioneering research on domain wall motion in ferrites and on cyclotron resonance in metals, and for technical leadership in the application of physics to communications devices"; Frederick Philips Award, IEEE, 1984.

GAMBLE, GEORGE P., AIEE Fellow 1945. Born: September 20, 1899, Richmond, Va. Degrees: E.E., 1922, University of Virginia.

Other Awards: Honorary Member, Engineers' Club of St. Louis, 1961.

GANDHI, OM P., Fellow 1979. Born: September 23, 1934, Multan, India. Degrees: B.S.(Physics, with Honors), 1952, University of Delhi, India; M.S.E.E., 1957, Sc.D., 1961, University of Michigan.

Fellow Award "For contributions to the understanding of nonionizing radiation effects, to the development of electron devices, and to engineering education"; Special Award, Utah Section, IEEE, 1975; Chairman, IEEE Committee on Man and Radiation, 1980-1982. Other Awards: Member, Diagnostic Radiology Study Section, National Institutes of Health, 1978-1981; Distinguished Research Award, University of Utah, 1979-1980; Member, Board of Directors, Bioelectromagnetics Society, 1979-1982; Member, ANSI Committee C95.4, 1978-present.

GANNETT, ELWOOD K., IRE Fellow 1962. Born: August 26, 1923, New York, N.Y. Degrees: B.S., 1944, University of Michigan.

Fellow Award "For unremitting and effective management of the procurement, editing, and production of the many publications of the Institute of Radio Engineers"; IEEE Centennial Medal, 1984. Other Awards: Fellow, Society for Technical Communication.

GARCIA, OSCAR N., Fellow 1984. Born: Sept. 10, 1936, Havana, Cuba. Degrees: B.S.E.E., 1961, M.S.E.E., 1964, North Carolina State University; Ph.D., 1969, University of Maryland.

Fellow Award "For contributions to arithmetic coding theory and leadership in computer science education"; Special Group Award for Outstanding Work in the Model Curricula of the IEEE Computer Society, 1977. Other Awards: ASEE Outstanding Young Engineering Teacher for the Southeastern Region, 1972; NSF Science Faculty Fellow, 1968-9; Member, ACM, DPMA, ASEE, Eta Kappa Nu, Tau Beta Pi, Sigma Xi, Phi Kappa Phi.

GARCIA-SANTESMASES, JOSE, Fellow 1983. Born: May 2, 1907, Barcelona, Spain. Degrees: E.E., 1930, Ecole Superieure d'Electricite, Paris, France; Licenciate in Physics, 1935, Barcelona University, Spain; Ph.D., 1943, Madrid University, Spain.

Fellow Award "For leadership in establishing research in computers and related topics in Spain." Other Awards: Echegaray Gold Medal, Royal Academy of Sciences (Madrid), 1979; Honorary Fellow, World Organisation of General Systems and Cybernetics (WOGSC), 1978; Silver Core Holder, International Federation of Information Processing (IFIP), 1974; Physics Medal, Royal Spanish Society of Physics and Chemistry, 1968; Professor Emeritus, Complutense University of Madrid; Member, Royal Academy of Sciences,

Madrid.

GARDINER, ROBERT A., Fellow 1971. Born: 1919, Ohio. Degrees: B.S.E.E., 1940, Case Institute of Technology.

Fellow Award "For contributions and leadership in the development of manned spacecraft guidance, navigation, and control systems." Other Awards: World War II Service Award, National Defense Research Council, 1945; Exceptional Service Medal, NASA, Apollo 8, 1968; Exceptional Service Medal, NASA, 1969; Electrical Engineering Visiting Committee, University of Texas, 1969-1975.

GARDNER, B. CLIFFORD, Fellow 1967. Born: Jensen, Utah.

Fellow Award "For his contribution to the television and microwave tube arts, and his technical leadership of projects in these fields."

GARDNER, FLOYD M., Fellow 1980. Born: October 20, 1929, Chicago, IL. Degrees: B.S.E.E., 1950, Illinois Institute of Technology; M.S.E.E., 1951, Stanford University; Ph.D., 1953, University of Illinois.

Fellow Award "For contributions to the understanding and applications of phase-lock loops."

GARDNER, LEONARD B., II, Fellow 1981. Born: February 16, 1927, Lansing, MI. Degrees: B.Sc.(Applied Physics), 1951, University of California; M.Sc.(E.E.), 1953, Sc.D.(E.E.), 1954, Golden State University; Certificate in Industrial Relations, 1956, University of California; M.Sc. Applied (Computer) Science, 1977, Augustana College.

Fellow Award "For pioneering concepts and leadership in the development of computerized control systems used in process simulators and in automated production lines"; Chairman, Los Angeles Chapter, IEEE Professional Group on Biomedical Electronics, 1955; Chairman, Los Angeles Chapter, IEEE Professional Group on Nuclear Science, 1960; Chairman, Los Angeles Chapter, IEEE Professional Group on Instrumentation, 1961; Member, WESCON Future Engineers Show, 1955-1966; Chairman, Education Committee, Los Angeles Section, IEEE, 1965; Chairman WESCON Future Engineers Show, 1966; Member, NECON, 1979; Chairman, North New Jersey Chapter, IEEE Control Systems Society, 1978; Chairman, North New Jersey Section, Education Committee, 1978; Cochairman, New Jersey SILA Committee, 1979; Cofounder and Publicity Chairman, New York/New Jersey Chapter, IEEE Engineering Management Society, 1979; Member-at-Large, Executive Committee, North Jersey Section, IEEE, 1979; Chairman, San Diego Chapter, IEEE Computer Society, 1981; Chairman, San Diego Chapter 299 Robotics International of SME, 1983-1984. Other Awards: Registered Professional Engineer; Certified Manufacturing Engineer; Member, Research Society of America; Sigma Xi; Society for Computer Simulation, Numerical Control Society.

GARDNER, MASTON L., AIEE Fellow 1951. Born: May 5, 1905, Mt. Gileod, N.C. Degrees: E.E., 1929, Westinghouse School of Technology.

GARFINKEL, DAVID, Fellow 1978. Born: May 18, 1930, New York, N.Y. Degrees: A.B., 1951, University of California, Berkeley; Ph.D., 1955, Harvard University; M.A.(Hon.), University of Pennsylvania, 1971.

Fellow Award "For contributions to computer simulation of biological systems." Other Awards: Senior Member, Society for Computer Simulation, 1979.

GARLAN, HERMAN, Fellow 1978. Born: December 5, 1907, New York, N.Y. Degrees: B.S., 1929, M.S.Ed., 1932, City College of New York; E.E., 1936, Columbia University.

Fellow Award "For the development of regulations to control radio frequency interference"; Certificate of Appreciation, 1967, 1968, Certificate of Gratitude, 1977, IEEE Electromagnetic Compatibility Society.

GARMIRE, ELSA M., Fellow 1981. Born: November 9, 1939. Degrees: A.B.(Physics), 1961, Harvard University; Ph.D.(Physics), 1965, Massachusetts Institute of Technology.

Fellow Award "For contributions in nonlinear and guided wave optics."

GARNER, HARVEY L., Fellow 1975. Born: December 23, 1926. Degrees: B.S., 1949, M.S., 1951, University of Denver; Ph.D., 1958, University of Michigan; Magistri Artium(Hon.), 1971, University of Pennsylvania.

Fellow Award "For contributions to the development of residue codes, and for leadership in engineering education." Other Awards: Member, Association for Computing Machinery; Member, American Society for Engineering Education; Member, American Association for Advancement of Science; Eta Kappa Nu; Sigma Xi; Sigma Pi Sigma; Fellowship, Ford Faculty Development, 1961; National Lecturer, Association for Computing Machinery, 1965.

GAROFF, KENTON, Fellow 1968. Born: June 9, 1918. Degrees: B.A., 1940, Brooklyn College.

Fellow Award "For leadership in planning and executing electron devices research and development programs in the area of high power tubes for military equipment."

GARR, DONALD E., AIEE Fellow 1951. Born: October 23, 1914, Kans. Degrees: B.S.E.E., 1936, Kansas State College.

Other Awards: New York State Society of Professional Engineers; Massachusetts Society of Professional Engineers; Coffin Award, General Electric.

GARRETT, P.B., AIEE Fellow 1949. Born: October 15, 1900, Ft. Collins, Colo. Degrees: B.S.E.E., 1922, Colorado State University.

GARVER, LEONARD L., Fellow 1977. Born: September 11, 1932, Columbus, Ohio. Degrees: B.S.E.E., 1956, M.S.E.E., 1957, Ph.D., 1961, Northwestern University.

Fellow Award "For development of advanced computer techniques in the solution of transmission expansion problems."

GARVER, ROBERT V., Fellow 1981. Born: June 2, 1932, Minneapolis, MN. Degrees: B.S.(Physics), 1956, University of Maryland; M.S. (Eng. Admin.), 1968, George Washington University.

Fellow Award "For contributions to the development and understanding of the limitations of broad-band solid-state microwave switches."

GARY, CLAUDE H., Fellow 1982. Born: October 22, 1926, France. Degrees: B.A.(Math.), 1945; Ingenieur Diplome, 1949, Ecole Superieure D'Electricite, Paris.

Fellow Award "For contributions to the understanding of corona and field effects in high-voltage power transmission." Other Awards: Medaille Blondel, 1968; Laureate Member, French Society of Electrical and Electronics Engineers, 1960, 1970.

GATTI, EMILIO C., Fellow 1973. Born: March 18, 1922. Degrees: Dr.E.E., 1946, Padova University.

Fellow Award "For contributions to research in nuclear instrumentation, and in the statistical limits of time and amplitude measurements;" IEEE Centennial Medal, 1984. Other Awards: Gold Medal "Bianchi", AEI, 1956; Gold Medal "Righi", AEI, 1972; Foundation "Angelo Della Riccia" Award, 1968; Gold Medal, Ministry of Public Education, 1971; Instituto Lombardo di Scienze e Lettre, 1980; Gold Medal "Pubblicazioni", HEI 1982; National Academy of Sciences "dei XL", 1982-.

GAUDERNACK, LEIF L., Fellow 1968. Born: June 16, 1897, Oslo, Norway. Degrees: E.E., 1920, Norges Tekniske Hoiskole.

Fellow Award "For his contributions to the development of circuits, methods and components in communication engineering, and his leadership of engineering projects." Other Awards: Knight, The Royal Norwegian Order of Saint Olav, 1968.

GAUSTER, WILHELM F., Fellow 1965. Born: January 6, 1901, Vienna, Austria. Degrees: Dipl.Ing., 1923, Dr.Tech., 1924, Dr.Habil., 1927, Technische Hochschule Vienna, Austria; Honorary Doctor's Degree, University of Technology, Graz, 1982.

Fellow Award "For contributions in the fields of engineering research and teaching." Other Awards: Fellow, American Physical Society; Member, Austrian Academy of Sciences, 1968; PRECHTL Medal, University of Technology, Vienna, 1981; Golden Stefan Medal, Austrian Institute of Electrical Engineers, Vienna, 1983.

GAYER, JOHN H., Fellow 1965. Born: September 20, 1919, Lincoln, Nebr. Degrees: B.S.E.E., 1942, University of Nebraska; Advanced Study, Harvard University,

1942; MIT, 1943.

Fellow Award "For contributions to the development of international radio communications"; Chairman, Geneva Section, Region 8, IEEE. Other Awards: World's First International Festival of Television Arts and Science Citation, 1961; Haile Selassie I Prize, Trust Development of Telecommunications in Africa, 1963; Member, International Frequency Reg. Board, International Telecommunications Union, Geneva, Switzerland, 1953-66.

GAYLORD, THOMAS K., Fellow 1983. Born: September 22, 1943, Casper, Wyoming. Degrees: B.S. (Physics), 1965, M.S.(E.E.), 1967, University of Missouri-Rolla; Ph.D.(E.E.), 1970, Rice University.

Fellow Award "For contributions to the fields of grating diffraction, optical data storage and processing, and engineering education." Other Awards: Fellow, Optical Society of America, 1982; Curtis W. McGraw Research Award, American Society for Engineering Education, 1979; Outstanding Young Engineer Award, Georgia Society of Professional Engineers, 1977; Research Awards, Sigma Xi, 1976,77,78,83; Outstanding Teacher Award, Georgia Tech 1976.

GAZZANA-PRIAROGGIA, PAOLO, Fellow 1972. Born: December 21, 1917, Casale Monferrato, Italy. Degrees: Dr. E.E., 1940, Politecnico di Milano.

Fellow Award "For contributions to high-voltage power cables and systems." Other Awards: Premium Pugno-Vanoni, Italian Electrotechnical Association, 1949-1951; Fellow, IEE, 1961; Premium, Institution of Electrical Engineers, 1962; Contributor to ANIAI Premium, 1964.

GEAR, C. WILLIAM, Fellow 1984. Born: February 1, 1935, London, England. Degrees: B.A.(Mathematics), 1956, Cambridge University; M.S.(Math), 1957, University of Illinois; M.A.(Math), 1960, Cambridge University; Ph.D.(Math/E.E.), 1960, University of Illinois.

Fellow Award "For contributions to the numerical solution of stiff and other differential equations and to computer science education." Other Awards: Fulbright Fellowship, 1956-1960; ACM SIGNUM George E. Forsythe Memorial Lecturer Award, 1979.

GEDDES, LESLIE A., Fellow 1977. Born: May 24, 1921, Port Gordon, Scotland. Degrees: B.Eng.(E.E.), 1945, M.Eng.(E.E.), 1953, McGill University; Ph.D.(Physiology), 1959, Baylor Medical College; D.Sc.(honoris causa), 1971, McGill University.

Fellow Award "For contributions to biomedical instrumentation, including techniques for measuring pulmonary function from the body's electrical impedance."

GEFFE, PHILIP R., Fellow 1981. Born: October 22, 1920, Napa, CA.

Fellow Award "For contributions to computer-aided filter design."

GEHRIG, EDWARD H., Fellow 1981. Born: October 31, 1925, Portland, OR. Degrees: B.A.(Physics), 1948, Reed College; B.S.E.E., 1949, Stanford University; M.S.E.E., 1951, Oregon State University.

Fellow Award "For contributions to the planning and design of high-voltage ac and dc transmission systems"; First Prize Paper Award, Portland Section, AIEE, 1959. Other Awards: Meritorious Service Award, Dept. of Interior, 1977; EPRI Certificate of Appreciation, 1978-1981; Chairman, Radio Noise S/C of T&D Com., 1967-8; Chairman, Portland Section Power Chapter, 1968-9; Licensed Engineer, Oregon.

GEIGES, KARL S., AIEE Fellow 1951. Born: August 24, 1908, Providence, R.I. Degrees: B.S., 1928, New Jersey Institute of Technology; M.S., 1943, Stevens Institute of Technology.

Fellow Award "For outstanding work in the field of electrical safety standards and the techniques for making them effective." Other Awards: Astin-Polk International Standards Medal, 1978.

GELB, ARTHUR, Fellow 1976. Born: September 20, 1937, Brooklyn, N.Y. Degrees: B.E.E., 1958, City College of the City University of New York; S.M.(Applied Physics), 1959, Harvard University; Sc.D.(Systems Engineering), 1961, Massachusetts Institute of Technology.

Fellow Award "For contributions to the application of modern control and estimation theory to integrated navigation and guidance systems." Other Awards: Special Fiftieth Anniversary Medal for Outstanding Contribution to Society as a Young Engineer, City College of the City University of New York, 1968.

GELLER, SEYMOUR, Fellow 1980. Born: March 28, 1921, New York, NY. Degrees: B.A. (Math. and Chemistry), 1941, Ph.D. (Physical Chemistry), 1949, Cornell University.

Fellow Award "For contributions to the crystalline structure of materials and for co-discovery of yttrium iron garnet." Other Awards: Fellow, American Physical Society, 1963; Fellow, American Mineralogical Society, 1970; University of Colorado Faculty Fellowship, 1977-1978; Research Award, College of Engineering and Applied Science, University of Colorado, 1979; Croft Research Professor, College of Engineering and Applied Science, University of Colorado, 1980; National Science Foundation Creativity Award, 1983.

GELNOVATCH, VLADIMIR G., Fellow 1982. Born: New York, NY. Degrees: B.S.E.E., 1963, Monmouth College; M.S.E.E., 1966, New York University.

Fellow Award "For contributions to the field of microwave circuit design and optimization"; Popov Society USSR/IEEE exchange delegation, 1974; IEEE/MTT ADCOM, 1979-present; Program Committee, Microwave Monolithic GaAs Symposium, 1981-present. Other Awards: U.S. Army Research and Development Award, 1972; Army Finalist, Arthur S. Flemming Award, 1973-1974.

GENIN, YVES V., Fellow 1984. Born: September 15, 1938, Etalle, Belgium. Degrees: S.B., S.M.(E.E.), 1962, University of Louvain, Belgium; Ph.D., 1963, University of Liege, Belgium.

Fellow Award "For contributions to circuit and system theory, especially applications of network theoretic concepts to multidimensional systems"; IEEE Guillemin-Cauer Paper Award, 1974. Other Awards: Fulbright-Hays Foundation Award, 1973.

GENS, RALPH S., Fellow 1969. Born: November 25, 1924. Degrees: B.S.E.E., 1949, Oregon State University.

Fellow Award "For significant contributions to the electric power industry in the fields of high-voltage transmission of dc and ac power"; William M. Habirshaw Award, IEEE, 1984. Other Awards: Distinguished Service Award, U.S. Dept. of Interior, 1978.

GENTRY, FINIS E., Fellow 1968. Born: June 5, 1929, Norene, Tenn. Degrees: B.S.E.E., 1951, University of Tennessee; M.E.E., 1958, Syracuse University.

Fellow Award "For his contributions to the development and understanding of high power semiconductor devices."

GERBER, EDUARD A., IRE Fellow 1958. Born: April 3, 1907, Fuerth, Bavaria, Germany. Degrees: Dipl.Phys., 1930, Dr.Ing., 1934, Technical University, Munich, Germany.

Fellow Award "For his many contributions to piezoelectricity and frequency control." Other Awards: Outstanding Achievement Award, Army Science Conference, West Point, 1964; Meritorious Civilian Service Award, U.S. Dept of Army, 1965, 1970; C.B. Sawyer Memorial Award, 1981.

GERBER, WALTER E., Fellow 1965. Born: September 2, 1902, Berne, Switzerland. Degrees: Dipl.Ing., 1925, Dr.Ing., 1930, Prof.Dr.Ing., 1970, Swiss Federal Institute of Technology, Zurich, Switzerland.

Fellow Award "For contributions to the development of television in Europe." Other Awards: Fernseh-Technische Gesellschaft, Germany, 1954; Fondation des Freres Lumieres, 1957; Gold Medal, Istituto P.T., Rome, 1959; Citation, INTERNATIONAL TELECOMMUNICATION JOURNAL, 1964; Honorary Chairman and Citation, International Television Symposium, Montreux, Switzerland, 1976; Honorary Chairman, Swiss National Committee of U.R.S.I.

GERHARDT, LESTER A., Fellow 1979. Born: January 28, 1940, Bronx, N.Y. Degrees: B.E.E., 1961, City College of New York; M.S.E.E., 1964, Ph.D., 1969, State University of New York at Buffalo.

Fellow Award "For research in digital signal processing, communications and adaptive systems, and pattern recognition, and for leadership in engineering education."

GERKS, IRVIN H., IRE Fellow 1958. Born: October 30, 1905, New London, Wis. Degrees; B.S., 1927, University of Wisconsin; M.S., 1932, Georgia Institute of Technology.

Fellow Award "For his contributions to the knowledge of tropospheric and ionospheric propagation."

GERLACH, HORST W.A., Fellow 1982. Born: June 23, 1912, Dresden, Germany. Degrees: M.E.E., 1938, Technical College, Dresden, Germany; 1940, Technical University, Berlin, Germany; USA: 1958-61 Grad. Courses at NBS Campus, Dept. of Commerce (G.W.U.; U of Md.).

Fellow Award "For design and development of microwave tubes and solid-state sources for missile fuze systems"; IEEE Centennial Medal, 1984. Other Awards: Special Act Award, H.D.L., Dept. of Army, 1973; Citation for service in AGED, USDR&E, Dept. of Defense, 1982; 14 Patents.

GERMESHAUSEN, KENNETH J., Fellow 1964. Born: May 12, 1907, Woodland, Calif. Degrees: B.S.E.E., 1931, Massachusetts Institute of Technology; D.Sc.(Hon.), Franklin Pierce College, 1976.

Fellow Award "For contributions to the technology of gaseous discharge flash lamps and stroboscopic lighting equipment." Other Awards: Fellow, American Academy of Arts and Sciences, 1969; Holley Medal, American Society of Mechanical Engineers, 1973.

GERSHO, ALLEN, Fellow 1982. Born: January 18, 1940, Quebec, Canada. Degrees: B.S.(E.E.), 1960, Massachusetts Institute of Technology; M.S.(E.E.), 1961, Ph.D., 1963, Cornell Univesity.

Fellow Award "For contributions to the theory of signal processing in communications"; Guillemin-Cauer Prize Paper Award, 1980; Member, Board of Governors, IEEE Communications Society.

GESELOWITZ, DAVID B., Fellow 1978. Born: May 18, 1930, Philadelphia, Pa. Degrees: B.S.E.E., 1951, M.S.E.E., 1954, Ph.D., 1958, University of Pennsylvania.

Fellow Award "For contributions to the application of electromagnetic theory in electrocardiography"; IEEE Centennial Medal, 1984. Other Awards: Fellow, American College of Cardiology, 1975; Guggenheim Fellow, 1978-79.

GETSINGER, WILLIAM J., Fellow 1980. Born: January 24, 1924, Waterbury, CT. Degrees: B.S.E.E., 1949, University of Connecticut; M.S.E.E., 1959, Engineer, 1961, Stanford University.

Fellow Award "For contributions to modeling and computer applications of microwave circuit design."

GETTING, IVAN A., IRE Fellow 1954. Born: January 18, 1912, New York, N.Y. Degrees: S.B., 1933, Massachusetts Institute of Technology; Dr.Phil., 1935, University of Oxford; D.Sc.(Hon.), 1956, Northeastern University.

Fellow Award "For his contributions to the development of automatic tracking radar systems"; Pioneer Award, 1975; President, IEEE, 1978. Other Awards: Presidential Medal for Merit, 1944; Naval Ordnance Development Award, 1944; Exceptional Service Award, U.S. Dept. of Air Force, 1960; Fellow, American Physical Society; Fellow, American Academy of Arts and Sciences; Member, National Academy of Engineering; Fellow, American Institute of Aeronautics and Astronautics; Kitty Hawk Award, 1975.

GETTING, MILAN P., JR., AIEE Fellow 1962. Born: November 19, 1908, Pittsburgh, Pa. Degrees: B.S., 1931, University of Pittsburgh.

Fellow Award "For contributions to the art of manufacturing and testing of transformers." Other Awards; American Society for Testing Materials; American Radiographic Society.

GEWARTOWSKI, JAMES W., Fellow 1982. Born: November 10, 1930, Chicago, IL. Degrees: B.S.E.E., 1952, Illinois Institute of Technology; M.S.E.E., 1953, Massachusetts Institute of Technology; Ph.D.(E.E.), 1958, Stanford University.

Fellow Award "For contributions to microwave solid-state circuits"; Browder J. Thompson Memorial Prize Award, IEEE, 1960.

GHANDHI, SORAB K., Fellow 1965. Born: January 1, 1928, Allahabad, India. Degrees: B.Sc., 1947, Benares Hindu University; M.S., 1948, Ph.D., 1951, University of Illinois.

Fellow Award "For contributions to circuit theory and the application of semiconductor devices." Other Awards: J.N. Tata Scholar, 1947-1950; Eta Kappa Nu; Pi Mu Epsilon; Phi Kappa Phi; Sigma Xi; Distinguished Teacher Award, Rensselaer Polytechnic Institute, 1976.

GHAUSI, MOHAMMED S., Fellow 1973. Born: February 16, 1930, Kabul, Afghanistan. Degrees: B.S.(highest honors), 1956, M.S., 1957, Ph.D., 1960, University of California at Berkeley.

Fellow Award "For contributions in active and distributed circuits, and education in network theory"; Regional Outstanding Lecturer, IEEE, 1974-1976. Other Awards: Outstanding Foreign Student Award in Bay Area, Mayor of Oakland, 1958; John F. Dodge Professor of Engineering, Oakland University, 1978-1983; Phi Beta Kappa.

GHOSE, RABINDRA N., Fellow 1970. Born: September 1, 1925, Howrah, India. Degrees: B.E.E., 1946, Jadavpur University; DIIS, 1948, Indian Institute of Sciences; M.S., 1952, University of Washington; M.A., 1954, Ph.D., 1954, E.E., 1956, University of Illinois.

Fellow Award "For contributions in the field of microwave theory and techniques, and for novel antenna concepts." Other Awards: Fellow, AAAS; Fellow, Washington Academy of Science; Fellow, Institution of Electrical Engineers; Fellow, American Physical Society; Fellow, Institute of Physics (London); Executive and Professional Hall of Fame; Charter Engineer, British Commonwealth; Certificate of Commendation, California State Legislature; Engineer of the Year, San Fernando Engineering Council, 1972; Registered Professional Engineer, California; Member, State Bar of California; Member, American Bar Association.

GIACOLETTO, L. J., IRE Fellow 1958. Born: November 14, 1916, Clinton, Ind. Degrees: B.S., 1938, Rose-Hulman Institute of Technology; M.S., 1939, State University of Iowa; Ph.D., 1952, University of Michigan.

Fellow Award "For his contributions to the understanding of transistors." Other Awards: Fellow, American Association for the Advancement of Science.

GIALLORENZI, THOMAS G., Fellow 1984. Born: February 28, 1943, New York, NY. Degrees: B.Eng.(Physics), 1965, M.Eng.(Physics), 1966, Ph.D., 1969, Cornell University.

Fellow Award "For leadership in the development of optical fiber systems for military applications." Other Awards: Research Society of America Award for Applied Sciences, 1973; Optical Society of American Fellows; Navy Meritorious Civilian Service Medal, 1976; Meritorious Presidential Senior Executive Rank, 1983.

GIANOLA, U. F., Fellow 1969. Born: October 29, 1927, Birmingham, England. Degrees: B.Sc., 1948, Ph.D, 1951, University of Birmingham, U.K.

Fellow Award "For outstanding contributions in the field of computer memories and magnetic logic, and in the activities of professional societies." Other Awards: 1981 Thomas Alva Edison Patent Award, Research and Development Council of New Jersey; Sigma Xi; AAAS.

GIBBONS, JAMES F., Fellow 1972. Born: September 19, 1931, Leavenworth, Kans. Degrees: B.S., 1953, Northwestern University; M.S., 1954, Ph.D., 1956, Stanford University.

Fellow Award "For contributions to solid-state electronics and engineering education"; Jack A. Morton Award, IEEE, 1980. Other Awards: Esbach Award for Most Outstanding Engineering Graduate, Northwestern University, 1953; Sigma Xi; Tau Beta Pi; Pi Mu Epsilon; Eta Kappa Nu; Phi Eta Sigma; Western Electric Fund Award for Excellence in Teaching, 1971; Member, National Academy of Engineering, 1974; Award for Outstanding Achievement, Northern California Solar Energy Association, 1975; Award for Outstanding Teaching of Undergraduate Engineering Studies, Tau Beta Pi, Stanford, 1976; Founder's Prize, Texas Instruments Foundation, 1983; Reid Weaver Dennis Professorship in Electrical Engineering, 1983.

GIBSON, J. JAMES, Fellow 1979. Born: March, 12, 1923, St. Albans, Great Britain. Degrees: Civilingenjor

(B.S.E.E.), 1947, Royal Institute of Technology, Stockholm, Sweden; Tekn. licentiat (M.S.E.E.), 1971, Chalmer Institute of Technology, Goteborg, Sweden.

Fellow Award "For contributions to consumer electronic systems and solid-state circuits." Other Awards: RCA Laboratory Achievement Award, 1961, 1962, 1968, 1979; Fellow, Technical Staff, RCA Laboratories, 1969; Fellow, Audio Engineering Society, 1980.

GIBSON, JOHN E., Fellow 1977. Born: June 11, 1926, Providence, R.I. Degrees: B.S.E.E., 1950, University of Rhode Island; M.S.E.E., 1952, Ph.D., 1956, Yale University.

Fellow Award "For leadership in electrical engineering education and contributions to automatic control and systems engineering"; First Prize Paper, AIEE, Connecticut Section, 1955, for "A Note on the Speed Torque Curves of AC Adjustable Speed Drives"; First Prize Paper, AIEE, Great Lakes District No. 5, 1961, for "A Set of Standard Specifications for Linear Automatic Control Systems" (co-authors Rekasius, McVey, Sridhar, Leedham); First Prize Paper in Industrial Division, AIEE, 1962, for "The Variable Gradient Method for Generating Liapunov Functions" (co-author D. G. Schultz).

GILBERT, BARRIE, Fellow 1984. Born: June 5, 1937, Bournemouth, England. Degrees: H.N.C.(Applied Physics), 1962, Bournemouth Muncipal College.

Fellow Award "For the discovery of the translinear principle and invention of the translinear multiplier"; IEEE Achievement Award, 1970.

GILBERT, EDGAR N., Fellow 1974. Born: July 25, 1923. Degrees: B.S., 1943, City College of The City University of New York; Ph.D., 1948, Massachusetts Institute of Technology.

Fellow Award "For contributions to information theory, and for applications of probability theory and combinatory analysis to electrical engineering." Other Awards: Member, Society for Industrial and Applied Mathematics (SIAM).

GILBERT, ELMER G., Fellow 1979. Born: March 29, 1930, Joliet, Ill. Degrees: B.S.E.E., 1952, M.S.E.E., 1953, Ph.D.(Instrum. Eng.), 1957, University of Michigan.

Fellow Award "For contributions to multivariable and optimal control systems." Other Awards: D. Hugo Schuck Best Paper Award, American Automatic Control Council, 1977; IEEE Control Systems Society, Best Paper Award, 1978.

GILBERT, ROSWELL W., IRE Fellow 1960. Born: July 21, 1909, Brooklyn, N.Y.

Fellow Award "For application of electronics to measurement techniques."

GILLETTE, ROBERT W., AIEE Fellow 1962. Born: October 29, 1905, Holyoke, Mass. Degrees: B.S., 1927, Worcester Polytechnic Institute.

Fellow Award "For contributions in the field of power cable engineering."

GILLIES, DONALD A., Fellow 1977. Born: November 22, 1920, South Bend, Wash. Degrees: B.S.E.E., 1949, Washington State College.

Fellow Award "For contributions to high-voltage equipment reliability through evaluation and maintenance." Other Awards: Distinguished Service Award, Gold Medal (highest honor), U.S. Dept. of Interior, 1973.

GILMAN, GEORGE W., IRE Fellow 1950. Born: August 26, 1901, Nova Scotia, Canada. Degrees: B.S., M.S., 1923, Massachusetts Institute of Technology.

Fellow Award "For contributions to the communication art and for his direction of important developments in the field of radio transmission systems." Other Awards: Order of the Rising Sun, Japan, 1934.

GILMER, BEN S., Fellow 1969. Born: March 5, 1905, Savannah, Ga. Degrees: B.S.E.E., 1926, D.Sc.(Hon.), 1958, Auburn University.

Fellow Award "For demonstrated managerial and engineering skills, and for outstanding contributions in directing the growth and development of the communications network."

GILSON, WALTER J., AIEE Fellow 1949. Born: March 17, 1898, Napa, Calif. Degrees: Dipl., 1924, University of Manchester, England. Awards: Vice President, District 10, AIEE, 1941-1943. Other Awards: Registered Professional Engineer, Ontario, Canada.

GINSBURG, CHARLES P., Fellow 1965. Born: July 27, 1920, San Francisco, Calif, Degrees: B.A., 1948, San Jose State College.

Fellow Award "For contributions to video recording techniques and equipment"; Vladimir K. Zworykin Award, "For pioneering contributions to the development of video magnetic recording." Other Awards: David Sarnoff Gold Medal, Society of Motion Picture and Television Engineers, 1957; Fellow, Society of Motion Picture and Television Engineers, 1958; Valdemar Poulsen Gold Medal, Danish Academy of Technical Sciences, 1960; Howard N. Potts Medal, Franklin Institute, 1969; Life Fellow, Franklin Institute, 1969; John Scott Medal, City of Philadelphia, 1970; Member, National Academy of Engineering, 1973; Honorary Member, Society of Motion Picture and Television Engineers, 1976; Honorary Fellow, Royal Television Society.

GINSBURG, SEYMOUR, Fellow 1973. Born: December 12, 1927, Brooklyn, N.Y. Degrees: B.S., 1948, City College of The City University of New York; M.S., 1949, Ph.D., 1952, University of Michigan.

Fellow Award "For contributions to switching theory, and work in automata and formal languges." Other Awards: Guggenheim Fellowship, 1974-1975.

GINZTON, EDWARD L., IRE Fellow 1951. Born: December 27, 1915, Russia. Degrees: B.S., 1936, M.S., 1937, University of California; E.E., 1938, Ph.D., 1940, Stanford University.

Fellow Award "For numerous contributions to microwave art, especially in the development of high-power klystrons"; Morris N. Liebmann Award, IRE, 1958, "For his creative contribution to the generation and useful application of high energy at microwave frequencies"; IEEE, Medal of Honor Award, 1969, "For his outstanding contributions in advancing the technology of high power klystrons and their application, especially to linear particle accelerators." Other Awards: National Academy of Engineering, 1965; National Academy of Science, 1967; Member, American Academy of Arts and Sciences, 1971; California Manufacturer of the Year, 1974.

GIORDANO, ANTHONY B., IRE Fellow 1958. Born: February 1, 1915, New York, N.Y. Degrees: B.E.E., 1937, M.E.E., 1939, D.E.E., 1947, Polytechnic Institute of Brooklyn.

Fellow Award "For his contributions to microwave measurements and modern communication curricula"; Meritorious Service Award, IEEE Communications Society, 1976; Honorable Mention Paper Citation, IEEE Education Society, 1981; Achievement Award, IEEE Education Society, 1982; IEEE Centennial Medal, 1984. Other Awards: Certificate of Distinction, Polytechnic Institute of Brooklyn, 1957; Distinguished Alumnus Award, Polytechnic Institute of Brooklyn, 1967; Fellow, American Association for the Advancement of Science, 1977; Distinguished Citation, Graduate Studies Division, American Society for Engineering Education, 1980; Certificate of Appreciation, Engineering Foundation of the United Engineering Trustees, 1980; Western Electric Fund Award, American Society for Engineering Education, 1981; Fellow, American Society for Engineering Education, 1983.

GIORDMAINE, JOSEPH A., Fellow 1978. Born: April 10, 1933, Toronto, Canada. Degrees: B.A., 1955, University of Toronto; A.M., 1957, Ph.D., 1960, Columbia University.

Fellow Award "For pioneering contributions to nonlinear optics"; IEEE MTT-S Distinguished Lecturer Award, 1983. Other Awards: Fellow, American Physical Society, 1967; Fellow, Optical Society of America, 1976; Fellow, New York Academy of Sciences, 1969; Fellow, American Association for the Advancement of Science, 1981.

GLASER, EDWARD L., Fellow 1969. Born: October 7, 1929, Evanston, Ill. Degrees: B.A., 1951, Dartmouth College; D.Sc.(Hon.), 1980, Heriot-Watt University, Scotland.

Fellow Award "For contributions to the design of computers and information systems." Other Awards: Phi Beta Kappa, 1951; Computer Man of the Year,

1974. National Academy of Engineering, 1977.

GLASFORD, GLENN M., Fellow 1964. Born: November 4, 1918, Arcola, Tex. Degrees: B.S., 1940, University of Texas; M.S., 1942, Iowa State University.

Fellow Award "For contributions to engineering education and leadership in engineering activities."

GLASS, ALEXANDER J., Fellow 1982. Born: January 4, 1933, Pittsfield Twp, NY. Degrees: B.S.(Physics), 1954, Rensselaer Polytechnic Institute; M.S.(Physics), 1955, Ph.D.(Physics), 1963, Yale University.

Fellow Award "For contributions to quantum electronics and to the design of glass lasers for fusion research." Other Awards: Director, Laser Institute of America, 1974-1980; Fellow, Optical Society of America, 1976; Special Citation, DOE and DARPA, 1978; Director, Optical Society of America, 1980-1982.

GLICKSMAN, MAURICE, Fellow 1978. Born: October 16, 1928, Toronto, Canada. Sc.M., 1952, Ph.D., 1954, University of Chicago.

Fellow Award "For contributions to the understanding of transport, optical, and plasma phenomena in semiconductors." Other Awards: RCA Award for Outstanding Achievement in Research, 1956, 1962.

GLOGE, DETLEF C., Fellow 1976. Born: February 2, 1936, Germany. Degrees: Dipl.Ing., 1961, Dr.Ing., 1964, Technical University of Braunschweig, Germany.

Fellow Award "For contributions in the field of optical transmission in fibers and in lens waveguides."

GLOMB, WALTER L., Fellow 1970. Born: February 7, 1925, Glen Ridge, N.J. Degrees: B.S.E.E., 1946, M.S.E.E., 1948, Columbia University.

Fellow Award "For contributions to communication transmission systems, particularly those utilizing satellite repeaters."

GLOVER, ALAN M., IRE Fellow 1956. Born: September 26, 1909, Rochester, N.Y. Degrees: B.A., 1930, M.A., 1932, Ph.D., 1935, University of Rochester.

Fellow Award "For contributions to the development of photo tubes."

GLUYAS, THOMAS M., JR., Fellow 1971. Born: 1913, Pennsylvania. Degrees: B.S.E.E., 1935, Pennsylvania State University.

Fellow Award "For technical leadership in the development and design of television transmitters and radar equipment."

GODSEY, FRANK W., JR., AIEE Fellow 1945. Born: September 14, 1906, Beaumont, Tex. Degrees: B.S., 1927, Rice University; M.S., 1929, E.E. 1933, Yale University.

Other Awards: Award for Advancement of Basic and Applied Science, Yale Engineering Association, 1947; Best Paper Award, Electrochemical Society; U.S. Navy Commendations, 1955, 1960; Westinghouse Order of Merit.

GOELL, JAMES E., Fellow 1979. Born: October 13, 1939, New York City, N.Y. Degrees: B.E.E., 1962, M.S., 1963, Ph.D., 1965, Cornell University.

Fellow Award "For technical contributions and leadership in the fields of optical fibers, integrated optical circuits, and millimeter waveguides." Other Awards: Honorable Mention, Eta Kappa Nu Outstanding Young Electrical Engineer of the United States for 1973.

GOETZBERGER, ADOLF, Fellow 1978. Born: November 29, 1928, Munich, Germany. Degrees: Dr. rer. nat. (in Physics), 1955, University of Munich.

Fellow Award "For contributions to the understanding of semiconductor/oxide interfaces"; J.J. Ebers Award, IEEE Electron Devices Society, 1983.

GOFF, KENNETH W., Fellow 1979. Born: June 14, 1928, Salem, W. Va. Degrees: B.S.E.E., 1950, West Virginia University; M.S.E.E., 1952, Sc.D.E.E., 1954, Massachusetts Institute of Technology.

Fellow Award "For technical contributions to and leadership in the design and development of hardware and software for digital computer-based process-control systems." Other Awards: Fellow, Instrument Society of America, 1973.

GOLAY, MARCEL J. E., IRE Fellow 1960. Born: May, 1902, Neuchatel, Switzerland. Degrees: Lic.E.E., 1924, Federal Institute of Technology, Zurich, Switzerland; Ph.D., 1931, University of Chicago; Dr.sc.techn.-(honoris causa), Federal Institute of Technology, Lausanne, Switzerland, 1977.

Fellow Award "For contributions to the fields of communications and infrared technology"; Harry Diamond Award, IRE, 1951, "For his many contributions in the over-all Signal Corps research and development program and particularly for his accomplishments leading toward a reduction in the infrared-radio gap." Other Awards: Sargent Award, American Chemical Society, 1961; Distinguished Achievement Award, Instrument Society of America, 1961; Jimmie Hamilton Award, American Society of Naval Engineers, 1971; M.S. Tswett Chromatography Medal, 1976; Tswett Memorial Medal, All-Union Chromatography Council, Academy of Sciences of the U.S.S.R., 1979; 1981 ACS Award in Chromatography, American Chemical Society; Dal Nogare Award, Chromotography Forum of Delaware Valley, 1982.

GOLD, BERNARD, Fellow 1972. Born: 1923, New York, N.Y. Degrees: B.E.E., 1944, City College of The City University of New York; M.S.E.E., 1946, Ph.D., 1948, Polytechnic Institute of Brooklyn.

Fellow Award "For contributions to speech communication and digital signal processing." Other Awards: Fulbright Fellow, 1954-55; Fellow, Acoustical Society of America; National Academy of Engineering, 1981.

GOLD, S. HAROLD, Fellow 1976. Born: September 4, 1923, Tucson, Ariz. Degrees: B.S.E.E., 1945, University of New Mexico.

Fellow Award "For contributions to the development, testing, application, and standardization of improved electrical power utility apparatus." Other Awards: Fellow, Institute for the Advancement of Engineering, 1971.

GOLDBERG, DANIEL L., Fellow 1984. Born: April 5, 1922, Philadelphia, PA. Degrees: B.S., M.S., New Jersey Institute of Technology.

Fellow Award "For innovative engineering of large transportation facilities"; IEEE Standards Board Medallion, 1975. Other Awards: Port Authority of New York & New Jersey Distinguished Service Award, 1979; I&CPS Committee Achievement Award, 1974.

GOLDBERG, HAROLD, IRE Fellow 1955. Born: January 31, 1914, Milwaukee, Wis. Degrees: B.S.E.E., 1935, M.S.E.E., 1936, Ph.D.(E.E.), 1937, Ph.D.(Physiology), 1941, University of Wisconsin.

Fellow Award "For his contributions to the field of guided missile armament." Other Awards: Exceptional Service Award, U.S. Dept. of Commerce, 1953; Distinguished Service Citation, University of Wisconsin, 1964.

GOLDBERG, HAROLD S., Fellow 1982. Born: January 22, 1925, New York, NY. Degrees: B.E.E., 1944, Cooper Union; M.E.E., 1949, Polytechnic Institute of Brooklyn.

Fellow Award "For contributions to the design and introduction of new instrumentation, especially medical instrumentation"; USAB Citation of Honor, IEEE, 1978; IEEE Centennial Medal, 1984. Other Awards: Award of Distinction, Polytechnic Institute of New York, 1980.

GOLDEY, JAMES M., Fellow 1969. Born: July 3, 1926, Wilmington, DE. Degrees: B.S., 1950, University of Delaware; Ph.D., 1955, Massachusetts Institute of Technology.

Fellow Award "For contributions to the science and technology of thyristors, transistors, and semiconductor integrated circuits."

GOLDMAN, STANFORD, IRE Fellow 1956. Born: November 14, 1907, Cincinnati, Ohio. Degrees: B.A., 1926, M.A., 1928, University of Cincinnati; Ph.D., 1933, Harvard University.

Fellow Award "For contributions to signal and circuit theory and to electro-medical research." Other Awards: Fellow, American Association for the Advancement of Science, 1981.

GOLDSMITH, LESTER M., AIEE Fellow 1947. Born: July 1, 1893, Pottsville, Pa. Degrees: D.Sc.(Hon.), 1942, Drexel Institute of Technology.

Other Awards: Mellville Medal, American Society of Mechanical Engineers; U.S. Army Medal for Exceptional Civilian Service.

GOLDSMITH, THOMAS T., JR., IRE Fellow 1949. Born: January 9, 1910, Greenville, S.C. Degrees: B.S., 1931, Furman University; Ph.D., 1936, Cornell University.

Fellow Award "For his contributions in the development of cathode ray instruments and in the field of television." Other Awards: Member, National Television System Committee; Fellow, Society of Motion Picture and Television Engineers; Fellow, Radio Club of America.

GOLDSTEIN, ALEXANDER, Fellow 1981. Born October 6, 1916, Zurich, Switzerland. Degrees: Engineer Diploma, 1940, Dr.(Tech.Sc.), 1948, Swiss Federal Institute of Technology ETHZ, Zurich, Switzerland.

Fellow Award "For contributions in the design, construction, and testing of EHV transformers."

GOLDSTEIN, EDWARD, Fellow 1974. Born: January 28, 1923, Stanislawow, Poland. Degrees: B.S.E.E., 1949, University of Minnesota.

Fellow Award "For leadership in the development of communications facilities for military and civilian applications."

GOLDSTEIN, LADISLAS, IRE Fellow 1956. Born: February 6, 1906, Dombrad, Hungary. Degrees: B.A., 1925; M.S., 1928, D.Sc., 1937, University of Paris.

Fellow Award "For contributions in the field of microwave gaseous electronics"; Microwave Prize, 1958, "For significant contribution to the field of endeavor of the Professional Group on Microwave Theory and Techniques" in IRE TRANSACTIONS, "Nonreciprocal Electromagnetic Wave Propagation in Ionized Gaseous Media." Other Awards: John Simon Guggenheim Fellow, 1957.

GOLOMB, SOLOMON W., Fellow 1982. Born: May 31, 1932, Baltimore, MD. Degrees: B.A., 1951, Johns Hopkins Univesity; M.A., 1953, Ph.D., 1957, Harvard University.

Fellow Award "For pioneering contributions to the development of digital communications and information theory, embodying unique and novel applications of techniques from discrete mathematics"; Member, Board of Governors, IEEE Information Theory Society, 1967-1973. Other Awards: USC Associates Award for Research, 1960; National Academy of Engineering, 1976; Archimedes Circle Award, 1978.

GONZALEZ, RAFAEL C., Fellow 1984. Born: August 26, 1942, Havana, Cuba. Degrees: B.S.E.E., 1965, University of Miami; M.E., 1967, Ph.D., 1970, University of Florida (all in E.E.).

Fellow Award "For contributions to the theory and application of pattern recognition and digital image processing." Other Awards: UTK Chancellor's Research Scholar Award, 1978; Magnavox Engineering Professor Award, 1980; M.E. Brooks Distinguished Professor Award, 1980; IBM Professorship, 1981.

GOOD, WILLIAM E., IRE Fellow 1959. Born: April 25, 1916, Hillsdale, Mich. Degrees: B.A., 1937, Kalamazoo College; M.S., 1939, University of Illinois; Ph.D., 1946, University of Pittsburgh.

Fellow Award "For pioneering development in microwave spectroscopy and contributions to color television receivers." Other Awards: Alumni Citation, Kalamazoo College, 1951; Life Fellow, Society for Information Display, 1975; Distinguished Alumni Award, Kalamazoo College, 1977, Steinmetz Award, General Electric, 1977.

GOODALL, WILLIAM M., IRE Fellow 1951. Born: September 7, 1907, Washington, D.C. Degrees: B.S., 1928, California Institute of Technology.

Fellow Award "For his researches in the fields of ionospheric phenomena and pulse communications." Other Awards: Life Member, New York Academy of Sciences, 1975; "Who's Who in Electrical Engineering," 1976.

GOODELL, PAUL H., AIEE Fellow 1959. Born: May 25, 1908, Wickliffe, Ohio. Degrees: E.E., 1931, University of Cincinnati.

Fellow Award "For contributions to applications of electricity to industrial heating"; Conference Chairman, Industrial Process Heating, 1952. Other Awards: Engineering Society of Detroit; National Society of Professional Engineers; Fellow, Illuminating Engineering Society, 1947; Fellow, AAAS, 1967; Honorary Member, Australian Institute of Metal Finishing, 1974; Recipient, National Fire Protection Association "Distinguished Service Award", 1982; 30 year Member, Safety Standards Committee on Industrial Ovens and Furnaces.

GOODHEART, CLARENCE F., AIEE Fellow 1962. Born: January 24, 1916, Porterville, Calif. Degrees: B.S., 1936, California Institute of Technology; M.S., 1938, Ohio State University.

Fellow Award "For contributions to the teaching of electrical engineering." Other Awards: Sigma Xi; Tau Beta Pi; Eta Kappa Nu.

GOODMAN, JOSEPH W., Fellow 1974. Born: February 8, 1936, Boston, Mass. Degrees: A.B., 1958, Harvard University; M.S., 1960, Ph.D., 1963, Stanford University.

Fellow Award "For contributions to the theory of imaging." Other Awards: Fellow, Optical Society of America, 1970; F.E. Terman Award, American Society for Engineering Education, 1971; Fellow, Society for Photo-Optical Instrumentation Engineers, 1975; Max Born Award, Optical Society of America, 1983.

GORDON, BERNARD M., Fellow 1972. Born: 1927, Mass. Degrees: B.S.E.E., 1948, M.S.E.E., 1948, Massachusetts Institute of Technology.

Fellow Award "For contributions to the analog-digital interface." Other Awards: National Telemetry Prize for

Best Technical Presentation in Telemetry, 1958; One of Ten Outstanding Men of the Year, Boston Chamber of Commerce, 1958; Bausch and Lomb Science Award; Outstanding Living Engineer of the Year, New England Award, Engineering Societies of New England, 1971; Tau Beta Pi; Eta Kappa Nu; Sigma Xi; Hex Alpha (M.I.T.).

GORDON, EUGENE I., Fellow 1968. Born: September 14, 1930, New York, N.Y. Degrees: B.S., 1952, City College of The City University of New York; Ph.D., 1957, Massachusetts Institute of Technology.

Fellow Award "For his scientific contributions in the fields of electro-optics and quantum electronics"; Vladimir K. Zworykin Award, 1975, "For the invention and leadership in the development of the silicon target camera tube, and in the extension of electronic television into new applications." Other Awards: National Academy of Engineering, 1979.

GORDON, WALTER S., AIEE Fellow 1954. Born: November 17, 1904, Seattle, Wash. Degrees: B.S.E.E., 1926, University of Washington.

Fellow Award "For contributions in the application of high intensity airfield lighting and in the adoption of higher distribution voltages in heavy western industries." Other Awards: Legion of Merit, U.S. Dept. of Army, 1947; Engineer of Year Award, Washington Society of Professional Engineers, Tacoma Chapter; Electric League of the Pacific Northwest, 1973; Registered Professional Engineer, Washington, Oregon, California, Pennsylvania.

GORDON, WILLIAM E., IRE Fellow 1961. Born: January 8, 1918, Paterson, N.J. Degrees: B.A., 1939, M.A., 1942, Montclair State College; M.S., 1946, New York University; Ph.D., 1953, Cornell University.

Fellow Award "For contributions to the understanding of radio scattering"; IEEE Centennial Medal, 1984. Other Awards: Balth, van der Pol Award, International Scientific Radio Union, 1966; Walter R. Read Professor of Engineering, Cornell University, 1965-1966; National Academy of Sciences, 1968; American Meteorological Society Medal, 1969; Guggenheim Fellow, 1972-73; National Academy of Engineering, 1975; National Academy of Sciences Arctowski Medal, 1984.

GORDY, THOMAS D., AIEE Fellow 1956. Born: October 8, 1915, High Point, N.C. Degrees: B.S.E.E., 1936, University of North Carolina; M.E.E., 1948, Ph.D., 1955, Rensselaer Polytechnic Institute.

Fellow Award "For contributions to the theory, design and development of transformer magnetic circuits."

GORHAM, ROBERT C., AIEE Fellow 1955. Born: April 11, 1893, Smith Center, Kans. Degrees: B.A., 1917, Nebraska Wesleyan University; E.E., 1924, Cornell University; M.S., 1941, University of Pittsburgh.

Fellow Award "For diversified development work as

an adjunct to electrical engineering education." Other Awards: Pennsylvania Society for Professional Engineers, 1954; Sigma Tau, 1954; Distinguished Service Award, National Council of State Boards of Engineering Examiners, 1958.

GOUDET, GEORGES, IRE Fellow 1960. Born: June 2, 1912, Dijon, France. Degrees: Dr.Phys., 1943, University of Paris.

Fellow Award "For contribution to engineering education and for the administration of research." Other Awards: Officier de la Legion d'Honneur.

GOULD, GERALD G., Fellow 1974. Born: November 17, 1913, Budapest, Hungary. Degrees: B.S.E.E., 1935, City College of The City University of New York; M.S.E.E., 1936, Purdue University;.

Fellow Award "For leadership in underwater systems development, and for contributions to the design of a major underwater tracking range"; Outstanding Engineer for Region 3, IEEE, 1973. Other Awards: Rear Admiral William S. Parsons Award for Scientific and Technical Progress, Navy League of the U.S., 1975; U.S. Navy Superior Service Award, 1981.

GOULD, LEONARD A., Fellow 1973. Born: November 20, 1927, Brooklyn, N.Y. Degrees: S.B., 1948, Sc.D., 1953, Massachusetts Institute of Technology.

Fellow Award "For contributions to the theory and design of process control systems and engineering education." Other Awards: Fulbright Research Fellowship, Denmark, 1957.

GOULD, ROY W., Fellow 1965. Born: April 25, 1927, Los Angeles, Calif. Degrees: B.S., 1949, California Institute of Technology; M.S., 1950, Stanford University; Ph.D., 1956, California Institute of Technology.

Fellow Award "For contributions to the theory of microwave tubes." Other Awards: Member, National Academy of Engineering, 1971; Member, National Academy of Sciences, 1974; American Academy of Arts and Sciences, 1982.

GOWEN, RICHARD J., Fellow 1982. Born: July 6, 1935, New Brunswick, NJ. Degrees: B.S., 1957, Rutgers University; M.S., 1961, Ph.D., 1962, Iowa State University.

Fellow Award "For contributions to aerospace biomedical engineering technology and to education"; IEEE President, 1984. Other Awards: Outstanding Engineer Award, Rutgers University, 1977; Professional Achievement Citation in Engineering, Iowa State University, 1983.

GRADY, ROBERT E., AIEE Fellow 1962.

Fellow Award "For contributions to industrial and utility construction."

GRAHAM, BEARDSLEY, Fellow 1965. Born: April 24, 1914, Berkeley, Calif. Degrees: B.S., University of California at Berkeley.

Fellow Award "For promotion of satellite communi-

cations systems."

GRAHAM, DUNSTAN, Fellow 1969. Born: August 17, 1922, Princeton, N.J. Degrees: B.S.E., 1943, M.S.E., 1947, Princeton University.

Fellow Award "For many pioneering contributions to automatic flight control theory and systems, and to related education."

GRAHAM, MARTIN, Fellow 1968. Born: July 12, 1926. Degrees: B.E.E., 1947, D.E.E., 1952, Polytechnic Institute of Brooklyn; M.Sc., 1948, Harvard University.

Fellow Award "For contributions to digital computers and biomedical electronics."

GRAHAM, ROBERT C., Fellow 1968. Born: October 13, 1907, Schenectady, N.Y. Degrees: B.S.E.E., 1930, Union College.

Fellow Award "For contributions to advancement of cable technology for underground distribution systems"; Chairman, IEEE Insulated Conductors Committee, 1961-1962. Other Awards: Listed in "Who's Who in Engineering," 1977.

GRAHAM, ROBERT E., IRE Fellow 1959. Born: October 4, 1916, Kansas City, Mo. Degrees: B.S.E.E., 1937, Purdue University.

Fellow Award "For contributions in the field of radar tracking systems and television research."

GRAMMER, GEORGE, IRE Fellow 1959. Born: January 25, 1905, Philadelphia, Pa. Degrees: 1926, Drexel Institute.

Fellow Award "For contributions in single sideband radio telephony and for publication of technical information."

GRANGER, JOHN V., IRE Fellow 1956. Born: September 14, 1918, Cedar Rapids, IA. Degrees: A.B., 1941, Cornell College, Iowa; M.S.(Comm. Eng.), 1942, Ph.D.(Applied Physics), 1948, Harvard University.

Fellow Award "For leadership in research and contributions to aircraft radiation systems"; President, IEEE, 1970. Other Awards: Member, National Academy of Engineering, 1974.

GRANLUND, JOHN, Fellow 1974. Born: May 17, 1924. Degrees: S.B., 1944, S.M., 1947, Sc.D., 1950, Massachusetts Institute of Technology.

Fellow Award "For developments in communications and radar systems and related electronic circuitry, and for leadership of scientific efforts in satellite communications." Other Awards: Eta Kappa Nu; Sigma Xi.

GRANQVIST, C. E., IRE Fellow 1955. Born: March 2, 1910, Ljungby, Sweden. Degrees: M.A., 1933, Royal University of Technology, Stockholm.

Fellow Award "For his contributions to air navigation systems and devices, and for his leadership in the engineering of electronic apparatus."

GRASSELLI, ANTONIO, Fellow 1972.

Fellow Award "For contributions to switching circuits, pattern recognition, and computer science education."

GRAY, AUGUSTINE H., JR., Fellow 1982. Born: August 18, 1936, Long Beach, CA. Degrees: B.S.E.E., M.S.E.E., 1959, Massachusetts Institute of Technology; Ph.D.(Eng.Sc.), 1964, California Institute of Technology; M.B.A., 1981, Pepperdine University.

Fellow Award "For contributions to the theory of linear prediction and its applications to speech processing"; Senior Award(co-author), IEEE Acoustics, Speech, and Signal Processing Society, 1974; Achievement Award, IEEE Acoustics, Speech, and Signal Processing Society, 1977.

GRAY, JAMES P., Fellow 1982. Born: June 4, 1943, Los Angeles, CA. Degrees: B.E.E., 1965, M.S., 1966, Ph.D.(Comm.Theory), 1970, Yale University.

Fellow Award "For creative contributions to computer network architecture." Other Awards: IBM Outstanding Contribution Award, 1975; IBM Outstanding Innovation Award, 1981; IBM Outstanding Innovation Award, 1983.

GRAY, PAUL E., Fellow 1972. Born: February 7, 1932, Newark, N.J. Degrees: S.B., 1954, S.M. 1955, Sc.D., 1960, Massachusetts Institute of Technology.

Fellow Award "For contributions to solid-state electronics and engineering education." Other Awards: Ford Foundation Postdoctoral Fellowship, 1961-63; Member, National Academy of Engineering, 1975; Member: Sigma Xi, Eta Kappa Nu, Tau Beta Pi; Fellow, American Academy of Arts and Sciences.

GRAY, PAUL RUSSELL, Fellow 1981. Born: December 8, 1942, Jonesboro, AR. Degrees: B.S., 1963, M.S., 1965, Ph.D., 1969, University of Arizona at Tucson.

Fellow Award "For contributions to development of integrated circuits"; Best Paper Award, International Solid-State Circuits Conference, 1975; Baker Prize, 1980; Morris N. Liebmann Memorial Award, IEEE, 1983.

GRAY, RICHARD E., Fellow 1969. Born: February 15, 1902, Dorsetshire, England. Degrees: Diploma, 1923, Farady House, London.

Fellow Award "For pioneering advances in radio communication, and for contributions in the field of radio wave propagation.

GRAY, ROBERT M., Fellow 1980. Born: November 1, 1943, San Diego, CA. Degrees: B.S., M.S., 1966, Massachusetts Institute of Technology; Ph.D., 1969, University of Southern California.

Fellow Award "For contributions to information and communication theory"; Corecipient, Prize Paper Award, IEEE Information Theory Group, 1976; Editor, IEEE Transactions on Information Theory, 1980-1983; IEEE Centennial Medal, 1984. Other Awards:

Fellow, Japan Society for the Promotion of Science, 1981; Fellow, John Simon Guggenheim Foundation, 1981.

GRAY, TRUMAN S., AIEE Fellow 1945. Born: May 3, 1906, Spencer, Ind. Degrees: B.S., 1926, B.A., 1927, University of Texas; S.M., 1929, Sc.D., 1930, Massachusetts Institute of Technology.

Other Awards: Tau Beta Pi, 1926; Phi Beta Kappa, 1927; Charles A. Coffin Fellow, 1928-1929; Henry Saltonstall Fellow, 1929-1930; Fellow, Massachusetts Institute of Technology (honorary), 1929-1930; Sigma Xi, 1937; Eta Kappa Nu, 1956.

GRAYBILL, HOWARD W., Fellow 1975. Born: August 2, 1915, Bareville, Pa. Degrees: B.S.E.E., 1937, Drexel University.

Fellow Award "For contributions in the development and design of compact high-voltage gas-insulated substations." Other Awards: Eta Kappa Nu, 1936; Tau Beta Pi, 1937; Outstanding Engineering Achievement of 1971, National Society of Professional Engineers, 1972; Gould Scientific Achievement Award, 1976.

GRAYSON, LAWRENCE P., Fellow 1976. Born: May 16, 1937, Brooklyn, N.Y. Degrees: B.E.E., 1958, M.E.E., 1959, Ph.D.(E.E.), 1962, Polytechnic Institute of Brooklyn.

Fellow Award "For contributions to the application of technology to education"; Achievement Award, IEEE Education Society, 1979; IEEE Centennial Medal, 1984. Other Awards: U.S. Army Achievement Award, 1962; Special Award on Social Justice, National Association for Public Continuing and Adult Education, 1975; Distinguished Service Award, Dept. of Health, Education, and Welfare, 1976; Distinguished Service Award, American Society for Engineering Education, 1977; Fellow, American Association for the Advancement of Science, 1977; William Elgin Wickenden Best Paper Award, American Society for Engineering Education, 1979; Second Century Award, Polytechnic Institute of New York, 1980; Distinguished Service Citation, American Society for Engineering Education, 1981; Certificate of Appreciation, ASEE Continuing Professional Development Division, 1983; Fellow, American Society for Engineering Education, 1983; Certificate of Merit, ASEE Continuing Professional Development Division, 1984.

GREATBATCH, WILSON, Fellow 1971. Born: September 6, 1919, Buffalo, N.Y. Degrees: B.E.E., 1950, Cornell University; M.S.E.E., 1957, University of Buffalo; Sc.D.(Hon.), 1970, Houghton College.

Fellow Award "For vital contributions in biomedical engineering"; William J. Morlock Award, IEEE/GMB, 1968, "For the invention of the implantable cardiac pacemaker"; IEEE Centennial Medal, 1984. Other Awards: 125th Anniversary Award, State University of New York at Buffalo, 1970; Professional Engineer of the Year Award, New York State Society of Professional Engineers, 1974; Laufman Award, AAMI, 1982; Wunsch Award, Brooklyn Polytechnic Institute, 1980; Fellow, Royal Society of Health, FRSH, (Great Britain), 1974; Fellow, American College of Cardiology, FACC, 1975; Senior Scientist Award, North American Society for Pacing and Electrophysiology, 1983.

GREBENE, ALAN B., Fellow 1979. Born: March 13, 1939, Istanbul, Turkey. Degrees: B.Sc.E.E., 1961, Robert College, Istanbul, Turkey; M.S.E.E., 1963, University of California, Berkeley; Ph.D., 1968, Rensselaer Polytechnic Institute.

Fellow Award "For contributions to the development of monolithic linear integrated circuits and leadership in engineering administration and management;" IEEE Centennial Medal, 1984.

GREEFKES, JOHANNES A., Fellow 1973. Born: April 26, 1912, Amsterdam, Netherlands. Degrees: Electronics degree, 1933, Technical Highschool, Amsterdam.

Fellow Award "For contributions to communication systems and circuitry"; IEEE Field Award, 1972, "In international communication in honor of Hernand and Sosthes Behn"; Vice President, Comsoc International Affairs, 1979, 1980. Other Awards: Veder Award, 1958.

GREEN, PAUL E.,JR., IRE Fellow 1962. Born: January 14, 1924, Durham, N.C. Degrees: A.B., 1944, University of North Carolina; M.S., 1948, North Carolina State College; Sc.D., 1953, Massachusetts Institute of Technology.

Fellow Award "For contributions to radar astronomy and communication through multipath channels"; Pioneer Award, IEEE Aerospace and Electronics Systems Society, 1980; IEEE Centennial Medal, 1984. Other Awards: National Academy of Engineering, 1981; Distinguished Engineering Alumnus, N.C. State Univ., 1983.

GREENE, DONALD L., Fellow 1966. Born: October 14, 1901, Ellicottville, N.Y. Degrees: B.S., 1924, Tri-State University.

Fellow Award "For contributions to the electric power field, both domestic and foreign." Other Awards: Letter of Appreciation, Matsuoka Electric Power Co., 1964; Letter of Appreciation, Okinawa Association of Power Distributing Companies, 1965; Letter of Appreciation, Government of the Ryukyus, 1965; Letter of Appreciation, Transmission Section, Gilbert Pacific, Inc., Okinawa Island Power Systems, 1965.

GREENFIELD, EUGENE W., AIEE Fellow 1951. Born: November 27, 1907, Baltimore, Md. Degrees: B.E., 1924, D.Eng., 1934, The Johns Hopkins University.

Awards: First Paper Prize, AIEE, District 9, 1956; Achievement Award, Region 6, 1964; Engineer of the Year, IEEE Spokane Section, 1973; Citation, AIEE, as Vice President, District 9, 1962-1964; IEEE Centennial Medal, 1984. Other Awards: Citations: State of Washington Air Pollution Control Board, 1966, State of Washington Energy Policy Council, 1974, National Particle Board Association, 1970, Engineering Foundation, 1972; Chief Technical Engineer Peer Panel-- Bonneville Power Administration, 1978, 1981; Visiting Scientist, Los Alamos National Laboratory, Los Alamos, NM, 1975-76; Visiting Scientist, Brookhaven National Laboratory, Long Island, NY, 1979-.

GREENWOOD, ALLAN, Fellow 1969. Born: Leeds, England. Degrees: B.A., 1943, M.A., 1948, Cambridge University; Ph.D. 1952, Leeds University.

Fellow Award "For outstanding contributions to the development of vacuum interrupters and to the understanding of the interaction between power systems and circuit interrupting devices"; First Prize Paper, Middle Eastern District No. 2, IEEE, 1960-61, 1961-62; Philip Sporn Professor of Engineering, Rensselaer Polytechnic Institute, 1979.

GREGORY, BOB L., Fellow 1977. Born: September 30, 1938, Allen, Okla. Degrees: B.S., 1960, M.S., 1961, Ph.D., 1963, Carnegie Institute of Technology.

Fellow Award "For contributions to the field of radiation-tolerant semiconductor devices and integrated circuits."

GREIG, JAMES, Fellow 1968. Born: April 24, 1903, Edinburgh. Degrees: E.E., 1923, Heriot-Watt College, Edinburgh; B.Sc., 1928, University College; M.Sc., 1933, University of London; Ph.D., 1938, University of Birmingham.

Fellow Award "For his contributions in the educational field, and research in electronics and electrical measurements especially on the behavior of ferromagnetic materials." Other Awards: William Siemens Professor of Electrical Engineering, King's College, University of London, 1940-1970.

GREW, LAWRENCE B., AIEE Fellow 1961. Born: December 12, 1906, Brockton, Mass. Degrees: B.S., 1927, M.S., 1928, Massachusetts Institute of Technology.

Fellow Award "For contributions to communication transmission and application of radio techniques to a communication system." Other Awards: Eta Kappa Nu, 1953.

GRIEG, DONALD D., IRE Fellow 1961. Born: February 26, 1916, London, England. Degrees: E.E., 1938, City College of the City University of New York; M.E.E., 1941, New York University.

Fellow Award "For contributions to pulse modulation systems."

GRIFFITH, M. SHAN, Fellow 1984. Born: December 25, 1940, Pontiac, MI. Degrees: B.S.(Science Engineering), Univ. of Michigan, 1962.

Fellow Award "For technical leadership and contributions to application practices in the field of industrial power system analysis and protection"; Industrial Power Systems Dept. Prize Paper Award, 1977; IAS Prize Paper Award, 1978; Industrial Power Systems Dept. Prize Paper Award, 1978; IAS Prize Paper Award, 1979; Industrial Power Systems Dept. Achievement Award, IEEE Industry Applications Society, 1983.

GRIVET, PIERRE AUGUSTE, Fellow 1971. Life Member 1982. Born: November 14, 1911, France. Degrees: Licence, 1935, Ecole Normale Superieure; Ph.D.(Physics), 1941.

Fellow Award "For research in electronics and for contributions to engineering education." Other Awards: Member, "Academie des Sciences de Paris, 1972; Member, Poland Academy of Science, 1977.

GROGAN, WILLIAM R., Fellow 1983. Born: August 2, 1924, Pittsfield, Massachusetts. Degrees: B.S.(E.E.), 1945, M.S.(E.E.), 1949, Worcester Polytechnic Institute.

Fellow Award "For contributions to engineering educational programs emphasizing the relationship of science and technology to societal concerns." Other Awards: William E. Wickenden Award, ASEE, 1980; Chester F. Carlson Award, ASEE, 1979; Scientific Achievement Award, Worcester Engineering Society, 1979.

GROGINSKY, HERBERT L., Fellow 1982. Born: July 10, 1930, Newark, NJ. Degrees: B.E.E.(summa cum laude), 1952, Polytechnic Institute of Brooklyn; M.S.E.E., 1954, D.Sc., 1959, Columbia University.

Fellow Award "For contributions to radar signal processing for detection, spectral estimation, and Doppler weather processing." Other Awards: Tau Beta Pi, 1952; Eta Kappa Nu, 1952; Sigma Xi, 1959; "Outstanding Young Men in America," 1966; "American Men of Science," 1972; "Who's Who in the East," 1974; Member, Technical Committee on Atmospheric Environment, American Institute of Aeronautics and Astronautics, 1978-1980.

GROH, GEORGE H., AIEE Fellow 1959. Born: January 17, 1901, Myerstown, Pa. Degrees: B.S.E.E., 1923, Pennsylvania State University.

Fellow Award "For contributions to the development of a rapidly expanding electric power system."

GROSS, AL, Fellow 1982. Born: February 22, 1918, Toronto, Canada.

Fellow Award "For pioneering work in VHF and UHF mobile radios"; IEEE Power Generation Committee; Excitation Subcommittee and Working Group. Other Awards: Fellow, Radio Club of America; Honorary Life Member, World CB Union and European CB Federation, Belgium; Honorary Life Member, Federazione Italiana Ricetrasmissioni CB, Milano; Honorary Life Member, National CB Council, Republic of Ireland; Honorary Life Member, Netherland CB Federation; Honorary Life Member, Deutscher CB Dachverband e.V., West Germany; Honorary Life Member, La Federazione CB Ticino, Switzerland; Honorary Member, Association, Francaise des Amateuro Radio (AFA), France; Honorary Member, Radio Club Austria; Honorary Member, Citizens Band Association, U.K.; Presidential Commendation for affirmative contributions in the telecommunications field, 1981; Who's Who in America, 43rd Edition.

GROSS, ERIC T. B., AIEE Fellow 1948. Born: May 24, 1901, Vienna, Austria. Degrees: E.E., 1923, D.Sc., 1932, Technical University Vienna.

Awards: Best Paper Award, AIEE, Chicago Section, 1953; IEEE, Brazil Council, Silver Plaque, 1974, "In recognition and appreciation of valued services and contributions to the education of power systems engineers in Brazil"; Award for Major Contributions in Electrical Engineering Management, IEEE Region 1, 1976, conferred "For contributions to furthering the objectives of IEEE through managerial excellence and achievement in organizing the Electric Power Engineering Operation and Curriculum at Rensselaer Polytechnic Institute, together with Faraday's Law Machines Laboratory"; Citation and Medal, IEEE Power Engineering Society, 1977, "For distinguished personal service to the Power Generation Committee"; and 1978, "For distinguished personal service to the Power Engineering Education Committee"; IEEE Centennial Medal, 1984. Other Awards: Fellow, Institution of Electrical Engineers, U.K., 1943; Sigma Xi, 1943; Tau Beta Pi, 1945; Fellow, AAAS, 1945; Fellow, N.Y. Academy of Sciences, 1967; Citation, American Power Conference, 1972; Distinguished Faculty Award, Rensselaer Polytechnic Institute, 1972; Western Electric Fund Award, ASEE, 1972; Eminent Member, Eta Kappa Nu, 1975; Special Citation, Edison Electric Institute, 1976; National Academy of Engineering, 1978; Austrian Cross of Honor in Science and the Arts, First Class, 1980.

GROSS, FRITZ A., Fellow 1966. Born: October 8, 1910, Germany. Degrees: Dipl., 1930, Lowell Institute, Massachusetts Institute of Technology; Dr.Eng.(Hon.), 1975, Northeastern University.

Fellow Award "For pioneering in the development of early microwave radar and continuing contributions and leadership in that field." Other Awards: Certificate of Commendation, U.S. Dept. of Navy, 1947; Reginald A. Fessenden Award, National Marine Electronics Assoc., 1981.

GROSZKOWSKI, JANUSZ, Fellow 1964. Born: March 21, 1898, Warsaw, Poland. Degrees: M.El.Eng., 1922, D.Techn.Sci., 1928, Dr.habil., 1929, Technical University, Warsaw; Dr. honoris causa of Technical Universities of Warszawa, Lodz, Gdansk.

Fellow Award "For contributions in the field of frequency stability and electronic technology, and leadership in engineering education and research." Other

Awards: Member, Academies of Science in Poland, Bulgaria, Czechoslovakia, Cuba, Hungary, Rumania, Soviet Union; Honorary Member, Societe Francaise des Electroniciens et Radio-techniciens; State Prize (First Grade): For Scientific Research, 1951, 1968, 1979, For Technical Research, 1955; Award of Honors, International Television Symposium, Montreux, 1963.

GROTZINGER, JOHN, AIEE Fellow 1957. Born: December 22, 1891, Schaffhausen, Switzerland. Degrees: Ing., 1913, Technical Institute of West Switzerland, Bienne.

Fellow Award "For his contributions to the technology of power systems and power supplies for a large industry."

GROVE, ANDREW S., Fellow 1972. Born: 1936, Hungary. Degrees: B.Ch.E., 1960, City College of The City University of New York; Ph.D., 1963, University of California at Berkeley.

Fellow Award "For contributions to metal-oxide-silicon devices and engineering education"; Region 6 Achievement Award, IEEE, 1969, "For advancing the state-of-the-art in semiconductor devices through his leadership, teaching, publications, and research on the metal-oxide-silicon system"; J.J. Ebers Award, IEEE, 1974, "For outstanding technical contributions to electron devices." Other Awards: Certificate of Merit, Franklin Institute, 1975; National Academy of Engineering, 1979; Townsend Harris Medal, City College of New York, 1980.

GROVER, FREDERICK W., IRE Fellow 1949. Born: September 3, 1876, Lynn, Mass. Degrees: S.B., 1899, Massachusetts Institute of Technology; M.S., 1901, Wesleyan University; Ph.D., 1907, George Washington University; Ph.D., 1908, Ludwig Maximilians Universitat, Munich, Germany.

Fellow Award "For his long activities and contributions in the field of electrical units and measurements, and for his publications." Other Awards: Fellow, American Association for the Advancement of Science; American Physics Society.

GROW, RICHARD W., Fellow 1973. Born: October 31, 1925, Lynndyl, Utah. Degrees: B.S.E.E., 1948, M.S.E.E., 1949, University of Utah; Ph.D., 1955, Stanford University.

Fellow Award "For contributions to electron devices and engineering education"; IEEE Utah Section Award, 1971, "In recognition of outstanding technical achievement." Other Awards: Member: Tau Beta Pi, Phi Kappa Phi, Eta Kappa Nu, Sigma Xi; Distinguished Research Award, University of Utah, 1982-83.

GUARRERA, JOHN J., fellow 1974. Born: March 4, 1922. Degrees: B.S.E.E., 1943, Massachusetts Institute of Technology.

Fellow Award "For contributions to microwave technology"; Chairman, San Fernando Valley Section, IEEE,

1957; Chairman, Los Angeles Council, IEEE, 1962; Chairman, Wescon Board, 1968; Director, Region Six, IEEE, 1971, 1972; Vice President, IEEE, 1973; President, IEEE, 1974; Vice President Professional Activities, 1977; U.S. Activities Board, IEEE, 1977; Engineering Professionalism Award, U.S. Activities Board, IEEE, 1978. Los Angeles, IEEE, Council Leadership Award, 1958; San Fernando Valley Sub-Section Service Award, 1958; Los Angeles, IEEE, Council Recognition Award, 1962. Other Awards: Service Award, Office of Scientific Research and Development , USA, 1945; Pi Mu Epsilon, 1947; Wescon Awards, 1960, 1966, 1968; WINCON Awards, 1960, 1963, 1966;Special Recognition Award, San Fernando Valley Engineers Council, 1964; Engineering Merit Award, 1971; Engineer of the Year Award, 1973; Fellow, Institute for the Advancement of Engineering; Member, American Society for Engineering Education; National Society of Professional Engineers, California Society of Professional Engineers; President-Elect, National Computer Graphics Association; Director of Research, School of Engineering and Computer Science, California State University, Northridge; Commissioner, Board of Water and Power Commissioners, Los Angeles; Director, Healthwest; Director Northridge Hospital Foundation; Director, San Fernando Valley Child Guidance Clinic; Commissioner, Citizens Commission on Pension Policy, Registered Licensed Professional Engineer in New York & California; Engineers Joint Council Recognition Award, 1964; Fellow, Institute for Advancement of Engineering, 1972; OSRD Award for World War II Work at Massachusetts Institute of Technology Radiation Laboratory; Tau Beta Pi, 1973; Distinguished Service Citation, Dept. of Consumer Affairs, California, 1973.

GUENTHER, ARTHUR H., Fellow 1979. Born: April 20, 1931, Hoboken, N.J. Degrees: B.S., 1953, Rutgers University; Ph.D., 1957, Pennsylvania State University; D.Sc.(Honoris Causa), 1973, University of Albuquerque.

Fellow Award "For contributions to high-power laser technology and pulsed-power engineering"; Harry Diamond Award, 1971. Other Awards: Fellow, Optical Society of America, 1974; Decoration for Exceptional Civilian Service, DoD, 1976; Distinguished Scientist of the Year, New Mexico Academy of Science, 1977; New Mexico Distinguished Public Service Award, 1982; Ad Com, Quantum Electronics and Applications Society, IEEE, 1983. Other Awards: Fellow, Optical Society of America; Arthur L. Schawlow Medal, Laser Institute of America, 1983; Distinguished Senior Executive, U.S. Government, 1983.

GUENTHER, RICHARD, Fellow 1965. Born: September 9, 1910, Lodz, Poland. Degrees: Dipl.Ing., 1934, Dr.Ing., 1937, Technische Hochschule Danzig.

Fellow Award "For contributions to advanced communications systems." Other Awards: Siemens Ring Stiftung, 1935.

GUERNSEY, ERNEST W., Fellow 1964. Born: May 1, 1896, Harrison Twp., Knox County, Ind. Degrees; B.S., 1918, University of Illinois; M.A., 1922, American University; Ph.D., 1924, George Washington University.

Fellow Award "For contributions to generation, transmission, and distribution of electric power." Other Awards: Fellow, American Association for the Advancement of Science, 1933; Certificate of Recognition, American Institute of Chemical Engineers, 1945; Award of Merit, American Gas Association, 1955.

GUMMEL, HERMANN K., Fellow 1973. Born: July 6, 1923, Hanover, Germany. Degrees: M.S., 1952, Ph.D., 1957, Syracuse University.

Fellow Award "For contributions to mathematical modeling of electronic devices"; Guillemin-Cauer Prize, IEEE Circuits and Systems Society, 1977, for paper "MOTIS - An MOS Timing Simulator" (co-authors B.R. Chawla and P. Kozak); David Sarnoff Award, IEEE, 1983.

GUNDLACH, FRIEDRICH W., Fellow 1974. Born: February 2, 1912, Berlin, Germany. Degrees: Dipl.-Ing., 1935, Dr.-Ing., 1938, Technische Hochschule Berlin, Germany.

Fellow Award "For contributions to the development of microwave tubes."

GUNN, J. B., Fellow 1968. Born: May 13, 1928, Cairo, Egypt. Degrees: B.A., 1948, Cambridge University.

Fellow Award "For his discovery and elucidation of a bulk phenomenon in III-V semiconductors now known as the Gunn Effect, which has made possible a new source of high frequency oscillations in the microwave region"; Morris N. Liebmann Prize, 1969, "For contributions to solid-state microwave power generation." Other Awards: Fellow, American Physical Society; John Scott Award and Medal, City of Philadelphia, 1971; Valdemar Poulsen Medal, Danish Academy of Technical Sciences, 1972; Fellow, American Academy of Arts and Sciences, 1977; Foreign Associate, National Academy of Engineering, 1978.

GUNTHER, CLARENCE A., IRE Fellow 1954. Born: June 29, 1903, Los Angeles, Calif. Degrees: B.S.E., 1926, Princeton University.

Fellow Award "For his contributions to the design and his leadership in the development of military electronic equipment." Other Awards: Award of Merit, Radio Corporation of America, 1952.

GUNTHER, FRANK A., Fellow 1964. Born: February 3, 1908, New York, N.Y.

Fellow Award "For contributions to the fields of UHF and VHF communications and leadership in communications practice." Other Awards: Fellow, Radio Club of America, 1951; Distinguished Service Award, U.S. Armed Forces Communications and Electronics As-

sociation, 1963; Meritorious Service Award, U.S. Armed Forces Communications and Electronics Association, 1967; DeForest Audion Medal, Veteran Wireless Operators Association, 1969; Edwin H. Armstrong Medal, Radio Club of America, 1970; Medal of Honor, U.S. Armed Forces Communications and Electronics Association, 1971.

GUSSETT, N. BERNARD, AIEE Fellow 1948. Born: December 27, 1899, Corpus Christi, Tex. Degrees: B.S.M.E., 1921, University of Texas.

GUTTON, HENRI, Fellow 1968. Born: February 22, 1905, Nancy, France. Degrees: D.Sc., 1929, University of Nancy.

Fellow Award "For his contributions to the development of UHF wave generation, its application, particularly to radar."

GUTZWILLER, F. WILLIAM, Fellow 1966. Born: December 8, 1926, Chicago, Ill. Degrees: B.E.E., 1949, Marquette University.

Fellow Award "For contributions to the development and application of the silicon controlled rectifier and other power semiconductors." Other Awards: Cordiner Award, General Electric, 1963; Steinmetz Award, General Electric, 1977; Professional Achievement Award, Marquette University, 1979.

GUY, ARTHUR W., Fellow 1977. Born: December 10, 1928, Helena, Mont. Degrees: B.S.E.E., 1955, M.S.E.E., 1957, Ph.D.(E.E.), 1966, University of Washington.

Fellow Award "For contributions to the understanding of microwave interactions with living tissues, and to methods of tissue dosimetry in microwave irradiation." Other Awards: Westinghouse Achievement Award, 1954; Ph.D. Program Award, Boeing, 1964-1965.

GUYKER, WILLIAM C., JR., Fellow 1982. Born: August 21, 1933, Donora, PA. Degrees: B.S.E.E., 1959, Massachusetts Institute of Technology.

Fellow Award "For contributions to the engineering and design of 500 kV transmission systems"; Power Group Award, 1970; Chairman, Pittsburgh Section, IEEE, 1973-1974; Area Chairman, Region 2, IEEE, 1974-1976; IEEE Centennial Medal, 1984. Other Awards: FCC-First Class Radio Telephone License, 1953; Registered Professional Engineer, 1963; Adjunct Member, EPRI Fusion Committee, 1976; Electric Power Research Institute Distribution Task Force, 1979-1982; Member, ECPD (now ABET) Engineering Technology Review Committee (Nuclear and Power Engineering), 1977 to present; Accreditation for National Association of Trade and Technical Schools (NATTS), 1974-present; Listed in the 43rd Edition of Marquis' Who's Who, Inc; Instructor and Lecturer at West Virginia University, Penn State University, and University of Pittsburgh.

GWYN, CHARLES W., Fellow 1982. Born: September 3, 1936, Sterling, CO. Degrees: B.S.E.E., 1961, Kansas University; M.S.E.E., 1963, Ph.D., 1968, University of New Mexico.

Fellow Award "For contributions to the fundamental analysis of radiation effects in semiconductor devices and circuits and for innovative developments of computer aids for integrated circuit design"; Outstanding Conference Paper Award, 1967, 1978. Other Awards: Tau Beta Pi; Eta Kappa Nu; Sigma Tau; Phi Kappa Phi; American Men and Women of Science; Registered Professional Engineer, New Mexico; Panel Member, Board of Assessment of National Bureau of Standards Programs.

H

HABER, FRED, Fellow 1978. Born: July 1, 1921, New York, N.Y. Degrees: B.S.E.E., 1948, Pennsylvania State University; M.S.E.E., 1953, Ph.D., 1960, University of Pennsylvania.

Fellow Award: "For contributions to electromagnetic compatibility measurement techniques."

HABERL, H. W., AIEE Fellow 1962. Born: June 3, 1902, Denver, Colo. Degrees: Duquesne University; Carnegie Institute of Technology.

Fellow Award "For contributions to electric utility design and system engineering." Other Awards: Engineering Institute of Canada; Professional Engineer, Province of Quebec.

HACHTEL, GARY D., Fellow 1980. Born: August 21, 1936. Degrees: B.Sc.(E.E.), 1959, California Institute of Technology; Ph.D.(E.E.), 1965, University of California at Berkeley.

Fellow Award "For contributions in computer-aided circuit design."

HADDAD, ABRAHAM H., Fellow 1982. Born: January 16, 1938, Baghdad, Iraq. Degrees: B.Sc.(E.E.), 1960, M.Sc.(E.E.), 1963, Technion-Israel Institute of Technology; M.A.(E.E.), 1964, Ph.D.(E.E.), 1966, Princeton University.

Fellow Award "For contributions to stochastic systems analysis and electrical engineering education." Other Awards: Sustained Superior Performance Award, National Science Foundation, 1982.

HADDAD, GEORGE I., Fellow 1968. Born: April 7, 1935, Aindara, Lebanon. Degrees: B.S.E., 1956, M.S.E., 1958, Ph.D., 1963, University of Michigan.

Fellow Award "For contributions to solid-state and quantum electronic devices and engineering education"; Distinguished Service Award, IEEE, Microwave Theory and Techniques Society, 1976. Other Awards: Curtis W. McGraw Research Award, American Society for Engineering Education, 1970.

HADDAD, JERRIER A., Fellow 1968. Born: July 17, 1922, New York, N.Y. Degrees: B.E.E., 1945, Cornell University; Honorary Science Degree, 1971, Union College; Honorary Science Degree, 1978, Clarkson College.

Fellow Award "For creative contributions to and engineering management direction of electronic data processing systems and their applications." Other Awards: National Academy of Engineering, 1968; Order of the Cedars Medal, Republic of Lebanon, 1970.

HADDOCK, FRED T., IRE Fellow 1962. Born: May 31, 1919, Independence, Mo. Degrees: S.B., 1941, Massachusetts Institute of Technology; M.S., 1950, University of Maryland; Sc.D.(Hon.), 1965, Southwestern at Memphis; Sc.D.(Hon.), 1966, Ripon College.

Fellow Award "For contributions to radio astronomy." Other Awards: Letter of Commendation, Secretary of the Navy, 1945.

HAEFF, ANDREW V., IRE Fellow 1949. Born: December 30, 1904, Moscow, Russia. Degrees: E.E., M.E., 1928, Polytechnic Institute, Harbin, China; M.S., 1929, Ph.D., 1932, California Institute of Technology.

Fellow Award "For contributions to ultra-high-frequency radio tubes and electronics"; Harry Diamond Award, IRE, 1950, "For contribution to the study of the interaction of electrons and radiation, and for contributions to the storage tube art." Other Awards: Fellow, American Physical Society, 1953.

HAERTLING, GENE H., Fellow 1979. Born: March 15, 1932, Old Appleton, MO. Degrees: B.S., 1954, University of Missouri at Rolla; M.S., 1960, Ph.D., 1961, University of Illinois.

Fellow Award "For contributions to ferroelectric and electrooptic ceramic materials and devices." Other Awards: Outstanding Paper Award, Electronics Division, American Ceramic Society, 1960-1970; Fellow, American Ceramic Society, 1971. PACE Award, NICE, 1972.

HAFNER, ERICH, Fellow 1978. Born: July 26, 1928. Degrees: Ph.D.(Physics), 1952, University of Graz, Austria.

Fellow Award "For contributions to the improvement of piezoelectric crystals and frequency control devices."

HAFSTAD, LAWRENCE R., IRE Fellow 1953. Born: June 18, 1904, Minneapolis, Minn. Degrees: B.E.E., 1926, University of Minnesota; Ph.D., 1933, The John Hopkins University.

Fellow Award "In recognition of his basic research work in electronics and nuclear science, and his administration of important scientific programs." Other Awards: AAAS Award, 1931; Medal of Merit, U.S. Dept of Navy, 1946; King's Medal in Defense of Freedom, Great Britain, 1946; Distinguished Service Award, Atomic Energy Commission, 1954; William Proctor Prize, Scientific Research Society of America, 1956; Regents Outstanding Achievement Award, University of Minnesota, 1962; Charles F. Kettering Award, Patent, Trademark and Copyright Institute, George Washington University, 1967; Sesquicentennial Award, University of Michigan, 1967; Citation, U.S. Atomic Energy Commission, 1968; Stevens Honor Award, Stevens Institute of Technology, 1968; Distinguished Civilian Service Award, DOD, 1972; U.S. Congress Citation, 1980.

HAGEN, JOHN P., IRE Fellow 1954. Born: July 1908, Amherst, Nova Scotia, Canada. Degrees: A.B., 1929, Boston University; M.A., 1931, Wesleyan University; Ph.D., 1949, Georgetown University; Sc.D.(Hon.), 1958, Boston University; Sc.D.(Hon.), 1959, Fairfield University; Sc.D.(Hon.), 1959, Adelphi College; Sc.D.(Hon.), 1960, Mt. Allison University, New Brunswick.

Fellow Award "For contributions to the development of military electronic equipment, and for measurements of solar microwave radiation."

HAGLER, MARION O., Fellow 1980. Born: September 7, 1939, Temple, TX. Degrees: B.A., 1962, B.S.E.E., 1963, Rice University; M.S.E.E., 1964, Ph.D., 1967, University of Texas at Austin.

Fellow Award "For contributions to plasma science and optical signal processing." Other Awards: Spencer A. Wells Award, 1978; P.W. Horn Professorship, 1980; Fellow, Optical Society of America, 1981.

HAGN, GEORGE H., Fellow 1979. Born: September 15, 1935, Houston, Tex. Degrees: B.S.E.E., 1959, M.S.E.E., 1961, Stanford University.

Fellow Award "For contributions to spectrum management and electromagnetic compatibility."

HAIMES, YACOV Y., Fellow 1980. Born: June 18, 1936. Degreees: B.S., 1964, Hebrew University, Jerusalem; M.S., 1967, Ph.D.(with distinction), 1970, University of California at Los Angeles.

Fellow Award "For contributions to the theory of large-scale systems." Other Awards: Sigma Xi Distinguished Research Award, 1976; First American Geophysical Union Congressional Science Fellow, 1977-1978; Fellow, American Society of Civil Engineers, 1979; Case Centennial Scholar, Case Institute of Technology, 1980; Fellow, American Association for the Advancement of Science, 1981; Registered Professional Engineer, Ohio.

HAKIMI, S. LOUIS, Fellow 1972. Born: December 16, 1932, Meshed, Iran. Degrees: B.S, 1955, M.S., 1957, Ph.D., 1959, University of Illinois (Urbana).

Fellow Award "For contributions in applications of graph theory to circuit analysis, analysis and design of communications networks, and coding theory."

HALACSY, ANDREW A., Fellow 1970. Born: May 15, 1907, Budapest, Hungary. Degrees: B.Sc., 1929, Ph.D., 1941, Engineering University, Budapest.

Fellow Award "For contributions to the design of transformers and electric power systems." Other Awards: Zipernowski Medal, Hungarian Electrotechnical Society, 1940; Engineer of the Year, Reno, Nevada Chapter of the NSPE, 1971; Fellow, Alexander v. Humboldt Foundation, Germany, 1973.

HALFMANN, EDWARD S., Fellow 1969. Born: June 11, 1914, New York, N.Y. Degrees: B.S.E.E., 1936, M.S.E.E., 1937, Massachusetts Institute of Technology.

Fellow Award "For significant contributions to the science of control and instrumentation and for the application of analog and digital computers leading to the more efficient operation of electric power systems."

HALL, ALBERT C., AIEE Fellow 1950, IRE Fellow 1958. Born: June 27, 1914, Port Arthur, Tex. Degrees: B.S., 1936, Texas A and M University; M.S., 1938, Sc.D., 1943, Massachusetts Institute of Technology.

IRE Fellow Award "For his contributions of theory and practice of servomechanisms." Other Awards: Ordnance Award, U.S. Dept. of Navy, 1946; Outstanding Young Engineer Award, Eta Kappa Nu, 1947; Meritorious Civilian Service Award, Secretary of Defense, 1965; Fellow, American Institute of Aeronautics and Astronautics; Exceptional Civilian Service Medal, Defense Intelligence Agency, 1972; Distinguished Public Service Medal, 1973, 1975; Member, Cosmos Club, 1961; Member, National Academy of Engineering, 1968; Exceptional Service Award, U.S. Air Force, 1977.

HALL, HENRY P., Fellow 1979. Born: June 28, 1928, East Orange, N.J. Degrees: A.B., Williams College, 1952; B.S., M.S.E.E., Massachusetts Institute of Technology, 1952.

Fellow Award "For contributions to electrical measurement technology, including the development of impedance bridges and standards, and the application of microprocessors to impedance measurements."

HALL, JAMES A., Fellow 1972. Born: April 25, 1920, Rhode Island. Degrees: B.S.E.E., 1942, Brown University; Ph.D., 1971, University of Rhode Island.

Fellow Award "For contributions to the development of cathode ray tubes and television camera tubes." Other Awards: Most Meritorious Patent Disclosure Award, Westinghouse Electric; Member, Optical Society of America; Member, American Vacuum Society.

HALL, ROBERT N., IRE Fellow 1957. Born: December 25, 1919, New Haven, Conn. Degrees: B.S., 1942, Ph.D., 1948, California Institute of Technology.

Fellow Award "For contributions in the field of semiconductor devices"; David Sarnoff Award, 1963, "For meritorious achievement in the field of electronics"; Jack A. Morton Award, 1976, "For outstanding achievement in solid-state physics and chemistry and the invention and development of semiconductor devices." Other Awards: Fellow, American Physical Society, 1962; William D. Coolidge Fellowship Award, General Electric Co., 1970; Member, Bomische Physical Society, 1976; Award in Solid State Science and Technology, Electrochemical Society, 1977; Member, National Academy of Engineering, 1977; National Academy of Sciences, 1978.

HALL, WILLIAM M., Fellow 1966. Born: July 10, 1906, Burlington, Vt. Degrees: S.B., 1928, S.M., 1932, Sc.D., 1935, Massachusetts Institute of Technology.

Fellow Award "For contributions to the development of radar through fundamental calculation and prediction of performance." Other Awards: Certificate of Commendation, U.S. Dept. of Navy, Bureau of Ships, 1947.

HALLER, CHARLES E., Fellow 1972. Born: 1924, Conn. Degrees: B.E.E., 1948, Rensselaer Polytechnic Institute.

Fellow Award "For contributions to engineering management and the use of computer technology in international communications."

HALLER, GEORGE L., IRE Fellow 1950, AIEE Fellow 1962. Born: May 8, 1907, Pittsburgh, Pa. Degrees: B.S., 1927, E.E., 1932, M.S., 1933, Ph.D., 1942, Pennsylvania State University; LL.D.(Hon.), 1962, Syracuse University; D.Sc.(Hon.), 1963, Stetson University; LH.D.(Hon.), 1963, Susquehanna University.

IRE Fellow Award "For his work on aircraft antennas and for his diversified radio effort during the war"; AIEE Fellow Award "For contributions to education, to communications, and to engineering management of a large defense enterprise."

HALLIGAN, WILLIAM J., IRE Fellow 1962. Born: December 8, 1898, South Boston, Mass. Degrees; B.S., 1923, Tufts University; 1922, United States Military Academy; D.Sc., 1976, Tufts University.

Fellow Award "For leadership in the communications industry."

HALLMAN, LUDLOW B., JR., IRE Fellow 1952. Born: July 21, 1907, Sneads, Fla. Degrees: B.S., 1929, E.E., 1934, Auburn University.

Fellow Award "For his accomplishments in the development of electronic equipment for airborne communication and navigation." Other Awards: Meritorious Civilian Service Award, War Dept., 1946; Exceptional Civilian Service Award, U.S. Dept. of Air Force, 1963.

HALPERIN, HERMAN, AIEE Fellow 1945. Born: June 21, 1898, Toledo, Ohio. Degrees: M.E., E.E., 1920, Cornell University.

Awards: National Best Paper Prize in Engineering Practice, AIEE, 1942; First Paper Prize, AIEE, 1953, Great Lakes District; William M. Habirshaw Award, AIEE, 1962, "For contributions in the fields of high-voltage cable systems and of allowable economic loading of cable systems and electrical equipment under varying environments." Other Awards: Sigma Xi, 1947; Eta Kappa Nu, 1952; Tau Beta Pi, 1953; Golden Acorn Award, Chamber of Commerce, Menlo Park, California, 1975.

HAM, JAMES M., Fellow 1967. Born: September 21, 1920, Coboconk, Ontario, Canada. Degrees: B.A.Sc., 1943, University of Toronto; S.M., 1947, Sc.D., 1952, Massachusetts Institute of Technology; D.esSc.A.(Hon.), 1973, University of Montreal; D.Sc.(Hon.), 1974, Queen's University; D.Sc.(Hon.), 1979, University of New Brunswick, McGill University; D.Eng.(Hon.), 1980, McMaster University; D.Eng.(Hon.), 1980, Technical University of Nova Scotia; LL.D.(Hon.), University of Manitoba, 1980; D.Eng.(Hon.), 1981, Memorial University; D.Eng.(Hon.), Hanyang University, Korea, 1981; LL.D.(Hon.) Concordia University, 1983; D.S.L. Wycliffe College, 1983.

Fellow Award "For leadership in engineering education and research, and for contributions to the field of automatic control systems"; McNaughton Medal, IEEE, 1977. Other Awards: Fellow, New College, University of Toronto, 1963; Centennial Medal of Canada, 1967; Engineering Alumni Medal, 1973; Engineering Medal, Association of Professional Engineers, Ontario, 1974; Fellow, Engineering Institute of Canada, 1977; Silver Jubilee Medal of Canada; Officer, Order of Canada, 1980; Sir John Kennedy Medal, Engineering Institute of Canada, 1983.

HAMBLEY, J. MERVYN, AIEE Fellow 1961. Born: May 26, 1905, Copper Cliff, Ontario, Canada. Degrees: B.Sc., 1929, Queen's University, Ontario; D.Eng.(Hon.), 1965, University of Waterloo; D.Sc.(Hon.), 1967, Queen's University.

Fellow Award "For contributions to the expansion of a large electric utility."

HAMBURGER, FERDINAND, JR., AIEE Fellow 1948, IRE Fellow 1953. Born: July 5, 1904, Baltimore, Md. Degrees: B.E., 1924, D.Eng., 1931, The Johns Hopkins University.

IRE Fellow Award "For his leadership as a teacher and author in the electronics and electrical engineering fields."

HAMER, WALTER J., AIEE Fellow 1962. Born: November 5, 1907, Altoona, Pa. Degrees: B.S., 1929, Juniata College; Ph.D., 1932, Yale University; D.Sc.(Hon.), 1966, Juniata College.

Fellow Award "For contributions to the field of electrochemistry"; First Paper Prize, AIEE, Industry Division, 1954, for "Some Aspects of the Charge and Discharge Processes in Lead Acid Storage Batteries." Other Awards: Certificates of Merit, Manhattan Project and Office of Scientific Research and Development, 1945; Superior Accomplishment Awards, U.S. Dept. of Commerce, 1954, 1962, 1965; Gold Medal, U.S. Dept. of Commerce, 1965; Life Fellow, New York Academy of Sciences, 1972; Fellow: Washington Academy of Sciences; American Institute of Chemists; American Association for the Advancement of Science; Honorary Member, Electrochemical Society, 1980.

HAMID, MICHAEL, Fellow 1979. Born: June 7, 1934, Tulkarm, Palestine. Degrees: B.Eng.(Electrical), 1960, M.Eng.(Electrical), 1962, McGill University; Ph.D.(Electrical Engineering), 1966, University of Toronto.

Fellow Award "For contributions to electromagnetic scattering and diffraction and development of dielectric-loaded waveguides, resonators, and antennas"; Past Chairman, Educational Activities Board, IEEE. Other Awards: Phi Eta Sigma, 1953; President, Board Director, International Microwave Power Institute, 1969-1973; Associate Editor, Journal of Microwave Power, 1969-1977; Member, Canadian Delegation to International Union of Radio Science, 1970-1981; Registered Professional Engineer, Province of Manitoba, Canada, 1971; Visting Professor, Defense Research Council of Canada Associate Committee on Bird Hazards to Aircraft 1972-1977; Associate Editor, J. of Microwave Power (JMP); Member, Manitoba Research Council and Chairman of Electrical Products Research Committee, 1972-1975; Invited Speaker, AGARD E.M. Wave Propagation Panel, Netherlands, April 1974; Invited Speaker, 25th Annual Meeting of Brazilian Society for Advancement of Science, 1973; Session Organizer and Invited Speaker, URSI General Assembly, Commission V1, Warsaw, Poland; General Chairman, 1971 Microwave Power Symposium, Monterey, California; Member, Editorial Boards of Microwave Journal, Journal of Microwave Power, and IEEE Transactions on Microwave Theory and Techniques 1969-; Invited Speaker, Session Chairman, and Member of International Organizing Committee of Colloquium on Microwave Communication, Budapest, Hungary 1974-1984; Author of over 155 Technical Articles, 90 Conference Papers and 25 Patents. Board of Canada, 1972; Member, Manitoba Research Council, 1972-1975; Centex Committee's Appreciation Award, Province of Manitoba, 1972; Fellow, Institution of Electrical Engineers, 1974; 25 Patents; Chartered Engineer, United Kingdom, 1974; Visting Professor, Naval Postgraduate School, 1979-1981. Adjunct Professor, University of Manitoba, 1970-1977; Member, Policy Committee of University of Manitoba Centre for Transportation Studies, 1972-1979.

HAMILTON, DOUGLAS JAMES, Fellow 1978. Born: December 6, 1930. Degrees: Ph.D.(E.E.), 1959, Stanford University; M.S.E.E., 1956, University of California at Los Angeles; B.S.E.E., 1953, Case Institute of Technology.

Fellow Award "For contributions to education in sol-

id-state electronics."

HAMILTON, HOWARD B., Fellow 1980. Born: October 28, 1923, Augusta, KS. Degrees: B.S.E.E., 1949, University of Oklahoma; M.S.E.E., 1955, University of Minnesota; Ph.D., 1962, Oklahoma State University.

Fellow Award "For contributions to electrical machinery and applications of power systems technology." Other Awards: "Professor Distinguido Honoris Causa," University Tecnica, Valparaiso, Chile; Listed in Marquis "Who's Who, Inc"; "American Men of Science"; "Directory of International Biography"; Registered Professional Engineer, Pennsylvania and Ohio; Fellow, American Association for the Advancement of Science.

HAMILTON, J. L., AIEE Fellow 1921. Born: September 24, 1883, Morrisville, Mo. Degrees: B.S., 1904, University of Missouri.

HAMILTON, WILLIAM H., Fellow 1966. Born: April 2, 1918, Greenville, Pa. Degrees: B.S., 1940, Washington and Jefferson College; M.S., 1948, University of Pittsburgh.

Fellow Award "For contributions to the development of control systems for nuclear reactor plants." Other Awards: Order of Merit, Westinghouse Electric, 1958.

HAMM, WILLIAM J., IRE Fellow 1960. Born: July 26, 1910, Belleville, Ill. Degrees: B.S., 1930, University of Dayton; M.S., 1935, Catholic University of America; Ph.D., 1942, Washington University.

Fellow Award "For service in teaching and research." Other Awards: Minnie-Stevens-Piper Teaching Award, 1958; National Institutes of Health Extramural Associate, 1978; National Research Council Senior Postdoctoral Associate, 1979.

HAMMING, RICHARD W., Fellow 1968. Born: February 11, 1915, Chicago, Ill. Degrees; B.S., 1937, University of Chicago; M.A., 1939, University of Nebraska; Ph.D. 1942, University of Illinois.

Fellow Award "For contributions to numerical analysis, information coding, and improved operation of computing centers"; Emanuel R. Piore Award, 1979. Other Awards: Turing Prize, 1968; Member, National Academy of Engineering, 1980; Harold Pender Award, University of Pennsylvania, 1981.

HAMMOND, DONALD L., Fellow 1977. Born: August 7, 1927, Kansas City, Mo. Degrees: B.S.(Physics), 1950, M.S.(Physics), 1952, D.Sc.(Hon.), 1974, Colorado State University.

Fellow Award "For contributions to the technology of quartz crystals and their applications as transducers." Other Awards: Sawyer Award, 1971.

HANCOCK, JOHN C., Fellow 1972. Born: October 21, 1929, Martinsville, Ind. Degrees: B.S.E.E., 1951, M.S.E.E., 1955, Ph.D., 1957, Purdue University.

Fellow Award "For contributions to engineering education and the field of communications through research, teaching, and administration." Other Awards:

Member, National Academy of Engineering, 1974; Lamme Award, American Society for Engineering Education, 1980; Fellow, American Association for the Advancement of Science, 1981; President, American Society for Engineering Education, 1983; Charter Fellow, American Society for Engineering Education, 1983.

HANNA, CLINTON R., AIEE Fellow 1944. Born: December 17, 1899. Degrees: B.S.E.E., 1922, E.E., 1926, D.Eng., 1945, Purdue University.

Awards: Lamme Medal, AIEE, 1955, "For his fundamental calculations and developments in the field of electrodynamics and particularly for his achievements in the design of generator voltage regulators, automatic rolling mill controls, and tank gun stabilizers." Other Awards: Order of Merit, Westinghouse Electric, 1940; Presidential Citation, 1942; Howard N. Potts Medal, Franklin Institute, 1949.

HANNA, WILLIAM M., AIEE Fellow 1962. Born: January 16, 1902, Washington, D.C. Degrees: B.S., 1924, E.E., 1925, Princeton University.

Fellow Award "For contributions to electric power system planning." Other Awards: Tau Beta Pi, 1972.

HANNAN, PETER W., Fellow 1968. Born: February 19, 1927, New York, N.Y. Degrees: B.S.M.E., 1948, Stevens Institute of Technology; M.S.E.E., 1950, University of Pennsylvania.

Fellow Award "For fundamental contributions to the theory and development of antennas, notably the Cassegrain, monopulse, and phased-array types."

HANO, ICHIRO, Fellow 1972. Born: October 2, 1898, Toyama-Ken, Japan. Degrees: Dr.Eng., 1937, Waseda University.

Fellow Award "For contributions to power systems engineering and in the development of power system stabilizing method." Other Awards: Power Engineering Prize, 1955, Achievement Prize, 1965, Authorship Prize, 1970, Institute of Electrical Engineers of Japan; Founder's Memorial Academic Award, Waseda University, 1968; Third Rising-Sun Prize, Japanese Government, 1969.

HANSEL, PAUL G., IRE Fellow 1957. Born: June 22, 1917, Grand Island, Nebr. Degrees: B.S., 1946, University of Kansas.

Fellow Award "For contributions in the fields of radio navigation and direction finding"; Pioneer Award, Aerospace and Electronic Systems Society , 1970, "For invention of Doppler VOR and contributions toward its development." Other Awards: Meritorious Civilian Service Award, War Dept., 1946; Distinguished Service Award, University of Florida, College of Engineering, 1976.

HANSEN, ALBERT, JR., AIEE Fellow 1958. Born: January 10, 1916. Degrees: E.E., 1936, Pratt Institute.

Fellow Award "For his contributions to design and improvement of electrical instruments." Other Awards:

Charles A. Coffin Award, General Electric, 1948.

HANSEN, ROBERT C., IRE Fellow 1962. Born: August 3, 1926, St. Louis, Mo. Degrees: B.S., 1949, University of Missouri; M.S., 1950, Ph.D., 1955, University of Illinois; D.Eng.(Hon.), 1975, University of Missouri.

Fellow Award "For contributions to the theory of antenna systems." Other Awards: Tau Beta Pi, Eta Kappa Nu, Sigma Xi, Phi Kappa Phi; Member, Commission B International Scientific Radio Union.

HANSEN, V. GREGERS, Fellow 1980. Born: September 30, 1934, Odense, Denmark. M.Sc., 1959, Technical University of Denmark.

Fellow Award "For contributions to the theory of false alarm control in radar systems."

HANSON, D. WAYNE, Fellow 1983. Born: February 9, 1937, Denver, Colorado. Degrees: B.S.E.E., 1959, University of Colorado; M.S.E.E., 1961, Stanford University.

Fellow Award "For the development and implementation of improved time and frequency dissemination services using satellite techniques"; IEEE M. Barry Carlton Award, best technical paper, 1975, Aerospace and Electronic Systems. Other Awards: IR-100 Award, Industrial Research Magazine, 1977.

HANSON, FREDRICK E., Fellow 1964. Born: February 4, 1901, Chicago, Ill. Degrees: New York University, 1923-1927; LL.D.(Hon.), 1962, Muhlenberg College.

Fellow Award "For contributions to manufacturing, engineering, and administration in the field of electron devices." Other Awards: Distinguished Service Award, American Standards Association, 1946.

HANSTEEN, HENRY B., AIEE Fellow 1951. Born: April 28, 1904, Brooklyn, N.Y. Degrees: E.E., 1924, Polytechnic Institute of Brooklyn; M.A., 1929, Ph.D., 1942, Columbia University.

Other Awards: Sigma Xi; Eta Kappa Nu; Tau Beta Pi.

HAPP, HARVEY HEINZ, Fellow 1973. Born: June 27, 1928, Berlin, Germany. Degrees: B.S.E.E., 1953, Illinois Institute of Technology; M.S.E.E., 1958, Rensselaer Polytechnic Institute; D.Sc., 1962, University of Belgrade.

Fellow Award "For the development of technical and economic analysis of electric power systems"; First Prize Paper, Midwest District, 1962; First Prize Paper, Power System Engineering Committee of the IEEE Power Engineering Society, 1977; Region I Award, IEEE, 1980. Other Awards: Eta Kappa Nu, 1952; Tau Beta Pi, 1953; Sigma Xi, 1963.

HARADA, KOOSUKE, Fellow 1981. Born: November 10, 1929, Fukuoka, Japan. Degrees: B.S.E.E., 1953, M.S.E.E., 1955, Dr. Eng., 1958, Kyushu University.

Fellow Award "For contributions to the development of combined magnetic and semiconductor devices for power control." Other Awards: Prize Paper Award, Institute of Electrical Engineers, Japan, 1965; Inoue

Harushige Prize, Research Development Corporation, Japan, 1980; Award for Outstanding Technology, Society of Instrument and Control Engineers, Japan, 1982.

HARADA, TATSUYA, Fellow 1981. Born: October 5, 1924, Yamanashi-ken, Japan. Degrees: B.E., 1947, Ph.D.(Eng.), 1962, Tokyo Institute of Technology.

Fellow Award "For contributions to high-voltage testing techniques and to the understanding of lightning and switching impulse flashover phenomena." Other Awards: Thesis Award, Institute of Electrical Engineers, Japan, 1966; Progress Award, Institute of Electrical Engineers, Japan, 1968, 1971; Progress Award with Fukuda Prize, Institute of Electrical Engineers, Japan, 1973; Electric Power Award, Institute of Electrical Engineers, Japan, 1975.

HARALAMPU, GEORGE S., Fellow 1977. Born: March 20, 1925, Lynchburg, Va. Degrees: B.S.E.E., 1952, Tufts University; M.S.(Electric Power Engineering), 1960, Northeastern University.

Fellow Award "For contributions in the field of power system protection and for achievements in the applications of computers to the analysis, design, and operation of large electric power systems."

HARALICK, ROBERT M., Fellow 1984. Born: September 30, 1943, New York, N.Y. Degrees: B.A.(Math), 1964, B.S.(E.E.), 1966, M.S.(E.E.), 1967, Ph.D.(E.E.), 1969, University of Kansas.

Fellow Award "For contributions in image processing and computer vision." Other Awards: Dow Chemical Young Outstanding Faculty Award, 1975; Mideast Region, ASEE, Recognition of Outstanding Young Electrical Engineers, Honorable Mention, 1975; Best Paper Award, AIPR, 1975; Eta Kappa Nu.

HARDER, EDWIN L., AIEE Fellow 1948. Born: April 28, 1905, Buffalo, N. Y. Degrees: E.E., 1926, Cornell University; M.S., 1930, Ph.D., 1946, University of Pittsburgh.

Awards: Lamme Medal, AIEE, 1962, "For meritorious achievement in the development of electrical machinery." Other Awards: Order of Merit, Westinghouse Electric, 1945; Distinguished Service Award, AFIPS, 1971; National Academy of Engineering, 1976.

HARDIN, CLYDE D., Fellow 1977. Born: May 26, 1925, Fort Worth, Tex. Degrees: B.S.(Physics, Math), 1948, Wake Forest College.

Fellow Award "For contributions to and leadership in the development of radar and ordnance electronics." Other Awards: Commendation, Secretary of the Army, 1956; R & D Achievement Award, U.S. Army, 1969; Decoration for Meritorious Civilian Service, Dept. of Army, 1971; Secretary of Defense Medal for Meritorious Civilian Service, 1973; Republic of Korea Cheonsu (National Security) Medal, 1976; AOC Medal for Electronic Warfare Management, 1976.

HARGENS, CHARLES W., 3RD, Fellow 1968. Born: October 21, 1918, Philadelphia, Pa. Degrees: S.B., 1941, Massachusetts Institute of Technology.

Fellow Award "For contributions in the application of electronics to bio-medical instrumentation." Other Awards: Award, Office of Scientific Research and Development, 1945; Sigma Xi; P.E., Pennsylvania; 12 Patents; Adjunct Associate Professor, Drexel University.

HARGER, ROBERT, Fellow 1979. Born: September 15, 1932, Flint, Mich. Degrees: B.S.(Eng.), 1955, M.S.(E.E.), 1959, Ph.D.(E.E.), 1961, University of Michigan, Ann Arbor.

Fellow Award "For contributions to radar systems, including synthetic aperture and harmonic radar, to optical communication theory, and to graduate and continuing education"; H. Barry Carlton Paper Award, IEEE Aerospace and Electronic Systems Society, 1977.

HARLOW, JAMES H., AIEE Fellow 1958. Born: October 1, 1901, Pittsburgh, Pa. Degrees: B.S., 1923, E.E., 1933, University of Maryland.

Fellow Award "For notable contributions to improvement in economy of electric power generation." Other Awards: Fellow, American Society of Mechanical Engineers, 1958; Prime Movers Award, American Society of Mechanical Engineers, 1960; Philadelphia Engineer of Year, 1962; Newcomen Medal, Franklin Institute, 1963; Pennsylvania Engineer of Year, 1966; President, ASME, 1966-1967.

HARMAN, GEORGE G., Fellow 1982. Born: December 7, 1924, Norfolk, VA. Degrees: B.S.(Indust. Physics), 1949, Virginia Polytechnic Institute; M.S.(Physics), 1959, University of Maryland.

Fellow Award "For development of process control and screening procedures for microelectronic welding and bonding"; IEEE Centennial Medal, 1984. Other Awards: Silver Medal, U.S. Dept. of Commerce, 1972; Gold Medal, U.S. Dept. of Commerce, 1979; Technical Achievement Award, International Society for Hybrid Microelectronics, 1981.

HARMON, RALPH N., IRE Fellow 1952. Born: August 29, 1905, California, Pa. Degrees: B.S.E.E., 1928, Carnegie Institute of Technology.

Fellow Award "For his contributions in the fields of broadcast antennas and transmitters and to electronic development in World War II." Other Awards: Outstanding Contribution Award, War Dept., U.S. Dept. of Navy; Certificate of Effective Service, Office of Scientific Research and Development; Silver W Award, Order of Merit, Westinghouse Electric, 1957; Engineering Achivement Award, National Association of Broadcasters, 1962.

HARMON, ROBERT W., Fellow 1978. Born: October, 22, 1929, Winchester, Ind. Degrees: B.S.E.E.(with Highest Distinction), 1951, M.S.E.E., 1955, Purdue

University.

Fellow Award "For developments in UHV insulation and suspension systems." Other Awards: Phi Eta Sigma; Eta Kappa Nu; Tau Beta Pi; Sigma Xi.

HARNDEN, JOHN D., JR., Fellow 1976. Born: September 20, 1928. Degrees: B.S.E.E., 1950, Union College.

Fellow Award "For contributions by leadership in the development of new electric power controls." Other Awards: Management Award, General Electric Co., 1963.

HARNESS, GEORGE T., AIEE Fellow 1949. Born: March 14, 1906, Galveston, Ind. Degrees: B.S., 1928, Ph.D., 1933, California Institute of Technology.

Other Awards: San Fernando Valley Engineer of the Year Award, 1963.

HARPER, JOHN D., AIEE Fellow 1950. Born: April 6, 1910, Louisville, Tenn. Degrees: E.E., 1933, University of Tennessee; LL.D.(Hon.), University of Evansville; Dr.Eng.(Hon.): Lehigh University, Maryville College, Rennesselaer Polytechnic Institute; Dr. of Commercial Science(Hon.), Widener College; D.Sc.(Hon.), Clarkson College of Technology; Dr. of Laws (Hon.), Carnegie-Mellon University, 1982.

Other Awards: Member, American Society of Mechanical Engineers; Nathan W. Dougherty Award, University of Tennessee; Professional Engineers Distinguished Service Award, Pittsburgh Chapter, Pennsylvania Society of Engineers; William Metcalf Award, Engineers' Society of Western Pennsylvania; Gold Medal, Pennsylvania Society; Invest-In-America "Spokesman for Business Award"; Knight's Cross, Order of St. Olav (Norway); National Academy of Engineering; First Bryce Harlow Award (Government/Business Recognition, 1982.

HARRINGTON, DEAN B., Fellow 1973. Born: Schenectady, N.Y. Degrees: B.S.E.E., 1944, Massachusetts Institute of Technology.

Fellow Award "For contributions to the design and performance analysis of large steam turbine driven generators"; Nikola Tesla Award, IEEE, 1981. Other Awards: National Academy of Engineering, 1981.

HARRINGTON, EVERETT J., Fellow 1965. Born: Febuary 1, 1911, Marshfield, Oreg. Degrees: B.S., 1934, Oregon State University.

Fellow Award "For contributions to power circuit breakers and the application of high voltage capacitors"; First Student Paper Prize, Region, 1934; First Paper Prize, AIEE, Northwest District, 1954, for "Shunt Capacitors in Large Transmission Networks"; William M. Habirshaw Award, IEEE, 1975, "For contributions in the application of high-voltage power circuit breakers, high-voltage series and shunt capacitors, and high-voltage direct-current transmission technology." Other Awards: Eta Kappa Nu; Tau Beta Pi; Sigma Tau; Special Act Award, Bonneville Power Administration,

1965; Distinguished Service Award, Gold Medal, Bonneville Power Administration, 1966; CIGRE Attwood Award, 1983.

HARRINGTON, JOHN V., IRE Fellow 1962. Born: May 9, 1919, New York, N.Y. Degrees: B.E.E., 1940, Cooper Union; M.E.E., 1948, Polytechnic Institute of Brooklyn; Sc.D., 1958, Massachusetts Institute of Technology.

Fellow Award "For contribution to radar detection." Other Awards: Exceptional Civilian Service Award , U.S. Dept. of Air Force, 1952; Citation for Distinguished Professional Achievement, Cooper Union, 1965; Fellow, American Association for the Advancement of Science, 1976; Fellow, American Institute of Aeronautics and Astronautics, 1977.

HARRINGTON, ROGER F., Fellow 1968. Born: December 24, 1925, Buffalo, N.Y. Degrees: B.S.E.E., 1948, M.S.E.E., 1950, Syracuse University; Ph.D., 1952, Ohio State University.

Fellow Award "For contributions to electromagnetic field theory and engineering education"; Best Paper Award, Antennas and Propagation Society, 1971; Best Paper Award, Syracuse Section IEEE, 1973, 1974, 1977, 1980; Distinguished Lecturer, IEEE Group on Antennas and Propagation, 1973-1975; Best Paper Award, IEEE Electromagnetic Compatability Society, 1978; Kurt Schlesinger Award, Syracuse Section, IEEE, 1981; IEEE Centennial Medal, 1984. Other Awards: Fulbright Lecturer, Technical University of Denmark, 1969; Distinguished Alumni Award, Ohio State University, 1970; Visiting Scientist, Yugoslavian Academies of Science, 1972.

HARRIS, CYRIL M., IRE Fellow 1961. Born: June 20, 1917, Detroit, Mich. Degrees: B.A., 1938, M.A., 1940, University of California at Los Angeles; Ph.D., 1945, Massachusetts Institute of Technology; D.Sc.(Hon.), 1981, New Jersey Institute of Technology.

Fellow Award "For contributions to the science of acoustics." Other Awards: Honorary Member, Audio Engineering Society, 1973; Member, National Academy of Engineering, 1975; Honorary Award, U.S. Institute for Theatre Technology, 1977; Franklin Medal, 1977; W.C. Sabine Medal, Acoustical Society of America, 1979; Member, National Academy of Sciences, 1980; American Institute of Architects Medal, 1980; Charles Batchelor Professor of Electrical Engineering and Professor of Architecture, Columbia Univeristy, New York, N.Y.

HARRIS, FOREST K., AIEE Fellow 1960. Born: August 26, 1902, Gibson County, Ind. Degrees: B.A., 1921, M.S., 1923, Oklahoma University; Ph.D., 1932, The Johns Hopkins University.

Fellow Award "For contributions to the art of electrical instrumentation"; Morris E. Leeds Award, 1972, "For a lifetime of making outstanding advances in the science of high-accuracy electrical measurements, and of stimulating further advances through his teaching, authorship, and committee activity." Other Awards, Fellow, Washington Academy of Sciences; Medal, U.S. Department of Commerce, 1956; Rosa Award, National Bureau of Standards, 1967; Si Fluor Award, Instrument Society of America, 1974; William A. Wildhack Award, National Conference of Standards Laboratories, 1981.

HARRIS, H., JR., Fellow 1966. Born: December 18, 1917, New York, N.Y. Degrees: M.S.E.E., 1939, Rensselaer Polytechnic Institute.

Fellow Award "For contributions to the theory and design of electronic control systems and for military systems engineering development."

HARRIS, JAY H., Fellow 1981. Born: June 3, 1936, Newark, NJ. Degrees: B.E.E. (summa cum laude), 1958, Polytechnic Institute of New York; M.S., 1959, California Institute of Technology; Ph.D., 1965, University of California at Los Angeles.

Fellow Award "For pioneering contributions to the understanding and application of wave effects in integrated and thin-film optical devices." Other Awards: New York State Scholar 1954-1958; Sigma Xi, 1958; Hughes Master's (1958) and Doctoral Fellow (1962-1965); Fulbright-Hayes Fellow, 1965; Stanford Ames Fellow, 1967; Traveling Lecturer, Optical Society of America, 1971; Special Achievement Award, National Science Foundation, 1980; Tau Beta Pi, 1981.

HARRIS, ROY H., Fellow 1983. Born: December 17, 1928, Madison, Georgia. Degrees: B.S.E.E., 1951, Georgia Inst. of Technology; M.E.E., 1956, Polytechnic Inst. of N.Y.

Fellow Award "For the development of highly reliable systems for outer space and underwater applications"; Outstanding Service Award, Region 3, IEEE, 1981.

HARRIS, STEPHEN E., Fellow 1972. Born: November 29, 1936, New York, N.Y. Degrees: B.S., 1959, Rensselaer Polytechnic Institute; M.S., 1961, Ph.D., 1963, Stanford University.

Fellow Award "For contributions in lasers, nonlinear optics, optical parametric devices, and acousto-optics"; David Sarnoff Award, 1978. Other Awards: Alfred Noble Prize, 1965; Fellow, Optical Society of America, 1968; Curtis W. McGraw Research Award, 1973; Fellow, American Physical Society, 1975; Guggenheim Fellowship, 1977; National Academy of Engineering, 1977; National Academy of Sciences, 1981.

HARRIS, WALTER R., AIEE Fellow 1957. Born: June 3, 1915, Crockett, Va. Degrees: B.S.E.E., 1937, West Virginia University; M.S.E.E., 1941, University of Pittsburgh.

Fellow Award "For contributions to the design and application of control systems for the precise regulation of speed, tension and position"; Outstanding

Achievement Award, IEEE Industry Applications Society, 1975; Tau Beta Pi; Eta Kappa Nu.

HARRISON, CHARLES W., JR., Fellow 1969. Born: September 15, 1913. Degrees: B.S.E., 1939, E.E., 1940, University of Virginia; S.M., 1942, M.E., 1952, Ph.D., 1954, Harvard University.

Fellow Award "For numerous and fundamental contributions to antenna theory, and for the application of electromagnetic principles to nuclear weapon and missile design"; Region 6, IEEE, Electronics Achievement Award, 1966, Citation: same as fellow award; IEEE EMC Best Paper Award (with R.W.P. King), 1972.

HARRISON, GORDON R., Fellow 1975. Born: December 14, 1931, Wister, Okla. Degrees: B.S., 1952, University of Central Arkansas; M.S., 1954, Ph.D., 1958, Vanderbilt University.

Fellow Award "For contributions and technical leadership in the development of microwave ferrimagnetic compounds and their application in microwave components and integrated circuits"; Florida West Coast Section Special Citations, 1964, 1965, "For perseverance and creativity as Editor of the Suncoast Signal, resulting in a timely and informative publication of great value to the members" 1968, 1969, "For services rendered in furthering the objectives of the IEEE and the Section." Other Awards: Best Tutorial Paper, National Electronics Conference, 1969.

HARRISON, THOMAS J., Fellow 1979, Born: May 13, 1935, Wausau, Wis. Degrees: B.S.E.E., 1957, M.S.E.E., 1958, Carnegie Institute of Technology; Ph.D., 1964, Stanford University.

Fellow Award "For contributions to computer systems for use in process-control applications and to the development of standards for process-control computer systems." Other Awards: Fellow, Instrument Society of America, 1981.

HARROLD, RONALD T., Fellow 1982. Born: April 4, 1933, London, England. Degrees: H.N.C.(E.E.), 1955, Twickenham College of Technology, England; H.N.C.(Electron.), 1961, Chelmsford College of Technology, England.

Fellow Award "For development of diagnostic measurement techniques for locating discharges in high-voltage apparatus." Other Awards: IR-100 Award, 1981; IR-100 Award, 1983; Engineering Achievement Award, Westinghouse Electric Corporation, 1983.

HART, HAROLD M., Fellow 1977. Born: November 19, 1913. Degrees: B.A.(magna cum laude), 1934, University of Buffalo; M.S.(with honors), 1936, University of Oklahoma.

Fellow Award "For contributions to missile fire control radars, air traffic control, and landing systems."

HART, PETER E., Fellow 1983. Born: February 27, 1941, Brooklyn, New York. Degrees: B.E.E., 1962, Rensselaer Polytechnic Institute; M.S., 1963, Ph.D.,

1966, Stanford University.

Fellow Award "For advances in pattern recognition and artificial intelligence." Other Awards: Eta Kappa Nu; Tau Beta Pi; Sigma Xi; "Citation Classic" recognition, Institute for Scientific Information, 1982; Distinguished Computer Science Lecturer, Carnegie-Mellon University, 1981; Distinguished Computer Science Lecturer, MIT, 1983.

HARTH, WOLFGANG, Fellow 1984. Born: February 7, 1932, Straubing, Bavaria, W. Germany. Degrees: Dipl.-Phys., Dr.rer.nat., Dr.-Ing.habil.

Fellow Award "For contributions to optoelectronic and high-frequency semiconductor devices."

HARTIG, ELMER O., Fellow 1964. Born: January 28, 1923, Evansville, Ind. Degrees: B.S.E.E., 1946, University of New Hampshire; S.M., 1947, Ph.D., 1950, Harvard University.

Fellow Award "For contributions to the field of coherent high resolution radar"; Electronics Award, IRE, Phoenix Section, 1960.

HARTUNG, ARTHUR F., AIEE Fellow 1956. Born: January 17, 1905, Kansas City, Mo. Degrees: B.S., 1926, Kansas University.

Fellow Award "For his contributions to the design and development of electric power systems."

HARTY, EDGAR A., AIEE Fellow 1953. Born: November 13, 1897, Constantinople, Turkey. Degrees: B.S., 1920, Robert College.

Fellow Award "For outstanding research on metallic (power semiconductor) rectification and fundamental contributions in the field of rectifier engineering"; Initial Paper Prize AIEE, 1937, for "Aging in Copper Oxide Rectifiers"; Certificate of Appreciation, AIEE-IEEE, 1959, as Chairman Power Semiconductor Committee, 1967, as Chairman Columbus Section. Other Awards: Charles A. Coffin Award, General Electric, 1946; Naval Ordnance Development Award, U.S. Navy, 1946; Certificate of Appreciation, U.S. Dept. of Defense, 1947; President, Massachusetts Society of Professional Engineers, 1949-1950; Establishment of the Edgar A. Harty Scholarship at Wittenberg University, by Dr. Stanley M. Hanley, 1973; Registered Professional Engineer, Massachusetts.

HARVEY, NORMAN L., IRE Fellow 1954. Born: October 9, 1911, Ossian, Iowa. Degrees: B.S.(E.E.), 1934, Iowa State University.

Fellow Award "For his contributions in the fields of electronic navigation and communication."

HASEGAWA, AKIRA, Fellow 1979. Born: June 17, 1934, Tokyo, Japan. Degrees: B.Eng., 1957, M.Eng., 1959, Osaka University; Ph.D., 1964, University of California, Berkeley; Dr.Sci., 1966, Nagoya University.

Fellow Award "For contributions to linear and nonlinear wave dynamics in laboratory and space plasmas, to nonlinear optics, and to nonlinear fluid dynamics."

Other Awards: Fulbright Scholar, 1961-64; Fellow, American Physical Society; Distinguished Technical Staff, ATT Bell Laboratories, 1983.

HASTINGS, HARRIS F., IRE Fellow 1962. Born: April 14, 1896, Fullerton, Nebr.

Fellow Award "For contributions to radio instrumentation in the field of precision time and frequency measurement." Other Awards: Distinguished Civilian Service Award, U.S. Dept of Navy; Award of Merit, Fleet Ballistic Missile Program; Group Achievement, U.S. Dept. of Navy.

HATCH, MAURICE F., AIEE Fellow 1961. Born: April 17, 1909, Spokane, Wash. Degrees: B.S., 1932, University of Washington.

Fellow Award "For contributions to coordination and integration of regional generating resources and transmission facilities." Other Awards: Phi Beta Kappa; Sigma Xi; Tau Beta Pi; Northwest Electric Light and Power Association.

HATCH, PHILIP H., AIEE Fellow 1946. Born: May 25, 1899, Albany, N.Y. Degrees: S.B., 1921, Massachusetts Institute of Technology.

Other Awards: Registered Professional Engineer, 1936.

HATCHER, CHARLES T., AIEE Fellow 1954. Born: October 4, 1897, Chester, Va. Degrees: E.E., 1920, University of Virginia.

Fellow Award "For service and leadership in research and engineering of electric power cables and their application." Other Awards: Charter Member, Tau Beta Pi.

HATOYAMA, GEORGE M., Fellow 1984. Born: March 31, 1911, Tokyo, Japan. Degrees: B.S., 1933, Ph.D.(E.E.), 1960, University of Tokyo.

Fellow Award "For leadership of industrial semiconductor research and development." Other Awards: Excellent Books Award for "Solid State Physics for Electronics Engineers", IEECE Japan, 1972.

HATSOPOULOS, GEORGE N., Fellow 1975. Born: January 7, 1927, Athens, Greece. Degrees: B.S., 1949, M.S., 1950, M.E., 1954, Ph.D., 1956, Massachusetts Institute of Technology; Dr.Sc.,(Honoris Causa), New Jersey Institute of Technology, 1982.

Fellow Award "For contributions to classical thermodynamic theory, and for research and development on the thermionic conversion of heat to electrical energy." Other Awards: Gold Medal Award for Outstanding Achievement in the Field of Engineering, Pi Tau Sigma, 1950-60; Gold Plate Award, Academy of Achievement, 1961; Fellow, American Academy of Arts and Sciences, 1966; National Academy of Engineering, 1978.

HAUS, HERMANN A., IRE Fellow 1962. Born: August 8, 1925, Ljubljana, Yugoslavia. Degrees: B.Sc., 1949, Union College; M.E.E., Rensselaer Polytechnic Institute;

Sc.D., Massachusetts Institute of Technology.

Fellow Award "For contributions to noise theory in electron-beam devices and electric networks." Other Awards: Guggenheim Fellow, 1959-60; Teaching Award, American Society of Engineering Education, 1972; National Academy of Engineering, 1976; Fellow, American Academy of Arts and Sciences, 1981.

HAUSPURG, ARTHUR, Fellow 1970. Born: 1925, New York, N.Y. Degrees: B.S., 1945, M.S., 1947, Columbia University.

Fellow Award "For contributions in the field of complex interconnected power systems reliability and UHV transmission systems." Other Awards: Member, National Academy of Engineering.

HAUSZ, WALTER, IRE Fellow 1954. Born: May 17, 1917, New York, N.Y. Degrees: B.S., 1937, M.S.E.E., 1938, Columbia University.

Fellow Award "For his contributions in the fields of radar and telemetry"; Chairman, Santa Barbara Section, IRE, 1958; Chairman, Los Angeles Section, 1960. Other Awards: Charles A. Coffin Award, General Electric, 1952; Fellow, AAAS.

HAVENS, BYRON L., Fellow 1969. Born: October 16, 1914, Spokane, Wash. Degrees: B.S.E.E., 1938, University of Washington; M.S.E.E., 1939, California Institute of Technology.

Fellow Award "For outstanding technical contributions to modern digital computers, and for sustained leadership in developing technical capabilities of European data processing laboratories." Other Awards: Presidential Certificate of Merit, 1948.

HAVILAND, ROBERT P., Fellow 1969. Born: October 18, 1913, Cleveland, Ohio. Degrees: B.S.E.E., 1939, Missouri School of Mines.

Fellow Award "For contributions to promote the use of satellites for communications, particularly direct broadcasting." Other Awards: Fellow, American Astronautical Society; Fellow, British Interplanetary Society.

HAYAKAWA, SHIGERU, Fellow 1984. Born: August 11, 1925, 55 Suibun, Chihaya-Akasakamura, Minamikawashi-gun, Osaka, Japan. Degrees: B.Sc., 1950, M.Sc., 1959, Ph.D., 1963, Osaka University.

Fellow Award "For leadership in consumer electronics industries and for development of piezoelectric ceramics." Other Awards: Mainichi Industrial Technology Award, 1967, 1979; Outstanding Research Award, Japan Society of Powder and Powder Metallurgy, 1973; Book Honours, The Institute of Electronics and Communication Engineers of Japan, 1976; Prizes, Director of Science and Technology Agency, 1981; Medal with Honours, Purple Ribbon, 1983.

HAYASHI, IZUO, Fellow 1976. Born: May 1, 1922. Degrees: M.S., 1946, D.Sc., 1962, University of Tokyo.

Fellow Award "For contributions to the development

and understanding of room temperature (Al-Ga) As laser diodes." Other Awards: Fujiwara Award, 1962; Ichimura Award, 1974; Award, 1975, Institute of Electronics and Communications Engineers of Japan.

HAYCOCK, OBED C., Fellow 1964. Born: October 5, 1901, Panguitch, Utah. Degrees: B.S., 1925, University of Utah; M.S., 1931, Purdue University.

Fellow Award "For contributions to research on the upper atmosphere and to engineering education." Other Awards: Outstanding Engineer Award, Utah Joint Engineering Council, 1962.

HAYES, J. EDMUND, IRE Fellow 1954. Born: February 14, 1910, Arcola, Saskatchewan, Canada. Degrees: B.Sc., 1935, Queen's University, Ontario; M.Eng., 1947, McGill University.

Fellow Award "For contributions in Canada to the development of outstanding shortwave antennas and switching designs." Other Awards: Fellow, Society of Motion Picture and Television Engineers, 1966.

HAYES, JEREMIAH F., Fellow 1983. Born: July 8, 1934, New York, New York. Degrees: B.E.E., Manhattan College; M.S., New York University; Ph.D., University of California, Berkeley.

Fellow Award "For contributions to the theory of local distribution in computer communications"; Special Acknowledgement Prize Paper Award, IEEE Information Theory Group, 1981; Magazine Prize Paper Award, IEEE Communication Society, 1982.

HAYES, THOMAS B., Fellow 1971. Born: August 25, 1912, Pendleton, Oreg. Degrees: B.S.E.E., Oregon State University; S.M.E.E., Massachusetts Institute of Technology.

Fellow Award "For contributions in planning and development of energy systems."

HAYES, VINCENT J., AIEE Fellow 1959. Born: September 11, 1902, Winsted, Conn. Degrees: B.S., 1925, Yale University.

Fellow Award "For contributions to planning, designing and construction of a statewide power system." Other Awards: Honorary Member, Eta Kappa Nu, 1956.

HAYKIN, SIMON, Fellow 1982. Born: January 6, 1931. Degrees: B.Sc.(first class honors), 1953, Ph.D.(E.E.), 1956, D.Sc.,1967, University of Birmingham, England.

Fellow Award "For contributions to signal processing, communication theory, and electrical engineering education." Other Awards: Fellow, Royal Society of Canada, 1980.

HAYT, WILLIAM H., JR., Fellow 1964. Born: July 1, 1920, Wilmette, Ill. Degrees: B.S.E.E., 1942, M.S.E.E., 1948, Purdue University; Ph.D., 1954, University of Illinois.

Fellow Award "For contributions to electrical engineering education." Other Awards: ASEE Western Electric Fund Award for Excellence in Instruction of Engineering Students, 1973; Sagamore of the Wabash, State of Indiana, 1969; Who's Who in America.

HEAFER, HAROLD P., AIEE Fellow 1962. Born: December 28, 1899, Georgetown, Tex.

Fellow Award "For contributions to the development of communication plant facilities in rapidly growing communities while Divison Engineer, Southwestern Bell Telephone Co. - Period 1920-1965."

HEATON-ARMSTRONG, L. JOHN, Fellow 1967. Born: April 26, 1905, Limerick, Ireland. Degrees: B.Sc., 1926, D.I.C., 1927, University of London.

Fellow Award "For contribution over many years to the design and manufacture of radio communication equipment for both civil and military use." Other Awards: Diploma of Membership, Imperial College of Science and Technology, University of London; Associate, City and Guilds College, University of London; Fellow, Institution of Electrical Engineers, U.K., 1967.

HEBERLEIN, GUSTAVE E., Fellow 1964. Born: July 12, 1902, Great Falls, Mont. Degrees: B.S., 1926, Washington State University.

Fellow Award "For creative leadership in the switchgear field"; IEEE, Philadelphia Special Section Award, 1970, "For noteworthy and creative contributions and inspiring leadership in the art and science of system protection"; Power and Life Award, IEEE Switchgear Committee, 1977. Other Awards: Award, National Electrical Manufacturers Association.

HECHTEL, J. RICHARD, Fellow 1971. Born: May 21, 1913, Schwabach, Germany. Degrees: Diplomphysiker, 1938, Dr.rer.nat., 1940, Technische Hochschule Munich.

Fellow Award "For contributions to electron optics, microwave tubes, and the development of the electrostatically focused power amplifier klystron."

HECKENDORN, HAROLD R., AIEE Fellow 1961. Born: Cedar Point, Kans. Degrees: B.S.E.E., 1934, Kansas State University.

Fellow Award "For contributions in the engineering of manufacture for the rapidly advancing communications art."

HEDMAN, DALE E., Fellow 1977. Born: August 22, 1935, Albert City, Iowa. Degrees: B.S.E.E., 1958, M.S.E.E., 1959, University of Nebraska.

Fellow Award "For contributions to the development of solutions for wave progagation and electrical transients on extra-high-voltage transmission systems"; Award for Noteworthy Paper in IEEE TRANSACTIONS on Power Apparatus and Systems, 1969, entitled "Field and Laboratory Tests of Contaminated Insulators for the Design of the State Electricity Commission of Victoria's 500 kV System." Other Awards: Member, Secretary IEC-TC-37, Surge Protective Devices Committee; CIGRE; Member: Sigma Tau, Pi Mu Epsilon, Eta

Kappa Nu.

HEGBAR, HOWARD R., IRE Fellow 1957. Born: February 22, 1915, Valley City, N.D. Degrees: B.S.E.E., 1937, North Dakota State College; M.S.E.E., 1938, Ph.D., 1941, University of Wisconsin.

Fellow Award "For contributions to the development of analog computers and their applications."

HEILMEIER, GEORGE H., Fellow 1973. Born: 1936, Pennsylvania. Degrees: B.S.E.E., 1958, University of Pennsylvania; M.S.E., 1960, M.A., 1961, Ph.D., 1962, Princeton University.

Fellow Award "For fundamental discoveries of electrooptic effects in liquid crystals, and basic liquid crystal device concepts"; David Sarnoff Award, IEEE, 1976, "For combining science with engineering to create the dynamic scattering liquid crystal display." Other Awards: RCA Laboratories Achievement Awards, 1960, 1962, 1965, 1969; Outstanding Young Electrical Engineer, Eta Kappa Nu, 1968; IR-100 Award, 1968, 1969; Outstanding Young Electrical Engineer in the United States of America, Eta Kappa Nu, 1969; White House Fellow Award, 1970; 26th Annual Arthur Flemming Award as Outstanding Young Man in Government, 1974; Secretary of Defense Civilian Meritorious Service Medal, 1974; Special Recognition Award, Society for Information Display, 1975, 1977; Dept. of Defense Distinguished Civilian Service Medal, 1975,-1977; Tau Beta Pi; Sigma Xi; Eta Kappa Nu; National Academy of Engineering; Defense Science Board; Air Force Scientific Advisory Board; Visitors Committee of Stanford University, Princeton University, University of Pennsylvania.

HEKIMIAN, NORRIS C., Fellow 1981. Born: January 14, 1926, Washington, D.C. Degrees: B.E.E., 1949, George Washington University; M.S., 1951, Ph.D., 1969, University of Maryland.

Fellow Award "For contributions to circuit design in the field of telecommunications test instrumentation"; Patron Award, Washington Section, IEEE, 1979. Other Awards: Sigma Tau, 1948; Alumni Achievement Award, George Washington University, 1976; Eta Kappa Nu, 1979.

HELLER, DOUGLAS M., Fellow 1966. Born: February 23, 1918, Dover, England. Degrees: B.Sc., 1939, Royal College of Science, Imperial College, University of London.

Fellow Awards "For engineering contributions to VHF air ground communications, and technical direction of commercial and military radio systems." Other Awards: Fellow, Institute of Physics, U.K., 1947; Associate, Royal College of Science, University of London; Associate Fellow, AIAA, 1955.

HELLER, GERALD S., Fellow 1969. Born: September 5, 1920, Detroit, Mich. Degrees: Sc.B., 1942, Sc.M., 1946, Brown University; Ph.D., 1948, Brown University.

Fellow Award "For fundamental contributions to ferrite and antiferromagnetic devices at microwave and millimeter-wave frequencies."

HELLIWELL, ROBERT A., IRE Fellow 1960. Born: September 2, 1920, Red Wing, Minn. Degrees: A.B., 1942, M.A., 1943, E.E., 1944, Ph.D., 1948, Stanford University.

Fellow Award "For contributions to ionospheric radio propagation." Other Awards: Antarctica Service Medal, 1966; National Academy of Sciences, 1967; Appleton Prize, 1972.

HELLMAN, MARTIN E., Fellow 1980. Born: October 2, 1945, New York, NY. Degrees: B.E. (E.E.), 1966, New York University; M.S., 1967, Ph.D., 1969, Stanford University.

Fellow Award "For contributions to cryptography"; Prize Paper Award, IEEE Information Theory Group, 1978; Donald G. Fink Award, IEEE, 1981. Other Awards: California State Psychological Association Award, 1978.

HELLMANN, REINHARD K., IRE Fellow 1960, AIEE Fellow 1961. Born: August 25, 1909, Aachen, Germany. Degrees: Dipl.Ing., 1932, Dr.Ing., 1937, Technische Hochschule Aachen, Germany.

IRE Fellow Award "For work in military electronics"; AIEE Fellow Award "For contributions in the field of civilian and military electronics"; IEEE Communication Society, Achievement Award, 1970.

HELLWARTH, ROBERT W., Fellow 1971. Born: December 10, 1930, Ann Arbor, Mich. Degrees: B.Sci.E., 1952, Princeton University; D.Phil., 1955, St. John's College, Oxford University.

Fellow Award "For discovery, invention, leadership, and teaching in the field of quantum electronics"; Student Award, Princeton Section of IRE, 1952, "In recognition of his professional development and accomplishments." Other Awards: Sigma Xi, 1951; Phi Beta Kappa, 1951; Rhodes Scholar, 1952; Honoree and Fellow, American Academy of Achievement, 1967; Lawrence A. Hyland Patent Award, 1968; National Science Foundation Senior Postdoctoral Fellow, St. Peter's College, Oxford University, 1970-71; Outstanding Educators of America, 1973; National Academy of Engineering, 1977; Fellow, American Physical Society, 1982; Fellow, Am. Ass. Advancement of Sci., 1982; 1983 Charles Hard Townes Award of the Optical Society of America.

HELSTROM, CARL W., Fellow 1970. Born: February 22, 1925, Easton, Pa. Degrees: B.S., 1947, Lehigh University; M.S., 1949, Ph.D., 1951, California Institute of Technology.

Fellow Award "For contributions to communication and detection theory"; Editor, IEEE Transactions on Information Theory, 1967-71; IEEE Centennial Medal, 1984. Other Awards: Fellow, Optical Society of America, 1975.

HENDERSON, J. T., IRE Fellow 1951. Born: December 9, 1905, Montreal, Quebec, Canada. Degrees: B.S., 1927, M.S., 1928, McGill University, Montreal; Ph.D., 1932, University of London, England.

Fellow Award "For his contributions in the field of radio direction-finding systems, and in particular, for his work in the development of Canadian radar during the war"; First McNaughton Award, Canadian Region, IEEE, 1969. Other Awards: Member of the Order of the British Empire, 1943; Fellow, Royal Society of Canada, 1944.

HENDRICKS, CHARLES D., Fellow 1974. Born: December 5, 1926, Lewiston, Utah. Degrees: B.S., 1949, Utah State University; M.S., 1951, University of Wisconsin; Ph.D., 1955, University of Utah.

Fellow Award "For contributions to electric propulsion and small particle technology." Other Awards: Fellow, American Physical Society; Fellow, American Association for the Advancement of Science; Associate Fellow, American Institute of Aeronautics and Astronautics; Member, American Association of Physics Teachers; Charter Member, Electrostatics Society of America; Phi Beta Kappa; Phi Kappa Phi; Sigma Xi; Sigma Pi Sigma; Tau Beta Pi; Eta Kappa Nu.

HENLE, ROBERT A., Fellow 1966. Born: April 27, 1924, Virginia, Minn. Degrees: B.S.E.E., 1949, M.S.E.E., 1951, University of Minnesota.

Fellow Award "For his contributions to the development of solid-state switching circuits." Other Awards: Fellow, IBM, 1964; Twelfth IBM Invention Achievement Award, 1978; National Academy of Engineering, 1982.

HENNEY, KEITH, IRE Fellow 1943. Born: October 28, 1896, McComb, Ohio. Degrees: B.A., 1921, Western Reserve University; M.A., 1925, Harvard University.

Fellow Award "In recognition of his accomplishments in obtaining publication of the technical information essential to the radio engineer"; Plaque, IRE, Rochester Fall Meeting, 1944, "For his many years of unselfish service to the radio and electronic industries through the technical press." Other Awards: Pioneer Award, Radio Club of America, 1982.

HENNING, RUDOLF E., Fellow 1965. Born: August 3, 1923, Hamburg, Germany. Degrees: B.S., 1943, M.S., 1947, D.Eng.Sc., 1954, Columbia University.

Fellow Award "For contributions to microwave instrumentation and measurement techniques."

HENNINGER, G. ROSS, AIEE Fellow 1943. Born: May 22, 1898, Hamilton, Ohio. Degrees: B.S., 1922, University of Southern California.

Other Awards: Eta Kappa Nu, 1953; Arthur L. Williston Award, American Society for Engineering Education, 1964; James H. McGraw Award, American Society for Engineering Education, 1973; Life Member, American Society for Engineering Education; Life Member, National Society of Professional Engineers.

HENRY, RAYMOND T., AIEE Fellow 1933. Born: November 17, 1889, Stronghurst, Ill.

HERBSTREIT, JACK W., IRE Fellow 1958. Born: September 7, 1917, Cincinnati, Ohio. Degrees: E.E., 1939, University of Cincinnati.

Fellow Award "For his leadership of a group studying radiowave scattering by atmospheric turbulence"; Harry Diamond Award, IRE, 1959, "For original research and leadership in radiowave propagation." Other Awards: Gold Medal, U.S. Dept. of Commerce, 1966; Director, Consultative Committee on International Radio, International Telecommunications Union, Geneva, Switzerland, 1966-74.

HERMACH, FRANCIS L., Fellow 1965. Born: January 8, 1917, Bridgeport, Conn. Degrees: B.E.E., 1943, George Washington University.

Fellow Award "For contributions to the field of low frequency electrical standards and measurement techniques"; Morris E. Leeds Award, IEEE, 1976, "For research and development of extremely accurate ac-dc transfer standards and for outstanding service on standards committees"; IEEE Centennial Medal, 1984. Other Awards: Meritorious Service Award, U.S. Dept. of Commerce, 1954; Distinguished Service Award, National Bureau of Standards, 1963.

HERMAN, ELVIN E., Fellow 1975. Born: March 17, 1921, Sigourney, Iowa. Degrees: B.S.E.E., 1942, State University of Iowa.

Fellow Award "For contribution to radar signal processing." Other Awards: Meritorious Civilian Service Award, U.S. Navy, 1946; L.A. Hyland Patent Award, Hughes Aircraft Co., 1971.

HERMAN, FRANK, Fellow 1964. Born: March 21, 1926, New York, N.Y. Degrees: B.S.E.E., 1945, M.S.E.E., 1949, Ph.D., 1953, Columbia University.

Fellow Award "For contributions in the field of the energy band structure of solids." Other Awards: Outstanding Research Awards, Radio Corporation of America Laboratories, 1950, 1952, 1955; Outstanding Young Man Award, New Jersey Chamber of Commerce, 1959; Tau Beta Pi; Sigma Xi; Fellow, American Physical Society; Visiting Professor: Technion, Haifa, 1971, University of Paris, 1973, Chalmers University, Gothenburg, 1977; Lecturer, Stanford University, 1964-present.

HERNQVIST, KARL G., Fellow 1969. Born: September 19, 1922, Boras, Sweden. Degrees: Doctor of Technology, 1959, Royal Institute of Technology, Stockholm.

Fellow Award "For invention and outstanding technical work in the field of gas discharge devices, thermionic energy conversion, and laser technology." Other

Awards: IR-100 Award, Industrial Research, 1967; David Sarnoff Award for Outstanding Technical Achievement, 1974 and 1982.

HEROLD, EDWARD W., IRE Fellow 1948. Born: October 15, 1907, New York, N.Y. Degrees: B.Sc., 1930, University of Virginia; M.Sc., 1942, D.Sc.(Hon.), 1961, Polytechnic Institute of Brooklyn.

Fellow Award "For his contributions to theory and design in the field of vacuum tubes"; Founders Medal, IEEE, 1976, "For his outstanding contributions to the electrical engineering profession at large, and in particular his insight and leadership in the development of color television." Other Awards: Phi Beta Kappa, 1930; Sigma Xi, 1942.

HERSHBERGER, W. DELMAR, IRE Fellow 1954. Born: May 10, 1903, Wellman, Iowa. Degrees: A.B., 1927, Goshen College; M.A., 1930, George Washington University; Ph.D., 1937, University of Pennsylvania.

Fellow Award "For his early contributions to the development of radar, and to frequency stabilization using microwave absorption lines"; IEEE Awards Committee, 1962-1964. Other Awards: RCA Laboratories Award, 1948; Chairman WESCON, 1954; Fulbright Research Scholar, Leyden University, 1955-1956; Visiting Professor, Technical University, Munich, 1968; University of Hawaii, 1975.

HERSHENOV, BERNARD, Fellow 1974. Born: September 22, 1927, New York, N.Y. Degrees: B.S., 1950, M.S., 1952, Ph.D., 1959, University of Michigan.

Fellow Award "For contributions to microwave devices." Other Awards: RCA Outstanding Achievement Award, 1964, 1967.

HERSKIND, C. CURTIS, AIEE Fellow 1948. Born: June 22, 1903, Hellerup, Denmark. Degrees: B.S.(E.E.), 1925, University of Denver.

Other Awards: Licensed Professional Engineer, New York State, 1938.

HERWALD, SEYMOUR W., IRE Fellow 1960, AIEE Fellow 1960. Born: January 17, 1917, Cleveland, Ohio. Degrees: B.S.M.E., 1938, Case Institute of Technology; M.S.M.E., 1940, Ph.D., 1944, University of Pittsburgh.

AIEE Fellow Award "For contributions to the theory and development of servomechanisms"; IRE Fellow Award "For contributions to servomechanism technology." Other Awards: Order of Merit, Westinghouse Electric Corp., 1958; Member, National Academy of Engineering, 1967; Exceptional Civilian Service Award, Dept. of Defense, 1974.

HERZ, ERIC, Fellow 1983.

Fellow Award "For contributions to the development and management of information systems for testing aerospace vehicles and for valuable services to the Institute."

HERZOG, GERALD B., Fellow 1966. Born: August 19, 1927, Minneapolis, Minn. Degrees: B.S.E.E., 1950,

M.S.E.E., 1951, University of Minnesota.

Fellow Award "For contributions to solid-state television and computer circuits." Other Awards: RCA Achievement Awards, 1953, 1960; David Sarnoff Outstanding Technical Achievement Awards, 1962, 1972; Outstanding Achievement Award, University of Minnesota, 1972; RCA Fellowship, University of Cambridge, England.

HERZOG, GERHARD, IRE Fellow 1955. Born: April 30, 1904, Zurich, Switzerland. Degrees: M.M.E., 1926, Ph.D., 1929, Eid. Technische Hochschule, Zurich.

Fellow Award "For his contributions to radioactive instrumentation for geological survey and medical applications." Other Awards: Fellow, American Physical Society.

HESS, HOWARD M., AIEE Fellow 1959. Born: September 11, 1908, Akron, Mich. Degrees: B.S.E.E., 1934, Wayne State University; M.S.E.E., 1937, University of Michigan.

Fellow Award "For contributions to the organization of electrical engineering education." Other Awards: Contribution and Leadership Award, Engineeering Society of Detroit, 1965; Fellow Award, Engineering Society of Detroit, 1974; Faculty Service Award, Wayne State University, 1976; Distinguished Service Award, Engineering Society of Detroit, 1978; Distinguished Engineering Alumni Achievement Award, Wayne State University, 1983.

HESSEL, ALEXANDER, Fellow 1976. Born: October 19, 1916, Vienna, Austria. Degrees: M.Sc., 1943, Hebrew University, Jerusalem, Israel; D.E.E., 1960, Polytechnic Institute of Brooklyn.

Fellow Award "For contributions to the theory of periodic structures as applied to phased array antennas"; Best Paper Award (corecipient), IEEE Transactions on Antennas and Propagation, 1974. Other Awards: Citation for Distinguished Research in the Field of Electromagnetics, Sigma Xi, 1977.

HEUMANN, GERHART W., AIEE Fellow 1960. Born: April 26, 1900, Luckenwalde, Germany. Degrees: Dipl.Ing., 1923, Technische Hochschule Dresden, Germany.

Fellow Award "For contributions to design of magnetic controls."

HEWLETT, WILLIAM R., IRE Fellow 1948, AIEE Fellow 1959. Born: May 20, 1913, Ann Arbor, Mich. Degrees: B.A., 1934, Stanford University; M.S., 1936, Massachusetts Institute of Technology; E.E., 1939, Stanford University; LL.D., 1966, University of California at Berkeley; LL.D., 1976, Yale University; D.Sc., 1978, Kenyon College; D.Sc., 1978, Polytechnic Institute of Technology; D.Eng.(Hon.), 1980, University of Notre Dame; D.Eng.(Hon.), Utah State University, 1983, Dartmouth College Eng.D(Hon.); Mills College, LL.D(Hon.).

IRE Fellow Award "For his initiative in the development of special radio measuring techniques"; AIEE Fellow Award "For technical and administrative achievements in the development and manufacture of precision electrical measuring equipment"; Founders Medal (with D. Packard), IEEE, 1973. Other Awards: Honorary Life Member, Instrument Society of America; National Academy of Engineering, 1965; Honorary Trustee, California Academy of Sciences, 1969; California Manufacturer of the Year, California Manufacturer's Assn., 1969; Business Statesman of the Year, Harvard Business School of Northern California, 1970; Medal of Achievement, WEMA (Western Electronic Manufacturers Assn.), 1971; Industrialist of the Year (with David Packard), California Museum of Science and Industry and California Museum Foundation, 1973; Award (with David Packard), Scientific Apparatus Makers Association, 1975; Vermilye Medal (with David Packard), Franklin Institute, 1976; Life Fellow Membership, Franklin Institute, 1976; Corporate Leadership Award, Massachusetts Institute of Technology, 1976; Medal of Honor, City of Boeblingen, Germany, 1977; National Academy of Sciences, 1977; Herbert Hoover Medal for Distinguished Service, Stanford University Alumni Association, 1977; American Philosophical Society, 1981.

HEYWANG, WALTER, Fellow 1979. Born: October 1, 1923, Schweinfurt, West Germany. Degrees: Master degree, 1948, Ph.D., 1950, University of Wurzburg; Dr.Ing.(h.c.), 1981, University of Lund.

Fellow Award "For contributions to the development of semiconductor and piezoelectric materials and devices." Other Awards: Wilhelm-Exner Medal, 1983, Austria.

HIATT, RALPH E., Fellow 1967. Born: April 12, 1910, Portland, Ind. Degrees: A.B., Indiana Central University; M.A., 1939, Indiana University.

Fellow Award "For research, development, design and construction of radar antennas and for fundamental efforts in radar cross section measurements"; President, Administrative Committee, IEEE Antenna and Propagation Society, 1970. Other Awards: Fellow, American Association for the Advancement of Science, 1969; Eta Kappa Nu, Sigma Xi, Tau Beta Pi; Professor Emeritus, Electrical and Computer Engineering, University of Michigan; Certificate of Service Award, Office of Scientific Research and Development, 1945.

HIBSHMAN, NELSON S., AIEE Fellow 1941. Born: January 20, 1902, Harrisburg, Pa. Degrees: B.S., 1924, Pennsylvania State University; M.S., 1927, Lehigh University.

HICKOK, ROBERT L., Fellow 1980. Born: February 25, 1929. Degrees: B.S., 1951, Rensslaer Polytechnic Institute; M.A., 1953, Dartmouth University; Ph.D., 1956, Rensslaer Polytechnic Institute.

Fellow Award "For contributions to plasma diagnostics."

HIEBERT, ALVIN L., Fellow 1971. Born: July 14, 1918, Mt. Lake, Minn. Degrees: B.A., 1942, Willamette University; E.E., 1943, Texas A&M College.

Fellow Award "For research and leadership on electromagnetic compatibility analysis techniques and spectrum engineering." Other Awards: Effective Service Award, Office of Scientific Research and Development, National Research Committee, 1945; American Man of Science, 1958; Listed in "Who's Who in Engineering," 1978.

HIERONYMUS, T. GALEN, AIEE Fellow 1950. Born: November 21, 1895, Valpariso, Ind.

Other Awards: Honorary Doctor of Philosophy, The Universal Society of Pathometrists in Atomic-Medicine, 1939; Silver Beaver Award for Distinguished Service to Boyhood, Boy Scouts of America, 1941; Annual Achievement Award, U.S. Radionics Congress, 1976; Professional Engineer, Missouri; patents in three countries.

HIGGINS, THOMAS J., AIEE Fellow 1957. Born: July 4, 1911, Charlottesville, Va. Degrees: E.E., 1932, M.A., 1937, Cornell University; Ph.D., 1941, Purdue University.

Fellow Award "For his researches in circuit theory and allied fields and for his leadership in electrical engineering education." Other Awards: George Westinghouse Award, American Society for Engineering Education, 1954; Benjamin Smith Reynolds Award, College of Engineering, University of Wisconsin, 1963; Engineer of the Year Award, Wisconsin Society of Professional Engineers, 1964; Donald P. Eckman Education Award, Instrument Society of America, 1964.

HIGGINS, WILLIAM H. C., IRE Fellow 1960. Born: April 26, 1908, Montclair, N.J. Degrees: B.S.E.E., 1929, E.E., 1934, D.Eng.(Hon.), 1978, Purdue University.

Fellow Award "For contributions to military electronics"; Edward Howard Armstrong Achievement Award, IEEE Communications Society, 1977. Other Awards: Naval Ordnance Development Award, U.S. Navy Bureau of Ordnance, 1945.

HIGHT, STUART C., IRE Fellow 1955. Born: July 28, 1906, Berkeley, Calif. Degrees: B.S., 1930, University of California; M.S., 1934, Columbia University.

Fellow Award "For his contributions to communication and weapon systems development." Other Awards: Outstanding Young Electrical Engineer, Eta Kappa Nu, 1940.

HIGINBOTHAM, WILLIAM A., IRE Fellow 1959. Born: October 25, 1910, Bridgeport, Conn. Degrees: A.B., 1932, D.Sc.(Hon.), 1963, Williams College.

Fellow Award "For contributions to pulse circuits and nuclear instrumentation"; First Annual Award, IEEE Nuclear Science Group, 1972, "In recognition of professional leadership in nuclear science and instrumentation and the quality and excellence of his technical contributions." Other Awards: Fellow, American Physical Society; Fellow, American Nuclear Society; Fellow, American Association for the Advancement of Science; First Annual Distinguished Service Award, Institute of Nuclear Materials Management, 1979.

HILEMAN, ANDREW R., Fellow 1970. Born: August 28, 1926, New Bethlehem, Pa. Degrees: B.S.E.E., 1951, Lehigh University; M.S.E.E., 1954, University of Pittsburgh.

Fellow Award "For contributions to the theory of lightning and switching surges and for the application of insulation concepts to EHV and UHV transmission systems design."

HILIBRAND, JACK, Fellow 1980. Born: September 15, 1930, New York, NY. Degrees: B.E.E., 1951, City College of New York; Sc.D., 1957, Massachusetts Institute of Technology.

Fellow Award "For contributions to the development of integrated circuits"; Outstanding Chapter Award, Philadelphia Section, IEEE, 1975-1976. Other Awards: Fellow, National Science Foundation, 1952-1954; Registered Professional Engineer, N.Y. State.

HILL, ALBERT G., IRE Fellow 1950. Born: January 11, 1910. Degrees: B.S, 1930, M.S., 1934, Washington University; Ph.D., 1937, University of Rochester.

Fellow Award "For work in the utilization of electronics to research in physics and his contribution in the conversion of wartime development laboratories to peacetime fundamental research." Other Awards: Presidential Certificate of Merit, 1948; Civilian Service Medal, U.S. Dept. of Air Force, 1955; American Ordnance Association Award, 1956; Distinguished Civilian Service Award, U.S. Dept. of Defense, 1959; Designer, Missile Viewing Radar, 1954-1955; Architecht, Designer, SHAPE Communication System, 1955-1959.

HILL, JAMES STEWART, Fellow 1979. Born: December 2, 1912, Washington, D.C. Degrees: B.S.E.E., 1934, Case Western Reserve University.

Fellow Award "For leadership in the international exchange of electromagnetic compatibility technology and for contributions to the field of measurement of the electromagnetic environment"; Certificate of Apprecitation, IEEE Electromagnetic Compatibility Society, 1969; Citation, Montreux Electomagnetic Compatibility Symposium, 1977; Chairman, PGRFI Symposium, 1960; Electromagnetic Compatibility Society Certificate of Appreciation, 1969; Electromagnetic Compatibility Society Honorary Life Member, 1982; Laurence G. Cumming Award, IEEE Electromagnetic Compatibility Society, 1983; IEEE Centennial Medal, 1984. Other Awards: Listed in "Who's Who in the World", "Who's Who in Engineering", Who's Who in Finance and Industry", "Men of Achievement".

HILL, LUTHER W., AIEE Fellow 1948. Born: August 9, 1892, Darlington, S.C. Degrees: B.S., 1915, C.E., 1916, University of South Carolina.

HILL, W. SCOTT, AIEE Fellow 1950. Born: September 6, 1902, Vicksburg, Mich. Degrees: B.S.E.E., 1923, University of Nevada.

HILLIARD, JOHN K., IRE Fellow 1952. Born: October 22, 1901, Wyndmere, N.D. Degrees: B.S., 1925, Hamline University; D.Sc.(Hon.), 1951, University of Hollywood.

Fellow Award "For his contributions in the field of motion-picture and audio engineering, and in the advancement of standards"; Audio Achievement Award, IRE, 1963. Other Awards: John H. Potts Award, Audio Engineering Society, 1961; Fellow, Society of Motion Picture Television Engineers; Fellow, Acoustical Society of America; Fellow, Audio Engineering Society.

HILLIER, JAMES, IRE Fellow 1957. Born: August 22, 1915, Brantford, Ontario, Canada. Degrees: B.A., 1937, M.A., 1938, Ph.D., 1941, D.Sc.(Hon.), University of Toronto; D.Sc.(Hon.), New Jersey Institute of Technology.

Fellow Award "For contributions in the field of electron optics, particularly electron microscopy"; IEEE David Sarnoff Medal, 1967; Founders Medal, IEEE, 1981. Other Awards: Albert Lasker Award, 1960; Industrial Research Institute Medal, 1975; Distinguished Contribution Award, Electron Microscopy Society of America; National Inventors Hall of Fame, 1980.

HILLS, HENRY W., AIEE Fellow 1951. Born: November 16, 1899, Melrose, Mass. Degrees: B.S., 1920, Massachusetts Institute of Technology.

Other Awards: Henry L. Doherty Medal, National Electric Light Association, 1926; Who's Who in America, 1952-1964; M.I.T. Sustaining Fellow Award, 1980.

HILSUM, CYRIL, Fellow 1984. Born: May 17, 1925, London, England. Degrees: B.Sc., Ph.D.

Fellow Award "For pioneering theoretical predictions of negative resistance effects due to hot electrons in gallium arsenide"; David Sarnoff Award, IEEE, 1981. Other Awards: Faraday Medal, 1982; GaAs Symposium Award, 1978; Rank Prize for Opto-Electronics, 1980; F.Inst.P.; F.I.E.E.; F.Eng., F.R.S. Foreign Associate, N.A.E.

HIMMEL, LEON, Fellow 1970. Born: January 24, 1921, New York, N.Y. Degrees: B.E.E., 1942, City College of The City University of New York.

Fellow Award "For contributions and leadership in the development of aerial navigation." Other Awards: Honorary Vice President, American Society for Zero Defects; Tau Beta Pi.

HINES, MARION E., Fellow 1968. Born: November 30, 1918, Bellingham, Wash. Degrees: B.S.(Applied Physics), 1940, M.S.E.E., 1946, California Institute of Technology.

Fellow Award "For contributions to the theory and application of solid-state microwave devices"; Outstanding Paper Award, IEEE, Solid State Circuits Conference, 1967; IEEE-S-MTT, Microwave Prize, 1972, 1977; J.J. Ebers Award, IEEE-S-ED, 1976; Microwave Prize, 1978; IEEE Lamme Medal, 1983; Microwave Career Award, IEEE S-MTT, 1983; IEEE Centennial Medal, 1984.

HINGORANI, NARAIN G., Fellow 1975. Born: June 15, 1931. Degrees: B.E., 1954, Baroda University, India; M.Sc., 1957, Ph.D., 1960, University of Manchester, England.

Fellow Award "For contributions through teaching, research, writing, and practice to the art of high-voltage direct-current power transmission"; Best Paper Prize, Portland IEEE Section, 1969, for "Transient Overvoltages on a Bipolar HVDC Line Cause by d-c Line Faults"; Prize Paper, T D Committee Award, 1980; Prize Paper Award, IEEE Power Engineering Society, 1980. Other Awards: Certificate of Recognition, Minnesota Society of Professional Engineers, 1978.

HINMAN, WILBUR STANLEY, JR., IRE Fellow 1957. Born: March 14, 1906, Washington, D.C. Degrees: B.S., 1926, Virginia Military Institute.

Fellow Award "For contributions in the fields of meteorographic instrumentation, aircraft navigation, and ordnance electronics"; Harry Diamond Award, IRE, 1956, "For his contributions to the electronic art in the fields of meteorology and proximity fuses." Other Awards: Exceptional Civilian Service Award, U.S. Dept. of Army, 1958; Distinguished Civilian Service Award, U.S. Dept. of Defense, 1960; Presidential Award for Distinguished Civilian Service, 1961.

HINTON, RAYMOND P., Fellow 1971. Born: January 18, 1925, Hereford, England. Degrees: B.Sc., 1944, University of Bristol.

Fellow Award "For contributions and leadership in the development of electronic switching and communication systems."

HIRAYAMA, HIROSHI, Fellow 1982. Born: January 29, 1922, Hyogo Perfecture. Degrees: B.E.E., 1944, Dr.Eng., 1960, Waseda University.

Fellow Award "For contributions to active network theory and leadership in graduate education." Other Awards: Best Paper Award, Institute of Electronics and Communication Engineers, Japan, 1960; Distinguished Services Award, Institute of Electronics and Communication Engineers, Japan, 1981.

HITCHCOCK, HARRY W., AIEE Fellow 1937. Born: January 1, 1890, Lincoln, Nebr. Degrees: B.S., 1911, Pomona College.

Awards: AIEE, 1912; National Director, Vice President District 8, IEEE. Other Awards: Medal of Freedom, 1944-1945; NATO Staff, 1953-1954; San Marino

City Council, 1955-1969; Mayor, San Marino, 1960-1969; Armed Forces Communications and Electronics Association; American Association for the Advancement of Science; Phi Beta Kappa; Tau Beta Kappa; Tau Beta Phi; Registered and Certified Electrical Engineer, California.

HITCHCOCK, LEON W., AIEE Fellow 1950. Born: June 21, 1886, West Medway, Mass. Degrees: B.S.E.E., 1908, Worcester Polytechnic Institute.

Other Awards: Charter Member, Phi Kappa Phi, University of New Hampshire, 1922; Charter Member, Tri-County Electrical Associates, 1953; Establishment of Leon W. Hitchcock Scholarship Award, Tri-County Electric Association; Dedication of Hitchcock Hall, University of New Hampshire, 1958; Member: New Hampshire Academy of Science, New Hampshire Society of Professional Engineers, American Society for Engineering Education, Listed in: "American Men of Science," "Who's Who in the East," "Who's Who in American Education"; Robert H. Goddard Award, Worcester Polytechnic Institute, 1967; Profile of Service Award, University of New Hampshire, 1978.

HITTINGER, WILLIAM C., Fellow 1967. Born: November 10, 1922, Bethlehem, Pa. Degrees: B.S., 1944, D.Eng.(Hon.), 1973, Lehigh University.

Fellow Award "For his contributions and leadership in electron device research and development." Other Awards: National Academy of Engineering, 1976.

HO, YU-CHI, Fellow 1973. Born: March 1, 1934, China. Degrees: S.B.E.E., 1953, S.B.E.E., 1955, Massachusetts Institute of Technology; Ph.D., 1961, Harvard University.

Fellow Award "For contribution to control theory and differential games"; Best Paper, IEEE TRANSACTIONS on Automatic Control, 1972-1973, "Team Decision Theory and Information Structures in Optimal Control Problems" (co-author K.C. Chu). Other Awards: Guggenheim Fellow, 1970.

HOADLEY, GEORGE B., AIEE Fellow 1962. Born: June 24, 1909, Swarthmore, Pa. Degrees: B.S., 1930, Swarthmore College; M.S., 1932, D.Sc., 1937, Massachusetts Institute of Technology.

Fellow Award "For contributions in the field of electrical engineering education." Other Awards: Outstanding Engineer Award, Raleigh Engineers Club, 1973; Certificate of Appreciation, North Carolina State University Chapter of Eta Kappa Nu, 1974.

HOAGLAND, ALBERT S., Fellow 1966. Born: September 13, 1926, Berkeley, Calif. Degrees: B.S., 1947, M.S., 1948, Ph.D., 1954, University of California at Berkeley.

Fellow Award "For contributions to the fields of magnetic recording and data storage for electronic computers"; Best Paper Prize, AIEE, 1956, for "Magnetic Data Recording Theory: Head Design"; President, IEEE

Computer Society, 1972-1973; Director (Board Member), IEEE, 1974-1975. Other Awards: Phi Beta Kappa; Sigma Xi; Eta Kappa Nu; Tau Beta Pi; President, AFIPS, 1978-1980; Trustee, Charles Babbage Foundation, 1979-1981.

HOAGLAND, JACK C., Fellow 1980. Born: November 14, 1918, Los Angeles, CA. Degrees: B.S., 1942, California Institute of Technology; Graduate studies, UCLA, UCI, 1972-75.

Fellow Award "For contributions to space communications"; M. Barry Carlton, IEEE, 1967; Outstanding Leadership, Orange County, IEEE, 1977, Outstanding Leadership, Southern Area Region 6, Los Angeles Council, IEEE, 1980. Other Awards: Citation for Professional Achievement, MacDonnell Douglas, 1965-1969; Engineer of the Month, Rockwell International, 1979; Listed in "Who's Who in America," "Who's Who in Technology," "Who's Who in California", 1980; WESCON Award, 1981; "American Men and Women of Science," 1982; Who's Who in Engineering, 1982; Engineer of the Year, Rockwell State, 1983.

HOBART, EDWARD A., Fellow 1964. Born: 1888, Middletown, Ohio. Degrees: E.E., 1912, Ohio State University.

Fellow Award "For contributions in the fields of battery charging and welding."

HOBBS, LINDER C., Fellow 1970. Born: February 4, 1925, Anniston, Ala. Degrees: B.E.E., 1948, Georgia Institute of Technology; M.S.E.E., 1952, M.B.A., 1956, University of Pennsylvania.

Fellow Award "For contributions to the planning, analysis, and forecasting of computer systems and technology"; Leadership Award, Los Angeles District of IEEE, 1965, "For excellence in leadership both as Professional Coordinator and Section Coordinator for the Los Angeles District of the IEEE." Other Awards: Tau Beta Pi, 1947; Eta Kappa Nu, 1947.

HODGES, DAVID A., Fellow 1977. Born: August 25, 1937, N.J. Degrees: B.E.E., 1960, Cornell University; M.S.E.E., 1961, Ph.D.(E.E.), 1966, University of California at Berkeley.

Fellow Award "For design and evaluation of semiconductor integrated circuits, and for education in digital electronic circuits and devices"; Outstanding Paper Award, IEEE International Solid-State Circuits Conference, 1963, 1969; Morris N. Liebmann Memorial Award, (co-recipient), IEEE, 1983.

HOEPPNER, CONRAD H., IRE Fellow 1958. Born: March 12, 1928, Spooner, Wis. Degrees: B.S.E.E., 1949, M.S.E.E., 1950, Professional E.E., 1957, University of Wisconsin; Graduate Studies, Massachusetts Institute of Technology.

Fellow Award "For his contributions to the field of telemetry and remote control." Other Awards: Fellow,

Tau Beta Pi; Fellow, Massachusetts Institute of Technology; Distinguished Service Citation, University of Wisconsin; Associate Fellow, AIAA; Pioneer, Missile, Range and Space; 72 patents; 64 publications.

HOERNI, JEAN A., Fellow 1970. Born: September 26, 1924, Geneva, Switzerland. Degrees: Ph.D.(Physics), 1950, University of Geneva; Ph.D.(Physics), 1952, University of Cambridge.

Fellow Award "For the invention of the planar transistor, and for other contributions to the theory and technology of semiconductor devices"; McDowell Award, IEEE Computer Society, 1972. Other Awards: Sigma Xi, 1953; John Scott Award, City of Philadelphia, 1966; Edward Longstreth Medal, Franklin Institute, 1969.

HOFF, MARCIAN E., Fellow 1982. Born: October 28, 1937. Degrees: B.E.E., 1958, Rensselaer Polytechnic Institute; M.S., 1959, PhD., 1962, Stanford University.

Fellow Award "For the conception and development of the microprocessor"; Cledo Brunetti Award, IEEE, 1980. Other Awards: Stuart Ballantine Medal, Franklin Institute, 1979.

HOFFMAN, CHARLES H., Fellow 1965. Born: September 5, 1917, Allentown, Pa. Degrees: B.S., 1938, Lehigh University; M.S., 1939, Massachusetts Institute of Technology.

Fellow Award "For contributions to the economic planning of large power systems."

HOFFMANN, JEAN A. J. L., Fellow 1966. Born: October 16, 1902, Marcinelle, Belgium. Degrees: E.E., M.E., 1927, Faculte des Sciences Appliquees, Brussels University.

Fellow Award "For contributions to engineering, teaching, and analog computation in Belgium." Other Awards: Doyen d'Honneur du Travail Scientifique, 1950; Professor Emeritus, 1973.

HOFT, RICHARD G., Fellow 1980. Born: December 4, 1926, Carroll, IA. Degrees: B.S.E.E., 1948, Iowa State University; M.E.E., 1954, Rensselaer Polytechnic Institute; Ph.D., 1965, Iowa State University.

Fellow Award "For contributions to power conversion systems"; William E. Newell Power Electronics Award (first recipient), IEEE, 1977.

HOGAN, C. LESTER, IRE Fellow 1960. Born: February 8, 1920, Great Falls, Mont. Degrees: B.S., 1942, Montana State College; M.S., 1947, Ph.D., 1950, Lehigh University; A.M.(Hon.), 1954, Harvard University; D.Engr.(Hon.), 1968, Montana State University; D.Sc.(Hon.),1969, Worcester Polytechnic Institute; D.Eng.(Hon.), 1971, Lehigh University.

Fellow Award "For pioneering in the application of ferrites"; Frederik Philips Award, 1975, "For outstanding accomplishments in the management of research and development resulting in important innovation in the semiconductor industry." Other Awards: National

Academy of Engineering, 1977; Honorary Fellow, IEE (London), 1978; Medal of Achievement, American Electronics Association, 1978; Fellow, American Association for the Advancement of Science, 1978; The Berkeley Citation, University of California at Berkeley, 1981.

HOGAN, JAMES J., Fellow 1982. Born: October 31, 1937, Cleveland, OH. Degrees: B.S.E.E., 1959, University of Dayton; M.S.E.E., 1961, Ph.D.(E.E.), 1964, University of Illnois.

Fellow Award "For contributions to terrain-sensing correlation guidance"; IRE Outstanding Student Award, Dayton Section, 1959. Other Awards: Award of Excellence in Electrical Engineering, University of Dayton, 1959; Dr. Karl Arnstein Award for Technical Achievement, Goodyear Aerospace Corporation, 1979.

HOGAN, JOSEPH C., Fellow 1984. Born: May 26, 1922, St. Louis, MO. Degrees: B.S.(E.E.), 1943, Washington University (St. Louis, Mo.); M.S.(E.E.), 1949, University of Missouri; Ph.D., 1953, University of Wisconsin.

Fellow Award "For contributions to the advancement of engineering education"; First Prize, Prize Paper Competition, AIEE Section, 1953. Other Awards: Fellowship, University of Wisconsin, 1951-53; Honor Award for Distinguished Service, Univ. of Missouri, 1973; Board of Governors, American Ass'n of Eng. Societies, 1982-83; Board of Directors, ASEE, 1973-75, 1981-84; Board of Directors, Engineers Council for Professional Development, 1974-80; Alumni Achievement Award, Washington University, 1979; Engineer of the Year, Indiana Society of Professional Engineers, 1981; Special Presidential Award, University of Notre Dame, 1981; Dean Emeritus of Engineering, University of Notre Dame, 1981; President, American Society for Engineering Education, 1982-83; Fellow, American Society for Engineering Education, 1983.

HOGG, DAVID C., Fellow 1965. Born: September 5, 1921, Vanguard, Saskatchewan, Canada. Degrees: B.Sc., 1949, University of Western Ontario; M.Sc., 1950, Ph.D., 1953, McGill University, Montreal, P.Q., Canada.

Fellow Award "For contributions in the field of tropospheric radio wave propagation and low-noise microwave antennas." Other Awards: National Academy of Engineering, 1978; Silver Medal Award, U.S. Department of Commerce, 1983.

HOHKI, SHIRO, Fellow 1975. Born: August 4, 1912, Toyko, Japan. Degrees: Dr.Eng., 1946, Tokyo University.

Fellow Award "For contributions to lightning protection of power transmission lines and to standardization of impulse voltage testing, and for leadership in research and development of power cables." Other Awards: Outstanding Paper Prize, 1946, Outstanding

Technical Achievement Prize, 1951, Outstanding Book Prize, 1953, Prize of Merit in Electrical Power Field, 1954, Institute of Electrical Engineers of Japan.

HOLBEN, WILMER P., AIEE Fellow 1942. Born: September 28, 1892, Allentown, Pa. Degrees: B.S.E.E., 1916, Pennsylvania State University.

Other Awards; Tau Beta Pi; Eta Kappa Nu; Phi Kappa Phi; American Society for Testing Materials; American Standards Association; Silver Beaver Award, Boy Scouts of America, 1938; "Who's Who in Engineering," 1977, 1980; "Who's Who in Eng. Encl.", 1980; 7 technical papers, 1937-1957.

HOLLANDER, GERHARD L., Fellow 1971. Born: February 27, 1922, Berlin, Germany. Degrees: B.S.E.E., 1947, Illinois Institute of Technology; M.S.E.E., 1948, Washington University; E.E., 1953, Massachusetts Institute of Technology.

Fellow Award "For contributions to the design of computing equipment and to evaluation methodology for large-scale information systems"; IEEE Centennial Medal, 1984. Other Awards: Commendation, U.S. Navy, 1944; Fellow, Institute for the Advancement of Engineering, 1970; Fellow, National Contract Management Association, 1978.

HOLLEY, CHARLES H., Fellow 1970. Born: April 15, 1919, Pittsburgh, Pa. Degrees: B.S.E.E., 1941, Duke University.

Fellow Award "For contributions to the design of large turbine-driven synchronous generators"; District Prize Paper Award, AIEE, for "Water Cooling of Turbine-Generator Stator Windings"; Award Citation, IEEE Region 1, 1975, "Through his outstanding leadership and management of large-scale research, development and production organizations in the electrical power field"; Nikola Tesla Award, 1978; IEEE Centennial Medal, 1984. Other Awards: National Academy of Engineering, 1976.

HOLLINGSWORTH, PHILIP M., Fellow 1966. Born: February 17, 1906, St. Helens, Lancashire, England. Degrees: B.Eng., 1928, M.Eng., 1936, University of Liverpool.

Fellow Award "For contributions to the development and application of underground cables." Other Awards: Fellow, Institution of Electrical Engineers, U.K., 1946.

HOLLYWOOD, JOHN M., IRE Fellow 1960. Born: February 4, 1910, Red Bank, N.J. Degrees: B.S., 1931, S.M., 1932, Massachusetts Institute of Technology.

Fellow Award "For contributions to electronic countermeasures and color television." Other Awards: Army-Navy Certificate of Appreciation, 1946.

HOLMES, D. BRAINERD, Fellow 1964. Born: May 24, 1921, New York, N.Y. Degrees: B.S.E.E., 1943, Cornell University; D.Sc.(Hon.), 1963, University of New Mexico, D.Sc.(Hon.), 1978, Worcester Polytechnic Institute.

Fellow Award "For contributions to electronic early warning systems and leadership of manned space flight programs." Other Awards: Paul T. Johns Award, Arnold Air Society, 1963; Outstanding Leadership, NASA, 1963; Eta Kappa Nu; Tau Beta Pi; Chi Psi; Aerospace Industries Association; American Institute of Aeronautics and Astronautics; National Academy of Engineering; National Aeronautic Association; National Space Club; The Navy League.

HOLMES, LYNN C., IRE Fellow 1949, AIEE Fellow 1951. Born: April 10, 1904, Brookfield, N. Y. Degrees: E.E., 1925, M.E.E., 1932, Rensselaer Polytechnic Institute.

IRE Fellow Award "For contributions to theory and practice in the field of magnetic recording." Other Awards: Leo H. East Memorial Award, Engineer of the Year, City of Rochester, NY, 1967.

HOLONYAK, NICK., JR., Fellow 1967. Born: November 3, 1928, Zeigler, Ill. Degrees: B.S., 1950, M.S., 1951, Ph.D., 1954, University of Illinois.

Fellow Award "For his contributions to solid state device applications including generation of visible coherent radiation from a semiconductor junction and discovery of phonon-assisted tunneling"; IEEE Morris Liebmann Award, 1973, "For outstanding contribution to the field of visible light emitting diodes and diode lasers"; Jack A. Morton Award, IEEE, 1981. Other Awards: Cordiner Award, General Electric, 1962; Texas Instruments Fellowship, University of Illinois, 1953-54; Member, National Academy of Engineering, 1973; John Scott Award, City of Philadelphia, 1975; GaAs Conference Award with Welker Medal, 1976; Center of Advanced Study, University of Illinois, 1977; Electrotechnical Society Solid State and Technology Award, 1983.

HOLT, ARTHUR W., Fellow 1969. Born: May 15, 1921. Degrees: B.A., 1942, M.A., 1949, Williams College.

Fellow Award "For major contributions to electronic digital computer technology and particularly, for the invention and development of a wide variety of optical character recognition equipment." Other Awards: Silver Medal, U.S. Commerce Dept.

HOLTON, WILLIAM C., Fellow 1982. Born: July 24, 1930, Washington, DC. Degrees: B.S.(Physics), 1952, University of North Carolina; M.S.(Physics), 1957, Ph.D.(Physics), 1960, University of Illnois.

Fellow Award "For technical leadership in semiconductor research and development." Other Awards: Phi Eta Sigma, 1949; Phi Beta Kappa, 1951; Fellow, American Physical Society, 1965; J. Daniels Award (1st in graduating NROTC class), 1952.

HONEY, RICHARD C., Fellow 1968. Born: March 9, 1924, Portland, Oreg. Degrees: B.S., 1945, California Institute of Technology; Ph.D., 1953, Stanford University.

Fellow Award "For his contributions to the fields of microwave antennas and laser applications." Other Awards: Fellow, Optical Society of America.

HONG, SE JUNE, Fellow 1983. Born: May 14, 1944, Seoul, Korea. Degrees: B.Sc.(Electr. Eng.) 1965, Seoul National University; M.S.(Elec. Eng.) 1967, Ph.D.(Elec. Eng.) 1969, University of Illinois.

Fellow Award "For contributions to fault-tolerant computing, design automation, and error-correcting codes." Other Awards: Honorable Mention Award for Outstanding Young Electrical Engineer, 1975; Three Outstanding Innovations Awards, IBM.

HONNELL, MARTIAL A., Fellow 1977. Born: October 23, 1910, Lyons, France. Degrees: B.S.E.E., 1934, M.S.E.E., 1940, E.E., 1945, Georgia Institute of Technology.

Fellow Award "For contributions to electrical engineering education and electronic instrumentation." Other Awards: M. A. Ferst Sigma Xi Research Award First Prize, Georgia Institute of Technology, 1951; Distinguished Graduate Faculty Lecturer, Auburn University, 1976-1977; Nine Certificates of Recognition, NASA, 1970-1981; Sigma Xi; Phi Kappa Phi; Tau Beta Pi; Eta Kappa Nu; Professor Emeritus, Auburn University 1981.

HONNELL, PIERRE M., AIEE Fellow 1957, IRE Fellow 1958. Born: January 28, 1908, Paris, France. Degrees: B.Sc., 1930, E.E., 1938, Texas A and M University; M.Sc., 1939, Massachusetts Institute of Technology; M.Sc., 1940, California Institute of Technology; Ph.D., 1950, St. Louis University.

AIEE Fellow Award "For his contributions to electrical engineering in the fields of research and education"; IRE Fellow Award "For distinguished teaching and research in applied electronics"; Paper Prize, AIEE, District 7, 1958. Other Awards: Legion of Merit, United States Military Academy, 1946; Fellow, American Association for the Advancement of Science, 1955; Marconi Premium, Institution of Radio Engineers, U.K., 1956; Colonel, USAR-Retired 1968; Professor Emeritus, Washington University, 1976.

HONSINGER, VERNON BERTRAM, Fellow 1979. Born: January 23, 1917, Saginaw, MI. Degrees: Assoc. Engrg., 1938, Bay City Junior College; B.S.E.E., 1940, University of Michigan, Ann Arbor.

Fellow Award "For contributions to the theory and design of electrical machinery, including reluctance machines." Other Awards: Eminence in Science and Engineering, 1962; Engineering Award of Merit, 1964, 1965.

HOOVER, CHARLES W., JR., Fellow 1980. Born: October 7, 1925, Akron, OH. Degrees: B.E.(Mech. Eng.), 1946, Yale University, B.S.(E.E.), 1947, Massachusetts Institute of Technology; M.S.(Physics), 1951,

Ph.D.(Physics), 1954, Yale University.

Fellow Award "For contributions to interconnection technology and memory components."

HOOVER, PAUL L., AIEE Fellow 1944. Born: December 20, 1898, Mansfield, Ohio. Degrees: B.S., 1921, Carnegie Institute of Technology; M.A., 1923, D.Sc., 1926, Harvard University.

Other Awards: Sigma Xi; Tau Beta Pi; Eta Kappa Nu.

HOOVER, WILLIAM G., AIEE Fellow 1962. Born: November 10, 1906, Pueblo, Colo. Degrees: A.B., 1928, E.E., 1929, Ph.D., 1936, Stanford University.

Fellow Award "For contributions to coordinated electronics and high- voltage engineering, and to engineering education." Other Awards: Phi Beta Kappa; Tau Beta Pi; Sigma Xi.

HOPFER, SAMUEL, Fellow 1982. Born: November 21, 1914, Rexingen, Germany. Degrees: B.A., 1944, West Virginia University; M.A., 1945, Cornell University; Ph.D., 1954, Polytechnic Institute of Brooklyn.

Fellow Award "For contributions to microwave components and instruments, especially to wide-band power measuring devices." Other Awards: New York Academy of Sciences; Sigma Psi, American Physical Society.

HOPKIN, ARTHUR M., Fellow 1973. Born: 1919. Degrees: B.S., 1942, Georgia Institute of Technology; M.S., 1947, Ph.D., 1950, Northwestern Technological Institute.

Fellow Award "For contributions to nonlinear control systems and engineering education"; Best Paper Prize, AIEE, Industrial Group, 1951. Other Awards: Member, American Society for Engineering Education; Phi Eta Sigma; Sigma Xi; Eta Kappa Nu; Tau Beta Pi.

HOPKINS, RICHARD J., Fellow 1971. Born: 1912, Vancouver, Canada. Degrees: B.S.E.E., 1934, Washington State University.

Fellow Award "For contributions to power capacitor design and application in transmission systems"; Member, IEEE Capacitor Subcommittee 1957-1975. Other Awards: Member, CIGRE; U.S. Representative, CIGRE Study Committee 18, Capacitors, 1960-1967; Member, NEMA Capacitor Tech. Comm., 1956-1975; Chairman, NEMA Capacitor Section, 1964-1967. Sigma Tau; Tau Beta Pi; Phi Kappa Phi.

HOPPER, FRANCIS L., IRE Fellow 1960. Born: February 24, 1901, Chicago, Ill. Degrees: B.S., 1922, California Institute of Technology.

Fellow Award "For contributions in underwater sound search and sound recording"; IEEE Committees on Audio, Stereo, Sound Recording, and Acoustics. Other Awards: Society of Motion Picture and Television Engineers; American Standards Association; Citation, Office of Scientific Research and Development.

HOPPER, GRACE M., IRE Fellow 1962. Born: December 9, 1906, New York, N.Y. Degrees: B.A., 1928, Vassar College; M.A., 1930, Ph.D., 1934, Yale University; D.Eng.(Hon.), 1972, Newark College of Engineering; D.Sc.(Hon.), 1973, C.W. Post College of Long Island University; D.Laws(Hon.), 1974, University of Pennsylvania; D.Sc.(Hon.), 1976, Pratt Institute; D.Sc.(Hon.), 1980, Linkoping University, Sweden; D.Sc.(Hon.), 1980, Bucknell University; D.Sc.(Hon.), 1980, Acadia University; D.Sc.(Hon.), 1981, Loyola University; D.Sc.(Hon.), 1981, Southern Illinois; D.Public Service(-Hon.), 1981, George Washington University, 1982; Marquette University, D.Sc.(Hon.), 1982; Seton Hill College, Dr. Humane Letters (Hon.) 1982.

Fellow Award "For contributions in the field of automatic programming"; IEEE, Philadelphia Section Achievement Award, 1968; W. Wallace McDowell Award, IEEE Computer Society, 1979. Other Awards: Phi Beta Kappa, 1928; Sigma Xi, 1930, Naval Ordnance Development Award, 1946; Achievement Award, Society of Women Engineers, 1964; Fellow, American Association for the Advancement of Science, 1964; Computer Sciences Man-of-the-Year Award, Data Processing Management Association, 1969; Harry Goode Memorial Award, AFIPS, 1970; Wilbur Lucius Cross Medal, Yale University, 1972; Fellow, ACPA, 1972; Member, National Academy of Engineering, 1973; Distinguished Fellow, British Computer Society, 1973; Legion of Merit, U.S. Navy, 1973; Distinguished Member Award, Washington, D.C. Chapter, Association for Computing Machinery, 1976; W. Wallace McDowell Award, IEEE Computer Society 1979; U.S. Navy Meritorious Service Medal, 1980; Fellow, Institute for Certification of Computer Professionals, 1981; The Gold Medal, Armed Forces Communications & Electronics Association, 1982; Honorary Professor, Defense Systems Management College, 1982; Pioneer Award, IEEE Computer Society, 1982; The Franklin Institute; International Oceangraphic Foundation; Armed Forces Communications and Electronics Association; Data Processing Management Association; Society of Women Engineers; British Computer Society; Association of Computer Programmers and Analysts; The Planetary Society; New York Academy of Sciences; The Oceanic Society; American Institute of Industrial Engineers; CODASYL Executive Committee; Management Systems Division Council, American Management Association; U.S. Naval Institute; "Who's Who"; "Who's Who in American Women"; "Who's Who in American Scientists";

HORGAN, JAMES D., Fellow 1968. Born: May 21, 1922, Grand Rapids, MI. Degrees: B.S.E.E., 1947, M.S., 1951, Marquette University; Ph.D., 1957, University of Wisconsin.

Fellow Award "For contributions to the understanding of biological control systems and to curriculum development." Other Awards: Engineer of the Year Award, Engineers and Scientists of Milwaukee, 1981.

HORNBECK, JOHN A., IRE Fellow 1962. Born: November 4, 1918, Minnesota. Degrees: A.B., 1939, Oberlin College; Ph.D., 1946, Massachusetts Institute of Technology.

Fellow Award "For contributions to electron devices."

HORNE, CHARLES F., JR., IRE Fellow 1958. Born: January 3, 1906, New York, N.Y. Degrees: B.S., 1926, United States Naval Academy; M.S., 1935, Harvard University.

Fellow Award "For his contributions to the field of communications in both military and civilian applications." Other Awards: Combat Legion of Merit; National Regional Medal, Civil Aeronautics Administration, 1952; Medal of Honor, Electronic Industries Association, 1964; Gold Knight Award, National Management Association, 1962; Governor's Award, California Advisory Council on Vocational Education Services (CACVE), 1975; Member, American Association for the Advancement of Science.

HOROWITZ, ISAAC M., Fellow 1972. Born: December 15, 1920, Safed, Israel. Degrees: B.Sc., 1944, University of Manitoba; B.S.E., 1948, Massachusetts Institute of Technology; M.E.E., 1953, D.E.E., 1956, Polytechnic Institute of Brooklyn.

Fellow Award "For contributions to active network theory and feedback control theory." Other Awards: Best Paper, National Electronics Conference, 1956.

HOROWITZ, STANLEY H., Fellow 1979. Born: August 5, 1925, New York, N.Y. Degrees: B.S.E.E., 1949, City College of New York.

Fellow Award "For contributions to power-system integrity through protective relaying and to education through industry/education research programs."

HORRELL, MAURICE W., IRE Fellow 1962. Born: January 9, 1912, Baldwin City, Kans. Degrees: B.S., 1935, M.S., 1938, Kansas State University.

Fellow Award "For contributions to military electronics and computer engineering."

HORTON, BILLY M., Fellow 1969. Born: December 27, 1918, Bartlett, Tex. Degrees: B.A., 1941, University of Texas; M.S., 1949, University of Maryland.

Fellow Award "For the technical leadership of a large government laboratory and, specifically for the invention of fluid amplification." Other Awards: Arnold O. Beckman Award, Instrument Society of America, 1960; U.S. Army Decoration for Exceptional Civilian Service, 1965; John Scott Award, City of Philadelphia, 1966; Distinguished Civilian Service Award, Department of Defense, 1967; Inventor of the Year, PTC Research Institute, George Washington University, 1971; National Academy of Engineering, 1979.

HOTH, DANIEL F., Fellow 1969. Born: December 4, 1913, Arlington, N.J. Degrees: M.E., 1935, Stevens Institute of Technology.

Fellow Award "For contributions to the engineering of pulse code modulation and satellite communication systems."

HOUCK, HARRY W., IRE Fellow 1957. Born: April 11, 1896, New Cumberland, Pa.

Fellow Award "For pioneering contributions in the field of broadcast reception." Other Awards: Armstrong Medal, Radio Club of America, 1941; Marconi Medal, Veteran Wireless Operators Association, 1955.

HOUGH, RICHARD R., Fellow 1964. Born: December 13, 1917, Trenton, N.J. Degrees: B.S.E., 1939, E.E., 1940, Princeton University; D.Sc.(Hon.), 1977, Susquehanna University.

Fellow Award "For leadership in military electronics development associated with guided missiles"; Alexander Graham Bell Medal, IEEE, 1980. Other Awards: Phi Beta Kappa; Sigma Xi; Outstanding Young Electrical Engineer Award; Eta Kappa Nu, 1947; Member, National Academy of Engineering, 1983.

HOUGH, WILLIAM R., AIEE Fellow 1948. Born: July 9, 1907, Kalamazoo, Mich. Degrees: B.S., 1929, University of Michigan.

Other Awards: Engineer of the Year Award, Cleveland Society of Professional Engineers, 1967; Distinguished Service Award, Cleveland Technical Societies Council, 1972.

HOUSE, HAZEN E., Fellow 1971. Born: September 15, 1906, Knoxville, Tenn. Degrees: S.B., 1929, S.M., 1932, Massachusetts Institute of Technology.

Fellow Award "For contributions to the determination of electrical and thermal characteristics of stranded and bare conductors." Other Awards: Life Member, American Society of Mechanical Engineers; Member Emeritus, Sigma Xi.

HOUSE, ROBERT W., Fellow 1979. Born: May 31, 1927, Wellsville, Ohio. Degrees: B.S., 1949, M.S., 1952, Ohio University; Ph.D., 1959, Pennsylvania State University; P.M.D., 1966, Harvard Business School.

Fellow Award "For contributions to the logical design of computers." Other Awards: Commendation, President's Commission, 1974-75; Letter of Commendation, U.S. Air Force.

HOUSTON, JOHN M., Fellow 1984. Born: December 26, 1924, Cincinnati, OH. Degrees: B.S.(Physics), 1949, Brown Univ.; Ph.D.(Physics), 1955, M.I.T.

Fellow Award "For research and engineering in electron physics."

HOWARD, WILLIAM G., JR., Fellow 1981. Born: November 6, 1941, MA. Degrees: B.E.E.(with distinction), 1964, M.S., 1965, Cornell University; Ph.D., 1967, University of California at Berkeley.

Fellow Award "For contributions to the advancement of semiconductor circuits and devices, especially analog and digital LSI technology."

HOWE, HARLAN G., JR., Fellow 1980. Born: 1936, New York, NY. Degrees: B.S.(Optics), 1957, University of Rochester.

Fellow Award "For contributions to microwave integrated circuits."

HOWE, ROBERT M., Fellow 1976. Born: August 28, 1925, Oberlin, Ohio. Degrees: B.S.E.E., 1945, California Institute of Technology; A.B.(Physics), 1947, Oberlin College; M.S.(Physics), 1948, University of Michigan; Ph.D.(Physics), 1950, Massachusetts Institute of Technology.

Fellow Award "For contributions to analog and hybrid computer technology."

HOWELL, IRVIN N., JR., Fellow 1975. Born: January 3, 1922, Chickasaw, Ala. Degrees: B.S.E.E., 1951, The Citadel, Charleston, S.C.

Fellow Award "For contributions in the field of communications, and for leadership in meeting quality control objectives and standards"; Outstanding Achievement Award, IEEE Industry Applications Society, 1977; President, IEEE Industry Applications Society, 1972; Elected Chairman, IEEE Standards Board, 1980-1982; Elected Director, IEEE Division II, 1983-1984; Member, IEEE Board of Directors, 1980-1984; IEEE Centennial Medal, 1984. Other Awards: Listed in: "Who's Who in the South and the Southwest," 1975, 1978, "Men of Achievement," 1976, "Who's Who in Engineering," 1977-82; "Who's Who in Industry and Finance," 1979-1980; "Who's Who in the World," 1981; Tau Beta Pi; Registered Professional Engineer, Tenn. & Miss.

HOWES, FREDERICK S., IRE Fellow 1950. Born: July 25, 1896, Paris, Ontario, Canada. Degrees: B.Sc., 1924, M.Sc., 1926, McGill University, Quebec; D.I.C., 1928, Ph.D., 1929, University of London.

Fellow Award "For his contributions as a teacher in the field of communication engineering"; Lecturer, McGill Univ., Dept. of Elec. Eng., 1929-43; Professor Emeritus of Electrical Engineering, 1964.

HOYT, THERON C., AIEE Fellow 1962. Born: September 9, 1916, Jamaica, N.Y. Degrees: B.S., 1938, Pennsylvania State University.

Fellow Award "For contributions to engineering of domestic and foreign utility generation and transmission facilities."

HSIAO, MU-YUE, Fellow 1974. Born: July 17, 1933, Hunan, China. Degrees: B.S.E.E., 1956, Taiwan University; M.S., 1960, University of Illinois; Ph.D., 1967, University of Florida.

Fellow Award "For contribution to computer reliability, error-correcting codes, and the detection and testing of faults." Other Awards: Outstanding Invention Awards, IBM, 1972, 1979.

HUANG, THOMAS S., Fellow 1979. Born: June 26, 1936, Shanghai, China. Degrees: B.S., 1956, National

Taiwan University; M.S., 1960, Sc.D., 1963, Massachusetts Institute of Technology.

Fellow Award "For contributions to the theory and application of image processing and digital filtering." Other Awards: Guggenheim Fellow, 1971-72; U.S. Senior Scientist Award, Humboldt Foundation, 1976-77.

HUBBARD, DAVID C., AIEE Fellow 1952. Born: August 11, 1908, Baltimore, Md. Degrees: B.S.E.E., 1930, Purdue University.

Fellow Award "For fifteen years of major contributions to live line maintenance equipment." Other Awards: Anchor Hall of Fame, A.B. Chance, 1972. 57 U.S. Patents, 1941-1974; International Merit Award for Creative Thinking, 1975; International Inventors Incorporated.

HUBBARD, MALCOLM M., AIEE Fellow 1950. Born: December 12, 1906, New York, N.Y. Degrees: B.S.E.E., 1929, Massachusetts Institute of Technology.

Other Awards: Presidential Certificate of Merit, 1946.

HUELSMAN, LAWRENCE P., Fellow 1978. Born: January 22, 1926, Chicago, Ill. Degrees: B.S.E.E., 1950, Case Institute of Technology; M.S.E.E., 1956, Ph.D., 1960, University of California.

Fellow Award "For contributions to electrical engineering education and circuit theory." Other Awards: National Electronics Conference Award, 1961; National Aeronautics and Space Administration Award, 1970; Anderson Prize, 1975.

HUENING, WALTER C., JR., Fellow 1981. Born: February 10, 1923, Boston, MA. Degree: B.S. (E.E.) (magna cum laude), 1944, Tufts University.

Fellow Award "For contributions to the standardization of short-circuit current calculation practices in the industrial power field"; First Prize Paper, IEEE Industry and General Applications Group, 1970; Second Prize Paper, IEEE Industry Applications Society, 1982. Other Awards: Tau Beta Pi, 1943.

HUGGINS, WILLIAM H., IRE Fellow 1957. Born: January 11, 1919, Rupert, Idaho. Degrees: B.S., 1941, M.S., 1942, Oregon State University; Sc.D., 1953, Massachusetts Institute of Technology.

Fellow Award "For contributions in the field of circuit theory"; Browder J. Thompson Award, IRE, 1948; Education Medal, 1966, "For his creative approach to and extraordinary effectiveness in teaching, and for his abiding interest in and rapport with students." Other Awards: Exceptional Civilian Service Award, U.S. Dept. of Air Force, 1954; Annual Award, National Electronics Conference, 1955; Fellow, Acoustical Society of America, 1959; Western Electric Fund Award, American Society for Engineering Education, 1965; National Academy of Engineering, 1970.

HUGHES, GEORGE W., Fellow 1973. Born: 1929, Michigan. Degrees: B.S., 1952, M.S., 1952, E.E., 1955, Sc.D., 1960, Massachusetts Institute of Technology.

Fellow Award "For contributions to understanding of speech." Other Awards: IR 100 Award, 1981; Special Fellow, National Institute of Neurological Diseases and Blindness, Istituto di Fisiologica, University of Pisa, Italy, 1963.

HUGHES, RUSSELL H., AIEE Fellow 1941. Born: 1891, Decatur, Ala. Degrees: B.Eng., 1913, Vanderbilt University.

HUGHES, WILLIAM L., IRE Fellow 1962. Born: December 2, 1926, Rapid City, S.D. Degrees: B.S., 1949, South Dakota School of Mines and Technology; M.S., 1950, Ph.D., 1952, Iowa State University.

Fellow Award "For contributions to color television." Other Awards: Man of the Year, Stillwater, OK Junior Chamber of Commerce, 1963; Outstanding Alumni, College of Engineering, Iowa State University, 1980.

HULL, DAVID R., IRE Fellow 1947. Born: October 29, 1903, Newton, N.J. Degrees: B.S., 1925, United States Naval Academy; M.S., 1933, Harvard University.

Fellow Award "For electronic contributions to the military services." Other Awards: Commendation Medal, U.S. Dept. of Navy, 1940; Legion of Merit, 1946; Fellow, Acoustical Society of America, 1952; Medal of Honor, Electronic Industries Association, 1960.

HUMPHREY, FLOYD B., Fellow 1974. Born: May 20, 1925, Greeley, Colo. Degrees: B.S., 1950, Ph.D., 1956, California Institute of Technology.

Fellow Award "For contributions to magnetic device flux reversal mechanisms, and for improving engineering education using laboratory instruction techniques."

HUMPHREY, WATTS S., Fellow 1972. Born: July 4, 1927, Battle Creek, Mich. Degrees: B.S.(Physics), 1949, University of Chicago; M.S.(Physics), 1950, Illinois Institute of Technology; M.B.A., 1951, University of Chicago.

Fellow Award "For his contributions to the development of specialized digital computers and communications systems for military application and leadership in the area of advanced computer programming"; Member, Editorial Board, IEEE Spectrum; Member, ACM; 5 U.S. patents (computer field).

HUNG, JAMES CHEN, Fellow 1984. Born: February 18, 1929, Foochow, China. Degrees: B.S.E.E., 1953, National Taiwan University; M.E.E., 1956, Eng.Sc.D., 1961, New York University.

Fellow Award "For contributions to industrial electronics and control automation." Other Awards: NASA Technology Award, NASA, 1969; Certificate for Contribution to Technology, NASA/MSFC, 1970; M.E. Brooks Engineering Professor Award, Univ. of Tenn., 1973; Letter of Commendation, U.S. Department of

Army, MICOM, 1981; Distinguished Service Professorship, Univ. of Tenn., 1983.

HUNT, A. BREWER, Fellow 1968. Born: June 15, 1902. Degrees: B.A.Sc., 1928, University of Toronto; D.Eng.(Hon.), 1967, Carleton University.

Fellow Award "For contributions to the organization of telecommunications research and development."

HUNT, BOBBY R., Fellow 1983. Born: August 24, 1941, McAlester, Oklahoma. Degrees: B.Sc., 1964, Wichita State University; M.Sc., 1965, Oklahoma State University; Ph.D., 1967, University of Arizona.

Fellow Award "For contributions to the theory and practice of digital image processing."

HUNT, LLOYD F., AIEE Fellow 1938. Born: June 17, 1897, San Jose, Calif. Degrees: B.S., 1919, E.E., 1929, D.Eng.(Hon.), 1967, University of Southern California.

Other Awards: U.S. Dept. of Navy; Modern Pioneer Award, National Association of Manufacturers, 1940; First Distinguished Alumnus Award, University of Southern California, School of Engineering, 1966; Lloyd F. Hunt Chair in Electric Power Engineering, University of Southern California, 1968; Sigma Xi; Eta Kappa Nu; CIGRE.

HUNTER, LLOYD P., Fellow 1967. Born: February 11, 1916, Wooster, Ohio. Degrees: B.A., 1939, College of Wooster; B.S., 1939, Massachusetts Institute of Technology; M.S., 1940, D.Sc., 1942, Carnegie Institute of Technology.

Fellow Award "For contributions to semiconductor research and publications." Other Awards: Fellow, American Physical Society; Outstanding Invention Award, IBM, 1964.

HUNTLEY, EDSON D., Fellow 1964. Born: March 31, 1902, New Woodstock, N.Y. Degrees: B.S., 1924, Union College.

Fellow Award "For contributions to the design of large turbine generatiors."

HUNTOON, ROBERT D., IRE Fellow 1954. Born: July 20, 1909, Waterloo, Iowa. Degrees: B.A., 1932, Iowa State Techers College; M.S., 1935, Ph.D., 1938, State University of Iowa.

Fellow Award "For his contributions in the fields of electronic ordnance and electron physics." Other Awards: Certificate of Appreciation, Office of Scientific Research and Development, 1945; Certificate of Appreciation, War Dept., Secretary of War, 1945; Naval Ordnance Development Award, 1945; Presidential Certificate of Merit, 1948; Award in Physical Sciences, Washington Academy of Sciences, 1948; Exceptional Service Award, U.S. Dept. of Commerce, 1949; Achievement Award, Iowa State Teachers College, 1957.

HUSE, RAYMOND A., Fellow 1977. Born: May 8, 1916. Degrees: B.S.E.E.(special honors), 1938, University of New Hampshire; M.S.E.E.(with honors), 1939, Harvard University.

Fellow Award "For technical leadership in developing innovative energy conversion and transmission technology for bulk power systems." Other Awards: Senior Member, Instrument Society of America; Phi Kappa Phi; Phi Lambda Phi.

HUSKEY, HARRY D., Fellow 1967. Born: January 19, 1916, Whittier, N.C. Degrees: B.S., 1937, University of Idaho; M.A., 1940, Ph.D., 1943, Ohio State University.

Fellow Award "For outstanding contributions in education and in the design of computing machinery"; Pioneer Award, IEEE Computer Society, 1982. Other Awards: U.S. Senior Scientist Awardee, Alexander von Humboldt Foundation, 1974-1975; Fellow, AAAS; Fellow, British Computer Society; Honored at Pioneer Session, National Computer Conference, 1978.

HUSSON, SAMIR S., Fellow 1983.

Fellow Award "For contributions to microprogramming, design automation, and computer system development methodology."

HUTCHESON, JOHN A., IRE Fellow 1948, AIEE Fellow 1951. Born: January 21, 1905, Park River, N.D. Degrees: B.Sc., 1926, D.Sc., 1943, University of North Dakota; D.Sc., 1963, Geneva College.

IRE Fellow Award "For his contributions and technical direction of work in radio research." Other Awards: Certificate of Commendation, U.S. Dept. of Navy, 1946.

HUTCHISON, JOHN A., AIEE Fellow 1962. Born: March 6, 1911, Baird, Tex. Degrees: B.S., 1932, California Institute of Technology.

Fellow Award "For progressive planning, system design and management of electric utility operation."

HUTTER, RUDOLF G. E., IRE Fellow 1954. Born: February 12, 1910, Berlin, Germany. Degrees: Ph.D., 1944, Stanford University.

Fellow Award "For his contributions in the field of electron ballistics and electron optics."

HYDE, MERRITT A., AIEE Fellow 1959. Born: July 1, 1896, Owego, N.Y. Degrees: A.B., 1919, Marietta College; B.S., 1922, E.E., 1944, Case Institute of Technology; Sc.D., 1948, Marietta College.

Fellow Award "For contributions to electrical application and pipeline pumping." Other Awards: Award of Merit, Westingthouse Electric, 1953.

HYLAND, LAWRENCE A., IRE Fellow 1955. Born: August 26, 1897, Nova Scotia, Canada. Degrees: D.E.(Hon.), 1954, Lawrence Institute of Technology.

Fellow Award "For his contributions to aircraft radio direction finding, and his effective direction of research"; Pioneer Award, IRE, 1957; Founders Medal, IEEE, 1974, "For leadership and management in the field of electronics." Other Awards: Distinguished Public Service Medal, U.S. Dept. of Navy, 1950; Medal of Achievement, Western Eletronic Manufacturers Association, 1967; Robert J. Collier Trophy, National Aeronautic Association of the United States, 1968; Gold Medal for Meritorious Service, Armed Forces Communications and Electronics Association, 1968; Barnstormer Trophy, Antelope Valley Aero Museum; Member, American Nuclear Society; Member, Scientific Research Society of America; Howard R. Hughes Memorial Award, 1981.

HYLTEN-CAVALLIUS, NILS R., Fellow 1982. Born: Malmo, Sweden. Degrees: M.Sc.(Elec.), 1943, Royal Technical University, Stockholm, Sweden.

Fellow Award "For leadership in ultra-high-voltage laboratory design and realization"; Prize Paper Award, Power System Instrumentation and Measurements Committee, IEEE, 1978; Prize Paper Award, Power System Instrumentation and Measurements Committee, IEEE, 1982; W.R.G. Baker Award, IEEE, 1983. Other Awards: Member, Swedish Academy of Engineering Science.

HYNES, LEE P., AIEE Fellow 1943. Born: October 15, 1881, Old Ripley, Ill.

Other Awards: American Society of Mechanical Engineers; American Society of Heating and Air Conditioning Engineers; National Society of Professional Engineers; Institution of Mechanical Engineers, U.K.; Engineering Institute, Canada.

I

IAMS, HARLEY, IRE Fellow 1950. Born: March 13, 1905, Lorentz, W.Va. Degrees: B.A.(Mechanical Engineering), 1927, Stanford University.

Fellow Award "For the development of electronic apparatus for converting images formed by electromagnetic waves to electrical signals, first in the television field, and later, using new principles, in the realm of short radio waves." Other Awards: Phi Beta Kappa; Sigma Xi, 1932; Modern Pioneer Award, National Association of Manufacturers, 1940.

IBUKA, MASARU, Fellow 1972. Born: April 11, 1908, Nikko City, Japan. Degrees: B.Sc., 1933, Waseda University; Dr.Sci., 1974, Plano University; Dr.Eng.(Hon.), 1976, Sophia University, Tokyo; Dr. Sc. (Hon.), 1979, Waseda University, Tokyo; D.Hum.(Hon.), 1982, Mindanao State University.

Fellow Award "For contributions to the development and application of solid-state devices in consumer and computer electronics"; IEEE Founders Medal, 1972, "For outstanding administrative leadership in applying solid-state devices in consumer electronics, thereby enhancing industry growth and bringing distinction to the profession." Other Awards: Medal of Honor with Blue Ribbon, H.M. Emperor of Japan, 1960; Niwa-Takayanagi Memorial Award, Institute of Television Engineers, Japan, 1973; NHK Broadcast Cultural Award,

Japan Broadcasting Corp., 1977; Imperial Order of the Sacred Treasure I, H.M. Emperor of Japan, 1978; Humanism and Technology Award, Aspen Institute for Humanistic Studies, 1981.

IDE, JOHN M., Fellow 1971. Born: 1907, New York, N.Y. Degrees: B.A., 1927, Pomona College; M.S.E., 1929, M.S., 1930, Sc.D., 1931, Harvard University; D.Sc.(Hon.), 1965, Pomona College.

Fellow Award "For achievements in the management of engineering research, and for contributions to sonar and other underwater devices." Other Awards: Harvard Engineering School Prize Scholarship, 1927; Civilian Meritorious Service Award, 1945; National Civil Service League Award, 1958; Fellow, Acoustical Society of America; Phi Beta Kappa; Tau Beta Pi; Delta Sigma Rho.

ILGENFRITZ, LESTER M., AIEE Fellow 1961. Born: October 8, 1897, Monroe, Mich. Degrees: B.S., 1920, University of Michigan.

Fellow Award "For contributions to design and development of submarine cable repeaters and systems."

IMMEL, RALPH B., Fellow 1968. Born: February 28, 1915, LaPorte, Ind. Degrees: B.S.E.E., 1936, Purdue University; M.S.E.E., 1941, University of Pittsburgh.

Fellow Award "For his inventiveness in developing electrical and electronic control devices for industry." Other Awards: Bronze Patent Award Plaque, Westinghouse Electric Corp., 1966; 54 U.S. Patents, 1941-1980.

IMMICH, HARRY R., AIEE Fellow 1961. Born: June 11, 1906, New York, N.Y. Degrees: B.S.E.E., 1931, University of Kansas.

Fellow Award "For contributions to storm-proofing of inter-city and rural communications."

INCE, A. NEJAT, Fellow 1972. Born: November 16, 1928, Bodrum, Turkey. Degrees: B.Sc.(Hon.), 1952, University of Birmingham, England; Ph.D., 1955, Cambridge University, England.

Fellow Award "For technical contributions and pioneering leadership in the development of international communication systems"; IEEE Award in International Communication, 1979; Science Award, Turkish Scientific and Technical Research Council, 1982; Secretary General, The Turkish Science and Engineering Research Council, Ankara.

INCULET, ION I., Fellow 1978. Born: February 11, 1921, Iasi, Romania. Degrees: Diplom. Engineer(Electrical), 1944, Politechnica, Bucharest; M.S.(Electrical Engineering), 1962, Laval, Quebec, Canada.

Fellow Award "For contributions to the applications of electrostatics in mineral separation and gas cleaning"; Prize Paper Award, Honorable Mention, IEEE Industry Applications Society, 1978; Outstanding Achievement Award, IEEE Industry Applications Society, 1983. Other Awards: Award for Visiting Scientists to France, National Research Council of Canada, 1975; Award for Visiting Scientists to Japan, National Research Council of Canada, 1976.

INGERSLEV, FRITZ H. B., Fellow 1970. Born: July 6, 1912, Aarhus, Denmark. Degrees: M.Sc., 1936, D.Sc., 1954, Technical University of Denmark.

Fellow Award "For contributions to electroacoustics and noise control." Other Awards: Danish Esso Prize, 1952; Prize, Christian Moller Sorensen and Marie Christine Sorensen Foundation, 1961; Prize, Finn Henriksen Foundation, 1969; Honorary Fellowship, British Institute of Acoustics, 1981; Medaille d'argent du groupement des acousticiens de langue Francaise, 1981; Foreign Associate, National Academy of Engineering, 1982; "DIN Ehrennadel", DIN Deutsches Institut fur Normung, 1982.

INGRAM, SYDNEY B., AIEE Fellow 1951. Born: October 8, 1903, Bottineau Co., N.D. Degrees: B.A., 1925, University of British Columbia; Ph.D.(Physics), 1928, California Institute of Technology.

INMAN, DONOVAN E., AIEE Fellow 1951. Born: April 25, 1901, Ft. Dodge, Iowa. Degrees: B.S., 1925, Colorado State University.

INOSE, HIROSHI, Fellow 1978. Born: January 5, 1927. Degrees: B.E., 1948, Ph.D., 1955, University of Tokyo.

Fellow Award "For contributions to the development of digital switching, digital modulation, and road traffic control systems"; IEEE Award in International Communication in Honor of Hernand and Sosthenes Behn, 1982. Other Awards: Marconi International Fellowship, 1976; Japan Academy Prize, 1979; Foreign Associate, National Academy of Sciences, 1977; Member, American Philosophical Society, 1979.

IPPEN, ERICH P., Fellow 1984. Born: March 29, 1940, Fountain Hill, PA. Degrees: S.B., 1962, M.I.T.; M.S., 1965, Ph.D., 1968, University of California, Berkeley.

Fellow Award "For contributions to picosecond optics and optical instrumentation"; Morris Leeds Award, IEEE, 1983. Other Awards: R.W. Wood Prize, Optical Society of America, 1981; Edward Longstreth Medal, Franklin Institute, 1982.

IPPONMATSU, TAMAKI, Fellow 1973. Born: April 29, 1901, Hiroshima-ken, Japan. Degrees: Dr.Phil., 1945, Osaka University.

Fellow Award "For contribution to power systems engineering, and pioneering leadership in nuclear electric power plants." Other Awards: Honorary Member, Society of Atomic Energy of Japan, 1967; Honorary Member, Institute of Electrical Engineers of Japan, 1971; Foreign Associate, National Academy of Engineering, 1978.

IRELAND, GLEN, AIEE Fellow 1953. Born: September 4, 1895, Independence, Iowa. Degrees: B.S.E.E., 1917, University of Iowa.

Fellow Award "For outstanding contributions in the planning and basic design of the nationwide telephone system, in particular the transmission and switching aspects of long distance service." Other Awards: Tau Beta Pi, 1915; Sigma Xi, 1917.

IRWIN, JOHN DAVID, Fellow 1982. Born: August 9, 1939, Minneapolis, MN. B.E.E., 1961, Auburn University; M.S., 1962, Ph.D., 1967, University of Tennessee.

Fellow Award "For contributions to industrial electronics, control instrumentation, and engineering education"; IEEE Centennial Medal, 1984.

ISBERG, REUBEN A., Fellow 1971. Born: December 11, 1913, Chugwater, Wyo. Degrees: A.B., 1935, University of Northern Colorado.

Fellow Award "For contributions to the engineering aspects of television broadcasting, and for leadership in demonstrating the important application of television techniques to university-level teaching." Other Awards: Fellow, Audio Engineering Society, 1965; Fellow, Society of Motion Picture and Television Engineers, 1972.

ISBISTER, ERIC J., IRE Fellow 1953. Born: June 11, 1912, Brooklyn, N.Y. Degrees: B.S.E.E., 1934, Massachusetts Institute of Technology; M.E.E., 1940, Polytechnic Institute of Brooklyn.

Fellow Award "For his contributions in the development of equipment for radio navigation, aircraft radar and marine radar"; Certificate of Appreciation, IRE, Long Island Section, 1957. Other Awards: Certificate of Commendation, U.S. Dept. of Navy, 1947; Certificate of Distinction, Polytechnic Institute of Brooklyn, 1956.

ISHIMARU, AKIRA, Fellow 1973. Born: March 16, 1928, Fukuoka, Japan. Degrees: B.S., 1951, University of Tokyo; Ph.D., 1958, University of Washington.

Fellow Award "For contributions to the theory of antennas and wave propagation, and to electrical engineering education"; IEEE Region VI Achievement Award, 1968, "In recognition of his advancement of the state of the art in the field of antennas and radio propagation"; Distinguished Lecturer, IEEE, Antennas and Propagation Society, 1976-1977; IEEE Centennial Medal, 1984. Other Awards: Fellow, Optical Society of America.

ISOM, WARREN REX, Fellow 1965. Born: June 8, 1910, Mitchell, Ind. Degrees: A.B., 1931, Butler University; A.M., 1933, George Washington University; A.M., 1937, Harvard University

Fellow Award "For contributions to electromechanics and wideband magnetic recording." Other Awards: Fellow, Society of Motion Picture and Television Engineers, 1956; Fellow, Audio Engineering Society, 1971; Engineer of Distinction, Engineers Joint Council, 1974.

ITOH, TATSUO, Fellow 1982. Born: May 5, 1940, Tokyo, Japan. Degrees: B.S., 1964, M.S., 1966, Yokohama National University; Ph.D., University of Illinois.

Fellow Award "For contributions to dielectric and printed waveguide technology for millimeter-wave intergrated circuits."

IVEY, HENRY F., Fellow 1975. Born: June 16, 1921, Augusta, Ga. Degrees: A.B., 1940, M.S., 1941, University of Georgia; Ph.D., 1944, Massachusetts Institute of Technology.

Fellow Award "For contributions and leadership in the advancement of the science of luminescence." Other Awards: Honorary Member, Electrochemical Society, 1980.

IWASAKI, SHUN-ICHI, Fellow 1984.

Fellow Award "For contributions to the development of magnetic high-density recording technology."

IWERSEN, J. E., Fellow 1973. Born: April 2, 1928, Staten Island, N.Y. Degrees: B.S., 1949, Wagner College; M.A., 1951, Ph.D., 1955, The Johns Hopkins University.

Fellow Award "For contributions and leadership in the development of semiconductor devices."

IZEKI, NOBORU, Fellow 1980. Born: March 14, 1925, Amagasaki, Japan. Degrees: B.E., 1949, Dr.Eng., 1962, Kyoto University.

Fellow Award "For contributions to corona measurement techniques." Other Awards: OHM Award, OHM Award Association, Japan, 1967; Progress Award, Japanese Electrical Manufacturing Association, 1968; Invention Award, Tokyo Chapter, Japan Invention Association, 1973.

J

JACKSON, LELAND B., Fellow 1981. Born: July 23, 1940, Atlanta, GA. Degrees: S.B., 1963, S.M., 1963, Massachusetts Institute of Technology; Sc.D., 1970, Stevens Institute of Technology.

Fellow Award "For contributions to the finite word length design and hardware implementation of digital filters."

JACKSON, WILLIAM D., Fellow 1983. Born: May 20, 1927, Edinburgh, Scotland. Degrees: B.Sc., 1947, University of Glasgow; A.R.C.S.T., 1948, University of Strathclyde, Glasgow; Ph.D., 1960, University of Glasgow.

Fellow Award "For contributions to novel energy conversion systems and for the promotion of international cooperation in energy research and development." Other Awards: United Kingdom Fulbright Scholar, 1955-57; Fellow, Institution of Electrical Engineers, United Kingdom, 1974; U.S. Energy Research and Development Administration, Special Achievement Award, 1976; Societe d'encouragement pour la re-cherche et l'invention, SERI Award, 1978.

JACOBAEUS, A. CHRISTIAN, Fellow 1977. Born: June 15, 1911, Stockholm, Sweden. Degrees: M.E.E., 1933, D.E.E., 1950, Royal Institute of Technology, Stockholm, Sweden; D.Sc.(honoris causa), 1978, University of Lund.

Fellow Award "For pioneering in the theory of switching systems and technical leadership in the development of telecommunication systems"; Alexander Graham Bell Medal, 1979. Other Awards: Polhem Award, Swedish Association of Engineers and Architects, 1955; Member, Swedish Academy of Engineering Science, 1957; Member, Royal Academy of Science, 1974; Major Gold Medal, Swedish Academy of Engineering Science, 1976.

JACOBS, DONALD H., Fellow 1979. Born: August 26, 1915, New York, N.Y. Degrees: B.Sc.(Physics), 1936, Rutgers University; M.A.(Physics), 1937, Duke University.

Fellow Award "For contributions to and technical leadership in electronic computers, aircraft navigation systems, guided-missile range instrumentation, aviation fire control, and missile guidance"; IEEE Centennial Medal, 1984. Other Awards: National Honor Society, 1932; Phi Beta Kappa, 1936; Pi Mu Epsilon, 1936; Psi Chi, 1936; Sigma Pi Sigma, 1937; Listed in "Who's Who in Engineering."

JACOBS, GEORGE, Fellow 1969. Born: July 16, 1924, New York, N.Y. Degrees: B.E.E., 1949, Pratt Institute; M.S.E.E., 1960, University of Maryland.

Fellow Award "For outstanding engineering contributions in the development of international broadcasting and telecommunications." Other Awards: Telepersonality Award, International Telecommunication Union, 1964; Superior Honor Award, U.S. Information Agency, 1976; Marconi Memorial Gold Medal for Achievement, Veterans Wireless Operators Association, 1977; Fellow, Radio Club of America, 1977; Outstanding Performance Award, U.S. Board for International Broadcasting, 1980; Appointed to Presidential Commission for Broadcasting to Cuba, 1982; Registered Professional Engineer, Maryland and District of Columbia.

JACOBS, HAROLD, Fellow 1968. Born: November 21, 1917, Port Chester, N.Y. Degrees: B.A., 1938, The Johns Hopkins University; M.A., 1942, Ph.D., 1945, New York University.

Fellow Award "For his many contributions in the field of solid-state and semiconductor devices"; Harry Diamond Award, 1973, "For identification of new bulk semiconductor effects at millimeter-waves, with applications to the fields of imaging and surveillance"; Guest Editor, Special Issue on Millimeter Waves, IEEE Transactions on Microwave Theory and Techniques, 1983. Other Awards: Decoration for Exceptional Civilian Service, Dept. of the Army, 1969; U.S. Army Research and Development Achievement Award, 1978; State of New Jersey, Science Advisory Committee, 1981.

JACOBS, IRA, Fellow 1981. Born: January 3, 1931. Degrees: B.S.(Physics), 1950, City College; M.S.(Physics), 1952, Ph.D.(Physics), 1955, Purdue University.

Fellow Award "For contributions to and leadership in the development of lightwave communications systems."

JACOBS, IRWIN M., Fellow 1974. Born: October 18, 1933, New Bedford, Mass. Degrees: B.E.E., 1956, Cornell University; M.S., 1957, Sc.D., 1959, Massachusetts Institute of Technology.

Fellow Award "For contributions to information and communication theory and its applications"; Small Company of the Year Award, Region Six (Awarded to Linkabit Corp.), 1974, "For their outstanding support of IEEE activities"; Paper Award with J.A. Heller, Communications Society, 1971, for Paper "Viterbi Decoding for Satellite and Space Communication." Other Awards: Outstanding Paper Award, National Electronics Conference, 1962; NASA Patent Award; National Academy of Engineering, 1982; American Institute of Aeronautics & Astronautics (AIAA) Biannual Award, (with A. J. Viterbi), 1980.

JACOBSEN, WILLIAM E., AIEE Fellow 1951. Born: December 30, 1912, Brooklyn, N.Y. Degrees: B.E.E., Polytechnic Institute of Brooklyn.

Other Awards: Certificate of Commendation, U.S. Dept. of Navy, 1947.

JACOBSON, A. WALTER, Fellow 1966. Born: April 25, 1913, New Haven, Conn. Degrees: B.S., 1934, University of Maryland; E.E., 1940, Yale University.

Fellow Award "For contributions to the field of industrial instrumentation and control."

JACOBSON, DAVID H., Fellow 1980. Born: February 23, 1943, Johannesburg, South Africa. Degrees: B.Sc. (E.E.) (cum laude), 1963, University of Witwatersrand; Ph.D. (Eng.), 1967, Imperial College, University of London; Diploma of Imperial College, 1967.

Fellow Award "For contributions to dynamic programming and singular optimal control." Other Awards: South African Cable Maker's Association Award, 1965; Jaycee's Four Outstanding Young South Africans Award, 1974; Registered Professional Engineer and Registered Natural Scientist in South Africa; Fellow, South African Institute of Electrical Engineers (SAIEE), 1982; Formerly Post Doctoral Fellow, Assistant Professor, Associate Professor at Harvard University, Cambridge, Massachusetts; Deputy President, Council for Scientific & Industrial Research, Pretoria, South Africa.

JACOBY, GEORGE V., Fellow 1981. Born: February 26, 1918, Esztergom, Hungary. Degrees: Dipl. Ing., 1941, Royal Hungarian University of Technical Sciences.

Fellow Award "For contributions to coding and signal processing for digital magnetic recording." Other Awards: ISS/Sperry Univac Outstanding Contributor's Award, 1978; Registered Professional Engineer, Pennsylvania, 1954; 20 U.S. Patents.

JAEGER, RICHARD J., JR., Fellow 1981. Born: July 28, 1923. Degrees: B.A.(Math.) (cum laude), 1951, Hofstra University.

Fellow Award "For contributions to the development and deployment of electronic operator assistance telephone switching systems."

JAFFE, LEONARD, Fellow 1971. Born: 1926, Ohio. Degrees: E.E., 1948, Ohio State University.

Fellow Award "For contributions to the science of space telecommunications"; IEEE Award in International Communication, 1967. Other Awards: Exceptional Service Award, NASA, 1964; Arthur S. Flemming Award for Scientific and Technical Achievement, 1965; Fellow, American Astronautical Society; Member, International Astronautical Federation; Eta Kappa Nu; Tau Beta Pi; Lloyd V. Berkner Space Utilization Award, AAS, 1967; NASA Medal for Outstanding Leadership, 1972; NASA Distinguished Service Medal, 1977; Aerospace Communications Award, American Institute of Aeronautics and Astronautics, 1978; President's Commission on Accident at Three Mile Island, Exceptional Service Award, 1979.

JAKES, WILLIAM C., JR., IRE Fellow 1962. Born: May 15, 1922, Milwaukee, Wis. Degrees: B.S.E.E., 1944, M.S.E.E., 1947, Ph.D.E.E., 1949, Northwestern University; Ph.D.(Hon.), 1961, Iowa Wesleyan University.

Fellow Award "For contributions to microwave antenna research and earth-satellite communications"; Annual Paper Award, IEEE, 1971. Other Awards: Alumni Award of Merit, Northwestern University, 1962.

JALONACK, HAROLD M., AIEE Fellow 1950. Born: March 16, 1896, Chicago, Ill. Degrees: E.E., 1917, Syracuse University.

Other Awards: Charles A. Coffin Award, General Electric, 1931; Citation, Office of Price Stabilization, U.S. Government, 1952; Citation, National Electrical Manufacturers Association; Chairman, Lamme Gold Medal Committee; Chairman, ASA C57 Committee; Member, IEEE Standards Committee; Member, Transformer Standards Committee.

JAMES, IVANHOE J. P., Fellow 1975. Born: February 25, 1911, Plymouth, Devon, England. Degrees: B.Sc., 1933, Dipl.Educ., 1934, University of Bristol.

Fellow Award "For contributions and leadership in the development of television systems and color television cameras." Other Awards: British IRE Marconi Award, 1961; British IRE Rediffusion Premium (Shared), 1962; Electronic Engineering Premium, Television Society, 1962; IERE Rediffusion Premium

(Shared), 1970; Fellowship, Royal Television Society, 1973.

JAMES, JACK N., Fellow 1970. Born: November 22, 1920, Tex. Degrees: B.S.E.E., 1942, Southern Methodist University; M.S.E.E., 1948, Union College, Albany.

Fellow Award "For leadership in directing space flight projects." Other Awards: Associate Fellow, American Institute of Aeronautics and Astronautics; Sigma Tau; Sigma Xi; Member, Caltech Management Club; Louis W. Hill Space Transportation Award (with corecipient), 1962; Astronautics Engineer Award, National Rocket Club, 1963; Public Service Award, NASA, 1963; Medal for Exceptional Scientific Achievement, Presented at White House, 1965; Distinguished Engineering Alumnus, Southern Methodist University, 1966; Stuart Ballantine Medal (with co-recipient), Franklin Institute, 1967.

JAMISON, JOHN S., JR., AIEE Fellow 1959. Born: March 18, 1904, Emlenton, Pa. Degrees: B.S.E.E., 1926, Virginia Military Institute; M.S.E.E., 1934, University of Pittsburgh.

Fellow Award "For contributions to engineering education." Other Awards: Distinguished Service Certificate, National Council of State Boards of Engineering Examiners.

JANCKE, GUNNAR, Fellow 1966. Born: December 26, 1913, Uppsala, Sweden. Degrees: 1935, Royal Institute of Technology, Stockholm; Dr.(Hon.), 1961, Technical University of Dresden; Dr.(Hon.), 1966, Chalmers Institute of Technology, Gothenburg.

Fellow Award "For contributions to the development of extra high voltage power systems"; IEEE William M. Habirshaw Award, 1971, "For leadership in the creation of the world's first 400 KV extra-high voltage transmission system, including the application of series capacitors at that voltage." Other Awards: Gold Medal, Royal Swedish Academy of Engineering Sciences, 1954; Member, Royal Swedish Academy of Engineering Sciences, 1956; Gold Medal, Swedish Standards Association, 1967; Foreign Associate, National Academy of Engineering, 1977; President, CIGRE, 1972-1978.

JANES, LEONARD R., AIEE Fellow 1949. Born: January 13, 1896, LaFayette, Ill. Degrees: B.S., 1917, Northwestern University; S.B., 1921, S.M., 1923, Massachusetts Institute of Technology.

JANISCHEWSKYJ, WASYL, Fellow 1977. Born: January 21, 1925, Prague, Czechoslovakia. Degrees: B.A.Sc., 1952, M.A.Sc., 1954, University of Toronto.

Fellow Award "For contributions in power systems analysis, long-distance transmission, and high-voltage corona research."

JATLOW, JACK L., Fellow 1964. Born: April 7, 1903, Poland. Degrees: E.E., 1924, Rensselaer Polytechnic Institute.

Fellow Award "For contributions in the fields of FM carrier telegraphy and tropospheric scatter communication systems"; Chairman, Admission and Advancement Committee, IEEE, 1981; Vice Chairman, Admission & Advancement Committee, 1982, 1983, 1984; Vice Chairman, IEEE Group/Society Awards Committee, 1981; Member ADCOM, IEEE Engineering Management Society, 1981, 1982, 1983; Member of ADCOM, IEEE Society on Social Implications of Technology, 1981-1982, 1983, 1984; Region I Award, IEEE, 1983. Other Awards: Fellow, American Association for the Advancement of Science; Member, New York Academy of Sciences.

JAYANT, NUGGEHALLY S., Fellow 1982. Born: January 9, 1946, Bagalore, India. Degrees: B.Sc., 1962, University of Mysore, India; B.E., 1965, Ph.D., 1970, Indian Institute of Science, Bangalore, India.

Fellow Award "For contributions to adaptive quantization and digital speech communication"; Browder J. Thompson Award, IEEE, 1975.

JELINEK, FREDERICK, Fellow 1974. Born: November 18, 1932, Prague, Czechoslovakia. Degrees: S.B., 1956, S.M., 1958, Ph.D., 1962, Massachusetts Institute of Technology.

Fellow Award "For contributions to the theory of information transmission"; Prize Paper Award, Information Theory Group, 1969-70, "For his outstanding contribution to information theory research as author of paper entitled 'Tree Encoding of Memoryless Time-Discrete Sources with a Fidelity Criterion.'"

JENKINS, FRANK A., Fellow 1967. Born: January 5, 1920, Stanley, N.C. Degrees: B.E.E., 1941, North Carolina State University.

Fellow Award "For contributions to the design and construction of high and extra high voltage transmission systems." Other Awards: Distinguished Service Award, Charlotte Engineers Club, 1976; Decorated by Yugoslavia with Order of the Yugoslav Flag With Gold Wreath, 1977, for, "Promoting the name and work of Nikola Tesla".

JENNY, HANS K., Fellow 1966. Born: September 14, 1919, Switzerland. Degrees: M.S.E.E., 1943, Eid. Technische Hochschule, Zurich, Switzerland.

Fellow Award "For contributions to the development of microwave devices."

JENSEN, ARTHUR S., IRE Fellow 1961. Born: December 24, 1917, Trenton, N.J. Degrees: B.S., 1938, M.S., 1939, Ph.D., 1941, University of Pennsylvania; Dipl.Adv.Eng., 1972, Dipl. Computer Science, 1977, Westinghouse School of Applied Engineering Science.

Fellow Award "For contributions to the development of electronic storage tubes and special electron devices." Other Awards: Naval Reserve Medal, 1951; Armed Forces Reserve Medal, 1961; Patent Award, Westinghouse Electric, 1965, 1976, 1982; Professional Member, Maryland Academy of Sciences, 1966; Special Corporate Patent Award, Westinghouse Electric, 1972; Fellow, Washington Academy of Science, 1976; Captain, U.S.N.R., 1960; Member, Maryland State Board of Registration for Professional Engineers, 1979-1984.

JEPSEN, ROBERT L., IRE Fellow 1962. Born: June 16, 1920, Valley, Wash. Degrees: B.S.E.E., 1944, Washington State College; Ph.D., 1951, Columbia University.

Fellow Award "For contributions to microwave tubes and ultrahigh vacuum."

JESPERS, PAUL G.A., Fellow 1982. Born: September 30, 1929. Degrees: Ingenieur civil electricien-mecanicien, 1953, Universite Libre de Bruxelles, Belgium; Dr.(Applied Sc.), 1958, Universite Catholique de Louvain, Belgium.

Fellow Award "For leadership in microelectronics research, development, and education."

JESSEL, JOSEPH J. A., AIEE Fellow 1961. Born: December 20, 1907, Lawrence, Mass. Degrees: B.S., 1931, M.S., 1932, D.Sc., 1934, Harvard University.

Fellow Award "For contributions to the coordination of power generation and transmission"; Patron of Washington, D.C. Section, 1967. Other Awards: Meritorious Service Award, Federal Power Commission, 1967; Special Award, Federal Power Commission, 1969; Distinguished Service Award, Federal Power Commission, 1972.

JESTY, LESLIE C., IRE Fellow 1956. Born: August 14, 1907, Alresford, Hants, United Kingdom. Degrees: B.Sc.(First Class Honors), 1928, D.Sc., 1966, University of London.

Fellow Award "For leadership and personal contributions in the development and evaluation of television systems." Other Awards: Fellow, Institution of Electrical Engineers, U.K., 1945; Fellow, Royal Television Society, 1945; Fellow, British Kinematograph, Sound and Television Society, 1954.

JEWELL, RICHARD G., AIEE Fellow 1953. Born: October 17, 1906, Hillsboro, Wis. Degrees: B.S.E.E., 1929, University of Wisconsin.

Fellow Award "For accomplishments in the design and invention of electric measuring instruments, especially as applied to the aircraft field."

JEYNES, PAUL H., Fellow 1964. Born: June 17, 1898, Ansonia, Conn. Degrees: Ph.B., 1918, M.E., 1920, Yale University.

Fellow Award "For contributions to the economic analysis of power distribution"; Second Paper Prize, Power Division, 1952, For "Evaluation of capacity differences in the economic comparison of alternative facilities." Other Awards: Sigma Xi; McCarter Medal, 1940.

JOEL, AMOS E., JR., AIEE Fellow 1962. Born: March 12, 1918, Philadelphia, Pa. Degrees: B.S.E.E., 1940, M.S.E.E., 1942, Massachusetts Institute of Technology.

Fellow Award "For contributions to telephone switching"; Achievement Award of the Communications Society, 1972, "In recognition of his inventiveness and leadership in the field of telephone switching"; Achievement Award, IEEE Communications Society, 1972; Alexander Graham Bell Medal, 1976, "For the conception and development of Electronic Switching Systems and their effective introduction into a nation-wide telephone system" (with W. Keister and R.W. Ketchledge). Other Awards: Sigma Xi, 1940; Outstanding Patent Award (corecipient), New Jersey Research and Development Council, 1972; Stuart Ballantine Medal, Franklin Institute, 1981; National Academy of Engineering, 1981; 2nd Centenary Award, International Telecommunication Union, 1983.

JOHANNESEN, JOHN D., Fellow 1976. Born: July 7, 1918, Brooklyn, N.Y. Degrees: B.Sc.E., 1943, Bucknell University; M.S.E.E., 1948, Ph.D.(E.E.), 1953, Case Institute of Technology.

Fellow Award "For the introduction of electronic switching into telephone systems and for contributions to digital switching."

JOHANSEN, OLAV S., Fellow 1982. Born: December 9, 1917, Grimstad, Norway. Degrees: M.S., 1942, Norwegian Institute of Technologi.

Fellow Award "For leadership and contributions to international research and standardization of high-voltage equipment and systems." Other Awards: Best Contribution, Elektroteknisk Tidsskrift, 1955.

JOHNSON, ALVA A., AIEE Fellow 1950. Born: July 25, 1909, Mineral, Va. Degrees: B.S.E.E., 1930, University of Virginia; A.M.P., 1956, Harvard University.

Other Awards: Tau Beta Pi; Raven Society, University of Virginia.

JOHNSON, E. CALVIN, Fellow 1967. Born: April 18, 1926, Tampa, Fla. Degrees: B.E.E., 1947, Georgia Institute of Technology; S.M., 1949, Sc.D., 1951, Massachusetts Institute of Technology.

Fellow Award "For contributions to the field of control systems technology, numerically controlled machine tool systems, and automation." Other Awards: Outstanding Young Engineer Award, Engineering Society of Detroit, 1955; Outstanding Young Alumnus Award, Georgia Institute of Technology, 1961; Outstanding Michigan Inventor Award, Michigan Patent Law Association, 1963.

JOHNSON, EDWARD O., IRE Fellow 1960. Born: December 21, 1919, Hartford Conn. Degrees: B.S.E.E., 1948, Pratt Institute.

Fellow Award "For contributions to gaseous electronics"; Editor's Award, IRE, 1953. Other Awards: Eta

Kappa Nu.

JOHNSON, FRANCIS S., Fellow 1973. Born: July 20, 1918, Omak, Wash. Degrees: B.Sc., 1940, University of Alberta; M.A., 1942, Ph.D., 1958, University of California at Los Angeles.

Fellow Award "For numerous contributions in space research"; IEEE Outstanding Engineer, Dallas Section, 1971. Other Awards: Bronze Star Medal, U.S. Air Force, 1944; Space Science Award, American Institute of Aeronautics and Astronautics, 1966; Henry Arctowski Medal, National Academy of Sciences, 1972; Exceptional Scientific Achievement Medal, NASA, 1973; John Adam Fleming Medal, American Geophysical Union, 1977; Meritorious Civilian Service Award, U.S. Air Force, 1979.

JOHNSON, FRITHIOF V., Fellow 1977. Born: February 10, 1909, Helena, Mont. Degrees: B.S.E.E., 1930, Montana State University.

Fellow Award "For contributions in inertial sensors and avionic systems." Other Awards: Charles A. Coffin Award, 1946, Charles Proteus Steinmetz Award, 1973, General Electric Co.

JOHNSON, H. RICHARD, IRE Fellow 1962. Born: April 26, 1926, Jersey City, N.J. Degrees: B.E.E., 1947, Cornell University; Ph.D., (Physics), 1952, Massachusetts Institute of Technology.

Fellow Award "For contributions to voltage-tuned microwave oscillators." Other Awards; Member, National Academy of Engineering, 1973; Cofounder, Watkins-Johnson Company, 1957.

JOHNSON, I. BIRGER, AIEE Fellow 1958. Born: September 29, 1913, Brooklyn, N.Y. Degrees: B.E.E., 1937, M.E.E., 1939, Polytechnic Institute of New York.

Fellow Award "For leadership and contributions in the field of electric power system protection"; William M. Habirshaw Award, 1966, "For his contribution to the reliability and economy of electric power transmission, particularly as affected by insulation requirements, and to the understanding of surge voltage phenomena"; First Paper Prize, Power Division, 1956, for "Switching Surges and Arrester Performance on High-Voltage Stations"; First Paper Prize, Pittsfield Section, 1960, for "Project EHV-Insulation Design"; Prize Paper Award, Surge Protective Devices Committee, IEEE Power Engineering Society, 1979; Distinguished Service Award, IEEE, 1980. Other Awards: Graduate Fellow, Polytechnic Institute of New York, 1937-1939; Eta Kappa Nu, 1938; Tau Beta Phi, 1938; "Who's Who in the East," 1959; Sigma Xi, 1962; Registered Professional Engineer, New York, 1967; Steinmetz Award, General Electric Co., 1975; "Who's Who in Engineering," 1977; Outstanding Alumnus Award, Polytechnic Institute of New York, 1980; Member, National Academy of Engineering, 1980; Distinguished Alumnus Award, Polytechnic Institute of

New York (Brooklyn), 1983.

JOHNSON, J. KELLY, IRE Fellow 1949. Born: January 27, 1903, Oskaloosa, Iowa. Degrees: A.B., B.S., 1923, Penn College; A.B., 1924, B.S., 1926, E.E., 1927, Columbia University.

Fellow Award "In recognition of his leadership in the design and manufacture of radio broadcast receivers."

JOHNSON, J. STUART, AIEE Fellow 1959. Born: May 8, 1912, Gower, Mo. Degrees: B.S., 1932, M.S., 1934, University of Missouri; Ph.D., 1937, Iowa State University; Sc.D.(Hon.), 1963, Lawrence Institute of Technology.

Fellow Award "For contribution to engineering education." Other Awards: Honor Award, University of Missouri, 1966.

JOHNSON, J.S., AIEE Fellow 1962. Born: September 10, 1913, Englewood, N.J. Degrees: B.S., 1935, Newark College of Engineering.

Fellow Award "For contributions to the development of electrical insulation, for high voltage generators"; Second Paper Prize, 1963, for "Insulation Failure Mechanisms on Large Rotating Machines."

JOHNSON, LOERING M., Fellow 1977. Born: September 22, 1926, Dickinson, N.Dak. Degrees: B.S.E.E., 1952, University of North Dakota; M.S.Eng.Sci., 1961, Rensselaer Polytechnic Institute.

Fellow Award "For contributions to nuclear power and to professional engineering practice"; IEEE Standards Medallion, 1975. Other Awards: Engineer-of-the-Year--Industry, Award, Connecticut Society of Professional Engineers, 1983.

JOHNSON, MAJOR A., Fellow 1971. Born: August 16, 1926, Council Bluffs, Iowa. Degrees: B.S.E.E., 1948, Purdue University; M.E.E., 1962, Syracuse University.

Fellow Award "For contributions to phased-array radar theory and its applications."

JOHNSON, PAUL L., AIEE Fellow 1943.

JOHNSON, RICHARD C., Fellow 1975. Born: May 9, 1930. Degrees: B.S., 1953, M.S., 1958, Ph.D., 1961, Georgia Institute of Technology.

Fellow Award "For contributions to and leadership in antenna measurements and rapid-scan microwave antennas." Other Awards: Sigma Xi Research Award, Georgia Institute of Technology Chapter, 1974.

JOHNSON, ROBERT R., Fellow 1970. Born: 1928, Wis. Degrees: B.S.E.E., 1950, University of Wisconsin; M.E., 1951, Yale University; Ph.D., 1956, California Institute of Technology.

Fellow Award "For leadership and management of pioneering technical projects in computerized banking, and system architecture"; Outstanding Contribution to Electronics, IRE Annual Award, Phoenix Section, 1962. Other Awards: Sigma Xi; Tau Beta Pi; Eta Kappa Nu; Outstanding Young Electrical Engineer (Honorary Mention), Eta Kappa Nu, 1961; Fellow, Engineering Society of Detroit, 1980.

JOHNSON, WALTER C., AIEE Fellow 1955, IRE Fellow 1962. Born: January 6, 1913, Weikert, Pa. Degrees: B.S., 1934, E.E., 1942, Pennsylvania State University.

AIEE Fellow Award "For contributions as author, teacher and research worker in the field of electrical engineering"; IRE Fellow Award "For contributions to electrical engineering education"; Best Initial Paper Prize, AIEE, 1939. Other Awards: Western Electric Fund Award for Excellence in Engineering Education, American Society for Engineering Education, 1967.

JOHNSON, WAYNE N., AIEE Fellow 1949. Born: May 8, 1900, Riverside, Calif.

JOLLY, ROBERT M., AIEE Fellow 1962. Degrees: B.S., 1932, University of Texas.

Fellow Award "For contributions to the design of regulation and control equipment and to the operation of a utility system."

JONES, ARTHUR A., AIEE Fellow 1951. Born: July 31, 1907, Yorkshire, England. Degrees: B.Sc.E.E., M.Sc.E.E., 1930, Massachusetts Institute of Technology.

Awards: Second Paper Prizes, AIEE, Power Group, 1948, 1949. Other Awards: Certificate of Commendation, U.S. Dept. of Navy, 1948; Award of Merit, American Society for Testing and Materials, 1962; Fellow, IEE(London), 1966; Fellow, ASTM, 1962; Honorary Member, American Society for Testing and Materials, 1980; Registered Professional Engineer, New York; Chartered Engineer (Council of Engineering Institutions), London.

JONES, EDWARD M. T., Fellow 1973. Born: August 19, 1924, Topeka, Kans. Degrees: B.S., 1944, Swarthmore College; Ph.D., 1950, Stanford University.

Fellow Award "For contributions in the analysis and development of radio frequency circuit components and radiating structures." Other Awards: Sigma Xi, 1944; Sigma Tau.

JONES, EDWARD MONRAD, Fellow 1966. Born: August 4, 1921, Hanover, N.H. Degrees: B.S., 1944, Massachusetts Institute of Technology.

Fellow Award "For development of opto-electronic systems producing accurate and intricate circular patterns." Other Awards: Professional Accomplishment in Industry, Technical and Scientific Societies Council of Cincinnati, Ohio.

JONES, EDWIN C., AIEE Fellow 1950. Born: July 20, 1903, Smithburg, W.Va. Degrees: B.S.E.E., 1925, West Virginia University; M.S.E.E., 1929, University of Illinois.

JONES, HOWARD S., JR., Fellow 1970. Born: August 18, 1921, Richmond, Va. Degrees: B.S., 1943, D.Sc., 1971, Virginia Union University; M.S.E.E, 1973, Bucknell University.

Fellow Award "For contributions to microwave an-

tenna design, with particular emphasis on antenna arrays." Other Awards: Outstanding Graduate in Science, Virginia Union University, 1956; Seven Letters of Commendation from Government and Private Industry, 1956-1975; Sustained Superior Performance Award, Harry Diamond Laboratories, 1956, 1970, 1975; Special Act Award, Harry Diamond Laboratories, 1968; Inventor of the Year Award, Harry Diamond Laboratories, 1972; Secretary of the Army Fellowship for Research and Study, 1972; Army R&D Achievement Award, 1975; Army Civilian Meritorious Service Award, 1976, 1980. Listed in: "Who's Who in America," "Who's Who in Engineering," "International Who's Who in Engineering," 1984, and "Leaders in Electronics"; 1977; Fellow, American Association for the Advancement of Science; Fellow, Washington Academy of Sciences; Registered Professional Engineer.

JONES, JESSE E., AIEE Fellow 1953. Born: December 20, 1902, Gilberts, Ohio. Degrees: B.E.E., 1928, Ohio State University.

Fellow Award "For contributions made to the field of control and regulating systems as found in industrial, naval and marine applications."

JONES, LOREN F., Fellow 1971. Born: July 4, 1905, St. Louis, Mo. Degrees: B.S., 1926, Washington University; Stanford Graduate School of Business.

Fellow Award "For contributions to AM and TV broadcast engineering, air navigation, computers, and evaluations of new technology." Other Awards: Sigma Xi, 1927; Presidential Certificate of Merit, 1946; Associate Fellow, American Institute of Aeronautics and Astronautics, 1946.

JONES, MORTON E., Fellow 1982. Born: April 12, 1928, Alhambra, CA. Degrees: B.S.(Chem.), 1949, University of California at Berkeley; Ph.D.(Chem.), 1953, California Institute of Technology.

Fellow Award "For technical leadership and for pioneering contributions to compound semiconductor material and processing technology."

JONES, RICHARD W., Fellow 1965. Born: May 9, 1904, St. Paul, Minn. Degrees: B.S., 1926, University of Minnesota; M.S., 1941, Northwestern University.

Fellow Award "For contributions in the fields of physiological control systems and biomedical engineering education."

JONES, TREVOR O., Fellow 1975. Born: November 3, 1930, Maidstone, Kent, United Kingdom. Degrees: HNC(Electrical), 1952, Aston Technical College; ONC(Mechanical), 1957, Liverpool Technical College.

Fellow Award "For leadership in the application of electronics to the automobile"; Edison Lecture Medal, Cleveland Section, IEEE, 1983. Other Awards: Silver Medal, Electrical Contractors Association, U.K., 1950; IEE Hooper Memorial Prize, 1951; SAE Arch T. Colwell Merit Award, 1974, 1975; Fellow, IEE, 1974; SAE Vincent Bendix Automotive Electronics Engineering Award, 1976; Safety Award for Engineering Excellence, U.S. Department of Transportation, 1978; Fellow, Society of Automotive Engineers, 1982; National Academy of Engineering, 1982.

JORDAN, ANGEL G., Fellow 1977. Born: September 19, 1930, Pamplona, Spain. Degrees: M.S.(Applied Physics), 1952, University of Zaragoza, Spain; M.S.E.E., 1959, Ph.D.(E.E.), 1959, Carnegie-Mellon University.

Fellow Award "For educational leadership in the field of solid-state device research." Other Awards: Nato Senior Scientist Award, 1976; U.A. and Helen Whitaker Professor of Electronics and Electrical Engineering, 1972-1980; Member of the Board, Pennsylvania Science and Engineering Foundation 1981-.

JORDAN, EDWARD C., IRE Fellow 1953. Born: December 31, 1910, Edmonton, Alberta, Canada. Degrees: B.Sc., 1934, M.Sc., 1936, University of Alberta; Ph.D., 1940, Ohio State University.

Fellow Award "For his work in radio direction-finding and antenna development"; Education Medal, IEEE, 1968, "For leadership in bringing new technological developments into electrical engineering education, and in creating an integration of research with education in a major department"; Honorary Life Member, Administrative Committee, IEEE Antennas and Propagation Society. Other Awards: Centennial Award, Ohio State University, 1970; Eminent Member, Eta Kappa Nu.

JORDAN, JOHN F., IRE Fellow 1954. Born: November 23, 1908, Pittston, Pa. Degrees: E.E., 1932, University of Cincinnati.

Fellow Award "For his contributions in the application of electronics to musical instruments." Other Awards: Engineer of the Year Award, Technical and Scientific Society of America, 1960.

JORDAN, KENNETH L., JR., Fellow 1980. Born: May 10, 1933, Portland, ME. Degrees: B.S.E.E., 1954, Rensselaer Polytechnic Institute; M.S.(E.E.), 1956, Sc.D.(E.E.), 1960, Massachusetts Institute of Technology.

Fellow Award "For contributions to military satellite systems communications." Other Awards: Fulbright Scholar, 1956.

JOSHI, ARAVIND K., Fellow 1977. Born: August 5, 1929, Poona, India. Degrees: B.E., 1950, University of Poona, India; Dipl., 1951, Indian Institute of Science, Banglore, India; M.S., 1958, Ph.D., 1960, University of Pennsylvania.

Fellow Award "For contributions to man-machine communication through work in natural language processing by computer and mathematical theory of languages"; "Distinguished Visitor," IEEE Computer Group, 1970-1971. Other Awards: Guggenheim Fellow, Guggenheim Memorial Foundation, 1971-1972; Henry S. Salvatori Professor of Computer and Cognitive Sciences, 1983.

JOUBERT, LLOYD P., AIEE Fellow 1941. Born: June 9, 1892, Malvern, Iowa. Degrees: B.S., 1915, University of Washington.

JOY, HENRY M., AIEE Fellow 1949. Born: September 14, 1889. Degrees: A.B., 1911, Harvard University.

JULIEN, KENNETH S., Fellow 1983. Born: September 2, 1932, Arouca, Trinidad, Trinidad & Tobago, West-Indies. Degrees: B.Sc.(Hon.)(Elec. Eng.), University of Nottingham; Ph.D.(Elec. Eng.), University of British Colombia.

Fellow Award "For the development of power engineering educational programs and the application of engineering principles to power systems and industrial developments in the Caribbean region." Other Awards: Chaconia Medal of Trinity, Government of Trinidad & Tobago; Fellow, Association of Professional Engineers.

JURY, ELIAHU I., Fellow 1968. Born: May 23, 1923, Baghdad, Iraq. Degrees: E.E., 1947, Israel Institute of Technology; M.S., 1949, Harvard University; Eng.Sc.D., 1953, Columbia University.

Fellow Award "For his teaching and comprehensive research contributions in Sampled-data Systems, and writing of authoritative books." Other Awards: Scholar Award, Israel Institute of Technology; Higgins Fellow, Columbia University; National Science Senior Postdoctoral Fellow, Washington, D.C.; Honorary Member, Sigma Xi, 1976-1977; Senior Fulbright-Hays Fellow; Centennial Medal, American Society of Mechanical Engineers; Dr.Sc.Techn.(honoris causa), Swiss Federal Institute of Technology, Zurich, Switzerland.

K

KAAR, IRA J., IRE Fellow 1941, AIEE Fellow 1951. Born: October 17, 1902, Dunsmuir, Calif. Degrees: B.S.E.E., 1924, University of Utah.

Awards: Citation, IRE, Professional Group on Broadcasting and Television Receivers, 1959, "In recognition of his outstanding contribution to the group's progress and as an expression of our high esteem"; Director, IRE, 1939, 1948. Other Awards: U.S. Navy Commendation, World War II, 1947; Award, Radio and Television Manufacturers Association, 1957; Fellow, American Association for the Advancement of Science, 1959; Theta Tau; Tau Beta Pi; Chairman, Panel 6 NTSC (black & white)(color); Licensed Engineer, Calif., Conn., N.Y.; Chairman NTSC; Chairman, Television Committee IRE; Fellow, Amer. Inst. for the Advancement of Science; Hall of Fame, University of Utah.

KADOTA, T. THEODORE, Fellow 1980. Born: November 14, 1930, Ehime-ken, Japan. Degrees: B.S.(E.E.), 1953, Yokohama National University, Japan; M.S.(E.E.), 1956, Ph.D.(E.E.), 1960, University of California at Berkeley.

Fellow Award "For contributions to the statistical theory of detection and estimation."

KAENEL, REG A., Fellow 1975. Born: October 22, 1929, Berne, Switzerland. Degrees: Dipl.El.Ing., 1955, Sc.D., 1958, Swiss Federal Institute of Technology.

Fellow Award "For contributions to audio, electroacoustics, and signal processing." Other Awards: Moorehead-Patterson Highest Merit Award, 1977.

KAGAYA, SEIICHI, Fellow 1980. Born: January 22, 1925. Degrees: B.E., 1947, Ph.D.(Eng.), 1961, Tokyo University, Japan.

Fellow Award "For contributions to power cables."

KAHN, ALAN R., Fellow 1973. Born: March 1, 1932. Degrees: M.D., 1959, University of Illinois Medical School.

Fellow Award "For technical administrative achievements in the application of engineering principles to the fields of biology and medicine." Other Awards: Charter Member, Biomedical Engineering Society; Founding Member, Neuroelectric Society; Member, International Federation for Medical and Biological Engineering; Member, American Medical Association; Member, Aerospace Medical Association; Member, Biophysical Society; Member, Association for the Advancement of Medical Instrumentation; Alliance for Engineering in Medicine and Biology; International Institute for Medical and Biological Engineering.

KAHN, LEONARD R., Fellow 1964. Born: June 16, 1926, New York, N.Y. Degrees: B.E.E., 1951, Polytechnic Institute of Brooklyn.

Fellow Award "For contributions to single sideband transmitting and receiving systems." Other Awards: Armstrong Medal, Radio Club of America, 1980; Member, Tau Beta Pi; Eta Kappa Nu; Sigma Xi.

KAHN, ROBERT E., Fellow 1981. Born: December 23, 1938. Degrees: B.E.E., 1960, City College of New York; M.A., 1962, Ph.D., 1964, Princeton University.

Fellow Award "For original work in packet switching mobile radio telecommunications technology.

KAHN, WALTER K., Fellow 1967. Born: March 24, 1929, Mannheim, Germany. Degrees: B.E.E., 1951, Cooper Union; M.E.E., 1954, D.E.E., 1960, Polytechnic Institute of Brooklyn (now Polytechnic Institute of New York).

Fellow Award "For significant original contributions to microwave networks theory and measurement"; Certificate of Achievement for Outstanding Contribution, IEEE, TRANSACTIONS on Antennas and Propagation, 1970, for "Theory of Mutual Coupling Among Minimum-Scattering Antennas." Other Awards: Member, American Association for the Advancement of Science; Member, Optical Society of America; Member, Commissions B and C, International Scientific Radio Union; Sigma Xi; Eta Kappa Nu; Tau Beta Pi; Member, The

Philosophical Society of Washington; Member, The Society of Photo-Optical Instrumentation Engineers; NATO Senior Fellow in Science, 1973; U.S. Navy Research Publication Award, 1976.

KAHNE, STEPHEN J., Fellow 1979. Born: April 5, 1937, New York, N.Y. Degrees: B.E.E., 1960, Cornell University; M.S., 1961, Ph.D., 1963, University of Illinois.

Fellow Award "For contributions to systems engineering and engineering education"; Distinguished Member, IEEE Control Systems Society, 1983. Other Awards: John A. Curtis Lecture Award, Computers in Education Division, American Society for Engineering Education, 1975; Fellow, American Association for the Advancement of Science, 1980; Case Centennial Scholar, 1980.

KAHNG, DAWON, Fellow 1972. Born: May 4, 1931, Republic of Korea. Degrees: B.S., 1955, Seoul National University; M.Sc., 1956, Ph.D., 1959, Ohio State University.

Fellow Award "For his contributions to semiconductor electronics, especially in the development of insulated-gate field-effect transistors and the high-speed Schottky barrier diodes." Other Awards: Stuart Ballantine Medal, Franklin Institute, 1975.

KAHRILAS, PETER J., Fellow 1980. Born: July 3, 1922, New York, NY. Degrees: B.E.E., 1944, City College of New York; M.E.E., 1946, Polytechnic Institute of Brooklyn.

Fellow Award "For contributions to electronic scanning radar systems." Other Awards: Author, Electronic Scanning Radar Systems Design Handbook, 1976; Registered Professional Engineer, New York and California; 40 technical papers; listed in Who's Who in Massachusetts, California, and Engineering.

KAILATH, THOMAS, Fellow 1970. Born: June 7, 1935, Poona, India. Degrees: B.E., 1956, University of Poona; S.M., 1959, Sc.D., 1961, Massachusetts Institute of Technology.

Fellow Award "For inspired teaching of and contributions to information, communication, and control theory"; Outstanding Paper Award, Information Theory Group, 1965-66. Other Awards: Guggenheim Fellowship, 1969-70; Churchill Fellow, Cambridge, England, 1977.

KAISEL, STANLEY F., Fellow 1964. Born: August 2, 1922, St. Louis, Mo. Degrees: B.S.E.E., 1943, Washington University; M.A., 1946, Ph.D., 1949, Stanford University.

Fellow Award "For contribution to the development of traveling wave tubes."

KAISER, JAMES F., Fellow 1973. Born: December 10, 1929, Piqua, Ohio. Degrees: E.E., 1952, University of Cincinnati; S.M., 1954, Sc.D., 1959, Massachusetts Institute of Technology.

Fellow Award "For contributions in digital signal processing, and the synthesis of digital filters"; Technical Achievement Award, IEEE Acoustics, Speech and Signal Processing Society, 1977; Meritorious Service Award, IEEE Acoustics, Speech, and Signal Processing Society, 1978; Special Award, IEEE Acoustics, Speech, and Signal Processing Society, 1980; Society Award, IEEE Acoustics, Speech, and Signal Processing Society, 1981; IEEE Centennial Medal, 1984. Other Awards: Distinguished Engineering Alumnus Award, University of Cincinnati, 1970; Eta Kappa Nu Award of Merit, University of Cincinnati, 1980; Distinguished Technical Staff Award, Bell Laboratories, 1982.

KAISER, WOLFGANG A., Fellow 1980. Born: February 22, 1923, Schoental, Germany. Degrees: Diplomingenieur, 1951, Dr. Ingenieur, 1955, University of Stuttgart, Germany.

Fellow Award "For contributions to advanced communication systems."

KAKIZAKI, KENJI, Fellow 1977. Born: March 2, 1919, Korea. Degrees: B.E., 1942, D.Eng., 1959, Waseda University, Tokyo, Japan.

Fellow Award "For development of traveling-wave tubes and for leadership in development of semiconductor memories for computers." Other Awards: Award, Japan Society for Promotion of Machine Industry, 1969; Special Award, from the President of Toshiba Corp., 1970.

KALB, JOHN W., AIEE Fellow 1959. Born: June 16, 1918, Columbus, Ohio. Degrees: B.S., 1940, Swarthmore College.

Fellow Award "For contributions to lightning protection of electric power systems, particularly the development of the magnetic gap arrester"; First Paper Prize, AIEE, Power Division, 1962; Power Group Award, 1968; Prize Paper, Power Engineering Society, 1971. Other Awards: Member, National Academy of Engineering, 1980.

KALB, ROBERT M., Fellow 1968. Born: April 30, 1904, Fostoria, Ohio. Degrees: B.E.E., 1927, Ohio State University.

Fellow Award "For pioneering contributions in real-time information-processing applications, and for sustained service to the profession in this field"; Certificate of Appreciation, Communication Technology Group, 1968, "For services rendered in furthering the objectives of the IEEE." Other Awards: Fellow, American Association for the Advancement of Science.

KALBACH, JOHN FREDERICK, Fellow 1970. Born: January 2, 1914, Seattle, Wash. Degrees: B.S.E.E., 1937, University of Washington.

Fellow Award "For contributions to the design and production of commercial data processing systems." Other Awards: Fellow, Institute for Advancement of Engineering, 1973.

KALBFELL, DAVID C., Fellow 1971. Born: August 20, 1914, Indiana. Degrees: A.B., 1934, University of California at Los Angeles; M.S., 1936, Ph.D., 1939, University of California at Berkeley.

Fellow Award "For contributions to instrumentation design and production."

KALMAN, R. E., Fellow 1964. Born: May 19, 1930, Budapest, Hungary. Degrees: D.Sc., 1957, Columbia University.

Fellow Award "For contributions to the theory of system control." IEEE Medal of Honor, 1974. Other Awards: Outstanding Young Scientist Award, Maryland Academy of Sciences, 1962; Rufus Oldenburger Medal, American Society of Mechanical Engineers, 1976; Honorary Member, Hungarian Academy of Sciences, 1976.

KAMAL, ADITYA K., Fellow 1975. Born: July 5, 1927, Kairana (M.Nagar), India. Degrees: B.Sc., 1946, Benares Hindu University, India; M.Sc., 1948, Allahabad University, India; D.I.I.Sc., 1951, Indian Institute of Science, India Dr.Ing., 1957, University of Paris, France.

Fellow Award "For leadership in establishing postgraduate teaching and research facilities, and research contributions in microwave and quantum electronics." Other Awards: Fellow, Institution of Engineers, India; Fellow, Academy of Sciences, India.

KAMINOW, IVAN P., Fellow 1974. Born: March 3, 1930, Union City, N.J. Degrees: B.S.E.E., 1952, Union College; M.S.E., 1954, University of California at Los Angeles; A.M., 1957, Ph.D., 1960, Harvard University.

Fellow Award "For contributions to the science and application of electro-optic materials"; IEEE Quantum Electronics Award, 1983. Other Awards: Fellow, American Physical Society, 1975; Fellow, Optical Society of America, 1976; Distinguished Technical Staff Award, Bell Lab, 1982.

KANAL, LAVEEN N., Fellow 1972. Born: September 29, 1931, Dhond, India. Degrees: B.S.E.E., 1951, M.S.E.E., 1953, University of Washington; Ph.D., 1960, University of Pennsylvania.

Fellow Award "For contributions to the theory of statistical pattern classification." Other Awards: Fellow, American Association for the Advancement of Science, 1972.

KANDOIAN, ARMIG G., IRE Fellow 1951. Born: November 28, 1911, Van, Armenia. Degrees: B.S., 1934, M.S., 1935, Harvard University; D.Eng.(Hon.), 1967, New Jersey Institute of Technology.

Fellow Award "For his important contributions to the development of antennas and navigation systems"; Achievement Award, Group on Communication Technology, 1965, "For his leadership in the planning and engineering of vital, large-scale communication systems and for promoting and managing the organization of important research and development teams in the field of systems engineering for global networks'"; IEEE Award in International Communication in Honor of Hernand and Sosthenes Behn, 1980; Chairman, IEEE Telecommunications Policy Committee; Chairman, Editorial Board, Reference Data for Radio Engineers. Other Awards: Tau Beta Pi, 1933; Honorable Mention Award, Outstanding Young Electrical Engineer, Eta Kappa Nu, 1943; Professor Dikran Kabakjian Science Award, Armenian Students Association, 1962.

KANEKO, HISASHI, Fellow 1981. Born: November 19, 1933. Degrees: B.S. (E.E.), 1956, University of Tokyo; M.S. (E.E.), 1962, University of California at Berkeley; Dr. Eng., 1967, University of Tokyo.

Fellow Award "For contributions to and leadership in digital communication systems." Other Awards: Inada Memorial Prize, Institute of Electronics and Communication Engineers, Japan, 1962; Best Paper Award, Institute of Electronics and Communication Engineers, Japan, 1967; Kajii Memorial Prize, Electrical Commmunication Association, Japan, 1979.

KANNO, MAKOTO, Fellow 1984. Born: April 1, 1925, Sendai, Japan. Degrees: Ph.D., 1961, Tohoku University.

Fellow Award "For contributions and leadership in the development of precise capacitance standards and research in precision electrical measurements." Other Awards: Prize of Progress, The Institute of Electrical Engineers of Japan, 1960.

KANOUSE, EDGAR L., AIEE Fellow 1958. Born: August 23, 1910, Amarillo, Tex. Degrees: B.S., 1932, University of Oklahoma; M.S., 1934, California Institute of Technology; E.E., 1938, Ph.D., 1941, Stanford University.

Fellow Award "For contributions to the art of high voltage power transmissions." Other Awards: Eminent Member, Eta Kappa Nu, 1973.

KANTEBET, S. R., IRE Fellow 1949. Born: December 9, 1900, Kanara, Bombay, India. Degrees: B.A., 1921, St. Xavier's College, Bombay.

Fellow Award "For his services as an educator, engineer and administrator in the fields of radio and cable communication in India." Other Awards: Fellow, Institution of Telecommunication Engineers, 1964; Institution of Electrical Engineers, U.K., 1967.

KAO, CHARLES K., Fellow 1979. Born: November 4, 1933, Shanghai, China. Degrees: B.Sc., 1957, Ph.D., 1965 (both in Electrical Engineering), University of London.

Fellow Award "For contributions to the practical use of optical waveguides for communications"; Morris Liebmann Award, IEEE, 1978. Other Awards: Morey Award, American Ceramic Society, 1976; Stewart Ballantine Medal, Franklin Institute, 1977; Rank Prize, 1978; L.M. Ericsson Prize Award, 1979; Gold Medal, Armed Forces Communications and Electronics Association, 1980; Achievement Award, Chinese Engineers and Scientists Association of Southern California, 1981; U.S.-Asia Institute 1983 Achievement Technology Award.

KAPLAN, SAMUEL H., Fellow 1979. Born: January 16, 1915, Chicago, Ill. Degrees: B.S., 1937, Armour Institute of Technology.

Fellow Award "For contributions to the theory and development of shadow-mask color TV picture tubes"; Vladimir K. Zworykin Award, 1978. Other Awards: IR-100 Award, 1970; Eugene McDonald Award, Zenith Radio Corporation, 1970; Electrochemical Society, Society for Information Display, Society of Motion Picture and Television Engineers.

KAPPEL, FREDERICK R., AIEE Fellow 1950. Born: January 14, 1902, Albert Lea, Minn. Degrees: B.S.E., 1924, University of Minnesota; LL.D.(Hon.), 1958, Lehigh University; LL.D.(Hon.), 1959, Knox College; D.C.L.(Hon.), 1959, Union College; LL.D.(Hon.), 1960, Ohio Wesleyan University; D.Eng.(Hon.), 1961, Rensselaer Polytechnic Institute; LL.D.(Hon.), 1962, Columbia University; LL.D.(Hon.), 1962, Williams College; LL.D.(Hon.), 1963, University of New Hampshire; LL.D.(Hon.), 1963, Michigan State University; LL.D.(Hon.), 1965, Pace College; D.Eng.(Hon.), 1966, University of Minnesota; LL.D.(Hon.), 1966, Butler University; LL.D.(Hon.), 1967, Colgate University; LL.D.(Hon.), 1967, Yale University.

Other Awards: Outstanding Achievement Award, University of Minnesota, 1954; Business Leaders Citation, FORBES Fifty Foremost, 1957; Gold Medal, Economic Club of New York, 1959; Annual Achievement Award, Deafness Research Foundation, 1961; Cross of the Commander of the Postal Award, Republic of France, 1962; Distinguished Life Member, American Society for Metals, 1962; Citation, SATURDAY REVIEW, 1962; Silver Quill Award, National Business Publications, 1963; Presidential Medal of Freedom, 1964; Award of Merit, Institute of Consulting Engineers, 1964; Gold Medal Award, National Institute of Social Sciences, 1964; John Fritz Medal, 1965; Man in Management Award, Pace College, 1966; Michael I. Pupin Medal, Columbia University, 1966; Business Leadership Award, University of Michigan, 1966; Equal Opportunity Medal, National Urban League, 1966; Distinguished Public Service Award, Tax Foundation, 1967; Chair endowed at University of Minnesota, 1967; Henry L. Gantt Memorial Gold Medal, American Management Association and American Society of Mechanical Engineers, 1970; Herbert Hoover Medal, 1972; Captain of Industry Award, WALL STREET JOURNAL, 1973; Eta Kappa Nu; Phi Sigma Phi; Tau Beta Pi; Beta Gamma Sigma.

KAPRIELIAN, ELMER F., Fellow 1979. Born: January 12, 1922, Fresno, Calif. Degrees: B.E.(Electrical Engineering), 1942, University of Southern California.

Fellow Award "For leadership and management of power-system operations"; IEEE Centennial Medal, 1984. Other Awards: Eminent Engineer Award, Tau Beta Pi, 1977.

KARADY, GEORGE, Fellow 1978. Born: August 17, 1930. Degrees: E.E., 1952, Technical University, Budapest, Hungary; Candidate of Technical Sciences, 1958, Hungarian Academy of Sciences; Ph.D.E.E., 1960, Technical University, Budapest, Hungary.

Fellow Award "For advancement in the art of high-voltage dc transmission."

KARAKASH, JOHN J., Fellow 1966. Born: June 14, 1914. Degrees: B.S.E.E., 1937, Duke University, M.Sc.E.E., University of Pennsylvania; Eng.D.(Hon.), Lehigh University.

Fellow Award "For contributions to engineering education and the achievement of effective college-industry cooperation in research"; IEEE Centennial Medal, 1984. Other Awards: Alfred Noble Robinson Award, 1948; Distinguished Professor of Electrical Engineering, 1961; Hillman Award, Lehigh University, 1962, 1981; Distinguished Engineer Award, Professional Engineers Society, Outstanding Teaching Award, 1968; Faculty Award for Outstanding Service, 1975.

KARNAUGH, MAURICE, Fellow 1976. Born: October 4, 1924, New York N.Y. Degrees: B.S., 1948, City College of the City University of New York; M.S., 1950, Ph.D., 1952, Yale University.

Fellow Award "For contributions to the understanding and application of digital techniques in telecommunications."

KARNI, SHLOMO, Fellow 1976. Born: June 23, 1932, Poland. Degrees: B.S.E.E.(cum laude), 1956, Technion, Israel Institute of Technology, Israel; M.Eng., 1957, Yale University; Ph.D., 1960, University of Illinois, Urbana.

Fellow Award "For contributions to circuit theory and engineering education"; "Teacher of the Year," IEEE Student Branch, University of New Mexico, 1965-66, 1967-68, 1977; "Teaching Excellence Award," Univ. of New Mexico Engineering Alumni, 1983.

KARP, SHERMAN, Fellow 1982. Born: October 12, 1935, Brooklyn, NY. Degrees: B.S.E.E., 1960, M.S.E.E., 1962, Massachusetts Institute of Technology; Ph.D., 1967, University of Southern California.

Fellow Award "For contributions to the advancement of optical communication theory and the development of the Blue-Green optical satellite communication system for submarine communcations." Other Awards: Scientist of the Year, Naval Ocean Systems Center, 1976; Tau Beta Pi; Sigma Xi; Eta Kappa Nu.

KARPLUS, WALTER J., Fellow 1971. Born: April 23, 1927, Vienna, Austria. Degrees: B.E.E., 1949, Cornell University; M.S., 1951, University of California at Berkeley; Ph.D., 1955, University of California at Los Angeles.

Fellow Award "For contributions to the theory and application of analog and hybrid computers and to engineering education." Other Awards: Senior Scientific Simulation Award, Society for Computer Simulation, 1972; Silver Core Award, International Federation of Information Processing Societies, 1980.

KARRER, LAWRENCE E., AIEE Fellow 1962. Born: December 22, 1904, Roslyn, Wash. Degrees: B.S.(magna cum laude), 1927, University of Washington.

Fellow Award "For contributions to planning and directing programs for utility expansion." Other Awards: Award, Seattle World's Fair, 1962; Award, Washington State Research Council, 1970; Award, Northwest Kidney Center, 1971; Member, Tau Beta Pi; Sigma Xi.

KASAMI, TADAO, Fellow 1975. Born: April 12, 1930, Kobe, Japan. Degrees: B.E., 1958, M.E., 1960, Ph.D., 1963, Osaka University.

Fellow Award "For contributions to the theory of error-correcting codes, automata theory, and formal languages, and to engineering education."

KASHYAP, RANGASAMI L., Fellow 1980. Born: March 28, 1938, Mysore, India. Degrees: D.I.I.Sc., 1960, M.E., 1962, Indian Institute of Science; Ph.D., 1966, Harvard University.

Fellow Award "For contributions to pattern recognition, and to the optimization of finite state systems." Other Awards: Best Research Paper Award, National Electronic Conference, 1966.

KASSON, JAMES M., Fellow 1981. March 19, 1943, Muncie, IN. Degrees: B.S.E.E., 1964, Stanford M.S.E.E., 1965, University of Illinois.

Fellow Award "For contributions to and leadership in the design of digital switching systems for customer premises business communications systems."

KATAOKA, SHOEI, Fellow 1984. Born: April 5, 1929, Fukushima, Japan. Degrees: B.Eng., 1953, University of Tokyo; M.Sc.(Eng.), 1957, University of London; Dr.Eng., 1966, University of Tokyo; D.Sc.(Eng.), 1978, University of London.

Fellow Award "For contributions in the field of compound semiconductor devices." Other Awards: Kelvin Premium, IEE(London), 1964; Inada Prize, IECE of Japan, 1963; The First Prize of the Association for the Promotion of Machine Industry of Japan, 1965; Distinguished Invention Prize, Japanese Government, 1970, 1972, 1974.

KATZ, ISADORE, Fellow 1970. Born: October 21, 1916. Degrees: B.S., 1937, Temple University.

Fellow Award "For contributions and research leadership in radar reflection from the sea and from the atmosphere."

KATZEFF, KURT, Fellow 1979. Born: June 22, 1926, Stockholm, Sweden. Degrees: Master of Electrical Engineering (Civilingenjor), 1950, Royal Institute of Technology (Kungl-Tekniska Hogskolan), Stockholm.

Fellow Award "For leadership in the development of telecommunications switching systems." Other Awards: Gold Medal, The Swedish Academy of Engineering Sciences, 1967; Member, The Swedish Academy of Engineering Sciences, 1983.

KATZENELSON, JACOB, Fellow 1980. Born: January 14, 1935, Afula, Israel. Degrees: B.Sc.(E.E.) (summa cum laude), 1957, M.Sc.(E.E.), 1959, Technion-Israel Institute of Technology, Haifa; Sc.D., 1962, Massachusetts Institute of Technology.

Fellow Award "For contributions to simulation and computer-aided design." Other Awards: Joseph and Sadie Riesman Professorship of Electrical Engineering, Technion- Israel Institute of Technology, 1978.

KATZER, HERIBERT J., AIEE Fellow 1962. Born: January 5, 1904, Vienna, Austria. Degrees: E.E., 1926, Technische Hochschule Vienna.

Fellow Award "For contributions in design and operation of electric utility systems in Brazil."

KAUFMAN, IRVING, Fellow 1973. Born: January 11, 1925. Degrees: B.E., 1945, Vanderbilt University; M.S., 1949, Ph.D., 1957, University of Illinois.

Fellow Award "For contributions to microwave electronics, and to education"; Annual Achievement Award, Phoenix Section, 1968, "For outstanding contributions to electrical engineering." Other Awards: Fulbright Senior Research Fellow, 1964-65, 1973-74; Alumni Award for Distinguished Career in Electronics, Electro-Physics Laboratory, University of Illinois, 1976; Sigma Xi; Eta Kappa Nu; Tau Beta Pi; Pi Mu Epsilon.

KAUFMAN, RICHARD B., AIEE Fellow 1962. Born: November 15, 1912, Baltimore, Md. Degrees: B.S.E.E., 1934, The Johns Hopkins University.

Fellow Award "For contributions to improved insulation and design of large transformers"; Member Emeritus, IEEE Power Engineering Society Transformers Committee.

KAUTZ, WILLIAM H., Fellow 1971. Born: 1924, Wash. Degrees: B.S.E.E., 1948, M.S.E.E., 1948, M.A.(Math.), 1949, Sc.D., 1951, Massachusetts Institute of Technology.

Fellow Award "For contributions to the fields of switching networks and error-correcting codes, and to engineering education."

KAWAHASHI, TAKESHI, Fellow 1983. Born: February 21, 1921, Nagasaki, Japan. Degrees: B.S.(Elec. Eng.), 1943, Dr. Eng., 1959, University of Tokyo.

Fellow Award "For developments in microwave communications and in satellite communications earth stations"; Leonard G. Abraham Prize Paper Award, IEEE, 1977. Other Awards: Achievement Award, Institute of Electronics and Communication Engineers of Japan (IECE), 1979; Technical Paper Prize, IECE of Japan, 1959; Purple Ribbon Medal, His Majesty the Emperor of Japan, 1971; Minister of Posts and Telecommunication Award, Minister of Posts and Telecommunication, 1979; Director-General Prize, Science and Technology Agency of Japan, 1965; Asahi News Paper's Prize of Invention, Japanese Association of Invention, 1965.

KAWASHIMA, MASAO, Fellow 1984. Born: January 14, 1929, Yokohama, Japan. Degrees: B.Eng.(Electronic Communications), 1952, Ph.D.Eng., 1952, Osaka University, Japan.

Fellow Award "For contributions to the development of digital transmission systems"; IEEE Centennial Medal, 1984. Other Awards: The Fourth OHM Technical Prize, The Foundation of Electrical Science and Technology of Japan, 1961; The Excellent Paper Prize, 1967, The Excellent Book Prize, 1975, The Institute of Electronics and Communication Engineers of Japan; The Excellent Invention Prize, 1974, The Invention Society of Japan; The Industrial Achievement Prize for the PCM-100M System Development, 1976, The FUJITSU LIMITED.

KAYTON, MYRON, Fellow 1981. Born: April 26, 1934. Degrees: B.S.(Mech.Eng.), 1955, Cooper Union; S.M.(E.E.), 1956, Harvard University; Ph.D.(Instrumentation), 1960, Massachusetts Institute of Technology.

Fellow Award "For contributions to the design of avionic systems for spacecraft and aircraft." Other Awards: Fellow, National Science Foundation, 1957-1960; Outstanding Young Men of America, 1965; Tau Beta Pi; Sigma Xi; Gano Dunn Medal, 1975; Cooper Union Citation, 1979.

KAZAN, BENJAMIN, IRE Fellow 1962. Born: May 8, 1917, New York, N.Y. Degrees: B.S., 1938, California Institute of Technology; M.A., 1940, Columbia University; Dr.Rer.Nat., 1961, Technical University of Munich, Germany.

Fellow Award "For contributions to the field of storage and display tubes." Other Awards: Silver Medal, American RoentgenRay Society, 1957; Coolidge Award, General Electric, 1958.

KAZDA, LOUIS F., Fellow 1965. Born: September 21, 1916, Dayton, Ohio. Degrees: E.E., 1940, M.S.E., 1943, University of Cincinnati; Ph.D., 1962, Syracuse University.

Fellow Award "For contributions to automatic control systems and engineering education"; First Paper Prize, AIEE 1960; Distinguished Member Award, IEEE Control Systems Society, 1983; IEEE Centennial Medal, 1984. Other Awards: Distinguished Alumnus Award, University of Cincinnati, 1981.

KEAR, FRANK G., IRE Fellow 1953. Born: October 18, 1903, Minersville, Pa. Degrees: E.E., 1926, Lehigh

University; S.M., 1926, Sc.D., 1933, Massachusetts Institute of Technology.

Fellow Award "In recognition of his contributions in the fields of radio aids to air navigation and radio and television broadcasting." Other Awards: Certificate of Commendation, U.S. Dept. of Navy, 1945.

KEARNEY, JOSEPH W., Fellow 1964. Born: April 6, 1922, Denver, Colo. Degrees: B.S.E.E., 1943, University of Colorado.

Fellow Award "For contributons to the development of electronic reconnaissance systems and to student activities."

KEIGLER, JOHN E., Fellow 1984. Born: July 10, 1929, Baltimore, MD. Degrees: B.E., 1950, M.S., 1951, Johns Hopkins University; Ph.D., 1958, Stanford University.

Fellow Award "For leadership and contributions to communications satellite technology." Other Awards: David Sarnoff Award, 1976.

KEISER, BERNHARD E., Fellow 1980. Born: November 14, 1928, Richmond Heights, MO. Degrees: B.S.E.E., 1950, M.S.E.E., 1951, D.Sc.E.E., 1953, Washington University, Missouri.

Fellow Award "For contributions to spacecraft electromagnetic compatibility."

KEISTER, JAMES E., IRE Fellow 1957. Born: July 11, 1914, Coburg, Iowa. Degrees: E.E., 1935, Cornell University.

Fellow Award "For contributions in the field of television transmitters and military electronic equipment." Other Awards: Eta Kappa Nu, 1934; Tau Beta Pi, 1934; Navy E, U.S. Dept. of Navy, 1944; Certificate of Commendation, U.S. Dept. of Navy, 1947; Apollo Achievement Award, NASA, 1969; Public Service Award, NASA, 1969.

KEISTER, WILLIAM, Fellow 1966. Born: June 6, 1907, Montgomery, Ala. Degrees: B.S.E.E., 1930, Auburn University.

Fellow Award "For fundamental contributions to switching theory"; Alexander Graham Bell Medal, 1976, "For the conception and development of Electronic Switching Systems and their effective introduction into a nation-wide telephone system" (with A.E. Joel, Jr. and R.W. Ketchledge); Ben S. Gilmer Award, Auburn University School of Engineering, 1976.

KEITHLEY, JOSEPH F., IRE Fellow 1962. Born: August 3, 1915, Peoria, Ill. Degrees: S.B., 1937, S.M., 1938, Massachusetts Institute of Technology.

Fellow Award "For contributions to electronic instrumentation." Other Awards: Distinguished Civilian Service Award, U.S. Dept. of Navy, 1945.

KELL, RAY D., IRE Fellow 1947. Born: June 7, 1904, Kell, Ill. Degrees: B.S., 1926, University of Illinois.

Fellow Award "For his extensive contributions over many years in television for both civilian and military use"; Vladimir K. Zworykin Award, 1966, "For his extensive and significant contributions, papers, and inventions which have been fundamental in the development of both black-and-white and color television." Other Awards: Modern Pioneer Award, National Association of Manufacturers, 1940; Certificate of Commendation, U.S. Dept. of Navy, 1946; Stuart Ballantine Medal, Franklin Institute, 1948; David Sarnoff Gold Medal, Society of Motion Picture and Television Engineers, 1954; David Sarnoff Outstanding Achievement Award, Radio Corporation of America, 1962; Certificate of Appreciation, National Association of Broadcasters, 1976. Sigma Xi, 1939; Certificate of Commendation, Bureau of Ships for Outstanding Service During WWII, 1947; College of Engineering Alumni Honor Award, 1974; Distinguised Alumnus Award, The Electrical Engineering Alumni Association, University of Illinois, 1973; N.Y. Academy of Science Recognition, 1979.

KELLEHER, JOHN J., Fellow 1980. Born: October 14, 1914, Lynbrook, NY.

Fellow Award "For contributions to international radio regulations." Other Awards: Department of Army Decoration for Meritorious Civilian Service, 1960; NASA Exceptional Service Medal, 1969.

KELLER, ARTHUR C., IRE Fellow 1961. Born: August 18, 1901, New York, N.Y. Degrees: B.S., 1923, Cooper Union; M.S., 1925, Yale University; E.E., 1926, Cooper Union; Columbia University, 1926-1930.

Fellow Award "For contributions to the recording and reproduction of stereophonic sound and to telephone switching"; First Prize Paper AIEE, Communications Division, 1954, for "A New General Purpose Relay for Telephone Switching Systems." Other Awards: Award, U.S. Dept. of Navy, Bureau of Ships, 1947; Award, U.S. Dept. of Navy, Bureau of Naval Ordnance, 1947; Emile Berliner Award, Audio Engineering Society, 1962; Achievement Award, First Paper Prize, Fellow, National Association of Relay Manufacturers, 1966; Fellow, Acoustical Society of America; Gold Medal, Audio Engineering Society, 1981; 40 U.S. patents; 150 Worldwide patents; Am. Physical Soc.; Audio Eng. Soc.; Yale Eng. Assoc.

KELLEY, ALBERT J., Fellow 1967. Born: July 27, 1924, Boston, Mass. Degrees: B.S., 1945, United States Naval Academy; B.S., 1948, Sc.D., 1956, Massachusetts Institute of Technology.

Fellow Award "For contributions to military and space guidance and control systems, and for leadership in space electronics research and technology." Other Awards: Fellow, American Institute of Aeronautics and Astronautics; Armed Forces Communications and Electronics Association; Exceptional Service Medal, NASA, 1967.

KELLEY, WENDELL J., Fellow 1972. Born: May 2, 1926, Champaign, Ill. Degrees: B.S.E.E., 1949, University of Illinois.

Fellow Award "For contributions to the electrical utility industry with special emphasis on reliability of service and protection of the environment"; Chairman, Central Illinois Section, IEEE. Other Awards: President, Electrical Engineering Alumni Association, University of Illinois, 1969-1971; Distinguished Alumnus Award, Electrical Engineering Dept., University of Illinois, 1973; College of Engineering Award for Distinguished Service in Engineering, University of Illinois, 1974; Alex Van Praag, Jr. Distinguished Engineering Service Award, Central Illinois Chapter of Illinois Society of Professional Engineers, 1983; Member, Eta Kappa Nu; National Society of Professional Engineers; Illinois Society of Professional Engineers;

KELLING, LEROY U. C., Fellow 1967. Born: August 12, 1913, Milwaukee, Wis. Degrees: B.S.E.E., 1941, University of Wisconsin.

Fellow Award "For contribution to the development and design of numerical control for machine tools."

KELLY, KENNETH C., Fellow 1972. Born: March 6, 1928, New York, N.Y. Degrees: B.S.E.E., 1953, Brooklyn Polytechnic Institute; M.S., 1963, University of California at Los Angeles.

Fellow Award "For his contributions to antenna theory and design and to the profession in guiding youth and furthering equal opportunities."

KELSO, JOHN M., Fellow 1974. Born: March 12, 1922, Punxsutawney, Pa. Degrees: B.A., 1943, Gettysburg College; M.S., 1945, Ph.D., 1949, Pennsylvania State University.

Fellow Award "For contributions in the fields of radio wave propagation and ionospheric physics." Other Awards: Pi Mu Epsilon, 1944; Sigma Pi Sigma, 1944; Sigma Xi, 1949.

KENNEDY, DAVID P., Fellow 1974. Born: November 15, 1923, Boston, Mass. Degrees: 1941, Massachusetts Institute of Technology; 1943, University of Chicago; 1946, University of Oklahoma.

Fellow Award "For contributions to computer-aided design of semiconductor devices and to the quantitative design tools for devices and processes."

KENNEDY, ROBERT S., Fellow 1975. Born: 1933, Augusta, Kans. Degrees: B.S., 1955, University of Kansas; S.M., 1959, Sc.D., 1963, Massachusetts Institute of Technology.

Fellow Award "For contributions to the analysis of fading and dispersive communications channels and to the field of optical communication through the atmosphere."

KENT, HARRY E., AIEE Fellow 1950. Born: August 22, 1901, Providence, R.I. Degrees: B.S.E.E., 1923, M.S.E.E., 1924, Massachusetts Institute of Technolo-gy.

KEONJIAN, EDWARD, Fellow 1965. Born: August 14, 1909, Tiflis, Russia. Degrees: M.S., Ph.D.(equivalent), 1932, Leningrad Institute of Electrical Engineering, U.S.S.R.

Fellow Award "For contributions to transistor circuitry and to micro-electronics." Other Awards: Distinguished Colleague Award, Aerospace Industries Association, 1974; Grant, Visiting Professor, Cairo University; U.S. National Science Foundation, 1976; Professional Engineer in the State of New York, 1953; Member, N.Y. Academy of Science, 1978; The Explorers, 1982; The Circumnavigators, 1981.

KERCHNER, RUSSELL M., AIEE Fellow 1957. Born: July 29, 1899, Belleville, Ill. Degrees: B.S., 1922, University of Illinois; M.S., 1927, Kansas State College.

Fellow Award "For distinction as a teacher of electrical engineering and as an author of textbooks in this field"; Award, IEEE, Kansas City Section, 1964, "For outstanding service and contributions toward the advancement of science and technology." Other Awards: Tau Beta Pi; Phi Kappa Phi; Eta Kappa Nu.

KERR, ANTHONY R., Fellow 1984. Born: August 30, 1941, England. Degrees: B.E.(Hon.), 1964, M.Eng.Sc.(Hon.), 1967, Ph.D., 1969, University of Melbourne, Australia.

Fellow Award "For contributions to millimeter-wave receivers"; 1978 IEEE Microwave Prize, 1978. Other Awards: NASA Exceptional Engineering Achievement Medal, 1983.

KERTZ, HUBERT L., AIEE Fellow 1960. Born: July 11, 1910, San Francisco, Calif. Degrees: A.B., 1934, E.E., 1936, Stanford University.

Fellow Award "For contributions to the design, construction and administration of a large complex communication network."

KERWIN, WILLIAM JAMES, Fellow 1982. Born: September 27, 1922. Degrees: A.A., 1942, San Bernardino Valley College; B.S., 1948, Redlands University; M.S., 1954, Ph.D., 1967, Stanford University.

Fellow Award "For contributions to the theory and practical realization of active RC circuits."

KESSELMAN, WARREN A., Fellow 1983. Born: April 8, 1927, Newark, N.J. Degrees: B.Sc.(E.E.), 1950, New Jersey Institute of Technology.

Fellow Award "For contributions to electromagnetic compatibility and interference measurement techniques"; Laurence G. Cumming Award, IEEE EMCS, 1981.

KETCHLEDGE, RAYMOND W., Fellow 1965. Born: December 8, 1919, Harrisburg, Pa. Degrees: B.S., 1942, M.S., 1942, Massachusetts Institute of Technology.

Fellow Award "For his inventions and leadership in electronic telephone switching"; Alexander Graham

Bell Medal, 1976, "For the conception and development of Electronic Switching Systems and their effective introduction into a nation-wide telephone system" (with A.E. Joel, Jr. and W. Keister). Other Awards: Member, National Academy of Engineering, 1970.

KETO, JOHN E., IRE Fellow 1948. Born: June 9, 1909, Maynard, Mass. Degrees: E.E., 1932, M.S., 1935, University of Cincinnati, D.Sc. (Hon.), 1961, Bradley University.

Fellow Award "In recognition of his contributions to the development of airborne radar equipment for the armed services during World War II." Other Awards: Presidential Medal of Merit, 1946; Exceptional Civilian Service Award, U.S. Dept. of Air Force, 1959; Award of Merit, Eta Kappa Nu; Sigma Xi; Tau Beta Pi.

KEY, EDWIN LEE, Fellow 1979. Born: May 7, 1926. Degrees: B.S.E.E., 1951, St. Louis University.

Fellow Award "For contributions to the theory of modern radar and leadership in the development of military electronic systems."

KEYES, ROBERT W., Fellow 1976. Born: December 2, 1921, Chicago, Ill. Degrees: B.S., 1942, Ph.D., 1953, University of Chicago.

Fellow Award "For contributions to solid-state physics, and for the development of a theory defining the physical limits in digital computer devices"; W. R. G. Baker Prize Award, IEEE, 1976, for paper "Physical Limits in Digital Electronics." Other Awards: National Academy of Engineering, 1976.

KIDD, KEITH H., Fellow 1966. Born: February 16, 1921, Toronto, Ontario, Canada. Degrees: B.A.Sc., 1942, M.A.Sc., 1948, University of Toronto.

Fellow Award "For contributions to electrical system planning and engineering."

KIDDER, ALLAN H., AIEE Fellow 1950. Born: December 25, 1899, Proctor, Vt. Degrees: B.S., 1922, M.S., 1923, Massachusetts Institute of Technology.

Awards: First Paper Prize, IRE, Power Division, 1955, for "Power Distribution System Parameters." Other Awards: James H. McGraw Prize, Edison Electric Institute, 1939.

KIGHT, MAX H., AIEE Fellow 1959. Born: October 26, 1900, Bluffdale, Tex. Degrees: B.S.E.E., 1923, University of Illinois.

Fellow Award "For contributions to the establishment of large scale hydro plants and associated power transmission." Other Awards: Distinguished Service Award, U.S. Dept. of Interior, 1966.

KIHARA, NOBUTOSHI, Fellow 1983. Born: October 14, 1926, Tokyo, Japan. Degrees: B.S., 1947, Waseda University, Japan.

Fellow Award "For contributions to magnetic video tape"; Outstanding Papers Award, IEEE CE-Group, 1977; Eduard Rhein Prize, 1978; Outstanding Papers Award, IEEE CE-Group, 1981; David Sarnoff Award,

IEEE, 1982. Other Awards: Nikkan Kogyo Shinbun Award, 1965; Prize of State Minister for Science and Technology, Japan, 1967; 10th Annual Wescon Industrial Design Awards, 1968.

KIHN, HARRY, IRE Fellow 1962. Born: January 24, 1912, Tarnow, Austria. Degrees: B.S.E.E., 1934, Cooper Union; M.S.E.E., 1952, University of Pennsylvania.

Fellow Award "For contributions to color television circuitry"; IEEE Centennial Medal, 1984. Other Awards: RCA Research Awards, 1952, 1956; Miniaturization Award, Miniature Precision Bearings, Inc., 1958.

KIKUCHI, HIROSHI, Fellow 1984. Born: May 6, 1926, Sendai, Japan. Degrees: B.S.(Eng.Sc.), 1949, Ph.D.(Eng.Sc.), 1959, University of Tokyo.

Fellow Award "For contributions and leadership in plasma studies." Other Awards: Prize for the Promotion of Science and Paper Award, Inst. Elec. Eng. Japan, 1956; Inada Memorial Prize Award, Inst. Elec. Commun. Eng. Japan, 1959.

KIKUCHI, MAKOTO, Fellow 1983. Born: December 6, 1925, Tokyo, Japan. Degrees: M.Sci., 1948, Ph.D., 1959, Tokyo University.

Fellow Award "For contributions to amorphous semiconductor research and semiconductor devices."

KILBOURNE, CHARLES E., AIEE Fellow 1947. Born: April 16, 1906, Ft. Sam Houston, Tex. Degrees: B.S., 1927, M.S., 1932, Virginia Military Institute.

Other Awards: Certificate of Commendation, U.S. Dept. of Navy, 1947.

KILBY, JACK S., Fellow 1966. Born: November 8, 1923, Jefferson City, Mo. Degrees: B.S.E.E., 1947, University of Illinois; M.S.E.E., 1950, University of Wisconsin.

Fellow Award "For contributions to the field of integrated circuits through basic concepts, inventions and development"; Outstanding Electrical Engineer Award, Dallas Section, 1965; David Sarnoff Award, 1966; Cledo Brunetti Award, 1979; IEEE Consumer Electronics Award, 1980. Other Awards: Daedalian Minuteman Electronics Award, 1965; Stuart Ballantine Medal, Franklin Institute, 1966; National Academy of Engineering, 1967; National Medal of Science, 1969; Alumni Achievement Award, University of Illinois, 1975; Zworykin Medal, National Academy of Engineering, 1975; Holley Award, ASME, 1982; National Inventors Hall of Fame, 1982.

KILGORE, G. ROSS, IRE Fellow 1957. Born: January 31, 1907, Fremont, Nebr. Degrees: B.S.E.E., 1928, University of Nebraska; M.S.E.E., 1931, University of Pittsburgh.

Fellow Award "For pioneering work in high frequency electron tubes and for leadership in the field of military electronic devices."

KILGORE, LEE A., AIEE Fellow 1945. Born: August 10, 1905, Leavitt, Nebr. Degrees: B.S.E.E., 1927, E.E., 1932, University of Nebraska; M.A., 1929, University of Pittsburgh. D.Eng.(Hon.), 1956, University of Nebraska.

Awards: Lamme Medal, AIEE, 1960, "For meritorious achievement in the design of electrical machinery; for analysis of synchronous machine reactances; for invention of special armature windings, and for inventions and designs related to large adjustable speed ac motors." Other Awards: Order of Merit, Westinghouse Electric, 1952; Member, National Academy of Engineering.

KILLGORE, CECIL L., AIEE Fellow 1961. Born: June 3, 1910, Hollywood, Calif. Degrees: B.S., 1932, California Institute of Technology at Pasadena.

Fellow Award "For contributions to investigation, planning and design of hydroelectric systems."

KILLIAN, STANLEY C., AIEE Fellow 1958. Born: July 4, 1906, Chicago, Ill. Degrees: B.S., 1934, University of Michigan.

Fellow Award "For contributions to design of high voltage bus and switching apparatus"; Paper Prize, AIEE, 1951, for "Induced Currents in High Capacity Bus Enclosures." Other Awards: Member, U.S. Delegation to International Electrotechnical Congress, Stockholm, Sweden, 1958.

KILLINGSWORTH, HENRY T., AIEE Fellow 1951. Born: September 16, 1900, Fort Gaines, Ga. Degrees: B.S., 1919, Auburn University.

KIMBALL, CHARLES N., IRE Fellow 1952. Born: April 21, 1911, Boston, Mass. Degrees: B.E.E., 1931, D.Eng.(Hon.), 1955, Northeastern University; S.M., 1932, Sc.D., 1934, Harvard University; Sc.D.(Hon.), 1958, Park College; D.P.S.(Hon.), 1974, Rockhurst College; Litt.D.(Hon.), Westminster College, 1978.

Fellow Award "In recognition of his contributions and leadership in applying electronic techniques to a wide variety of industrial uses"; IRE, Kansas City Section, 1960, "For outstanding service and contributions toward the advancement of science and technology." Other Awards: Mr. Kansas City, Greater Kansas City Chamber of Commerce, 1973; Chancellor's Medal, University of Missouri at Kansas City, 1976; Annual Trustees' Citation, Midwest Research Institute, 1975; Life Trustee, Menninger Foundation, 1976; Life Trustee, Committee for Economic Development, 1977; Chairman, Advisory Committee, Office of Technology Assessment, U.S. Congress, 1980; Greater KC Hall of Fame, 1982.

KIMBARK, EDWARD W., AIEE Fellow 1948. Born: September 21, 1902, Chicago, Ill. Degrees: B.S., 1924, E.E., 1925, Northwestern University; S.M., 1933, Sc.D., 1937, Massachusetts Institute of Technology.

Awards: First Paper Prize, AIEE, Northeastern District, 1934; Paper Prize, 1977; William M. Habirshaw Award, IEEE, 1980. Other Awards: Distinguished Service Award, U.S. Dept. of Interior, 1974.

KIMEL, HARRY, Fellow 1974. Born: January 24, 1912. Degrees: B.S.E.E., 1934, M.S.E.E., 1938, Harvard University.

Fellow Award "For leadership in the advancement of reliability technology." Other Awards: Honorary Member of Faculty, Defense Weapons Systems Management Center, Wright-Patterson Air Force Base, Dayton, Ohio.

KIMURA, HISAO, Fellow 1975. Born: November 27, 1911, Tokyo, Japan. Degrees: B.E., 1934, Ph.D., 1945, University of Tokyo.

Fellow Award "For contributions to the development of surge-proof transformers, field researches of abnormal voltages of power circuits, and computer analysis of power network problems." Other Awards: Seventh "Electric Power Prize," Institute of Electrical Engineers of Japan, 1958.

KINARIWALA, BHARAT, Fellow 1974. Born: October 14, 1926, Abmedabad, India. Degrees: B.S., 1950, Benares Hindu University, India; M.S., 1954, Ph.D., 1957, University of California at Berkeley.

Fellow Award "For contributions to research in circuit and system theory and to engineering education." Other Awards: IBM Fellowship, 1952-53.

KIND, DIETER HANS, Fellow 1978. Born: October 5, 1929, Reichenberg/Boehmen. Degrees: Dr. Ing.(summa cum laude), 1957, Techn. Hochschule Munchen.

Fellow Award "For development of high-voltage measurement and test methods." Other Awards: Honorary Professor, Technische Universitat Braunschweig; Verband deutscher Elekrotechniker (VDE); Comite International des Poids et Mesures; International Electrotechnical Commission Technical Committee 42.

KING, DONALD D., IRE Fellow 1959. Born: August 7, 1919, Rochester, N.Y. Degrees: A.B., 1942, A.M., 1944, Ph.D., 1946, Harvard University.

Fellow Award "For contributions to microwave technology and electronic countermeasures"; President-Elect, IEEE, 1985.

KING, RONOLD W. P., IRE Fellow 1953. Born: September 19, 1905, Williamstown, Mass. Degrees: A.B., 1927, S.M., 1929, University of Rochester; Ph.D., 1932, University of Wisconsin.

Fellow Award "For his many contributions to the theory of radiating systems." Other Awards: Guggenheim Fellow, 1937, 1958; Fellow, American Academy of Arts and Sciences, 1947; Corresponding Member, Bavarian Academy of Science, 1962; Distinguished Service Award, University of Wisconsin, 1973.

KING, SING-YUI, Fellow 1981. Born: September 25, 1919. Degrees: B.Sc. (Eng.) (with first class Honors), 1940, University of Hong Kong; Ph.D., 1947, University of London; D.Sc. (Hon.), 1981, University of Hong Kong.

Fellow Award "For contributions to studies of electric fields and thermal environments near bundled conductors in power systems and to electrical engineering education." Other Awards: Hon. Fellow Hong Kong Institution of Engineers, 1982.

KINGHORN, JOHN H., AIEE Fellow 1962. Born: August 26, 1904, Wheeling, W. Va. Degrees: B.S., 1926, Carnegie Institute of Technology.

Fellow Award "For contributions to the application of high speed protective relay systems and automatic controls to large power systems."

KINGSLEY, CHARLES, JR., AIEE Fellow 1959. Born: October 14, 1904, New York, N.Y. Degrees: S.B., 1927, S.M., 1932, Massachusetts Institute of Technology.

Fellow Award "For contributions to the teaching of electrical engineering and to the theory of electrical machinery."

KINGSTON, ROBERT H., IRE Fellow 1962. Born: February 13, 1928, Somerville, Mass. Degrees: B.S., M.S., 1948, Ph.D., 1951, Massachusetts Institute of Technology.

Fellow Award "For contributions to solid state devices, particularly the solid state maser." Other Awards: Fellow, American Physical Society, 1958; Fellow, Optical Society of America, 1980.

KINO, GORDON S., Fellow 1966. Born: June 15, 1928, Melbourne, Australia. Degrees: B.Sc.(Math)(Hon.), 1948, M.Sc.(Math), 1950, University of London; Ph.D.(E.E.), 1955, Stanford University.

Fellow Award "For contributions to the design of electron guns and to high frequency behavior of gaseous plasmas." Other Awards: American Physical Society; Guggenheim Fellow, University College, London, England, 1967, 1968; Fellow, American Institute of Physics, 1967; Member, National Academy of Engineering, 1976; Fellow, American Association for the Advancement of Science, 1982.

KIRBY, RICHARD C., Fellow 1970. Born: November 22, 1922, Galesburg, Ill. Degrees: B.E.E., 1951, University of Minnesota.

Fellow Award "For leadership in telecommunications research, and for contributions to radio propagation." Other Awards: Gold Medals, U.S. Department of Commerce, 1955, 1967; Fellow, Radio Club of America, 1978; IEEE Communications Society, Don McLellan Meritorious Service Award, 1979; IEEE Award in International Communications in Honor of Hernand and Sosthenes Behn, 1981; Outstanding Achievement Award, University of Minnesota, 1983.

KIRCHMAYER, LEON K., Fellow 1965. Born: July 24, 1924, Milwaukee, Wis. Degrees: B.E.E., 1945, Marquette University; M.S., 1947, Ph.D., 1950, University of Wisconsin.

Fellow Award "For advances in the economic operation of large electric power systems"; IEEE Centennial Medal, 1984. Other Awards: Honorable Mention: Outstanding Young Electrical Engineer in the U.S.A., Eta Kappa Nu, 1954. Engineer of the Year Award, Schenectady Professional Engineering Society, 1966; Tau Beta Pi; Pi Mu Epsilon; Eta Kappa Nu; Sigma Xi; Fellow, American Society of Mechanical Engineers, 1969; Distinguished Service Citation, University of Wisconsin, 1972; National Academy of Engineering, 1979; ASME Centennial Medal, 1980.

KIRCHNER, FRANCOIS F., Fellow 1974. Born: 1924. Degrees: Eng., 1946, Ecole Polytechnique, Paris; Eng., 1948, Ecole des Ponts et Chaussees, Paris; Lic., 1949, Faculte des Sciences, University of Paris; Eng., 1950, Ecole Superieure d'Electricite, Paris.

Fellow Awards: "For contributions to the field of very-high-voltage electric energy transmission." Other Awards: Member, Societe Francaise des Electriciens (S.F.E.); Member, Association Francaise de Cybernetique Economique and Technique (AFCET); Permanent Member, CIGRE.

KIRKPATRICK, GEORGE M., Fellow 1965. Born: August 18, 1919, Roseville, Ill. Degrees: B.S.E.E., 1941, University of Illinois; M.S.E., 1963, Syracuse University.

Fellow Award "For developments in monopulse radar antennas and leadership in engineering."

KIRSCHBAUM, HERBERT S., Fellow 1971. Born: February 6, 1920, Ohio. Degrees: B.S.E.E., 1942, Cooper Union Institute of Technology; M.S.E.E., 1946, University of Pittsburgh; Ph.D., 1953, Carnegie Institute of Technology.

Fellow Award "For contributions in the development and application of analog computers, helical transmission-line theory, and to control of reactors and radio telescopes." Other Awards: Member, Instrument Society of America.

KISHI, GENYA, Fellow 1983. Born: March 23, 1928, Tokyo, Japan. Degrees: B.Eng., 1951, Dr. Eng., 1960, Tokyo Institute of Technology.

Fellow Award "For contributions to the theory of lumped-constant networks." Other Awards: Achievement Award, Inst. Electronics & Communication Engineers of Japan, 1981; Best Paper Awards, Inst. E. C. E. Japan, 1968, 1969, 1973, 1983.

KITAGAWA, KAZUE, Fellow 1970. Born: July 26, 1904, Nishinomiya City, Japan. Degrees: B.E.E., 1927, Dr. Eng., 1944, Tokyo University.

Fellow Award "For contributions to the fields of oil-filled capacitors and high-voltage cables." Other Awards: Prize for Extraordinary Achievement, Society of Electrical Engineering, 1974.

KITAMURA, ZEN'ICHI, Fellow 1976. Born: December 10, 1911, Kyoto, Japan. Degrees: Ph.D., 1959, Osaka University.

Fellow Award "For contributions to electronic exchange systems and digital computer technology." Other Awards: Award, Minister of Postal Services, Japan, 1976; Award, Institute of Electronics and Communications Engineers of Japan, 1977.

KITSUREGAWA, TAKASHI, Fellow 1976. Born: May 16, 1917, Osaka, Japan. Degrees: B.S., 1941, D.Eng., 1959, Osaka University.

Fellow Award "For contributions to the development of high-performance microwave antennas"; Citation, IEEE Antennas and Propagation Society, 1974. Other Awards: Progress Award, Japan Electrical Manufacturers Association, 1959; Achievement Award, Institute of Electronics and Communications Engineers of Japan, 1967; Twelve Outstanding Patent of the Year Awards, Invention Association of Japan, 1962-1977; Medal of Honor with Purple Ribbon, Japanese Government, 1971; Distinguished Services Award, IECE of Japan, 1978; Honorary Member, IECE of Japan, 1982; Distinguished Services Medal, JEMA, Japan, 1983; Honorary Member, The Society of Instrument and Control Engineers of Japan, 1983.

KLASS, PHILIP J., Fellow 1973. Born: November 8, 1919, Des Moines, Iowa. Degrees: B.S.E.E., Iowa State University.

Fellow Award "For keeping the technical community accurately informed of significant new avionics developments and business trends."

KLEEN, WERNER J., IRE Fellow 1957. Born: October 29, 1907, Hamburg, Germany. Degrees: Dr.Phil.Nat., 1931, Dr.Habil., 1936, University of Heidelberg, Germany.

Fellow Award "For contributions to electron tube theory and microwave techniques"; Microwave Career Award, IEEE Microwave Theory and Techniques Society, 1980; Frederick Philips Award, IEEE, 1982. Other Awards: Medal of Svenska Teknolog Foereningen, Stockholm, Sweden, 1950; Gauss-Weber Medal, University of Goettingen, 1955; Honorary Professor, Technical University of Munich, 1956; Ring of Honour, Verband Deutscher Elektrotechniker, 1978.

KLEINROCK, LEONARD, Fellow 1973. Born: June 13, 1934, New York, N.Y. Degrees: B.E.E., 1957, City College of the City University of New York; M.S.E.E., 1959, Ph.D., 1963, Massachusetts Institute of Technology.

Fellow Award "For contributions in computer-communication networks, queueing theory, time-shared computer systems, and engineering education"; Leonard G. Abraham Prize Paper Award, IEEE Communications Society, 1959. Other Awards: Guggenheim Fellow, 1971-72; Lanchester Prize Paper Award, 1976; Prize Paper Award, ICC, 1978; Member,

National Academy of Engineering, 1980; L.M. Ericsson Award, co-recipient, 1982; Townsend Harris Medal, 1982.

KLINGER, ALLEN, Fellow 1984. Born: April 2, 1937, New York, N.Y. Degrees: B.E.E., 1957, Cooper Union; M.S., 1958, California Institute of Technology; Ph.D., 1966, University of California, Berkeley.

Fellow Award "For contributions to image analysis by means of computers." Other Awards: National Academy of Science and U.S.S.R. Academy of Science Exchange Program, 1982.

KLIPSCH, PAUL W., AIEE Fellow 1962. Born: March 9, 1904, Elkhart, Ind. Degrees: B.S.E.E., 1926, Dr.(Hon.Caus.), 1981, New Mexico State University; E.E., 1934, Stanford University.

Fellow Award "For contributions to electrical reproduction of sound." Other Awards: Tau Beta Pi, 1934; Sigma Xi, 1934; Fellow, Audio Engineering Society, 1957; Distinguished Alumnus Award, N. Mex. State U., 1966; Silver Medal, Audio Engineering Society, 1978; Eta Kappa Nu, 1981.

KNAPP, FLOYD H., AIEE Fellow 1945. Born: August 12, 1903, Susquehanna, Pa. Degrees: E.E., 1927, Rensselaer Polytechnic Institute.

KNERR, REINHARD H., Fellow 1980. Born: February 18, 1939, Pirmasens Federal Republic of Germany. Degrees: 1960, Cand. Ing.-Technische Hochschule, Aachen, Germany; 1962, Ing. Dipl.-Ecole Nat. Sup. D'Electrotechnique, D'Electronique et D'Hydraulique, Toulouse, France; M.S.(E.E.), 1964, Ph.D.(E.E.), 1968, Lehigh University.

Fellow Award "For contributions to lumped element microwave circulators." Other Awards: NATO Scholar, 1964.

KNIGHT, C. RAYMOND, Fellow 1970. Born: September 25, 1918, Salt Lake City, Utah. Degrees: B.S., 1940, University of Utah; M.S., 1959, George Washington University.

Fellow Award "For contributions to increasing reliability and effectiveness in electronic and electromechanical systems; Annual Reliability Group Award, 1975, "For significant technical and management contributions to the reliability and maintainability of electronic equipment through extensive and continuing activity in the development and application of new techniques - as both an individual leader and team contributor"; President, Reliability Group, 1973-74; IEEE Centennial Medal, 1984.

KNOBLAUGH, ARMAND F., IRE Fellow 1958. Born: June 10, 1903, Cincinnati, Ohio. Degrees: B.S.E. 1925, M.A., 1927, Ph.D., 1929, University of Cincinnati.

Fellow Award "For contributions to the field of electronic generation and the reproduction of music." Other Awards: Tau Beta Pi, 1925; Sigma Xi, 1957; Distinguished Engineering Alumnus Award, University of Cincinnati, 1974.

KNOWLES, HUGH S., Fellow 1969. Born: September 23, 1904, Hynes, Iowa. Degrees: A.B., 1928, Columbia University; Honorary Doctor of Science, Northwestern University, 1982.

Fellow Award "For engineering leadership in the field of acoustics and its radio applications"; Achievement Award, IEEE Audio and Electroacoustics Group, 1966, "For audio technology." Other Awards: Fellow, Acoustical Society of America; Fellow, Audio Engineering Society; National Academy of Engineering, 1969; Honorary Member, American National standards Institute (formerly ASA), 1972; Silver Medal in Engineering Acoustics, Acoustical Society of America, 1976; Gold Medal, Audio Engineering Society, 1978; First American Recipient, Alexander Graham Bell Award, Horgerate-Akustiker Kongress, 1978.

KNOWLTON, ALFRED D., AIEE Fellow 1950. Born: October 9, 1898, Haverford, Pa. Degrees: B.Sc., Haverford College.

KNOX, JAMES B., IRE Fellow 1951. Born: November 10, 1910, Winnipeg, Manitoba, Canada.

Fellow Award "For contributions to the development and design of radio transmitters and communication systems, and radio aids to air navigation in Canada."

KNUDSEN, NIELS HERVARD, Fellow 1979. Born: August 24, 1911, Randers, Denmark. Degrees: Electrical Engineer, 1936, Polytechnical Institute, Copenhagen; Doctor of Technology, 1953, Royal Technical Institute, Stockholm.

Fellow Award "For contributions to the development of extra-high-voltage transmission systems and equipment and to engineering education."

KO, HSIEN C., Fellow 1971. Born: April 28, 1928, Taiwan, Formosa. Degrees: B.S., 1951, National Taiwan University; M.S., 1953, Ph.D., 1955, Ohio State University.

Fellow Award "For contributions to radio astronomy and radio astronomical measurements." Other Awards: Distinguished Teaching Award, Eta Kappa Nu, Ohio State University, 1975; Charles E. MacQuigg Award, Ohio State University, 1977.

KO, WEN H., Fellow 1978. Born: April 12, 1923, Shang-Hong Fukien, China. Degrees: B.S.E.E., 1946, National Amoy University; M.S.E.E., 1956, Ph.D.E.E., 1959, Case Institute of Technology; Post Ph.D., 1968, National Institutes of Health Postdoctoral Fellowship, Stanford, Calif.

Fellow Award "For leadership and contributions in the field of microelectronics for biomedical instrumentation." Other Awards: Best Technical Paper Award, CECON, 1970; Achievement Award, Chinese Engineer in America, 1978; Visiting Professor, Katholic University of Leuven, Leuven, Belguim, 1983.

KOBAYASHI, HISASHI, Fellow 1977. Born: June 13, 1938, Tokyo, Japan. Degrees: B.S., 1961, M.S., 1963, University of Tokyo; M.A., 1966, Ph.D., 1967, Princeton University.

Fellow Award "For contributions to data transmission and to modeling and performance analysis of computer communication systems"; Honor Roll, IEEE Computer Society, 1980. Other Awards: David Sarnoff RCA Scholarship Award, 1960; Orson Desaix Munn Fellowship, 1965; IBM Invention Achievement Award, 1971, 1973; IBM Outstanding Contribution Award, 1975; Senior U.S. Scientist Award of the Alexander von Humboldt Foundation, the Federal Republic of Germany, 1979; Silver Core, International Federation of Information Processing, 1980; "Who's Who in America," 1982; "Who's Who in the World," 1983.

KOBAYASHI, KOJI, Fellow 1968. Born: February 17, 1907, Yamanashi Prefecture, Japan. Degrees: B.S., 1929, Dr.Eng., 1939, Tokyo Imperial University; LL.D.(Hon.), 1968, Monmouth College; Dr.Eng.(Hon.), 1971, Polytechnic Institute of New York; Dr.(Hon.), 1980, Autonomous University of Guadalajara, Mexico.

Fellow Award "For contributions to multichannel carrier techniques and to leadership in the growth of the electronics industry"; Frederik Philips Award, IEEE, 1976. Other Awards: Distinguished Service Award, Institute of Electrical Communication of Japan, 1957; Purple Ribbon Medal, 1957; Meritorious Services Award for Export Promotion, Prime Minister of Japan, 1964; Blue Ribbon Medal, 1964; Mainichi Press' Prize for Development of Industrial Technology, 1965; Golden Plate Award, American Academy of Achievement, 1966; Grand Cross of the Peruvian Government, 1970; Medal of Honor from the Paraguayan Government, 1971; Jordan Star with Order of the Third Class from His Majesty The King of Jordan, 1972; The First Class Order of the Sacred Treasure from His Majesty the Emperor of Japan, 1978; "Orden Nacional del Merito" en el grado de Comendador, the Government of the Republic of Paraguay, 1978; The Order of Merit of the First Grade, the Government of the Arab Republic of Egypt, 1979; The National Order of the Southern Cross in the Grade of Commander, the Government of the Federative Republic of Brazil, 1979; Commander (Third Class) of the Most Exalted Order of the White Elephant from His Majesty the King of the Kingdom of Thailand, 1981; "Shevalier du Merite de Madagascar" from President of the Malagasy Democratic Republic, 1981.

KOBER, CARL L., IRE Fellow 1960. Born: November 22, 1913, Vienna, Austria. Degrees: Ph.D.(Theoretical Physics), 1935, University of Vienna; Ph.D.(Electron.), 1939, Technische Hochschule Vienna.

Fellow Award "For application of electronic techniques to defense systems." Other Awards: Fellow, American Association for the Advancement of Science,

1967; Fellow, Explorers Club, 1978.

KOCH, WINFIELD R., IRE Fellow 1954. Born: December 23, 1903, Omaha, Nebr. Degrees: B.S., 1925, University of Minnesota.

Fellow Award "For his creative inventions and developments in radio and television circuitry." Other Awards: Modern Pioneer Award, National Association of Manufacturers; Tau Beta Pi.

KODALI, V. PRASAD, Fellow 1980. Born: November 1, 1939. Degrees: B.E.(with honors), 1961, University of Madras, India; M.S., 1963, Case Institute of Technology; Ph.D., 1967, University of Leeds, England.

Fellow Award "For leadership in the planning of radar development"; IEEE Centennial Medal, 1984. Other Awards: Fellow, Institution of Engineers, India, 1977; Fellow, Institution of Electronics and Telecommunication Engineers, India, 1979.

KOENIG, HERBERT W., Fellow 1979. Born: May 26, 1908, Neufeld, Austria. Degrees: Ph.D., 1932, University of Vienna.

Fellow Award "For contributions to the understanding of amplification and noise in electron-drift tubes and leadership in engineering education." Other Awards: VDE-Ehrenring des Verbandes Deutscher Elektrotechniker, 1972; Schrodinger-Preis der Osterreichischen Akademie der Wissenschaften, 1976; Silberne Ehrennadel des Verbandes Deutscher Elektrotechniker, 1979.

KOEPFINGER, JOSEPH L., Fellow 1976. Born: May 6, 1925, Sewickley, Pa. Degrees: B.S.E., 1949, M.S.E., 1953, University of Pittsburgh.

Fellow Award "For contributions to the application of relay protection in power systems."

KOFOID, MELVIN J., Fellow 1967. Born: July 16, 1910, Portland, Oreg. Degrees: B.S., 1933, M.S., 1935, Oregon State University; Ph.D., 1942, Stanford University.

Fellow Award "For his original contributions to the understanding of spark, arc, plasma, and dielectric phenomena."

KOGELNIK, HERWIG, Fellow 1973. Born: June 2, 1932, Graz, Austria. Degrees: Dipl.Ing., 1955, Dr.Techn., 1958, Technische Hochschule Wien; D.Phil., 1960, Oxford University.

Fellow Award "For contributions to the understanding of the structure and coupling of modes in optical beams and resonators." Other Awards: Fellow, Optical Society of America, 1972; National Academy of Engineering.

KOGO, HIROSHI, Fellow 1984. Born: August 16, 1921, Tokyo, Japan. Degrees: B.Sc.(E.E.), 1944, Ph.D., 1958, Tokyo Institute of Technology.

Fellow Award "For contributions to closed-circuit television." Other Awards: Achievement Award, Institute of Television Engineers of Japan, 1978.

KOHL, WALTER H., IRE Fellow 1958. Born: January 22, 1905, Kitzingen, Germany. Degrees: Dipl.Ing., 1928, Dr.Ing., 1930, Technische Hochschule Dresden, Germany.

Fellow Award "For his research on methods of constructing vacuum tubes individually and in mass production." Other Awards: Fellow, American Ceramic Society, 1967.

KOHLI, FAQIR C., Fellow 1976. Born: February 28, 1924. Degrees: B.A., 1943, B.Sc.(Hons.)(Physics), 1944, Government College, Punjab University, Lahore, West Pakistan; B.Sc.(E.E.)(with honors), 1948, Queen's University, Canada; S.M.E.E., 1950, Massachusetts Institute of Technology.

Fellow Award "For leadership, planning, control technology, and management of efficient and reliable power systems." Other Awards: Gold Medal and Cash Award for Best Paper in Power System, Institution of Engineers, India, 1964, 1972; Gold Medal for Best Paper on Power System, Central Board of Irrigation and Power, 1970; Fellow, Institution of Engineers, India; Fellow, Institution of Electrical Engineers, U.K.; Fellow, Computer Society of India; Associate Member, Sigma Xi, MIT Chapter.

KOJIMA, SATOSHI, Fellow 1975. Born: February 28, 1912, Tokyo, Japan. Degrees: B.E., 1935, Tokyo Imperial University; D.E., 1950, Tokyo University.

Fellow Award "For development of traffic theory in communication and the nationwide dialing system in Japan." Other Awards: Mainichi Industry Award, Mainichi Press, 1963; Advancement Award, IEEJ, 1963; Distinguished Achievement Award, IECEJ, 1969; Medal of Honor with Purple Ribbon, Japanese Government, 1971.

KOKOTOVIC, PETAR V., Fellow 1980. Born: March 18, 1934. Degrees: Dipl. Eng., 1958, Magistar (E.E.), 1963, University of Belgrade, Yugoslavia; Kandidat of Technical Sciences, 1965, USSR Academy of Sciences, Moscow.

Fellow Award "For contributions to sensitivity analysis and singular perturbation theory."

KONCEL, EDWARD F., JR., Fellow 1977. Born: August 3, 1927, Chicago, Ill. Degrees: B.S.E.E., 1949, M.S.E.E., 1950, Illinois Institute of Technology.

Fellow Award "For innovations in technical and economic development of a major electric utility and for pioneering the computerized accounting of nuclear fuel"; Charles LeGeyt Fortescue Fellow of AIEE, 1949. Other Awards: Member: Eta Kappa Nu, Tau Beta Pi, Sigma Xi.

KONISHI, YOSHIHIRO, Fellow 1981. Born: September 24, 1928, Nara, Japan. Degrees: B.E., 1951, Dr.Eng., 1963, Kyoto University.

Fellow Award "For contributions to microwave component technology of particular significance to satellite broadcasting." Other Awards: Institute of Television Engineering, Japan, 1974; Institute of Electronics and Communication Engineers, Japan, 1975; Minister, Post Office Ministry, Japan, 1978; Ministry, Patent Bureau, Japan, 1977; Minister of State, Science and Technology, 1979; Medal with Purple Ribbon, Emperor, Japan, 1982.

KOOPMAN, RICHARD J. W., AIEE Fellow 1953. Born: June 24, 1905, St. Louis, Mo. Degrees: B.S., 1928, University of Missouri; M.S., 1933, Yale University; Ph.D., 1942, University of Missouri.

Fellow Award "For contributions to the theory and practice of servomechanisms and telemetering as well as his diversity of interest in the several fields of electrical engineering, through which he has become a well-rounded and inspiring teacher and educational administrator"; First Paper Prize, AIEE, 1937; Award of Honor, St. Louis Section IEEE, 1982. Other Awards: Tau Beta Pi; Sigma Xi; Eta Kappa Nu; First Samuel C. Sachs Professor of Electrical Engineering, 1972; Honorary Member, Engineers' Club of St. Louis, 1972; Distinguished Service Award, St. Louis Electrical Board of Trade, 1973; Faculty Award, Alumni Association, Washington University, 1973.

KOPPER, JOHN M., AIEE Fellow 1961. Born: September 15, 1912, New York, N.Y. Degrees: B.E., 1933, D.Eng., 1944, The Johns Hopkins University.

Fellow Award "For contributions to the solution of problems of electromagnetic warfare."

KORMAN, N. I., IRE Fellow 1956. Born: February 23, 1916, Providence, R.I. Degrees: B.S., 1937, Worcester Polytechnic Institute; S.M., 1938, Massachusetts Institute of Technology; Ph.D., 1958, University of Pennsylvania.

Fellow Award "For contributions in the field of radar fire control and missile guidance." Other Awards: Sigma Xi, 1937; Charles A. Coffin Fellowship, General Electric, 1937; Award of Merit, RCA Victor, 1951.

KORN, FRANKLIN A., Fellow 1969. Born: July 10, 1905, Elizabeth, N.J.

Fellow Award "For contributions to the development of crossbar switching technology"; IEEE, Columbus (Ohio) Technical Council Award, "Technical Man of the Year," 1970.

KORN, GRANINO A., Fellow 1979. Born: May 7, 1922, Berlin, Germany. Degrees: B.A., 1942, Ph.D., 1948, Brown University; M.A., 1943, Columbia University.

Fellow Award "For technical contributions to the field of computer simulation and to electrical engineering education." Other Awards: Senior Prize for Advanced Simulation, 1968; Humboldt Foundation Prize, 1976.

KORNEI, OTTO, Fellow 1968. Born: April 29, 1903, Vienna, Austria. Degrees: M.S.E.E., 1925, Institute of Technology, Vienna; Dipl.Ing., 1931, Institute of Technology, Berlin, Germany.

Fellow Award "For fundamental and extensive contributions to magnetic recording technology for video and audio applications, as well as creative contributions to the beginnings of xerography." Other Awards: Honorary Member, Audio Engineering Society, 1960; "Who's Who in Engineering," 1977; "Who's Who in Technology Today," 1981; "International Who's Who in Engineering," 1982.

KORPEL, ADRIANUS, Fellow 1975. Born: February 18, 1932, Rotterdam, Holland. Degrees: B.S.E.E., 1953, M.S.E.E, 1955, Ph.D., 1969, University of Delft, Netherlands.

Fellow Award "For contributions to the understanding and applications of acoustooptic interactions."

KOSMAHL, HENRY G., Fellow 1979. Born: December 14, 1919, Wartha, Germany. Degrees: Dipl. Physicist, 1943, University of Dresden; Doctor of Science, 1949, University of Darmstadt.

Fellow Award "For contributions to the theory of traveling-wave tubes and klystrons leading to increased efficiencies"; Technology Advancement Award, IEEE/DoD, 1977; Associate Editor, IEEE Transactions on Electron Devices, 1983. Other Awards: Exceptional Scientific Achievement Award, NASA, 1974; NASA Major Inventor Award, 1980.

KOSONOCKY, WALTER F., Fellow 1976. Born: December 15, 1931, Sieradz, Poland. Degrees: B.S.E.E., 1955, M.S.E.E., 1957, Newark College of Engineering; Eng.Sc.D., 1965, Columbia University.

Fellow Award "For contributions to solid-state logic, memory, and imaging." Other Awards: David Sarnoff Fellowship, 1958-1959; Achievement Award, RCA Laboratories, 1959, 1963, 1979; Honor Roll Award, NJIT Alumni Association, 1960; Fellow, Technical Staff, RCA Laboratories, 1979; David Sarnoff Award, RCA, 1981.

KOTELNIKOV, V.A., Fellow 1964.

Fellow Award "For contributions to the theory and practice of radio communications and radar astronomy."

KOUYOUMJIAN, ROBERT G., Fellow 1976. Born: April 26, 1923. Degrees: B.Sc.(Physics), 1948, Ph.D.(Physics), 1953, The Ohio State University.

Fellow Award "For contributions to the geometrical theory of diffraction of electromagnetic waves." Other Awards: Member, URSI Commission B; Sigma Xi; Eta Kappa Nu; Sigma Pi Sigma.

KOVALY, JOHN J., Fellow 1981. Born: June 12, 1928, McKeesport, PA. Degrees: B.S., 1950, Muskingum College; M.S., 1953, University of Illinois.

Fellow Award "For contributions to synthetic aperture radar." Other Awards: Outstanding Author Award, Raytheon Company, 1976; "Who's Who in Engineering"; "Who's Who in America."

KRAMER, ANDREW W., Fellow 1971. Born: April 10, 1893, Chicago, Ill. Degrees: B.S.E.E., 1916, Armour Institute of Technology (now Illinois Institute of Technology.)

Fellow Award "For contributions in the field of technical literature, particularly in the interpretation of highly technical material into simple understandable language." Other Awards: Certificate of Appreciation for Outstanding Service to the Power Industry and the American Power Conference, Illinois Institute of Technology, 1975; Special Award for Services to the American Power Conference, American Nuclear Society, 1975.

KRAUS, C. RAYMOND, AIEE Fellow 1962. Born: Philadelphia, Pa. Degrees: B.S.(E.E.), Pennsylvania State University.

Fellow Award "For contributions to radio and wire telephony"; Communications Switching Committee, 1969-1983; Chairman, AIEE Radio Committee, 1960; Philadelphia Section Award, IEEE, 1969. Other Awards: Fellow, Radio Club of America, 1978; Phi Kappa Phi; Eta Kappa Nu; Tau Beta Pi.

KRAUS, JOHN D., IRE Fellow 1954. Born: June 28, 1910, Ann Arbor, Mich. Degrees: B.S., 1930, M.S., 1931, Ph.D., 1933, University of Michigan.

Fellow Award "For his leadership in antenna development, and his research contributions in radio astronomy"; IEEE Centennial Medal, 1984. Other Awards: Sullivant Medal, 1970; National Academy of Engineering, 1972; Outstanding Achievement Award, Univ. of Michigan, 1981.

KRAUSE, PAUL C., Fellow 1977. Born: January 27, 1932, Reynolds, Nebr. Degrees: B.S.E.E., 1956, B.S.M.E., 1957, M.S.E.E., 1958, University of Nebraska; Ph.D.(E.E.), 1961, University of Kansas.

Fellow Award "For contributions to electric machine theory and hybrid computer simulation of power system components." Other Awards: Edward and Rosa Uhrig Foundation Award for Excellence in Teaching, University of Wisconsin, Milwaukee, 1964; Eta Kappa Nu Teaching Award, Purdue University, 1975.

KRENDEL, EZRA S., Fellow 1967. Born: March 5, 1925, New York, N.Y. Degrees: A.B., 1945, Brooklyn College; Sc.M., 1947, Massachusetts Institute of Technology; M.A., 1949, Harvard University.

Fellow Award "For pioneering in and extending the techniques of electrical engineering to the study of a wide range of human behavior-manual control dynamics, power production, vision and bioelectric potentials such as brain waves." Other Awards: Louis E. Levy Gold Medal (corecipient), Franklin Institute; 1959; NATO Lecturer, 1968, 1971; Member, Philadelphia Mayor's Science and Technology Advisory Council.

KRESGE, JAMES S., Fellow 1984. Born: December 7, 1924, Alberta, Canada. Degrees: B.S.E.E., 1950, Univ.

of California.

Fellow Award "For contributions to the commercial development of metal-oxide-surge arresters"; Three Prize Paper Awards, IEEE Power Engineering Society; W.R.G. BAKER Prize Award, IEEE, 1978.

KRESSEL, HENRY, Fellow 1974. Born: January 24, 1934, Vienna, Austria. Degrees: B.A., 1955, Yeshiva College; M.S., 1956, Harvard University; M.B.A., 1959, Ph.D., 1965, University of Pennsylvania.

Fellow Award "For contributions to the development of semiconductor devices"; IEEE Centennial Medal, 1984. Other Awards: David Sarnoff Award for Outstanding Technical Achievement, RCA, 1973; Bernard Revel Award, Yeshiva University Outstanding Alumni, 1981.

KRESSER, JEAN V., AIEE Fellow 1960. Born: September 26, 1904, Noumea, New Caledonia. Degrees: B.S., 1930, Massachusetts Institute of Technology.

Fellow Award "For contributions to the design, control and literature of electric power apparatus"; Best District Paper Prize, AIEE, 1946. Other Awards: Outstanding Engineering Performance Award, Westinghouse Electric Corp., 1968.

KRETZMER, ERNEST R., Fellow 1977. Born: December 24, 1924, Germany. Degrees: B.S.E.E., 1945, Worcester Polytechnic Institute; M.S.E.E., 1946, Sc.D.(E.E.), 1949, Massachusetts Institute of Technology.

Fellow Award "For contributions to the understanding of video signal transmission, and for leadership in the development of data communications systems."

KREUZER, BARTON, Fellow 1973. Born: February 18, 1909, New York, N.Y. Degrees: E.E., 1928, Polytechnic Institute of Brooklyn.

Fellow Award "For contributions in applying electronics to sound motion pictures, satellite meteorology, lunar exploration, and communications." Other Awards: Life Fellow, Society of Motion Picture and Television Engineers.

KREZDORN, ROY R., AIEE Fellow 1962. Born: January 30, 1910, Shreveport, La. Degrees: B.S., 1932, Texas A and M University; M.S., 1951, University of Texas; E.E., 1954, Texas A and M University.

Fellow Award "For contributions to electrical engineering education."

KRISTIANSEN, MAGNE, Fellow 1978. Born: April 14, 1932, Elverum, Norway. Degrees: B.S.E.E., 1961, Ph.D.(E.E.), 1967, University of Texas at Austin.

Fellow Award "For contributions to plasma technology and pulsed power." Other Awards: Spencer A. Wells Award, 1972; NATO Senior Fellowship in Science, 1975; P.W. Horn Professorship, 1977; Japan Society for Promotion of Science Fellowship, 1978; Distinguished Faculty Research Award, 1980; Fellow, American Physical Society, 1982.

KROEMER, HERBERT, Fellow 1970. Born: August 25, 1928, Weimar, Germany. Degrees: Diplom-Physiker, 1951, Dr.rer.nat., 1952, Gottingen University.

Fellow Award "For the invention of the drift transistor and other semiconductor devices"; J.J. Ebers Award, 1973, "For outstanding technical contributions to electron devices including conception of the drift transistor, of wide-gap emitters, and of the double-hetero junction laser, and for contributions to the understanding of the Gunn effect"; GaAs Symposium Award, 1982. Other Awards: Fellow, American Physical Society, 1973.

KRONEBERG, ALEX A., AIEE Fellow 1948. Born: July 23, 1899, Ekaterinburg, Russia. Degrees: B.S.E.E., 1926, California Institute of Technology.

KRONMILLER, CHARLES W., AIEE Fellow 1962. Born: September 14, 1903, Fort Wayne, IN. Degrees: B.S.E.E., 1927, Purdue University.

Fellow Award "For contributions to the art of lighting including the development of ballasts for fluorescent lamps." Other Awards: Eta Kappa Nu, 1926; Charles A. Coffin Award, General Electric, 1946; Citizen-Engineer Award, Fort Wayne Area, 1962; Numerons Awards, ARRL.

KRUTTER, HARRY, IRE Fellow 1960. Born: March 17, 1911, Boston, Mass. Degrees: S.B., 1932, S.M., 1933, Ph.D., 1935, Massachusetts Institute of Technology.

Fellow Award "For contributions to defense electronics." Other Awards: Distinguished Civilian Service Award, U.S. Dept. of Navy, 1956; Distinguished Civilian Service Award, U.S. Dept. of Defense, 1957.

KSIENSKI, AHARON A., Fellow 1977. Born: June 23, 1924, Warsaw, Poland. Degrees: M.Sc.(E.E.), 1952, Ph.D.(E.E.), 1958, University of Southern California.

Fellow Award "For contributions to signal processing antennas." Other Awards: Fellow, United States Steel, 1957-59; Lord Brabazon Award, Institution of Electronics and Radar Engineers, 1967, 1976; R. W. Thompson Award for Meritorious Achievement, Ohio State University, 1969; Commendation, Ohio House of Representatives, 1977.

KU, WEI SHING, Fellow 1982. Born: July 13, 1923, Wusih, Kiangsu, China. Degrees: B.S.E.E., 1944, Tatung University, Shanghai, China; M.S.(Eng.), 1948, Cornell University.

Fellow Award "For technical leadership in the application of advanced planning concepts to the development of high-voltage networks."

KU, YU H., AIEE Fellow 1945, IRE Fellow 1961. Born: December 24, 1902, Wusih, Kiangsu, China. Degrees: S.B., 1925, S.M., 1926, Sc.D., 1928, Massachusetts Institute of Technoloy; M.A., LL.D.(Hon.), University of Pennsylvania.

IRE Fellow Award "For contributions to nonlinear circuit analysis"; IEEE, Lamme Medal Award, 1972. Other

Awards: Professional Accomplishment Award, Chinese Institute of Engineers, 1959; Gold Medal, Ministry of Education, Republic of China, 1960; Professional Accomplishment Award, Chinese Engineers and Scientists Association of Southern California, 1969; Gold Medal, Chinese Institute of Electrical Engineers, 1972; Fellow, Institution of Electrical Engineers, London; Fellow, Academia Sinica; Personal Member of General Assembly, International Union of Theoretical and Applied Mechanics (IUTAM); Gold Medal, Pro Mundi Beneficio, Brazilian Academy of Humanities, 1975; Honorary Member, United Poets Laureate International, 1977; American Society for Engineering Education; Sigma Xi; Eta Kappa Nu; Phi Tau Phi; U.S. National Committee of Theoretical and Applied Mechanics.

KUBO, TOSHIHIKO, Fellow 1979. Born: December 8, 1909, Tottori, Japan. Degrees: Bachelor of Physics, 1932, Doctor of Engineering, 1945, University of Tokyo.

Fellow Award "For leadership in the communications and electronics industries of Japan and contributions to the development of mercury rectifiers." Other Awards: Medal of Honor with Purple Ribbon from Prime Minister of Japan, 1973; Third Class Order of the Rising Sun, Emperor of Japan, 1980.

KUEHL, HANS H., Fellow 1980. Born: March 16, 1933, Detroit, MI. Degrees: B.S., 1955 Princeton University; M.S., 1956, Ph.D., 1959, California Institute of Technology.

Fellow Award "For contributions to the theory of antennas in plasmas." Other Awards: Teaching Excellence Award, University of Southern California Associates, 1964; Faculty Service Award, Archimedes Circle, University of Southern California, 1970; Outstanding Electrical Engineering Faculty Member Award, Eta Kappa Nu, 1977; Halliburton Award for Exceptional Service, University of Southern California, 1980.

KUEHN, RALPH E., Fellow 1971. Born: March 22, 1918, Belleville, Ill. Degrees: B.S.E.E., 1941, M.S.E.E., 1948, University of Illinois.

Fellow Award "For contributions to the reliability of military and space electronic systems, and leadership in related professional activities." Other Awards: National Reliability Award.

KUH, ERNEST S., Fellow 1965. Born: October 2, 1928, Peking, China. Degrees: B.S., 1949, University of Michigan; M.S., 1950, Massachusetts Institute of Technology; Ph.D., 1952, Stanford University.

Fellow Award "For contributions to active and passive circuit theory and engineering education"; Guillemin-Cauer Award, Circuits and Systems Society, 1973; IEEE Education Medal, 1981. Other Awards: Senior Postdoctoral Fellow, National Science Foundation, 1962-1963; Award, National Electronics Conference, 1966; Distinguished Alumni Award, University of Michigan, 1970; Member, National Academy of Engineering, 1975; Member, Academia Sinica, 1975; Alexander von Humboldt Senior Scientist Award, 1977; Honorary Professor, Shanghai Jiao-tong University, 1979; Lamme Award, ASEE, 1981; Fellow, AAAS, 1981.

KUHLMANN, JOHN H., AIEE Fellow 1956. Born: November 26, 1893, Waverly, Iowa. Degrees: B.A., 1913, Warburg College; B.E., 1917, E.E., 1921, University of Iowa.

Fellow Award "For creative contributions to the teaching of electrical apparatus design."

KUHN, MATTHEW, Fellow 1982. Born: March 19, 1936, Sacalaz, Rumania. Degrees: B.Sc.(Eng. Physics), 1962, Queen's University, Canada; M.Sc.(Elect.), 1964, Ph.D.(Elect.), 1967, University of Waterloo, Canada.

Fellow Award "For technical leadership and contributions to solid-state device technology and its application in telecommunications systems."

KUMAGAI, NOBUAKI, Fellow 1981. Born: May 19, 1929, Ryojun, Japan. Degrees: B.E., 1953, Dr. Eng., 1959, Osaka University, Japan.

Fellow Award "For contributions to the study of wave propagation in electromagnetics, optics, and acoustics." Other Awards: Special Award for outstanding contributions, Laser Society of Japan; Member, Laser Society of Japan; Member, Institute of Electronics and Communication Engineers, Japan; Member, Institute of Electrical Engineers, Japan; President, Microwave Theory and Techniques Society, Institute of Electronics and Communication Engineers, Japan, 1979-1981.

KUMMER, WOLFGANG H., Fellow 1974. Born: October 10, 1925, Stuttgart, Germany. Degrees: B.S., 1946, M.S., 1947, Ph.D., 1954, University of California at Berkeley.

Fellow Award "For contributions to adaptive array antenna systems." Other Awards: Phi Beta Kappa; Tau Beta Pi; Eta Kappa Nu; Sigma Xi; Alpha Mu Gamma.

KUNO, H. JOHN, Fellow 1977. Born: March 27, 1938, Osaka, Japan. Degrees: B.S.(with honors), 1961, M.S., 1963, Ph.D., 1966, University of California at Los Angeles.

Fellow Award "For contributions to the field of solid-state generation and modulation of millimeter-wave signals."

KUNZE, ALLAN A., Fellow 1965. Born: December 23, 1912, East Lansing, Mich. Degrees: B.S., 1934, Michigan State University; M.S., 1935, University of Michigan.

Fellow Award "For contributions to the advancement of global communications systems." Other Awards: Tau Beta Pi; Sigma Xi; Eta Kappa Nu; Phi Kappa Phi; Armed Forces Communication and Electronics Association; Distinguished Alumnus, College of Engineering, University of Michigan, 1970.

KUO, FRANKLIN F., Fellow 1972. Born: April 22, 1934, China. Degrees: B.S., 1955, M.S., 1956, Ph.D., 1958, University of Illinois.

Fellow Award "For contributions to computer-aided design and engineering education."

KUO, YEN-LONG, Fellow 1982. Born: November 18, 1936, Taipei, Taiwan. Degrees: M.S.E.E., 1961, Oklahoma State University; Ph.D.(E.E.), 1966, University of California, Berkeley.

Fellow Award "For contributions to the application of analysis and computer-aided design to communication circuits and systems"; Darlington Award, IEEE Circuits and Systems Society, 1975.

KUPER, J. B. HORNER, IRE Fellow 1954. Born: November 5, 1909, New York, N.Y. Degrees: B.A., 1930, Williams College; Ph.D., 1938, Princeton University.

Fellow Award "For his contributions to nucleonic instrumentation and health physics."

KUROKAWA, KANEYUKI, Fellow 1974. Born: August 14, 1928. Degrees: B.S., 1951, Ph.D., 1958, University of Tokyo.

Fellow Award "For contributions to microwave solid-state circuits, oscillators, and devices." Other Awards: Okabe Memorial Prize, IECE of Japan, 1956; Progress Award, Institute of Electrical Engineers of Japan, 1959; Certificate of Appreciation, ISSCC, 1965; Certificate of Outstanding Contributor, ICCD, 1983.

KURTH, CARL F., Fellow 1979. Born: May 11, 1928, Germany. Degrees: Grad. Ing., 1951, Polytechnikum Mittweida, Germany.

Fellow Award "For contributions to the practical use of active filters and leadership in the application of digital signal processing to telecommunications systems"; Darlington Prize Paper Award (corecipient), IEEE Circuits and Systems Society, 1977; President, IEEE Circuits and Systems Society, 1980; IEEE Centennial Medal, 1984. Other Awards: Best Paper Award, NTG (German Communications Society), 1964.

KURTH, HENRY R., AIEE Fellow 1961. Born: April 26, 1900, Cambridge, Mass. Degrees: B.S., 1921, Massachusetts Institute of Technology.

Fellow Award "For contributions to power system operating practices and safety standards."

KUSHNER, HAROLD J., Fellow 1974. Born: July 29, 1933, New York, N.Y. Degrees: B.E.E., 1955, City College of The City University of New York; M.S., 1956, Ph.D., 1958, University of Wisconsin.

Fellow Award "For contributions to stochastic control theory."

KUSKO, ALEXANDER, Fellow 1977. Born: April 4, 1921, New York, N.Y. Degrees: B.S.E.E., 1942, Purdue University; S.M.E.E., 1944, Sc.D., 1951, Massachusetts Institute of Technology.

Fellow Award "For contributions to education in power electronics and to the application of solid-state power devices to industry"; William E. Newell Power Electronics Award, IEEE, 1980.

KUSSY, FRANK WERNER, Fellow 1979. Born: October 13, 1910. Degrees: Diplom. Ingenieur (Master of Science), 1934, Munich T.H.; Doktor T.W. (Ph.D.), 1936, University of Vienna.

Fellow Award "For leadership in the development and application of advanced devices and protection systems for motor control." Other Awards: Gould Scientific Achievement Award, 1979; Professional Engineer, Pa, Md, La.

KWAKERNAAK, HUIBERT, Fellow 1983. Born: March 18, 1937, Rijswijk (Z.H.), The Netherlands. Degrees: Diploma (Eng. Physics), 1960, Delft University of Technology; M.Sc.(E.E.), 1962, Ph.D.(E.E.), 1963, University of California at Berkeley.

Fellow Award "For contributions to the theory of control systems and for educational leadership."

KYLE, WILLIAM D., JR., AIEE Fellow 1956. Born: May 18, 1915, Milwaukee, Wis. Degrees: A.E.E.E., 1936, Cornell University.

Fellow Award "For contributions to the development of components for electric power distribution."

L

LACK, FREDERICK R., IRE Fellow 1937, AIEE Fellow 1948. Born: July 18, 1895, Eastbourne, East Sussex, England. Degrees: B.S.(with great distinction), 1925, Harvard University; Ph.D.(Hon.), 1956, Albright College.

Other Awards: Order of the Rising Sun, 1922; Presidential Certificate of Merit, 1947; Medal of Honor, Electronic Industries Association, 1959; Tau Beta Phi.

LACY, PETER, Fellow 1967. Born: December 6, 1920, Jacksonville, Fla. Degrees: B.S.E.E., 1942, University of Florida; M.S., 1947, Ph.D., 1952, Stanford University.

Fellow Award "For contributions in electronics measurements by the design of phase and delay distortion instrumentation and by furthering the development of microwave signal sources and their application in electronic instruments." Other Awards: Phi Kappa Phi, 1942; Sigma Xi, 1947; Sperry Fellow, 1948; Listed in American Men and Women of Science and Who's Who in America.

LAFFERTY, JAMES M., IRE Fellow 1954. Born: April 27, 1916, Battle Creek, Mich. Degrees: B.S., 1939, M.S., 1940, Ph.D., 1946, University of Michigan.

Fellow Award "For his research contributions to microwave tubes and high-current density cathodes"; Lamme Medal, 1979. Other Awards: Naval Ordnance Development Award; Distinguished Alumnus Citation, University of Michigan, 1953; IR-100 Award (Triggered Vacuum Gap), 1968; National Academy of Engineering, 1981.

LAFFERTY, RAYMOND E., Fellow 1980. Born: July 12, 1918.

Fellow Award "For contributions to loss measurements of reactive components." Other Awards: Fellow, Radio Club of America, 1957.

LAGRONE, ALFRED H., IRE Fellow 1961. Born: September 25, 1912, DeBerry, Tex. Degrees: B.S.E.E., 1938, M.S.E.E., 1948, Ph.D., 1954, University of Texas at Austin.

Fellow Award "For contributions to the theory and application of tropospheric radio wave propagation"; Scott Helt Award, IRE, Professional Group on Broadcasting, 1959, "For the most outstanding paper presented or published in the field of broadcast engineering."

LAITHWAITE, ERIC R., Fellow 1969. Born: June 14, 1921, Atherton, Lancashire, England. Degrees: B.Sc., 1949, M.Sc., 1950, Ph.D., 1957, D.Sc., 1964, Manchester University.

Fellow Award "For outstanding work on electrical machines and especially for development of the linear induction motor." Other Awards: S.G. Brown Award and Gold Medal, Royal Society, 1966; Premium Awards, Institution of Electrical Engineers, London, 1955-1969.

LAKATOS, EMORY, IRE Fellow 1961. Born: August 5, 1905, Losoncz, Hungary. Degrees: M.E., 1926, Stevens Institute of Technology; LL.B., 1939, New York Law School.

Fellow Award "For contributions in communications and weapon systems design." Other Awards: Prize Essayist, Edison Lighting Institute, 1927; Certificate of Appreciation, Office of Scientific Research and Development, 1945.

LAKER, KENNETH R., Fellow 1983. Born: September 10, 1946, Newburgh, New York. Degrees: B.E. (Elec. Eng.), 1968, Manhattan College; M.S. (Elec. Eng.), 1970, Ph.D., 1973, New York University.

Fellow Award "For contributions to filter design and microcircuit implementation"; President, IEEE Circuits and Systems Society, 1983; IEEE Centennial Medal, 1984. Other Awards: Eta Kappa Nu; Sigma Xi.

LALLIER, W. C., AIEE Fellow 1960. Born: July 6, 1900, Denison, Tex. Degrees: B.S.,1922, University of Wisconsin.

Fellow Award "For contributions to engineering of a state-wide communication system."

LAMB, JAMES J., IRE Fellow 1954. Born: November 30, 1900, Michigan, N.D. Degrees: B.S.E.E., 1922, Catholic University of America.

Fellow Award "For his technical contributions to amateur radio activities, and for direction of radio and electronic circuitry development." Other Awards: Fellow, Radio Club of America, 1958; Technical Merit

Award, American Radio Relay League, 1959; Certificate of Achievement, U.S. Dept. of Army, 1965, 1968; Meritorious Civilian Service Medal, U.S. Dept. of Army, 1966; Pioneer Award, Radio Club of America, 1981.

LAMBERT, ROBERT F., Fellow 1970. Born: March 14, 1924, Warroad, Minn. Degrees: B.E.E., 1948, M.S.E.E., 1949, Ph.D., 1953, University of Minnesota.

Fellow Award "For contributions in theoretical and applied acoustics with special reference to noise and system dynamics." Other Awards: Fellow, Acoustical Society of America, 1974; NASA, Recognition Award, 1979; Progressive Architecture Design Awards, 1968, 1969, 1982.

LAMM, A. UNO, AIEE Fellow 1962. Born: May 22, 1904, Gothenburg, Sweden. Degrees: M.S., 1927, Dr.Tech., 1943, Royal Institute of Technology, Stockholm; Dr.(Hon.), 1965, University of Technology, Copenhagen.

Fellow Award "For contributions to the development of high-voltage dc transmission and mercury-arc valves"; Lamme Medal, 1965, "For outstanding achievement in developing the high-power high-voltage mercury arc valve and a unique system of control and protection for its application as rectifier and inverter in high-voltage dc power transmission"; Uno Lamm High Voltage DC Award, IEEE Power Engineering Society, established 1981. Other Awards: Gold Medal, Royal Swedish Association of Engineers and Architects, 1940; Arnberg Award, Royal Swedish Academy of Science, 1947; John Ericsson Medal, American Society of Swedish Engineers, 1962; Medal, Swedish Society of Inventors, 1961; Honorary Fellow, Manchester University, England, 1962; Order of Merit, Societe pour la Recherche et l'Invention, Paris, France, 1965; Member, Royal Academy of Science, Sweden; Member, Royal Academy of Engineering Sciences, Sweden; Member, Royal Society of Science, Uppsala; Foreign Associate, National Academy of Engineering, 1976; Member, National Academy of Engineering, 1979; Howard N. Potts Award, Franklin Institute, 1981.

LAMM, RALPH A., Fellow 1967. Born: July 13, 1908, Bisbee, Ariz.

Fellow Award "For contribution to the development of missile guidance systems and leadership in missile program management"; Leadership Citation, IRE, 1960; Leadership Citation, IRE, Los Angeles Section, 1962; Certificate of Appreciation, WESCON, 1966; Certificate of Appreciation, Los Angeles Council, 1967, "For his farsighted and distinguished leadership in technical and management responsibilities"; Citation, "For services rendered in furthering the objectives of the IEEE, Vice Chairman Region 6, 1981-82"; Director, Region 6, IEEE, 1983-84; IEEE Centennial Medal, 1984. Other Awards: Diploma, U.S. War Manpower Commission, 1943; Certificate of Appreciation, Office of Scientific Research and Development, 1945; Naval

Ordnance Development Award, 1945; Presidential Certificate of Merit, 1948; Meritorious Service Award, U.S. Dept. of Commerce, 1949; Bronze Medallion, American Ordnance Association, 1969; Engineer of the Year, San Fernando Valley Engineers Council, 1970; All Ahead Bendix Award, Bendix Corporation, Electrodynamics Division, 1972; Founding Director, Electronic Conventions, Inc., 1973-1981; Founding Director & President, Missile Technology Historical Association, 1980-1981.

LAMPARD, DOUGLAS G., IRE Fellow 1962. Born: May 4, 1927, Sydney, Australia. Degrees: B.Sc., 1951, M.Sc., 1952, Sydney University; Ph.D., 1954, University of Cambridge.

Fellow Award "For contributions to noise theory and absolute capacitance measurement." Other Awards: Heaviside Premium, Institution of Electrical Engineers, U.K., 1957; Albert F. Sperry Award, Instrument Society of America, 1965; Fellow, Australian Academy of Science; Honorary Fellow, Faculty of Anaesthetists of the Royal Australasian College of Surgeons.

LAMPE, JOHN HAROLD, AIEE Fellow 1948. Born: December 1, 1896, Baltimore, Md. Degrees: B.E., 1918, M.E.E., 1926, D.Eng., 1931, The Johns Hopkins University; D.Sc.(Hon.), 1953, Clarkson College of Technology; D.Hum.(Hon.), 1965, North Carolina State University; D.Sc.(Hon.), 1981, Old Dominion University.

Awards: Citation, IEEE Student Section, Southeastern Region III, 1977. Other Awards: Sigma Xi, 1931; Citation, North Carolina Society of Engineers, 1954; Fellow, American Association for the Advancement of Science, 1958; Congressional Citation, 1958; Citation, Phi Kappa Phi, 1960; Citation, U.S. Dept. of Army, 1960; Distinguished Alumni Award, The Johns Hopkins University, 1963; Citation, Virginia Society of Professional Engineers, 1965; Citation, American Society for Engineering Education, 1966; Distinguished Alumni Award, Baltimore Polytechnic Institute, 1976.

LAMPERT, MURRAY A., Fellow 1971. Born: November 29, 1921, New York, N.Y. Degrees: B.A., 1942, M.A., 1945, Harvard University.

Fellow Award "For contributions in understanding of injected currents in insulators and semiconductors." Other Awards: Fellow, APS; Fellow, American Association for the Advancement of Science.

LAMSON, JOSEPH V., AIEE Fellow 1958. Born: March 26, 1902, Evanston, Ill. Degrees: B.S., 1926, M.S., 1936, University of Washington.

Fellow Award "For contributions to transmission design." Other Awards: Meritorious Service Silver Medal, U.S. Dept. of Interior, 1960.

LAND, CECIL E., Fellow 1974. Born: January 8, 1926, Lebanon, Mo. Degrees: B.S., 1949, Oklahoma State University; D.Sc.(Hon.), 1978, Oklahoma Christian

College.

Fellow Award "For contributions in ferroelectric materials and devices"; IEEE Centennial Medal, 1984. Other Awards: Engineer of the Year, Albuquerque Chapter, New Mexico Society of Professional Engineers, 1973; Frances Rice Darne Memorial Award, Society for Information Display, 1976; Distinguished Scientist Award, New Mexico Academy of Science, 1976; Fellow Award, Society for Information Display, 1980; Fellow Award, American Ceramic Society, 1981; Japan Society for the Promotion of Science Fellow, 1983; Distinguished Member, Technical Staff Award, Sandia National Laboratories, 1983.

LANDAUER, ROLF, Fellow 1972. Born: February 8, 1927, Suttgart, Germany. Degrees: B.S., 1945, M.S., 1947, Ph.D., 1950, Harvard University.

Fellow Award "For theoretical contributions to solid-state physics and technology and for technical management of research." Other Awards: National Academy of Engineering, 1978.

LANDGREBE, DAVID A., Fellow 1977. Born: April 12, 1934, Huntingburg, Ind. Degrees: B.S.E.E., 1956, M.S.E.E., 1958, Ph.D., 1962, Purdue University.

Fellow Award "For leadership in the application of remote sensing to enhance knowledge of the earth's resources." Other Awards: Medal for Exceptional Scientific Achievement, NASA, 1973; Certificate of Recognition and Cash Award, NASA, 1976.

LANDGREN, GEORGE L., Fellow 1982. Born: July 22, 1919, Duluth, MN. Degrees: B.E.E., 1941, University of Minnesota; M.S.E.E., 1955, Northwestern University.

Fellow Award "For contributions in the analysis of generator dynamic stability, interconnection capability, and transmission reliability"; Prize Paper Award, IEEE Power Engineering Society, 1973.

LANE, F. H., AIEE Fellow 1937. Born: June 7, 1882, St. Louis, Mo. Degrees: M.E., 1904, Illinois Institute of Technology.

LANE, FRANCIS J., Fellow 1967. Born: September 28, 1902, Bradford, England. Degrees: M.Sc.(Electrical Engineering), 1925, University of Leeds.

Fellow Award "For outstanding contributions in the field of high-voltage power transmission, especially in the areas of protective gear, system design, and high-voltage direct current"; William M. Habirshaw Award, 1976, "For his international leadership in development and application of high-voltage ac and dc transmission." Other Awards: Institution of Electrical Engineers, U.K., Premiums: Student Premium, 1926, Institution Premium, 1952, John Hopkinson Premium, 1954, Blumlein-Brown-Willans Premium, 1958; Officer of the Order of the British Empire (OBE), 1951; Fellow, Institution of Electrical Engineers, U.K., 1942; Fellowship of Engineering, U.K., 1976.

LANE, L. JUBIN, Fellow 1983. Born: February 19, 1927, Morganton, North Carolina. Degrees: B.E.E., 1950, North Carolina State College; M.S.E.E., 1972, University of Virginia.

Fellow Award "For the development of static excitation systems for large electric generators." Other Awards: Eta Kappa Nu; Tau Beta Pi; Phi Eta Sigma; Phi Kappa Phi.

LANG, WILLIAM W., Fellow 1971. Born: 1926, Mass. Degrees: B.S., 1946, Iowa State University; M.S., 1949, Massachusetts Institute of Technology; Ph.D., 1958, Iowa State University.

Fellow Award "For contributions in engineering acoustics, noise control, digital processing of acoustic signals, and in establishing acoustical standards"; Achievement Award, IEEE Group on Audio and Electroacoustics, 1970, "For contributions to audio technology as documented in the publications of the IEEE." Other Awards: Fellow, Acoustical Society of America, 1967; Fellow, Audio Engineering Society, 1969; National Academy of Engineering, 1978; Fellow, American Association for the Advancement of Science, 1982; P.E., New York State.

LANGLEY, RALPH H., IRE Fellow 1929. Born: January 5, 1889, New York, N.Y. Degrees: E.E., 1913, Columbia University.

LANGMUIR, DAVID B., IRE Fellow 1962. Born: December 14, 1908, Los Angeles, Calif. Degrees: B.S., 1931, Yale University; Sc.D., 1935, Massachusetts Institute of Technology.

Fellow Award "For contributions to the field of electrostatic propulsion."

LANTZ, MARTIN J., AIEE Fellow 1962. Born: February 17, 1905, Pendleton, Oreg. Degrees: B.S., 1927, M.S.(Electrical Engineering), 1932, Oregon State University.

Fellow Award "For contributions to the technology of power system protection." Other Awards: Kappa Kappa Psi, 1926; Eta Kappa Nu, 1927; Silver Medal, U.S. Dept. of Interior, 1965.

LAPOSTOLLE, PIERRE M., Fellow 1964. Born: May 29, 1922, Vanves, Seine, France. Degrees: Ing., 1941, Ecole Polytechnique, Paris; D.Sc., 1947, University of Paris.

Fellow Award "For contributions in the field of electron optics, traveling-wave tubes, and particle acceleration." Other Awards: Medaille Blondel, 1958; Prix Petit d'Ormoy de l'Academie des Sciences, 1975.

LA ROSA, RICHARD, Fellow 1982. Born: October 4, 1925. Degrees: B.E.E., 1947, M.E.E., 1947, D.E.E., 1953, Polytechnic Institute of Brooklyn.

Fellow Award "For contributions to the development of electroacoustic signal processing devices."

LARSEN, PHILLIP N., Fellow 1974. Born: February 27, 1929, Montrose, Colo. Degrees: B.S.E.E., 1950,

Colorado State University; M.S.E.E., 1953, Ph.D., 1956, University of Illinois; M.S.(Business Administration), 1967, George Washington University.

Fellow Award "For leadership in research and development of military electronics systems." Other Awards: Distinguished Service Medal; Silver Star; Legion of Merit with Oak Leaf Cluster; Distinguished Flying Cross; Meritorious Service Medal; Air Medal with Eleven Oak Leaf Clusters; Air Force Commendation Medal; Vietnamese Cross of Gallantry with Gold Star; Fellow, American Association for the Advancement of Science; Sigma Xi; Eta Kappa Nu; Sigma Tau; Pi Mu Epsilon; Kappa Mu Epsilon; Registered Professional Engineer; Brigadier General, USAF (Retired).

LARSON, ARVID G., Fellow 1984. Born: July 26, 1937, Chicago, IL. Degrees: B.S.E.E., 1959, Illinois Institute of Technology; M.S.E.E., 1966, Ph.D., 1973, Stanford University.

Fellow Award "For contributions to real-time digital processing techniques applied to the spectrum analysis of radar"; IEEE Centennial Medal, 1984. Other Awards: Sigma Xi, 1974.

LARSON, HARRY T., Fellow 1971. Born: October 16, 1921, Berkeley, Calif. Degrees: B.S.E.E.(with highest honors), 1947, University of California at Berkeley; M.S.E.E., 1954, University of California at Los Angeles.

Fellow Award "For contributions to the applications and social implications of computer systems"; IEEE Centennial Medal, 1984. Other Awards: Eta Kappa Nu; Tau Beta Pi; Sigma Xi.

LARSON, ROBERT E., Fellow 1973. Born: September 19, 1938, Stockton, Calif. Degrees: S.B., 1960, Massachusetts Institute of Technology; M.S., 1961, Ph.D., 1964, Stanford University.

Fellow Award "For contributions in the application of modern control and estimation theory and algorithms to defense and industrial problems"; Best Paper Award, IEEE Automatic Control Group, 1966, "In recognition of his contribution toward furthering the objectives of the IEEE Automatic Control Group, as exemplified by his outstanding technical paper entitled 'Dynamic Programming with Reduced Computational Requirements'"; President, IEEE Control Systems Society, 1975-1976; Division I Director, IEEE, 1978-1979; Vice President, Technical Activities, IEEE, 1980-1981; President, IEEE, 1982. Other Awards; Donald P. Eckman Award, American Automatic Control Council, 1968; Outstanding Young Electrical Engineer, Eta Kappa Nu, 1969.

LASSEN, EIVIND U., AIEE Fellow 1946. Born: September 30, 1900, Glemmen, Norway. Degrees: E.E., 1923, Eid. Technische Hochschule Zurich, Switzerland.

Other Awards: Citation of Honor, Purdue University, 1957; Award, National Electrical Manufacturers Association, 1965; Award, National Fire Protection Association, 1966.

LATHROP, JAY W., Fellow 1976. Born: September 6, 1927, Bangor, Maine. Degrees: B.S., 1948, M.S. 1949, Ph.D., 1952, Massachusetts Institute of Technology.

Fellow Award "For contributions to the technology of monolithic and hybrid integrated circuits." Other Awards: Meritorious Civilian Service Award, U.S. Dept. of Army, 1959.

LAUDER, ARTHUR H., AIEE Fellow 1949. Born: December 27, 1900, Evanston, Wyo. Degrees: B.S.E.E., 1922, University of Wyoming.

Other Awards: Charles A. Coffin Award, General Electric Co., 1938; Member: National Society of Professional Engineers, New York Society of Professional Engineers; Phi Kappa Phi.

LAURENT, PIERRE G., Fellow 1970. Born: February 12, 1904, Brest, France. Degrees: 1924, Ecole Polytechnique, Paris.

Fellow Award "For achievements in the development and design of electrical apparatus and their application to transmission systems." Other Awards: Eric Gerard Medal, Institut Montefiore, Liege, Belgium, 1937-39; Andre Blondel Medal, Paris, 1948; John Hopkinson Premium, London, 1965; Willans Premium, London, 1965; Laureate Member, Societe Francaise des Electriciens.

LAUVER, ROBERT M., Fellow 1978. Born: August 25, 1934. Degrees: B.S., 1956, University of Connecticut; M.S., 1961, New York University.

Fellow Award "For contributions to ocean acoustic surveillance systems."

LAVERICK, ELIZABETH, Fellow 1971. Born: November 25, 1925, Amersham, England. Degrees: B.Sc., 1946, Ph.D., 1950, Durham University.

Fellow Award "For contributions in the field of millimetric measurements, and for leadership of microwave research and development facilities." Other Awards: Member, Women's Engineering Society; Fellow, Institute of Electrical Engineers (U.K.); Fellow, Institute of Physics; Honorary Fellow, UMIST.

LAVI, YESHAYAHU, Fellow 1971. Born: June 21, 1926, Berlin, Germany. Degrees: E.E., 1953, Columbia University.

Fellow Award "For contributions to the advancement of communication and electronics"; IEEE Centennial Medal, 1984. Other Awards: Israel Defense Prize, 1968.

LAW, HAROLD B., IRE Fellow 1952. Born: September 7, 1911, Douds, Iowa. Degrees: B.S., 1934, Kent State University; M.S., 1936, Ph.D., 1941, Ohio State University.

Fellow Award "For his development of techniques and structural methods leading to practical storage-

and television-tube designs"; Vladimir K. Zworykin Award, IRE, 1955, "For development of techniques and processes resulting in a practical form of shadow-mask tri-color kinescope"; Outstanding Contribution Award, Consumer Electronics Groups, 1966; IEEE Lamme Award, 1975, "For outstanding contributions in developing color picture tubes, including the fabrication techniques which made color television practical." Other Awards: Award, Television Broadcasters Association, 1946; David Sarnoff Outstanding Team Award, 1961; Frances Rice Darne Award, Society for Information Display, 1975; National Academy of Engineering, 1979.

LAWRENCE, ROBERT F., Fellow 1964. Born: May 2, 1921, Watertown, N.Y. Degrees: B.E.E., 1943, Pratt Institute.

Fellow Award "For contributions in the field of power transmission and distribution"; Paper Prize, 1946. Other Awards: Order of the Yugoslav Flag with Gold Wreath, presented by the President of Yugoslavia in New York City, 1977.

LAWRENSON, PETER J., Fellow 1976. Born: March 12, 1933, Prescot, Lancashire, England. Degrees: B.Sc., 1954, M.Sc., 1956, D.Sc., 1971, University of Manchester.

Fellow Award "For contributions to electromagnetic field analysis and to the theory and design of electrical machines." Other Awards: Duddell Scholarship, 1951-1954, Crompton Premium, 1957, John Hopkinson Premium, 1965, Crompton Premium, 1967, The Institution Premium, 1981, Institution of Electrical Engineers, U.K.; Fellow, Institution of Electrical Engineers, U.K., 1974; Fellowship of Engineering, U.K., 1980; Fellow, Royal Society, 1982; James Alfred Ewing Medal, Institution of Civil Engineers & Royal Society, 1983.

LAWRIE, DUNCAN H., Fellow 1984. Born: April 26, 1943, Chicago, IL. Degrees: B.A., 1966, DePauw University; B.S.E.E., 1966, Purdue; M.S., 1969, Ph.D., 1973, University of Illinois.

Fellow Award "For contributions to very-high-speed computing systems." Other Awards: Most Original Paper Award, 1980 International Conference on Parallel Processing; Editor, Communications of the ACM (1982); General Chairman, Fourth International Conference on Distributed Computing Systems (1984); Excellence in Teaching, University of Illinois, 1975-1979, 1982-1983; Eta Kappa Nu; Tau Beta Pi; Sigma Xi.

LEADABRAND, RAY L., Fellow 1970. Born: 1927, California. Degrees: B.S., 1950, San Jose State College; M.S., 1953, Stanford University.

Fellow Award "For contributions to the electromagnetic measurements of the aurora and of the radar characteristics of nuclear explosions." Other Awards: Member, RESA (Scientific Research Society of America); Member, American Association for the Advancement of Science; Member, American Geophyscal Union; Sigma Xi; Board Member, AUVS (Association for Unmanned Vehicle Systems).

LEAL, ALEXANDRE H., Fellow 1976. Born: June 20, 1910, Rio de Janeiro, Brazil. Degrees: E. and C.E., 1932, Escola Politecnica, Rio de Janeiro; S.M., 1935, Massachusetts Institute of Technology; 1959, Brazilian War College, Rio de Janeiro.

Fellow Award "For leadership in planning and development of power systems."

LEARNED, VINCENT R., IRE Fellow 1955. Born: January 21, 1917, San Jose, Calif. Degrees: B.S.E.E., 1938, University of California; Ph.D., 1943, Stanford University.

Fellow Award "For his contributions to research and development of microwave electron tubes."

LEAS, J. WESLEY, Fellow 1965. Born: June, 14, 1916, Delaware, Ohio. Degrees: B.S.E.E., 1938, Ohio State University.

Fellow Award "For leadership in the development of electronic data processing and communications." Other Awards: Outstanding Engineering Alumnus Award, Ohio State University, 1960; Distinguished Service Award, Ohio State University, 1975; Distinguished Engineering Alumnus Award, Ohio State University, 1983.

LEBENBAUM, MATTHEW T., Fellow 1964. Born: November 29, 1917, Portland, Oreg. Degrees: B.A., 1938, Stanford University; M.S., 1945, Massachusetts Institute of Technology; M.S.E.E., 1975, Stanford University.

Fellow Award "For contributions and leadership in the development and design of low-noise amplifiers"; Outstanding Engineering Manager Award, Long Island Section, IEEE, 1979. Other Awards: Tau Beta Pi; Phi Beta Kappa; Sigma Xi.

LEBENS, JOHN C., AIEE Fellow 1962. Born: July 31, 1911, St. Louis, Mo. Degrees: B.S., 1932, M.S., 1933, Washington University.

Fellow Award "For contributions to the field of low voltage electrical protection"; Outstanding Young Electrical Engineer Award, AIEE, St. Louis Section, 1947; Paper Prize, AIEE, St. Louis Section, 1952, 1958.

LEBOW, IRWIN L., Fellow 1978. Born: April 27, 1926, Boston, Mass. Degrees: S.B., 1948, Ph.D., 1951, Massachusetts Institute of Technology.

Fellow Award "For contributions to satellite communications technology." Other Awards: Fellow, American Physical Society, 1957; Meritorious Executive, Senior Executive Service, 1980.

LECHNER, BERNARD J., Fellow 1977. Born: January 25, 1932, New York, N.Y. Degrees: B.S.E.E., 1957, Columbia University.

Fellow Award "For contributions to flat panel displays and two-way cable television systems." Other Awards: David Sarnoff Team Award in Science, RCA, 1962; Achievement Award, RCA, 1962, 1967; Best Paper Award, International Solid-State Circuits Conference, 1965; Fellow, Society for Information Display, 1970; Frances Rice Darne Award, Society for Information Display, 1971; Beatrice Winner Award, 1983.

LEE, CHESTER C.Y., Fellow 1971. Born: December 10, 1926. Degrees: B.E.E., 1947, Cornell University; M.S.E.E., 1949, Ph.D., 1954, University of Washington.

Fellow Award "For contribution to switching and automata, and for leadership in the use of computers for scientific and commercial applications."

LEE, E. BRUCE, Fellow 1983. Born: February 1, 1932, Brainerd, Minnesota. Degrees: B.S.M.E., 1955, M.S.M.E., 1956, Ph.D., 1960, Univ. of Minnesota.

Fellow Award "For contributions to the foundations of optimal control theory and to engineering education." Other Awards: Gold Medal Award, A.S.M.E.; Fellow, Sigma Tau; Tau Beta Pi; Sigma Xi.

LEE, FRANCIS F., Fellow 1971. Born: January 28, 1927, Nanking, China. Degrees: S.B., 1950, S.M., 1951, Ph.D., 1966, Massachusetts Institute of Technology.

Fellow Award "For contributions to research, engineering, and education in the field of digital systems." Other Awards: Hertz Fellow, 1964-66.

LEE, LOW K., IRE Fellow 1959. Born: February 12, 1916, Oakland, Calif. Degrees: B.S., 1937, M.S., 1939, University of California at Berkeley; Ph.D., 1979, California Western University.

Fellow Award "For contributions in the development of subminiature components and production techniques."

LEE, MARVIN, AIEE Fellow 1960. Born: March 31, 1901, New York, N.Y. Degrees: E.E., 1922, Polytechnic Institute of New York.

Fellow Award "For contributions to engineering development of electrical connectors and connector tools."

LEE, RALPH H., Fellow 1971. Born: April 6, 1911, Las Vegas, Nev. Degrees: B.S.E.E., 1934, University of Alberta.

Fellow Award "For contributions to reliability and safety of electric industrial distribution and utilization systems and components"; Achievement Award, IEEE Industrial and Commercial Power Systems Committee of the Industry Applications Society, 1975, "For leadership, counseling and technical contributions made through his long dedication to this Committee"; Outstanding Achievement Award, IEEE Industry Applications Society, 1976; IEEE Centennial Medal, 1984.

LEE, REUBEN, IRE Fellow 1954. Born: November 8, 1902, Shirland, Derby, England. Degrees: B.S.E.E.,

1924, West Virginia University.

Fellow Award "For his contributions to the design and development of inductive components." Other Awards: Order of Merit, Westinghouse Electric, 1960.

LEE, SHUNG-WU, Fellow 1981. Born: December 15, 1940. Degrees: B.S., 1961, Cheng Kung University, Taiwan; M.S., 1964, Ph.D., 1966, University of Illinois at Urbana.

Fellow Award "For contributions to phased array antennas, geometrical theory of diffraction, and engineering education.

LEE, T. H., Fellow 1965. Born: May 11, 1923, Shanghai, China. Degrees: B.S.M.E., 1946, National Chiao Tung University; M.S.E.E., 1950, Union College; Ph.D., 1954, Rensselaer Polytechnic Institute.

Fellow Award "For contributions in the field of high temperature arc plasma and in research and development of high power circuit interrupters"; President, IEEE Power Engineering Society, 1964-1965; Haraden Pratt Award, IEEE, 1983; Power Life Award, IEEE. Other Awards: Member, National Academy of Engineering; Philip Spurn Professor of Energy Processing, Massachusetts Institute of Technology.

LEE, TIEN PEI, Fellow 1984. Born: September 8, 1933, Nanking, China. Degrees: B.S., 1957, National Taiwan University; M.S., 1959, Ohio State University; Ph.D., 1963, Stanford University.

Fellow Award "For contributions to semiconductor elements for lightwave communications." Other Awards: Distinguished Staff Award, Bell Laboratories 1983.

LEE, WILLIAM CHIEN-YEH, Fellow 1982. Born: July 20, 1932, London, England. Degrees: B.Sc., 1954, Chinese Naval Academy, Taiwan; M.S., 1960, Ph.D., 1963, Ohio State University.

Fellow Award "For contributions to the analysis, modeling, and conceptual and practical design of mobile telephone systems."

LEEDS, WINTHROP M., AIEE Fellow 1948. Born: August 18, 1905, Moorestown, N.J. Degrees: B.S., 1926, Haverford College; M.S., 1930, Ph.D., 1945, University of Pittsburgh.

Award: Lamme Scholarship, 1937; National Paper Prize, 1943; Paper Prize, District 2, 1952; Chairman, Research Committee, 1958-1960; Lamme Medal, 1971. Other Awards: Special Patent Awards, Westinghouse Electric, 1941, 1952, 1966.

LEEDY, HALDON A., AIEE Fellow 1959. Born: April 15, 1910, Fremont, Ohio. Degrees: B.A., 1933, North Central College; M.A., 1935, Ph.D., 1938, University of Illinois.

Fellow Award "For contributions to the organization and administration of an engineering research center devoted to community betterment." Other Awards: Fellow, Acoustical Society of America; Fellow, American

Association for the Advancement of Science; American Physical Society; Sigma Xi; Tau Beta Pi.

LEFKOWITZ, IRVING, Fellow 1981. Born: July 8, 1921, New York, NY. Degrees: B.Ch.E., 1943, Cooper Union School of Engineering; M.S. (Instrum. Eng.), 1955, Ph.D., 1958, Case Institute of Technology.

Fellow Award "For contributions to computer control of industrial process systems." Other Awards: NATO Post-Doctoral Fellow, 1962; American Automatic Control Council's Control Heritage Award, 1982; Fellow, AAAS, 1982.

LEHAN, FRANK W., IRE Fellow 1962. Born: January 26, 1923, Los Angeles, Calif. Degrees: B.S.E.E., 1944, California Institute of Technology.

Fellow Award "For contributions to space electronics and telemetry"; Annual Award, IRE, Professional Group on Space Electronics and Telemetry, "For valuable contribution to the field of space electronics and telemetry." Other Awards: Outstanding Merit Award, Mohawk Valley Engineers Executive Council, 1966; Member, National Academy of Engineering.

LEHMANN, GERARD J., Fellow 1964. Born: April 6, 1909, Paris, France. Degrees: Ing. Arts et Manufactures, 1931, Ecole Centrale, Paris.

Fellow Award "For his origination and development of techniques and theory in the fields of communication, radio navigation, radar, and engineering education." Other Awards: President, Societe Francaise des Electroniciens et des Radioelectriciens, 1963; Prix de L'Academie des Sciences, Paris, 1965; President, Societe des Ingenieurs Civils de France, 1967.

LEHOVEC, KURT, IRE Fellow 1956. Born: June 12, 1918, Ladowitz, Czechoslovakia. Degrees: Ph.D., 1941, University of Prague.

Fellow Award "For basic work in the field of semiconduction." Other Awards: Fellow, American Physical Society, 1953.

LEIB, FRANCIS E., Fellow 1968. Born: June 8, 1905, Rushville, Ill. Degrees: B.S.E.E., 1926, University of Illinois.

Fellow Award "For contributions to overhead line materials and construction practices."

LEIFER, MEYER, IRE Fellow 1955. Born: April 18, 1914, New York, N.Y. Degrees: B.S., 1933, Brooklyn College; A.M., 1935, Columbia University.

Fellow Award "For his contributions to the fields of electronic navigation and information theory." Other Awards: Pi Mu Epsilon, 1933; Sigma Pi Sigma; Sigma Xi; Scientific Research Society of America.

LEITH, EMMETT N., Fellow 1971. Born: March 12, 1927, Detroit, Mich. Degrees: B.S., 1949, M.S., 1952, Ph.D., 1978, Wayne State University.

Fellow Award "For contributions to holography and the field of optical data processing"; Morris N. Liebman Award, IEEE, 1967. Other Awards: Gordon Memorial

Award, S.P.I.E., 1965; Man of the Year, Industrial Research Award, 1966; U.S. Camera Achievement Award, U.S. CAMERA and TRAVEL MAGAZINE, 1967; Daedalion Minuteman Award, U.S. Air Force, 1968; Fellow, Optical Society of America; National Medal of Science, 1979; National Academy of Engineering, 1982.

LEMPEL, ABRAHAM, Fellow 1982. Born: February 10, 1936, Poland. Degrees: B.Sc., 1963, M.Sc., 1965, D.Sc., 1967, Technion-Israel Institute of Technology.

Fellow Award "For contributions to the theory of data complexity, the practice of data compression, and the algebraic analysis and synthesis of digital sequences"; Paper Award (corecipient), IEEE Information Theory Society, 1979.

LEMPERT, JOSEPH, Fellow 1968. Born: July 3, 1913, North Adams, Mass. Degrees: B.S., 1935, Massachusetts Institute of Technology; M.S., 1942, Stevens Institute of Technology.

Fellow Award "For advances in the concept, design and fabrication of x-ray tubes, image tubes, and electron beam welders."

LENDARIS, GEORGE G., Fellow 1983. Born: April 2, 1935, Helper, Utah. Degrees: B.S., 1957, M.S., 1958, Ph.D., 1961 (all in Electrical Engineering), Univ. of Calif., Berkeley.

Fellow Award "For conceiving, developing, and implementing the use of sampled optical diffraction patterns for image analysis"; District First Prize, Best Paper, AIEE, 1959.

LENDER, ADAM, Fellow 1974. Born: 1921, Germany. Degrees: B.S.E.E., 1954, M.S.E.E., 1956, Columbia University; Ph.D., 1972, Stanford University.

Fellow Award "For contributions to data communications"; Paper Award, 1968, "In recognition of his contribution to the technical literature as author of a paper entitled 'Correlative Data Transmission with Coherent Recovery Using Absolute Reference.'" Other Awards: Stuart Ballantine Medal, Franklin Institute, 1983.

LENEHAN, BERNARD E., Fellow 1964. Born: February 18, 1901, Deep Valley, Pa. Degrees: B.S., 1921, Carnegie Institute of Technology.

Fellow Award "For contributions in the field of electrical instruments and measurments"; Morris E. Leeds Award, AIEE, 1962, "For meritorious achievement in the field of electrical measurements."

LENT, STANLEY B., AIEE Fellow 1960. Born: January, 28, 1902, Dorchester, Mass.

Fellow Award "For contributions to transit system electrification and engineering management." Other Awards: Fellow, A.A.A.S., 1967.

LEON, BENJAMIN J., Fellow 1970. Born: March 20, 1932, Austin, Tex. Degrees: B.S., 1954, University of Texas; S.M., 1957, Sc.D., 1959, Massachusetts Institute of Technology.

Fellow Award "For contributions to circuit theory and engineering education"; IEEE Vice President, 1979-1980.

LEONDES, CORNELIUS T., Fellow 1969. Born: July 21, 1927, Philadelphia, Pa. Degrees: B.S.E.E., 1949, M.S.E.E., 1951, Ph.D., 1954, University of Pennsylvania.

Fellow Award "For contributions to the theory and application of computer and control systems technology, and to the advancement of education in the field"; Baker Prize Award, 1970; Barry Carlton Honorable Mention Award, 1973. Other Awards: Guggenheim Fellow, 1962-63; Fulbright Research Scholar, 1962-63.

LEPAGE, WILBUR R., IRE Fellow 1959, AIEE Fellow 1962. Born: November 16, 1911, Kearny, N.J. Degrees: B.S., 1933, Cornell University; M.S., 1939, University of Rochester; Ph.D., 1941, Cornell University.

IRE Fellow Award "For contributions to electronic engineering education and literature"; AIEE Fellow Award "For contributions to electrical engineering education."

LEPSELTER, MARTIN P., Fellow 1972. Born: 1929, New York, N.Y. Degrees: B.M.E., 1951, City College of The City University of New York.

Fellow Award "For contributions to the advancement of transistor and integrated circuit arts"; Daniel C. Hughes, Jr., Memorial Award, ISHM; Jack A. Morton Award, IEEE, 1979. Other Awards: Member, American Society of Mechanical Engineers, 1948.

LEROY, GERARD L., Fellow 1979. Born: July 15, 1928, France. Degrees: Electrical Engineer, 1948, Ecole Breguet, Paris.

Fellow Award "For contribution to the scientific knowledge of long air-gap discharges and to the improvement of testing and measuring techniques of transient phemomena." Other Awards: Societe Francaise des Electriciens, 1963; Blondel Medal, France, 1974.

LESK, I. ARNOLD, Fellow 1964. Born: January 26, 1927, Regina, Saskatchewan, Canada. Degrees: B.Sc., 1948, University of Alberta; M.S., 1949, Ph.D., 1951, University of Illinois.

Fellow Award "For contributions to the advancement of semiconductor technology and device development." Other Awards: American Physical Society; Electrochemical Society; Sigma Xi; Eta Kappa Nu; Phi Kappa Phi; Pi Mu Epsilon.

LESLIE, JOHN R., Fellow 1976. Born: June 9. 1917, Toronto, Ontario, Canada. Degrees: B.A.Sc.(Eng. Physics), 1938, University of Toronto.

Fellow Award "For leadership in the development of electronic devices and systems for the power industry." Other Awards: Wason Medal for Noteworthy Research, American Concrete Institute, 1950.

LESTER, GEORGE N., Fellow 1983. Born: April 7, 1928, Atlanta, Georgia. Degrees: B.E.E., 1950,

Georgia Institute of Technology.

Fellow Award "For contributions to the standardization, design, and application of high-voltage power circuit breakers and to the understanding of transient switching of capacitor banks."

LEVERENZ, HUMBOLDT W., IRE Fellow 1956. Born: July 11, 1909, Chicago, Ill. Degrees: B.A., 1930, Stanford University.

Fellow Award "For contributions to the field of luminescence as applied to electronic devices." Other Awards: Fellow, American Physical Society, 1941; Frank P. Brown Medal, Franklin Institute, 1954; Fellow, Optical Society of America, 1961; Fellow, American Association for the Advancement of Science, 1966; Member, National Academy of Engineering, 1970.

LEVERTON, WALTER F., Fellow 1975. Born: December 24, 1922, Imperial, Saskatchewan, Canada. Degrees: B.A., 1946, M.A., 1948, University of Saskatchewan; Ph.D., 1950, University of British Columbia.

Fellow Award "For leadership in the development of military space systems, and for contributions to semiconductor technology."

LE VESCONTE, LESTER B., AIEE Fellow 1951. Born: December 4, 1903, Scott Co., Minn. Degrees: B.A., 1926, Macalester College; B.S.E.E., 1926, University of Minnesota.

Other Awards: Initial Chairman, Computer Application Subcommittee and PICA; Chairman, Power System Engineering Committee; Eta Kappa Nu; Tau Beta Pi; Prize Paper, "Lightning Protection in Extra-High-Voltage Stations".

LEVINE, ARNOLD M., IRE Fellow 1962. Born: August 15, 1916, Preston, Conn. Degrees: B.S., 1939, Tri-State College; M.S., 1940, University of Iowa; D.Sc.(Hon.), 1960, Tri-State College.

Fellow Award "For contributions to missile guidance." Other Awards: American Rocket Society; San Fernando Valley, Engineer of the Year Award, Engineering Council, 1968; Institute for the Advancement of Engineering Fellow; "Who's Who in America"; "Who's Who in Finance and Industry"; Distinguished Service Award, ITT; 50 patents.

LEVINE, JULES D., Fellow 1980. Born: June 24, 1937, New York, NY. Degrees: B.S.M.E., 1959, Columbia University; Ph.D., 1963, Massachusetts Institute of Technology.

Fellow Award "For contributions to the physics of semiconductor surfaces"; Founder and Chairman, IEEE Power Engineering Society, Princeton Section, 1974-1978; Chairman, Princeton Section, IEEE, 1977-1978. Other Awards: RCA Outstanding Achievement Award, 1968, 1972, 1975.

LEVINE, SAMUEL, Fellow 1968. Born: May 2, 1916, New York, N.Y. Degrees: B.E.E., 1938, City College of the City University of New York.

Fellow Award "For contributions to the applications of real time data processing systems"; IEEE Centennial Medal, 1984. Other Awards: Fellow, American Association for the Advancement of Science, 1968.

LEVOY, LOUIS G., JR., AIEE Fellow 1955. Born: March 18, 1907, Webster, S.D. Degrees: B.S.E.E., 1929, University of California at Berkeley; Advanced Engineering Program, General Electric.

Fellow Award "For his contributions to and development of electronic control, welding circuits and electric equipment for aircraft." Other Awards: 22 technical papers.

LEVY, RALPH, Fellow 1973. Born: April 12, 1932, London, England. Degrees: B.A., 1953, M.A., 1957, St. Catharine's College, Cambridge; Ph.D., 1966, Queen Mary College, University of London.

Fellow Award "For contributions to the theory and design of transmission line networks."

LEWIN, LEONARD, Fellow 1981. Born: July 22, 1919, Essex, England. Degrees: D.Sc.(Hon.), 1967, University of Colorado.

Fellow Award "For contributions to the theory of waveguides, antennas, and microstrip transmission lines"; Microwave Prize, 1963; W. G. Baker Award, 1963. Other Awards: Premium Award, Institution of Electrical Engineers, 1952, 1960; Fulbright Scholar, 1982.

LEWIS, BERNARD L., Fellow 1984. Born: December 19, 1923, Storm Lake, IA. Degrees: B.Sc., 1946, M.Sc.(Physics), 1948, Tulane University.

Fellow Award "For contributions to radar signal processing." Other Awards: Navy Distinguished Civilian Service Award: Six Best Papers of the Year Awards, Radar Division of NRL; 180 Patent Awards.

LEWIS, EDWIN R., Fellow 1976. Born: July 14, 1934, Los Angeles, Calif. Degrees: A.B.(Biological Sciences), 1956, M.S.E.E., 1957, E.E., 1959, Ph.D.(E.E.), 1961, Stanford University.

Fellow Award "For contributions to the modeling of neurological processes and the understanding of neurological transducers." Other Awards: Fellow, Neurosciences Research Program, 1966; Distinguished Teaching Citation, University of California, 1972.

LEWIS, FRANK M., AIEE Fellow 1944. Born: May 13, 1892, Portland, Oreg. Degrees: B.S.E.E., 1917, University of Washington.

LEWIS, GEORGE A., AIEE Fellow 1961. Born: February 24, 1909, Cuba, Ill. Degrees: B.S.E.E., 1932, South Dakota School of Mines and Technology.

Fellow Award "For contributions to the planning and design of a large hydro-electric power generation and transmission system." Other Awards: Meritorious Service Award, Dept. of Interior, 1969; Privileged Member, Montana Society of Engineers, 1972.

LEWIS, JOHN B., Fellow 1972. Born: May 28, 1922, Poughkeepsie, N.Y. Degrees: B.S., 1944, B.S.E.E., 1947, Duke University; M.S., 1951, University of Tennessee; Ph.D., 1961, Purdue University.

Fellow Award "For contributions to sampled data control systems, optimal control, and digital and hybrid simulation."

LEWIS, MELVIN G., AIEE Fellow 1959. Born: August 9, 1912, Santa Cruz, Calif. Degrees: B.S.E.E., 1935, Santa Clara University.

Fellow Award "For contributions to design and construction of large electrical plants and installations."

LEWIS, PHILIP M., II, Fellow 1979. Born: May 30, 1931. Degrees: B.E.E., 1952, Rensselaer Polytechnic Institute; M.S., 1954, Sc.D., 1956, Massachusetts Institute of Technology.

Fellow Award "For contributions to switching theory, compiler design, and computer science and for leadership in professional activities."

LEWIS, W. DEMING, IRE Fellow 1954. Born: January 6, 1915, Augusta, Ga. Degrees: A.B., 1935, Harvard University; B.A., 1938, University of Oxford; M.A., 1939, Ph.D., 1941, Harvard University; M.A., 1945, University of Oxford; LL.D.(Hon.), 1965, Lafayette College; LL.D.(Hon.), 1966, Rutgers University; LL.D.(Hon.), 1966, Hahnemann Medical College; LH.D.(Hon.), 1966, Moravian College; LL.D.(Hon.), 1968, Muhlenberg College; D.Sc.(Hon.), 1972, Medical College of Pennsylvania; D.Eng.(Hon.), 1974, Lehigh University.

Fellow Award "For his contributions to research, particularly in the fields of microwave filters and switching systems." Other Awards: Member, National Academy of Engineering (Council 1972-1978).

LEWIS, WILLIAM A., AIEE Fellow 1945. Born: January 21, 1904, Harriman, Tenn. Degrees: B.S., 1926, M.S., 1927, Ph.D.(summa cum laude), 1929, California Institute of Technology.

Awards: IEEE Power Engineering Society, Power System Relaying Committee Award, First Award, 1977, "For distinguished service to the Committee"; Member, AIEE Board of Directors, 1956-1961; IEEE Centennial Medal, 1984.

LEYDORF, GEORGE F., IRE Fellow 1962. Born: April 24, 1908, Perrysburg, Ohio. Degrees: B.E.E., 1931, M.Sc., 1933, Ohio State University.

Fellow Award "For contributions to high power radio broadcasting." Other Awards: Public Service Certificate, American Radio Relay League, 1937; Honorable Mention, Eta Kappa Nu Recognition, 1942; Naval Ordnance Development Award, 1946.

LEYLAND, SIMEON C., AIEE Fellow 1951. Born: July 8, 1902, Fall River, Mass. Degrees: B.S.E.E., 1924, Worcester Polytechnic Institute.

Other Awards: Order of Merit, Westinghouse Electric, 1966.

LI, CHING-CHUNG, Fellow 1978. Born: March 30, 1932, Changshu, Kiangsu, China. Degrees: B.S.E.E., 1954, National Taiwan University; M.S., 1956, Ph.D., 1961, Northwestern University.

Fellow Award "For contributions to biocybernetics."

LI, TINGYE, Fellow 1972. Born: July 7, 1931, Nanking, China. Degrees: B.Sc., 1953, University of Witwatersrand, South Africa; M.S., 1955, Ph.D., 1958, Northwestern University.

Fellow Award "For contributions to laser resonator theory and the application of digital computation to the analysis of lasers and microwave antennas"; W.R.G. Baker Prize Award, 1975; David Sarnoff Award, 1979. Other Awards: Achievement Award, Chinese Institute of Engineers(U.S.), 1978; Member, National Academy of Engineering, 1980; Merit Award, Northwestern University Alumni Association, 1981; Achievement Award, Chinese American Academic and Professional Association, 1983.

LIBOIS, LOUIS JOSEPH, Fellow 1969. Born: July 22, 1921, Simandre, Ain, France. Degrees: Engr., 1941, Ecole Polytechnique; Engr., 1945, Ecole Nationale Superieure des Telecommunications.

Fellow Award "For contributions to the application of computers to telecommunications and to the progress of microwave links." Other Awards: Prix General FERRIE, France, 1953; Medaille d'or de e'industrie nationale, France, 1968.

LIECHTI, CHARLES A., Fellow 1979. Born: March 12, 1937, Basel, Switzerland. Degrees: Diploma Physics, 1962, Ph.D. E.E., 1967, Swiss Federal Institute of Technology.

Fellow Award "For contributions to the development of gallium- arsenide field-effect transistors, microwave amplifiers, and gallium-arsenide digital integrated circuits"; Two Outstanding Contributed Paper Awards, IEEE International Solid-State Circuits Conference, 1973, 1974; Microwave Prize, 1975, National Lecturer, IEEE Microwave Theory and Techniques Society, 1979.

LIED, FINN, Fellow 1968.

Fellow Award "For distinguished technical leadership and contribution in ionospheric research."

LIEN, JESSE R., Fellow 1981. Born: November 15, 1917, Portland, OR. Degrees: B.A. (Physics), 1941, Reed College; M.A. (Physics), 1953, Boston University.

Fellow Award "For leadership in electronics research and engineering." Other Awards: Member, Signal Corp., Research and Development Advisory Council, 1959-1962; Member, Security Advisory Board, NSA, 1962-1971; Honorary Scientific Advisor, USA Electronics Command, 1965.

LIMB, JOHN O., Fellow 1978. Born: August 19, 1939. Degrees: B.E.E., 1963, Ph.D.(Electrical Engineering),

1967, University of Western Australia.

Fellow Award "For contributions to efficient coding of color and monochrome video signals"; Leonard G. Abraham Prize Paper Award, IEEE Communications Society, 1973; Donald G. Fink Prize Award (with A.N. Netravali), 1982; Board of Governors, IEEE Communications Society, 1983-1985.

LIN, HUNG C., Fellow 1969. Born: August 8, 1919, Shanghai, China. Degrees: B.S.E.E., 1941, Chiao Tung University; M.S.E., 1948, University of Michigan; D.Eng., 1956, Polytechnic Institute of Brooklyn.

Fellow Award "For contributions to semiconductor electronics and circuits and pioneering of integrated circuits"; J. J. Ebers Award, IEEE Electron Devices Society, 1978. Other Awards: Sigma Xi, 1956; Patent Awards, Westinghouse Electric Corp., 1965, 1966, 1971; Phi Tau Phi Scholastic Honor Society, 1966; NASA Certificate, Technology Utilization Program, 1970; Achievement Award, Chinese Institute of Engineers, 1974.

LIN, PEN-MIN, Fellow 1981. Born: October 17, 1928, Liaoning, China. Degrees: B.S.E.E., 1950, National Taiwan University; M.S.E.E., 1956, North Carolina State University; Ph.D.(E.E.), 1960, Purdue University.

Fellow Award "For contributions to computer-aided analysis and applications of graph theory to circuits and systems."

LIN, SHU, Fellow 1980. Born: May 20, 1936, Nanking, China. Degrees: B.S.E.E., 1959, National Taiwan University; M.S., 1964, Ph.D., 1965 Rice University.

Fellow Award "For contributions in coding theory and engineering education."

LIND, ANTHONY HOWARD, Fellow 1978. Born: April 28, 1921. Degrees: B.S.E.E., 1946, M.S.E.E., 1949, University of Wisconsin, Madison.

Fellow Award "For technical leadership in the design and product development of video tape recorders and color television cameras." Other Awards: Fellow, Society of Motion Picture and Television Engineers, 1964.

LINDER, CLARENCE H., AIEE Fellow 1957. Born: January 18, 1903, Ogema, Wisc. Degrees: B.S., 1924, M.S., 1927, University of Texas; D.E.(Hon.), 1955, Worcester Polytechnic Institute; D.E.(Hon.), 1956, Clarkson College; LL.D.(Hon.), 1965, Union College; LL.D.(Hon.), 1972, Lehigh University.

Fellow Award "For his contributions in coordinating electrical development and cultivating engineering talent"; The Haraden Pratt Award, IEEE, 1976. Other Awards: Fellow, American Society of Mechanical Engineers; Fellow, AAAS, Distinguished Alumnus Award, University of Texas, 1962; Founding Member, National Academy of Engineering.

LINDERS, JOHN R., Fellow 1967. Born: September 30, 1908, Cleveland, Ohio. Degrees: B.S.E.E.(cum laude), 1930, Lafayette College.

Fellow Award "For major contributions to the advancement of the philosophy and techniques used for the protection of electrical systems and apparatus." Other Awards: Military Commendation, 1945; Tau Beta Pi; Eta Kappa Nu; Distinguished Service Award, Power System Relaying Committee, 1983.

LINDHOLM, FREDRIK A., Fellow 1978. Born: February 26, 1936, Tacoma, Wash. Degrees: B.S., 1958, M.S., 1960, Stanford University; Ph.D., 1963, University of Arizona (Tucson).

Fellow Award "For contributions to transistor and solar cell modeling"; Outstanding Paper Awards, IEEE International Solid-State Circuits Conference, Philadelphia, 1963, 1965; Certificate of Merit, IEEE Photovoltaic Specialists Conference, Washington, D.C. Other Awards: American Men and Women of Science, 1966; Outstanding Faculty Award, University of Florida, 1975. Outstanding Research Award, American Society of Engineering Education, 1978; Florida Blue Key Distinguished Faculty Award, 1978.

LINDSEY, WILLIAM C., Fellow 1974. Born: December 6, 1936. Degrees: B.S.E.E., 1958, University of Arkansas; M.S.E.E., 1959, Ph.D., 1962, Purdue University.

Fellow Award "For leadership in the development of communications systems." Other Awards: Senior Scholar, University of Arkansas, 1958; Philip Sporn Award, 1961; Outstanding Performance Award, Mariner Mars Project, 1969; Certificate of Recognition, NASA, 1971; Outstanding and Dedicated Service Award, International Telemetering Foundation, 1972; Tau Beta Pi; Eta Kappa Nu; Pi Mu Epsilon; Sigma Pi Sigma; Sigma Xi.

LINDVALL, FREDERICK C., AIEE Fellow 1943. Born: May 29, 1903, Moline, Ill. Degrees: B.S., 1924, University of Illinois; Ph.D., 1928, California Institute of Technology; D.Sc.(Hon.), 1963, National University of Ireland; D.Eng.(Hon.), 1966, Purdue University.

Other Awards; Lamme Medal, Honorary Member, American Society for Engineering Education, 1966; National Academy of Engineering.

LINK, FRED M., Fellow 1973. Born: October 11, 1904, York, Pa. Degrees: B.S.E.E., 1927, Pennsylvania State University.

Fellow Award "For contributions to mobile radio communications"; Member, Board of Directors, IEEE Vehicular Technology Society; IEEE Centennial Medal, 1984. Other Awards: Fellow and President, Radio Club of America, 1968-date; Honorary Life Member, Associated Public Safety Communications Officers, 1970; DeForest Audion Award, DeForest Pioneers Association, 1971; DeForest Award, Veteran Wireless Operators Association, 1973; Sarnoff Citation, Radio Club of America, 1976; Honorary Life Member IMSA, 1980; Honorary Life Member, ENTELEC Association, 1982; Allen B. DuMont Citation, Radio Club of Ameri-

ca, 1983; Honorary Life Member, Society of Wireless Pioneers, 1983; Member: Tau Beta Pi; Eta Kappa Nu; Pi Kappa Alpha.

LINVILL, JOHN G., IRE Fellow 1960. Born: August 8, 1919, Kansas City, Mo. Degrees: A.B., 1941, William Jewell College; S.B., 1943, S.M., 1945, Sc.D., 1949, Massachusetts Institute of Technology; Sc.D.(Hon.), 1966, University of Louvain, Belgium.

Fellow Award "For contributions to network theory and transistor circuits"; IEEE Education Medal, 1976, "For leadership as a teacher, author and administrator, and for contributions in solid-state electronics and technology." Other Awards: Member, National Academy of Engineering, 1971; Member, American Academy of Arts and Sciences, 1974; John Scott Award, City of Philadelphia, 1980; Medal of Achievement, American Electronics Association, 1983.

LINVILLE, THOMAS M., AIEE Fellow 1947. Born: March 3, 1904, Washington, D.C. Degrees: E.E., 1926, University of Virginia; A.M.P., 1950, Harvard University.

Fellow Award "For contributions to theory, design, and application of electrical motors, generators, and drive systems." Other Awards: Charles A. Coffin Award, General Electric, 1946; Citation, U.S. Dept. of Navy, 1946; Engineer of Year Award, Schenectady Professional Engineers Society, 1960; Eminent Member, Tau Beta Pi, 1967; Fellow, American Society of Mechanical Engineers; President, Tau Beta Pi Association, 1974-76, National Society of Professional Engineers, 1966-67.

LIOU, MING-LEI, Fellow 1979. Born: January 6, 1935, Tinghai, Chekiang, China. Degrees: B.S.E.E., 1956, National Taiwan University; M.S.E.E., 1961, Drexel University; Ph.D., 1964, Stanford University.

Fellow Award "For contributions to the application of analysis and computer-aided design to communications circuits and systems"; Special Prize Paper Award, IEEE Circuits and Systems Society 1973; Darlington Prize Paper Award, IEEE Circuits and Systems Society, 1977.

LISCHER, LUDWIG F., Fellow 1967. Born: March 1, 1915, Darmstadt, Germany. Degrees: B.S.E.E., 1937, Dr.Eng.(Hon.), 1976, Purdue University.

Fellow Award "For developing new technical approaches in planning generating station and major transmission line additions, also for pioneering extra high voltage interconnections between large integrated electric power systems." Other Awards: Distinguished Alumnus, Purdue University, 1965; National Academy of Engineering, 1978.

LITCHFORD, GEORGE B., Fellow 1964. Born: August 12, 1918, Long Beach, Calif. Degrees: B.A., 1941, Reed College.

Fellow Award "For contributions in the field of air

navigation and traffic control"; Pioneer Award, S-AES, IEEE, 1974, "For pioneering contributions to the design of precision omnidirectional microwave beacons; Lamme Medal, IEEE, 1981. Other Awards: RTCA-Collier Trophy, 1950; Fellow, American Institute of Aeronautics and Astronautics, 1970; Wright Brothers Medal/Lectureship, American Institute of Aeronautics and Astronautics, 1978.

LITTLE, CHARLES G., Fellow 1965. Born: November 4, 1924, Liuyang, Hunan, China. Degrees: B.S., 1948, Ph.D., 1952, University of Manchester.

Fellow Award "For contributions to the fields of radio propagation and atmospheric research." Other Awards: Fellow, Royal Astronomical Society; American Meteorological Society; American Association for the Advancement of Science; Member, National Academy of Engineering.

LITTLE, D.G., IRE Fellow 1929, AIEE Fellow 1950.

LIU, BEDE, Fellow 1972. Born: September 25, 1934, Shanghai, China. Degrees: B.S.E.E., 1954, National Taiwan University; M.E.E., 1956, D.E.E., 1960, Polytechnic Institute of Brooklyn.

Fellow Award "For contributions to discrete-time signal and system theory and engineering education."

LIU, CHAO-HAN, Fellow 1981. Born: January 3, 1939, Kwangsi, China. Degrees: B.S., 1960, National Taiwan University; Ph.D., 1965, Brown University.

Fellow Award "For contributions to wave propagation in the ionosphere and to engineering education"; Outstanding Achievement Award, IEEE Antennas and Propagation Group, 1968. Other Awards: National Science Council "Special Chair" as Visiting Professor at National Taiwan University, Republic of China, 1981.

LIU, MING-TSAN (MIKE), Fellow 1983. Born: August 30, 1934, Peikang, Taiwan. Degrees: B.S.E.E., 1957, National Cheng Kung University, Tainan, Taiwan; M.S.E.E., 1961, Ph.D., 1964, Univ. of Pennsylvania.

Fellow Award "For contributions to distributed processing, computer networking and education in computer science." Other Awards: Engineering Research Award, Ohio State University, 1982.

LIU, RUEY-WEN, Fellow 1981. Born: March 18, 1930. Degrees: B.S., 1954, M.S., 1955, Ph.D, 1960, University of Illinois.

Fellow Award "For contributions to the analysis of nonlinear circuits and systems."

LIVINGSTON, ORRIN W., AIEE Fellow 1949. Born: July 11, 1905, Roselle, N.J. Degrees: B.S., 1927, Rutgers University.

Other Awards: Charles A. Coffin Award, General Electric, 1932; Outstanding Engineer Award, Rutgers Engineering Society, 1973.

LIVINGSTON, ROBERT S., Fellow 1968. Born: September 20, 1914, Summerland, Calif. Degrees: A.B.,

1935, Pomona College; M.A., Ph.D., 1941, University of California at Berkeley.

Fellow Award "For contributions to cyclotron focusing developments." Other Awards: Fellow, American Physical Society; Fellow, American Association for the Advancement of Science.

LLOYD, C. G., Fellow 1965. Born: December 8, 1913, Toronto, Ontario, Canada. Degrees: B.A.Sc., 1935, University of Toronto.

Fellow Award "For contributions to electrical engineering standards."

LO, ARTHUR W., Fellow 1966. Born: May 21, 1916, Shanghai, China. Degrees: B.S., 1938, Yenching University, China; M.A., 1946, Oberlin College; Ph.D., 1949, University of Illinois.

Fellow Award "For contributions to the field of digital devices and circuits, and to education."

LO, YUEN T., Fellow 1969. Born: January 31, 1920, Hankow, China. Degrees: B.S.E.E., 1942, National Southwest Associated University; M.S., 1949, Ph.D., 1952, University of Illinois.

Fellow Award "For research and contributions to the theory of antenna arrays"; Best Paper Award, IEEE Antennas and Propagation Society, 1964, 1979; IEEE Centennial Medal, 1984. Other Awards: John T. Bolljahn Memorial Award, 1964.

LOANE, EDWARD S., Fellow 1965. Born: March 14, 1908, Baltimore, Md. Degrees: B.E., 1928, The Johns Hopkins University.

Fellow Award "For contributions to electric power systems development." Other Awards: Fellow, American Society of Civil Engineers, 1960; Rickey Medal, American Society of Civil Engineers, 1976.

LOAR, HOWARD H., Fellow 1969. Born: May 23, 1923, Wilsonville, Nebr. Degrees: A.B., 1946, Nebraska Wesleyan University; M.A., 1949, Ph.D., 1952, Columbia University.

Fellow Award "For contributions to and supervision of the development of new communication systems and concepts."

LOCKIE, ARTHUR M., Fellow 1959. Born: September 13, 1909, Buffalo, N.Y. Degrees: E.E., 1931, M.E.E., 1935, Rensselaer Polytechnic Institute.

Fellow Award "For contributions to thermal analysis of performance and life of loaded distribution transformers." Other Awards: Special Patent Award, Westinghouse Electric, 1957; Product Engineering Master Design Award, 1963.

LODGE, WILLIAM B., IRE Fellow 1949. Born: August 17, 1907, Whitemarsh, Pa. Degrees: B.S., 1930, M.S., 1931, Massachusetts Institute of Technology.

Fellow Award "For his many contributions to broadcast engineering and in particular for his work in the field of frequency allocations." Other Awards: Society of Motion Picture and Television Engineers.

LOFERSKI, JOSEPH J., Fellow 1973. Born: August 7, 1925, Hudson, Pa. Degrees: B.S., 1948, University of Scranton; M.S., 1949, Ph.D. 1953, University of Pennsylvania.

Fellow Award "For contributions to radiation damage in semiconductors, solar cell technology, and education"; IEEE William E. Cherry Award for Outstanding Contributions to Photovoltaic Science and Technology, 1981. Other Awards: Fellow, American Association for Advancement of Science; Freeman Award, Providence Engineering Society, 1974; Fellow, U.S.-Poland Academies of Science Exchange Program, 1974-75.

LOGAN, EDWIN W., AIEE Fellow 1956. Born: April 21, 1912, Monroe, La. Degrees: B.S.E.E., 1937, E.E.(Hon.), 1966, University of Missouri at Rolla.

Fellow Award "For his contributions to the development of automatic equipment for the direction of military weapons." Other Awards: Charter Member, Academy of Electrical Engineers, UMR.

LOGUE, JOSEPH C., Fellow 1966. Born: December 20, 1920, Philadelphia, Pa. Degrees: B.E.E., 1944, M.E.E., 1949, Cornell University.

Fellow Award "For contributions to advanced solid-state techniques in the field of digital computation"; Certificate of Appreciation, IRE, 1957; Computer Society Honor Roll, 1974, "For his long, dedicated service as Chairman of the IEEE Computer Society Awards Committee." Other Awards: Eta Kappa Nu, 1944; Sigma Xi, 1946; Phi Kappa Phi, 1949; IBM Fellow, 1971; Fifth Invention Achievement Award, IBM, 1974.

LOHMAN, ROBERT D., Fellow 1972. Born: August 19, 1924, Chicago, Ill. Degrees: B.S.E.E., 1949, Norwich University; M.S.E.E., 1951, North Carolina State College.

Fellow Award "For contributions to transistor technology applied to television and computers."

LOHMANN, CARLOS ALBERTO J., Fellow 1981. Born: October 19, 1934, Rio de Janeiro, Brazil. Degrees: B.S.(E.E.), 1957, B.S.(Civil Eng.), 1957 Mackenzie University, Brazil.

Fellow Award "For contributions to management techniques in power distribution for electrical utilities in Latin America"; Eminent Engineer, Region 9, IEEE, 1980; Member, Fellow Committee, 1982-1984; IEEE Centennial Medal, 1984. Other Awards: Engineer of Distinction, Engineers Joint Council, 1973.

LOHSE, EDWARD, Fellow 1972. Born: January 3, 1918, Brooklyn, N.Y. Degrees: B.E.E., 1938, City College of The City University of New York.

Fellow Award "For leadership in the application of solid-state technology to computers and the establishment of data communications and computer standards." Other Awards: Associate Fellow, American Institute of Aeronautics and Astronautics; Fellow, American Association for the Advancement of Science, 1981;

Astin-Polk medalist, American National Standards Institute, 1982.

LOKAY, HENRY E., Fellow 1979. Born: July 8, 1927, Oak Park, Ill. Degrees: B.S.E.E., 1951, Illinois Institute of Technology; M.S.E.E., 1956, University of Pittsburgh.

Fellow Award "For contributions to new methods of the simulation of turbine generators, power-system dynamics, and distribution engineering." Other Awards: Eta Kappa Nu, 1950; Prime Movers Committee Award, American Society of Mechanical Engineers, 1979; U.S. Representative SC #11, CIGRE.

LOMBARDINI, PIETRO PAOLO, Fellow 1978. Born: June 27, 1913. Degrees: Ph.D.(Nautical Sciences), 1937, University of Naples; Ph.D.(Physics), 1950, University of Pisa.

Fellow Award "For contributions to radar astronomy, artificial ion cloud communication, and far infrared radiometry."

LOMBARDO, PETER P., Fellow 1980. Born: October 11, 1927. Degrees: B.E.E., 1953, Polytechnic Institute of Brooklyn.

Fellow Award "For contributions in the field of low-noise wide-band receiver technology."

LONG, LEONARD W., Fellow 1980. Born: January 8, 1920, Forest City, NC. Degrees: B.E.E., 1943, M.S.E.E., 1950, North Carolina State University.

Fellow Award "For leadership in the development and application of power transformer standards"; Distinguished Service, Transformers Committee, 1979; Charlotte Section Chairman, IEEE, 1965-66; Transformers Committee, 1969; PES Awards Committee, 1983; PES Fellows Committee, 1983-; numerous Working Groups.

LONG, MAURICE W., Fellow 1969. Born: April 20, 1925, Madisonville, Ky. Degrees: B.E.E., 1946, Georgia Institute of Technology; M.S.E.E., 1948, University of Kentucky; M.S., 1957, Ph.D., 1959, Georgia Institute of Technology.

Fellow Award "For sustained contributions to and leadership in advancing the technology of radar systems and the understanding of electromagnetic scattering." Other Awards: ABOU Research Award, Georgia Institute of Technology, 1960, 1971.

LONGINI, RICHARD L., Fellow 1968. Born: March 11, 1913, Chicago, Ill. Degrees: S.B., 1940, University of Chicago; M.S., 1944, Ph.D., 1948, University of Pittsburgh.

Fellow Citation "For contributions to research and development in solid-state metallurgy." Other Awards: Fellow, American Physical Society, 1958.

LONGO, THOMAS A., Fellow 1974. Born: January 11, 1927. Degrees: B.S., 1947, M.S., 1953, Ph.D., 1957, Purdue University.

Fellow Award "For development of transistortransitor-logic integrated circuits, and other contributions to semiconductor technology." Other Awards: Member, American Physical Society; Sigma Pi Sigma; Sigma Xi.

LONNGREN, KARL E., Fellow 1982. Born: August 8, 1938, Milwaukee, WI. B.S.E.E., 1960, M.S.E.E., 1962, Ph.D., 1964, University of Wisconsin.

Fellow Award "For contributions to the understanding of nonlinear effects in plasmas and related areas." Other Awards: Fellow, American Physical Society, 1980.

LOPER, WILLIAM B., AIEE Fellow 1958. Born: November 2, 1896, Montrose, Colo. Degrees: B.S.E.E., 1918, University of Colorado.

Fellow Award "For contributions to the development of a rapidly expanding electric power system." Other Awards: Citation, Bureau of Ships, U.S. Dept. of Navy, 1946.

LOPEZ, ALFRED R., Fellow 1983. Born: November 10, 1930, New York, N.Y. Degrees: B.E.E., 1958, Manhattan College; M.S.E.E., 1963, Polytechnic Institute of Brooklyn.

Fellow Award "For contributions to the design and application of electronically scanned antennas."

LORD, HAROLD W., AIEE Fellow 1951. Born: August 20, 1905, Eureka, Calif. Degrees: B.S.E.E., 1926, California Institute of Technology.

Other Awards: Charles A. Coffin Award, General Electric, 1933.

LORENZEN, HOWARD O., Fellow 1970. Born: June 24, 1912, Atlantic, Iowa. Degrees: B.S., 1935, Iowa State College.

Fellow Award "For leadership and contributions to countermeasures technology." Other Awards: U.S. Navy Meritorious Civilian Service Award, 1948; U.S. Navy Distinguished Civilian Service Award, 1960; Distinguished Engineering Alumni Award, Iowa State University, 1970; U.S. Navy Capt. Robert Dexter Conrad Award for Outstanding Scientific Contributions, 1973.

LORRAINE, RICHARD G., AIEE Fellow 1951. Born: December 26, 1904.

LOTZ, WALTER E., JR., Fellow 1968. Born: August 21, 1916, Johnsonburg, Pa. Degrees: B.S., 1938, United States Military Academy; M.S.E.E., 1947, University of Illinois; Ph.D., 1953, University of Virginia; D.Sc.(Hon.), 1974, Florida Institute of Technology.

Fellow Award "For contributions to research in military communications and planning for world-wide implementation." Other Awards: Bronze Star Medal, 1945; Army Commendation Medal, with Oak Leaf Cluster, 1946, 1960; Distinguished Service Medal, 1966, 1974; Legion of Merit, with Three Oak Leaf Clusters 1946, 1963, 1966, 1971; Airmans Medal, 1966; Distinguished Electrical Engineering Alumnus, University of Illinois, 1972; Secretary of Defense

Meritorious Civilian Service Award, 1977.

LOUGHLIN, BERNARD D., IRE Fellow 1955. Born: May 19, 1917, New York, N.Y. Degrees: B.E.E., 1939, E.E., 1945, Cooper Union; M.S.E.E., 1946, Stevens Institute of Technology.

Fellow Award "For contributions to color television, frequency modulation, and superregeneration"; Vladimir K. Zworykin Award, IRE, 1952, "For his outstanding contribution to the theory, the understanding, and the practice of color television"; Award, Professional Group on Broadcast and Television Receivers, 1957, "For his significant contributions in the field of color television circuitry"; Consumer Electronics Award, 1972. Other Awards: David Sarnoff Gold Medal, Society of Motion Picture and Television Engineers, 1955; Modern Pioneer Award, National Association of Manufacturers, 1965; National Academy of Engineering, 1967; Cooper Union Professional Achievement Citation, 1969; Gano Dunn Medal, Cooper Union, 1970; International Television Symposium Citation, 1973; Special Commendation, Society of Motion Pictures and Television Engineers, 1977.

LOUGHREN, ARTHUR V., IRE Fellow 1944, AIEE Fellow 1957. Born: September 15, 1902, Rensselaer, N.Y. Degrees: A.B., 1923, E.E., 1925, Columbia University.

IRE Fellow Award "For his many valuable contributions to broadcast and television engineering and his untiring efforts to advance this profession"; AIEE Fellow Award, "For technical leadership in the electronic industry with major contributions toward standardization of color television"; Morris N. Liebmann Award, IRE, 1955, "For his leadership and technical contributions in the formulation of the signal specifications for compatible color television"; President, IRE, 1956; IEEE Consumer Electronics Group, Consumer Electronics Award, 1974, "For outstanding contributions to consumer electronics." Other Awards: Certificate of Commendation, U.S. Dept. of Navy, 1947; David Sarnoff Gold Medal, Society of Motion Picture and Television Engineers, 1953; Egleston Medal, Columbia University, 1964; Fellow, Society of Motion Picture and Television Engineers; Edwin H. Armstrong Medal, Radio Club of America, 1981.

LOUISELL, WILLIAM H., Fellow 1978. Born: August 22, 1924, Mobile, Ala. Degrees: B.S., 1948, M.S., 1949, Ph.D., 1953 (all in Physics), University of Michigan.

Fellow Award "For contributions to quantum statistical properties of radiation." Other Awards: Alexander von Humboldt U.S. Senior Scientist Award, 1979-80; Fellow, American Physical Society.

LOVE, ALLAN W., Fellow 1980. Born: May 28, 1916, Toronto, Ontario, Canada. Degree: B.A. (Math. and Physics), 1938, M.A. (Physics), 1939, Ph.D. (Physics),

1951, University of Toronto.

Fellow Award "For contributions to the theory and practice of spherical reflector antennas and radiometer systems"; Best Paper Award, IEEE Antennas and Propagation Society, 1973; Member, Antennas and Propagation Society AdCom, 1980-1982, Vice President, 1983; President, 1984; Associate Editor, IEEE Transactions on Antennas and Propagation, 1977-1982; Distinguished Lecturer, IEEE Antennas and Propagation Society, 1979-80; Editor, IEEE Press Reprint Books "Electromagnetic Horn Antennas", 1976, and "Reflector Antennas", 1978. Other Awards: Certificate of Appreciation, Department of Energy, NASA, 1978-1980; Engineer of Year Award, Rockwell International, 1981.

LOVE, CARLOS O., Fellow 1978. Born: June 30, 1920, Corsicana, Tex. Degrees: B.S.E.E., 1947, Texas A&M University.

Fellow Award "For contributions to utilizing alternate fuel resources and to creating an effective interconnected system"; Electrical Engineer of the Year, Dallas Section, IEEE, 1974. Other Awards: Commended by Federal Aviation Administration for emergency service following Hurricane Beulah, 1967; Citation by Secretary Rogers Morton for outstanding service during Mississippi, Louisiana, and Texas hurricanes, 1971.

LOVETT, ISRAEL H., AIEE Fellow 1950. Born: March 1, 1890, Council Bluffs, Iowa. Degrees: S.B., 1914, Massachusetts Institute of Technology; E.E., 1924, University of Missouri at Rolla; M.S.E., 1928, University of Michigan.

Awards: AIEE, Student Branch, "For dedicated service to the electrial engineering profession and engineering education." Other Awards: Award, Missouri Society of Professional Engineers, 1961; Outstanding Service to College Youth, Missouri School of Mines and Metallurgy, 1973.

LOZIER, JOHN C., Fellow 1966. Born: February 5, 1912, Bayside, N.Y. Degrees: A.B., 1934, Columbia University.

Fellow Award "For contributions in the field of automatic controls and the use of digital computers in real time control systems." Other Awards: Chevalier of the Legion of Merite Postal, France, 1962.

LUBCKE, HARRY R., IRE Fellow 1951. Born: August 25, 1905, Alameda, Calif. Degrees: B.S., 1929, University of California.

Fellow Award "For courageous pioneering work in television production and transmission"; Best Paper Prize, AIEE, Branch, 1930; Other Awards: Certificate, War Dept., 1946; Certificate of Commendation, U.S. Dept. of Navy, 1947; Founder Member, Society of Television Engineers (Los Angeles), 1938, President, 1942, Past President, Academy of Television Arts and Sciences, 1949; Award, Academy of Television Arts

and Sciences, 1951; Fellow, American Association for the Advancement of Science, 1958; Pacific Pioneer Broadcasters Achievement Award, 1966; Fellow, Society of Motion Picture and Television Engineers, 1967; Phi Beta Kappa; Tau Beta Pi; Eta Kappa Nu; Sigma Xi; Diamond Circle Award, Pacific Pioneer Broadcasters, 1980.

LUBORSKY, FRED E., Fellow 1978. Born: May 14, 1923, Philadelphia, Pa. Degrees: B.S., 1947, University of Pennsylvania; Ph.D., 1952, Illinois Institute of Technology.

Fellow Award "For contributions toward the theoretical understanding of magnetic properties and the practical utilization of this knowledge"; Distinguished Lecturer, IEEE Magnetics Society, 1979; Awarded the first IEEE Magnetics Society Achievement Award, 1981; IEEE Centennial Medal, 1984. Other Awards: Outstanding Achievement in Industrial Science Award, American Association for the Advancement of Science, 1956; Fellow, New York Academy of Science, 1968; Fellow, American Institute of Chemists; British Science Research Council Fellowship, 1977; Collidge Fellow, General Electric Corporate Research and Development, 1978.

LUCAS, PIERRE M., Fellow 1984. Born: July 13, 1924, 51330 Givry en Argonne, France. Degrees: Ecole Polytechnique, 1944, Paris; Ecole Nationale Superieure des Telecommunications, 1950, Paris.

Fellow Award "For contributions to electronic switching systems."

LUCCIO, FABRIZIO, Fellow 1983. Born: June 20, 1938, Tripoli, Libya. Degrees: Dr. Ing., 1962, Politecnico of Milano; Libera Docenza, 1968, Italian University System.

Fellow Award "For contributions to the theory of sequential switching networks and for leadership in computer science education."

LUCKY, ROBERT W., Fellow 1972. Born: January 9, 1936, Pittsburgh, Pa. Degrees: B.S.E.E., 1957, M.S.E.E., 1959, Ph.D., 1961, Purdue University.

Fellow Award "For contributions to the theory and practice of data communications"; Achievement Award, IEEE Communications Society, 1975. Other Awards: Honorable Mention, Outstanding Young Electrical Engineer, Eta Kappa Nu, 1967; Distinguished Engineering Alumnus, Purdue University, 1969.

LUDEKENS, LOUIS E., Fellow 1966. Born: May 29, 1904, Pine Grove, Calif.

Fellow Award "For contributions in microwave VHF mobile radio and carrier relaying in the utility field." Other Awards: Commendation, U.S. Navy Submarine Service; Commendation, Office of Scientific Research and Development - National Defense Research Council; Commendation, Columbia University, 1945.

LUENBERGER, DAVID G., Fellow 1975. Born: September 16, 1937, Los Angeles, Calif. Degrees: B.S., 1959, California Institute of Technology; M.S., 1961, Ph.D., 1963, Stanford University.

Fellow Award "For contributions to modern control theory." Other Awards: Honorable Mention, Outstanding Paper Award, Joint Automatic Control Conference, 1970.

LUMLEY, CHARLES S., AIEE Fellow 1945. Born: December 14, 1897, Belper, Derby, England.

LUMNITZER, PAUL L., Fellow 1968. Born: June 21, 1912, Johnstown, Pa. Degrees: B.S.E.E., 1933, University of Pittsburgh.

Fellow Award "For pioneering contributions in planning and design of extra-high-voltage transmission systems."

LUNAS, LAWRENCE J., AIEE Fellow 1957. Born: September 2, 1905, Balfour, N.D. Degrees: B.S., 1926, University of North Dakota; M.S., 1939, University of Pittsburgh; E.E., 1941, University of North Dakota.

Fellow Award "For his contributions in the field of design and development of electrical instruments"; IEEE-North Jersey Section Award, 1970. Other Awards: Sigma Xi; Sigma Tau; Tau Beta Pi.

LUNDSTROM, ALEXIS A., Fellow 1970. Born: November 26, 1905, Portland, Oreg. Degrees: B.S., 1928, Oregon State University.

Fellow Award "For original contributions to systems for telephone dialing, and for satellite tracking and guidance." Other Awards: Tau Beta Pi, 1928; Phi Kappa Phi; Eta Kappa Nu; Naval Ordnance Development Award, 1945; Certificate of Appreciation, Officers and Men of the USS Northampton, 1952; French Post Office Medal for Work on Telstar Earth Station at Pleumeur Bobou, 1962.

LUNSFORD, JESSE B., AIEE Fellow 1947. Born: August 14, 1890, Hickman, Ky. Amarillo Business College; Highland Park College of Engineering; Univeristy of Arkansas, 1910.

Awards: Standards Committee, AIEE. Other Awards: Listed in "Who's Who in Engineering," 1931, 1980, 1981; Certificate, American Standards Association, 1946; Commendation, Chief Signal Officer, U.S. Dept of Army, 1946; Commendations, Chief of Bureau of Ships, U.S. Dept of Navy, 1946, 1951; Listed in: "Who's Who in Technology Today", 1982; "International Who's Who in Engineering", 1982; in categories of "Technology" and "Intellectuals", International Biogeographical Center, Cambridge, England, 1983.

LUSIGNAN, JOSEPH T., AIEE Fellow 1950. Born: June 30, 1902, Stockton, Calif. Degrees: S.B., 1924, S.M., 1925, Massachusetts Institute of Technology; Ph.D., 1928, Stanford University.

LUSTED, LEE B., IRE Fellow 1959. Born: May 22, 1922, Mason City, Iowa. Degrees: B.A., 1943, Cornell College; M.D., 1950, Harvard University; D.Sc.(Hon.), 1963, Cornell College.

Fellow Award "For technical achievements and leadership in relating medicine and electronics." Other Awards: Memorial Fund Lecture, 1959; Magna Cum Laude Award, Radiological Society of North America, 1964; Fellow, American College of Radiology; Fellow, New York Academy of Science; Fellow, AAAS; Sigma Xi. Listed in "Who's Who in America," 1979; Honorary Member, Japan Radiological Society, 1970; Editor, Medical Decision Making, 1980-.

LUSTGARTEN, MERRILL NATHANIEL, Fellow 1979. Born: February 15, 1922. Degrees: A.B.(Chemistry), 1941, Brooklyn College; B.S.E.E., 1945, North Carolina State College; M.S.E.E., 1947, Columbia University.

Fellow Award "For contributions to electromagnetic compatibility technology."

LUTHER, ARCH C., JR., Fellow 1974. Born: December 5, 1928, Philadelphia, Pa. Degrees: B.S.E.E., 1950, Massachusetts Institute of Technology.

Fellow Award "For engineering contributions to the design of color television cameras and video tape recorders." Other Awards: SMPTE Fellow, 1969; SMPTE David Sarnoff Gold Medal, 1973; RCA David Sarnoff Award, 1975; Sigma Xi; Eta Kappa Nu.

LYNCH, EDWARD E., AIEE Fellow 1956. Born: August 20, 1903, Easthampton, Mass. Degrees: B.S., 1925, M.S., 1926, Massachusetts Institute of Technology.

Fellow Award "For his contribution in the fields of electrical instrumentation and control."

LYNCH, WILLIAM A., IRE Fellow 1960. Born: July 11, 1904, Brooklyn, N.Y. Degrees: M.E., 1925, M.E.E., 1948, Polytechnic Institute of Brooklyn.

Fellow Award "For contributions to engineering education."

LYON, JOHN A. M., AIEE Fellow 1959. Born: October 9, 1912, Saginaw, Mich. Degrees: B.S.E., 1933, M.S., 1934, Ph.D., 1936, University of Michigan.

Fellow Award "For contributions to engineering education." Other Awards: Phi Kappa Phi, 1933; Sigma Xi, 1936; Eta Kappa Nu, 1950; Distinguished Alumnus Citation, University of Michigan, 1953; Tau Beta Pi, 1954; Fellow, American Association for the Advancement of Science, 1954; Honor Alumnus, Arthur Hill High School, Saginaw, Michigan, 1969; Western Electric Fund Award, 1970; Certificate of Appreciation, U.S. Army, 1970; S. S. Attwood Distinguished Engineering Achievement Award, University of Michigan, 1978; Prince Visiting Scholar, Arizona State University, 1982-83; Professor Emeritus, Electrical and Computer Engineering, University of Michigan, 1983.

M

MA, MARK TSU-HAN, Fellow 1982. Born: March 21, 1933, China. Degrees: B.S.E.E., National Taiwan University; M.S.E.E., University of Illinois at Urbana; Ph.D.(E.E.), Syracuse University.

Fellow Award "For contributions to the analysis and synthesis of antenna arrays"; Best Original Paper Award, IRE-NEC, Chicago, 1960; Best Paper Award, IRE Syracuse Section, 1961. Other Awards: Commerce Science and Technology Fellowship Award, 1971; Special Achievement Award, Office of Telecommunications, U.S. Dept. of Commerce, 1974, 1975; Superior Accomplishment Award, Office of Telecommunications, U.S. Dept. of Commerce, 1978.

MACADAM, WALTER K., AIEE Fellow 1959. Born: November 16, 1913, New York, N.Y. Degrees: S.M., 1937, Massachusetts Institute of Technology.

Fellow Award "For contributions to the growth and improvement of telephone communications"; President, IEEE, 1967.

MACCARTHY, DONNELL D., AIEE Fellow 1958. Born: January 31, 1905, Raleigh, N. C. Degrees: E.E., 1926, M.E.E., 1930, Cornell University.

Fellow Award "For his contributions to overvoltage protective apparatus for electric power systems."

MACDONALD, J. ROSS, IRE Fellow 1959. Born: February 27, 1923, Savannah, Ga. Degrees: B.A., 1944, Williams College; S.B., 1944, S.M. 1947, Massachusetts Institute of Technology; D.Phil., 1950, D.Sc., 1967, Oxford University, England.

Fellow Award "For contributions to the theory of electronic circuits and electrical properties of solids"; Senior Paper Prize, IRE, Professional Group on Audio, 1957; Achievement Award, IRE, Professional Group on Audio, 1962; Meritorious Services Award, IEEE, Acoustics, Speech and Signal Processing Group, 1974. Other Awards: Rhodes Scholar, 1948-50; Member, National Academy of Engineering, 1970; Member, National Academy of Sciences, 1973.

MACFARLANE, ALISTAIR G.J., Fellow 1982. Born: May 9, 1931, Edinburgh, Scotland. Degrees: B.Sc.(first class honors), 1953, D.Sc., 1969, University of Glasgow; Ph.D., 1964, University of London; M.Sc., 1973, University of Manchester; M.A., 1974, Sc.D., 1979, University of Cambridge.

Fellow Award "For contributions to control engineering and to the development of frequency domain techniques for linear multivariable feedback systems." Other Awards: Institution Premium, Institution of Electrical Engineers, London; Control and Automation Premium, Institution of Electrical Engineers, London; I.C.I. Prize, Institute of Measurement and Control; Centennial Medal, American Society of Mechanical Engineers, 1980; Sir Harold Hartley Medal, Institute of Measurement and Control, London, 1982.

MACHLER, R. C., AIEE Fellow 1961. Born: February 9, 1906. Degrees: B.S., 1929; M.A., 1933; Ph.D., 1936, Northwestern University.

Fellow Award "For contributions in the field of electrical measurements." Other Awards: Sigma Xi, 1929; Phi Beta Kappa, 1936; Merit Award, Office of Scientific Research and Development, 1945; Naval Ordnance Award, 1945.

MACINTYRE, JOHN R., AIEE Fellow 1956. Born: June 29, 1908, England. Degrees: B.S.E.E., 1934, University of Michigan.

Fellow Award "For his contributions in the field of instrumentation and control." Other Awards: Fellow, Instrument Society of America, 1975.

MACKAY, R. STUART, IRE Fellow 1962. Born: January 3, 1924, San Francisco, Calif. Degrees: B.A., 1944, Ph.D., 1952, University of California at Berkeley.

Fellow Award "For contributions to the field of medical electronics." Other Awards: Guggenheim Fellow, Karolinska Institute, Stockholm, 1957, 1958; Erskine Fellow, University of Canterbury, Christchurch, New Zealand, 1969, 1980; Fulbright Lecturer, University of Cairo, Egypt, 1960-1961; Phi Beta Kappa; Sigma Xi; Pi Mu Epsilson; Apollo Award, American Optometric Association, 1962; Senior Scientist on Galapagos International Scientific Project, 1964; Distinguished Visitor, Antarctica, 1969; Opening Lecture at the 9th International Conference on Medical and Biological Engineering, Melbourne, 1971; Keynote Address at Biotelemetry Congress, Pretoria, 1971; Keynote Address at 20th Australasian Conference on Physical Science and Engineering in Medicine and Biology, Christchurch, 1980.

MACKINNON, KEITH A., IRE Fellow 1949. Born: September 10, 1905, Kingston, Ontario, Canada. Degrees: B.Sc., 1926, B.Sc., 1928, M.Sc., 1929, Queen's University, Ontario.

Fellow Award "For his technical contributions in Canada to the theory and design of transmitting antennas and the development of a coverage plan for a national network."

MACLEAN, JAMES B., AIEE Fellow 1947. Born: November 4, 1899, New York, N.Y. Degrees: B.A., 1922, M.A., 1923, Wesleyan University; B.S., 1924, M.S., 1925, Massachusetts Institute of Technology.

Other Awards: 3 Patents.

MACNEE, ALAN B., Fellow 1964. Born: September 19, 1920, New York, N.Y. Degrees: S.B., 1943, Sc.D., 1948, Massachusetts Institute of Technology.

Fellow Award "For contributions to research and teaching in the field of circuit theory"; Browder J. Thompson Award, IRE, 1951. Other Awards: Western Electric Fund Award, American Society for Engineering Education, 1968-69; Fellow, American Association for the Advancement of Science, 1980.

MACNEILL, JOHN B., AIEE Fellow 1942. Born: June 21, 1888, Summerside, Prince Edward Island, Canada. Degrees: S.B., 1913, Massachusetts Institute of

Technology.

Awards: Lamme Medal, AIEE, 1947. Other Awards: Award of Merit, Westinghouse, 1936.

MACNICHOL, EDWARD F., JR., Fellow 1964. Born: October 24, 1918, Toledo, Ohio. Degrees: A.B., 1941, Princeton University; Ph.D., 1952, The Johns Hopkins University.

Fellow Award "For contributions to medical electronics and to the physiology of vision"; Morlock Award, Professional Group on Biomedical Electronics, 1965. Other Awards: Member, American Physical Society; Member, Biophysical Society; Member, American Physiological Society; Honorary Life Member, Association for Research in Vision and Ophthalmology.

MACOVSKI, ALBERT, Fellow 1976. Born: May 2, 1929, New York. Degrees: B.E.E., 1950, City College of the City University of New York; M.E.E., 1953, Polytechnic Institute of Brooklyn; Ph.D., 1968, Stanford University.

Fellow Award "For contributions to television engineering"; Award IRE Professional Group on Broadcast and Television Receivers, 1957, "For his contributions in the area of color television demodulators"; Vladimir K. Zworykin Award, IEEE, 1973, "For contributions to single tube encoded color cameras and color television receiving circuits." Other Awards: Award for Outstanding Work in Research, RCA Laboratories, 1952, 1954; Special Fellow, NIH, 1972; Fellow, OSA, 1977.

MAC RAE, ALFRED URQUHART, Fellow 1979. Born: April 14, 1932, New York, N.Y. Degrees: Ph.D.(Physics), 1960, Syracuse University.

Fellow Award "For leadership in the development of ion implantation technology and its application to semiconductor device fabrication"; Outstanding Paper Award, IEEE Solid-State Circuits Conference, 1969; Meetings Chairman, ADCOM, IEEE Electron Devices Society. Other Awards: Fellow, American Physical Society, 1965; Scientific Fellow, Bohmische Physical Society, 1978.

MAILLOUX, ROBERT J., Fellow 1978. Born: June 20, 1938, Lynn, Mass. Degrees: B.S.(Electrical Eng.), 1961, Northeastern University; S.M.(Applied Physics), 1962, Ph.D.(Applied Physics), 1965, Harvard University.

Fellow Award "For contributions to enhancing the performance of phased array antennas"; Special Achievement Paper Award, IEEE Antennas and Propagation Society, 1969; Best Paper Award, Honorable Mention, IEEE Antennas and Propagation Society, 1976; Boston Chapter Chairman, AdCom Member, IEEE Antennas and Propagation Society; Meetings Chairman, Distinguished Lecturer, Vice President and 1983 President, AP-S. Other Awards: Marcus O'Day Memorial Award for best paper published in a recog-

nized journal, Air Force Cambridge Research Laboratories, 1972; URSI Commission B Technical Activities Chairman, 1979-82.

MAISSEL, LEON I., Fellow 1983. Born: May 31, 1930, Cape Town, South Africa. Degrees: B.Sc., 1949, M.Sc., 1951, University of Cape Town; Ph.D., 1955 Imperial College (London).

Fellow Award "For contributions to computer design languages and thin film deposition."

MAKHOUL, JOHN I., Fellow 1980. Born: September 19, 1942. Deirmimas, Lebanon. Degrees: B.E., 1964, American University of Beirut; M.Sc., 1965, Ohio State University; Ph.D., 1970, Massachusetts Institute of Technology.

Fellow Award "For contributions to the theory of linear prediction and its applications to spectral estimation, speech analysis, and data compression"; Senior Award, IEEE Acoustics, Speech, and Signal Processing Society, 1978; Technical Achievement Award, IEEE Acoustics, Speech, and Signal Processing Society, 1982; Proc. IEEE paper on Linear Prediction (April 1975) was named a "Citation Classic" by the Institute for Scientific Information, 1982. Other Awards: Fellow, Acoustical Society of America.

MALEWSKI, RICHARD A., Fellow 1981. Born: September 3, 1935, Warsaw, Poland. Degrees: Dipl.Eng.(E.E.), 1956, M.Sc.(E.E.), 1958, Technical University of Warsaw, Poland; D.Sc., 1967, Electrotechnical Institute of Warsaw, Poland.

Fellow Award "For contributions to high-voltage measuring techniques and to the study of the transient skin effect in massive conductors"; Prize Paper Award, Power System Instrumentation and Measurement Committee, IEEE, 1978, 1979, 1980, 1981; W.R.G. Baker Prize Award, IEEE, 1983.

MALEY, GERALD A., Fellow 1983. Born: March 5, 1929, Buffalo, N.Y. Degrees: B.S.E.E., Univ. of Buffalo.

Fellow Award "For contributions to logic design using gate arrays and to sequential switching circuits"; Outstanding Section Award, IEEE.

MALING, GEORGE C., JR., Fellow 1974. Born: February 24, 1931. Degrees: A.B., 1954, Bowdoin College; S.B., 1954, S.M., 1954, E.E., 1958, Ph.D., 1963, Massachusetts Institute of Technology.

Fellow Award "For contributions to noise control engineering and to standardized methods of sound power measurements." Other Awards: Fellow, Acoustical Society of America; Fellow, Audio Engineering Society; Member, Institute of Noise Control Engineering; Member, American Association for the Advancement of Science; Eta Kappa Nu; Tau Beta Pi; Sigma Xi.

MALLARD, STEPHEN A., Fellow 1981. Born: September 15, 1924, Jersey City, NJ. Degrees: M.E., 1948, M.S. (E.E.), 1951, Stevens Institute of Technology.

Fellow Award "For the development and application

of simulation methods for reliability evaluations of complex electric power systems"; Section Chairman Certificate of Appreciation, IEEE, 1967; Energy Committee, IEEE. Other Awards: Member, CIGRE; Member, Pennsylvania-New Jersey-Maryland (PJM) Management Committee; Executive Board, Mid Atlantic Council (MAAC); Advisor, Gas Research Institute; Member, National and New Jersey Society of Professional Engineers; Advisor, Energy Group, Stevens Institute of Technology; Tau Beta Pi, 1948; Eta Kappa Nu, 1970; Executive Board of the Onsite Fuel Cell Users Group; Electric Power Research Institute, Research Advisory Committee; American Red Cross, Board of Trustees; Ethics Committee, District V - Essex County Bar Association.

MALLINSON, JOHN C., Fellow 1982. Born: January 30, 1932, Bradford, Yorkshire, England. Degrees: B.A.(Physics), 1953, M.A.(Physics), 1953, University College, Oxford, England.

Fellow Award "For contributions to the theory of magnetic recording"; Distinguished Lecturer, IEEE Magnetics Society, 1983-84.

MALSBARY, JAMES S., AIEE Fellow 1948. Born: June 17, 1905, Montgomery Co., Ind. Degrees: B.S.E.E., 1927, Purdue University.

MALTHANER, WILLIAM A., IRE Fellow 1962. Born: July 9, 1915, Columbus, Ohio. Degrees: B.E.E., 1937, Rensselaer Polytechnic Institute.

Fellow Award "For contributions to electronic switching."

MANDELBROT, BENOIT B., Fellow 1969. Born: November 20, 1924, Warsaw, Poland. Degrees: Engineer's Diploma, 1947, Ecole Polytechnique, Paris; M.S., 1948, California Institute of Technology; Ph.D.(Mathematics), 1952, University of Paris.

Fellow Award "For contributions to information theory and its application, and to the understanding of 1/f and random noise processes." Other Awards: Fellow, American Academy of Arts and Sciences, 1982; Fellow, John Simon Guggenheim Memorial Foundation, 1968; Fellow, Econometric Society, 1968; Fellow, Institute of Mathematical Statistics, 1968; Fellow, American Statistical Association, 1972; Elected Member, International Statistical Institute; IBM Fellow, T.J. Watson Research Center; Professor of the Practice of Mathematics, Harvard.

MANINGER, R. CARROLL, Fellow 1978. Born: December 24, 1918, Harper, Kans. Degrees: B.S., California Institute of Technology, 1941.

Fellow Award "For contributions to the development of measurement techniques used in the nuclear and environmental sciences."

MANLEY, JACK M., IRE Fellow 1960. Born: March 9, 1909, Farmington, Mo. Degrees: B.S., 1930, University of Missouri.

Fellow Award "For contributions to the theory of modulators and parametric amplifiers"; David Sarnoff Award (with H.E. Rowe), IEEE, 1977, "For their work on the properties of nonlinear devices resulting in the well-known Manley-Rowe relations." Other Awards: Visiting Professor, University of Wisconsin at Madison, 1974-1981.

MANN, ROBERT W., Fellow 1979. Born: October 6, 1924, Brooklyn, N.Y. Degrees: S.B., 1950, S.M., 1951, Sc.D., 1957 (all in Mechanical Engineering), Massachusetts Institute of Technology.

Fellow Award "For leadership in biomedical engineering research and education and in the application of technology to the problems of the handicapped." Other Awards: Tau Beta Pi, 1949; Pi Tau Sigma, 1949; Sigma Xi, 1950; Alfred P. Sloan Award, 1957; Talbert Abrahms Photogrammetry Award, 1962; Citation for Sensory Aids, Assoc. Blind of Massachusetts, 1969; Germeshausen Professor, M.I.T., 1970; Institute of Medicine, National Academy of Science, 1971; IR-100 Innovation Award, 1972; Fellow, American Academy of the Arts and Sciences, 1972; National Academy of Engineering, 1973; Whitaker Professor of Biomedical Engineering, M.I.T., 1974; M.I.T. Alumni Assoc., Bronze Beaver Award, 1975; Goldenson Award, 1976, H.R. Lissner Award, 1977; Gold Medal Award, American Society of Mechanical Engineers, 1977; Sigma Xi National Lecturer, 1979-1981; 37th New England Award, Engineering Societies of New England, 1979; ALZA Distinguished Lecturer, 1980; Fellow, American Association for The Advancement of Science, 1982; National Academy of Sciences, 1982; James R. Killian, Jr. Faculty Achievement Award, M.I.T., 1983; President, M.I.T. Alumni Assoc., 1983-1984.

MANNING, ERIC G., Fellow 1984. Born: August 4, 1940, Windsor, Canada. Degrees: B.Sc., M.Sc., Waterloo; Ph.D.(E.E.), Illinois at Urbana.

Fellow Award "For contributions to fault diagnosis and simulation in digital systems."

MANNING, LAURENCE A., IRE Fellow 1962. Born: April 28, 1923. Degrees: A.B., 1944, M.Sc., 1948, Ph.D., 1949, Stanford University.

Fellow Award "For techniques in ionospheric measurements."

MANNING, MELVIN L., AIEE Fellow 1948. Born: November 26, 1900, Miller, S.D. Degrees: B.S., 1927, Dr.Eng., 1978, South Dakota State University; M.S., 1932, University of Pittsburgh.

Awards: First Paper Prize, AIEE-NEMA Conference on Electrical Insulation, 1960, "In recognition of his efforts expended in the preparation and presentation of a technical paper designed to meet the high standard of AIEE relating to electrical insulation materials"; Member, IEEE Transformer Committee and WG's, IEEE SCC 4.0 and 4.1 Committees; ANSI C57 Committee;

Other Awards: Sigma Xi; Sigma Tau; Eta Kappa Nu; Phi Kappa Phi; Life Member, American Society for Engineering Education; Award of Merit, "Dictionary of International Biography," 1968; Award of Merit, National Council of Engineering Examiners, 1971; Award, "Distinguished Engineer," Alumni, South Dakota State University, 1977; Certificate, American Society for Testing and Materials, 1980; Licensed Professional Engineer Pa. and S.D.; Prof. and Dean Emeritus, College of Engineering, S.D. State Univ. Harding Hall, E.E. Dept, Brookings, S.D., 57007.

MANO, KUNIO, Fellow 1983. Born: September 29, 1910, Aichi Prefecture, Japan. Degrees: B.Sc.(E.E.), 1934, Tohoku Imperial University; Ph.D., 1960, Tohoku University.

Fellow Award "For contributions to the analysis and measurement of electrical contact phenomena and the development of electromechanical components." Other Awards: Violet Ribbon Medal from His Majesty the Emperor, 1973; Ragnar Holm Scientific Achievement Award, 1975.

MANSPEAKER, EDWIN D., AIEE Fellow 1961. Born: February 27, 1900, Elwood, Nebr. Degrees: B.S.E.E., 1924, University of Kansas.

Fellow Award "For contributions to service engineering of aircraft electrical systems."

MAPES, CHARLES M., AIEE Fellow 1951. Born: March 14, 1901, N.Y. Degrees: B.S., 1923, Massachusetts Institute of Technology.

MARBURGER, THOMAS E., AIEE Fellow 1960. Born: January 28, 1901, Point Patience, Md. Degrees: B.S., 1943, The Johns Hopkins University.

Fellow Award "For contributions to the management of construction, operation and engineering of a major electric utility system." Other Awards: Tau Beta Pi, 1960.

MARBURY, RALPH E., AIEE Fellow 1945. Born: February 3, 1897, Commerce, Ga.

Other Awards: Order of Merit, Westinghouse Electric, 1942.

MARCATILI, ENRIQUE A. J., Fellow 1967. Born: August 1, 1925, Argentina. Degrees: A.E., 1947, E.E., 1948, University of Cordoba, Argentina.

Fellow Award "For contributions to microwave circuitry; specifically for novel multimode, millimeter-wave filters"; W.R.G. Baker Prize Award (with S.E. Miller and T. Li), 1975, for paper, "Research Toward Optical-Fiber Transmission Systems." Other Awards: Member, National Academy of Engineering.

MARCHAND, NATHAN, Fellow 1964. Born: June 20, 1916, Shawinigan Falls, Quebec, Canada. Degrees: B.S.E.E., 1937, City College of the City University of New York; M.S.E.E., 1941, Columbia University.

Fellow Award "For contributions in communication principles, aerial navigation, and medical electronics."

Other Awards: Fellow, American Association for the Advancement of Science; Eta Kappa Nu; Tau Beta Pi; Sigma Xi

MARCUM, CHARLES R., Fellow 1981. Born: September 17, 1919, Clay County, KY. Degrees: B.S.E.E., 1942, University of Kentucky; M.S.(Ind.Mgmt.), 1954, Massachusetts Institute of Technology.

Fellow Award "For technical and management contributions to the fields of large rotating machines and static power devices."

MARCUSE, DIETRICH, Fellow 1973. Born: February 27, 1929, Koenigsberg, Germany. Degrees: Diplom-Physiker, 1954, Freie Universitaet, Berlin; D.Ing., 1962, Technische Hochschule Karlsruhe, Germany.

Fellow Award "For contributions to the understanding of light guidance through gas lenses and imperfect optical fibers."

MARCUVITZ, N., IRE Fellow 1958. Born: December 29, 1913, Brooklyn, N.Y. Degrees: B.E.E., 1935, M.E.E., 1941, D.E.E., 1947, Polytechnic Institute of Brooklyn.

Fellow Award "For fundamental contribution to the solution of microwave field problems." Other Awards: Member, National Academy of Engineering, 1978.

MARCY, H. TYLER, Fellow 1967. Born: September 14, 1918, Rochester, N.Y. Degrees: B.S.E.E., 1940, M.S.E.E., 1941, Massachusetts Institute of Technology.

Fellow Award "For leadership in feedback control, in professional society activities and significant contribution to management of engineering enterprises." Other Awards: Eta Kappa Nu, 1940; Tau Beta Pi, 1940; Sigma Xi, 1941; Member, Instrument Society of America, 1963; Fellow, Instrument Society of America, 1977.

MARIHART, DONALD JOSEPH, Fellow 1982. Born: November 13, 1926. Degrees: B.S.E.E., 1949, Washington State University.

Fellow Award "For contributions to the application of modern communication techniques to the control and protection of power systems."

MARKEL, JOHN D., Fellow 1981. Born: January 20, 1943. Degrees: B.S.E.E., 1965, Kansas State University; M.S.E.E., 1968, Arizona State University; Ph.D., 1970, University of California at Santa Barbara.

Fellow Award "For contributions to the theory and applications of linear prediction of speech."

MARKO, HANS, Fellow 1978. Born: February 24, 1925, Kronstadt/Siebenburgen. Degrees: Dipl. Ing., 1950, Dr. Ing., 1953, Technical University of Stuttgart; Professor, 1962, Technical University of Munchen.

Fellow Award "For contributions to digital communication transmission systems." Other Awards: Prize, Nachrichtentechnische Gesellschaft, 1957.

MARNER, GENE R., Fellow 1967. Born: November 8, 1923, Johnson Co., Iowa. Degrees: B.A., 1947, Ph.D., 1956, University of Iowa.

Fellow Award "For contributions to the fields of radio-celestial navigation, radio meteorology, microwave propagation and radiometric techniques." Other Awards: Phi Beta Kappa, 1948; Sigma Xi, 1953; Eta Kappa Nu, 1956; Distinguished Service Award, ION, 1968; Hays Award, 1969; Fellow Award, American Association for the Advancement of Science, 1968; Distinguished Alumni Award, Westmar College, 1970.

MARQUIS, VERNON M., AIEE Fellow 1954. Born: March 30, 1898, Garfield County, Okla. Degrees: B.A., 1921, E.E., 1922, Stanford University; M.S., 1925, Union College.

Fellow Award "For his contributions to the integration, interconnections, stable and economical operation of power systems." Other Awards: War Production Board, Office of War Utilities.

MARSHALL, CHARLES J., IRE Fellow 1955. Born: March 27, 1912, San Antonio, Tex. Degrees: B.Sc., 1939, E.E., 1946, University of Cincinnati.

Fellow Award "For his contributions to airborne television and radar research and development." Other Awards: Distinguished Alumnus Award, College of Engineering, Univ. of Cincinnati, 1983.

MARSTEN, JESSE, IRE Fellow 1957. Born: January 1, 1897, Budapest, Hungary. Degrees: B.Sc., 1917, City College of the City University of New York.

Fellow Award "For contributions to the design and manufacture of electronic components."

MARSTEN, RICHARD B., Fellow 1971. Born: 1925, New York, N.Y. Degrees: S.B., 1946, S.M., 1946, Massachusetts Institute of Technology; Ph.D., 1951, University of Pennsylvania.

Fellow Award "For contributions to the development of radar and communications satellite systems." Other Awards: Tau Beta Pi, Sigma Xi, 1946; Member, New York Academy of Sciences, 1967; White House Citation for Outstanding Sustained Performance, 1972; Exceptional Service Medal, NASA, 1974; Group Achievement Award, NASA, ATS-6 Project Team, 1974; Citation for Outstanding Achievement in Research and Educational Development, Committee on Social Justice, International Women's Year, 1975; Apollo-Soyuz Project Space Medal, 1976; Professional Member, Eta Kappa Nu, 1976; National Advisory Board, National Energy Foundation; Licensed Professional Engineer, New Jersey.

MARTIN, DANIEL W., IRE Fellow 1957. Born: November 18, 1918, Georgetown, Ky. Degrees: A.B., 1937, Georgetown College, Ky.; M.S., 1939, Ph.D., 1941, University of Illinois; D.Sc., 1981, Georgetown College.

Fellow Award "For contributions in the field of acoustics"; Audio Senior Award, IRE, 1958; Audio Achieve-

ment Award, IRE, 1959. Other Awards: Fellow, Acoustical Society of America, 1950; Fellow, Audio Engineering Society, 1963; Audio Engineering Society Award, 1969; Engineer of the Year, Technical and Scientific Societies Council of Cincinnati, 1972; Ohio Technical Engineer of the Year, OSPE, 1972.

MARTIN, EDGAR T., Fellow 1971. Born: May 1, 1918, Princeton, W.Va. Degrees: B.S., 1938, M.S., 1953, Virginia Polytechnic Institute; M.A., 1957, American University; J.D., 1957, George Washington University; L.L.M., 1959, Georgetown University; Graduate, 1963, U.S. Army Command and General Staff College; M.Forensic Sc., 1977, George Washington University.

Fellow Award "For leadership, management, and technical contributions in the development of international broadcasting and telecommunications."

MARTIN, JOHN A., Fellow 1976. Born: March 12, 1924, La Grande, Oreg. Degrees: B.S.E.E., 1948, University of Tennessee.

Fellow Award "For contributions to heavy ion cyclotron developments"; Special Merit Award, IEEE Nuclear and Plasma Sciences Society, 1977. Other Awards: Tau Beta Pi; Phi Kappa Phi; Sigma Xi.

MARTIN, THOMAS L., JR., Fellow 1966. Born: September 26, 1921, Memphis, Tenn. Degrees: B.S.E.E., 1942, M.E.E., 1948, Rensselaer Polytechnic Institute; Ph.D., 1951, Stanford University; D.Eng.(Hon.), 1967, Rensselaer Polytechnic Institute.

Fellow Award "For leadership in graduate engineering education"; Electronic Achievement Award, IRE, Region 7, 1956, "For his leadership in engineering education and many contributions to the technical literature"; Engineer of the Year, IEEE, Dallas, Texas, 1968. Other Awards: Bronze Star, U.S. Dept. of Army, 1946; Outstanding Service Award, Florida Engineering Society, 1966; Member, National Academy of Engineering, 1972.

MARTIN, WILLIAM G., III, AIEE Fellow 1962. Born: September 3, 1915, Athens, Ala. Degrees: B.E., 1938, Vanderbilt University; M.Sc., 1940, Rutgers University.

Fellow Award "For contributions to the development of electric rotating machinery, magnetic clutch, and arc welding equipment."

MARTINDALE, ROY, AIEE Fellow 1940. Born: February 2, 1884, Pine Village, Ind. Degrees: B.S.E.E., 1906, Purdue University.

MARTINO, JOSEPH P., Fellow 1976. Born: July 16, 1931, Warren, Ohio. Degrees: A.B.(Physics), 1953, Miami University, Ohio; M.S.E.E., 1955, Purdue University; Ph.D.(Mathematics), 1961, Ohio State University.

Fellow Award "For contributions to engineering management, R&D planning, and R&D resource allocation"; Prize Paper Award, IEEE Engineering Management Society, 1973; IEEE Centennial Medal, 1984. Other

Awards: Air Force Systems Command Technical Achievement Award, 1971; Air Force Commendation Medal with Two Oak Leaf Clusters.

MARTZLOFF, FRANCOIS D., Fellow 1983. Born: June 9, 1929, Strasbourg, France. Degrees: Ecole Speciale de Mecanique et d'Electricite, 1951; M.S.E.E., 1952, Georgia Tech; M.S.I.A., 1971, Union College.

Fellow Award "For contributions to reliability, safety, and cost-effectiveness of electronic equipment subjected to surge voltages"; Surge Protective Devices Committee Paper Award, 1977; Surge Protective Devices Committee Paper Award, 1982. Other Awards: Industrial Research IR-100 Award, 1972.

MARUHASHI, TORU, Fellow 1983. Born: July 26, 1924, Kyoto, Japan. Degrees: B.E., 1947, Ph.D.(Eng.), 1962, Osaka University, Japan.

Fellow Award "For research on the analysis and simulation of high frequency inverters and other power electronic circuits."

MARUVADA, SARMA P., Fellow 1983. Born: January 1, 1938, Rajahmundry, India. Degrees: B.E.(Hon.), 1958, Andhra University; M.E., 1959, Indian Institute of Science; M.A.Sc., 1966, Ph.D., 1968, University of Toronto.

Fellow Award "For contributions to the theoretical and experimental aspects of corona and radio interference performance of high-voltage ac and dc systems."

MARYSSAEL, GUSTAVE J. C., AIEE Fellow 1961. Born: October 1, 1903, Verona, Italy. Degrees: Civil and Mining Eng., 1926, University of Brussels; E.E., 1927, University of Ghent.

Fellow Award "For contributions to the design and construction of a large public utility in Mexico."

MASON, WARREN P., IRE Fellow 1941. Born: September 28, 1900, Colorado Springs, Colo. Degrees: B.S., 1921, University of Kansas; M.A., 1924, Ph.D., 1928, Columbia University.

Awards: Lamme Medal, 1967, "For outstanding contributions in the fields of sonics and ultrasonics and for his original work in designs of and applications for electro-mechanical transducers." Other Awards: Honorary Member, Audio Engineering Society, 1957; Arnold O. Beckman Award, Instrument Society of America, 1964; Distinguished Alumni Award, University of Kansas, 1965; C. B. Sawyer Award, Signal Corps Symposium, 1966; Gold Medal, Acoustical Society of America, 1971; Honorary Fellow (First Honorary Member, 1974), British Institute of Acoustics; Honorary Fellow, Institute of Acoustics of India; Dedication of First International Conference on Fatigue and Corrosion Fatigue up to Ultrasonic Frequencies to Warren P. Mason, Engineering Foundation Conference, 1981.

MASSA, FRANK, IRE Fellow 1959. Born: April 10, 1906, Boston, Mass. Degrees: B.S., 1927, M.S., 1928, Massachusetts Institute of Technology.

Fellow Award "For pioneering in electroacoustical engineering." Other Awards: Swope Fellow, Massachusetts Institute of Technology, 1927-1928; Fellow, Acoustical Society of America.

MASSEY, JAMES L., Fellow 1971. Born: February 11, 1934, Wauseon, Ohio. Degrees: B.S., 1956, University of Notre Dame; M.S.E.E., 1960, Ph.D., 1962, Massachusetts Institute of Technology.

Fellow Award "For contributions to coding theory, particularly to threshold decoding and its application, and to engineering education"; Annual Paper Award, IEEE Group on Information Theory, 1963, "For his valuable contribution of a practical yet efficient decoding algorithm as demonstrated in his monograph entitled 'Threshold Decoding' included as part of the Press Series of the Massachusetts Institute of Technology"; IEEE Centennial Medal, 1984. Other Awards: Best Tutorial Paper Award, National Electronics Conference, 1963; Thomas P. Madden Award, University of Notre Dame, 1969; Faculty Award, University of Notre Dame, 1970; ASEE Western Electric Fund Award, 1975; International Telemetry Conference Award, 1975; Special Presidential Award, University of Notre Dame, 1976.

MASTERS, R. WAYNE, IRE Fellow 1962. Born: May 25, 1914, Fort Wayne, Ind. Degrees: B.S., 1938, University of Alabama; M.S., 1941, Ohio State University; Ph.D., 1957, University of Pennsylvania.

Fellow Award "For contributions in the field of antennas and RF transmission systems."

MASUDA, SENICHI, Fellow 1978. Born: April 12, 1926, Imchon, South Korea. Degrees: Ph.D., 1961, University of Tokyo.

Fellow Award "For contributions to the understanding and application of electrostatic precipitation technology." Other Awards: Achievement Award, Institute of Electrical Engineers of Japan; Paper Award, Institute of Electrical Engineers of Japan, 1974; Invention Award, Minister of Science and Technology, 1975; Technical Award, Association of Painting Technology, 1977; Contribution Award, Society of Industrial Pollution Control, 1977; Paper Award, Institute of Electrostatics Japan, 1978; Paper Award, Society of Industrial Pollution Control, 1980; Corresponding Member, Verein Deutscher Ingeniuere, 1984.

MATARE, HERBERT F., Fellow 1973. Born: September 22, 1912, Aachen, Germany. Degrees: B.S., 1933, Dipl.Ing., 1939, Aachen; Dr.Ing., 1942, Technical University, Berlin; Dr.Sc.Phys., 1950, Ecole Normale Superieure, Paris.

Fellow Award "For contributions to understanding of high-frequency semiconductor devices, and crystal properties, and electronic effects of dislocations." Other Awards: Electrochemical Society; Over 100 publications; 60 patents; Member, American Physical Society;

Member, Thin-Film Division, American Vacuum Society; Member, American Association for the Advancement of Science; Member, New York Academy of Science; Listed in "American Men of Science," "Who's Who in the West," "Leaders in American Science," "World's Who's Who in Science," "Who's Who in California," "Dictionary of International Biography"; "Who's Who in Technology Today"; Visit. Prof., Electronics Engineering Dpt., UCLA, 1968-1969; Internat. Solid State Conf., Brussels, 1959; Solar Cell Meeting, Dallas, 1976, Conf. on Polycrystalline Semiconductors, Perpignan, France, 1982. Visiting Prof., Physics Dpt., Calif. State University, Fullerton, Calif., 1969-1970; Keynote Speaker and Conference Chairman; Internat. Meetings: Electrochem. Society, Chicago, 1955; New York, 1958; New York, 1969; Internat. Solid State Conf., Brussels, 1959; Solar Cell Meeting, Dallas, 1976; Conf. on Polycrystalline Semiconductors, Perpignan, France, 1982.

MATHES, KENNETH N., AIEE Fellow 1962. Born: June 30, 1913, Schenectady, N.Y. Degrees: B.S., 1935, Union College.

Fellow Award "For achievements in the field of insulation materials"; Distinguished Service Award, IEEE Electrical Insulation Society, 1981; IEEE Centennial Medal, 1984. Other Awards: Engineer of the Year Award, Schenectady County National Society of Professional Engineers, 1969; Arnold H. Scott Award, American Society for Testing and Materials, 1972; Award of Merit and Fellow, American Society for Testing and Materials, 1976; Fellow, American Association for the Advancement of Science, 1980.

MATHES, R. E., IRE Fellow 1957. Born: April 1902, Minneapolis, Minn. Degrees: B.S., 1924, University of Minnesota.

Fellow Award "For contributions to the development of terminal apparatus for point-to-point radio service."

MATHEWS, MAX V., Fellow 1975. Born: November 13, 1926, Columbus, Nebr. Degrees: B.S., 1950, California Institute of Technology; M.S., 1952, Sc.D., 1954, Massachusetts Institute of Technology.

Fellow Award "For advances in the analysis, synthesis, and recognition of speech and the generation of musical sounds by computer and electronic methods"; Fellow, Acoustical Society of America, 1961; IEEE David Sarnoff Gold Medal Award, 1973, "For leadership in applying electronics to art and for his contribution to the production of musical sounds by computer." Other Awards: Member, National Academy of Sciences, 1975; National Academy of Engineering, 1979; Fellow, American Academy of Arts & Sciences, 1982; Fellow, Audio Engineering Society, 1983.

MATHEWS, WARREN E., Fellow 1973. Born: November 10, 1921, Osborne, Kans. Degrees: B.A., 1942, Ohio Wesleyan University; B.S., M.S., 1944, Massachusetts Institute of Technology; Ph.D., 1953, California Institute of Technology.

Fellow Award "For contributions to the fields of systems and aerospace electronics." Other Awards: Howard Hughes Fellowship in Science and Engineering, 1950-52; Honorary Member, Beta Gamma Sigma, 1969.

MATISOO, JURI, Fellow 1980. Born: February 10, 1937, Estonia. Degrees: B.S.E.E., 1959, M.S.E.E., 1960, Massachusetts Institute of Technology; Ph.D., 1964, University of Minnesota.

Fellow Award "For the invention and development of Josephson computer circuits"; Jack A. Morton Award, IEEE, 1978.

MATSCH, LEANDER W., AIEE Fellow 1950. Born: January 23, 1902, New Ulm, Minn. Degrees: B.S., 1926, Lewis Institute; M.S., 1948, Illinois Institute of Technology.

MATSUMOTO, AKIO, Fellow 1974. Born: October 13, 1908, Hokkaido, Japan. Degrees: B.E., 1931, Hokkaido University, Japan; D.Eng., 1942, Tohoku University, Japan.

Fellow Award "For contributions to the theory and design of electrical networks and microwave filters." Other Awards: Science and Technology Award, Hokkaido Shimbunsha, 1967; Contribution Award, The Institute of Electronics and Communication Engineers, Japan, 1972; Second Grade National Medal, Japanese Government, 1981; 1982 Microwave Career Award, IEEE, 1982.

MATSUMOTO, YOSHIHIRO, Fellow 1982. Born: February 21, 1932, Kobe, Japan. Degrees: B.S., 1954, D.Eng., 1974, University of Tokyo.

Fellow Award "For leadership in the application of computers in industrial control systems." Other Awards: New Invention Award, Japanese Association of Electrical Equipments Manufacturers, 1963; Special Contributor Award, Toshiba Corporation, 1980; All Japan Invention Award, Japan Institute of Invention and Innovation, 1981; Excellent Researcher Award, Japanese Government, 1982.

MATSUO, TADAYUKI, Fellow 1984. Born: January 1, 1924, Gosen City, Niigata Prefecture, Japan. Degrees: B.E., 1946, Dr.Eng., 1963, Tohoku University, Sendai, Japan.

Fellow Award "For the development of chemically sensitive solid-state devices and their applications to biomedicine." Other Awards: 17th Achievement Award, IECE of Japan, 1980.

MATTHAEI, GEORGE L., Fellow 1965. Born: August 28, 1923, Tacoma, Wash. Degrees: B.S., 1948, University of Washington; M.S., 1949, E.E., 1950, Ph.D., 1952, Stanford University.

Fellow Award "For contributions to the theory and design of microwave networks and parametric amplifiers"; Microwave Prize, IRE, 1961, "For very significant contribution to the field of endeavor of the IRE-PGMTT."

MATTHIAS, LYNN H., Fellow 1965. Born: July 11, 1904, Antigo, Wis. Degrees: B.S., 1926, University of Wisconsin.

Fellow Award "For contributions to the development and design of electric motor controls." Other Awards: Eta Kappa Nu; Phi Kappa Phi; Tau Beta Pi; American Association for the Advancement of Science; University of Wisconsin Citation, 1960.

MATTHYSSE, IRVING F., Fellow 1969. Born: April 25, 1912, New York, N.Y. Degrees: B.S.E.E., 1934, The Cooper Union; B.S.E.E., 1936, M.S., 1941, New York University.

Fellow Award "For orignal contributions and inventions of electrical connectors, precision fuses, and electrical switches, as well as numerous technical papers and publications concerning these products. Also for outstanding leadership of the engineering departments which he has headed."

MATTINGLY, ROBERT L., IRE Fellow 1962. Born: September 15, 1916, New York, N.Y. Degrees: B.S., 1938, University of Maryland; M.S., 1942, Stevens Institute of Technology.

Fellow Award "For contributions to radar and to military weapons systems". Other Awards: Fellow, Radio Club of America.

MATTSON, ROY HENRY, Fellow 1971. Born: December 26, 1927, Chisholm, Minn. Degrees: B.S.E.E., 1951, M.S.E.E., 1952, University of Minnesota; Ph.D., 1959, Iowa State University.

Fellow Award "For contributions to education and to solid-state and medical electronics research"; IEEE Education Society, Meritorious Service Award, 1982. Other Awards: Fellow, American Association for the Advancement of Science, 1972; Engineer of the Year Award, ASPE, 1978; Anderson Prize, University of Arizona, 1981.

MAURER, HANS A., Fellow 1976. Born: May 7, 1913, Frankfurt, Germany. Degrees: Ph.D.(Applied Physics), Frankfurt University.

Fellow Award "For contributions to radar and electrooptical technology."

MAWARDI, OSMAN K., Fellow 1967. Born: December 12, 1917, Cairo, Egypt. Degrees: B.Sc., 1940, M.Sc., 1946, Cairo University; A.M., 1947, Ph.D., 1948, Harvard University.

Fellow Award "For significant contributions in the field of acoustics and plasmas, and for leadership in the area of university education and research in plasma dynamics." Other Awards: Fellow, Acoustical Society of America, 1950; Biennial Award, Acoustical Society of America, 1952; Guggenheim Fellow, 1954; Fellow, American Association for the Advancement of Science,

1959; Research Award, Sigma Xi, 1964; Fellow, American Physical Society, 1966; Member, Institute for Advanced Study, Princeton, 1970-71; CECON Medal of Achievement, 1979.

MAXFIELD, JOSEPH P., AIEE Fellow 1927. Born: December 28, 1887, San Francisco, Calif. Degrees: B.S., 1910, Massachusetts Institute of Technology.

Other Awards: Meritorious Civilian Service Award, U.S. Dept. of Army, 1947; Meritorious Civilian Service Award, U.S. Dept. of Navy, 1953; John Potts Award, Acoustical Society of America, 1954.

MAXSON, RAYMOND D., AIEE Fellow 1955. Born: March 12, 1898, Rochelle, Ill. Degrees: B.S., 1921, University of Illinois.

Fellow Award "For contributions to the integration of the engineering practices of several companies into a combined power system." Other Awards: President's Award, Western Society of Engineers, 1960; Distinguished Service Award, American Nuclear Society, 1965; Honorary Member, Western Society of Engineers, 1969; Distinguished Alumnus Award, Electrical Engineering, University of Illinois, 1973.

MAXWELL, DONALD E., IRE Fellow 1961. Born: April 18, 1913, Lynn, Mass. Degrees: B.S.E.E., 1937, Tufts University; M.S., 1939, Harvard University.

Fellow Award "For contributions to radio-television broadcasting and standardization"; IEEE Centennial Medal, 1984.

MAXWELL, MARVIN V., AIEE Fellow 1951. Born: February 18, 1901, Carthage, Mo. Degrees: B.S., 1924, University of Missouri.

Other Awards: Life Fellow, American Society of Mechanical Engineers; Chairman, Chicago Section; Washington Awards Representative; Vice President and Trustee, Western Society of Engineers; President, Illinois Engineering Council; Eta Kappa Nu; Tau Beta Pi; Pi Mu Epsilon.

MAYER, FERDY P.M., Fellow 1982. Born: August 19, 1926, Grand Duchy of Luxembourg. Degrees: IEG, 1949, IRG(Radioelectricity and E.E.), 1950, Institut Polytechnique, Grenoble, France; Dr.Eng., 1957, University of Grenoble, France.

Fellow Award "For contributions to the theory of ferromagnetics and to the development of materials for the suppression of electromagnetic interference." Other Awards: Vice President, Alliance Francaise, France; Vice President, Association France-United Kingdom.

MAYER, HARRY F., IRE Fellow 1956. Born: January 19, 1912, Indianapolis, Ind. Degrees: B.S.E.E., 1932, M.S.E.E., 1933, Purdue University.

Fellow Award "For contributions in the development of airborne radar systems." Other Awards: Citation, U.S. Dept. of Navy, 1945.

MAYER, JAMES W., Fellow 1972. Born: April 24, 1930, Chicago, Ill. Degrees: B.S., 1952, Ph.D., 1960,

Purdue University.

Fellow Award "For contributions to the development of semiconductor nuclear radiation detectors and to the understanding of ion implantation processes in semiconductors." Other Awards: Fellow, American Physical Society; Scientific Member, Bohmische Physical Society, 1975; Von Hippel Award, Materials Research Society, 1981; 1000 most-cited contemporary scientists, 1965-78.

MAYER, ROBERT, Fellow 1981. Born: March 28, 1918. Degrees: B.S.E.E., 1940, M.S.E.E., 1949, University of Pennsylvania.

Fellow Award "For contributions to instrumentation and control systems in the petrochemical process industry"; Philadelphia Section Award, IEEE Industrial Electronics and Control Instrumentation Society, Achievement Award. Other Awards: Sigma Xi.

MAYES, PAUL E., Fellow 1975. Born: December 21, 1928, Frederick, Okla. Degrees: B.S.E.E., 1950, University of Oklahoma; M.S.E.E., 1952, Ph.D., 1955, Northwestern University.

Fellow Award "For contributions to the theory and development of the log-periodic antennas."

MAYNE, DAVID Q., Fellow 1981. Born: April 23, 1930, Germiston, South Africa. Degrees: B.Sc. (Eng.), 1950, M.Sc. (Eng.), 1956, University of the Witwatersrand, Johannesburg; Ph.D., 1967, D.Sc. (Eng.), 1981, University of London.

Fellow Award "For contributions to optimal control and dynamic programming." Other Awards: Research Fellowship, Harvard University, 1971; Research Engineer, University of California at Berkeley (summer), 1974, 1976, 1978, 1980; Senior Science Research Council Fellowship, 1979; FIEE, 1980.

MAYO, JOHN S., Fellow 1967. Born: February 26, 1930, Greenville, N.C. Degrees: B.S.E.E., 1952, M.S.E.E., 1953, Ph.D.E.E., 1955, North Carolina State University.

Fellow Award "For contributions to the development of pulse code modulation systems"; Alexander Graham Bell Award, 1978. Other Awards: Distinguished Alumnus Award, North Carolina State University, 1977; National Academy of Engineering, 1979.

MCADAM, WILL, Fellow 1981. Born: October 22, 1921. Degrees: B.S.E.E., 1942, Case Western Reserve University; M.S.E.E., 1959, University of Pennsylvania.

Fellow Award "For the development of reliable instrumentation for the measurement and control of industrial processes in high-noise environments."

MCADAMS, WILLIAM A., Fellow 1976. Born: February 24, 1918. Degrees: B.S.(Math, Physics), 1940, Washington College.

Fellow Award "For leadership in electrical and electronic standardization activities"; Charles Proteus Steinmetz Award, IEEE, 1983. Other Awards: Alumni

Citation in Science, Washington College, 1960; Gold Standards Medal, American National Standards Institute, 1965; Meritorious Service Award for International Standards, American National Standards Institute, 1974; President, U.S. National Committee of International Electrotechnical Commission, 1968-1974; Chairman, National Fire Protection Association, 1976-1978; President, Chairman, American Society for Testing and Materials, 1977-1978; Board of Governors, Washington College, 1980-1986; Leo B. Moore Award, Standards Engineers Society, 1977; General Electric Centennial Steuben Award, 1978; President, International Electrotechnical Commission, 1980-1983; James M. McGraw Award for Electrical Men, 1980; National Inventors Hall of Fame Medal, 1983.

MCARTHUR, ELMER D., IRE Fellow 1945. Born: May 3, 1903, Salamanca, N.Y. Degrees: B.S.E.E., 1925, M.S., 1948, Union College.

Fellow Award "For his developments in the field of ultrahigh-frequency electron tubes." Other Awards: Charles A. Coffin Award, General Electric, 1946; Certificate of Commendation, U.S. Dept. of Navy, 1947; Sigma Xi, 1948; Award, National Electronics Conference, 1954.

MCBEE, WARREN D., Fellow 1972. Born: March 18, 1925, Toledo, Ohio. Degrees: B.S., 1945, Marquette University; M.S., 1948, Ph.D., 1951, University of Michigan.

Fellow Award "For contributions to the development of microwave electron tubes and to the management of industrial research."

MCCANN, GILBERT D., AIEE Fellow 1950, IRE Fellow 1961. Born: January 12, 1912, Glendale, Calif. Degrees: B.S., 1934, M.S., 1935, Ph.D., 1939, California Institute of Technology.

IRE Fellow Award "For contributions to electrical engineering literature and education"; First Paper Prizes, AIEE, 1943, 1946. Other Awards: Sigma Xi; Tau Beta Pi; Eta Kappa Nu; American Society of Mechanical Engineers; American Society for Engineering Education; Outstanding Electrical Engineer Award, Eta Kappa Nu, 1942.

MCCLELLAN, LESLIE N., AIEE Fellow 1938. Born: March 27, 1888, Middletown, Ohio. Degrees: B.S., 1911, University of Southern California; D.Eng.(Hon.), 1949, University of Colorado.

Other Awards: Gold Medal, Colorado Engineering Council, 1951; Gold Medal, U.S. Dept. of the Interior, 1952; Distinguished Alumnus Award, School of Engineering, University of Southern California, 1967; Sigma Psi; Tau Beta Phi; Eminent Member, Eta Kappa Nu.

MCCLOSKA, FRED W., AIEE Fellow 1960. Born: July 8, 1908, Chicago Ill. Degrees: B.S., 1929, Illinois Institute of Technology.

Fellow Award "For contributions to design and application of electrical equipment to power systems, industrial and nuclear plants."

MCCLURE, GEORGE F., Fellow 1981. Born: February 9, 1933, Jacksonville, FL. Degrees: B.E.E., 1954, M.S. (Eng.), 1961, University of Florida.

Fellow Award "For contributions to mobile telephone communications systems engineering and the creation of new and more effective methods of spectrum utilization."

MCCLURE, J. BURNS, AIEE Fellow 1960. Born: December 6, 1900, Thamesford, Ontario, Canada. Degrees: B.Sc., 1923, Queen's University, Kingston, Ontario, Canada; M.S., 1927, Massachusetts Institute of Technology.

Fellow Award "For contributions in the field of electric utility generating stations and power transmission systems"; First Paper Prize, AIEE, 1953, for "Short Circuit Capabilities of Synchronous Machines for Unbalanced Faults."

MCCLUSKEY, EDWARD J., Fellow 1965. Born: October 16, 1929, New York, N.Y. Degrees: B.A., 1953, Bowdoin College; B.S., 1953, M.S., 1953, Sc.D., 1956, Massachusetts Institute of Technology.

Fellow Award "For contributions to switching theory and engineering education"; IEEE Centennial Medal, 1984. Other Awards: Fellow, American Association for the Advancement of Science.

MCCLYMONT, KENNETH R., Fellow 1976. Born: Novermber 13, 1924, Toronto, Ontario, Canada. Degrees: B.A.Sc., 1947, University of Toronto.

Fellow Award "For contributions and leadership in the planning and development of a high-voltage bulk power transmission system."

MCCONNELL, ANDREW J., AIEE Fellow 1960. Born: November 30, 1904, Atlantic City, N.J. Degrees: B.E.E., 1928, Cornell University.

Fellow Award "For contributions to design and application of power system protective relays." Other Awards: Tau Beta Pi; Eta Kappa Nu; Phi Kappa Phi.

MCCONNELL, LORNE D., Fellow 1976. Born: Rosetown, Saskatchewan, Canada. Degrees: B.S.E.E., 1948, University of Saskatchewan; 1950-1954, University of New Brunswick; Business Administration, 1962-1963, University of California at Los Angeles.

Fellow Award "For contributions to the research and development of power switchgear equipment."

MCCORMICK, GLENDON C., Fellow 1981. Born: September 20, 1915, Paradise, Nova Scotia. Degrees: B.Sc., 1935, King's-Dalhousie; M.Sc., 1945, Acadia; Ph.D., 1953, McGill University.

Fellow Award "For contributions to the understanding of electromagnetic wave propagation and scattering in precipitating atmospheres."

MCCOUBREY, ARTHUR O., Fellow 1973. Born: 1920, Canada. Degrees: B.S., 1943, California Institute of Technology; Ph.D., 1953, University of Pittsburgh.

Fellow Award "For contributions to atomic frequency standards and research management." Other Awards: Member, American Physical Society; Sigma Pi Sigma.

MCCOY, RAWLEY D., IRE Fellow 1959. Born: December 21, 1914, Nutley, N.J. Degrees: M.E., 1937, Stevens Institute of Technology.

Fellow Award "For contributions in the field of servo control systems and analog computers."

MCCREARY, RALPH L., IRE Fellow 1962. Born: December 23, 1916, Morning Sun, Ohio. Degrees: B.S. 1938, Miami University, Ph.D., 1942, University of Rochester.

Fellow Award "For contributions to the field of radio propagation." Other Awards: Fellow, American Physical Society.

MCCULLOUGH, JACK A., IRE Fellow 1953. Born: November 25, 1907, San Francisco, Calif.

Fellow Award "For his pioneering contributions to power tube design." Other Awards: Distinguished Public Service Award, U.S. Dept. of Navy, 1950.

MCCUNE, FRANCIS K., AIEE Fellow 1949. Born: April 10, 1906, Santa Barbara, Calif. Degrees: B.S., 1928, University of California at Berkeley.

Awards: Past Chairman, Committee on Management, AIEE; Operating Committee, AIEE. Other Awards: Member, National Academy of Engineering; Charter Member, American Nuclear Society; Member, American Society of Mechanical Engineers; Howard Coonley Medal, 1969; Past President and Honorary Director, Atomic Industrial Forum; Past President, American National Standards Institute; Member of the Corporation: Woods Hole Oceanographic Institute; Past Member, National Research Council.

MCCURLEY, JAMES B., AIEE Fellow 1950. Born: March 22, 1906, Baltimore, Md. Degrees: B.Eng., 1928, The Johns Hopkins University; M.S., 1930, Yale University; D.Eng., 1935, The Johns Hopkins University.

MCCUTCHEN, BRUNSON S., AIEE Fellow 1932. Born: November 20, 1892, North Plainfield, N.J. Degrees: Lit.B., 1915, E.E., 1917, Princeton University.

MCDANIEL, G. H., Fellow 1965. Born: September 1, 1911, Ligonier, Ind.

Fellow Award "For contributions to principles and practices for interconnected power systems operation." Other Awards: Certificate of Public Service, Federal Power Commission, 1964; Certificate of Outstanding Leadership, Defense Electric Power Administration, 1975

MCDONALD, HENRY S., Fellow 1972. Born: October 28, 1927, Carlisle Barracks, Pa. Degrees: B.E.E., 1950, The Catholic University of America; M.S.E.E., 1953, D.Eng., 1955, The Johns Hopkins University.

Fellow Award "For research on the digital processing

of speech and television, instrumention for simulation of digital communication systems, and for work on interactive graphics for computer"; Society Award, IEEE Acoustics, Speech, and Signal Processing Society, 1978.

MC DONALD, JOHN C., Fellow 1979. Born: January 23, 1936, San Bernardino, Calif. Degrees: B.S.(with distinction), 1957, M.S., 1959, Deg. of Engineer, 1964, Stanford University.

Fellow Award "For contributions to and leadership in the design of integrated digital transmission and switching systems." Other Awards: Tau Beta Pi, 1956; Sigma Xi, 1958; Teaching Fellowship, Stanford University, 1958, 1959.

MCEACHRON, KARL B., JR., AIEE Fellow 1951. Born: June 27, 1915, Ada, Ohio. Degrees: B.S.E.E., 1937, Purdue University.

Other Awards: Distinguished Alumnus Award, Purdue University, 1964.

MCELIECE, ROBERT J., Fellow 1984. Born: May 21, 1942, Washington, D.C. Degrees: B.S., 1964, Ph.D., 1967, Cal Tech.

Fellow Award "For contributions to information and coding theory research and applications."

MCELRATH, GEORGE, IRE Fellow 1956. Born: January 23, 1899, Marlboro, N.Y.

Fellow Award "For contributions in the development of broadcasting and television operating practices and techniques."

MCELROY, ALAN J., Fellow 1981. Born: September 1, 1934. Degrees: B.E.E., 1956, City College of New York; M.E.E., 1959, Polytechnic Institute of Brooklyn; Ph.D., 1969, Massachusetts Institute of Technology.

Fellow Award "For contributions to the field of failure history recording and to the coordination of power systems."

MCFARLAN, RONALD L., IRE Fellow 1961. Born: March 8, 1905, Cincinnati, Ohio. Degrees: A.B., 1926, University of Cincinnati; Ph.D., 1930, University of Chicago.

Fellow Award "For contributions to systems applications of electronic computers and for effective administrative activities during formative periods of technical advancement"; President, IRE, 1960; Member, IRE-AIEE Merger Committee, 1962; Director, IEEE, 1963.

MCFARLANE, MAYNARD D., AIEE Fellow 1951, IRE Fellow 1956. Born: August 12, 1895, London, England.

IRE Fellow Award "For contributions in the development of facsimile and radar." Other Awards: Fellow, American Association for the Advancement of Science, 1928; Certificate of Appreciation, War Dept., U.S. Dept. of Navy, 1947.

MCFEE, R., Fellow 1967. Born: January 24, 1925. Degrees: B.E.E., 1947, Yale University; M.S., 1949, Syra-

cuse University; Ph.D., 1954, University of Michigan.

Fellow Award "For contributions to the theory and instrumentation of electro- and magneto-cardiography."

MCGILLEM, CLARE D., Fellow 1975. Born: October 9, 1923, Clinton, Mich. Degrees: B.S., 1947, University of Michigan; M.S., 1949, Ph.D., 1955, Purdue University.

Fellow Award "For contributions in radar research"; IEEE Centennial Medal, 1984. Other Awards: Meritorious Civilian Service Award, U.S. Navy, 1954.

MC GRODDY, JAMES C., Fellow 1979. Born: April 6, 1937, New York, N.Y. Degrees: B.S., 1958, St. Joseph's College; Ph.D., University of Maryland, 1964.

Fellow Award "For contributions to the understanding of nonequilbrium transport and optical properties of semiconductors." Other Awards: Fellow, American Physical Society.

MCILVEEN, EDWARD E., Fellow 1983. Born: August 3, 1911, Passaic, N.J. Degrees: B.M.E. (cum laude), Polytechnic Institute of Brooklyn.

Fellow Award "For contributions in electric power cable design and testing." Other Awards: Sigma Xi, 1943.

MCINTYRE, ROBERT J., Fellow 1980. Born: December 19, 1928, Bathurst, New Brunswick, Canada. Degrees: B.Sc., 1950, St. Francis Xavier University, Nova Scotia; M.Sc., 1953, Dalhouse University, Nova Scotia; Ph.D., 1956, University of Virginia.

Fellow Award "For theoretical work on the noise properties of avalanche photodiodes, and for leadership in their commercial development." Other Awards: RCA David Sarnoff Award for Outstanding Technical Achievement, 1979.

MCKAY, KENNETH G., Fellow 1964. Born: April 8, 1917, Montreal, Quebec, Canada. Degrees: B.S., 1938, M.S., 1939, McGill University, Quebec; D.Sc., 1941, Massachusetts Institute of Technology; Dr.Eng.(Hon.), 1980, Stevens Institute of Technology.

Fellow Award "For fundamental advances in the physics and engineering of solid-state devices." Other Awards: Anne Molson Gold Medal for Mathematics and Natural Philosophy, 1938; NASA Public Service Award, 1969; NASA Public Service Group Achievement Award, 1969; Member, National Academy of Sciences, National Academy of Engineering; Fellow, American Physical Society.

MCKEAN, A. LAIRD, Fellow 1966. Born: April 25, 1916, New York, N.Y. Degrees: B.E.E., 1936, M.E.E., 1941, Polytechnic Institute of New York (Brooklyn).

Fellow Award "For contributions to research, design, fabrication and testing of power and communication cables."

MCKINLEY, D. W. R., IRE Fellow 1948. Born: September 22, 1912, Shanghai, China. Degrees: B.A., 1934,

M.A., 1935, Ph.D., 1938, University of Toronto.

Fellow Award "For his contributions to the development in Canada of radio aids to air navigation." Other Awards: Fellow, American Physical Society; Fellow, Royal Society of Canada; Order of the British Empire; Engineering Institute of Canada.

MCLEAN, CORBETT, AIEE Fellow 1958. Born: December 24, 1905, Portland, Oreg. Degrees: A.B.(E.E.), 1927, E.E., 1929, Stanford University.

Fellow Award "For his contributions to development of an extensive power system." Other Awards: Sigma Xi, 1929; Award, Electric Club of Oregon, 1941; Award, Portland Agenda Club, 1944; Award, Northwest Electric Light and Power Association, 1970.

MCLEAN, CYRUS H., Fellow 1969. Born: September 6, 1898.

Fellow Award "For planning, engineering, and implementing significant advancements in international telecommunications."

MCLEAN, FRANCIS C., Fellow 1967. Born: November 6, 1904, Birmingham, England. Degrees: B.Sc., 1925, Birmingham University.

Fellow Award "For his contributions to British broadcasting engineering, both radio and television." Other Awards: Fahie Premium, Institution of Electrical Engineers, U.K., 1946; Duddell Premium, Institution of Electrical Engineers, U. K., 1961.

MCLEAN, TRUE, Fellow 1965. Born: January 22, 1899, Staten Island, N.Y. Degrees: E.E., 1922, Cornell University; P.E., 1937, State University of New York.

Fellow Award "For contributions to engineering education and research in acoustics, communication, and electrical measurements." Other Awards: Sigma Xi, 1936; Eta Kappa Nu, 1952. Professor Emeritus, 1966, Cornell University.

MCLENNAN, MILES A., IRE Fellow 1960. Born: November 11, 1902, Bay City, Mich. Degrees: B.S.E.E., 1928, University of Michigan.

Fellow Award "For contributions in aero-medical electronics." Other Awards: Meritorious Civilian Service Award, U.S. Dept. of Air Force, 1956.

MCLUCAS, JOHN L., IRE Fellow 1962. Born: August 22, 1920, Fayetteville, N.C. Degrees: B.S., 1941, Davidson College; M.S., 1943, Tulane University; Ph.D., 1950, Pennsylvania State University; D.Sc.(Hon.), 1974, Davidson College.

Fellow Award "For contributions to the transmission and recording of radar presentations." Other Awards: Distinguished Public Service Award, 1964, Bronze Palm, 1973, Silver Palm, 1975, Dept. of Defense; National Academy of Engineering, 1969; Exceptional Civilian Service Award, Bronze Palm, U.S. Dept. of the Air Force, 1973, 1975; Distinguished Service Medal, NASA, 1975; Distinguished Alumnus Award, Tulane University, 1976, Pennsylvania State University,

1977; Secretary's Award for Outstanding Achievement, Dept. of Transportation, 1977; Fellow, American Institute of Aeronautics and Astronautics, 1979; AIAA Reed Aeronautics Award, 1982.

MCMAHON, EUGENE J., Fellow 1984. Born: April 22, 1917, Jersey City, N.J. Degrees: Assoc.(Chem.Eng.), 1938, Newark College of Engineering; Assoc.(Chemistry), 1941, Polytechnic Institute of Brooklyn.

Fellow Award "For contributions to understanding the mechanism of electrical breakdown in solid dielectrics called 'treeing'"; Editor, IEEE Transactions on Electrical Insulation, 1970-76. Other Awards: Arnold H. Scott Award, ASTM, 1977.

MCMILLAN, BROCKWAY, IRE Fellow 1962. Born: March 30, 1915, Minneapolis, Minn. Degrees: B.S., 1936, Ph.D., 1939, Massachusetts Institute of Technology.

Fellow Award "For contributions to information theory, circuit theory, and systems analysis." Other Awards: Member, National Academy of Engineering, 1969.

MCMORRIS, WILLIAM A., AIEE Fellow 1951. Born: June 26, 1906, Condon, Oreg. Degrees: B.S., 1928, Oregon State University.

MCMURRAY, WILLIAM, Fellow 1980. Born: August 15, 1929, Los Angeles, CA. Degrees: B.Sc.(Eng.), 1950, Battersea Polytechnic, London; M.S., 1956, Union College.

Fellow Award "For leadership in developing high-efficiency solid-state inverters, and for advancing the analysis and design of cycloconverters"; William E. Newell Power Electronics Award, IEEE, 1978; Lamme Medal, IEEE, 1984. Other Awards: Outstanding Technical Achievement Award, Group on Solid-State Applications and Measurements, General Electric, 1980.

MCMURTRY, BURTON J., Fellow 1969. Born: March 26, 1935, Houston, Tex. Degrees: B.A., 1956, B.S.E.E., 1957, Rice University; M.S.E.E., 1959, Ph.D., 1962, Stanford University.

Fellow Award "For contributions to the development of lasers and photodectors and for technical leadership in electrooptic research," Other Awards: Alfred Noble Prize, 1964.

MCNALLY, JAMES O., IRE Fellow 1957. Born: July 1, 1903, Fredericton, New Brunswick, Canada. Degrees: B.Sc., 1924, University of New Brunswick.

Fellow Award "For contributions in the field of industrial electronics."

MCNUTT, WILLIAM J., Fellow 1976. Born: August 31, 1927, Philadelphia, Pa. Degrees: B.S.E.E., 1950, Tufts College; M.S.E.E., 1952, Illinois Institute of Technology.

Fellow Award "For contributions to the design and standardization of test procedures of power transformers"; Chairman, Transformers Committee, 1981-1982.

MCRUER, DUANE T., Fellow 1967. Born: October 25, 1925, Bakersfield, Calif. Degrees: B.S., 1945, M.S., 1948, California Institute of Technology.

Fellow Award "For many pioneering contributions to flight control technology and to systems engineering, and for the creation of a systematic technique for the design of complex man-machine systems." Other Awards: Louis E. Levy Medal, Franklin Institute; Mechanics and Control of Flight Award, American Institute of Aeronautics and Astronautics, 1970; Alexander C. Williams Award, HFS, 1976.

MCVEY, EUGENE S., Fellow 1981. Born: December 6, 1927, Wayne, WV. Degrees: B.S.E.E., 1950, University of Louisville, M.S.E., 1955, Ph.D., 1960, Purdue University.

Fellow Award "For contributions to control theory and its application to switch power systems and image processing"; Eugene Mittelmann Award, IEEE Industrial Electronics Society, 1982. Other Awards: Eta Kappa Nu Award (first recipient), Virginia Chapter.

MCWHORTER, ALAN L., Fellow 1967. Born: August 25, 1930, Crowley, La. Degrees: B.S., 1951, University of Illinois; Sc.D., 1955, Massachusetts Institute of Technology.

Fellow Award "For outstanding contributions to the fundamental understanding and practical development of solid-state electronic devices"; David Sarnoff Award, 1971, "For outstanding contributions leading to a better understanding of semiconductor devices."

MEADOR, JACK R., AIEE Fellow 1952.

Fellow Award "For outstanding ingenuity and creative ability in devising insulation structures which have contributed substantially to improvements in the design of transformers, particularly of very large high-voltage units."

MEADOWS, HENRY E., JR., Fellow 1975. Born: May 27, 1931. Degrees: B.E.E., 1952, M.S.E.E., 1953, Ph.D., 1959, Georgia Institute of Technology.

Fellow Award "For contributions to the theory of linear time-variable networks." Other Awards: Ford Foundation Resident in Engineering Practice, 1968-1969.

MEAGHER, RALPH E., IRE Fellow 1962. Born: September 22, 1917, Chicago, Ill. Degrees: B.S., 1938, University of Chicago; M.S., 1939, Massachusetts Institute of Technology; Ph.D., 1949, University of Illinois.

Fellow Award "For contributions to the development of high speed computers." Other Awards: Predoctoral Fellow, National Research Council, 1945-1948; Presidential Certificate of Merit, 1946; Fellow, American Physical Society.

MEDITCH, JAMES S., Fellow 1976. Born: July 30, 1934, Indianapolis, Ind. Degrees: B.S.E.E., 1956, Purdue University; S.M., 1957, Massachusetts Institute of Technology; Ph.D., 1961, Purdue University.

Fellow Award "For contributions to the development

and application of estimation theory."

MEE, C. DENIS, Fellow 1970. Born: December 28, 1927, Loughborough, England. Degrees: Ph.D., 1951, D.Sc., 1967, Nottingham University.

Fellow Award "For contributions to the physics and technology of magnetic recording"; IEEE Audio Group Achievement Award, 1964, "For outstanding contributions to magnetic recording." Other Awards: IBM Corporate Recognition Award, 1980; IBM Fellow, 1983.

MEGLA, GERHARD K., Fellow 1968. Born: January 22, 1918. Degrees: B.S.E.E., 1939, University of Berlin; Ph.D., 1954, University of Dresden.

Fellow Award "For contributions to the development of microwave communications and optical processing techniques."

MEI, KENNETH K., Fellow 1979. Born: May 19, 1932. Degrees: B.S.E.E., 1959, M.S.E.E., 1960, Ph.D., 1962, University of Wisconsin, Madison.

Fellow Award "For contributions to the research and application of computer methods to electromagnetic theory"; Best Paper Award, IEEE Antenna and Propagation Society, 1967; Honorable Mention, Best Paper Award, IEEE Antennas and Propagation Society, 1974.

MEIKSIN, ZVI H., Fellow 1981. Born: July 9, 1926. Degrees: B.S.(E.E.), 1950, Dipl. Ing., 1951, Technion-Israel Institute of Technology; M.S.(E.E.), 1953, Carnegie Institute of Technology-Carnegie Mellon University; Ph.D., 1959, University of Pittsburgh.

Fellow Award "For contributions to piezoresistivity of thin films and leadership in electronics education."

MEINDL, JAMES D., Fellow 1968. Born: April 20, 1933, Pittsburgh, Pa. Degrees: B.S., 1955, M.S., 1956, Ph.D., 1958, Carnegie-Mellon University.

Fellow Award "For leadership and contribution in the field of microelectronics and integrated circuitry"; Outstanding Paper Award, ISSCC, 1970, 1975, 1976, 1977, 1978; J.J. Ebers Award, IEEE Electron Devices Society, 1980. Other Awards: Citation, Arthur S. Flemming Award Commission, 1967; National Academy of Engineering, 1978.

MEITZLER, ALLEN HENRY, Fellow 1981. Born: December 16, 1928, Allentown, PA. Degrees: B.S. (Physics), 1951, Muhlenberg College; M.S. (Physics), 1953, Ph.D. (Physics), 1955, Lehigh University.

Fellow Award "For developments in ferroelectric and piezoelectric devices and materials"; Best Paper Award, IEEE Sonics and Ultasonics Group, 1963. Other Awards: Fellow, Acoustical Society of America, 1963.

MELCHER, JAMES R., Fellow 1977. Born: July 5, 1936, Giard, Iowa. Degrees: B.S.E.E., 1957, M.S.(Nuclear Engineering), 1959, Iowa State University; Ph.D.(E.E.), 1962, Massachusetts Institute of Technology.

Fellow Award "For contributions to electrohydrodynamics and its practical application." Other Awards: First Mark Mills Award, American Nuclear Society; Western Electric Award, American Society for Engineering Education.

MELCHOR, JACK L., Fellow 1967. Born: July 6, 1925, Mooresville, N.C. Degrees: B.S.(Physics), 1948, M.S.(Physics), 1950, University of North Carolina; Ph.D.(Physics), 1953, University of Notre Dame.

Fellow Award "For contributions in the field of ferrite microwave devices and for leadership in industry exploitation of such devices." Other Awards: Science Centennial Award, University of Notre Dame, 1965.

MELSA, JAMES L., Fellow 1978. Born: July 6, 1938, Omaha, Nebr. Degrees: B.S.E.E., 1960, Iowa State University; M.S.E.E., 1962, Ph.D., 1965, University of Arizona.

Fellow Award "For educational leadership in the information and control sciences." Other Awards: Western Electric Award, Gulf Southwest Section, American Society for Engineering Education, 1973.

MELVIN, HOWARD L., AIEE Fellow 1931. Born: December 2, 1890, Latah, Wash. Degrees: B.S., 1911, E.E., 1919, Washington State University; M.S., 1917, Massachusetts Institute of Technology; M.S., 1917, Harvard University.

MENDEL, JERRY M., Fellow 1978. Born: May 14, 1938, New York, N.Y. Degrees: B.M.E., 1959, M.E.E., 1960, Ph.D.(E.E.), 1963, Polytechnic Institute of Brooklyn.

Fellow Award "For contributions to system identification, state estimation, and their application to aerospace technology"; Distinguished Member, IEEE Control Systems Society, 1983. Other Awards: Outstanding Presentation Award, Society of Exploration Geophysicists, 1976.

MENGEL, JOHN T., Fellow 1971. Born: 1918, Ballston Lake, N.Y. Degrees: B.S., 1939, Union College.

Fellow Award "For leadership and technical contributions to the advancement of space programs and world-wide satellite tracking networks." Other Awards: Outstanding Achievement Award, U.S. Naval Research Laboratory, 1951; Exceptional Service Award, NASA, 1965.

MERGLER, HARRY W., Fellow 1976. Born: June 1, 1924, Chillicothe, Ohio. Degrees: B.S., 1948, Massachusetts Institute of Technology; M.S., 1950, Ph.D., 1956, Case Institute of Technology.

Fellow Award "For contributions to engineering education and research in digital logic, numerical control, and aeronautical instrumentation"; Lamme Medal, 1978; Best Paper Award, IEEE Transactions on Industrial Electronics and Control Instrumentation, 1979. Other Awards: Leonard Case Professorial Chair in Electrical Engineering; Gold Medal For Scientific Achievement, Case Institute of Technology, 1980; National Academy of Engineering, 1980.

MERRIAM, CHARLES W., III, Fellow 1984. March 31, 1931, Birmingham, Al. Degrees: Sc.B.(Eng.), 1953, Brown University; M.S., 1955, Sc.D.(E.E.), 1953, M.I.T.

Fellow Award "For contributions to optimal control theory and its application." Other Awards: Best Paper Award, JACC(with F.J. Ellert), 1962.

MERRILL, GRAYSON, Fellow 1971. Born: 1912, California. Degrees: B.S., 1934, U.S. Naval Academy.

Fellow Award "For contributions in the field of guided missile systems, and in the management and technical development of automated test systems." Other Awards: G. Edward Pendray Award, American Rocket Society; Distinguished Achievement Award, L.I. Universities; Commendation, Secretary of the Navy; Legion of Merit, Polaris Program, Secretary of the Navy; Associate Fellow, Institute of Aeronautical Science; Member, American Ordnance Association; Senior Member, American Rocket Society.

MERRITT, M. STEPHENS, AIEE Fellow 1962. Born: November 2, 1903, Daviston, Ala. Degrees: B.S., 1924, University of Alabama.

Fellow Award "For contribution to the development of comprehensive relaying for a major power system"; Chairman, East Tennessee Section, AIEE, 1957-1958; Chairman, IEEE Line Relay Protection Subcommittee. Other Awards: U.S.A. National Committee, Technical Committee 41, International Electrotechnical Commission; Program Planning Board, Georgia Tech Annual Relay Conference, 1947-1982; State Director, Chapter Chairman, Tennessee Society of Professional Engineers, 1970.

MERSEREAU, RUSSELL M., Fellow 1983. Born: August 29, 1946, Cambridge, Massachusetts. Degrees: S.B., S.M. (Elec. Eng.), 1969, Massachusetts Institute of Technology; Sc.D. (Elec. Eng.), 1973, Massachusetts Institute of Technology.

Fellow Award "For contributions to multidimensional digital signal processing"; Browder J. Thompson Memorial Prize Award, IEEE, 1976. Other Awards: Research Unit Award, Southeastern Section, ASEE, 1977.

MESNER, MAX H., Fellow 1968. Born: April 16, 1912, Meadville, Mo. Degrees: B.S.E.E., 1940, University of Missouri.

Fellow Award "For major contributions to the design of television cameras for spacecraft."

MESSENGER, GEORGE C., Fellow 1976. Born: July 20, 1930, Bellows Falls, Vt. Degrees: B.S.(Physics), 1951, Worcester Polytechnic Institute; M.S.E.E., 1957, University of Pennsylvania.

Fellow Award "For contributions to the determination of radiation damage to semiconductors and advances in semiconductor device technology."

MESSERLE, HUGO K., Fellow 1983. Born: October 25, 1925, Haifa, Palestine. Degrees: B.E.E., 1951, M.Eng.Sc., 1952, University of Melbourne; Ph.D., 1957, University of Sydney; D.Sc., 1968, University of Melbourne.

Fellow Award "For leadership in education and research in energy conversion and electric energy systems." Other Awards: Dixon Scholarship, University of Melbourne, 1950; Argus Research Scholarship, University of Melbourne, 1951; Electrical Engineering Premium, Inst. of Engrs. Australia, 1953; Electrical Engineering Prize, Inst. of Engrs. Australia, 1961, 1970; Fellow, Australian Academy of Technological Sciences, 1978; Fellow, Inst. of Electr. Engrs.; Fellow, Inst. of Engrs., Australia.

MESSERSCHMITT, DAVID G., Fellow 1983.

Fellow Award "For contributions to the theory of transmitting digital waveforms on band-limited channels."

METCALF, GEORGE F., IRE Fellow 1942, AIEE Fellow 1948. Born: December 7, 1906, Milwaukee, Wis. Degrees: B.S., 1928, D.Eng., 1956, Purdue University.

IRE Fellow Award "For his development work on vacuum tubes, and vacuum tube circuits and the successful coordination of many technical projects requiring broad scientific knowledge and sound judgement."

METTLER, RUBEN F., IRE Fellow 1962. Born: February 23, 1924, Shafter, Calif. Degrees: B.S.E.E., 1947, Ph.D., 1949, California Institute of Technology; Doctor of Humane Letters (Hon.), 1980, Baldwin-Wallace College.

Fellow Award "For contributions to radar and ballistic missiles." Other Awards: Outstanding Young Electrical Engineer Award, Eta Kappa Nu, 1954; Ten Outstanding Young Men of America Award, National Junior Chamber of Commerce, 1955; Fellow, American Institute of Aeronautics and Astronautics, 1962; Southern California Engineer of the Year Award, 1964; Southern California Marketing Man of the Year Award, 1965; National Academy of Engineering, 1965; Alumni Distinguished Service Award, California Institute of Technology, 1966; Fellow, American Astronautical Society, 1967; Distinguished Public Service Medal, Department of Defense, 1969; National Human Relations Award, National Conference of Christians and Jews, 1979; Excellence in Management Award, Industry Week Magazine, 1979.

METZGER, SIDNEY, IRE Fellow 1962. Born: February 1, 1917. Degrees: B.S.E.E., 1937, New York University; M.E.E., 1950, Polytechnic Institute of Brooklyn.

Fellow Awards "For contributions to military communications"; IEEE EASCON Aerospace Electronics Award, 1975; IEEE Award in International Communication, 1976; Member, Joint Telecommunications Advisory Committee, 1971-1979; Chairman, 1975-1977.

Other Awards: Fellow, American Institute of Aeronautics and Astronautics, 1982; Member, National Academy of Engineering, 1976; Aeronautics and Space Engineering Board, National Academy of Engineering, 1981-; Tau Beta Pi; Sigma Xi.

MEYER, ROBERT G., Fellow 1981. Born: July 21, 1942, Melbourne, Australia. Degrees: B.E., 1963, M.Eng.Sc., 1965, Ph.D., 1968, University of Melbourne, Australia.

Fellow Award "For contributions to the analysis and design of high-frequency amplifiers."

MEYERAND, RUSSELL G., AIEE Fellow 1948. Born: April 6, 1902, Milwaukee, Wis. Degrees: B.S., 1925, M.S., 1926, Massachusetts Institute of Technology.

Awards: "Engineer of the Year", CIGRE, IEEE, St. Louis Section, 1968. Other Awards: Edison Electrical Institute; Association of Edison Illuminating Companies; Member, N.S.P.E.

MICHAEL, DONALD T., Fellow 1983. Born: July 13, 1908, Portsmouth, Ohio. Degrees: E.E., University of Cincinnati; LLB, Chase College, Cincinnati.

Fellow Award "For contributions to the development and administration of IEEE Standards on electrical power distribution"; Member Emeritus, IEEE Standards Board. Other Awards: Lt. Col. US Army Reserve Corps, Retired, Bronze Star Award.

MICHEL, ANTHONY N., Fellow 1982. Born: Rekasch, Romania. Degrees: B.S.E.E., 1958, M.S.(Math.), 1964, Ph.D.(E.E.), 1968, Marquette University; D.Sc.,(Appld.Math.), 1973, Technical University of Graz, Austria.

Fellow Award "For contributions in the qualitative analysis of large-scale dynamical systems"; Assoc. Editor, IEEE Circuits and Systems, 1977-1979; Best Paper Award, IEEE Control Systems Society, 1978; IEEE Ad Hoc Visitor for Accreditation Board for Engineering and Technology, 1980; Assoc. Editor, IEEE Transactions on Automatic Control, 1981; Editor, IEEE Transactions on Circuits and Systems, 1981-1983; Program Chairman, IEEE International Large Scale Systems Symposium, 1982; ADCOM Member, IEEE Circuits and Systems Society, 1984-1987; IEEE Centennial Medal, 1984; Program Chairman of the 1985 IEEE Conference on Decision and Control. Other Awards: Chairman, Midwest Symposium on Circuits and Systems, 1978; Listed in Who's Who in America, 1984.

MICHELSON, ERNEST L., AIEE Fellow 1955. Born: March 10, 1908, St. John, New Brunswick, Canada. Degrees: B.S.E.E., 1929, Illinois Institute of Technology.

Fellow Award "For his contributions to power system engineering."

MIDDELHOEK, SIMON, Fellow 1982. Born: September 12, 1931, Rotterdam, The Netherlands. Degrees:

M.(Physics), 1956, Delft University of Technology; Ph.D., 1961, University of Amsterdam.

Fellow Award "For contributions to the theory of magnetic thin films and to magnetic and semiconductor technologies and for leadership in engineering education." Other Awards: IBM Invention Achievement Awards, 1965, 1966, 1967, 1968; IBM Outstanding Contribution Award, 1968.

MIDDENDORF, WILLIAM H., Fellow 1968. Born: March 23, 1921, Cincinnati, Ohio. Degrees: B.E.E., 1946, University of Virginia; M.S., 1948, University of Cincinnati; Ph.D., 1960, Ohio State University.

Fellow Award "For contributions to product design and development and to engineering education in the field of design"; Second Paper Prize, AIEE, "An Approach to Induction Motor Synthesis," 1961, 1962; Outstanding Paper Award, IEEE Education Group, 1976, for "Methods for Improving Design Procedures." Other Awards: National Science Fellow, 1958-1959; Herman Schneider Distinguished Engineer Award, Technical and Scientific Societies Council of Cincinnati, 1978; Public Service Award, University of Cincinnati, 1981.

MIDDLEBROOK, R. DAVID, Fellow 1978. Born: May 16, 1929, England. Degrees: B.A., 1952, M.A., 1954, Cambridge University, England; M.S., 1953, Ph.D., 1955, Stanford University.

Fellow Award "For contributions to electronic circuit analysis"; IEEE William E. Newell Power Electronics Award for Outstanding Achievement in Power Electronics, 1982. Other Awards: IR 100 Award, Industrial Research and Development Magazine, 1980; Award for Excellence in Teaching, Associated Students of the California Institute of Technology, 1982.

MIDDLETON, DAVID, IRE Fellow 1959. Born: April 19, 1920, New York, N.Y. Degrees: A.B.(summa cum laude)(Physics), 1942, Harvard College; A.M., 1945, Ph.D.(Physics), 1947, Harvard University.

Fellow Award "For contributions to the theory of noise in electronic systems"; Prize Paper Awards, IEEE Electromagnetic Compatability Society, 1977, 1979. Other Awards: Phi Beta Kappa, 1941; Sigma Xi, 1945; Predoctoral Fellow, National Research Council, 1945-47; Fellow, American Physical Society, 1951; Award, National Electronics Conference, 1956; Fellow, American Association for the Advancement of Science, 1959; Member, Cosmos Club, Washington, D.C., 1965-; Wisdom Award of Honor, 1970; Naval Research Advisory Committee, 1980-1977; Outstanding Authorship Award, Institute of Telecommunications Sciences, U.S. Dept. of Commerce, 1978; Fellow, Explorers Club, 1978; Special Achievement Award, National Telecommunication and Information Administration, U.S. Dept. of Commerce, 1978; First Prize Paper, Third International Symposium on Electromagnetic Compatibility, Rotterdam, The Netherlands, 1979; Fellow, Acoustical Society of America, 1981. Listed in: "Who's Who in America"; "Who's Who in the World"; "Who's Who in Engineering" (AAES); "Leaders in Electronics"; "American Men and Women of Science."

MIDWINTER, JOHN E., Fellow 1983. Born: March 8, 1938, Newbury, Berkshire, UK. Degrees: B.Sc., 1961, Ph.D., 1968, London University.

Fellow Award "For leadership of a major research and development effort on optical-fiber transmission." Other Awards: FIEE, 1980; IEE Electronics Division Premium 1978; Assoc. of American Publishers "Best book on Technology published in USA in 1979"; "Bruce Prellar Prize" Lecture, Royal Society of Edinburgh, 1983; "Clifford Peterson Prize" Lecture, Royal Society (London), 1983.

MIEHER, WALTER W., IRE Fellow 1954. Born: October 30, 1916, St. Louis, Mo. Degrees: B.S., 1938, Washington University.

Fellow Award "For his engineering contributions and technical leadership in the development of precision radar systems."

MIERENDORF, ROBERT C., Fellow 1980. Born: December 12, 1917, Milwaukee, WI. Degrees: B.S.E.E., 1939, University of Wisconsin.

Fellow Award "For contributions and technical leadership in the development of products and standards for industrial control systems."

MIESSNER, BENJAMIN F., IRE Fellow 1937. Born: July 27, 1890, Huntingburg, Ind.

Other Awards: De Forest Audion Gold Medal, Veteran Wireless Operators Association, 1963.

MIHRAN, THEODORE G., Fellow 1964. Born: June 28, 1924, Detroit, Mich. Degrees: B.A., 1944, M.S., 1947, Ph.D., 1950, Stanford University.

Fellow Award "For contributions to the theory of wave phenomena in electron beams." Other Awards: Fortescue Fellow; Tau Beta Pi; Phi Beta Kappa; Sigma Xi.

MIKOS, JOHN J., Fellow 1966. Born: March 21, 1917, Cleveland, Ohio. Degrees: B.S., 1939, Case Institute of Technology; M.S., 1946, Carnegie Institute of Technology.

Fellow Award "For contributions to basic knowledge in power circuit interruption and for leadership in research and development of high voltage interrupter switchgear." Other Awards: Eta Kappa Nu; Tau Beta Pi; Fellow, Instrument Society of America, 1969.

MIKULSKI, JAMES J., Fellow 1983. Born: February 18, 1934, Chicago, Illinois. Degrees: B.S.E.E., 1955, Fournier Institute of Technology; M.S.E.E., 1956, California Institute of Technology; Ph.D., 1959, University of Illinois.

Fellow Award "For contributions to the development of cellular mobile radiotelephone systems." Other Awards: Tau Beta Pi; Sigma Xi; Dan Noble Fellow,

Motorola, Inc.

MILLAR, JULIAN Z., IRE Fellow 1956, AIEE Fellow 1962. Born: July 3, 1901, Mattoon, Ill. Degrees: B.S., 1923, University of Illinois.

IRE Fellow Award "For administrative contributions to military communication, and for commercial application of microwave systems"; AIEE Fellow Award "For contributions to record communications systems and for their application to national defense"; Achievement Award, Communication Technology Group, 1967, "For outstanding contributions to engineering and development in telegraph and radio relay systems"; IEEE North Jersey Section Award, 1972, "For distinguished service to the Section." Other Awards: Meritorious Service Citation, U.S. Dept. of Air Force; Distinguished Alumnus Award, Electrical Engineering Alumni of the University of Illinois, 1970; Fellow, Radio Club of America, 1973; Fellow, American Association for the Advancement of Science, 1980; Sarnoff Citation, Radio Club of America, 1982.

MILLER, EDMUND K., Fellow 1984. Born: December 24, 1935, Milwaukee, WI. Degrees: B.S.(E.E.), 1957, Michigan Technological University; M.S.(Nuclear Eng.), 1958, M.S.(E.E.), 1961, Ph.D.(E.E.), 1965, U. of Michigan.

Fellow Award "For contributions to the development and application of numerical methods in electromagnetic radiation and scattering"; Distinguished Lecturer, IEEE Antenna and Propagation Society, 1979-1981. Other Awards: AEC Fellow, 1957.

MILLER, GABRIEL L., Fellow 1977. Born: January 18, 1928, New York, N.Y. Degrees: B.Sc.(Physics), 1949, M.Sc.(Mathematics), 1952, Ph.D.(Physics), 1957, University of London.

Fellow Award "For contributions to nuclear instrumentation and its innovative extension to measurements in other scientific fields."

MILLER, KENNETH W., AIEE Fellow 1949. Born: April 24, 1898, Jacksonville, Ill. Degrees: B.S., 1919, E.E., 1929, University of Illinois; M.S., 1932, Union College.

Other Awards: American Mathematics Association; Sigma Xi; Tau Beta Pi.

MILLER, ORRIS J., AIEE Fellow 1950. Born: November 27, 1891, Dorset, Ohio. Degrees: B.S., 1914, Hiram College; B.C.E., 1916, Ohio University; C.E., 1920, Ohio State University.

Other Awards: Organizer, Chattanooga Engineers Club, 1925; Tau Beta Pi, 1950; Distinguished Service Award, Charlotte Engineers Club, 1966.

MILLER, RAYMOND E., Fellow 1970. Born: October 9, 1928, Bay City, Mich. Degrees: B.S.M.E., 1950, University of Wisconsin; B.S.E.E., 1954, M.S., 1955, Ph.D., 1957, University of Illinois.

Fellow Award "For contributions to the advancement of the theoretical understanding of computation

through work in switching theory and theoretical models."

MILLER, STEWART E., IRE Fellow 1958. Born: September 1, 1918, Milwaukee, Wis. Degrees: S.M., S.B., Massachusetts Institute of Technology.

Fellow Award "For leadership and invention in the waveguide art"; IEEE Morris N. Liebmann Award, 1972, "For pioneering research in guided millimeter and optical transmission systems"; W.R.G. Baker Prize Paper Award (With E.A.J. Marcatile and Tingye Li), 1975; Stuart Ballantine Medal, Franklin Institute, 1977; National Academy of Engineering, 1973.

MILLER, WILLIAM E., AIEE Fellow 1962. Born: December 18, 1917, Los Angeles, Calif. Degrees: B.S.E.E., 1939, University of California at Berkeley.

Fellow Award "For contributions to design of metal rolling mill electrical systems." Other Awards: Charles A. Coffin Award, General Electric, 1951; Vice President, Chairman Executive Board, International Federation of Automatic Control, 1981-1984; Secretary, American Automatic Control Council.

MILLER, WILLIAM F., Fellow 1980. Born: November 19, 1925, Vincennes, IN. Degrees: B.S., 1949, M.S., 1951, Ph.D. 1956, Purdue University.

Fellow Award "For contributions to education in computer science and to university administration." Other Awards: Faculty Citation Outstanding Alumnus, Vincennes University, 1962; PsC (honoris causa), Purdue University, 1972; Fellow, American Academy of Arts and Sciences, 1980.

MILLERMASTER, RALPH A., AIEE Fellow 1949. Born: July 27, 1905, Milwaukee, Wis. Degrees: B.S.E.E., 1927, University of Wisconsin.

Other Awards: Distinguished Service Citation, University of Wisconsin, 1967.

MILLMAN, GEORGE H., Fellow 1970. Born: June 2, 1919, Boston, Mass. Degrees: B.S., 1947, University of Massachusetts; M.S., 1949, Ph.D., 1952, Pennsylvania State University.

Fellow Award "For contributions to the understanding of environmental effects on radar propagation and to the field of ionospheric radio physics." Other Awards: Phi Kappa Phi, 1947; Sigma Pi Sigma, 1948; Pi Mu Epsilon,1949; Sigma Xi, 1952.

MILLMAN, JACOB, IRE Fellow 1958. Born: May 17, 1911, Russia. Degrees: B.S., 1932, Ph.D., 1935, Massachusetts Institute of Technology.

Fellow Award "For his contributions to the field of pulse circuits"; IEEE Education Medal, 1970. Other Awards: Fellow, American Physical Society, 1946; Citation, Office of Scientific Research and Development, 1947; Great Teachers Award, Columbia University, School of Engineering, 1967; Fulbright Lecture Grants, Italy, 1959-60, Uruguay, 1967-68, Israel, 1973-74; Author of eight textbooks on Electronics, 1941-79;

Charles Batchelor Professor of Electrical Engineering, Columbia University, Emeritus.

MILLMAN, SIDNEY, IRE Fellow 1959. Born: March 15, 1908, Dawid-Gorodok, Poland. Degrees; B.S., 1931, City College of the City University of New York; A.M., 1932, Ph.D., 1935, Columbia University; Sc.D.(Hon.), 1974, Lehigh University.

Fellow Award "For contributions in the field of magnetrons, traveling-wave amplifiers and backward-wave oscillators."

MILLS, ROBERT G., Fellow 1979. Born: January 20, 1924, Effingham, Ill. Degrees: B.S.E., 1944, Princeton University; M.A., 1947, University of Michigan; Ph.D., 1952, University of California, Berkeley.

Fellow Award "For contributions to research on nuclear fusion power reactors"; IEEE Centennial Medal, 1984. Other Awards: Phi Beta Kappa, 1944; Naval Ordnance Award, 1947; National Research Fellow, 1952; Sigma Xi, 1953.

MILNES, ARTHUR G., Fellow 1968. Born: July 30, 1922, Heswall, England. Degrees: B.Sc., 1943, M.Sc., 1947, D.Sc., 1956, University of Bristol.

Fellow Award "For contribution to graduate education and research in solid state electronics, 1967"; J.J. Ebers Award, IEEE Electron Devices Society, 1982. Other Awards: Paper Premiums, Institution of Electrical Engineers, 1954, 1957; Paper Prize, 1954, Institute of Physics, London; Fellow, Institution of Electrical Engineers, London, 1967; Fellow, American Physical Society, 1975.

MILTON, ROBERT M., AIEE Fellow 1960. Born: September 22, 1901, Baldwyn, Miss. Degrees: B.S., 1923, Mississippi State University.

Fellow Award "For contributions to design of high voltage substation, switchgear and heavy current buses."

MINARD, CLARENCE W., AIEE Fellow 1955. Born: February 14, 1893, Plattekill Township, N.Y. Degrees: E.E., 1916, Syracuse University.

Fellow Award "For contributions to planning and design of higher voltage distribution systems in extensive rural areas."

MINER, JOHN D., JR., AIEE Fellow 1949. Born: October 3, 1903, East Greenwich, R.I. Degrees: B.S., 1925, Brown University.

Other Awards: National Society of Professional Engineers; Ohio Society of Professional Engineers; Westinghouse Silver "W", 1952; Certificate; American Standards Association.

MINOZUMA, FUMIO, Fellow 1964. Born: September 28, 1916, Japan. Degrees: B.Sc., 1941, Ph.D., 1956, Kyoto University.

Fellow Award "For research in radio wave propagation and radio noise interference and leadership in the Radio Regulatory Bureau of Japan."

MINTER, JERRY B., Fellow 1969. Born: October 31, 1913. Degrees: B.S.E.E., 1934, Massachusetts Institute of Technology.

Fellow Award "For contribution to radio signal generating and measuring devices." Other Awards: Armstrong Medal, Radio Club of America, 1968; Past President, Life Fellow, Audio Engineering Society; Past President, Fellow, Radio Club of America; Life Member, SMPTE; Life Member, Amer. Soc. for Metals.

MITCHELL, DOREN, IRE Fellow 1960. Born: March 4, 1905, Columbus, Ohio. Degrees: B.S., 1925, Princeton Univeristy.

Fellow Award "For contributions to long-distance communication systems"; Chairman, Awards Committee, IEEE Communications Technology Group, 1965-1967. Other Awards: Governors Award, Ohio, 1963; Fellow Award, American Association for the Advancement of Science, 1981.

MITRA, SANJIT K., Fellow 1974. Born: November 26, 1935, Calcutta, India. Degrees: B.Sc., 1953, Utkal University, India; M.Sc., 1956, Calcutta University, India; M.S., 1960, Ph.D., 1962, University of California at Berkeley.

Fellow Award "For contributions to active network theory and to engineering education." Other Awards: F.E. Terman Award, American Society for Engineering Education, 1973; Visiting Professorship, Kobe University, Japan Society for Promotion of Science, 1973; Fellow, American Association for the Advancement of Science, 1982; Visiting Fellow, Australian National University, 1982.

MITTER, SANJOY K., Fellow 1979. Born: December 9, 1933. Degrees: B.Sc., 1954, Calcutta University; B.Sc., 1957, Ph.D., 1965, Imperial College, University of London.

Fellow Award "For contributions to optimization computation and control theory."

MITTRA, RAJ, Fellow 1971. Born: July 1, 1932, Banares, India. Degrees: M.S., University of Calcutta; Ph.D., University of Toronto.

Fellow Award "For contributions to electromagnetic theory as applied to the solution of problems in antennas and waveguides"; Best Paper Award, IEEE Antennas and Propogation Society, 1978; Past President, IEEE Antennas and Propagation Society; Past Editor, IEEE Transactions on Antennas and Propagation; IEEE Centennial Medal, 1984. Other Awards: Guggenheim Fellowship, 1965-66.

MIYAIRI, SHOTA, Fellow 1980. Born: April 24, 1917, Japan. Degrees: Dr.Eng., 1957, Tokyo Institute of Technology.

Fellow Award "For contributions to electrical machinery, power electronics, and leadership in electrical engineering education." Other Awards: Paper Prize, Institute of Electrical Engineers, Japan, 1962; Author

Prize, Institute of Electrical Engineers, Japan, 1966, 1976.

MIYAJI, KOH-ICHI, Fellow 1982. Born: October 31, 1914, Tokyo, Japan. Degrees: B.S.E.E., 1939, Tohoku Imperial University; Ph.D.(E.E.), 1951, Tohoku University.

Fellow Award "For contributions to the development of microwave tubes and solid-state image converters." Other Awards: Fellow, Society for Information Display, 1977.

MIYAOKA, SENRI, Fellow 1984. Born: February 16, 1937, Buenos Aires, Argentina. Degrees: B.Sc.(Phy.), 1959, Gakushuin University; Ph.D.(E.E.), Tokyo Institute of Technology.

Fellow Award "For contributions to the development of color television tubes"; IEEE Best Paper Award, 1968, 1973, 1974; Vladimir K. Zworykin Award, IEEE, 1974. Other Awards: Purple Ribbon Prize of Japan, 1973; The Prize of Minister of State for Science and Technology, 1973.

MOCHIZUKI, HITOSHI, Fellow 1984. Born: June 14, 1924, Sizuoka, Japan. Degrees: B.S.(E.E.), 1948, Ph.D.(Eng.), 1961, University of Tokyo.

Fellow Award "For contributions to maritime communications systems."

MOE, ROBERT E., IRE Fellow 1955. Born: April 2, 1912, Appleton, Wis. Degrees: B.S.E.E., 1933, University of Wisconsin.

Fellow Award "For his contributions to the field of electronics"; Annual Award, IRE, Professional Group on Reliability and Quality Control, 1961, "For his part in the development of the publication 'Parts Specification Management for Reliability.'" Other Awards: Certificate of Commendation, US. Dept. of Navy, 1946; Kentucky State Board of Registration for Professional Engineers, 1975-1982; Professional Engineer, New York 1948, Kentucky 1949.

MOHLER, RONALD R., Fellow 1982. Born: April 11, 1931, Ephrata, PA. Degrees: B.S., 1956, Pennsylvania State University; M.S., 1958, University of Southern California; Ph.D., 1965, University of Michigan.

Fellow Award "For contributions to automatic control theory of bilinear systems." Other Awards: NATO Senior Scientist Award, 1978; USSR Exchange Visitor, National Academy of Science, 1980.

MOHR, MILTON E., IRE Fellow 1961. Born: April 9, 1915, Milwaukee, Wis. Degrees: B.S., 1938, Dr.Eng.(Hon.), 1959, University of Nebraska.

Fellow Award "For contributions to military electronic systems." Other Awards: Eta Kappa Nu.

MOLL, JOHN L., IRE Fellow 1962. Born: December 21, 1921, Wauseon, Ohio. Degrees: B.S., 1943, Ph.D., 1952, Ohio State University; Doctor Honoris Causa, 1983, Katholieke Universiteir, Leuven, Belgium.

Fellow Award "For contributions to theory and devel-

opment of semiconductor devices"; Ebers Award, IEEE, 1971. Other Awards: Sigma Xi; American Physical Society; Postdoctoral Fellow, Guggenheim, 1964; Howard N. Potts Medal, Franklin Institute, 1967; Distinguished Alumnus Award, College of Engineering, Ohio State University, 1970; Member, National Academy of Engineering, 1974; 5 patents, 50 publications.

MONSETH, INGWALD T., AIEE Fellow 1945. Born: May 10, 1902, Minneapolis, Minn. Degrees: B.S., 1924, University of Minnesota; A.M.P., 1952, Harvard University.

Other Awards: American Society of Mechanical Engineers; Award of Merit, Westinghouse Electric Corp.

MONSHAW, VALDEMAR R., Fellow 1983. Born: April 27, 1926, Nesquehoning, Pennsylvania. Degrees: B.S.E.E., 1949, M.S.E.E., 1954, City College of New York.

Fellow Award "For contributions to the reliability engineering of space electronic systems"; Annual Reliability Award, IEEE Reliability Society, 1979; IEEE Centennial Medal, 1984.

MONTGOMERY, G. FRANKLIN, IRE Fellow 1962. Born: May 1, 1921, Oakmont, Pa. Degrees: B.S.E.E., 1941, Purdue University.

Fellow Award "For contributions to electronic instrumentation."

MONTGOMERY, LUKE H., IRE Fellow 1960. Born: January 18, 1907, Nashville, Tenn.

Fellow Award "For contributions to medical electronics." Other Awards: Fellow, American Association for the Advancement of Science; Sigma Xi.

MOORE, ARTHUR D., AIEE Fellow 1943. Born: January 7, 1895, Fairchance, Pa. Degrees: B.S.E.E., 1915, Carnegie Institute of Technology; M.S.E., 1923, University of Michigan.

Other Awards: Donald P. Eckman Education Award, Instrument Society of America, 1969; Stephen S. Attwood Distinguished Achievement Award, College of Engineering, University of Michigan, 1970; Outstanding Citizen Award, Institute of Gerontology, University of Michigan, 1977.

MOORE, GORDON E., Fellow 1968. Born: January 3, 1929, San Francisco, Calif. Degrees: B.S., University of California; Ph.D., 1954, California Institute of Technology.

Fellow Award "For contributions and leadership in research, development and production of silicon transistors and monolithic integrated circuits"; McDowell Award, IEEE Computer Society, 1978; Frederik Philips Award, 1979. Other Awards: National Academy of Engineering, 1976; Harry Goode Memorial Award, American Federation of Information Processing Societies, 1978.

MOORE, JOHN BARRATT, Fellow 1979. Born: April 4, 1941. Degrees: B.E., 1962, M.Eng.Sc., 1963, University of Queensland, Australia; Ph.D., 1967, University of Santa Clara.

Fellow Award "For contributions to optimal estimation and control and leadership in electrical engineering education."

MOORE, JOHN R., IRE Fellow 1958. Born: July 5, 1916, St. Louis, Mo. Degrees: B.S., 1937, Washington University; D.Sc.(Hon.), 1963, West Coast University.

Fellow Award "For his contributions to the field of automatic inertial navigation systems." Other Awards: Outstanding Mechanical Engineering Graduate, Washington University, 1960; Meritorious Public Service Citation, U.S. Dept. of Navy, 1961; Thurlow Award, Institute of Navigation, 1962; Distinguished Alumni Award, Washington University, 1964; Outstanding Achievement Award, American Institute of Industrial Engineers, 1967; WEMA Medal of Achievement, 1969; UCLA Executive Program Award, Graduate School of Business Administration, Executive Program Association, 1969; National Academy of Engineering, 1978; Fellow Award, AIAA, 1982.

MOORE, RICHARD K., IRE Fellow 1962. Born: November 13, 1923, St. Louis, Mo. Degrees: B.S.E.E., 1943, Washington University; Ph.D., 1951, Cornell University.

Fellow Award "For contributions to electromagnetic propagation and engineering education"; Outstanding Technical Achievement Award, IEEE Council on Oceanic Engineering, 1978; Distinguished Technical Achievement Award, IEEE Geoscience & Remote Sensing Society, 1982; IEEE Centennial Medal, 1984. Other Awards: Engineering Alumni Achievement Award, Washington University, 1978.

MOORE, WILLIAM J. M., Fellow 1976. Born: May 3, 1924, Edinburgh, United Kindom. Degrees: B.A.Sc., 1946, University of British Columbia; M.Eng., 1948, McGill University.

Fellow Award "For contributions to precise current comparators and the application of the comparator to industrial measurements."

MOOSE, LOUIS F., Fellow 1984.

Fellow Award "For contributions to microwave relay communication systems."

MORAN, JOHN H., JR., Fellow 1980. Born: September 22, 1923, Philadelphia, PA. Degrees: B.S.E.E., 1947, Case Western Reserve University.

Fellow Award "For contributions to the design, testing, and application of station and transmission line insulators"; Prize Paper Award (coauthor), Transmission and Distribution Committee, IEEE, 1981.

MOREHOUSE, STEPHEN B., Fellow 1976. Born: August 26, 1905, Sharon, Conn. Degrees: E.E., 1928, Rensselaer Polytechnic Institute.

Fellow Award "For contributions to generation control techniques for large-scale interconnected electric

power systems." Other Awards: Registered Professional Engineer.

MOREY, CHARLES V., AIEE Fellow 1960. Born: September 28, 1903, Fall River, Mass. Degrees: B.S., 1925, Worcester Polytechnic Institute.

Fellow Award "For contributions to the art and practice of electric power metering."

MORGAN, BERNARD S., JR., Fellow 1972. Born: June 30, 1927, Brooklyn, N.Y. Degrees: B.S., 1951, U.S. Naval Academy; M.S.E., 1958, Ph.D., 1963, University of Michigan.

Fellow Award "For contributions to aerospace research and development programs."

MORGAN, HOWARD K., IRE Fellow 1957. Born: February 22, 1906, New York, N.Y. Degrees: B.S.E.E., 1929, University of California at Berkeley.

Fellow Award "For engineering contributions in the fields of air communication and navigation." Other Awards: Collier Award, Radio Technical Commission for Aeronautics, 1948.

MORGAN, J. DERALD, Fellow 1984. Born: March 15, 1939, Hays, KS. Degrees: B.S.E.E., M.S.E.E., Ph.D.

Fellow Award "For contributions to engineering education"; IEEE Centennial Medal, 1984; Education Award, St. Louis Section, IEEE; Award of Honor, St. Louis Section, IEEE. Other Awards: Who's Who in America; Who's Who in Engineering; American Men and Women of Science; Who's Who Among Students in American Colleges & Universities; Outstanding Young Men of America; Community Leaders of America; Men of Achievement; Sigma Xi; Phi Kappa Phi; Tau Beta Pi; Omicron Delta Kappa; Eta Kappa Nu; Emerson Electric Professor of Electrical Engineering, University of MO-Rolla, 1976-present; Alcoa Foundation Professor of Electrical Engineering, University of MO-Rolla, 1973-75.

MORGAN, MILLETT G., Fellow 1971. Born: January 25, 1915, Hanover, N.H. Degrees: B.A., 1937, M.Sc.E., 1938, Cornell University; Eng., 1939, Ph.D., 1946, Stanford University.

Fellow Award "For contributions to international radio science, to engineering, and to engineering education."

MORGENTHALER, FREDERIC R., Fellow 1978. Born: March 12, 1933, Cleveland, Ohio. Degrees: S.B. and S.M., 1956, Ph.D., 1960, Massachusetts Institute of Technology.

Fellow Award "For contributions to the theory and applications of microwave magnetics."

MORISUYE, MASANOBU, AIEE Fellow 1962. Born: February 7, 1898, Japan. Degrees: B.S., 1921, University of California at Berkeley; M.E.E., 1923, Cornell University.

Fellow Award "For contributions to the development of high voltage instrument transformers." Other

Awards: Sigma Xi, 1923; Citation, American Legion, 1954; Silver Beaver Award, Boy Scouts of America, 1956; Citizenship Award, Westinghouse Electric, 1958; Citation, International Executive Service Corps, 1969; Citation, Shenango Valley Council of Churches, 1970; Distinguished Service Award, American Red Cross, 1976.

MORITA, KIYOSHI, Fellow 1967. Born: March 18, 1901, Tokyo, Japan. Degrees: D.Eng., 1932, Tokyo Imperial University.

Fellow Award "For contributions to engineering education." Other Awards: Asano Prize, Institute of Electrical Engineers of Japan, 1954; Medal of Meritorious Scientific Contribution, Institute of Electrical Communication Engineers of Japan, 1961; Honorary Member, Institute of Electrical Communication Engineers of Japan, 1965; Rising Sun Decoration of 3rd class, His Majesty the Emperor, 1971; Honorary Professor, Tokyo Institute of Technology.

MORITA, MASASUKE, Fellow 1977. Born: December 6, 1915, Japan. Degrees: B.Eng., 1939, D.Eng., 1951, University of Tokyo.

Fellow Award "For contributions to microwave communications and leadership in development, manufacture, and installation of radio communication systems." Other Awards: Medal of Honor with Purple Ribbon, Prime Minister of Japan, 1958; Imperial Invention Award, 1966; Medal of Honor, Institute of Electronic and Electrical Communication Engineers of Japan, 1967; Two Distinguished Paper Awards, Institute of Electronics and Communications Engineers of Japan; Commendations: Minister of Post and Telecommunication, Minister of Trade and Industry, Chairman of Nippon Telephone and Telegraph Public Corp., Japan Radio Association, Institute of Electrical Engineers of Japan; Honorable Member: Institute of Electronics and Electrical Engineers of Japan, Member, Institute of Electrical Engineers of Japan, Institute of Information Processing Engineers of Japan.

MORIWAKI, YOSHIO, Fellow 1975. Born: May 14, 1911, Tokyo, Japan. Degrees: B.E., 1933; D.Eng., 1947, University of Tokyo.

Fellow Award "For contributions to analyses of high-frequency and pulsed characteristics of electronic circuits."

MORONG, TRAFFORD M., Fellow 1982. Born: November 1, 1911, Danvers, MA. Degrees: B.S.E.E., 1934, University of New Hampshire; M.S.E.E., 1949, University of Pennsylvania.

Fellow Award "For leadership in transforming the staff and facilities of a 25 Hz rural power system into a modern public utility."

MORRILL, CHARLES D., Fellow 1964. Born: October 31, 1919, St. Louis, Mo. Degrees: B.S.E.E., 1941, University of Illinois.

Fellow Award "For contributions to the development of electronic analog computing techniques."

MORRILL, WAYNE J., AIEE Fellow 1957. Born: February 22, 1901, Lansing, MI. Degrees: B.S.E.E., 1923, Purdue University; Doctor of Engineering, 1974, Purdue University.

Fellow Award "For achievements in the design and manufacture of fractional horsepower motors." Other Awards: Sigma Xi, Purdue, 1923; Charles A. Coffin Award, General Electric, 1942.

MORRIN, THOMAS H., IRE Fellow 1960. Born: November 24, 1914, Woodland, Calif. Degrees: B.S., 1937, University of California at Berkeley; Graduate, 1941, Post Graduate School, United States Naval Academy.

Fellow Award "For the administration of research in electronics." Other Awards: Fellow, American Association for the Advancement of Science, 1961. Other Awards: Sigma Xi.

MORRIS, ALBERT J., IRE Fellow 1962. Born: January 3, 1919, New York, N.Y. Degrees: B.S., 1941, University of California at Berkeley; M.S., 1948, E.E., 1950, Stanford University.

Fellow Award "For contributions towards control systems, medical electronics and pulsed generators."

MORRIS, EDWIN W., AIEE Fellow 1948. Born: February 4, 1903, Blue Springs, Nebr. Degrees: B.Sc., 1925, University of Nebraska.

MORRIS, ROBERT M., IRE Fellow 1957. Born: January 18, 1902, Washington, D.C.

Fellow Award "For contributions in the field of radio and television broadcasting"; Scott Helt Award, Broadcasting Group, 1967. Other Awards: Exceptional Service Medal, War Dept., 1945; Engineering Achievement Award, NAB, 1967; Emile Berliner Award, Audio Engineering Society, 1967.

MORRIS, ROBERT M., Fellow 1978. Born: September 20, 1915, Trenton, Nova Scotia, Canada. Degrees: B.A., 1938, Mount Allison University; B.E., 1940, Nova Scotia Technical College; S.M., 1950, Massachusetts Institute of Technology.

Fellow Award "For contributions to the understanding of radio noise and corona on high-voltage transmission lines."

MORRISON, JOHN F., IRE Fellow 1951. Born: March 14, 1906, Buffalo, N.Y.

Fellow Award "In recognition of his contributions in the field of broadcast antenna and transmitter design."

MORRISON, W. C., Fellow 1964. Born: September 13, 1915, Sioux City, Iowa. Degrees: A.B., 1937, Morningside College; B.S., 1939, M.S., 1940, University of Iowa.

Fellow Award "For significant contributions to the fields of VHF, UHF, and color television."

MORROW, WALTER E., JR., Fellow 1966. Born: July 24, 1928. Degrees: B.S., 1949, M.S., 1951, Massachusetts Institute of Technology.

Fellow Award "For contributions to the development of space and tropospheric scatter communication systems"; IEEE Armstrong Award, 1976. Other Awards: Outstanding Achievement Award, President Julius A. Stratton, Massachusetts Institute of Technology, 1963; Member, National Academy of Engineering, 1978.

MORSE, A. STEPHEN, Fellow 1984. Born: June 18, 1939, Mt. Vernon, N.Y. Degrees: B.S.(E.E.), 1962, Cornell; M.S.(E.E.), 1964, U. of Arizona; Ph.D.(E.E.), 1967, Purdue.

Fellow Award "For contributions to the theory of multivariable control systems." Other Awards: Best Paper Prize, Joint Automatic Control Conference, 1970, 1972.

MORTENSON, KENNETH E., Fellow 1975. Born: December 14, 1926, Melrose, Mass. Degrees: B.S., 1947, B.E.E., 1948, M.E.E., 1950, Ph.D., 1954, Rennselaer Polytechnic Institute.

Fellow Award "For contributions to the field of microwave semiconductor devices and components, and to engineering education and management." Other Awards: Sigma Xi; Member, New York State Academy of Science.

MORTLOCK, JOSEPH R., Fellow 1967. Born: August 25, 1903, Quetta, Pakistan. Degrees: B.Sc., Ph.D., University of London.

Fellow Award "For contributions to high voltage transmission systems and components." Other Awards: Ayrton Premium, Institution of Electrical Engineers, U.K., 1946; Faraday Medal, Institution of Electrical Engineers, U.K., 1964.

MORTON, GEORGE A., IRE Fellow 1951. Born: March 24, 1903, New Hartford, N.Y. Degrees: B.S., 1926, M.S., 1928, Ph.D., 1932, Massachusetts Institute of Technology.

Fellow Award "For his contributions to research on electronic imaging and electronic applications in the field of nucleonics"; Vladimir K. Zworykin Award, IRE, 1962, "For his contribution to electronic television through the development of camera and imaging tubes"; Merit Award, IEEE Nuclear and Plasma Sciences Society, "In recognition of outstanding contributions towards the advancement of scintillation counting." Other Awards: Overseas Premiums, Institution of Electrical Engineers, U. K., 1937, 1938; David Richardson Medal, American Optical Society, 1937; Certificate of Commendation, U.S. Dept. of Navy, Bureau of Ships; Certificate of Commendation, National Defense Research Council; Certificate of Commendation, U.S. Dept. of Air Force.

MORTON, LYSLE W., AIEE Fellow 1951. Born: June 1, 1901, Blooming Prairie, Minn. Degrees: B.S.E.E., 1924, University of Minnesota.

Fellow Award "For the advancement of the theory and practice of electrical engineering and the allied arts and sciences, and the maintenance of high professional standards among its members"; Standards Board Award, IEEE, 1973. Other Awards: Citation, Eta Kappa Nu, 1952; Gold Key Award, General Electric, 1956; Citation, GOSAM, General Electric, 1966; Citation, I.E.C., 1970; Charles H. Tuttle Citation, Lake George Association, 1971; General Electric, Senior Elfun, Community Service Award, 1974-1975; Who's Who in Engineering.

MORTON, PAUL L., IRE Fellow 1955, AIEE Fellow 1962. Born: May 14, 1906, Silao, Mexico. Degrees: B.S., 1931, University of Washington; M.S., 1938, Massachusetts Institute of Technology; Ph.D., 1943, University of California at Berkeley; The Berkeley Citation, 1973, University of California at Berkeley.

IRE Fellow Award "For his contributions to the field of digital computers and the teaching of electronics"; AIEE Fellow Award "For contributions to the teaching of electrical engineering, to the direction of research and to the development of electronic digital computers."

MOSCHYTZ, GEORGE S., Fellow 1978. Born: April 18, 1934, Freiburg i. Breisgau, West Germany. Degrees: Dipl. Ing.(ETH), 1958, Dr.Tech.Sci.(ETH), 1960, Full Prof., 1973, Swiss Federal Institute of Technology (ETH), Zurich.

Fellow Award "For contributions to the theory and the development of hybrid-integrated linear communication networks"; Chairman, Swiss Section, IEEE, 1981. Other Awards: Outstanding Paper Award, Electronic Components Conference, 1969; Swiss Electrotechnical Society; Eta Kappa Nu.

MOSELEY, FRANCIS L., IRE Fellow 1957. Born: August 2, 1908, St. Paul, Minn.

Fellow Award "For contributions to the development of aircraft navigation systems and electronic instruments"; Pioneer Award, IRE, 1959. Other Awards: Flight Safety Foundation Award, 1955.

MOSTAFA, ABD EL-SAMIE, Fellow 1967. Born: April 27, 1913, Cairo, Egypt. Degrees: B.Sc., 1937, Ph.D., 1947, Faculty of Engineering, Cairo.

Fellow Award "For contributions to electrical engineering education in the United Arab Republic and for his work in analysis of nonlinear systems." Other Awards: Prize, Egypt, 1950; Prize, United Arab Republic, 1955; Top State Prize in Science, United Arab Republic, 1974.

MOUNTJOY, GARRARD, IRE Fellow 1950. Born: October 21, 1905, Boscobel, Wis. Degrees: B.S., 1929, Washington University.

Fellow Award "For his contributions to the design of radio and television broadcast receivers." Other Awards: Modern Pioneer Award, National Association of Manufacturers.

MOURADIAN, H., AIEE Fellow 1926.

MOURIER, GEORGES, Fellow 1979. Born: July 12, 1923, Paris, France. Degrees: Engineer, 1946, Ecole Sup. Physique et Chimie, Paris; Ph.D., 1971, Universite de Paris.

Fellow Award "For contributions to crossed-field microwave tubes and plasma science."

MOYER, ELMO E., AIEE Fellow 1951. Born: March 13, 1908, Chicago, Ill. Degrees: B.S., 1929, University of Notre Dame.

MUCHMORE, ROBERT B., IRE Fellow 1962. Born: July 8, 1917, Augusta, Kans. Degrees: B.S., 1939, University of California at Berkeley; E.E., 1942, Stanford University.

Fellow Award "For accomplishments in the field of microwave antennas and tubes."

MUELLER, CHARLES W., IRE Fellow 1959. Born: February 12, 1912, New Athens, Ill. Degrees: B.S., 1934, University of Notre Dame; M.S., 1936, Sc.D., 1942, Massachusetts Institute of Technology.

Fellow Award "For contributions to the development of electronic tubes and solid-state devices"; J.J. Ebers Award, 1972, "For outstanding technical contributions to electron devices, spanning the evolution of modern electronics from grid-controlled tubes through the alloy transistor, the thyrister, and MOS devices to silicon vidicons and silicon storage vidicons." Other Awards: David Sarnoff Outstanding Achievement Award, Radio Corporation of America, 1966; Engineering Honor Award, University of Notre Dame, 1975.

MUELLER, GEORGE E., IRE Fellow 1962. Born: July 16, 1918, St. Louis, Mo. Degrees: B.S.E.E., 1939, University of Missouri; M.S., 1940, Purdue University; Ph.D., 1951, Ohio State University; Honorary Doctorates: 1964, Wayne State University; 1964, New Mexico State University; 1964, University of Missouri; 1965, Purdue University; 1965, Ohio State University; 1979, Pepperdyne University.

Fellow Award "For accomplishments in the field of microwave antennas and tubes." Other Awards: Fellow, American Institute of Aeronautics and Astronautics, 1966; Fellow, American Physical Society, 1967; National Academy of Engineering, 1967; Fellow, American Astronautical Society; Fellow, Royal Aeronautical Society; Honorary Fellow, British Interplanetary Society; Recipient, A.A.S. Space Flight Award, 1968; Veterans of Foreign Wars Space Award, 1968; NASA Distinguished Service Medal, 1966, 1968, 1969; Eugen Sanger Award, 1970; Fellow, American Association for the Advancement of Science, 1970; National Medal of Science, 1970; Gold Plate Award, American Academy

of Achievement, 1972; National Transportation Award, 1979; Goddard Medal, AIAA, 1983.

MULLEN, JAMES A., Fellow 1974. Born: May 28, 1928, Malden, Mass. Degrees: B.S., 1950, Providence College; M.A., 1951, Ph.D., 1955, Harvard University.

Fellow Award "For contributions to statistical communication and oscillator noise theory and its application to the solution of a broad spectrum of problems."

MULLER, CHRISTIAAN A., Fellow 1973. Born: April 18, 1923, Alkmaar, The Netherlands. Degrees: M.Sc., 1950, Delft University.

Fellow Award "For contribution to the knowledge of the structure of galaxies." Other Awards: Veder Award, Veder Foundation, The Netherlands, 1956; Member, Netherlands Radio and Electronics Society.

MULLER, JEAN J., Fellow 1974. Born: September 6, 1910, Basel, Switzerland. Degrees: Eng., 1934, Ecole Centrale, Paris, France; Dr.(Tech.Sc.), 1937, Polytechnikum, Zurich, Switzerland.

Fellow Award "For contributions to microwave tubes and to television and radio-communications systems."

MULLER, MARCEL W., Fellow 1980. Born: November 1, 1922, Vienna, Austria. Degrees: B.S.E.E., 1949, A.M. (Physics), 1952, Columbia University; Ph.D. (Physics), 1957, Stanford University.

Fellow Award "For development of micromagnetic theory and applications to magnetic materials, and for contributions to noise theory of lasers and masers." Other Awards: Fellow, American Physical Society, 1965; Humboldt Award, 1976; Fulbright Fellowship (h.c.), 1976.

MULLIGAN, JAMES H., JR., IRE Fellow 1959. Born: October 29, 1920, Jersey City, N.J. Degrees: B.E.E., 1943, Cooper Union; M.S., 1945, Stevens Institute of Technology; E.E., 1947, Cooper Union; Ph.D., 1948, Columbia University.

Fellow Award "For contributions in the fields of network theory, feedback systems and engineering education"; Second Paper Prize, AIEE, District 3, 1961; Second Prize, AIEE, Communication Division, 1961; The Haraden Pratt Award, 1974.

MUMFORD, WILLIAM W., IRE Fellow 1952. Born: June 17, 1905, Vancouver, Wash. Degrees: A.B., 1930, Willamette University.

Fellow Award "In recognition of his contributions in the field of high-frequency propagation and in the development of microwave components"; Morris E. Leeds Award, 1966, "For his outstanding contribution to the theory and technique of microwave measurements, including his invention and application of standard noise sources and directional couplers"; Microwave Career Award, MTT-S, 1973, "For a career of meritorious achievement and outstanding technical contributions in the field of microwave theory and techniques." Other Awards: Burghardt Prize, Willamette University, 1926;

Mackay Visiting Professor of Electrical Engineering, University of California at Berkeley, 1955; Ford Foundation Visiting Professor of Electrical Engineering, University of Wisconsin, 1962; Honorary Life Member, MTTS, ADCOM, 1964; Alumni Citation Award, Willamette University, 1968; Engineer of Distinction, Engineers' Joint Council, 1970; Outstanding Achievement Award, COMAR, 1980.

MUNGALL, ALLAN G., Fellow 1980. Born: March 12, 1928, Vancouver, B.C., Canada. Degrees: B.A.Sc., 1949, M.A.Sc., 1950, University of British Columbia; Ph.D., 1954, McGill University.

Fellow Award "For contributions in the design, construction, and use of primary cesium frequency and time standards."

MUNSON, JOHN C., Fellow 1981. Born: October 9, 1926. Clinton, IA. Degrees: B.S.E.E., 1949, Iowa State University; M.S. (E.E.), 1952, Ph.D. (E.E.), 1962, University of Maryland.

Fellow Award "For leadership in acoustics research and development, especially in the area of sonar signal processing." Other Awards: Dept. of Navy Scholar, M.I.T.; Meritorious Civilian Service Awards, 1955, 1960; Superior Accomplishment Awards (Four Inventions); Major Invention Award (corecipient), PUFFS, 1970.

MUROGA, SABURO, Fellow 1977. Born: March 15, 1925, Numazu-shi, Japan. Degrees: Ph.D., 1958, Tokyo University.

Fellow Award "For contributions to switching theory, computer design, information theory, and engineering education." Other Awards: Inada Award, Institute of Electronics and Communications Engineers of Japan, 1955.

MURPHY, BERNARD T., Fellow 1977. Born: May 30, 1932, Hull, England. B.Sc.(Hons.)(Physics), 1953, Ph.D.(Medical Physics), 1959, Leeds University.

Fellow Award "For contributions to the field of integrated circuits"; Program Committee, IEDM, ISSCC, ESSCIRC; Outstanding Paper Award, ISSCC, 1968; Best Invited Paper Award, ESSCIRC, 1982; Editorial Board, IEEE Spectrum; IEEE Field Awards Committee; IEEE GIO Committee. Other Awards: Over 70 papers and patents in the fields of integrated circuits, microwave devices, electron beam systems and medical physics.

MURPHY, GORDON J., Fellow 1967. Born: February 16, 1927, Milwaukee, Wis. Degrees: BS., 1949, Milwaukee School of Engineering; M.S., 1952, University of Wisconsin; Ph.D., 1956, University of Minnesota.

Fellow Award "For contributions to education and research in the field of automatic control." Other Awards: Chicago's Ten Outstanding Young Men Award, 1961.

MURRAY, PETER R., Fellow 1969. Born: December 28, 1915, Hancock, Md. Degrees: B.S., 1938, Antioch College.

Fellow Award "For pioneering leadership in the guided missile program." Other Awards: Lawrence Sperry Award, AIAA, 1946; Air Force Scientific Advisory Board, 1980-present.

MUSA, SAMUEL A., Fellow 1980. Born: June 20, 1940, Degrees: B.S.(E.E.), 1961, B.A., 1961, Rutgers University; M.S.(Applied Physics), 1962, Ph.D.(Applied Physics), 1965, Harvard University.

Fellow Award "For contributions to nonlinear systems analysis as applied to communications, control, and military systems."

MUSHIAKE, YASUTO, Fellow 1977. Born: March 28, 1921, Okayama-ken, Japan. Degrees: B.S.E.E., 1944, D.Eng., 1954, Tohoku University.

Fellow Award "For contributions to linear antennas and self-complementary antennas." Other Awards: Literature Prize, Institute of Electrical Engineers, Japan, 1957; Niwa-Takayanagi Award, Institute of Television Engineers of Japan, 1964; Hattori Hohkoh Award, Hattori Foundation, 1971; Literature Prize, Institute of Electronics and Communication Engineers of Japan, 1980; Medal of Honor, IECE of Japan, 1982; Commendation for Distinguished Service, Minister of Science and Technology, 1982.

MYERS, EARL H., Fellow 1969. Born: February 28, 1914. Degrees: B.S.E.E., 1937, Kansas State University; M.S.E.E., 1944, University of Pittsburgh.

Fellow Award "For significant contributions in the fields of large direct current machinery design and specialized applications."

MYERS, PETER B., Fellow 1966. Born: April 24, 1926, Washington, D.C. Degrees: B.S.E.E., 1946, Worcester Polytechnic Institute; D.Phil.(Nuclear Physics), 1950, University of Oxford; D.Hum.Litt.(Hon.), 1973, College of Idaho.

Fellow Award "For contributions in solid-state circuits and for leadership in related professional activities." Other Awards: Rhodes Scholar, 1947-50; Sigma Xi, 1953; Fellow, American Association for the Advancement of Science, 1959.

MYLES, ASA H., AIEE Fellow 1959. Born: August 21, 1906, LaRue, Ohio. Degrees: B.S., 1929, Ohio University.

Fellow Award "For contributions by invention and development of electrical controls for heavy industry"; Coauthor, IEEE Standard Handbook of Electrical Engineering; Coauthor, Switchgear and Control Handbook (by Robert W. Smeaton).

N

NAGAI, KENZO, Fellow 1971. Born: March 21, 1901, Sendai, Japan. Degrees: B.S., 1925, D.Eng., 1933,

Tohoku University.

Fellow Award "For contributions to magnetic recording methods, the theory and design of electrical networks, and engineering education." Other Awards: Honorary Member, Audio Engineering Society, 1973; Japan Academy of Science, 1977.

NAGAI, KIYOSHI K.N., Fellow 1982. Born: March 29, 1929, Sendai, Japan. Degrees: B.Eng., 1951, D.Eng., 1959, Tohoku University, Japan.

Fellow Award "For contributions to the development of phased array antennas." Other Awards: Inada Memorial Scholarship Award, Institute of Electronics and Communication Engineers, Japan, 1954; Ichimura Prize, 1980; Niwa-Takayanagi Prize, Institute of Television Engineers, Japan, 1980; Achievement Award, Institute of Electronics and Communication Engineers, Japan, 1982. The Prize of Minister of State for Science and Technology, Science and Technology Agency, Japan, 1983.

NAGEL, THEODORE J., Fellow 1968. Born: December 20, 1913, Andes, N.Y. Degrees: A.B., 1936, Columbia Collge; B.S., 1937, M.S., 1938, Columbia University.

Fellow Award "For distinguished service through technical contributions and administrative leadership in the planning, development and operation of large power systems"; William M. Habirshaw Award, IEEE, 1979. Other Awards: Member, National Academy of Engineering, 1973; Eta Kappa Nu; Tau Beta Phi.

NAGLE, H. TROY, JR., Fellow 1983. Born: August 31, 1942, Booneville, Miss. Degrees: B.S.E.E., 1964, M.S.E.E., 1966, Univ. of Alabama; Ph.D., 1968, Auburn Univ.; M.D., 1981, Univ. of Miami.

Fellow Award "For contributions to industrial electronics, data acquisition, and control instrumentation." Other Awards: Prize Paper, Sigma Xi, 1969; Professional Engineer Award, State of Alabama, 1973.

NAGUMO, JIN-ICHI, Fellow 1981. Born: October 14, 1926, Tokyo, Japan. Degrees: B.Eng., 1948, M.Eng., 1950, Ph.D., 1953, University of Tokyo.

Fellow Award "For contributions to neuron modeling, medical electronics, human engineering, and nonlinear systems analyses."

NAHMAN, NORRIS S., Fellow 1977. Born: November 9, 1925, San Francisco, Calif. Degrees: B.S.(Electronics and Radio Engineering), 1951, California Polytechnic State University, San Luis Obispo; M.S.E.E., 1952, Stanford University; Ph.D.(E.E.), 1961, University of Kansas.

Fellow Award "For contributions to time domain metrology"; First Place, Student Paper Contest, IRE, South California Section, 1951; Recognition Award, IEEE Student Branch, Toledo Section, "For outstanding leadership as Chairman of Electrical Engineering, University of Toledo, 1973-75"; IEEE Instrumentation & Measurement Society: Adm. Committee 1982-1984; Tran-

sactions Editorial Review Committee, 1980-present. Other Awards: Dept. of the Army Commendation Pendant for research contributions, 1954; Sigma Tau, 1956; Eta Kappa Nu, 1958; Sigma Pi Sigma, 1959; URSI Commission A, 1966; Distinguished Alumnus Award, California Polytechnic State University, 1972; Tau Beta Pi, 1974; National Bureau of Standards Senior Scientist Fellowship, 1979-80.

NAKAGOME, YUKIO, Fellow 1979. Born: May 25, 1922, Yamanashi-Ken, Japan. Degrees: Bachelor of Engineering, 1949, Doctor of Engineering, 1960, Tokyo Institute of Technology.

Fellow Award "For contributions to electronic-communications switching systems and digital signal transmission." Other Awards: Paper Award, Institute of Electronics and Communications Engineers of Japan, 1957, 1973; Achievement Award, Institute of Electronics and Communications Engineers of Japan, 1976; Scientific Contribution Award, Minister of Science and Technology of Japan, 1978; Medal with Purple Ribbon, 1983.

NAKAHARA, TSUNEO, Fellow 1983. Born: August 29, 1930, Tokushima Pref., Japan. Degrees: B.S.E.E., 1953, D.Eng., 1961, University of Tokyo.

Fellow Award "For contributions to the development of microwave transmission lines, traffic control systems, and fiber optics"; Paper Award, IEEE Vehicular Technology Society, 1972. Other Awards: Inada Memorial Award, Institute of Electronics and Communication Engineers (IECE) of Japan, 1958; Okabe Memorial Award, IECE of Japan, 1959; Special Distinguished Service Award, Laser Society of Japan, 1983.

NAKANISHI, KUNIO, Fellow 1978. Born: July 4, 1922, Yokohama, Japan. Degrees: Bachelor of Engineering, 1944, Ph.D. of Engineering, 1957, Tokyo University.

Fellow Award "For contributions to the understanding of switching phemomena in high-power switchgear."

NAKANO, YOSHIEI, Fellow 1980. Born: February 21, 1910, Hiroshima, Japan. Degrees: B.S.(Eng.), 1934, Ph.D.(Eng.), 1950, Tokyo Institute of Technology.

Fellow Award "For contributions in the development and standardization of insulation systems for electrical locomotives and cars." Other Awards: Award, Govenor of Tokyo Metropolis, 1969; Award, Railway Electrification Association Incorporation, Japan, 1971.

NAMBA, SHOGO, IRE Fellow 1960. Born: July 14, 1904, Kyoto, Japan. Degrees: M.Eng., 1927, Dr.Eng., 1935, Kyoto University.

Fellow Award "For contributions to the understanding of ionospheric radio propagation." Other Awards: Asano Award, Institute of Electrical Engineers of Japan, 1936; Distinguished Service Award, Institute of Electronics and Communication Engineers of Japan, 1943.

NARENDRA, KUMPATI SUBRAHMANYA, Fellow 1979. Born: April 14, 1933. Degrees: B.E.(with honors), 1954, University of Madras; S.M., 1955, Ph.D., 1959, Harvard University; M.A.(Honorary), 1968, Yale University.

Fellow Award "For contributions to stability theory and its application to adaptive control systems."

NARUD, JAN A., Fellow 1969. Born: June 3, 1925. Degrees: Radio Eng., 1945, Oslo Institute of Technology; B.S.E.E., M.S.E.E., 1951, California Institute of Technology; Ph.D., 1955, Stanford University.

Fellow Award "For contributions to the basic knowledge and engineering of integrated circuits"; Outstanding Paper Award, International Solid-State Circuit Conference, 1963; Outstanding Paper Award, National Electronics Conference, 1967.

NASH, HAROLD E., Fellow 1967. Born: July 14, 1914, Lindsay, Calif. Degrees: B.A., 1938, University of California at Berkeley.

Fellow Award "For his contribution to research and development of sonar, both as a scientist and as a director of a major Navy laboratory." Other Awards: Fellow, Acoustical Society of America, 1957; Distinguished Civilian Service Award, Dept. of the Navy, 1975.

NASH, RAYMOND A., JR., Fellow 1983. Born: August 2, 1938, Pittsburgh, Pennsylvania. Degrees: B.S.E., 1960, Princeton University; M.S.E.E., 1961, Massachusetts Institute of Technology; Ph.D., 1966, Yale University.

Fellow Award "For contributions to the analysis of integrated navigation systems."

NASHMAN, ALVIN E., Fellow 1971. Born: December 16, 1926, New York, N.Y. Degrees: B.E.E., 1948, City College of The City University of New York; M.E.E., 1951, New York University; Sc.D.(Hon.), 1968, Pacific University.

Fellow Award "For contributions and leadership in the development of communication and geodetic satellite systems."

NASSER, ESSAM, Fellow 1976. Born: February 3, 1936, Cairo, Egypt. Degrees: B.E.E., 1952, Cairo University; Dipl.Ing., 1955, Dr.-Ing., 1959, Technical University of West Berlin.

Fellow Award "For contributions to the field of high-voltage engineering education and to the understanding of contaminated insulator breakdown." Other Awards: Award, Minister of Education, Cairo, 1947; Award, Senate of West Berlin, 1957; Award and Grant, Office of Naval Research, 1961; Grant, National Science Foundation, 1966-69, 1974-76.

NATHAN, MARSHALL I., Fellow 1971. Born: January 22, 1933, Lakewood, N.J. Degrees: B.S., 1954, Massachusetts Institute of Technology; M.A., 1956, Ph.D., 1958, Harvard University.

Fellow Award "For leadership and research in the development of the injection laser and the study of high field transport properties of semiconductors"; David Sarnoff Award, IEEE, 1980. Other Awards: Fellow, American Physical Society; IBM Outstanding Contribution Awards: 1963, Injection Laser; 1968, Microwave Oscillations.

NATHANSON, FRED E., Fellow 1984. Born: January 12, 1933, Baltimore, MD. Degrees: B.E., 1955, Johns Hopkins University; M.S.E.E., 1956, Columbia University.

Fellow Award "For contributions to radar-system signal processing." Other Awards: Eta Kappa Nu; URSI.

NATHANSON, HARVEY C., Fellow 1976 Born: October 22, 1936, Pittsburgh, Pa. Degrees: B.S., 1958, M.S., 1959, Ph.D.(E.E.), 1962, Carnegie-Mellon University.

Fellow Award "For contributions to solid-state devices through the combination of micromechanics and semiconductor physics." Other Awards: Outstanding Young Electrical Engineer Award (Honorable Mention), Eta Kappa Nu, 1966; Carnegie-Mellon Outstanding Alumnus Award, 1983.

NAU, ROBERT H., Fellow 1965. Born: April 21, 1913, Burlington, Iowa. Degrees: B.S., 1935, Iowa State University; M.S., 1937, Texas A and M University; E.E., 1941, Iowa State University.

Fellow Award "For contributions to engineering education"; "The Educational Award," IEEE, St. Louis Section, 1982. Other Awards: Breast Order of Yun Hui, President of China, 1946; Outstanding Chapter President, Missouri Society of Professional Engineers, 1963; Life Member, Radio Club, University of Missouri at Rolla, 1963; Outstanding President, Dept. of Missouri Reserve Officers' Association, 1964-65; Recipient of University of Missouri at Rolla, Outstanding Teacher Award, 1969; Sigma Xi; Phi Kappa Phi; Tau Beta Pi; Pi Mu Epilson; Eta Kappa Nu; Sigma Tau; Kappa Mu Epilson; Sigma Pi Sigma; Honorary Member, Blue Key National Honorary, UMR Student Body, 1983; "Honorary Professor of Military Science," UMR Military Dept., 1983; Professor Emeritus of Electrical Engineering, 1983, Board of Curators of University of Missouri; Award "ATA Hall of Fame Trapshooting Instructor," 1982, 83, 84; 5 textbooks, 64 technical papers.

NEILD, J.F., AIEE Fellow 1936.

NEIRYNCK, JACQUES, Fellow 1981. Born: August 17, 1931, Uccle, Belgium. Degrees: Ingenieur Civil Electricien, 1954, Docteur en Sciences Appliquees, 1958, University of Louvain, Belgium.

Fellow Award: "For contributions to circuit theory, especially in the area of filter design."

NELSON, A. L., AIEE Fellow 1929. Born: June 15, 1892, Charlottetown, Prince Edward Island, Canada. Degrees: B.S., 1915, Massachusetts Institute of Technology.

NERGAARD, LEON S., IRE Fellow 1952. Born: September 2, 1905, Battle Lake, Minn. Degrees: B.S.E.E., 1927, University of Minnesota; M.S.E.E., 1930, Union College; Ph.D. (Physics), 1935, University of Minnesota.

Fellow Award "In recognition of his contributions in the fields of ultra-high-frequency measurements, electron tubes and circuits"; Mervin J. Kelly Award, 1973, "For outstanding contributions and leadership in the introduction of very high and ultra high frequencies for telecommunications." Other Awards: David Sarnoff Outstanding Achievement Award in Science, 1960.

NESSMITH, JOSH T., JR., Fellow 1977. Born: June 18, 1923, Bulloch County, Ga. Degrees: B.E.E., 1947, Georgia School of Technology; M.S.E.E., 1957, Ph.D., 1965, University of Pennsylvania.

Fellow Award "For leadership in and technical management of radar systems engineering."

NEUBAUER, JOHN P., Fellow 1966. Born: November 16, 1904, Bridgeport, Conn. Degrees: E.E., 1926, Rensselaer Polytechnic Institute; M.S.E.E., 1928, Yale University.

Fellow Award "For contributions to the engineering design and operation of electrical systems for large metropolitan areas."

NEVITT, H. J. BARRINGTON, AIEE Fellow 1960. Born: June 1, 1908, St. Catharines, Ontario, Canada. Degrees: B.A.Sc., 1941, University of Toronto; M.Eng., 1945, McGill University, Quebec.

Fellow Award "For contributions to high quality international communication systems in Latin America." Other Awards: Fellow, American Association for the Advancement of Science, 1961; Fellow, Institution of Electrical Engineers, U.K., 1966; Fellow, Engineering Institute of Canada, 1972; Discoveries International Symposia, Honda Foundation, Tokyo, Japan.

NEWCOMB, ROBERT W., Fellow 1972. Born: June 27, 1933, Glendale, Calif. Degrees: B.S.E.E., 1955, Purdue University; M.S., 1957, Stanford University; Ph.D., 1960, University of California at Berkeley.

Fellow Award "For contributions to network synthesis, time-varying systems, and engineering education." Other Awards: Fulbright Fellow, Australia, 1963-64; Professor Invite, Louvain University, 1967-68; Fulbright-Hays Fellow, Malaysia, 1976.

NEWELL, ALLEN, Fellow 1974. Born: March 19, 1927, San Francisco, Calif. Degrees: B.S., 1949, Stanford University; Ph.D., 1957, Carnegie Institute of Technology.

Fellow Award "For contributions in computing through list processing language development, and for texts on computers, digital systems, and artificial intelligence"; Computer Pioneer Award, Charter Recipient, IEEE Computer Society, 1982. Other Awards: Harry

Goode Memorial Award, American Federation of Processing Societies, 1971; Member, American Academy of Arts and Sciences, 1972; Member, National Academy of Sciences, 1972; A.M. Turing Award (with H.A. Simon), ACM, 1975; Member, National Academy of Engineering, 1980; U.A. and Helen Whitaker University Professor of Computer Science, Carnegie-Mellon University.

NEWELL, HOBART H., AIEE Fellow 1949. Born: June 10, 1896, Cumberland, R.I. Degrees: B.S., 1918, D.Eng.(Hon.), 1957, Worcester Polytechnic Institute.

NEWHOUSE, RUSSELL C., IRE Fellow 1956. Born: December 17, 1906, Clyde, Ohio. Degrees: B.E.E., 1929, M.S., 1930, Ohio State University.

Fellow Award "For his work in the fields of terrain clearance indicators, airborne communications, and military weapons systems"; Pioneer Award, Aerospace and Electronic Systems Group, 1967, "In recognition of his contribution to radio altimetry." Other Awards: Guggenheim Fellow, 1929; Lawrence Sperry Award, 1938; Distinguished Alumnus Award, Ohio State University, 1959; Distinguished Public Service Plaque, Federal Aviation Agency, 1965.

NEWHOUSE, VERNON L., Fellow 1968. Born: January 30, 1928, Mannheim, W.Germany. Degrees: B.Sc.(Hon.), 1949, Ph.D., 1952, University of Leeds, England.

Fellow Award "For contributions to magnetic and superconducting memory elements and their applications in computers"; Honorable Mention Award, IEEE Sonics and Ultrasonics Group, 1976. Other Awards: Robert C. Disque Professor of Electrical and Computer Engineering, Drexel University; Adjunct Professor of Radiology, Jefferson University.

NEWMAN, MORRIS M., Fellow 1969. Born: September 7, 1909. Degrees: B.E.E., 1931, M.S., 1937, University of Minnesota.

Fellow Award "For contributions to aircraft lightning protection, and for development of techniques for the controlled study of natural lightning."

NEWMAN, ROGER, Fellow 1972. Born: August 16, 1925, New York, N.Y. Degrees: A.B., 1945, M.A., 1946, Ph.D., 1949, Columbia University.

Fellow Award "For contributions to semiconductor technology and for leadership in solid-date electronics research." Other Awards: Fellow, American Physical Society, 1959.

NEWMEYER, WILLIAM L., AIEE Fellow 1948. Born: April 19, 1894, Memphis, Tenn. Degrees: B.S., 1916, Case Institute of Technology; M.S., 1933, California Institute of Technology.

NICHOLAS, JAMES H., Fellow 1974. Born: March 11, 1916, Tarentum, Pa. Degrees: B.S.E.E., 1938, Purdue University.

Fellow Award "For contributions to the continuing

development, testing, and manufacture of devices for high-voltage power cable."

NICHOLS, CLARK, Fellow 1978. Born: August 9, 1914, Searsport, Me. Degrees: B.S. and M.S.(Electrical Engineering), 1935, Massachusetts Institute of Technology.

Fellow Award "For contributions to and leadership in the design of computer-control systems for electric power applications"; First Prize Paper, AIEE Power Division, 1952-53.

NICHOLS, MYRON H., Fellow 1974. Born: January 26, 1915, Nelson, Ohio. Degrees: A.B., 1936, Oberlin College; Ph.D., 1939, Massachusetts Institute of Technology.

Fellow Award "For contributions to physical electronics, radio telemetry, and research in upper atmosphere phenomena"; IRE-PGSET Award, 1957; IEEE, S-AES Pioneer Award, 1972, "For pioneering contributions to the field of high-speed electronic time-division multiplex radio telemetry and early FM/FM telemetry." Other Awards: Phi Beta Kappa, 1936; Sigma Xi, 1939; National Telemetry Conference Award, 1963.

NICHOLS, NATHANIEL B., Fellow 1968. Born: December 31, 1914, Mt. Pleasant, Mich. Degrees: B.S., 1936, Central Michigan University; M.S., 1937, University of Michigan; Sc.D.(Hon.), 1964, Central Michigan University; Sc.D.(Hon.), 1968, Case Western Reserve University.

Fellow Award "For contributions to the theory and applications of automatic control." Other Awards: Young Author's Prize, Electrochemical Society, 1938; Fellow, Instrument Society of America, 1960; Excellence in Documentation Award, Instrument Society of America, 1963; Honorary Member, Instrument Society of America, 1979; Fellow, American Association for the Advancement of Science, 1966; Rufus Oldenburger Award, American Society of Mechanical Engineers, 1969; Alfred F. Sperry Medal Award (with J.G. Ziegler), Instrument Society of America, 1974; Control Heritage Award, American Automatic Control Council, 1980.

NICKEL, DONALD L., Fellow 1981. Born: June 19, 1933, Hortonville, WI. Degrees: B.S.(Math. and Physics), 1957, University of Wisconsin.

Fellow Award "For leadership in planning, design, and equipment applications to distribution systems."

NICOLLIAN, EDWARD H., Fellow 1984. Born: May 23, 1927, New York, N.Y. Degrees: M.E., 1951, Stevens Institute of Technology; M.A., 1956, Columbia University.

Fellow Award "For contributions to metal-oxide-semiconductor interfaces."

NIELSEN, RUSSELL A., AIEE Fellow 1951. Born: January 19, 1912, Portland, Oreg. Degrees: A.B., 1933, A.M., 1934, Ph.D.(Physics), 1937, Stanford University.

NIKIFOROFF, BASIL, AIEE Fellow 1946. Born August 30, 1888, Russia. Degrees: E.E., 1913, Electrical Institute of Emperor Alexander III, Petrograd, Russia.

NILSSON, EINAR, AIEE Fellow 1962. Born: October 9, 1901, Sweden. Degrees: E.E., 1922, Federal Electrotechnical Institute, Vesteras, Sweden.

Fellow Award "For contributions to the design of electrical generating plants of many types and of metropolitan substations." Other Awards: Certificate of Appreciation, U.S. Army O.Q.M.G. Office, R&D Branch, 1944.

NISHINO, OSAMU, Fellow 1973. Born: 1912, Osaka, Japan. Degrees: B.E., 1936, D.Eng., 1950, University of Tokyo.

Fellow Award "For contributions to electrical and nuclear instrumentation and to engineering education." Other Awards: Distinguished Service Award, Association of Applied Electrical Technique, 1944; Commendation, Minister of International Trade and Industry, 1969; Blue Ribbon Medal, Japanese Government, 1977; Professor Emeritus, University of Tokyo; Member, Institute of Electrical Engineers of Japan; Member, Institute of Electronics and Communication Engineers of Japan; Member, Physical Society of Japan; Meritorious Member, Japan Society of Applied Physics; Member, Society for Instrument and Control Engineers (Japan); Member, Atomic Energy Society of Japan; Member, Japan Society for Medical Electronics and Biological Engineering.

NISHIZAWA, JUN-ICHI, Fellow 1969. Born: September 12, 1926, Japan. Degrees: B.S., 1948, D.Eng., 1960, Tohoku University.

Fellow Award "For technical contributions to solid state electronics, and leadership in related professional activities"; Jack A. Morton Award, IEEE, 1983. Other Awards: Director's Award, Japanese Science and Technology Agency, 1965, 1970; Imperial Invention Prize, 1966; Matsunaga Memorial Award, 1969; Okochi Memorial Technology Prize, 1971, 1980; Award of the Japan Academy, 1974; Meritorious Honor Award, Japanese Science and Technology Agency, 1975; Purple Ribbon Medal, Japanese Government, 1975; Achievement Award, Institute of Electronics and Communication Engineers of Japan, 1975; Inoue Harushige Award, 1982; Person of Cultural Merits, 1983.

NOIZEUX, P. J., IRE Fellow 1947. Born: 1900, France.

Fellow Award "For his leadership in development of radio communication services."

NOLL, PETER, Fellow 1982. Born: September 9, 1936, Oldenburg, Germany. Degrees: Dipl.Ing., 1964, Dr.Ing., 1969, Technical University of Berlin.

Fellow Award "For contributions to adaptive quantization and coding of speech signals"; Senior Award, IEEE Acoustics, Speech, and Signal Processing Society, 1977. Other Awards: NTG Award, 1976.

NOMURA, TATSUJI, Fellow 1971. Born: 1913, Tokyo, Japan. Degrees: B.E., 1935, University of Tokyo.

Fellow Award "For contributions and leadership in the field of broadcasting and television engineering." Other Awards: Niwa-Takayanagi Prize, Institute of Television Engineering, 1961; Minister's Award, Ministry of Science and Technology, 1963; Silver Prize, Emperor of Japan, 1964; Mainichi Industrial Engineering Prize, 1964; Member, Institute of Electrical Engineers of Japan; Member, Institute of Electronic and Communication Engineers of Japan; Member, Institute of Television Engineers of Japan; Blue Ribbon Medal, Emperor of Japan, 1970; Third Order Medal of Sacred Treasure, Emperor of Japan, 1983.

NONKEN, GORDON C., AIEE Fellow 1955. Born: July 1, 1908, Burns, Kans. Degrees: B.S., 1930, Kansas State University.

Fellow Award "For a diversity of contribution in the field of measurements and manufacturing processes; particulary as applied to transformers and capacitors." Other Awards: Charles A. Coffin Award, General Electric, 1948.

NOORDERGRAAF, ABRAHAM, Fellow 1975. Born: August 7, 1929, Utrecht, The Netherlands. Degrees: B.Sc., 1953, M.Sc., 1955, Ph.D., 1956, University of Utrecht, The Netherlands; M.A.(Hon.), 1971, University of Pennsylvania.

Fellow Award "For leadership in the utilization of electrical engineering principles in the solution of basic life science problems, especially hemodynamics." Other Awards: Research Award, Secretary of Education, Arts and Sciences (Dutch Government), 1952; Fellow, New York Academy of Sciences, 1976; Herman C. Burger Award, 1978; Fellow: American Association for the Advancement of Science, 1978, Explorer's Club, 1978, College of Physicians of Philadelphia, 1978; American College of Cardiology, 1979; Case Centennial Fellow, 1980.

NORBERG, HANS A., AIEE Fellow 1961. Born: August 19, 1901, Astoria, Oreg. Degrees: B.S.E.E., University of Minnesota.

Fellow Award "For contributions to electrical equipment for the petroleum industry."

NORTH, DWIGHT O., IRE Fellow 1943. Born: September 28, 1909, Hartford, Conn, Degrees: B.S., 1930, Wesleyan University; Ph.D., 1933, California Institute of Technology.

Fellow Award "For his contributions to the knowledge of the fundamentals of vacuum tube performance, especially with regard to fluctuation phenomena and the effects of electron transit time."

NORTH, HARPER Q., IRE Fellow 1957. Born: January, 24, 1917, Los Angeles, Calif. Degrees: B.S., 1938, California Institute of Technology; M.A., 1940, Ph.D., 1947, University of California at Los Angeles.

Fellow Award "For practical developments in the field of semi-conductors." Other Awards: Fellow, American Physical Society; Medal Of Honor, Electronic Industries Association, 1966; Honorary Director, Electronic Industries Association, 1966.

NORTH, JOHN R., AIEE Fellow 1941. Born: February 20, 1900, Cambridge, Mass. Degrees: B.S., 1923, California Institute of Technology.

Other Awards: Tau Beta Pi, 1938; Standards Council Award, American Standards Association (now ANSI), 1953.

NORTON, KENNETH A., IRE Fellow 1943, AIEE Fellow 1956. Born: February 27, 1907, Rockwell City, Iowa. Degrees: B.S., 1928, University of Chicago.

IRE Fellow Award "In recognition of his work in applying his conclusions from the theory of radio wave propagation to the problems of frequency allocation"; AIEE Fellow Award "For his contributions in the field of radio wave propagation"; Harry Diamond Award, IRE, 1960. Other Awards: Fellow, American Association for the Advancement of Science, 1943; Fellow, American Physical Society, 1945; Stuart Ballantine Medal, Franklin Institute, 1954; Exceptional Service Award, U.S. Dept. of Commerce, 1962; Silver Medal, U.S. Dept of Commerce, 1966.

NOUVION, FERNAND F. P., Fellow 1968. Born: November 2, 1905, Neuilly, Seine, France. Degrees: Ingenieur, 1927, Ecole Superieure d'Electricite, Paris.

Fellow Award "For important contribution to the development of modern electrical railroads." Other Awards: Prix Coignet, Societe des Ingenieurs Civils, 1946; Medaille de la Ville de Paris, 1955; Prix Giffard, l'Academie des Sciences, 1965; Grand Prix, la Societe des Ingenieurs Civils, 1968, Medaille Ampere, Societe Francaise des Electrciens; Paris Exhibition, 1881, Premium de l'Institution of Electrical Engineers of Great Britain; Commandeur de la Legion d'Honneur, 1975; Diploma and Medal, Pan American Railway Congress Association, Mexico, 1980.

NOVICK, ROBERT, Fellow 1967. Born: May 3, 1923, New York, N.Y. Degrees: M.E., 1944, M.S., 1949, Stevens Institute of Technology; Ph.D., 1955, Columbia University.

Fellow Award "For basic contributions to atomic and molecular physics." Other Awards: Alfred P. Sloan Research Fellow, 1958; American Astronomical Society; Fellow, American Physical Society; Tau Beta Pi; Sigma Xi; Fellow, American Association for the Advancement of Science; NASA Medal for Exceptional Scientific Achievement, 1980.

NOYCE, ROBERT N., Fellow 1966. Born: December 12, 1927, Burlington, Iowa. Degrees: B.A., 1949, Grinnell College; Ph.D, 1953, Massachusetts Institute of Technology.

Fellow Award "For leadership in research development and manufacture of semiconductor devices"; IEEE Medal of Honor, 1978; Cledo Brunetti Award, 1978. Other Awards: Stuart Ballantine Medal, Franklin Institute, 1966; National Academy of Engineering, 1969; Harry Goode Award, American Federation of Information Processing, 1978; Faraday Medal, Institution of Electrical Engineers, London, 1979; National Medal of Science, 1979; Harold Pender Award, Moore School of Electrical Engineering, University of Pennsylvania, 1980.

NUCCI, ELIDIO J., Fellow 1965. Born: October 30, 1919, New York, N.Y. Degrees: B.E.E., 1940, Manhattan College.

Fellow Award "For contributions to the advancement and implementation of reliability technology"; Annual Award, IRE, Professional Group on Reliability and Quality Control, 1961, "For his part in the development of the publication 'Parts Specification Management for Reliability,' as a member of the ad hoc study group on parts specification management for reliability."

NUSSBAUM, ALLEN, Fellow 1984. Born: August 22, 1919, Philadelphia, PA. Degrees: B.A., 1939, M.A., 1940, Ph.D., 1953, University of Pennsylvania.

Fellow Award "For contributions to the theory of semiconductor device physics." Other Awards: Fulbright Visiting Professor, Hebrew University of Jerusalem, 1971-72.

NUSSBAUMER, HENRI J., Fellow 1983. Born: December 22, 1931, Paris, France. Degrees: Ingenieur ECP, 1954, Ecole Centrale de Paris; Ph.D., 1977, University of Nice.

Fellow Award "For contributions to the theory and the development of line switching and data transmission systems." Other Awards: IBM Fellow, 1975.

O

OAKES, LESTER C., Fellow 1983. Born: October 11, 1923, Knoxville, Tennessee. Degrees: B.S.E.E., 1949, M.S., 1962, University of Tennessee.

Fellow Award "For contributions to nuclear power reactor control systems."

OATLEY, SIR CHARLES W., Fellow 1968. Born: February 14, 1904, Frome, Somerset, England. Degrees: M.A., Cambridge University.

Fellow Award "For contributions to research in radar and electron optics, especially for development of the scanning electron microscope." Other Awards: Fellow, Royal Society, 1969; Member, Fellowship of Engineering, 1976; Foreign Associate, National Academy of Engineering, 1979.

O'BRIEN, BRIAN, AIEE Fellow 1947. Born: January 2, 1898, Denver, Colo. Degrees: Ph.B.(E.E.), 1918, Ph.D., (Physics), 1922, Yale University.

Other Awards: Presidential Medal for Merit, 1948; Frederick Ives Medal, Optical Society of America,

1951; Exceptional Civilian Service Medal, USAF, 1969; Distinguished Public Service Medal, NASA, 1972; Exceptional Service Medal, U.S. Air Force, 1973; Member, National Academy of Sciences, American Philosophical Society, American Academy of Arts and Science, Connecticut Academy of Science and Engineering; National Academy of Engineering.

O'BRIEN, JOSEPH E., AIEE Fellow 1951. Born: December 16, 1905, Keokuk, Iowa. Degrees: B.S.(E.E.), 1927, M.A., 1928, Catholic University of America.

O'BRYAN, HENRY M., IRE Fellow 1961. Born: August 28, 1905, Chetopa, Kans. Degrees: A.B., 1926, Clark University; A.M., 1927, Northwestern University; Ph.D., 1930, The Johns Hopkins University.

Fellow Award "For contributions to the administration of scientific research."

OCCHINI, ELIO, Fellow 1982. Born: December 21, 1928, Viareggio, Italy. Degrees: E.E., 1952, Polytechnic Institute, Milan, Italy.

Fellow Award "For contributions to cable technology for EHV and UHV electric energy transmission." Other Awards: Fellow, Institution of Electrical Engineers, 1978; Charted Engineer, 1978.

O'CONNELL, J. D., IRE Fellow 1957. Born: September 25, 1899, Chicago, Ill. Degrees: B.S., 1922, United States Military Academy; M.S., 1930, Yale University.

Fellow Award "For distinguished leadership in the field of military electronics"; Chairman, Joint Technical Advisory Committee, IRE & EIA, IEEE and EIA JTAC. Other Awards: Order of Leopold, Degree of Crois de Guerre with Palm, Belgium, 1940; Legion of Merit, 1945; Oak Leaf Cluster, 1945; Croix de Guerre, Order of the Army with Palm, France, 1945; Gold Medal, Franklin Institute, 1954; Silver Medal, Poor Richard Club, 1956; Distinguished Service Medal, 1959; Distinguished Service Award, Executive Office of the President, 1969.

O'CONNOR, JAMES J., Fellow 1977. Born: August 3, 1917, New York, N.Y. Degrees: B.E.E., 1948, New York University.

Fellow Award "For contributions to the advancement of communication to the engineering profession." Other Awards: Engineering Award for Outstanding Leadership, American Society of Mechanical Engineers, 1972; Jesse H. Neal Editorial Achievement Award, American Business Press, Inc., 1974; Listed in "Who's Who in America," 1981; Listed in "Who's "Who in Engineering," 1980; Eta Kappa Nu; Licensed Professional Engineer.

OETINGER, H. W., AIEE Fellow 1952. Born: October 10, 1894, Philadelphia, Pa. Degrees: B.S., 1915, University of Pennsylvania.

Other Awards: Sigma Xi.

OETTINGER, ANTHONY G., Fellow 1970. Degrees: A.B., 1951, Harvard College; Ph.D., 1954, Harvard

University.

Fellow Award "For pioneering contributions to machine language translation, to information retrieval, and to the use of computers in education." Other Awards: Phi Beta Kappa; Sigma Xi; Fellow, American Association for the Advancement of Science; Fellow, American Academy of Arts and Sciences.

OFFNER, FRANKLIN F., Fellow 1966. Born: April 18, 1911, Chicago, Ill. Degrees: B.Chem., 1933, Cornell University; M.S., 1934, California Institute of Technology; Ph.D., 1938, University of Chicago.

Fellow Award "For the development of electronic equipment for medical research and for contributions to biomedical engineering education"; IEEE Centennial Medal, 1984. Other Awards: Lincoln Medal, 1966.

OGATA, KENJI, Fellow 1978. Born: October 9, 1917, Tokyo, Japan. Degrees: Ph.D.(Electrical Engineering), 1957, Tohoku University.

Fellow Award "For contributions to and leadership in telecommunications research and development."

OGUCHI, BUN-ICHI, Fellow 1977. Born: November 21, 1921, Nagano Prefecture, Japan. Degrees: D.E.E., 1951, University of Tokyo.

Fellow Award "For technical contributions to microwave and millimeter-wave transmission and for leadership in the research and the development of tele-communication." Other Awards: Award, President of Nippon Telegraph and Telephone Public Corporation, 1954; Twenty-seventh and twenty-ninth Mainichi Industrial Technology Special Awards, 1975, 1977; Achievement Award, Institute of Electronics and Communication Engineers of Japan, 1979; Achievement Award, Minister Award, Ministry of Post and Telecommunication, 1981; Achievement Award, Minister Award, Science and Engineering Agency, 1983.

OHASHI, KANICHI, Fellow 1964. Born: February 9, 1899, Kurashiki-shi, Okayama, Japan. Degrees: D.Eng., 1934, Tohoku University.

Fellow Award "For contributions to electrical communications and leadership in the theoretical, development, and manufacturing fields."

OHKOSHI, AKIO, Fellow 1983. Born: October 14, 1925, Saitama Prefecture, Japan. Degrees: B.S., 1953, Ph.D., 1981, Waseda University.

Fellow Award "For contributions to the invention of the Trinitron color cathode ray tube and the establishment of its manufacturing technology"; Outstanding Paper Awards, Chicago Spring Conference, BTR, 1968, 1973, 1974. Other Awards: Mainichi Industrial Technology Award, 1969; The Ohkohchi Memorial Technology Prize, 1973; Award for Contribution to Science and Technology, Science and Technology Agency, Japan, 1973; The Medal of Honor with Purple Ribbon, The Emperor of Japan, 1973.

OHL, RUSSELL S., IRE Fellow 1955. Born: January 31, 1898, Macungie, Pa. Degrees: B.S., 1918, Pennsylvania State University.

Fellow Award "For his contributions to the development of solid state point contact rectifiers." Other Awards: Distinguished Alumnus Award, Pennsylvania State University, 1976.

OHTSUKI, TATSUO, Fellow 1984. Born: June 16, 1940, Tokyo, Japan. Degrees: B.E., 1963, M.E., 1965, Dr.Eng., 1970, Waseda University.

Fellow Award "For contributions to circuit theory and computer-aided circuit analysis"; Guillmin-Causer Prize Award, IEEE Circuits and Systems Society, 1974. Other Awards: Best Paper Award, IECEJ (Institute of Electronics and Communication Engineers of Japan), 1969.

OIZUMI, JURO, Fellow 1975. Born: March 12, 1913, Sendai, Japan. Degrees: D.Eng., 1952, Tohoku University.

Fellow Award "For contributions to computer engineering." Other Awards: Mainichi Communication Prize, 1943; Minister Prize of International Trade and Industry, 1974; Honorary Member, Information Processing Society of Japan, 1980; Professor Emeritus, Tohoku University; Member, Board of Trustees, Pacific Telecommunications Council.

OKADA, MINORU, Fellow 1974. Born: October 31, 1907. Degrees: B.E., 1931, D.Eng., 1941, University of Tokyo.

Fellow Award "For contributions to air traffic control automation and navigational electronic engineering." Other Awards: Distinguished Service Award, Institute of Electronics and Communication Engineers of Japan, 1968; Medal of Honor with Purple Ribbon, 1968; Commendation, Minister of Transport, 1972; Life Member, Institute of Electrical Engineers of Japan; Fellow, Institute of Electronics and Communication Engineers of Japan, 1973; Commendation, Ministry of Posts and Telecommunications, 1978; Second Class Order of the Rising Sun, 1980.

OKAMOTO, HIDEO, Fellow 1979. Born: September 17, 1918, Tokyo, Japan. Degrees: Bachelor of Science, 1943, Ph.D., 1965, Osaka University.

Fellow Award "For research on corona measurements and the effects of corona on the deterioration of electrical insulation." Other Awards: Minister's Prize, Science and Technology Agency, Government of Japan, 1971.

OKAMURA, SOGO, Fellow 1973. Born: March 18, 1918, Mie Prefecture, Japan. Degrees: B.Eng., 1940, D.Eng., 1951, University of Tokyo.

Fellow Award "For contributions to microwave theory and techniques and to engineering education." Other Awards: Distinguished Service Award (Koseki-sho), Institute of Electronics and Communication Engineers of

Japan, 1974; Award for Scientific and Technological Development, Governor of Tokyo Metropolis, 1975; Broadcast Culture Prize, The Japan Broadcasting Corporation, 1979.

OKEAN, HERMAN C., Fellow 1975. Born: September 28, 1933, New York, NY. Degrees: B.A., 1955, Columbia College; B.S.E.E., 1956, Columbia School of Engineering; M.E.E., 1960, New York University; Eng. Sc.D., 1965, Columbia University.

Fellow Award "For contributions to microwave integrated circuit techniques." Other Awards: Cash Award for Invention, NASA; Phi Beta Kappa; Tau Beta Pi; Eta Kappa Nu.

OKOCHI, MASAHARU J. A., Fellow 1980. Born: March 30, 1916, Tokyo, Japan. Degrees: B.Eng., 1940, D.Eng., 1961, Tokyo Institute of Technology.

Fellow Award "For contributions to information and systems sciences, and to engineering education."

OKOSHI, TAKANORI, Fellow 1983. Born: September 16, 1932, Tokyo, Japan. Degrees: B.S., 1955, M.S., 1957, Ph.D., 1960, University of Tokyo.

Fellow Award "For contributions to lightwave and microwave engineering and, in particular, for the development of techniques for the analysis and synthesis of propagation in multimode fibers." Other Awards: Inada Memorial Award, IECE Japan, 1958; Okabe Memorial Award, IECE Japan, 1966; Hattori-Hoko Prize, Hattori Foundation, 1974; Paper Prize, IECE Japan, 1975; Book Prize, Institute of Television Engineers (ITE) of Japan, 1975; Book Prize, IECE Japan, 1978; Paper Prize, ITE Japan, 1979; Achievement Prize, IECE Japan, 1979; Achievement Prize, ITE Japan, 1981; Paper Prize, IECE Japan, 1983; Achievement Prize, IECE Japan, 1983.

OKRESS, ERNEST C., Fellow 1964. Born: March 9, 1910, Hamtramck, Mich. Degrees: B.E.E., 1935, University of Detroit; M.Sc., 1940, University of Michigan; Sc.D.(Thesis), 1974, Sussex Institute of Technology, England, with doctoral courses completed, summa cum laude, University of Michigan, New York University and Polytechnic Institute of Brooklyn.

Fellow Award "For contributions to microwave magnetron design." Other Awards: Certificate of Commendation, Office of Scientific Research and Development; Certificate of Recognition, NASA; Three Special Patent Awards for Outstanding Inventions, Westinghouse Electric Co.; Member at Large, National Science Honorary Society of Sigma Xi; Fellow Award, American Physical Society; National Council of Engineering Examiners' Award of Certificate; Registered Professional Engineer.

OKWIT, SEYMOUR, Fellow 1966. Born: August 31, 1929. Degrees: B.S., 1952, Brooklyn College; M.S., 1957, M.S., 1961, Adelphi University.

Fellow Award "For contributions to solid-state devices, particularly masers, parametric amplifiers, and ferrite devices"; Editor, IEEE Transactions on Microwave Theory and Techniques, 1966-1968; Member, GMTT Administrative Committee, IEEE, 1966-1971; President, Administrative Committee, IEEE Microwave Theory and Techniques Society, 1971; Member, IEEE Publications Board; Chairman, Finance Committee, IEEE Publications Board; Member, IEEE Budget Advisory Committee. National Lecturer, IEEE Microwave Theory and Techniques Society, 1974; Distinguished Service Award, IEEE Microwave Theory and Techniques Society, 1977.

OLDHAM, WILLIAM G., Fellow 1981. Born: May 5, 1938, Detroit, MI. Degrees: B.S., 1960, M.S., 1961, Ph.D., 1963, Carnegie Mellon University.

Fellow Award "For contributions to electron devices and large-scale integration technologies." Other Awards: Senior Postdoctoral Fellowship, National Science Foundation, 1970-1971.

OLINER, ARTHUR A., IRE Fellow 1961. Born: March 5, 1921, Shanghai, China. Degrees: B.A., 1941, Brooklyn College; Ph.D., 1946, Cornell University.

Fellow Award "For contributions to network representations of microwave structures"; Microwave Prize, 1966, "For many significant contributions to the field of endeavor of the IEEE-GMTT in numerous published papers, particularly the one entitled 'Equivalent Circuits for Discontinuities in Balanced Strip Transmission Line'"; National Lecturer, IEEE G-MTT, 1967; Honorary Life Member, IEEE Microwave Theory and Techniques Society, 1977; Microwave Career Award, IEEE, 1982. Other Awards: Institution Premium, Institution of Electrical Engineers, U.K., 1964; Guggenheim Fellow, 1965-66; Fellow, A.A.A.S., 1967; Outstanding Educators of America, 1973; Citation for Distinguished Research, Sigma Xi, 1974; Fellow, Institution of Electrical Engineers, U.K., 1980.

OLIVER, BERNARD M., IRE Fellow 1954. Born: May 27, 1916, Santa Cruz, Calif. Degrees: A.B., 1935, Stanford University; M.S.E.E., 1936, Ph.D., 1940, California Institute of Technology.

Fellow Award "For his contributions to communications, particularly in the field of information theory and coding systems"; Lamme Medal, IEEE, 1977, "For his contributions to the theory and practice of electronic instrumentation and measurements." Other Awards: Distinguished Alumnus Award, California Institute of Technology, 1972; Wright Prize, 1984.

OLSEN, KENNETH H., Fellow 1972. Born: February 20, 1926, Bridgeport, Conn. Degrees: B.S., 1950, M.S., 1952, Massachusetts Institute of Technology.

Fellow Award "For contributions to digital computer design, manufacturing, and marketing." Other Awards: Young Electrical Engineer of the Year, Eta Kappa Nu, 1960.

OLTMAN, H. GEORGE JR., Fellow 1983. Born: October 6, 1927, Brookhaven, Mississippi. Degrees: B.S., 1950, M.S., 1954, University of New Mexico.

Fellow Award "For contributions to antenna, microwave, and measurement technology"; Vice Chairman, Los Angeles Council, IEEE, 1977; President, IEEE Microwave Theory and Techniques Society, 1984. Other Awards: Lawrence A. Hyland Patent Award, Hughes Aircraft Company, 1981; Fellow, Institute for the Advancement of Engineering, 1974; Fellow, British Interplanetary Society, 1957; Registered Professional Engineer, New Mexico, 1956; Sigma Xi, 1965; Kappa Mu Epsilon, 1948.

OLYPHANT, MURRAY, JR., Fellow 1972. Born: May 2, 1923, New York, N.Y. Degrees: B.S., 1944, M.S.E., 1947, Princeton University.

Fellow Award "For contributions to the understanding of the breakdown mechanism of electrical insulation." Other Awards: 3M Carlton Society, 1976; Sigma Xi.

OMI, HANZO, Fellow 1966. Born: April 5, 1901, Ibaragi, Japan. Degrees: Eng., 1923, D.Eng., 1973, Tokyo Institute of Technology.

Fellow Award "For contributions to improvement of electrical communication devices and pioneering in manufacture of digital computers"; Founder's Medal, 1979. Other Awards: Mainichi Advancement Award, 1958; Distinguished Achievement Award, Institute of Electronics and Communications Engineers of Japan, 1964; Blue Ribbon Medal, Japanese Government, 1964; Third Order of Merit with Order of the Sacred Treasures, Japanese Government, 1971.

OMURA, JIM K., Fellow 1981. Born: September 8, 1940, San Jose, CA. Degrees: B.S., 1962, M.S., 1963, Massachusetts Institute of Technology; Ph.D., 1966, Stanford University.

Fellow Award "For contributions to information and communication theory as applied to communications systems design."

O'NEAL, RUSSELL D., IRE Fellow 1961. Born: February 15, 1914, Columbus, Ind. Degrees: A.B., 1936, DePauw University; M.S., Ph.D., 1941, University of Illinois.

Fellow Award "For leadership and technical administration in systems engineering." Other Awards: Fellow, American Physical Society, 1958; Associate Fellow, American Institute of Aeronautics and Astronautics, 1966; Fellow, AAAS, 1967; Michigan Man of Science, 1973; Engineering Society of Detroit.

O'NEIL, JAMES E., Fellow 1977. Born: April 1, 1929. Degrees: B.S.E.E., 1951, Washington State University; 1953-1959, University of Pittsburgh.

Fellow Award "For leadership in and management of innovative EHV projects." Other Awards: Member, Institute for Management Sciences; Tau Beta Pi.

O'NEILL, EUGENE F., Fellow 1969. Born: July 2, 1918, New York, N.Y. Degrees: B.S.E.E., 1940, M.S.E.E., 1941, Columbia University; D.Sc.(Hon.), 1963, Bates College; D.Eng.(Hon.), 1964, Politecnico di Milano, Italy; D.Sc.(Hon.), 1965, St. Johns University.

Fellow Award "For contributions and leadership in the engineering, development, and construction of international communication terminals and systems"; IEEE Award in International Communication in Honor of Hernand and Sosthenes Behn, 1971, "For outstanding technical innovations and management in the development of many key technologies underlying the present day international communications art, especially TELSTAR, the first operational telecommunications satellite, as well as his earlier contributions to transoceanic cable telephony." Other Awards: Tau Beta Pi; Sigma Xi; Member, National Academy of Engineering.

ONOE, MORIO, Fellow 1979. Born: March 28, 1926. Degrees: Engineer's Diploma, 1947, Ph.D., 1955, University of Tokyo.

Fellow Award "For contributions to the understanding of piezoelectric phemomena and the development of piezoelectric filters." Other Awards: C.B. Sawyer Award, Frequency Control Symposium, 1975; Award for Distinguished Achievement, Institute of Electronics and Communication Engineers of Japan, 1976; Niwa-Takayanagi Paper Award, Institute of Television Engineers of Japan, 1977.

OONO, YOSIRO, IRE Fellow 1962. Born: January 6, 1920, Fukuoka, Japan. Degrees: B.E., 1941, D.Eng., 1948, Kyushu University.

Fellow Award "For contributions to multiport network synthesis." Other Awards: Best Paper Award, Institute of Electronics and Communication Engineers of Japan, 1949, 1964, 1973, 1975; Asahi Press Prize, 1964.

OPPELT, WINFRIED, Fellow 1967. Born: June 5, 1912, Hanau, Germany. Degrees: Dipl.Ing., 1934, Dr.Ing., 1943, Darmstadt; Dr.Ing.(Hon.), 1965, Technische Hochschule Munich.

Fellow Award "For outstanding accomplishments in engineering education and contributions to the theory and application of automatic control.

OPPENHEIM, ALAN V., Fellow 1977. Born: November 11, 1937, New York, N.Y. Degrees: S.B., 1961, S.M., 1961, Sc.D., 1964, Massachusetts Institute of Technology.

Fellow Award "For contributions to digital signal processing and speech communications"; Senior Award, IEEE Group on Audio and Electroacoustics, 1969, "For authoring a paper of exceptional merit"; Technical Achievement Award, IEEE Group on Acoustics, Speech, and Signal Processing, 1977, "In recognition of fundamental contributions in the development of homomorphic processing techniques for speech and other signals"; Society Award, IEEE Acoustics, Speech, and Signal Processing Society, 1980, "In recognition of sustained leadership in the field of digital signal processing by your innovative research, writing of pioneering textbooks, and inspiring teaching". Other Awards: T. V. Shares Award for Excellence in Teaching, 1963; Guggenheim Fellow, 1972-1973; Cecil H. Green Distinguished Chair in Electrical Engineering, 1976-1978; M.I.T. Graduate Student Council Teaching Award, Dept. of EECS, 1981.

OPSAHL, ALERT M., AIEE Fellow 1948. Born: November 9, 1901, Seneca, Iowa. Degrees: A.B., 1924, Luther College.

Awards: Citation, 1966, "As a pioneer in the field of lightning and oscillography." Other Awards: Certificate, War Dept., 1945.

ORCHARD, H. J., Fellow 1968. Born: May 7, 1922, Oldbury, England. Degrees: B.Sc., 1946, M.Sc., 1951, London University.

Fellow Award "For contributions to the theory and engineering design of passive and active circuits"; Outstanding Paper Award, IEEE Group on Circuit Theory, 1969.

ORDUNG, PHILIP F., IRE Fellow 1959. Born: August 12, 1919, Luverne, Minn. Degrees: B.S., 1940, South Dakota State College; M.Eng., 1949, Yale University.

Fellow Award "For contributions to network theory and to electrical engineering education." Other Awards: Award, Yale Engineering Association, 1953; Fellow, Timothy Dwight College, Yale University, 1959; Fellow, American Association for the Advancement of Science, 1959; Sorenson Fellow, 1962; Member, New York Academy of Science, 1975; Regent, California Lutheran College, 1975.

OSATAKE, TONAU, Fellow 1975. Born: October 12, 1916. Degrees: B.E., 1941, D.Eng., 1953, University of Tokyo.

Fellow Award "For contributions to communication systems design." Other Awards: Akiyama and Shida Memorial Award (Prize Paper Award), Institute of Electronic and Communication Engineers of Japan, 1954; Member, Institute of Electronic Engineers of Japan; Honorary Member, Institute of Electronic and Communication Engineers of Japan; Member, Institute of Television Engineers of Japan; His Majesty Purple Ribbon Medal Award, Emperor of Japan, 1981.

OSBON, W. ORAN, AIEE Fellow 1948. Born: November 26, 1904, Indianapolis, Ind. Degrees: B.S.E.E., 1926, Purdue University.

OSBORNE, HAROLD S., AIEE Fellow 1921, IRE Fellow 1945. Born: August 1, 1887, Fayetteville, N.Y. Degrees: B.S., 1908, Eng.D., 1910, Massachusetts Institute of Technology; D.Litt.(Hon.), 1966, Montclair State College.

IRE Fellow Award "For his contributions in the electrical communication field including outstanding leadership and direction in the application of new techniques to telephony." Awards: Edison Medal, AIEE, 1960. Other Awards: Fellow, Acoustical Society of America; Fellow, American Physical Society; Fellow, American Association for the Advancement of Science; Fellow, American Society for Engineering Education; Honorary Member, Tau Beta Pi, 1947; Steinmetz Lecturer, Steinmetz Memorial Foundation, 1953; Eminent Member, Eta Kappa Nu, 1955; Knight, Royal Order of Vasa, Sweden, 1955; Howard Coonley Medal, American Standards Association, 1956; 75th Anniversary Medal, American Society of Mechanical Engineers, 1956; Special Award, Joint Council of Municipal Planning Boards, Essex County, New Jersey, 1957; Special Award, Engineers Joint Council, 1958; Award, Regional Plan Association of New York, 1959; Honorary Member, Montclair Society of Engineers, 1960; Medalist, First Golden Plate Banquet of Academy of Achievement, 1961; Silver Medal Award, American Society of Planning Officials, 1962; Honorary Life Fellow, Standards Engineers Society, 1963; Certificate of Appreciation, Mayor of Graz, Austria, 1965; Honorary Member, Institute of Electronics and Communications Engineers of Japan, 1965; Community Service Award, Montclair Chamber of Commerce, 1965; Citizenship Award, Service Clubs Council of Montclair, 1976.

OSCARSON, GERHARD L., AIEE Fellow 1955. Born: April 27, 1900, Wheaton, Minn. Degrees: B.S.E.E., 1922, E.E., 1955, University of Minnesota.

Fellow Award "For engineering development and coordination of the design of rotating electric machinery to the changing requirements of industry and national defense."

OSEPCHUK, JOHN M., Fellow 1978. Born: February 11, 1927. Degrees: A.B.(magna cum laude), 1949, A.M., 1950, Ph.D., 1957, Harvard University.

Fellow Award "For contributions to microwave technology and to microwave safety"; National Lecturer, IEEEMTT-S, 1977-1978. Other Awards: Fellow, International Microwave Power Institute, 1978.

OSHIMA, SHINTARO, Fellow 1976. Born: August 5, 1914, Kyushu, Japan. Degrees: B.E., 1940, D.E., 1962, Kyushu University.

Fellow Award "For contributions to magnetic thin-film memory, and for leadership in the field of electrical communications." Other Awards: Prize of Science and Technology Agency Minister, 1965; Cultural Decoration of the Purple Ribbon, 1966; Prize of the Governor of Tokyo, 1966; Achievements Prize of the Institute of Electronics and Communication Engineers, 1972; Services Prize of the Institute of Electronics and Communication Engineers, 1975.

OSTERLE, WILLIAM H., AIEE Fellow 1951. Born: August 1, 1898, Xenia, Ohio. Degrees: B.S., 1921, E.E., 1925, University of Santa Clara.

O'SULLIVAN, G. H., AIEE Fellow 1948. Born: September 21, 1909, New York, N.Y. Degrees: B.S., 1931, E.E., 1941, Cooper Union.

Other Awards: Illuminating Engineering Society; Institution of Electrical Engineers, U.K.

OUCHI, ATSUYOSHI, Fellow 1982. Born: October 10, 1919, Tokyo, Japan. Degrees: B.Eng., 1942, D.Eng., 1962, Tokyo, University.

Fellow Award "For leadership in the application of semiconductor devices and in the development of VLSI technology." Other Awards: Special invention Prize, Kanagawa Branch, Invention Society of Japan, 1962; All Japan Invention Prize, Invention Society of Japan, 1963; First Prize, Machine Development Society, Japan, 1966; Okochi Memorial Grand Production Prize, Okochi Memorial Foundation, 1969; Japan Industrial Technology Grand Prize, The Prime Minister, 1974.

OUIMET, J. ALPHONSE, IRE Fellow 1952. Born: June 12, 1908, Montreal, Quebec, Canada. Degrees: B.A., 1928, College Ste-Marie, Montreal; B.Eng., 1932, McGill University; D.A.Sc.(Hon.), 1957, University of Montreal; D.C.L.(Hon.), 1962, Acadia University; LL.D.(Hon.), 1962, University of Saskatchewan; LL.D.(Hon.), 1963, McGill University; D.Soc.Sc.(Hon.), 1967, University of Ottawa; D.Adm.(Hon.), 1967, University of Sherbrooke; LL.D.(Hon.), 1974, Royal Military College of Canada; D.es Arts, Universite Laval, 1978.

Fellow Award "For his engineering contributions in the development and direction of radio and television in Canada"; McNaughton Medal, 1972; International Public Service Award, IEEE Communications Society, 1978. Other Awards: Ross Medal, Engineering Institute of Canada, 1948; Archambault Medal, Association Canadienne-Francaise pour l'Avancement des Sciences, 1958; Julian C. Smith Medal, Engineering Institute of Canada, 1959; "Companion of the Order of Canada," 1969; Sir John Kennedy Medal, Engineering Institute of Canada, 1969; Special Award, Society of Motion Picture and Television Engineers, 1974; Canadian Engineers Gold Medal, 1975; International EMMY Directorate Award, 1976; The "Great Montrealer" Award, 1978; The Quebec Government Television Award, 1983.

OUYANG, MID, Fellow 1980. Born: May 14, 1921. Degrees: B.S.E.E., 1944, National Amoy University, China; Ph.D.(Electrical Engineering), 1952, Durham University.

Fellow Award "For contributions to high-voltage testing and measuring techniques, and to statistical insulation testing."

OWENS, C. DALE, Fellow 1966. Born: May 15, 1906, Wadesville, Ind. Degrees: A.B., 1928, Indiana University; M.A., 1936, Columbia University.

Fellow Award "For contributions in magnetic measurements, materials and components, and leadership in national and international standards on magnetics."

OWENS, JAMES B., Fellow 1969. Born: July 23, 1920. Degrees: B.S.E.E., 1941, Rice University.

Fellow Award "For significant contributions to the science of circuit protection, and for guidance and contributions on the formulation of U.S. and International Standards."

OZAKI, HIROSHI, Fellow 1975. Born: February 13, 1920, Osaka, Japan. Degrees: B.E., 1942, D.Eng., 1955, Osaka University.

Fellow Award "For contribution to circuit theory and engineering education." Other Awards: Paper Award, 1975, Achievement Award, 1976, Distinguished Services Awards, 1983, Institute of Electronics and Communications Engineers of Japan.

P

PACKARD, DAVID, IRE Fellow 1948. Born: September 7, 1912, Pueblo, Colo. Degrees: B.A., 1934, E.E., 1939, Stanford University; Sc.D.(Hon.), 1964, Colorado College; LL.D.(Hon.) 1966, University of California; LL.D.(Hon.), 1970, Catholic University; D.Litt.(Hon.), 1973, Southern Colorado State College; D.Eng.(Hon.), 1974, University of Notre Dame.

Fellow Award "For his initiative in the development of special radio testing and measuring techniques; Founders Medal (with Hewlett), IEEE, 1973. Other Awards: Herbert Hoover Medal, Stanford Alumni Association, 1966; Man of the Year, Peninsula Manufacturers' Association, 1969; Gantt Medal Award, American Management Association, 1970; Crozier Gold Medal, American Ordnance Association, 1970; Distinguished Alumnus of the Year, Alpha Delta Phi, Stanford University, 1970; Distinguished Citizen Award, Palo Alto (Calif.) Chamber of Commerce, 1972; James Forrestal Memorial Award, National Security Industrial Association, 1972; Very Distinguished Public Service Award, Federal City Club, Washington, D.C., 1972; Business Statesman of the Year Award, Harvard Business School of New York, 1972; The Silver Quill Award, American Business Press, Inc., 1972; Grand Cross of Merit, Federal Republic of Germany, 1972; Benjamin Fairless Memorial Medal, American Iron and Steel Institute, 1972; Business Statesman of the Year, Harvard Business School of Northern California, 1973; Industrialist of the Year (with William R. Hewlett), California Museum of Science and Industry and the California Museum Foundation, 1973; Silver Helmet Defense Award, AMVETS, 1973; Medal of Honor, Electronic Industries Association, 1974; The Washington Award, Western Society of Engineers, 1975; Hoover Medal, presented by the American Society of Civil Engineers, American Institute of Mining, Metallurgical and Petroleum Engineers, American Society of Mechanical Engineers and IEEE, 1975; Award (with William R. Hewlett), Scientific Apparatus Makers Association, 1975; Gold Medal Award, National Football Foundation and Hall of Fame, 1975; Vermilye Medal (with William R. Hewlett), Franklin Institute, 1976; Award for International Achievement, World Trade Club of San Francisco, 1976; Award of Merit, American Consulting Engineers Council Fellows, 1977; Achievement in Life Award, Encyclopedia Britannica, 1977; Founders Award, National Academy of Engineering, 1979; 1980 Engineering Award of Distinction, San Jose State University, 1980; Thomas D. White National Defense Award, United States Air Force Academy, 1981; Distinguished Information Sciences Award, Data Processing Management Association, 1981; "1981 Citizen of the Year," Police Activities League, San Jose; John Fritz Medal, Founders Society, WESCON, 1982; Sylvanus Thayer Award, United States Military Academy, West Point, New York, 1982; Environmental Leadership Award, Natural Resources Defense Council, 1983.

PACKER, LEWIS C., Fellow 1966. Born: February 21, 1893, Wellsville, Ohio. Degrees: B.E.E., Ohio State University.

Fellow Award "For contributions in the design and application of universal motors and low-loss induction motors"; Past Member, Rotating Machinery and Fractional Horsepower Committees. Other Awards: Order of Merit and Silver W Award, Westinghouse Electric Corp., 1955; Silver Bowl Citation, by Planning Board, City of East Longmeadow, Massachusetts; American Association of Engineering Societies; "Who's Who in Engineering," 1980, 1982; "Who's Who," by Cambridge, England Pub.

PAGANELLI, THOMAS I., Fellow 1969. Born: October 13, 1918, Cle Elum, Wash. Degrees: B.S.E.E., 1942, University of California.

Fellow Award "For engineering and management activities which led to the development of large-scale military electronic systems." Other Awards: Air Force Legion of Merit; Sigma Xi; Tau Beta Pi; Eta Kappa Nu.

PAGE, CHESTER H., IRE Fellow 1956. Born: November 13, 1912. Degrees: A.B., 1934, Sc.M., 1934, Brown University; Ph.D., 1937, Yale University.

Fellow Award "For contributions to military electronic research and development"; Harry Diamond Memorial Award, 1974, "For outstanding technical leadership and contributions to ordnance electronics, metrology and international standardization of electrical quantities and symbols." Other Awards: Exceptional Service Certificate, Bureau of Ordnance; Gold Medal Award, U.S. Dept. of Commerce; Certificate of Appreciation, War Dept., Office of Scientific Research and Development; Fellow, American Physical Society.

PAGE, ROBERT M., IRE Fellow 1957. Born: June 2, 1903, St. Paul, Minn. Degrees: B.S., 1927, Hamline University; M.A., 1932, George Washington University; D.Sc.(Hon.), 1943, Hamline University.

Fellow Award "In recognition of his pioneering achievements in solving some of the early problems of basic importance to radar"; Harry Diamond Award, IRE, 1953, "For outstanding contributions to the development of radar through pioneering work and through sustained efforts over the years"; Pioneer Award, IEEE Aerospace and Electronic Systems Society, 1977. Other Awards: Distinguished Civilian Service Award, U.S. Dept. of Navy, 1954; Presidential Certificate of Merit, 1946; Sigma Xi, 1950; Scientific Research Society of America, 1951; Fellow, American Association for the Advancement of Science, 1952; Stuart Ballantine Medal, Franklin Institute, 1957; Sigma Tau, 1959; Capt. Robert Dexter Conrad Award, U.S. Dept. of Navy, 1959; Centennial Award, Wheaton College, 1960; Federal Distinguished Civilian Service Award, 1960; Fellow, American Scientific Affiliation, 1960; Fellow, Washington Academy of Sciences, 1962; W. Randolf Lovelace II Award, 1967; Alumni Achievement Award, George Washington University, 1967; Bloomington (Minnesota) Bicentennial Distinguished Citizen Award, 1976; Book of Golden Deeds Award, National Exchange Club, 1979; Minnesota Inventors Hall of Fame, 1979.

PAKALA, WILLIAM E., AIEE Fellow 1951. Born: July 2, 1901, Red Lodge, Mont. Degrees: B.S., 1927, Montana State College.

Fellow Award "For research and development of ignitron tubes and their firing circuits, radio noise instrumentation and measurements of electrical apparatus and transmission lines, and prediction of radio noise from transmission lines"; Second Prize Paper, AIEE, District 6, "For Leadville High Altitude EHV Test Project, Part 3 Radio Influence Investigations." Other Awards: ANSI Radio-Electrical Coordination; Patent Awards, Westinghouse Electric Corporation, 1936-64.

PALMER, H. B., AIEE Fellow 1951. Born: August 9, 1900, Keota, Iowa. Degrees: B.S.E.E., 1923, M.S.E.E., 1927, E.E., 1927, University of Colorado.

Other Awards: Tau Beta Pi; Eta Kappa Nu; Sigma Xi; Sigma Tau.

PALMER, JAMES DANIEL, Fellow 1982. Born: March 8, 1930. Degrees: A.S., 1953, Fullerton Junior College; B.S.E.E., M.S.E.E., 1957, University of California at Berkeley; Ph.D., 1963, University of Oklahoma; Doctor of Public Service (honoris causa), 1977, Regis College.

Fellow Award "For contributions to applying systems engineering and systems sciences techniques to the study of large-scale socioeconomic systems."

PALMER, WINSLOW, IRE Fellow 1956. Born: July 8, 1912, New Brunswick, N.J. Degrees: B.S., 1937, University of Hawaii; E.E., 1939, Stanford University.

Fellow Award "For contributions to the theory and practice of radio navigation"; Pioneer Award, IEEE Aerospace and Electronic Systems Society, 1971. Other Awards: Certificate of Commendation, U.S. Dept. of Navy, Bureau of Ships; Thurlow Award, Institute of Navigation, 1967; BASIC Loran C System Patents 2,-811,717; 2,728,909.

PAN, WEN Y., IRE Fellow 1958. Born: July 15, 1912, Soochow, China. Degrees: B.S.E.E., 1935, Chiao Tung University; E.E., 1939, Ph.D., 1940, Stanford University.

Fellow Award "For his contributions to the advancement of radio broadcast and television receiver design." Other Awards: Professional Merit Awards, China Ministry of Communications, 1936, 1937; Merit Award, Office of Scientific Research and Development, 1944; Professional Achievements Award, C.I.E., 1961; Best Conference Paper Award, Electronic Components Conference, 1966, 1968; Best Paper Award, RCA, DCSD, 1968; Patent Recognition Award, RCA, 1974; Seminar Award on Engineering and Technology, 1976; Telecommunications Management Award, 1977; Contribution to Science Award, National Science Council, ROC, 1979; Outstanding Achievement Award, Industrial Technology Research Institute, ROC, 1980; Most Outstanding Service Award, United Microelectronics, 1982; Distinct Contributions Award, ERSO/ITRI, 1983.

PANKOVE, JACQUES I., Fellow 1965. Born: November 23, 1922, Russia. Degrees: B.S., 1944, M.S., 1948, University of California; Ph.D., 1960, University of Paris.

Fellow Award "For contributions to electrical and optical semiconductor research and development"; J.J. Ebers Award, IEEE, 1975. Other Awards: Achievement Awards, Radio Corporation of America, 1952, 1953, 1963; David Sarnoff Scholar, 1956; Fellow, American Physical Society, 1967.

PANNENBORG, ANTON EDUARD, Fellow 1980. Born: March 5, 1922, The Hague, The Netherlands. Degrees: Dr., 1952, Technical University of Delft, The Netherlands; Dr. Ing. (Hon.), 1977, Rheinisch-Westphalische Hochschule, Aachen, Germany.

Fellow Award "For leadership in the management of research and development."

PANSINI, ANTHONY J., AIEE Fellow 1954. Born: November 19, 1909, New York, N.Y. Degrees: B.S., 1931, E.E., 1938, Cooper Union.

Fellow Award "For contributions to the development of transmission and distribution systems capable of serving adequately a load of exceptionally rapid growth."

PANTELL, RICHARD H., Fellow 1977. Born: December 25, 1927, New York, N.Y. Degrees: B.S. and M.S., 1950, Massachusetts Institute of Technology; Ph.D., 1954, Stanford University.

Fellow Award "For contributions to electron devices and quantum electronics." Other Awards: Senior Member, Sigma Xi, 1955; Fulbright Award, 1965; Fellowship Grant, Science Research Council of the United Kingdom, 1976.

PAO, YOH-HAN, Fellow 1978. Born: July 17, 1922, Kiukiang, China. Degrees: B.S., 1945, Henry Lester Institute for Technical Education, Shanghai, China; Ph.D.(Applied Physics) 1952, Pennsylvania State University.

Fellow Award "For contributions to laser and electrooptic research and for leadership in engineering education." Other Awards: "Nuclear USA" Commendation, Atomic Energy Commission, 1970; Merit Award, CECON, Inc., 1978.

PAPIAN, WILLIAM N., IRE Fellow 1962. Born: July 27, 1916, New York, N.Y. Degrees: S.B., 1948, S.M., 1950, Massachusetts Institute of Technology.

Fellow Award "For pioneering the engineering development of the magnetic-core memory and for continuous leadership in the evolution of a series of advanced high-speed digital computers."

PAPOULIS, ATHANASIOS, Fellow 1967.

Fellow Award "For contributions to the understanding of probability and stochastic processes in engineering education."

PAPPENFUS, ERNEST W., IRE Fellow 1962. Born: September 23, 1917, Raymond, Mont. Degrees: B.E.E., 1941, University of Minnesota.

Fellow Award "For contributions to single sideband communications and to multichannel frequency control systems."

PARIS, DEMETRIUS T., Fellow 1975. Born: September 27, 1928, Stavroupolis, Thrace, Greece. Degrees: B.S.E.E., 1951, Mississippi State University; M.S.E.E., 1958, Ph.D., 1962, Georgia Institute of Technology.

Fellow Award "For leadership in engineering education and for the advancement of electromagnetic theory"; Achievement Award, IEEE Education Society, 1980; IEEE Centennial Medal, 1984. Other Awards: First Prize Paper by Member of Georgia Tech Faculty, Sigma Xi, 1965.

PARIS, LUIGI, Fellow 1981. Born: November 30, 1927, Pisa, Italy. Degrees: Dr. (Indust. Eng.), 1950, University of Pisa; Libera Docenza Impianti Elettrici, 1966.

Fellow Award "For contributions to the design of electric transmission systems and for leadership in their research." Other Awards: Premi "AEI-IEL" 1966, 1972; Premio "AEI Pubblicazioni" 1982.

PARK, ROBERT H., Fellow 1965. Born: March 15, 1902 (as U.S. citizen), Strassbourg, Germany. Degrees: B.S., 1923, Massachusetts Institute of Technology.

Fellow Award "For contributions to the theory of synchronous machinery and the stability of high voltage transmission systems"; National First Paper Prize, AIEE, 1930, for "Circuit Breaker Recovery Voltages"; Lamme Medal Award, 1972, "In recognition of his outstanding contributions to analysis of the transient behavior of ac machines and systems." Other Awards: Distinguished Civilian Service Award, U.S. Dept. of Navy, 1945.

PARKER, DON, Fellow 1982. Born: January 14, 1933, Ogden, UT. Degrees: B.S.E.E., 1956, Brigham Young University; M.S., 1957, Harvard University; D.Sc., 1964, Massachusetts Institute of Technology.

Fellow Award "For technical direction of the development of high-power solid-state sources."

PARKER, DONALD J., Fellow 1971. Born: 1926, Rochester, New York. Degrees: B.S., 1950, University of Rochester.

Fellow Award "For contributions to the field of electrooptical engineering." Other Awards: Member, Optical Society of America, 1949; Member, Society of Motion Picture and Television Engineers, 1950; Member, American Physical Society.

PARKER, NORMAN F., Fellow 1977. Born: May 14, 1923, Fremont, Nebr. Degrees: B.S., 1947, M.S., 1947, D.Sc., 1948, Carnegie-Mellon University.

Fellow Award "For achievements in inertial navigation and leadership in engineering management." Other Awards: Merit Award, Carnegie-Mellon Alumni Association, 1969; Member, National Academy of Engineering, 1976; Fellow, American Institute of Aeronautics and Astronautics, 1980.

PARKER, NORMAN W., Fellow 1974. Born: November 3, 1922, Brewster, Ohio.

Fellow Award "For contributions to the design of color television receivers"; IEEE Consumer Eelectronics Award, 1971, "For outstanding contributions to consumer electronics"; Two First Prize Papers, Consumer Electronics Group.

PARKER, SYDNEY R., Fellow 1975. Born: April 18, 1923, New York, N. Y. Degrees: B.E.E., 1944, City College of The City University of New York, M.S., 1948, Sc.D., 1964, Stevens Institute of Technology.

Fellow Award "For contributions to circuit and system theory"; Harry Diamond Memorial Award, IEEE, 1984. Other Awards: Carl E. Menneken Research Award, Sigma Xi, 1977; Tau Beta Pi; Eta Kappa Nu; Sigma Xi.

PARKS, ROBERT J., Fellow 1966. Born: April 1, 1922, Los Angeles, Calif. Degrees: B.S.E.E., 1944, California Institute of Technology.

Fellow Award "For contributions to the conception, design and development of missile and spacecraft systems." Other Awards: Man of the Year Award, La Canada, 1958; Public Service Award, NASA, 1963; Louis W. Hill Space Transportation Award, American Institute of Aeronautics and Astronautics, 1963; Fellow, American Institute of Aeronautics and Astronautics, 1965; Man of the Year Award, California Air Force Association Aerospace Industry, 1966; Exceptional Service Medal, NASA, 1967; Member, National Academy of Engineering, 1973; Distinguished Service Medal, NASA, 1980; AIAA Goddard Astronautics Award, 1980.

PARKS, THOMAS W., Fellow 1982. Born: March 16, 1939, Buffalo, NY. Degrees: B.E.E., 1961, M.S., 1964, Ph.D., 1967, Cornell University.

Fellow Award "For fundamental contributions to signal processing and digital filter design"; Technical Achievement Award, IEEE Acoustics, Speech, and Signal Processing Society, 1981. Other Awards: Senior Scientist Award, Alexander von Humboldt Foundation, 1973; Senior Fulbright Fellow, 1973-1974.

PARZEN, BENJAMIN, Fellow 1970. Born: April 5, 1913, Poland. Degrees: B.S.E., 1936, City College of The City University of New York.

Fellow Award "For contribution to the theory and practice of wide-range frequency synthesizers and to the precision measurement of signals of high spectral purity."

PATEL, ARVIND M., Fellow 1984. Born: October 19, 1937, Isnav (Gujarat), India. Degrees: B.E., 1959, Sardar Vallabh-bhai University, India; M.S., 1961, University of Illinois, Urbana; Ph.D., 1969, University of Colorado.

Fellow Award "For contributions to data encoding/decoding and error correction and their application to magnetic storage devices." Other Awards: Outstanding Technical Paper Award, AFIPS, 1970; Three Outstanding Innovation Awards, IBM, 1973, 1975, 1983.

PATEL, C. KUMAR N., Fellow 1975. Born: July 2, 1938, Baramati, India. Degrees: B.E., 1958, Poona University; M.S.E.E., 1959, Ph.D., 1961, Stanford University.

Fellow Award "For contributions to the development of the carbon dioxide and spin-flip Raman lasers"; Lamme Medal, IEEE, 1976, "For the invention and development of the carbon dioxide and spin-flip Raman lasers and for contributions to infrared spectroscopy of solids and gases." Other Awards: Ranade Prize, Poona University, 1956; P.T. Bastikar Memorial Prize, Poona University, 1958; Adolph Lomb Medal, Optical Society of America, 1966; Fellow, American Physical Society, 1966; Ballantine Medal, Franklin Institute, 1968; Coblentz Award, Coblentz Society, American Chemical Society, 1974; Member, National Academy of Sciences, 1974; Honor Award, Association of Indians in America, 1975; Zworykin Award, National Academy of Engineering, 1976; Fellow, American Academy of Arts and Sciences, 1976; Fellow, Optical Society of America, 1976; Member, National Academy of Engineering, 1978; Founders' Award, Texas Instruments, 1978; Fellow, American Association for the Advancement of Science, 1980; Honorary Member, Gynecologic Laser Society, 1981; Townes Award, Optical Society of America, 1982; NY Section Award, Society for Applied Spectroscopy, 1982; Honorary Fellow, Institution of Electronics and Telecommunications Engineers (India), 1983; Honorary Fellow, Indian National Science Academy, 1983.

PATTON, ALTON D., Fellow 1980. Born: February 1, 1935, Corpus Christi, TX. Degrees: B.S.(E.E.), 1957, University of Texas at Austin; M.S.(E.E.), 1961, University of Pittsburgh; Ph.D., 1972, Texas A & M University.

Fellow Award "For contributions to power system reliability analysis and assessment"; Prize Paper Award, Industrial & Commercial Power Systems Department, IEEE Industry Applications Society, 1975. Other Awards: Outstanding Young Engineer, Brazos Chapter, TSPE, 1970; Eta Kappa Nu; Tau Beta Pi; Sigma Xi.

PATTON, WILLARD T., Fellow 1979. Born: April 20, 1930, Schenectady, N.Y. Degrees: B.S.E.E., 1952, M.S., 1958, University of Tennessee; Ph.D., 1963, University of Illinois.

Fellow Award "For contributions to the development of phased-array antenna technology." Other Awards: Eta Kappa Nu, 1950; Tau Beta Pi, 1951; Phi Kappa Phi, 1952; Sigma Xi, 1962; RCA MSRD Annual Technical Excellence Award, 1966; David Sarnoff Outstanding Technical Achievement Award, RCA, 1975.

PATWARDHAN, PRABHAKAR K., Fellow 1976. Born: December 2, 1927, Indore, Madhya Pradesh, India. Degrees: M.Sc.(Physics), 1949, Ph.D. (Physics, Electronics and Instrumentation), Banaras Hindu University.

Fellow Award "For contributions to nuclear electronics, data-handling systems, and instrumentation." Other Awards: National Award for Engineering Technology (Citation and Cash Award), FIE Foundation, India, 1977; Dr. A. N. Khosla National Award (Citation, Gold Medal, and Cash Award), University of Roorkee, 1977; Federation of Indian Chambers of Commerce and Industries National Award for 1979; Fellow, Maharashtra Academy of Sciences, 1977; National Representative of the IERE (UK) Council in India; Member, National Council of Institution of Engineers (India) and Institution of Electronic and Telecommunication Engineers, (India); Chairman, Technology Transfer Group, Bhabha Atomic Research Centre and Head, Computer Centre, Bhabha Atomic Research Centre; Member, High Level Committee for Technological Research and

its Utilization of the Government of Maharashtra.

PAVLIDIS, THEDOSIOS, Fellow 1979. Born: September 8, 1934, Salonica, Greece. Degrees: Diploma, 1957, Tech. University of Athens; M.S., 1962, Ph.D., 1964, University of California, Berkeley.

Fellow Award "For contributions to the theory and application of pattern recognition."

PAYNE, LEONARD S., IRE Fellow 1949. Born: February 28, 1887, London, England.

Fellow Award "For his contributions in Canada to the field of International Communications." Other Awards: Institution of Electrical Engineers, (U.K.).

PEAIRS, MARSDEN H., AIEE Fellow 1962. Born: August 18, 1904, Lawrence, Kans. Degrees: UCLA pre-engineering.

Fellow Award "For contributions to development of multi-generator automatic-paralleling ac electrical power systems for large aircraft." Other Awards: Registered Electrical Engineer, California.

PEAKE, HAROLD J., Fellow 1971. Born: December 7, 1920, Norton, Va. Degrees: B.S., 1942, Virginia Polytechnic Institute; M.S., 1953, University of Maryland; M.Engrg.Adm., 1969, George Washington University.

Fellow Award "For contributions to spacecraft technology and electronic instrumentation." Other Awards: Eta Kappa Nu, 1942; Sigma Xi, 1953.

PEARCY, NOAH C., AIEE Fellow 1943. Born: December 5, 1898, Jeffersonville, Ind. Degrees: B.S.E.E., 1922, Purdue University.

PEARSON, GERALD L., Fellow 1964. Born: March 31, 1905, Salem, Oreg. Degrees: A.B., 1926, Willamette University; M.A., 1929, Stanford University; Sc.D., 1956, Willamette University.

Fellow Award "For contributions to semiconductor physics and to the development of the silicon solar cell." Other Awards: John Scott Award, City of Philadelphia, 1956; John Price Wetherill Medal, Franklin Institute, 1963; Gold Plate Award, American Academy of Achievement, 1963; Member, National Academy of Engineering, 1968; Member, National Academy of Sciences, 1970; Marian Smoluchowski Medal, Polish Physical Society, 1975; Fellow, American Physical Society; Sigma Xi; Solid State Science and Technology Award, Electrochemical Society, 1981; Gallium Arsenide Symposium Award, 1981; Alfried Krupp von Bohlen und Halbach Energy Research Prize, Krupp Foundation, 1981; Heinrich Welker Medallion, Siemens AG, 1983.

PEARSON, J. BOYD, JR., Fellow 1979. Born: June 3, 1930, McGehee, Ark. Degrees: B.S.E.E., 1958, M.S.E.E., 1959, University of Arkansas; Ph.D., 1962, Purdue University.

Fellow Award "For contributions to the theory of multivariable control systems."

PEASE, MARSHALL C., 3RD, IRE Fellow 1962. Born: July 30, 1920, New York, N.Y. Degrees: B.S., 1940, Yale University; M.A., 1943, Princeton University.

Fellow Award "For contributions to microwave tubes." Other Awards: Phi Beta Kappa; Sigma Xi.

PEATFIELD, ROBERT R., AIEE Fellow 1960. Born: November 1, 1906, Dorchester, Mass. Degrees: B.S., 1928, Massachusetts Institute of Technology.

Fellow Award "For contributions to coordinated design of thermal and hydraulic generation."

PECK, D. STEWART, Fellow 1966. Born: October 19, 1918, Grand Rapids, Mich. Degrees: B.S.E., 1939, M.S., 1940, University of Michigan.

Fellow Award "For contributions to gaseous electronics and to semiconductor device reliability."

PEDEN, IRENE C., Fellow 1974. Born: September 25, 1925, Topeka, Kans. Degrees: B.S.E.E., 1947, University of Colorado; M.S.E.E., 1958, Ph.D., 1962, Stanford University.

Fellow Award "For contributions to radioscience in the polar regions, and for leadership of women in engineering"; IEEE Antennas and Propagation Society "Man of the Year" (sic) 1979. Other Awards; Achievement Award, Society of Women Engineers, 1973; Distinguished Engineering Alumnus Award, University of Colorado, 1974; ABET Engineering Accreditation Commission Special Award, 1983; Who's Who in Engineering; Who's Who in America.

PEDERSEN, AAGE, Fellow 1984. Born: June 23, 1921, Odense, Denmark. Degrees: M.Sc.(E.E.), 1948, The Technical University of Denmark.

Fellow Award "For contributions to the breakdown mechanism of compressed gas insulation." Other Awards: Member, Danish Academy of Technical Sciences, 1970.

PEDERSON, DONALD O., Fellow 1964. Born: September 30, 1925, Hallock, Minn. Degrees: B.S., 1948, North Dakota State University; M.S., 1949, Ph.D., 1951, Stanford University; Dr. (Applied Science) (Hon.), 1979, Katholieke University of Leuven, Belgium.

Fellow Award "For contributions to circuit theory and engineering education"; Best Paper Prize, International Solid State Circuits Conference, 1963; Certificate of Appreciation, International Solid State Circuits Conference, 1966. Other Awards: Guggenheim Fellow, 1964; Education Medal, 1969; National Academy of Engineering, 1974; National Academy of Sciences, 1982.

PEIRCE, W. THEODORE, AIEE Fellow 1949. Born: July 20, 1900, Arlington, Mass. Degrees: A.B., 1921, Harvard University; B.S., 1922, Harvard Engineering School.

PEIXOTO, CARLOS, A. O., Fellow 1981. Born: June 26, 1933, Fortaleza-Ceara, Brazil. Degrees: E.E., 1959, Federal School of Engineering, Brazil.

Fellow Award "For leadership in the planning and design of electrical transmission systems for Brazil and for applications of analytic approaches in equipment specification."

PENDLETON, WESLEY W., AIEE Fellow 1962. Born: April 2, 1914, Providence, R.I. Degrees: B.S.E.E., 1936, University of Rhode Island; M.S.E.E., 1940, Massachusetts Institute of Technology.

Fellow Award "For contributions to the development of ultra high-temperature electrical insulation"; Paper Prize, Electrical Insulation Conference, 1963; Various Offices in E/EIC; 22 patents.

PENFIELD, PAUL, JR., Fellow 1972. Born: May 28, 1933, Detroit, Mich. Degrees: B.A., 1955, Amherst College, Sc.D., 1960, Massachusetts Institute of Technology.

Fellow Award "For contributions to varactor theory and to the electrodynamic theory of moving media."

PENN, WILLIAM B., Fellow 1975. Born: May 2, 1917, Lawrence, Mass. Degrees: B.S., 1937, M.S., 1938, Massachusetts Institute of Technology.

Fellow Award "For contributions to the development of very-high-temperature insulation systems"; President, IEEE Electrical Insulation Society, 1981, 1982.

PENNEY, GAYLORD W., AIEE Fellow 1945. Born: November 20, 1898, Stacyville, Iowa. Degrees: B.S., 1923, Iowa State University; M.S., 1929, University of Pittsburgh, Dr.Eng.(Hon.), 1980, Carnegie-Mellon University.

Other Awards: Order of Merit, Westinghouse Electric, 1937; John Price Wetherill Medal, Franklin Institute, 1951; Frank A. Chambers Award, Air Pollution Control Association, 1978; Professional Achievement Citation, Iowa State University, 1979.

PERKINS, JOSEPH R., Fellow 1966. Born: December 24, 1911, Richmond, Va. Degrees: B.S., 1933, University of Richmond; E.E., 1935, Princeton University.

Fellow Award "For leadership in research on the effects of electrical phenomena on insulating material." Other Awards: Fellow, American Society for Testing and Materials, 1976; Hall of Fame, Electrical/Electronics Insulation Conference, 1977.

PERKINS, WILLIAM R., Fellow 1972. Born: September 1, 1934, Council Bluffs, Iowa. Degrees: A.B., 1956, Harvard University; M.S., 1957, Ph.D., 1961, Stanford University.

Fellow Award "For contributions to sensitivity and feedback control theory."

PERPER, LLOYD J., Fellow 1971. Born: April 23, 1921, New York, N.Y. Degrees: S.B., 1941, Massachusetts Institute of Technology; M.Sc., 1948, Ohio State University.

Fellow Award "For contributions to air navigation and traffic control." Other Awards: Public Service Award, Arizona Council of Engineering and Scientific

Associations, 1973.

PERRY, JAMES S., Fellow 1972. Born: November 30, 1926, New York. Degrees: B.E.E., 1949, Rensselaer Polytechnic Institute.

Fellow Award "For his leadership in the development and application of radars to air traffic control." Other Awards: Pioneer Award, (co-recipient), 1983.

PERSONICK, STEWART D., Fellow 1983. Born: February 22, 1947, Brooklyn, New York. Degrees: B.E.E.-(E.E.), 1967, CCNY; S.M.(E.E.), 1968, Sc.D.(E.E.), 1969, M.I.T.

Fellow Award "For contributions to the theory and application of optical fiber transmission systems."

PETER, ROLF W., IRE Fellow 1958. Born: August 20, 1920, Zurich, Switzerland. Degrees: Diploma(E.E.), 1944, Dr.Sc.Tech., 1948, Eidg. Technische Hochschule, Zurich, Switzerland.

Fellow Award "For his contributions to the reduction of fluctuation noise in traveling wave tubes"; Founding Member, IEEE Society on Social Implications of Technology. Other Awards: American Physical Society; Sigma Xi; Founding Member, 1974, Honorary Member, Swiss Solar Energy Society, 1980.

PETERS, LEON, JR., Fellow 1981. Born: May 28, 1923, Degrees: B.E.E., 1950, M.Sc., 1954, Ph.D., 1959, Ohio State University.

Fellow Award "For contribution to the theory of microwave radiation and scattering."

PETERSON, A. M., IRE Fellow 1962. Born: May 21, 1922, Santa Clara, Calif. Degrees: B.S., 1948, M.S., 1949, Ph.D., 1952, Stanford University.

Fellow Award "For contributions to propagation aspects of communications." Other Awards: Scientific Research Society of America; Sigma Xi; Society for Industrial and Applied Mathematics.

PETERSON, ARNOLD P. G., IRE Fellow 1958. Born: August 7, 1914, Dekalb, Ill. Degrees: B.Eng., 1934, University of Toledo; S.M., 1937, Sc.D., 1941, Massachusetts Institute of Technology.

Fellow Award "For his contributions to the design of new audio frequency devices and acoustic measuring instruments." Other Awards: Fellow, Acoustical Society of America, 1951; John H. Potts Memorial Award, Audio Engineering Society, 1968.

PETERSON, HAROLD A., AIEE Fellow 1947. Born: December 28, 1908, Essex, Iowa. Degrees: B.S., 1932, M.S., 1933, University of Iowa.

Awards: IEEE Education Medal, 1978; IEEE Centennial Medal, 1984. Other Awards: Benjamin Smith Reynolds Award, University of Wisconsin, 1957; Citation of Merit, Wisconsin Utilities Association, 1965; Endowed Professorship, Wisconsin Utilities Association, 1967; Edward Bennett Professorship, 1974; National Academy of Engineering, 1978.

PETERSON, HAROLD O., IRE Fellow 1941. Born: November 3, 1899, Washington Co., Nebr. Degrees: B.S.E.E., 1921, D.Eng.(Hon.), 1951, University of Nebraska.

Other Awards: Modern Pioneer Award, National Association of Manufacturers, 1940.

PETERSON, W. WESLEY, IRE Fellow 1962. Born: April 22, 1924, Muskegon, Mich. Degrees: A.B., 1948, B.S.E., 1949, M.S.E., 1950, Ph.D., 1954, University of Michigan.

Fellow Award "For contributions to coding theory."

PETERSON, WALTER E., Fellow 1966. Born: January 5, 1921, Los Angeles, Calif. Degrees: B.S.E.E., 1943, University of California.

Fellow Award "For pioneering in aero-medical instrumentation and for technical direction of national defense projects." Other Awards: Northrop Patent Award, 1950; WESCON Award, 1960; Certicate of Appreciation, California State Board of Registration for Professional Engineers, 1971; Fellow, Institute for the Advancement of Engineering, 1974; Service Award for Conference Advisor to 1982 Auto Test Con.

PETRILLO, SALVATORE E., IRE Fellow 1956. Born: July, 1909, Waterbury, Conn. Degrees: B.S., 1930, Yale University.

Fellow Award "For contributions in the development, production, supply, and quality control of military signal equipment." Other Awards: Legion of Merit, U.S. Dept. of Army, 1945.

PETRITZ, RICHARD L., Fellow 1964. Born: October 24, 1922. Degrees: B.S., 1944, B.S.E.E., 1946, M.S.E.E., 1947, Ph.D.(Physics), 1950, Northwestern University.

Fellow Award "For contributions to semiconductor physics, fluctuation phenomena, and surface physics"; Browder J. Thompson Award, IRE, 1954. Other Awards: Meritorious Civilian Service Award, U.S. Dept. of Navy, 1954; Fellow, Texas Academy of Sciences; Fellow, American Physical Society.

PETROU, NICHOLAS V., Fellow 1971. Born: August 2, 1917, Springfield, Mass. Degrees: B.S., 1940, Northeastern University; M.S., 1942, Harvard University.

Fellow Award "For engineering management in military and space electronics, and incorporation of integrated electronics in aerospace systems." Other Awards: Order of Merit, Westinghouse, 1965; Society of Manufacturing Engineers, Region I Award of Commendation, 1982.

PETTEE, ALLEN D., AIEE Fellow 1946. Born: September 2, 1889, Andover, Mass. Degrees: B.A., 1911, Yale University; B.S.(Electrical Engineering), 1916, Massachusetts Institute of Technology; B.S., 1916, Harvard University.

PETTIT, JOSEPH M., IRE Fellow 1954. Born: July 15, 1916, Rochester, Minn. Degrees: B.S., 1938, University of California at Berkeley; E.E., 1940, Ph.D., 1942, Stanford University; Honorary Doctorate, Soong Jun University, Korea, 1975.

Fellow Award "For his outstanding work as engineer and educator in the field of high-frequency and microwave communications"; Electronic Achievement Award, IRE, Region 7, 1952; Founders Medal, IEEE, 1983. Other Awards: Presidential Certificate of Merit, 1949; Fellow, American Association for Advancement of Science, 1965; National Academy of Engineering, 1967; Engineering Educator of the Year, Engineers Week Committee of Atlanta, 1973; Honorary Member, American Society for Engineering Education, 1976; National Science Board, 1977-1982; Distinguished Engineering Alumnus Award, University of California, Berkeley, 1979; Distinguished and Unusual Service Award, American Society for Engineering Education, 1979; Archdiocesan Medal of St. Paul, Atlanta, 1980; Academia Nacional de Ingenieria, Mexico, 1981.

PFLUG, CHARLES E., AIEE Fellow 1961. Born: October 22, 1900, Tobias, Nebr. Degrees: B.S.E.E., 1926, University of Nebraska.

Fellow Awards "For contributions to electrified production of automobiles."

PHADKE, ARUN G., Fellow 1980. Born: August 27, 1938, Gwalior, India. Degrees: B.Sc., 1955, Agra University, India; B.Tech.(Hon.), 1959, Indian Institute of Technology, Khargpur; M.S., 1961, Illinois Institute of Technology; Ph.D., 1964, University of Wisconsin.

Fellow Award "For contributions to the application of digital computers to power systems"; Outstanding Paper Award, Power System Relaying Committee, IEEE, 1978. Other Awards: Gold Medalist, Madhya Pradesh, India, 1955.

PHELPS, JAMES D. M., Fellow 1981. Born: January 25, 1915, Bluefield, VA. Degree: B.S.E.E., 1938, Auburn University.

Fellow Award "For development of surge arrester applications and industry standards for EHV systems"; IEEE Standards Achievement Award, 1977. Other Awards: Surge Protective Devices Committee Award, IEEE, 1978.

PHELPS, PAUL L., JR., Fellow 1984. Born: March 2, 1922, Glendale, CA. Degrees: B.Sc.(E.E.), 1957, California Polytechnic State University.

Fellow Award "For contributions to the development and applications of instrumental methods in environmental radionuclide detection and behavior"; IEEE Centennial Medal, 1984.

PHILLIPS, ARTHUR H., AIEE Fellow 1957. Born: January 3, 1905, Reading, Pa. Degrees: E.E., 1927, Lehigh University.

Fellow Award "For his contributions to engineering in the fields of power generation and transmission."

PHILLIPS, SAMUEL C., Fellow 1974. Born: February 19, 1921, Springerville, Ariz. Degrees: B.S.E.E., 1942, University of Wyoming; M.S.E.E., 1950, University of Michigan; Dr. of Laws(Hon.), 1963, University of Wyoming.

Fellow Award "For management and leadership of aerospace vehicle and space programs"; Simon Ramo Medal, IEEE, 1984. Other Awards: Thomas D. White Space Trophy, National Geographic Society; Langley Medal, Smithsonian Institution; Astronautics Engineer Award, National Space Club; Flying Tiger Pilot Trophy; Distinguished Service Medal, awarded twice, National Aeronautics and Space Administration; Distinguished Service Medal with Oak Leaf Cluster; Legion of Merit; Distinguished Flying Cross with Oak Leaf Cluster; Air Medal with 7 Oak Leaf Clusters, U.S. Air Force; Army Commendation Medal; Croix de Guerre, France; Order of National Security Merit, Republic of Korea; Fellow, American Institute of Aeronautics and Astronautics; Fellow, American Astronautical Society; Member, National Academy of Engineering; Member, National Space Club; Member, Alpha Kappa Psi.

PHILLIPS, VIRGEL E., Fellow 1974. Born: September 17, 1918, Almota, Wash. Degrees: B.S.E.E., 1941, Washington State College.

Fellow Award "For contributions in the fields of power circuit breaker design, performance, and power system interaction."

PHISTER, MONTGOMERY, JR., Fellow 1969. Born: February 26, 1926, San Pedro, Calif. Degrees: B.S.E.E., 1949, M.S.E.E., 1950, Stanford University; Ph.D., 1953, Cambridge University.

Fellow Award "For contribution to logical design methods, to the advancement of computer education, and to the application of computers to real-time control of processes."

PICARD, DENNIS J., Fellow 1981. Born: August 25, 1932, Providence, RI. Degree: B.B.A. (Elect. Eng. & Mgmt.) (cum laude), 1961, Northeastern University.

Fellow Award "For leadership in the development and implementation of large phased array radars." Other Awards: Sigma Epsilon Rho Honor Fraternity.

PICKERING, WILLIAM H., IRE Fellow 1955. Born: December 24, 1910, Wellington, New Zealand. Degrees: B.S., 1932, M.S., 1933, Ph.D., 1936, California Institute of Technology.

Fellow Award "For his contributions as teacher of electronics, and his leadership in missile guidance, control and instrumentation", Annual Award, IRE, Professional Group on Reliability and Quality Control, 1958; Golden Omega Award, IEEE, 1968; Edison Medal, IEEE, 1972. Other Awards: Meritorious Civilian Service, U.S. Dept. of Army, 1945; James Wyld Memorial Award, American Rocket Society, 1957; Scientific

Achievement Award, Association of U.S. Army, 1958; Space Flight Achievement Award, National Industry Missile Conference, 1959; Distinguished Civilian Service Award, U.S. Dept. of Army, 1959; George Washington Achievement Award, Southern California Engineering Societies, 1963; Columbus Gold Medal, International Institute of Communications, Italy, 1964; Galabert Award, France, 1965; Special Award, British Interplanetary Society, 1965; Distinguished Service Medal, NASA, 1965; Robert H. Goddard Trophy, National Space Club, 1965; Crozier Medal, American Ordnance Association, 1965; Spirit of St. Louis Medal, American Society of Mechanical Engineers, 1965; William Proctor Prize, Scientific Research Society of America, 1965; Order of Merit, Italy, 1966; Louis W. Hill Transportation Award, AIAA, 1968; Gold Cup Man of Achievement Award, ARCS, 1969; Interprofessional Cooperation Award, Society of Manufacturing Engineers, 1970; Marconi Medal, Marconi Foundation, Bologna, Italy, 1974; Presidential Medal of Science, 1975; Fahrney Award, Franklin Institute, 1976; Honourary Knight Commander of the Civil Division of the Most Excellent Order of the British Empire, 1975; Herman Oberth Award, 1978.

PICKHOLTZ, RAYMOND L., Fellow 1982. Born: April 12, 1932, New York, NY. Degrees: B.S.E.E., 1954, M.S.E.E. 1958, City University of New York; Ph.D.(E.E.), 1966, Polytechnic Institute of New York.

Fellow Award "For contributions to the design of digital communications systems and to engineering education"; IEEE Centennial Medal, 1984. Other Awards: Research Award, RCA Laboratories, 1955.

PICKLES, SIDNEY, Fellow 1968. Born: April 23, 1909, Monterey, Calif. Degrees: B.S.(Physics), 1931, E.E., 1933, Stanford University;

Fellow Award "For outstanding contributions to navigational electronics," Other Awards: Citation, New Jersey Council for Research and Development; 38 patents and 2 pending.

PIERCE, HARVEY F., Fellow 1968. Born: October 18, 1909, Apalachicola, Fla. Degrees: B.S.E.E., 1932, University of Florida.

Fellow Award "For contributions to the engineering and direction of vital national projects, to the professional and educational furtherance of engineering." Other Awards: Engineer of the Year, FES, 1957, 1966; Outstanding Service to Society, FES, 1959; Outstanding Technical Achievement, FES, 1961; Citation for Work at Cape Canaveral, Dept. of the Army, 1960; University of Florida Significant Alumnus Award, 1966; Engineering Distinguished Service Award, University of Florida, 1967; Award, National Society of Professional Engineers, 1968.

PIERCE, JOHN ALVIN, IRE Fellow 1947. Born: December 11, 1907, Spokane, Wash. Degrees: A.B., 1937,

University of Maine.

Fellow Award "For his contributions to the development of Loran, a radio aid to navigation"; Morris N. Liebmann Award, IRE, 1953, "For his pioneering and sustained outstanding contributions to radio navigation, and his related fundamental studies of radio wave propagation"; Pioneer Award, Professional Group on Aeronautical and Navigation Electronics, 1961, "In recognition of his basic contributions to the early development and implementation of the Loran radio navigation system; his basic conceptions in allied types of systems including sky-wave synchronized Loran, Loran C. Radux and Omega; and his basic studies and experimental researches on radio propagation as related to radio navigation systems." Other Awards: Thurlow Award, Institute of Navigation, 1947; Presidential Certificate of Merit, 1948; Fellow, American Academy of Arts and Sciences, 1952; Honorary Member, Audio Engineering Society, 1956; Capt. Robert Dexter Conrad Award, U.S. Navy, 1975; Honorary Member, International Omega Association, 1977; Honorary Member, Institute of Navigation, 1979; Honorary Member, Wild Goose Association, 1980.

PIERCE, JOHN N., Fellow 1976. Born: July 9, 1933, Norwood, Mass. Degrees: S.B., 1955, S.M., 1955, E.E., 1959, Massachusetts Institute of Technology.

Fellow Award "For contributions to communication systems and to the theory of diversity combining and algebraic coding."

PIERCE, JOHN R., IRE Fellow 1948. Born: March 27, 1910, Des Moines, Iowa. Degrees: B.S., 1933, M.S., 1934, Ph.D., 1936, California Institute of Technology; D.Eng.(Hon.), 1961, Newark College of Engineering; D.Sc.(Hon.), 1961, Northwestern University; D.Sci.(Hon.), 1963, Yale University; D.Sc.(Hon.), 1963, Polytechnic Institute of Brooklyn; D.Eng.(Hon.), 1964, Carnegie Institute of Technology; D.Sc.(Hon.), 1965, Columbia University; D.Sc.(Hon.), 1970, University of Nevada; LL.D.(Hon.), 1974, University of Pennsylvania; E.El.Eng.(Hon.), 1974, University of Bologna; D.Sc.(Hon.), University of Southern California, 1977.

Fellow Award "For his many contributions to the theory and design of vacuum tubes"; Morris N. Liebmann Award, IRE, 1947, "For his development of a traveling-wave amplifier tube having both high gain and very great band width"; Edison Medal, 1963, "For his pioneer work and leadership in satellite communications and for his stimulus and contributions to electron optics, traveling wave tube theory and the control of noise in electron streams"; Medal of Honor, 1975, "For his pioneering concrete proposals and the realization of satellite communication experiments, and for contributions in theory and design of traveling wave tubes and in electron beam optics essential to this success." Other Awards: Stuart Ballantine Medal, Franklin Institute,

1960; H.H. Arnold Trophy, Air Force Association, 1962; Golden Plate Award, Academy of Achievement, 1962; Gen. Hoyt S. Vandenberg Trophy, Arnold Air Society, 1963; Valdemar Poulsen Medal, Danish Academy of Technical Sciences, 1963; National Medal of Science, 1963; H.T. Cedergren Medal, 1964; Distinguished Service Award, California Institute of Technology Alumni, 1966; John Scott Award, 1974; Marconi Award, 1974; Founders' Award, National Academy of Engineering, 1977; Marconi International Fellowship, 1979.

PIERCE, ROBERT E., AIEE Fellow 1947. Born: January 12, 1897, Indianapolis, Ind. Degrees: B.S., 1918, Purdue University.

PIERCE, ROGER J., Fellow 1969. Born: March 19, 1911, Des Moines, Iowa. Degrees: B.S.E.E., 1932, Iowa State University.

Fellow Award "For his contributions in the field of ground and aerospace telecommunications." Other Awards: 12 U.S. patents; Member, Sigma Xi; Charter Member, American Rocket and Marine Technical Societies.

PIETENPOL, WILLIAM J., IRE Fellow 1962. Born: July 15, 1921, Boulder, Colo. Degrees: B.A., B.S., 1943, University of Colorado; Ph.D., 1949, Ohio State University.

Fellow Award "For contributions to transistor technology."

PINDER, KENNARD, AIEE Fellow 1945. Born: November 9, 1900, Queen Ann Co., Md. Degrees: E.E., Drexel University; grad. work, power plant engineering, University of Pennsylvania.

Awards: section and IEEE national committees; chairman, former AIEE General Industry Practices Committee (now Industry Applications Society) in 1949-50; first chairman, Delaware Bay Section predecessor, the Wilmington Subsection, 1944-5; member, National Electric Code Panels No. 8 & 14 in 1963-5; and a Registered Professional Engineer in the State of Delaware; presented 30 papers at IEEE Section and National meetings; IEEE Red Book and Green Book; IEEE Centennial Medal, 1984. Other Awards: Tau Beta Pi.

PINGREE, GEORGE N., AIEE Fellow 1950. Born: April 28, 1898, New London, N.H. Degrees: B.S.E.E., 1920, University of New Hampshire.

PIORE, E. R., IRE Fellow 1950. Born: July 19, 1908, Wilno, Russia. Degrees: B.A., 1930, Ph.D., 1935, University of Wisconsin; D.Sc.(Hon.), 1962, Union University; D.Sc.(Hon.), 1966, University of Wisconsin.

Fellow Award "For his many contributions in the fields of engineering and physical sciences, and for outstanding service in enhancing the national effort in basic research." Other Awards: Distinguished Civilian Service Award, U.S. Dept. of Navy, 1955; Medal, Indus-

trial Research Institute, 1967; Morris J. Kaplun International Prize for Distinguished Research and Scholarship, Hebrew University of Jerusalem.

PIPPIN, JOHN E., Fellow 1971. Born: October 7, 1927, Kinard, Fla. Degrees: B.E.E., 1951, M.S.E.E., 1953, Georgia Institute of Technology; Ph.D., 1958, Harvard University.

Fellow Award "For contributions to the theory and development of microwave ferrite devices"; Outstanding Engineer, Region III, 1971. Other Awards: Engineer of the Year in Industry, Engineers of Greater Atlanta, Engineers Week Committee, 1973.

PLISKIN, WILLIAM A., Fellow 1981. Born: August 9, 1920, Akron, OH. Degrees: B.S., 1941, Kent State University; M.S. (Physics), 1943, Ph.D. (Physics), 1949, Ohio State University.

Fellow Award "For creative contributions to the development of thin-film insulators and glasses used in microelectronic devices." Other Awards: Annual Award, Mid-Hudson Section, American Chemical Society, 1964; IBM Corporate Invention Award, 1966; Annual Electronics Division Award, Electrochemical Society, 1973; IBM Corporate Recognition Award, 1979; IBM General Technology Division Invention Award, 1981; Sigma Pi Sigma, 1942; Sigma Xi, 1949; Pi Mu Epsilon, 1949.

PLONSEY, ROBERT, Fellow 1977. Born: July 17, 1924, New York, N.Y. Degrees: B.E.E., 1943, Cooper Union School of Engineering; M.E.E., 1948, New York University; Ph.D. 1957, University of California at Berkeley.

Fellow Award "For contributions to the application of electromagnetic field theory to problems in electrocardiography and electrophysiology"; Morlock Award, IEEE Engineering in Medicine and Biology Society, 1979. Other Awards: Sigma Xi, 1957; Eta Kappa Nu, 1963; NIH Senior Postdoctoral Fellow, 1980-1981.

POCH, WALDEMAR J., IRE Fellow 1956. Born: July 11, 1905, London, England. Degrees: B.S., 1928, University of Michigan; M.S., 1934, University of Pennsylvania.

Fellow Award "For contributions in the development and design of television studio equipment."

POHM, ARTHUR V., Fellow 1968. Born: January 11, 1927, Olmsted Falls, Ohio. Degrees: B.E.E., B.E.S., 1950, Fenn College; M.S., 1953, Ph.D., 1954, Iowa State University.

Fellow Award "For his significant contributions in the development of thin magnetic films as computer memory elements and for his inspired leadership as an effective researcher and teacher." Other Awards: Western Electric Award for Engineering Teaching, 1965; Cleveland State University Distinguished Service Award, 1967; Anston Marston Distinguished Professor of Engineering, 1971; NASA $1,000 Certificate of Recognition (Patent), 1980; Outstanding Engineering Alumnus Award, Cleveland State University, 1982.

POLAK, ELIJAH, Fellow 1977. Born: August 11, 1931, Bialystok, Poland. Degrees: B.E.E., 1956, University of Melbourne; M.S., 1959, Ph.D., 1961, University of California at Berkeley.

Fellow Award "For contributions to the theory and implementation of numerical algorithms in optimal control and nonlinear programming." Other Awards: Guggenheim Fellow, 1969; Senior Postdoctoral Fellow, United Kingdom Science Research Council, 1972, 1976, 1979, 1982.

POLKINGHORN, FRANK A., SR., IRE Fellow 1938. Born: July 23, 1897, Holbrook, Mass. Degrees: B.S., 1922, University of California at Berkeley.

Awards: North Jersey Section Award, 1961, 1969, 1975. Other Awards: Honorary Member, Institute of Electronics and Communication Engineers of Japan, 1952; Fellow, AAAS; Certificate of Achievement, Supreme Command Allied Powers, 1950; Appreciation, Institute of Electrical Engineers, Japan, 1950; Tau Beta Pi; Eta Kappa Nu; Sigma Xi; Phi Beta Kappa.

POLLACK, HERBERT W., Fellow 1980. Born: March 27, 1927, Brooklyn, NY. Degrees: B.E.E., 1950, City College of New York; M.E.E., 1953, New York University.

Fellow Award "For contributions to flexible printed circuits and their application." Other Awards: Tau Beta Pi, 1948; Eta Kappa Nu, 1948.

POLLACK, LOUIS, Fellow 1966. Born: November 4, 1920, New York, N.Y. Degrees: B.E.E., 1953, City College of the City University of New York.

Fellow Award "For the pioneering design and development of communication satellite earth terminals, employed in moon reflection relaying." Other Awards: Sigma Xi; Assoc. Fellow, Amer. Inst. Aero. and Astronautics.

POLLACK, MARTIN A., Fellow 1982. Born: December 25, 1937, Brooklyn, NY. Degrees: B.E.E., 1958, M.E.E., 1959, Polytechnic Institute of Brooklyn; Ph.D.(E.E.), 1963, University of California at Berkeley.

Fellow Award "For contributions to new molecular laser systems and for advances in mixed crystal semiconductor materials for optoelectronic applications." Other Awards: Westinghouse Fellow, 1958-1959; Ford Foundation Fellow, 1961-1962; NATO Postdoctoral Fellow, 1968-1969; Optical Society of America Fellow, 1978.

POLLARD, ERNEST I., AIEE Fellow 1951. Born: November 11, 1906, Nehawka, Nebr. Degrees: B.Sc., 1928, University of Nebraska; M.Sc., 1935, University of Pittsburgh.

Awards: AIEE, 1957-59, "For valued services and contributions as Chairman, Power Division." Other

Awards: Certificate of Commendation, U.S. Dept. of Navy; Fellow, ASME, 1981.

POLOUJADOFF, MICHEL E., Fellow 1982. Born: 1932, France. Degrees: Ingenieur Diplome, Ecole Superieure d'Electricite, France; M.S., Harvard University; Dr.Sc.Physiques, Sorbone, France.

Fellow Award "For contributions to the theory of electrical machinery, especially single-phase and linear induction motors, and for leadership in advancing engineering education." Other Awards: Member Laureate, Societe de Electriciens; Electroniciens et Radio-electriciens, France; Fellow, Institution of Electrical Engineers, London; Dr. Honoris Causa, Universite de Liege, Belgium, 1983.

POMERENE, JAMES H., Fellow 1971. Born: June 22, 1920, Yonkers, N.Y. Degrees: B.S.E.E., 1942, Northwestern University.

Fellow Award "For contributions to the design and development of computing systems." Other Awards: Tau Beta Pi, Sigma Xi, 1942; IBM Fellow, 1976.

PONTECORVO, PAUL J., Fellow 1964. Born: September 7, 1909, Pisa, Italy. Degrees: D.E.E., 1932, Polytechnical Institute, Turin, Italy.

Fellow Award "For contributions to microwaves communications systems." Other Awards: Certificate of Commendation, U.S. Dept. of Navy, 1960.

POPPELBAUM, WOLFGANG J., Fellow 1969. Born: August 28, 1924, Frankfurt, Germany. Degrees: M.S., 1948, Ph.D., 1953, University of Lausanne.

Fellow Award "For outstanding contributions to computer hardware and graduate education."

POPPELE, J. R., IRE Fellow 1950. Born: February 4, 1898, Newark, N.J.

Fellow Award "For his long and continued leadership in the broadcasting field, and in particular for his recent contributions to television broadcasting." Other Awards: Citation, American Television Society, 1951; Sarnoff Citation Award, Radio Club of America, 1974; Gold Medal Award, Veteran Wireless Operators Association, 1940, 1951; DeForest Service President Award, 1974; Chairman, Tele. Measurements, Electronics.

PORCELLO, LEONARD J., Fellow 1976. Born: March 1, 1934, New York, N.Y. Degrees: B.A.(Physics), 1955, Cornell Unversity; M.S.(Physics), 1957, M.S.E.(E.E.), 1959, Ph.D.(E.E.), 1963, University of Michigan.

Fellow Award "For contributions and leadership in coherent optics and synthetic aperture radar programs."

PORTNOY, WILLIAM M., Fellow 1983. Born: October 28, 1930, Chicago, Illinois. Degrees: B.S.(Eng. Physics), 1952, M.S.(Physics), 1952, Ph.D., 1959, University of Illinois.

Fellow Award "For contributions to solid-state electronics and component development." Other Awards: Western Electric Fund Award, American Society for Engineering Education, 1980; Fulbright Professor of Engineering, University of Warwick, UK, 1975-1976.

POSNER, EDWARD C., Fellow 1981. Born: August 10, 1933, New York, NY. Degrees: B.A., 1952, M.S., 1953, Ph.D., 1957, University of Chicago.

Fellow Award "For contributions to and leadership in the development of deep space communication systems."

POST, EDGAR A., IRE Fellow 1961. Born: May 20, 1914, Spokane, Wash. Degrees: B.S.E.E., 1936, University of Illinois.

Fellow Award "For contributions to aeronautical electronics." Other Awards: Outstanding Young Electrical Engineer, Honorable Mention, Eta Kappa Nu, 1945; Citation, Radio Technical Commission for Aeronautics, 1950, 1976; Commendation, Federal Aviation Administration, 1976.

POTTER, JAMES L., IRE Fellow 1957. Born: December 4, 1905, Carthage, Mo. Degrees: B.S., 1928, M.S., 1930, E.E., 1939, Kansas State University.

Fellow Award "For contributions as an engineer and educator"; First Paper Prize, AIEE, Science Division, 1954, "For development of magnettor current standard." Other Awards: Professor Emeritus, Rutgers University, 1975.

POTTS, JOHN A., AIEE Fellow 1957. Born: February 22, 1900, Chicago, Ill. Degrees: B.S.E.E., 1923, University of Wisconsin.

Fellow Award "For original engineering work and executive direction in a large electric utility."

POUNSETT, FRANK H. R., IRE Fellow 1947. Born: September 12, 1904, London, England. Degrees: B.A.Sc., 1928, University of Toronto.

Fellow Award "In recognition of his contributions to the engineering development and production design of radar apparatus in Canada."

POVEY, EDMUND H., Fellow 1965. Born: August 5, 1908, Boston, Mass. Degrees: B.E.E., 1929, M.S., 1955, Northeastern University.

Fellow Award "For development in instrumentation and test techniques for the evaluation of high-voltage insulation." Other Awards: Arnold H. Scott Award, American Society for Testing and Materials, 1976.

POWELL, A. HAMILTON, Fellow 1964. Born: December 24, 1914, Boston, Mass. Degrees: B.S.E.E., 1937, Worcester Polytechnic Institute.

Fellow Award "For development of high-power switching equipment and advancement of high voltage power systems." Other Awards: Tau Beta Pi, 1936; Associate Member, Sigma Xi, 1937; Certificate of Commendation, U.S. Dept. of Navy, 1947.

POWERS, EDWARD J., JR., Fellow 1983. Born: November 29, 1935, Winchester, Massachusetts. De-

grees: B.S., 1957, Tufts University; S.M., 1959, Massachusetts Institute of Technology; Ph.D., 1965, Stanford University (all in Elec. Eng.).

Fellow Award "For contributions to the analysis of data relating to nonlinear phenomena in materials such as controlled thermonuclear plasmas."

POWERS, KERNS H., Fellow 1969. Born: April 15, 1925, Waco, Tex. Degrees: B.S.E.E., 1951, University of Texas; Sc.D., 1956, Massachusetts Institute of Technology.

Fellow Award "For contributions to and supervision of the development of new communication systems and concepts." Other Awards: Achievement Award, RCA Laboratories, 1960; Forty-One for Freedom Award, U.S. Navy, 1967.

PRATT, MINOT H., AIEE Fellow 1953. Born: July 4, 1903. Degrees: B.S., 1926, University of Pennsylvania.

Fellow Award "For outstanding contributions in the development of improved methods and equipment essential to the efficient operation of a major electric distribution system." Other Awards: Outstanding Man in Engineering and Science, Syracuse HERALD JOURNAL, 1967; Engineer of the Year Award, Central New York Chapter, New York State Society of Professional Engineers, 1968.

PREISMAN, ALBERT, IRE Fellow 1953. Born: February 8, 1901, New York, N.Y. Degrees: A.B.(with honors in Mathematics), 1922, E.E., 1924, Columbia University.

Fellow Award "For contributions to the application of graphical constructions, video amplifier theory, and technical education"; Patron Award, Washington Section, IEEE, 1969. Other Awards: Engineering Vice President, Capitol Radio Engineering Institute, 1945-1959; Member, Sigma Xi.

PREIST, DONALD H., Fellow 1965. Born: January 18, 1916, Tunbridge Wells, England. Degrees: B.Sc., 1936, Dipl.Eng., University of London.

Fellow Award "For contributions in extending the limits on power and frequency in communications and radar." Other Awards: Fellow, Institution of Electrical Engineers (U.K.).

PREPARATA, FRANCO PAOLO, Fellow 1978. Born: December 29, 1935. Degrees: Dr. Eng., 1959, University of Rome.

Fellow Award "For contributions to coding theory." Other Awards: Remington-Rand Thesis Award, 1959; Libera Docenza, Italian University System, 1969.

PRESSLEY, JACKSON H., IRE Fellow 1931. Born: June 8, 1896, Rio Vista, Calif. Degrees: B.S., 1920, University of California at Berkeley.

PRESTON, KENDALL, JR., Fellow 1974. Born: October 22, 1927, Boston, Mass. Degrees: A.B., 1950, S.M., 1952, Harvard University.

Fellow Award "For contributions in the fields of synthetic aperture systems, optical computation, and pattern recognition." Other Awards: Cum Laude Society, 1945; Henry Warder Carey Prize, 1945.

PRESTON, WILLIAM F., AIEE Fellow 1962. Born: March 11, 1905, Abilene, Tex. Degrees: B.S., 1927, Yale University.

Fellow Award "For contributions to distribution standardization in a new area." Other Awards: Citation, Ministry of Labor, Industry, and Commerce, Brazil, 1954.

PREWITT, JUDITH M.S., Fellow 1981. Born: October 16, 1935. Degrees: B.A.(high honors), 1957, Swarthmore College; M.A., 1959, University of Pennsylvania; Ph.D., 1978, Uppsala University, Sweden.

Fellow Award "For contribution to applying image processing techniques to automated medical diagnostics"; IEEE Computer Society Certificate of Appreciation, 1979, 1981; Other Awards: Phi Beta Kappa, Sigma Xi, Eta Kappa Nu, Mortar Board; Alumnae Scholar, Leopold Shepp Foundation Scholar and Westinghouse Scholar, Swarthmore College, 1952-1956; University Scholar and Shohat Fellow in Mathematics, University of Pennsylvania, 1957-1959; Swedish Natural Science Coucil Fellow, 1976-1978; U.S. DHEW Public Health Service Award for Sustained High Quality Performance, 1973; National Cancer Institute Committee for Cytology Automation, 1972-1977, Diagnostic Radiology Advisory Committee, 1973-1977, Diagnostic Research Advisory Committee, 1977-1979; National Computer Graphics Association Certificate of Appreciation, 1981; SIAM National Lecturer, 1976-; Stocker Visiting Professor of Electrical Engineering, Ohio University, 1982-1983.

PRICE, ROBERT, IRE Fellow 1962. Born: July 7, 1929, West Chester, Pa. Degrees: A.B., 1950, Princeton University; Sc.D., 1953, Massachusetts Institute of Technology.

Fellow Award "For contributions to communication system theory and its use in radar contact with Venus"; Edwin Howard Armstrong Achievement Award, IEEE Communications Society, 1981.

PRICER, W. DAVID, Fellow 1978. Born: July 22, 1935, Des Moines, Iowa. Degrees: B.A.(Physics), 1959, Middlebury College; B.S., 1959, M.S., 1959 (both in Electrical Engineering), Massachusetts Institute of Technology.

Fellow Award "For contributions to the development of computer memory technology"; President, IEEE Solid-State Circuits Council, 1980-1981; Editor, IEEE Journal of Solid-State Circuits, 1983-. Other Awards: Sigma Xi.

PRINCE, DAVID C., AIEE Fellow 1926. Born: February 5, 1891, Springfield, Ill. Degrees: B.S.E.E., 1912, M.S.E.E., 1913, University of Illinois; Sc.D.(Hon.), 1943, Union College.

Awards: Lamme Medal, AIEE, 1946. Other Awards: Modern Pioneer Award, National Association of Manufacturers; Eta Kappa Nu; Tau Beta Pi; Sigma Xi; American Society of Mechanical Engineers; Society of Automotive Engineers; American Association for the Advancement of Science; American Helicopter Society; New York Society of Professional Engineers; Emeritus Member, American Nuclear Society.

PRINCE, M. DAVID, Fellow 1968. Born: March 27, 1926, Greensboro, N.C. Degrees: B.E.E., 1946, M.S.E.E., 1949, Georgia Institute of Technology.

Fellow Award "For outstanding leadership in the application and use of computers in the design of aerospace structures and systems." Other Awards: Citation, Institute of Aerospace Sciences, 1962; 4 Patents; Engineer/Scientist of the Year Award, Lockheed-Georgia Company, 1981; Chairman, CAM-I (Computer-Aided Manufacturing, Inc.) Electronics Automation Program, 1982-1983.

PRINCE, MORTON B., Fellow 1979. Born: April 1, 1924, Philadelphia, Pa. Degrees: A.B., 1947, Temple University; Ph.D., 1951, Massachusetts Institute of Technology.

Fellow Award "For contributions to semiconductor devices and leadership in the development of the silicon solar cell." Other Awards: Marconi Premium, British Institution of Radio Engineers, 1958.

PRINGLE, ARTHUR E., II, AIEE Fellow 1962. Born: October 31, 1899, Lansdowne, Pa. Degrees: B.S., 1921, Pennsylvania State University.

Fellow Award "For contributions in the field of electrical standardization." Other Awards: Fellow, Standards Engineers Society, 1961; James H. McGraw Award, Manufacturers Medal, 1962.

PRITCHARD, DALTON H., Fellow 1977. Born: September 1, 1921, Crystal Springs, Miss. Degrees: B.S.E.E., 1943, Mississippi State University.

Fellow Award "For contributions to the development and improvement of color television"; Vladimir K. Zworykin Award, IEEE, 1977, "For significant contributions to color television technology." Other Awards: Fellow Technical Staff, RCA Laboratories, 1975; Fellow, Society for Information Display, 1976; Eduard Rhein Prize, 1980; David Sarnoff Award, 1981; National Academy of Engineering, 1983.

PRITCHARD, R. L., IRE Fellow 1960. Born: September 8, 1924, N.J. Degrees: B.S., 1946, Brown University; M.S., 1947, Ph.D., 1950, Harvard University.

Fellow Award "For contributions in the field of transistor circuits." Other Awards: Sigma Xi, 1947; Tau Beta Pi, 1952; Award, Electronic Industries Association, 1963, 1967.

PRITCHARD, WILBUR L., Fellow 1968. Born: May 31, 1923, New York, N.Y. Degrees: B.S.E.E., 1943, City College of the City University of New York.

Fellow Award "For contributions and leadership in the development of military communication satellite systems." Other Awards: Systems Command Award for Outstanding Achievement, U.S. Air Force, 1967; 50th Anniversary Commemorative Medal, City College of the City University of New York, 1969; Fellow, American Institute of Aeronautics and Astronautics; Aerospace Communications Award, American Institute of Aeronautics and Astronautics, 1972; Townsend Harris Medal, City College of the City University of New York, 1976; Lloyd Berkner Award, American Astronautical Society.

PROAKIS, JOHN G., Fellow 1984. Born: June 10, 1935, Chios, Greece. Degrees: B.S.E.E., 1959, Univ. of Cincinnati; M.S.E.E., 1961, M.I.T.; Ph.D., 1967, Harvard University.

Fellow Award "For contributions to decision-directed measurement techniques and adaptive equalization techniques to digital communication over various channels."

PROEBSTER, WALTER E., Fellow 1977. Born: April 2, 1928, Manheim, Federal Republic of Germany. Degrees: Dipl.-Ing.(E.E.), 1951, Dr.-Ing.(E.E.), 1956, Technical University of Munich.

Fellow Award "For contributions to and leadership in the development of computer components and systems." Other Awards: Outstanding Paper Award, Nachrichtentechnische Gesellshaft (German Electrical Communication Society), 1956, 1960; Outstanding Paper Award, Solid-State Circuits Conference, Philadelphia, 1962; Patent Invention Achievement Award, IBM, 1963, 1969, 1971.

PRYWES, NOAH, Fellow 1969. Born: November 18, 1925. Degrees: B.S., 1949, Technion; M.S., 1951, Carnegie Institute of Technology; Ph.D., 1954, Harvard University.

Fellow Award "For contributions in developing large-scale high-speed computers, educational programs in computer and information sciences, and computer applications to information storage and retrieval."

PUCEL, ROBERT A., Fellow 1979. Born: December 27, 1926, Ely, Minn. Degrees: B.S. and M.S., 1951, Sc.D., 1955, Massachusetts Institute of Technology.

Fellow Award "For contributions to the modeling of microwave solid-state devices and leadership in their application"; Microwave Prize (corecipient), IEEE Microwave Theory and Techniques Society, 1976; National Lecturer, IEEE Microwave Theory and Techniques Society, 1980-1981.

PUENTE, JOHN G., Fellow 1977. Born: June 11, 1930, New York, N.Y. Degrees: B.S.E.E., 1957, Polytechnic Institute of Brooklyn; M.S.E.E., 1960, Stevens Institute of Technology.

Fellow Award "For contributions to the development of digital techniques and multiple access methods for satellite communications"; IEEE Award in International Communication in Honor of Hernand and Sosthenes Behn, 1975.

PUGH, EMERSON W., Fellow 1972. Born: May 1, 1929, Pasadena, Calif. Degrees: B.S., 1951, Ph.D., 1956, Carnegie-Mellon University.

Fellow Award "For contributions to information storage technology for digital computers"; Distinguished Lecturer, IEEE Magnetics Society, 1980. Other Awards: Fellow, American Physical Society, 1962; Fellow, American Association for the Advancement of Science, 1978; IBM Patent Achievement Awards 1966, 1977, 1978.

PUGSLEY, DONALD W., IRE Fellow 1952. Born: December 4, 1912, Salt Lake City, Utah. Degrees: B.S.E.E., 1935, University of Utah; M.S.E.E., 1954, Syracuse University.

Fellow Award "For his technical contributions and leadership in the development and design for production of commercial and military electronic equipment."

PULLEN, KEATS A., AIEE Fellow 1961. Born: November 12, 1916, Onawa, Iowa. Degrees: B.S., 1939, California Institute of Technology; D.Eng., 1946, The Johns Hopkins University.

Fellow Award "For contributions through research and literature to design of control and information-gathering systems for missiles." Other Awards: Marconi Memorial Award, VWOA, 1982.

PULLEY, ROBERT L., AIEE Fellow 1959. Born: October 19, 1901, Huntsville, Ala. Degrees: B.S.E.E., 1922, Auburn University.

Fellow Award "For contributions to design, operation and management of electric power systems." Other Awards: Florida Engineering Society.

PULVARI, CHARLES F., Fellow 1970. Born: July 19, 1907, Karlsbad, Austria-Hungary. Degrees: Dipl.Ing., 1929, Royal Hungarian University of Technical Sciences; Dr.Eng.(Hon.), 1981.

Fellow Award "For contributions to the theory of ferro electricity and its application to digital storage devices." Other Awards: Industrial Research 100 Award, 1963; Americanism Medal, DAR, 1975; Professor Emeritus of Electrical Engineering, Catholic University of America; Fellow, New York Academy of Sciences; Fellow, American Ceramic Society; Sigma Xi; Tau Beta Pi; Past Chairman, Chrystal Committee; Visiting Professor, Korea Advanced Institute of Sciences; 75 Patents.

PUMPHREY, FRED H., AIEE Fellow 1948. Born: July 31, 1898, Dayton, Ohio. Degrees: A.B., 1920, B.E.E., 1921, E.E., 1927, D.Sc.(Hon.), 1962, Ohio State University.

Other Awards: Listed in "Who's Who in America," 1951-1965; Regional Award, American Society of Mechanical Engineers, 1962; Letter of Commendation, Dr. Wernher von Braun, NASA, 1963; Letter of Commendation, U.S. Army Missile Command, 1963; Alabama Engineer of Year, A.S.P.E., 1969; Dean of Engineering, Emeritus, Auburn University, 1969.

PURINGTON, ELLISON S., Fellow 1964. Born: October 13, 1891, Mechanic Falls, Me. Degrees: A.B, 1912, Bowdoin College; A.M., 1913, Harvard University; Sc.D.(Hon.), 1965, Bowdoin College.

Fellow Award "For contributions to circuit design, radio control, and communication systems."

PURL, O. THOMAS, Fellow 1974. Born: June 5, 1924, East St. Louis, Ill. Degrees: B.S.E.E., 1951, M.S.E.E., 1952, Ph.D., 1955, University of Illinois.

Fellow Award "For contributions to high-power traveling-wave tubes, and for leadership of microwave electron device engineering."

PURSLEY, MICHAEL B., Fellow 1982. Born: August 10, 1945, Winchester, IN. Degrees: B.S., 1967, M.S., 1968, Purdue University; Ph.D.(E.E.), 1974, University of Southern California.

Fellow Award "For contributions to information theory and spread-spectrum communications"; President, IEEE Information Theory Group, 1983.

PUTNAM, RUSSELL C., AIEE Fellow 1951. Born: March 27, 1898, Jacksonville, Ill. Degrees: A.B., 1919, Butler University; B.S., 1923, E.E., 1928, University of Colorado; M.S., 1933, Case Institute of Technology.

Other Awards: Fellow Emeritus, Illuminating Engineering Society, 1965; Life Member, Cleveland Engineering Society, 1974.

Q

QUARLES, LAWRENCE R., IRE Fellow 1952, AIEE Fellow 1959. Born: January 26, 1908, Charlottesville, Va. Degrees: B.S.E., 1929, Ph.D.(Physics), 1935, University of Virginia.

IRE Fellow Award "For his contributions in the field of engineering education, particularly in the teaching of communication and circuit theory"; AIEE Fellow Award "For contributions to engineering education and nuclear engineering." Other Awards: Fellow, American Nuclear Society, 1965.

QUATE, C. F., Fellow 1965. Born: December 7, 1923, Baker, Nev. Degrees: B.S., 1944, University of Utah; Ph.D., 1950, Stanford University.

Fellow Award "For contributions to the theory and design of low noise amplifiers"; Morris N. Liebmann Memorial Award, IEEE, 1981. Other Awards: Tau Beta Pi; Sigma Xi; American Physical Society; Member, National Academy of Engineering, 1970; Member, National Academy of Sciences, 1975; Appointed Member, Editorial Board, Journal of Applied Physics and Applied Physics Letters, 1981-1984; Rank Prize, 1982; Fellow, American Academy of Arts and Sciences, 1982; Fellow, Acoustical Society of America, 1982.

QUILL, JOSEPH S., AIEE Fellow 1961. Born: December 22, 1918, Peterborough, N.H. Degrees: B.S., 1941, M.S., 1942, Massachusetts Institute of Technology.

Fellow Award "For contributions to application of gas turbines and associated electrical equipment to gas pipeline pumping." Other Awards: Charles A. Coffin Award, General Electric, 1955.

R

RABINER, LAWRENCE R., Fellow 1976. Born: September 28, 1943, Brooklyn, N.Y. Degrees: B.S., 1964, M.S., 1964, Ph.D., 1967, Massachusetts Institute of Technology.

Fellow Award "For contributions to digital signal processing and speech communication research"; Paper Award, Best Paper by an Author Under 30 Years of Age, IEEE Group on Audio and Electroacoustics, 1970; Achievement Award, IEEE Acoustics, Speech, and Signal Processing Society, 1979; Emanuel R. Piore Award, IEEE, 1980; Society Award, IEEE Acoustics, Speech, and Signal Processing Society, 1981. Other Awards: Outstanding Young Electrical Engineer, Honorable Mention, Eta Kappa Nu, 1973; Biennial Award, Acoustical Society of America, 1974; Fellow, Acoustical Society of America; Member, National Academy of Engineering.

RABINOW, JACOB, IRE Fellow 1956. Born: January 8, 1910, Kharkov, Russia. Degrees: B.S., 1933, E.E., 1934, City College of the City University of New York.

Fellow Award "For contribution in the fields of electronic ordnance and automatic control"; Harry Diamond Memorial Award, IEEE, 1977, "For important inventions in ordnance, computers, and Post Office automation." Other Awards: Naval Ordnance Development Award, 1945; Certificate of Commendation, National Defense Research Committee, 1945; Presidential Certificate of Merit, 1948; Exceptional Service Award, U.S. Dept. of Commerce, 1949; Certificate of Appreciation, War Dept., 1949; Edward Longstreth Medal, Franklin Institute, 1959; Robert C. Connelly Award, Miami Valley Computer Association, 1967; Jefferson Medal Award, New Jersey Patent Law Association, 1973; National Inventors Day Commendation, American Patent Law Association, 1973; Industrial Research and Development Magazine's Scientist of the Year, 1980; Doctor of Humane Letters, Towson State University, 1983.

RACE, HUBERT H., AIEE Fellow 1939. Born: February 12, 1899, Buffalo, N.Y. Degrees: E.E., 1922, Ph.D., 1927, Cornell University.

Awards: First Paper Prize, AIEE, 1930.

RADANT, MILTON E., Fellow 1984. Born: July 16, 1932, Evanston, IL. Degrees: B.S.E.E., 1954, University of Illinois; M.S., 1956, U.C.L.A.

Fellow Award "For leadership in development of pulse Doppler radar and digital signal processing." Other Awards: Bronze Tablet Honors, University of Illinois, 1954; Tau Beta Pi; Eta Kappa Nu; Phi Kappa Phi.

RADEKA, VELJKO, Fellow 1976. Born: November 21, 1930, Zagreb, Yugoslavia. Degrees: Dipl.Eng.(E.E.), 1955, D.Eng.Sci., 1961, University of Zagreb.

Fellow Award "For contributions to the theory and practice of nuclear detector signal processing."

RADER, CHARLES M., Fellow 1978. Born: June 20, 1939, Brooklyn, N.Y. Degrees: B.E.E., 1960, M.E.E., 1961, Polytechnic Institute of Brooklyn.

Fellow Award "For contributions to digital signal processing"; Technical Achievement Award, IEEE Acoustics, Speech, and Signal Processing Society, 1976; President, IEEE Acoustics, Speech, and Signal Processing Society, 1980, 1981; IEEE Centennial Medal, 1984.

RADER, LOUIS T., AIEE Fellow 1951. Born: Frank, Alberta, Canada. Degrees: B.S.E.E., 1933, University of British Columbia; M.S.E.E., 1935, Ph.D.(E.E.), 1938, California Institute of Technology; National Academy of Engineering, 1970; Professor Emeritus, University of Virginia, 1982.

Awards: IEEE Centennial Medal, 1984. Other Awards: Distinguished Service Award, California Institute of Technology, 1966; Virginia Engineering Foundation Award, 1982; Sigma Xi; Tau Beta Pi; Eta Kappa Nu; Beta Gamma Sigma.

RAE, JAMES R., Fellow 1968. Born: December 9, 1907, San Francisco, Calif. Degrees: B.S.E.E., 1929, M.S.E.E., 1929, Massachusetts Institute of Technology.

Fellow Award "For contributions to the improvement and expansion of world-wide communication services."

RAGALLER, KLAUS, Fellow 1983. Born: July 4, 1938, Furth, Germany. Degrees: Dipl.Ing., (Electrical Engineering), 1961, Ph.D.(Plasma Physics), 1966, Technical University, Munich.

Fellow Award "For contributions to the physical understanding of the electric arc and its interaction with gas flow in electric circuit breakers."

RAGAZZINI, JOHN R., IRE Fellow 1955, AIEE Fellow 1962. Born: January 3, 1912, New York, N.Y. Degrees: B.S., 1932, E.E., 1933, City College of New York; A.M., 1938, Ph.D., 1941, Columbia University.

IRE Fellow Award "For his contributions in the fields of computers and control systems and as a teacher of these subjects"; AIEE Fellow Award "For contributions to engineering education and to the theory of random noise and control systems"; AIEE Education Medal, 1962; IEEE Education Medal, 1979. Other Awards: Rufus Oldenburger Award, American Society of Mechanical Engineers, 1970.

RAINWATER, L. JAMES, Fellow 1966. Born: December 9, 1917, Council, Idaho. Degrees: B.S.(Physics), 1939, California Institute of Technology; M.A., 1941, Ph.D.(Physics), 1946, Columbia University.

Fellow Award "For basic contributions to nuclear science." Other Awards: Ernest Orlando Lawrence Award, Atomic Energy Commission, 1963; Nobel Prize in Physics, 1975; Fellow, American Physical Society; Fellow, N.Y. Academy of Sciences; Fellow, AAAS; Member, National Academy of Sciences; Optical Society of America; Fellow, 1982 Honorary Member, Swedish Royal Academy of Sciences.

RAISBECK, GORDON, Fellow 1969. Born: May 4, 1925, New York, N.Y. Degrees: B.A., 1944, Stanford University; Ph.D., 1949, Massachusetts Institute of Technology.

Fellow Award "For contributions and leadership in research on communication theory, transmission line theory, and transistor circuits." Other Awards: Rhodes Scholarship, 1947; Fellow, Acoustical Society of America, 1979.

RAJCHMAN, JAN A., IRE Fellow 1953. Born: August 10, 1911, London, England. Degrees: Dipl.Ing., 1934, Dr.Tech.Sc., 1938, Eid. Technische Hochschule, Zurich, Switzerland.

Fellow Award "For outstanding contributions in secondary emission multiplication and digital storage devices"; Morris N. Liebmann Award, IRE, 1960, "For his contributions to the development of magnetic devices for information processing"; Edison Medal, IEEE, 1974, "For a creative career in the development of electronic devices and for pioneering work in computer memory systems." Other Awards: Louis E. Levy Medal, Franklin Institute, 1947; National Academy of Engineering, 1966; Harold Pender Award, Moore School of Electrical Engineering, University of Pennsylvania, 1977; Francis Rice Darne Memorial Award, Society for Information Display, 1981.

RAMADANOFF, DIMITER, AIEE Fellow 1946. Born: October 8, 1900, Kustendil, Bulgaria. Degrees: B.S.E.E., 1924, University of Illinois; Ph.D.(Physics), 1932, Cornell University.

Awards: First Paper Prize, AIEE, for "High Altitude Brush Problem." Other Awards: McMullen Research Scholar, Cornell University, 1926-1937; Member, American Institute of Aeronautics and Astronautics, Society of Automotive Engineers, Society of Lubrication Engineers, Sigma Xi, Phi Kappa Phi; Listed in: "American Men of Science," "Who's Who in the Midwest," "Who's Who in Aviation," "Who's Who in Technology Today"; 15 Patents; 10 publications.

RAMAMOORTHY, C.V., Fellow 1978. Degrees: Ph.D., 1964, Harvard University.

Fellow Award "For contributions to computer architecture and software engineering"; Honor Roll Award,

IEEE Computer Society, 1975; Special Group Award, IEEE Computer Society, 1977. Other Awards: Faculty Award, College of Engineering, University of Texas, 1971.

RAMBERG, EDWARD G., IRE Fellow 1955. Born: June 14, 1907, Florence, Italy. Degrees: A.B., 1928, Cornell University; Ph.D., 1932, University of Munich.

Fellow Award "For his theoretical analyses of electronic devices"; IEEE David Sarnoff Award, 1972, "For meritorious achievement in electronics." Other Awards: Fellow, American Physical Society, 1957; Fulbright Fellow, 1960; David Sarnoff Outstanding Team Award, 1961; David Sarnoff Outstanding Individual Achievement Award in Science, 1972.

RAMBO, S. IVAN, Fellow 1970. Born: October 26, 1914, Granger, Washington. Degrees: B.S., 1937, Washington State College; M.S., 1950, The Johns Hopkins University.

Fellow Award "For contributions to the development of high-power high- frequency transmitters for radar systems."

RAMBO, WILLIAM R., IRE Fellow 1959. Born: September 3, 1916, San Jose, Calif. Degrees: A.B., 1938, Engr., 1941, Stanford University.

Fellow Award "For contributions to military electronics."

RAMEY, ROBERT A., IRE Fellow 1958. Born: December 14, 1918, Wayne, W.Va. Degrees: E.E., 1943, M.S., 1945, D.Sc., 1948, University of Cincinnati.

Fellow Award "For his contributions to magnetic amplifier theory and design."

RAMO, SIMON, IRE Fellow 1950. Born: May, 1913, Salt Lake City, Utah. Degrees: B.S., 1933, University of Utah; Ph.D., 1936, California Institute of Technology; D.Eng.(Hon.), 1960, Case Institute of Technology; D.Sc.(Hon.), 1961, University of Utah; D.Sc.(Hon.), 1963, Union College; D.Eng.(Hon.), 1966, University of Michigan; D.Sc.(Hon.), 1968, Worcester Polytechnic Institute; D.Sc.(Hon.), 1969, University of Akron; D.Law(Hon.), 1970, Carnegie-Mellon University; D.Eng.(Hon.), 1971, Polytechnic Institute of New York; D.Law(Hon.), 1972, University of Southern California; D.Sc.(Hon.), 1976, Cleveland State University; D. Law (Hon.), 1983, Gonzaga University.

Fellow Award "For his many contributions to the analysis of electromagnetic phenomena and for his leadership in research"; Electronic Achievement Award, IRE, 1953; Golden Omega Award, IEEE-MENA, 1975; IEEE Founders Medal, 1980. Other Awards: Paul T. Johns Award, Arnold Air Society, 1960; American Academy of Achievement Award, 1964; Air Force Association Award, 1964; Eminent Member, Eta Kappa Nu, 1966; Distinguished Service Gold Medal, AFCEA, 1970; Outstanding Achievement in Business Award, University of Southern California, 1971; Kayan

Medal, Columbia University, 1972; Award of Merit, American Consulting Engineers Council, 1974; Delmar S. Fahrney Award, The Franklin Institute, 1978; Honorary Fellow, American Institute of Aeronautics and Astronautics, 1979; Business Statesman, Harvard Business School Association, 1979; Member, National Academy of Sciences; Member, American Philosophical Society; Member, International Academy of Astronautics, France; Fellow, American Physical Society; Fellow, American Institute of Aeronautics and Astronautics; Fellow, American Association for the Advancement of Science; Fellow, Institute for the Advancement of Engineering; Fellow, American Academy of Arts and Sciences; Founding Member, National Academy of Engineering; Outstanding Achievement in Engineering Management Award, University of Southern California, 1979; National Medal of Science, 1979; Midwest Research Institute Board of Trustee's Citation, 1980; Distinguished Alumnus, University of Utah, 1981; The UCLA Medal, 1982; The Medal of Freedom, United States of America, 1983; Arthur M. Bueche Medal, National Academy of Engineering, 1983.

RAMSAY, JOHN F., Fellow 1965. Born: May 12, 1908, Milngavie, Dunbartonshire, Scotland. Degrees: M.A., 1934, University of Glasgow, Scotland.

Fellow Award "For contributions to the development of antennas and historical research into the origins of electronics technology." Other Awards: Ambrose Fleming Premium, Institution of Electrical Engineers, U.K., 1941; Premium, Institution of Electrical Engineers, U.K., 1953.

RANKIN, HARRY M., AIEE Fellow 1951. Born: June 17, 1888, Richland, Mich. Degrees: B.E.E., 1910, University of Michigan.

RANSOM, GLEN B., AIEE Fellow 1951. Born: April 29, 1898, Marengo, Iowa. Degrees: B.S., 1922, University of Minnesota.

RAO, THAMMAVARAM R. N., Fellow 1984. Born: June 5, 1933, Prakasam St., Andhra Pradesh, India. Degrees: B.S.(Physics), 1952, Andhra University, India; D.I.I.Sc.(E.E.), 1955, Indian Institute of Science, Bangalore; M.S.E.E., 1961; Ph.D., 1964, The University of Michigan, Ann Arbor.

Fellow Award "For contributions to the theory and applications of arithmetic codes, unidirectional error-correcting codes, and fault-tolerant computing." Other Awards: Tamma Sambiah Gold Medal, Andhra University, 1952.

RAPPAPORT, GEORGE, IRE Fellow 1956. Born: December 7, 1919, New York, N.Y. Degrees: B.S., 1941, City College of the City University of New York; M.S., 1947, Ph.D., 1949, Ohio State University; D.Sc.(Hon.), 1965, Parsons College.

Fellow Award "For contributions and leadership in the electronic countermeasures research and development." Other Awards: Pioneer Medal, Association of Old Crows, 1979.

RATCLIFFE, J. A., IRE Fellow 1953. Born: December 12, 1902, Bacup, Lancashire, England. Degrees: M.A., 1927, University of Cambridge.

Fellow Award "For his outstanding contributions to radio propagation research and to education." Other Awards: Fellow, Royal Society; Fellow, Institution of Electrical Engineers, U.K.

RAU, DAVID S., IRE Fellow 1960. Born: April 23, 1899, Little Egg Harbor, N.J. Degrees: B.S., 1922, United States Naval Academy.

Fellow Award "For contributions to international radio communications systems." Other Awards: Commendation Ribbon, U.S. Dept. of Navy, 1946; Achievement Award, Communications Technology Group, 1969.

RAUCH, LAWRENCE L., IRE Fellow 1960. Born: May 1, 1919, Los Angeles, Calif. Degrees: A.B., 1941, University of Southern California; A.M., 1948, Ph.D., 1949, Princeton University.

Fellow Award "For contributions to the theory and practice of radio telemetry"; Special Award, IRE, Professional Group on Telemetry and Remote Control, 1957, "For valuable contribution to the field of telemetry." Other Awards: War and Navy Dept.'s Award, 1947; National Telemetering Conference Award, 1960; Donald P. Eckman Education Award, Instrument Society of America, 1966.

RAWLINS, CHARLES B., Fellow 1981. Born: July 4, 1928, Annapolis, MD. Degrees: B.S. (M.E.), 1949, Johns Hopkins Univesity; M.S. (I.E.), 1965, Clarkson College.

Fellow Award "For contributions to the field of transmission line conductor dynamics."

RAWLS, JAMES A., AIEE Fellow 1961. Born: March 22, 1905, Holland, Va. Degrees: B.A., 1926, Virginia Polytechnic Institute.

Fellow Award "For contributions to the design and operation of electric power system"; William M. Habirshaw Award, IEEE, 1969, "For developing the first 500 kV. transmission line in the U.S. and the free world."

RAYMOND, RICHARD C., IRE Fellow 1958. Born: August, 1917, Spokane, Wash. Degrees: B.S., 1938, State College of Washington; Ph.D., 1941, University of California.

Fellow Award "For his contributions to communication theory."

REARDON, KENNETH N., AIEE Fellow 1962. Born: January 23, 1902, Boston, Mass. Degrees: B.S., 1923, Harvard University.

Fellow Award "For contributions to the development of power system relaying and control systems."

RECHTIN, EBERHARDT, IRE Fellow 1962. Born: January 16, 1926, N.J. Degrees: B.S., 1946, Ph.D.(cum laude), 1950, California Institute of Technology.

Fellow Award "For contributions to communication theory and space communication"; Electronic Achievement Award, IRE, Region, 1960, "For research in communications systems and leadership in setting up deep space communications stations"; Alexander Graham Bell Medal, IEEE, 1977, "For pioneering and lasting contributions to deep-space-vehicle communications technology and for leadership in defense telecommunications." Other Awards: Fellow, American Institute of Aeronautics and Astronautics, 1962; Exceptional Scientific Achievement Medal, NASA, 1965; Academician, International Astronautics Academy, 1965; Member, National Academy of Engineering, 1968; Aerospace Communications Award, American Institute of Aeronautics and Astronautics, 1970; Distinguished Public Service Medal, Dept. of Defense, 1973; Distinguished Public Service Medal, U.S. Navy, 1983.

REDDY, D. RAJ, Fellow 1983. Born: June 13, 1937, Katoor, India. Degrees: B.E., 1958, University of Madras; M.S., 1964, Ph.D., 1966, Stanford University.

Fellow Award "For contributions in automatic speech recognition, artificial intelligence, and robotics."

REDHEAD, PAUL A., IRE Fellow 1962. Born: May 25, 1924, Brighton, Sussex, England. Degrees: B.A., 1944, M.A., 1949, Ph.D., 1969, University of Cambridge.

Fellow Award "For contributions to the field of electron devices and to ultrahigh vacuum technology." Other Awards: Fellow, Royal Society of Canada, 1968; Honorary Member, American Vacuum Society, 1970; Medard W. Welch Award, American Vacuum Society, 1975.

REDIKER, ROBERT H., IRE Fellow 1962. Born: June 7, 1924, Brooklyn, N.Y. Degrees: B.S., 1947, Ph.D., 1950, Massachusetts Institute of Technology.

Fellow Award "For contributions to semiconductor device research"; David Sarnoff Award, IEEE, 1969. Other Awards: Fellow, American Physical Society; Sigma Xi.

REED, EUGENE D., Fellow 1972. Born: 1919, Vienna, Austria. Degrees: B.S.E.E., 1941, University of London; M.S.E.E., 1947, Ph.D., 1953, Columbia University.

Fellow Award "For contributions and leadership in the development of microwave tubes and solid-state components." Other Awards: Member, National Academy of Engineering, 1971.

REED, IRVING S., Fellow 1973. Born: 1923, Washington. Degrees: B.S., 1944, Ph.D., 1949, California Institute of Technology.

Fellow Award "For contributions to automatic detection and processing of radar data, multiple error-correcting communication codes, and digital computer design"; Shannon Lecturer, IEEE Information Theory Group, 1982. Other Awards: Sigma Xi; National Academy of Engineering, 1979; Charles Lee Powell Professor of Computer Engineering, University of Southern California, 1978.

REED, JOSEPH, Fellow 1977. Born: December 11, 1920, Bronx, N.Y. Degrees: B.E.E., 1944, The Cooper Union; M.E.E., 1951, Polytechnic Institute of Brooklyn; Ph.D.(E.E.), Polytechnic Institute of New York.

Fellow Award "For contribution to communication, automatic control, and navigational radar by the application of electronic, mechanical, and hydraulic technology." Other Awards: Sigma Xi, 1951, 1977; Tau Beta Pi, 1944; Oil Hydraulic Institute 1st Prize Paper, 1953.

REGGIA, FRANK, Fellow 1968. Born: October 30, 1921, Northumberland, Pa. Degrees: B.S.E.E., 1969, M.S.E.E., 1971, Bucknell University.

Fellow Award "For contributions to microwave ferrite technology." Other Awards: Engineer of Year Award, Washington Society of Engineers, 1952; Superior Accomplishment Award, National Bureau of Standards, 1953; Superior Accomplishment Award, Dept. of Army, 1958; Dept. of Army Scholarship Award, Bucknell University, 1969-70; National Engineering Honor Society (Tau Beta Pi), 1969; Fellow, Washington Academy of Science, 1971; Fellow, American Association for Advancement of Science, 1972; Research and Development Award, U.S. Dept. of Army, 1976; Inventor of the Year Award, Harry Diamond Labs. (Dept. of Army), 1977; Federal Retiree of the Year Award (For Outstanding Contributions to Public Service), National Association Retired Federal Employees, 1978; Awarded 22 U.S. and Canadian Patents, 1956-1979.

REGULINSKI, THADDEUS L., Fellow 1984. Born: October 9, 1932, Torun, Poland. Degrees: B.E.E., 1950, Manhattan College; M.S.E.E., 1953, New Jersey Institute of Technology; Ph.D., 1977, University of Bradford, Bradford, West Yorkshire, U.K.

Fellow Award "For contributions to research and education in reliability." Other Awards: Citizen of the Year Award, Omega Psi Phi, 1968; Outstanding Engineer Award, Ohio Society of Professional Engineers, 1972; Meritorious Civilian Service Award, USAF, 1979; Best Paper RAM Symposium, 1974, 1981.

REICH, BERNARD, Fellow 1973. Born: January 7, 1926, New York, N.Y. Degrees: B.S., 1948, City College of the City University of New York.

Fellow Award "For contributions to reliability improvement, standardization, and design of semiconductor devices and integrated circuits." Other Awards: R & D Annual Achievement Citation, U.S. Army Electronics Command, 1965; Charter Engineer and Fellow, Institution of Electrical Engineers, U.K.; Army Decoration for Meritorious Civilian Service.

REICH, HERBERT J., IRE Fellow 1949, AIEE Fellow 1954. Born: October 25, 1900, Staten Island, N.Y. Degrees: M.E., 1924, Ph.D., 1928, Cornell University.

IRE Fellow Award "For his contributions as a teacher and author in the radio and electronics fields"; AIEE Fellow Award "For his research in electronics, and his contributions to the technical literature"; IRE Director, 1944, 1948-1957; Board of Editors, 1943-1953; Member, Seven IRE Committees, 1941-1966; Chairman: Education Committee, 1945-1951, Electron Tube Standards Committee, 1964-1966, Electron Tube Conference Committee, 1951; Editor, IEEE Transactions on Electron Devices, 1952-1954. Other Awards: Technical Research Associate and Editor, Radio Research Laboratory, 1944-1946; Yale University Professor of Electrical Engineering, 1946-1962, and Engineering and Applied Science, 1962-1969; Emeritus Professor, 1969 to date; Author, three electronics textbooks, coauthor of five others; author of over 60 published technical papers; USA Member, Electron Tube Committee, International Electrotechnical Commission, 1960-1972; Chairman, Microwavetube Subcommittee, 1965-1972; USA Member, Working Group on Terms and Definitions, 1965-1972; Member, Advisory Group on Electron Tubes, Office of the Secretary of Defense, 1950-1958.

REID, JOHN M., Fellow 1984. Born: June 8, 1926, Minneapolis, MN. Degrees: B.S.E.E., 1950, M.S., 1957, University of Minnesota; Ph.D., 1965, University of Pennsylvania.

Fellow Award "For contributions to the field of medical diagnostic ultrasound." Other Awards: Fellow, Acoustical Society of America, 1975; Ernest N. Calhoun Chair of Biomedical Engineering, Drexel University, 1982-date.

REINGOLD, IRVING, Fellow 1975. Born: November 13, 1921, Newark, N.J. Degrees: B.S., 1942, Eng., 1949, New Jersey Institute of Technology.

Fellow Award "For leadership in the fields of display and microwave devices." Other Awards: Outstanding Technical Leadership Award, U.S. Army Electronics Command, 1962; Fellow, Society for Information Display, 1973; Frances Rice Darne Memorial Award, Society for Information Display, 1978.

REINTJES, J. FRANCIS, Fellow 1975. Born: February 19, 1912. Degrees: E.E., 1933, M.E.E., 1934, Rensselaer Polytechnic Institute.

Fellow Award "For contributions to the fields of radar and automation."

REMPT, HENRY F., AIEE Fellow 1958. Born: October 23, 1908, New York, N.Y.

Fellow Award "For contributions to proof tested applications of power and electronics for aircraft."

RENNER, JOHN J., Fellow 1976. Born: November 25, 1919, Boston, Mass. Degrees: S.B.E.E., 1941, Massachusetts Institute of Technology.

Fellow Award "For contributions in the application of systems engineering to telecommunications for government and industry." Other Awards: Bronze Star Medal, Citation Ribbon, World War II; Fellow, Radio Club of America, 1973.

RENNICK, JOHN L., Fellow 1967. Born: August 10, 1911, Chicago, Ill. Degrees: B.E.E., 1934, Marquette University.

Fellow Award "For his many contributions in the development of color television receivers." Other Awards: Outstanding Contribution Award, Consumer Electronics Groups, 1965.

RESWICK, JAMES B., Fellow 1974. Born: April 16, 1922, Ellwood City, Pa. Degrees: S.B., 1943, S.M., 1948, Sc.D., 1952, Massachusetts Institute of Technology; D.Eng.(Hon.), 1968, Rose Polytechnic Institute.

Fellow Award "For contributions to biomedical and rehabilitation engineering." Other Awards: Honors Award for Paper, American Society of Mechanical Engineers, Instruments and Regulators Division, 1955; Product Engineering Master Designer Award, 1969; Member, Institute of Medicine, National Academy of Science, 1972; Isabel and Leonard Goldenson Award, United Cerebral Palsy, 1974; Member, National Academy of Engineering, 1976.

RETTENMEYER, FRANCIS X., IRE Fellow 1944. Born: July 27, 1900, Kendrick, Okla. Degrees: B.S., 1922, University of Colorado; M.S., 1924, Columbia University.

Fellow Award "For advancement in the art of mobile communications."

REUDINK, DOUGLAS O., Fellow 1981. Born: May 6, 1939. Degrees: A.A.(Eng. Physics), 1959, Clark College; B.A.(Physics), 1962, Linfield College; Ph.D.(Math.), 1965, Oregon State University.

Fellow Award "For contributions to satellite communications and microwave mobile radio systems."

REYNOLDS, DONALD K., Fellow 1967. Born: December 9, 1919, Portland, Oreg. Degrees: B.A., 1941, M.A., 1942, Stanford University; Ph.D., 1948, Harvard University.

Fellow Award "For contributions to antenna theory and in electrical engineering education."

REZA, FAZLOLLAH, Fellow 1965. Born: January 1, 1915, Resht, Iran. Degrees: P.E., 1938, Teheran University; M.S., 1946, Columbia University; D.E.E., 1950, Polytechnic Institute of Brooklyn.

Fellow Award "For contributions to network and information theory." Other Awards: Special Award, Teheran University, 1962; Honorary Professor, Polytechnic Institute of New York, 1975; Honorary Professor, Carleton University, Ottawa, Canada, 1976; Fellow, AAAS, 1976; Life Member, New York Academy of Science; Member, American Mathematical Society; Member, Society for Industrial and Applied Mathematics; Honorary Professor, McGill University, 1978.

RHINEHART, ROBERT J., AIEE Fellow 1962. Born: August 20, 1900, Plymouth, Ind. Degrees: B.S.E.E., 1922, E.E., 1926, Purdue University.

Fellow Award "For contributions to the engineering and operation of a major electric utility system and to the advancement of the engineering profession"; IEEE Centennial Medal, 1984. Other Awards: Outstanding Service Award, Arkansas Society of Professional Engineers; Unselfish Service Award, National Society of Professional Engineers; Certificate of Merit, National Council of Engineering Examiners; Arkansas Traveler Award, Governor of Arkansas; Outstanding Engineer, Arkansas Society of Professional Engineers; Distinguished Service Award, Kiwanis International; Presbyterian Elder Emeritus; Member, Tau Beta Pi; President, National Society of Professinal Engineers, 1956-1957; Who's Who in Engineering; Who's Who in the South and Southwest; Governor, Mo-Kan-Ark District of Kiwanis International; Chairman, Last Man Club, Veterans of World War I; Commander, American Legion Post.

RHODES, CHARLES W., Fellow 1980. Born: August 1, 1929, Buffalo, NY.

Fellow Award "For contributions to measurement techniques and instrumentation for television." Other Awards: Fellow, Society of Motion Pictures and Television Engineers.

RHODES, DONALD R., Fellow 1967. Born: December 31, 1923, Detroit, Mich. Degrees: B.S.E.E., 1945, M.Sc., 1948, Ph.D., 1953, Ohio State University.

Fellow Award "For contributions to the theory and techniques of antenna synthesis"; John T. Bolljahn Award, IEEE, 1963. Other Awards: Fellow, American Association for the Advancement of Science, 1967; Benjamin G. Lamme Medal, Ohio State University, 1975.

RHODES, JOHN D., Fellow 1980. Born: October 9, 1943. Degrees: B.Sc., 1964, Ph.D., 1966, D.Sc., 1974, University of Leeds, England.

Fellow Award "For contributions to circuit theory, particularly distributed and multivariable filter networks"; Microwave Prize, IEEE Microwave Theory and Techniques Society, 1969; Browder J. Thompson Award, IEEE, 1970; Cauer-Guillimen Award, IEEE Circuits and Systems Society, 1974. Other Awards: J.J. Thompson Award, Institute of Electrical Engineers, London, 1971, 1973.

RIBLET, HENRY B., Fellow 1975. Born: May 20, 1911, Clayton, N.Mex. Degrees: B.A., 1934, Friends University, Wichita, Kansas.

Fellow Award "For contributions in the fields of guided missile telemetry and space telecommunications systems." Other Awards: Member, Society for Technical Communication, Washington, D.C. Chapter, 1972; Distinguished Alumni Award, Friends University, Wichita, Kansas, 1977.

RIBLET, HENRY J., IRE Fellow 1958. Born: July 21, 1913, Calgary, Alberta, Canada. Degrees: B.S., 1935, M.A., 1937, Ph.D., 1939, Yale University.

Fellow Award "For advances in the theory and design of microwave components."

RICARDI, LEON J., Fellow 1974. Born: March 21, 1924, Brockton, Mass. Degrees: B.S., 1949, M.S. 1952, Ph.D., 1969, Northeastern University.

Fellow Award "For contributions to the theory and design of microwave antenna systems for communication satellites and deep space radar applications."

RICE, REX, Fellow 1969. Born: January 27, 1918, Douglas, Arizona. Degrees: B.S.M.E., 1940, Stanford University.

Fellow Award: "For contributions to the organization and applications of digital systems"; IEEE International Solid State Circuits Conference, Outstanding Paper Award, 1964; IEEE Computer Society Special Award, 1975, "For contributions as architect of COMPCON and service as first chairman of COMPCON Standing Committee, 1972-75"; W.W. McDowell Award, IEEE Computer Society, 1982; IEEE Centennial Medal, 1984.

RICE, STEPHEN O., IRE Fellow 1953. Born: November 29, 1907, Shedds, Oreg. Degrees: B.S., 1929, D.Sc.(Hon.), 1961, Oregon State University.

Fellow Award "For his mathematical investigations of communication problems of noise, non-linear circuits, and efficient coding of information"; Mervin J. Kelly Award, 1965, "For his outstanding and fundamental contributions in the field of communications, particularly in the understanding of the effects of noise"; NTC '74 Award "For outstanding contributions to the theory and practice of communications"; Alexander Graham Bell Medal, IEEE, 1983.

RICH, EDWARD A. E., Fellow 1974. Born: March 30, 1916, Salt Lake City, Utah. Degrees: B.S.E.E., 1937, University of Utah.

Fellow Award "For contributions to automatic process control in the cement and glass industries"; Distinguished Service Award, Cement Industry Committee, IEEE Industry Applications Society, 1969; Achievement Award, IEEE Industry Applications Society, 1981; Editor, IEEE Transactions on Industry Applications, 1983-84; IEEE Centennial Medal, 1984. Other Awards: Phi Kappa Phi, 1937; Tau Beta Pi, 1937.

RICHARD, WILLIAM J., Fellow 1966. Born: February 20, 1901, Goldsboro, Md. Degrees: B.S., 1924, University of Maryland.

Fellow Award "For contributions to national and international standardization of insulated wire and cable

for power systems."

RICHARDSON, AVERY G., IRE Fellow 1956. Born: March, 1901, Brooklyn, N.Y. Degrees: E.E., 1924, Polytechnic Institute of Brooklyn.

Fellow Award "For contributions and leadership in the development of radio direction finders." Other Awards: Certificate of Commendation, U.S. Dept. of Navy; Award, National Defense Research Council.

RICHARDSON, JOHN M., Fellow 1965. Born: September 5, 1921, Rock Island, Ill. Degrees: B.A., 1942, University of Colorado; M.A., 1947, Ph.D., 1951, Harvard University.

Fellow Award "For contributions to atomic frequency and time interval standards." Other Awards: Gold Medal Award for Exceptional Service, U.S. Dept. of Commerce, 1964.

RICHARDSON, OLAN, AIEE Fellow 1956. Born: February 2, 1904, Gainesville, Ga.

Fellow Award "For contributions to protection and communication for electric power systems."

RICHMAN, DONALD, IRE Fellow 1961. Born: September 15, 1922, Brooklyn N.Y. Degrees: B.E.E., 1943, City College of the City University of New York; M.E.E., 1942, Polytechnic Institute of Brooklyn.

Fellow Award "For contributions to color television"; Vladimir K. Zworykin Award, IRE, 1957, "For contributions to the theory of synchronization, particularly that of color subcarrier reference oscillator synchronization in color television."

RICHMAN, PAUL, Fellow 1982. Born November 17, 1942, New York, NY. Degrees: B.S.E.E., 1963 Massachusetts Institute of Technology; M.S.E.E., 1964, Columbia University.

Fellow Award "For contributions to the theoretical analysis, development, and manufacture of large-scale integrated circuits"; Region 1 Award, IEEE, 1979. Other Awards: Annual Achievement Award, Electronics Magazine, 1978.

RICHMAN, PETER L., Fellow 1971. Born: November 7, 1927, New York, N.Y. Degrees: B.S., 1946, Massachusetts Institute of Technology; M.S., 1953, New York University.

Fellow Award "For many contributions to the precision measurement and generation of AC voltages and currents."

RICHMOND, JACK H., Fellow 1980. Born: July 30, 1922, Kalispell, MT. Degrees: B.S.E.E.(summa cum laude), 1950, Lafayette College; M.Sc.E.E., 1952, Ph.D.E.E., 1955, Ohio State University.

Fellow Award "For contributions to the theory of antennas and scattering." Other Awards: Tau Beta Pi; Phi Beta Kappa; Sigma Xi; Pi Mu Epsilon; Eta Kappa Nu.

RICHTER, WALTHER, AIEE Fellow 1942, IRE Fellow 1953. Born: May 17, 1900, Heidelberg, Germany. Degrees: B.S., 1923, Technische Hochschule Karlsruhe,

Germany.

IRE Fellow Award "In recognition of his contributions in the field of industrial electronics."

RIDEOUT, VINCENT C., IRE Fellow 1960. Born: May 22, 1914, Alberta, Canada. Degrees: B.Sc., 1938, University of Alberta; M.S., 1940, California Institute of Technology.

Fellow Award "For contributions in the fields of microwave and computer education." Other Awards: Reynolds Award, University of Wisconsin, 1960; Professor-Emeritus, Department of Electrical and Computer Engineering, Madison, Wisconsin.

RIDINGS, GARVICE H., Fellow 1965. Born: July 9, 1905, Buena Vista, Va. Degrees: B.S.E.E., 1926, Virginia Polytechnic Institute and State University.

Fellow Award "For achievements in the technology of facsimile transmission." Other Awards: F. E. d'Humy Award, Western Union, 1960.

RIEBMAN, LEON, Fellow 1966. Born: April 22, 1920. Degrees: B.S.E.E., 1943, M.S.E.E., 1947, Ph.D., 1951, University of Pennsylvania.

Fellow Award "For contributions to theory and practice in microwave solid-state devices"; Section Award, IEEE, 1971. Other Awards: Atwater Kent Award, University of Pennsylvania, 1943.

RIGAS, HARRIETT B., Fellow 1984. Born: April 30, 1934, Winnipeg, Manitoba, Canada. Degrees: B.Sc., 1956, Queens University, Ontario; M.S., 1959, Ph.D., 1963, University of Kansas.

Fellow Award "For contributions to programming of analog/hybrid computers and to the development of computer engineering curricula"; IEEE (Spokane Section) Engineer of the Year, 1980. Other Awards: FIER (Foundation for Instrumentation Education and Research) Fellow, 1961-63; SWE (Society of Women Engineers) Achievement Award, 1982; 1983 "Distinguished Engineering Service Award", The University of Kansas School of Engineering, Lawrence, Kansas.

RING, DOUGLAS H., Fellow 1966. Born: March 28, 1907, Butte, Mont. Degrees: A.B., 1929, E.E., 1930, Stanford University.

Fellow Award "For contributions to radar and microwave communication techniques."

RINGLAND, WILLIAM L., Fellow 1968. Born: September 14, 1914. Degrees: B.S.E.E., 1935, Purdue University.

Fellow Award "For significant contributions in the development and design of large rotating electrical machines." Other Awards: $5000 Engineering and Scientific Award, Allis-Chalmer, 1960.

RINGLEE, ROBERT J., Fellow 1974. Born: April 23, 1926, Sacramento, Calif. Degrees: B.S.E.E., 1946, M.S.E.E., 1948, University of Washington; Ph.D., 1964, Rensselaer Polytechnic Institute.

Fellow Award "For contributions to practical meth-

ods for assessing and improving reliability of power apparatus and systems"; Prize Paper Award (co-authors W.J. Degnan and G.G. Doucette), Power Apparatus and Systems, 1958, "An Improved Method of Oil Preservation and Its Effects on Gas Evolution"; Prize Paper Award (co-author M.W. Schulz, Jr.), Power Apparatus and Systems, 1960, "Some Characteristics of Audible Noise of Power Transformers and Their Relationship to Audibility Criteria and Noise Ordinances"; Prize Paper Award (co-author: R.J. Chambliss, G.A. Cucci, L.G. Leffler, N.D. Reppen), Power Apparatus and Systems, 1974, for "Operating Reserve and Generation Risk Analysis for the P-J-M Interconnection"; Fellow, American Association for the Advancement of Science, 1979.

RINIA, HERRE, IRE Fellow 1954. Born: March 30, 1905, Cornwerd, Friesland, Netherlands. Degrees: M.A., 1928, Dr.h.c., 1970, Technical University of Delft.

Fellow Award "For his creative contributions to radio engineering in Holland, and his leadership in the field of television"; IEEE Award in International Communication, 1970. Other Awards: Member, Koninklijke Nederlandse Akademie van Wetenschappen, 1947; Officer of the Order of Orange, Nassau, Netherlands, 1955.

RITCHIE, ALISTAIR E., Fellow 1979. Born: February 28, 1914, Lurgan, Ireland. Degrees: A.B., 1935, M.A., 1937, Dartmouth College.

Fellow Award "For contributions to telephone network switching and signaling development."

RITTENHOUSE, JOSEPH W., AIEE Fellow 1961. Born: January 22, 1917, Neosho, Mo. Degrees: B.S.(E.E.), 1939, Purdue University; M.S.(E.E.), 1949, University of Missouri at Rolla.

Fellow Award "For contributions to engineering education and to vacuum switching techniques."

RITTNER, EDMUND S., Fellow 1969. Born: May 29, 1919, Boston, Mass. Degrees: B.S., 1939, Ph.D., 1941, Massachusetts Institute of Technology.

Fellow Award "For contributions to photoconductors and semiconductors." Other Awards: Fellow, Massachusetts Institute of Technology, 1941, 1942; Fellow, American Physical Society, 1953.

RIVERS, ROBERT A., Fellow 1980. Born: September 5, 1923. Degrees: B.S.E.E., 1953, Massachusetts Institute of Technology.

Fellow Award "For leadership in the application of microwave technology and for contributions to the profession.

RIZK, FAROUK A.M., Fellow 1982. Born: July 6, 1934, Egypt. Degrees: B.Sc.E.E., 1955, M.Sc., 1958, Cairo University; Licentiate of Technology, 1960, Royal Institute of Technology, Sweden; Dr.Tech., 1963, Chalmers University of Technology, Sweden.

Fellow Award "For contributions to the science of

high-voltage technology and for technical leadership in the advancement of the electrical power industry." Other Awards: Egyptian National Prize in Engineering Science, 1971; Egyptian Order of Worthiness, Third Class, 1972; Egyptian Order of Science, First Class, 1973; Chairman, Technical Committee No. 28 "Insulation Coordination", International Electrotechnical Commission (IEC), 1983.

ROA, WILLIAM J., JR., Fellow 1965. Born: July 30, 1913, St. Louis, Mo. Degrees: B.S., 1934, M.S., 1936, Washington University.

Fellow Award "For contributions to the development, design, and construction of complex test facilities." Other Awards: Fellow, American Association for the Advancement of Science; National Society of Professional Engineers.

ROAD, RICHARD A., AIEE Fellow 1961. Born: August 17, 1907, Peru, Ind. Degrees: B.S.E.E., 1929, Purdue University.

Fellow Award "For contributions to invention, development and production of watt-hour meters."

ROBERTS, F. MORLEY, AIEE Fellow 1948. Born: May 26, 1901, Grappenhail-Chesire, England. Degrees: B.Sc.(E.E.), 1924, Queens University, Canada.

Other Awards: Engineer of the Year, Society of Professional Engineers, 1964.

ROBERTS, LOUIS W., Fellow 1964. Born: September 1, 1913, Jamestown, N.Y. Degrees: A.B., 1935, Fisk University; M.S., 1937, University of Michigan.

Fellow Award "For contributions and administrative leadership in the field of microwave tubes." Other Awards: Phi Beta Kappa, 1963, 1965; Certificate of Appreciation, NASA, 1970; Apollo Achievement Award, 1970, NASA; Secretary's Appreciation Award, 1975, Secretary's Silver Medal for Meritorious Achievement, 1976, U.S. Dept. of Transportation; Outstanding Achievement Award, 1978, University of Michigan.

ROBERTS, WALTER VAN B., IRE Fellow 1929. Born: November 13, 1893, Titusville, Pa. Degrees: B.S., 1915, E.E., 1917, Ph.D., 1924, Princeton University.

ROBERTSON, LAWRENCE M., AIEE Fellow 1945. Born: January 20, 1900, Denver, Colo. Degrees: B.S., 1922, E.E., 1927, University of Colorado; D.Law, 1930, University of Denver; M.S., 1938, D.Eng., 1955, University of Colorado.

Awards: Vice President, AIEE, 1945-1947; Director, AIEE, 1956-1960; Special Task Committee of five Societies, AIEE to Locate Headquarters, 1955-1956; Committee on Merger of AIEE with IRE-IEEE, 1962; William M. Habirshaw Award, 1963, "For his vigorous leadership and unique contributions in the field of high voltage transmission, particularly including lightning, corona and radio influence studies at high altitudes and at extra-high transmission voltages"; Power Group

Award, IEEE, 1969; Prize Paper Award, IEEE, 1969. Other Awards: Gold Medal, Colorado Engineering Council, 1954; Public Service Certificate, Federal Power Commission, 1964; Outstanding Service Certificate, Professional Engineers of Colorado, 1966; Fellow, New York Academy of Sciences, 1966; Distinguished Engineering Alumnus Award, University of Colorado, 1968; Alfred J. Ryan Award, Professional Engineers of Colorado, 1969; Distinguished Service Award, Denver Regional Council of Governments, 1971; Distinguished Service Award, National Council of Engineering Examiners, 1971; Exceptional Contributions to the Electrical Industry Award, Rocky Mountain Electrical League, 1977; "Who's Who in Engineering"; "Who's Who in the West"; College of Engineering and Applied Science Plague, University of Colorado, 1981; Attwood Associate in the U.S. National Committee CIGRE, 1982.

ROBINSON, ARTHUR S., Fellow 1970. Born: September 26, 1925, New York, N.Y. Degrees: B.S., 1948, Columbia University; M.S., 1951, New York University; D.Sc., 1957, Columbia University.

Fellow Award "For leadership in digital and analog computing and control systems, solid state radar, coherent electrooptical systems, and medical electronic instrumentation"; Most Significant Paper, IRE TRANSACTIONS on Electronic Computers, 1955. Other Awards: Tau Beta Pi; Chairman, Joint Automatic Control Conference, 1962.

ROBINSON, GLEN P., JR., Fellow 1977. Born: September 10, 1923, Jacksonville, Fla. Degrees: B.S.(Physics), 1948, M.S.(Physics), 1950, Georgia Institute of Technology.

Fellow Award "For leadership in microwave antenna measurement instrumentation"; Outstanding Engineer, Region 3, IEEE, 1978. Other Awards: Young President's Organization, 1960; Georgia's Outstanding Small Businessman, SBA, 1965; Georgia Entrepreneur Award, 1981.

ROBINSON, P. H., AIEE Fellow 1958. Born: April 24, 1904, Woods Co., Okla. Degrees: B.S., 1925, University of Oklahoma.

Fellow Award "For contributions to economical design and protection of an electric power system."

ROBSON, PETER N., Fellow 1981. Born: November 23, 1930, Bolton, England. Degrees: B.A., 1954, University of Cambridge; Ph.D., 1963, University of Sheffield.

Fellow Award "For contribution to the understanding of microwave solid-state devices." Other Awards: Fellowship of Engineering, 1983; OBE, 1983.

ROBUCK, JOHN B., Fellow 1971. Born: May 1, 1907, Helena, Tex. Degrees: B.S.(E.E.), 1929, University of Texas at Austin.

Fellow Award "For contributions to the planning, design, and development of modern electric utility systems"; Dallas Chapter, IEEE, National Engineers Week, 1970, "In recognition of outstanding achievements in the field of electrical engineering." Other Awards: Engineer of the Year, Dallas Chapter, Texas Society of Professional Engineers, 1975.

ROCHELLE, ROBERT W., Fellow 1970. Born: June 23, 1923, Nashville, Tenn. Degrees: B.S., 1947, University of Tennessee; M.S., 1949, Yale University; Ph.D., 1963, University of Maryland.

Fellow Award "For development of spacecraft information systems, in particular, pulse-frequency modulation; and for contributions to advancements in spacecraft electronics." Other Awards: La Medaille du CNES, France, 1965, 1972; Outstanding Engineering Alumnus, University of Tennessee, 1974.

ROCHESTER, NATHANIEL, IRE Fellow 1958. Born: January 14, 1919, Buffalo, N.Y. Degrees: B.S., 1941, Massachusetts Institute of Technology.

Fellow Award "For his work in the logical design of calculating machines." Other Awards: IBM Fellow, 1965.

ROCKEFELLER, GEORGE D., Fellow 1975. Born: December 31, 1927, Lancaster, Pa. Degrees: B.S., 1948, Lehigh University; M.S., 1968, New Jersey Institute of Technology; M.B.A., 1978, Fairleigh Dickinson University.

Fellow Award "For technical leadership in the application of protective relaying."

ROCKWELL, EDWARD W., AIEE Fellow 1948. Born: March 9, 1898, Syracuse, N.Y. Degrees: B.S., 1919, University of Southern California.

RODRIGUE, GEORGE P., Fellow 1975. Born: June 19, 1931, Paincourtville, La. Degrees: B.S., 1952, M.S., 1954, Louisiana State University; Ph.D., 1958, Harvard University.

Fellow Award "For contributions to the characterization and application of ferrimagnetic garnets in microwave devices."

ROGERS, ALDRED W., IRE Fellow 1960. Born: December 24, 1909, Laurel, Del. Degrees: B.S., 1930, University of Delaware.

Fellow Award "For contributions in the development of electronic components." Other Awards: Meritorious Civilian Service Award, U.S. Dept. of Army, 1966; Founder, Electronic Component Parts Group, 1953.

ROGERS, ELBERT W., AIEE Fellow 1962. Born: May 8, 1901, Celeste, Tex. Degrees: B.S.E.E., 1922, Texas A & M University.

Fellow Award "For planning, engineering and construction of electric utility system."

ROGERS, FREDERICK HELME, SR., Fellow 1964. Born: January 7, 1903, Shreveport, La. Degrees: B.S.E.E., 1925, University of Maryland.

Fellow Award "For contributions to the planning, engineering, and management of electric utility opera-

tions." Other Awards: Recognition Award, National Society of Professional Engineers, 1969; Distinguished Service Certificate, National Council of Engineering Examiners, 1973.

ROGERS, THOMAS F., Fellow 1964. Born: August 11, 1923, Providence, R.I. Degrees: B.Sc., 1945, Providence College; M.A., 1949, Boston University.

Fellow Award "For research on scatter propagation and for contributions to military communications systems." Other Awards: Civil Service Outstanding Performance Award, U.S. Dept. of Air Force, 1957; Certificate of Commendation, Secretary of the Navy, 1961; Meritorious Civilian Service Award, Secretary of Defense, 1967; Engineering News Record Award: Construction Man of the Year, 1969.

ROHDE, LOTHAR, IRE Fellow 1956. Born: October 4, 1906, Leverkusen, Germany. Degrees: Dr.phil., 1931, University of Jena; Dr.-Ing.E.h., 1953, Technische Hochschule Munich, Germany.

Fellow Award "For contributions to electronic measurement techniques and for technical administration."

ROHLFS, ALBERT F., AIEE Fellow 1962. Born: October 21, 1914, Barberton, Ohio. Degrees: B.S., 1937, Bucknell University.

Fellow Award "For contributions to improved high voltage testing and measuring techniques"; Second Paper Prize, AIEE, for "Sixty-Cycle and Impulse Sparkover of Large Gap Spacings," Second Paper Prize, AIEE, 1958, for "The Response of Resistance Voltage Dividers to Steep-Front Impulse Waves"; Third Paper Prize, AIEE, 1958, for "Impulse Flashover Characteristics of Long Strings of Suspension Insulators"; IEEE Power Group Award, 1969, for TRANSACTIONS Paper, "Influence of Air Density on Electrical Strength of Transmission Insulation."

ROHRER, RONALD A., Fellow 1980. Born: August 19, 1939, Oakland, CA. Degrees: B.S.E.E., 1960, Massachusetts Institute of Technology; M.S., 1961, Ph.D., 1963, University of California at Berkeley.

Fellow Award "For theoretical contributions and practical software for computer-aided circuit design"; Browder J. Thompson Award, IEEE, 1967; Prize Paper Award, IEEE, Circuit Theory Group, 1970, Other Awards: Outstanding Research Paper Award, National Electronics Conference, 1964; Frederick Emmons Terman Award, American Society for Engineering Education, 1978.

ROMAN, WALTER G., AIEE Fellow 1958.

Fellow Award "For contributions to development of nuclear reactors for naval and land power."

RONEY, ROBERT K., Fellow 1974. Born: August 5, 1922, Newton, Iowa. Degrees: B.S., 1944, University of Missouri; M.S., 1947, Ph.D., 1950, California Institute of Technology.

Fellow Award "For leadership and inspiration in the

exploration and applications of space technology." Other Awards: Missouri Honor Award, University of Missouri, 1979.

ROOT, WILLIAM L., Fellow 1965. Born: October 6, 1919, Des Moines, Iowa. Degrees: B.S., 1940, Iowa State University; S.M., 1943, Ph.D., 1952, Massachusetts Institute of Technology.

Fellow Award "For contributions to education and research in communications random processes, and radar detection theory." Other Awards: Senior Post-Doctoral Fellow, National Science Foundation, 1970.

ROSCH, SAMUEL J., Fellow 1977. Born: November 11, 1893, Austria. Degrees: B.S.E.E., 1911, Columbia University; D.Sc.(Hon.), 1939, University of Wisconsin.

Fellow Award "For contributions to the theory, development, manufacture, and application of high-voltage and specialty cable";Best Initial Paper Prize, AIEE, 1939, for "Current Carrying Capacity of Rubber-Insulated Electrical Conductors"; Citation, AIEE, 1961, "For Chairmanship (twelve years) of Subcommittee on Voltage Utilization Wiring." Other Awards: Three Citations, U.S. Navy Bureau of Ships; President, Insulator Power Cable Engineering Association, 1955-1958; Honorary Life Member, Insulated Power Cable Engineering Association, 1959; Honorary Life Member, American Society for Testing and Materials, 1959; Honorary Life Member, IAEI, Eastern Section, 1959; Honorary Life Member, Minnesota Electrical Association, 1959.

ROSE, ALBERT, IRE Fellow 1948. Born: March 30, 1910, New York, N.Y. Degrees: A.B., 1931, Ph.D., 1935, Cornell University.

Fellow Award "For his contributions in the field of television camera tubes and associated equipment"; Awards: Morris N. Liebmann Award, IRE, 1945; Edison Medal, 1979, "For contributions to television camera tubes, photoconductivity and human vision." Other Awards: Television Broadcasters Award, 1945; Journal Award, Society of Motion Picture and Television Engineers, 1946; David Sarnoff Gold Medal, Society of Motion Picture and Television Engineers, 1958; Mary Shepard Upson Distinguished Visiting Professor, Cornell University, 1967; Fairchild Distinguished Scholar, California Institute of Technology, 1975; Fellow, American Physical Society; Member; Societe Suisse de Physique; Member, National Academy of Engineering; SID Award, 1976; University Professor, Boston University, 1977, 1978; Leo Friend Award, American Chemical Society, 1982; Honorary Member, Society of Photographic Engineers and Scientists; Visiting Professor: Stanford University, 1976, Hebrew University (Jerusalem), 1976-77; Distinguished Visiting Professor: Polytechnic Institute, Mexico City, 1978, University of Delaware, 1979.

ROSE, HERBERT A., AIEE Fellow 1962. Born: December 14, 1899, Holden, Mo. Degrees: B.S.E.E., 1924, B.S.M.E., 1925, Kansas State University.

Fellow Award "For contributions to the utilization of electrical products in the lumber, pulp and paper industries." Other Awards: Fellow, American Society of Mechanical Engineers, 1970.

ROSEN, CHARLES A., Fellow 1974. Born: 1917. Degrees: B.E.E., 1940, Cooper Union Institute of Technology; M.Eng., 1950, McGill University; Ph.D., 1956, Syracuse University.

Fellow Award "For contributions to solid-state physics and the development of computer-controlled robots." Other Awards: Member, Scientific Research Society of America, 1956; Member, American Physical Society, 1958.

ROSEN, HAROLD A., Fellow 1970. Born: March 20, 1926, New Orleans, La. Degrees: B.E., 1947, Tulane University; M.S., 1948, Ph.D., 1951, California Institute of Technology; D.Sc.(Hon.), 1975, Tulane University.

Fellow Award "For contributions in missile guidance and satellite communications"; Mervin J. Kelly Award, IEEE, 1972; Alexander Graham Bell Award, IEEE, 1982. Other Awards: Tulane University Award, 1947; Astronautics Engineer Award, National Space Club, 1964; Golden Plate Award, American Academy of Achievement, 1965; Emmy Award, National Academy of Arts and Sciences, 1966; L.A. Hyland Patent Award, Hughes Aircraft Company, 1967; Communications Award, American Institute of Aeronautics and Astronautics, 1968; Inventor of the Year, Patent Law Association of Los Angeles, 1973; National Academy of Engineering, 1973; Spacecraft Design of the Year Award, American Institute of Aeronautics and Astronautics, 1973; L.M. Ericsson International Prize, 1976; Distinguished Alumni Award, California Institute of Technology, 1976; Lloyd V. Berkner Award, American Astronautic Society, 1976; Outstanding Technical Achievement Award, National Cable Television Association, 1980; 30 Patents.

ROSEN, PAUL, Fellow 1979. Born: January 4, 1922, Boston, Mass. Degrees: B.S.E.E., 1941, Tufts College; M.S.E.E., 1953, Northeastern University; Ed.M., 1977, Boston University.

Fellow Award "For leadership in and technical contributions to military satellite communications systems." Other Awards: Sigma Xi; American Psychological Association.

ROSENBACH, SAMUEL, AIEE Fellow 1948. Born: April 5, 1901, New York, N.Y. Degrees: B.S.E.E. (highest honor), 1921, University of New Mexico.

Other Awards: Graduate Fellow, Carnegie Institute of Technology, 1921-1922.

ROSENBAUM, FRED J., Fellow 1979. Born: February 15, 1937, Chicago, Ill. Degrees: B.S.E.E., 1959, M.S.E.E., 1960, Ph.D.E.E., 1963, University of Illinois.

Fellow Award "For contributions to the analysis and design of microwave ferrite and semiconductor components."

ROSENBERG, AARON E., Fellow 1984. Born: April 9, 1937, Malden, MA. Degrees: S.B.(E.E.), S.M.(E.E.), 1960, M.I.T.; Ph.D.(E.E.), 1964, Univ. of Penn.

Fellow Award "For contributions to speech and speaker recognition." Other Awards: Fellow, Acoustical Society of America.

ROSENBERG, LEON T., AIEE Fellow 1957. Born: November 9, 1904, South Bend, Ind. Degrees: B.Sc., 1926, University of Michigan; M.Sc., 1949, Illinois Institute of Technology.

Fellow Award "For contributions to the design and development of large steam turbine driven generators"; First Paper Prize, AIEE, District 5, 1959, for "A New Sator Coil Transposition for Large Machines"; First Nikola Tesla Award, 1976, "For his half century of development and design of large steam turbine driven generators and his important contributions to literature." Other Awards: Eta Kappa Nu, 1949; Tau Beta Pi, 1949; Sigma Xi, 1949; Yugoslav Nikola Tesla Medal, 1976.

ROSENBERG, PAUL, IRE Fellow 1961. Born: March 31, 1910, New York, N.Y. Degrees: B.A., 1930, M.A., 1933, Ph.D., 1941, Columbia University.

Fellow Award "For contributions in the field of electron physics." Other Awards: Fellow, American Association for Advancement of Science, 1943; Talbert Abrams Award, American Society of Photogrammetry, 1955; Associate Fellow, American Institute of Aeronautics and Astronautics, 1961; Fellow, American Institute of Chemists, 1968; Member, National Academy of Engineering, 1970; Fellow, Explorers Club, 1979; Honorary Member, Institute of Navigation, 1982.

ROSENFELD, AZRIEL, Fellow 1972. Born: February 19, 1931, New York, N.Y. Degrees: B.A., 1950, Yeshiva University; M.A., 1951, Columbia University; Ordination (Rabbi), 1952, M.H.L.(Hebrew Literature), 1953, M.S., 1954, D.H.L.(Hebrew Literature), 1955, Yeshiva University; Ph.D., 1957, Columbia University; D. Tech. (Hon.), 1980, Linkoping University, Sweden.

Fellow Award "For leadership and contributions to the field of computer processing of pictorial information."

ROSENSTEIN, ALLEN B., Fellow 1982. Born: August 25, 1920, Baltimore, MD. Degrees: B.S., 1940, University of Arizona; M.S., 1950, Ph.D., 1958, University of California at Los Angeles.

Fellow Award "For contributions to theory, design, and manufacture of power converters and for leader-ship and research in engineering education."

ROSENTHAL, JENNY E.(Mrs. A. Bramley), Fellow 1966. Born: July 31, 1909, Moscow, Russia. Degrees: Sc.B., 1926, University of Paris; Sc.M., 1927, Ph.D., 1929, New York University.

Fellow Award "For achievement in spectroscopy, optics, and mathematical techniques and their application to electronic engineering."

ROSS, CHARLES W., Fellow 1980. Born: November 2, 1920. Degrees: B.S.(E.E.), 1943, University of Alabama; S.M.(E.E.), 1949, D.Sc.(E.E.), 1952, Harvard University.

Fellow Award "For contributions to automatic generation control and the operation of interconnected power systems."

ROSS, GERALD F., Fellow 1972. Born: December 14, 1930, New York, N.Y. Degrees: B.E.E., 1952, City College of the City University of New York; M.E.E., 1955, Ph.D., 1963, Polytechnic Institute of Brooklyn.

Fellow Award "For his contributions in the field of picosecond pulse technology and its applications to microwave-distributed network theory and high-resolution electromagnetic sensors"; K.C. Black Best Paper Award, NEREM Meeting, 1974. Other Awards: Tau Beta Pi, 1951; Eta Kappa Nu, 1951; Sigma Xi, 1961; Member, Commission B, URSI.

ROSS, HUGH C., Fellow 1966. Born: December 31, 1923, Turlock, Calif. Degrees: B.S., 1950, Stanford University.

Fellow Award "For contributions in the development of high vacuum switching equipment."

ROSS, IAN M., Fellow 1966. Born: August 15, 1927, Southport, Lancashire, England. Degrees: B.A., 1948, M.A., 1952, Ph.D., 1952, University of Cambridge; Dr.Sc.(Hon.), Stevens Institute of Technology and the New Jersey Institute of Technology.

Fellow Award "For contributions to the design theory and development of transistors", Morris N. Leibmann Award, IRE, 1963, "For contributions to the development of the epitaxial transistor and other semiconductor devices." Other Awards: Certificate of Award, Pennsylvania Society of Professional Engineers, 1964; NASA Public Service Award, 1969; Public Service Group Achievement Award, 1969; National Academy of Engineering, 1973; National Academy of Science, 1982; President's Commission on Industrial Competitiveness, 1983.

ROSS, IRVINE E., JR., AIEE Fellow 1960. Born: September 28, 1908, Needham, Mass. Degrees: B.S., 1930, M.S., 1931, Massachusetts Institute of Technology.

Fellow Award "For contributions to the development of commutation at high altitudes." Other Awards: O.S.R.D. Citation, 1946; Fort Wayne Citizen Engineer, 1958; National Senior Elfun of the Year, G.E., 1979;

George B. Morgan Award, Massachusetts Institute of Technology, 1981.

ROSS, MONTE, Fellow 1975. Born: May 26, 1932, Chicago, Ill. Degrees: B.S.E.E., 1953, University of Illinois; M.S.E.E., 1962, Northwestern University.

Fellow Award "For leadership and contributions to the development of high-data-rate optical communications systems." Other Awards: "Technical Contributions" Award, American Institute of Aeronautics and Astronautics, St. Louis Section, 1976.

ROSSOFF, ARTHUR L., Fellow 1971. Born: April 4, 1922, New York, N.Y. Degrees: B.E.E., 1943, City College of the City University of New York; M.E.E., 1947, Polytechnic Institute of Brooklyn.

Fellow Award "For contributions to the understanding and application of the transistor, and to real-time topographic mapping"; Region 1 Citation 1974, "For services rendered in furthering the objectives of the IEEE through distinguished leadership in the area of professional activities in Region 1 and nationally"; USAB Award, IEEE, 1979; IEEE Centennial Medal, 1984. Other Awards: Registered Professional Engineer (by eminence), New York, 1972; Citation, American Institute of Aeronautics and Astronautics, Long Island Section, 1974; First Leaders in Long Island Technology Award, Long Island Forum for Technology (LIFT), 1979.

ROTH, J. PAUL, Fellow 1975. Born: December 16, 1922, Detroit, Mich. Degrees: B.M.E., 1946, University of Detroit; M.S., 1948, Ph.D., 1954, University of Michigan.

Fellow Award "For development of mathematical methods for computer design." Other Awards: ONR Fellowship, 1952-53; 141 papers; 8 U.S. patents, 64 patent publications.

ROTH, J. REECE, Fellow 1981. Born: September 19, 1937, Washington, PA. Degrees: S.B.(Physics), 1959, Massachusetts Institute of Technology; Ph.D.(Eng. Physics), 1963, Cornell University.

Fellow Award "For developments in superconducting magnet technology and discoveries in plasma instabilities and turbulence."

ROTH, WILFRED, Fellow 1968. Born: June 24, 1922. Degrees: B.S.E.E., 1943, Columbia University; Ph.D., 1948, Massachusetts Institute of Technology.

Fellow Award "For his contributions to ultrasonics and instrumentation." Other Awards: Fellow, Acoustical Society of America, 1959.

ROTHE, FREDERICK S., AIEE Fellow 1962. Born: August 21, 1913, Union City, Pa. Degrees: B.Sc., 1935, Drexel University; M.E.E., 1965, Rensselaer Polytechnic Institute.

Fellow Award "For contributions to the development and application of electrical system analysis."

ROTMAN, WALTER, Fellow 1969. Born: August 24, 1922, St. Louis, Mo. Degrees: B.S.E.E., 1947, M.S.E.E., 1948, Massachusetts Institute of Technology.

Fellow Award "For contributions to antenna technology and to the better understanding of the interaction of electromagnetic waves with plasmas"; Chairman, Boston Chapter, IEEE Microwave Theory and Techniques Group, 1956; IEEE Centennial Medal, 1984. Other Awards: Decoration for Exceptional Civilian Service, U.S. Air Force, 1948-80.

ROUBINE, ELIE, Fellow 1976. Born: 1913, Paris, France. Degrees: Agrege de l'Universite, 1935, Ecole Normale Superieure; Docteur es-Sciences, 1945, University of Paris.

Fellow Award "For contributions in teaching and research in the field of electromagnetic theory, communication theory, and applied mathematics." Other Awards: Blondel Medal, Societe Francaise des Electroniciens; Member, Societe Francaise de Physique; Blondel Medal, Societe Francaise Des Electriciens, 1953; Gen. Ferrie Prize, Academie Des Sciences, 1974.

ROWE, HARRISON E., Fellow 1971. Born: January 29, 1927, Chicago, Ill. Degrees: B.S., 1948, M.S., 1950, Sc.D., 1952, Massachusetts Institute of Technology.

Fellow Award "For contributions to the understanding of nonlinear microwave circuits and noise in communications systems and media"; Microwave Prize (with Dale T. Young), 1972, "For significant contributions to the theory of millimeter waveguides and optical fibers in their companion papers entitled 'Transmission Distortion in Multimode Random Waveguides, and 'Optimum Coupling for Random Guides with Frequency-Dependent Coupling'"; David Sarnoff Award (with J.M. Manley), 1977, "For their work on the properties of nonlinear devices resulting in the well-known Manley-Rowe relations."

ROWE, JOSEPH E., Fellow 1965. Born: June 4, 1927, Highland Park, Mich. Degrees: B.S.E., 1951, M.S.E., 1952, Ph.D., 1955, The University of Michigan.

Fellow Award "For contributions to electronics and engineering education." Other Awards: Curtis W. McGraw Research Award, American Society for Engineering Education, 1964; Fellow, American Association for the Advancement of Science, 1968; Distinguished Faculty Achievement Award, University of Michigan, 1970; Member, National Academy of Engineering, 1977.

ROYDEN, GEORGE T., IRE Fellow 1933. Born: June 20, 1895, Fort Clark, Tex. Degrees: A.B., 1917, Eng., 1924, Stanford University.

Awards: IRE Professional Group on Communications Systems, Special Award, 1961, "For his leadership and guidance as National Chairman of the Professional Group on Communications Systems"; IEEE Centennial

Medal, 1984. Other Awards: Sigma Xi, 1925; Silver Beaver, Boy Scouts of America, 1956.

ROYS, HENRY E., IRE Fellow 1955. Born: January 7, 1902, Beaver Falls, Pa. Degrees: B.S., 1925, University of Colorado.

Fellow Award "For his contributions to the improvement of disk and tape recording"; Achievement Award, IRE, Professional Group on Audio, 1956, "For contribution to technology as documented in publications of the IRE." Other Awards: Fellow, Audio Engineering Society, 1953; Fellow, Acoustical Society of America, 1954; Emile Berliner Award, Audio Engineering Society, 1957; Distinguished Alumnus Award, University of Colorado, 1967; Gold Medal Award, Audio Engineering Society, 1973.

RUBEL, JOHN H., Fellow 1964. Born: April 27, 1920, Chicago, Ill. Degrees: B.S.E.E., 1942, California Institute of Technology.

Fellow Award "For contributions to the management of research and engineering." Other Awards: Meritorious Civilian Service Award, 1960; Distinguished Civilian Service Award, 1963.

RUBENS, SIDNEY M., Fellow 1968. Born: March 21, 1910, Spokane, Wash. Degrees: B.S., 1934, Ph.D., 1939, University of Washington.

Fellow Award "For outstanding contributions in the applications of magnetics to computer technology." Other Awards: Sigma Xi, 1939; Meritorious Civilian Service Award, U.S. Navy Dept., 1945.

RUBIN, LAWRENCE G., Fellow 1982. Born: September 17, 1925, Brooklyn, NY. Degrees: B.S.(Physics), 1949, University of Chicago; M.A.(Physics), 1950, Columbia University.

Fellow Award "For contributions to cryogenic temperature and high-strength magnetic field measurement technology." Other Awards: Fellow, American Physical Society, 1972.

RUBIN, WILLIAM L., Fellow 1970. Born: December 12, 1926, Brooklyn, N.Y. Degrees: B.S.E.E., 1948, Columbia University; M.E.E., 1953, Sc.D.E.E., 1960, Polytechhnic Institute of Brooklyn.

Fellow Award "For contributions to the performance of radar systems by signal processing techniques"; M. Barry Carlton Award for Best Paper, 1962. Other Awards: Eta Kappa Nu, Sigma Xi.

RUBINOFF, MORRIS, IRE Fellow 1962. Born: August 20, 1917, Toronto, Ontario, Canada. Degrees: B.A., 1941, M.A., 1942, Ph.D., 1946, University of Toronto.

Fellow Award "For contributions to electronic computer developments and education"; Award, IRE, Electronic Computers Group, 1955, "For the most significant contribution to the electronics field for the year."

RUCKER, CHARLES T., Fellow 1982. Born: June 30, 1931, Augusta, GA. Degrees: B.E.E., 1957, Georgia Institute of Technology.

Fellow Award "For contributions to the development of techniques for power combining of microwave semiconductor devices"; Vice President, IEEE MTT-S, 1982; Member, IEEE MTT-S Administrative Committee, 1976-present; President, IEEE MTT-S, 1983; IEEE Centennial Medal, 1984.

RUDGE, ALAN W., Fellow 1983. Born: October 17, 1937, London, England. Degrees: Dip.E.E., 1964, London Polytechnic; Ph.D., 1968, Birmingham University UK.

Fellow Award "For contributions to the design and analysis of reflector antennas." Other Awards: Polytechnic Governors Medal, 1964; Sigma Xi, 1970; Electronics Letters Premium, 1975; Fellow, Institution of Electrical Engineers UK, 1979.

RUDIN, HARRY R., JR., Fellow 1983. Born: January 23, 1937, New Haven, Connecticut. Degrees: B.E., 1958, M.E., 1960, D.Eng., 1964, Yale University.

Fellow Award "For contributions to the analysis of computer communication network performance and protocols."

RUEGER, LAUREN J., Fellow 1984. Born: December 30, 1921, Archbold, OH. Degrees: B.Sc.(Engineering Physics), M.Sc.(Physics), Ohio State University.

Fellow Award "For contributions to and leadership in satellite navigation technology."

RUEHLI, ALBERT E., Fellow 1984. Born: June 22, 1937, Zurich, Switzerland. Degrees: Telecommunication Engineer, 1963, A.T.Z. Zurich; Ph.D., 1972, University of Vermont.

Fellow Award "For key contributions to the field of computer-aided design"; Guillemin-Cauer Price Paper Award, IEEE Grants and System Society, 1982. Other Awards: IBM Outstanding Contribution Awards 1975, 1978, 1982.

RUELLE, GILBERT L., Fellow 1977. Born: November 2, 1926, Eguilly, France. Degrees: Dipl.Eng.(E.E.), 1951, Ecole Nationale Superieure d'Electricite et Mecanique de Nancy, France; Dr. (honoris causa), 1980, Liege Universite.

Fellow Award "For contributions to the development and construction of very large electrical rotating machines." Other Awards: Ampere Medal and Mascart Medal, French Society of Electrical Engineers, 1953; Blondel Medal, France, 1966; Montefiore Medal, Belgium, 1976; Doctor Honoris Causa, Liege Universite, 1980.

RUGGLES, LEONARD L., AIEE Fellow 1950. Born: Chicago Heights, Ill. Degrees: B.S.E.E., 1918, Purdue University.

Other Awards: Talbot G. Martin Award.

RUINA, JACK P., Fellow 1966. Born: August 19, 1923. Degrees: B.E.E., 1944, City College of the City University of New York; M.E.E., 1949, D.E.E. 1951, Polytechnic Institute of Brooklyn.

Fellow Award "For contributions to the development of high resolution radar systems and outstanding technical direction of defense research and development programs." Other Awards: Flemming Award, Junior Chamber of Commerce, 1962; Meritorious Civilian Service Award, Secretary of Defense, 1963.

RUMSEY, VICTOR H., IRE Fellow 1960. Born: November 22, 1919, Devizes, Wilts, England. Degrees: B.A., 1941, University of Cambridge; D.Eng., 1962, Tohoku University, Japan; D.Sc., 1972, University of Cambridge.

Fellow Award "For contributions to antenna theory and practice"; Morris N. Liebmann Award, IRE, 1962, "For basic contributions to the development of frequency independent antennas." Other Awards: Guggenheim Fellow, 1964; Outstanding Teachers of America Award, 1971; National Academy of Engineering, 1980; George Sinclair Award, Ohio State University, 1982.

RUNGE, WILHELM T., Fellow 1974. Born: June 10, 1895, Hannover, Germany. Degrees: Dipl.Ing., Dr.Ing., 1923, Technische Hochschule Darmstadt.

Fellow Award "For early leadership in microwave techniques." Other Awards: Dr.Ing.E.H., Technische Universitaet Berlin, 1966; V.D.E. Ehrenring, 1968.

RUSCH, WILLARD V. T., Fellow 1976. Born: July 12, 1933. Degrees: B.S.E.E.(summa cum laude), 1954, Princeton University; M.S.E.E., 1955, Ph.D.(E.E.), 1959, California Institute of Technology.

Fellow Award "For contributions to reflector antennas and to engineering education." Other Awards: Class Valedictorian, Princeton University, 1954; National Science Fellow, California Institute of Technology, 1954-1958; Fulbright Scholar, Technische Hochschule Aachen, Germany, 1959-1960; Teaching Excellence Award, University of Southern California Associates, 1965; National Science Foundation Faculty Fellowship, 1966; Award for Contributions to Space Science, NASA, 1970; Distinguished Faculty Award, University of Southern California Engineering Alumni Association, 1970; Santa Fe International Professional Ethics Award, 1974; Senior U.S. Scientist Prize, Alexander von Humboldt Foundation, Max Planck Institut fuer Radioastronomie, Bonn, Germany, 1980-1981.

RUSHING, FRANK C., AIEE Fellow 1960. Born: July 11, 1906, Nordheim, Tex. Degrees: B.S., 1928, University of Texas; M.S., 1930, University of Pittsburgh.

Fellow Award "For contributions to dynamic balancing of rotating electrical apparatus." Other Awards: Fellow, American Society of Mechanical Engineers, 1967.

RUSSELL, FREDERICK A., Fellow 1976. Born: April 18, 1915, New York, N.Y. Degrees: B.S.E.E., 1935, E.E., 1939, Newark College of Engineering; M.S., 1941, Stevens Institute of Technology; Sc.D.(E.E.), 1953, Columbia University.

Fellow Award "For leadership in electrical engineering education." Other Awards: Senior Scientific Simulation Award, 1966; Member: Sigma Xi, Tau Beta Pi, Eta Kappa Nu; Allan R. Cullimore Distinguished Service Award, New Jersey Institute of Technology, 1981.

RUSSO, PAUL M., Fellow 1983.

Fellow Award "For developments in microprocessor technology and its applications to consumer and industrial automation products."

RUSSO, ROY L., Fellow 1983. Born: November 6, 1935, Kelayres, PA. Degrees: B.S.E.E., 1957, M.S.E.E., 1959, Ph.D., 1964, Pennsylvania State U.

Fellow Award "For contributions to design automation for large-scale integrated circuitry"; IEEE Computer Society, Certificate of Appreciation, 1978; IEEE Centennial Medal, 1984. Other Awards: IBM Invention Achievement Award, 1978; IBM Outstanding Contribution Award, 1968; Pennsylvania State University Outstanding Writing Award, 1967.

RUST, WILLIAM M., JR., AIEE Fellow 1953. Born: January 1907, Liberty, Tex. Degrees: B.A., 1928, M.A., 1929, Ph.D., 1931, Rice Institute.

Fellow Award "In recognition of his outstanding contributions to geophysics and communications engineering pertaining to the petroleum industry."

RUSTEBAKKE, HOMER M., Fellow 1980. Born: June 7, 1917, Scobey, MT. Degrees: B.S., 1941, Polytechnic College of Engineering; M.S., 1945, University of Pittsburgh; Ph.D., 1969, Illinois Institute of Technology.

Fellow Award "For developing a power engineering graduate education program, and for contributions to subsynchronous resonance phenomena in power systems." Other Awards: Sigma Xi; Tau Beta Pi; Eta Kappa Nu.

RUTHROFF, CLYDE L., Fellow 1974. Born: February 4, 1921, Sioux City, Iowa. Degrees: B.S.E.E., 1950, M.A., 1952, University of Nebraska.

Fellow Award "For contributions to microwave radio system design."

RUTTER, A. R., AIEE Fellow 1950. Born: June 19, 1894, Wellsburg, W.Va. Degrees: B.S.E.E., 1917, University of Pittsburgh.

RUZE, JOHN, IRE Fellow 1962. Born: May 24, 1916, New York, N.Y. Degrees: B.S., 1938, City College of the City University of New York; M.S, 1940, Columbia University; D.Sc., 1952, Massachusetts Institute of Technology.

Fellow Award "For contributions to the theory of antenna design." Other Awards: Tau Beta Pi; Eta Kappa Nu; Sigma Xi.

RYAN, FRANCIS M., IRE Fellow 1940, AIEE Fellow 1946. Born: February 4, 1895, Seattle, Wash. Degrees: B.S.E.E., 1919, University of Washington.

Other Awards: Registered Professional Engineer,

New York, 1924.

RYDER, JOHN D., AIEE Fellow 1951, IRE Fellow 1952. Born: May 8, 1907, Columbus, Ohio. Degrees: B.E.E., 1928, M.S., 1929, Ohio State University; Ph.D., 1944, Iowa State University; D.Eng., 1963, Tri-State College.

IRE Fellow Award "For his contributions in industrial applications of electronic circuits and to education in radio and allied fields"; The Haraden Pratt Award, IEEE, 1979. Other Awards: Man of the Year, National Electronics Conference, 1970.

RYDER, ROBERT M., IRE Fellow 1957. Born: March 8, 1915, Yonkers, N.Y. Degrees: B.S., 1937, Ph.D., 1940, Yale University.

Fellow Award "For contributions to the development of microwave tubes and applications of transistors."

RYERSON, JOSEPH L., Fellow 1967. Born: October 20, 1918, Goshen, N.Y. Degrees: B.E.E., 1941, Clarkson College of Technology; M.E.E., 1953, Ph.D., 1967, Syracuse University.

Fellow Award "For contributions in the field of earth satellite communications and for leadership in the field of military communications." Other Awards: Honorary Member, Sigma Pi Sigma, 1951; Outstanding Civil Service Award, 1963; Sigma Xi, 1967; NATO Star, 1967; Fellow, A.A.A.S., 1969; Special Achievement Awards, 1969, 1970; Civil Service (Outstanding) Award, 1972; Special Award, American National Metric Council, 1976; Golden Knight Award, Clarkson College, 1976; Listed in: "Who's Who in America," "Who's Who in Government" (1972), "Dictionary of International Biography of England," "Who's Who in Engineering" (1977); Professional Engineer, New York, Indiana.

RYNN, NATHAN, Fellow 1980. Born: December 2, 1923, New York, NY. Degrees: B.E.E., 1944, City College of New York; M.S., 1947, University of Illinois; Ph.D., 1956, Stanford University.

Fellow Award "For contributions to plasma engineering and the conception and the construction of the first Q machine." Other Awards: Fellow, American Physical Society; Senior Fulbright Fellow, France, 1978.

S

SAAD, THEODORE S., Fellow 1965. Born: September 13, 1920, Boston, Mass. Degrees: B.S.E.E., 1941, Massachusetts Institute of Technology.

Fellow Award "For contributions to the microwave field through technical publications and industrial leadership."

SAAL, RUDOLF, Fellow 1977. Born: March 17, 1920, Munich, F.R. Germany. Degrees: Dipl.-Ing., 1947, Dr.-Ing., 1963, Technische Universitaet Muenchen.

Fellow Award "For contributions to filter design and engineering education." Other Awards: Best Paper Award, Nachrichtentechnische Gesellschaft, 1959.

SABA, SHOICHI, Fellow 1979. Born: February 28, 1919, Tokyo, Japan. Degrees: B.E., 1941, University of Tokyo.

Fellow Award "For contributions to the development of fault locators and gas-insulated switch gears." Other Awards: The Development Award, Institute of Electrical Engineers of Japan, 1950.

SABATH, JAKOB, Fellow 1984. Born: July 5, 1925, Modosch, Banat, Jugoslavia. Degrees: E.E., Viena-Moedling, Austria; M.S.E.E., 1949, Technical University, Munich, Germany.

Fellow Award "For contributions to the development and advancement of economically and environmentally compatible power systems using EHV transmission."

SABY, JOHN S., Fellow 1968. Born: March 21, 1921, Ithaca, N.Y. Degrees: B.A., 1942, Gettysburg College; M.S., 1944, Ph.D., 1947, Pennsylvania State University; Sc.D.(Hon), 1969, Gettysburg College.

Fellow Award "For pioneering investigations in semiconductor junction devices"; IEEE Centennial Medal, 1984.

SACHDEV, MOHINDAR S., Fellow 1983. Born: April 1, 1928, Amritsar, India. Degrees: B.Sc.(E.E., M.E.), 1950, Benares Hindu University; M.Sc.(E.E.), 1965, Panjab University; M.Sc., 1967, Ph.D., 1969, University of Saskatchewan.

Fellow Award "For contributions to computer applications for power system analysis and protection." Other Awards: Fellow, Institution of Engineers, India, 1972; Fellow, Institution of Electrical Engineers, U.K., 1981.

SACK, EDGAR A., Fellow 1968. Born: January 31, 1930. Degrees: B.S., 1951, M.S., 1952, Ph.D., 1954, Carnegie Institute of Technology.

Fellow Award "For contributions to control systems and pattern recognition, and for his vigorous leadership in the integrated circuit field." Other Awards: Outstanding Young Electrical Engineer, Eta Kappa Nu, 1959; Fellow, Polytechnic Institute of New York, 1981; Alumni Merit Award, Carnegie Mellon University, 1981.

SADLER, ERNEST K., AIEE Fellow 1948. Born: June 4, 1896, Oakland, Calif. Degrees: B.S., 1916, Polytechnic College of Engineering.

Other Awards: Member, Electric Club of Greater Los Angeles, 1946; Life Member, Tau Beta Pi, Delta of California, 1953; Life Member, Pacific Coast Electrical Association, 1959.

SAEKS, RICHARD E., Fellow 1977. Born: November 30, 1941, Chicago, Ill. Degrees: B.S.E.E., 1964, Northwestern University; M.S.E.E., 1965, Colorado State University; Ph.D., 1967, Cornell University.

Fellow Award "For contributions to circuit and system theory." Other Awards: Member, Society for Industrial and Applied Mathematics; Member, American

Mathematical Society; Member, Sigma Xi; Distinguished Faculty Research Award, Texas Tech University, 1978; Paul Witfield Horn Professorship, Texas Tech University, 1979.

SAGE, ANDREW P., Fellow 1976. Born: August 27, 1933, Charleston, S.C. Degrees: B.S.E.E., 1955, The Citadel; S.M.E.E., 1956, Massachusetts Institute of Technology; Ph.D.(E.E.), 1960, Purdue University.

Fellow Award "For contributions to engineering education and to the theory of systems, identification, estimation, and control"; Barry Carlton Award, IEEE, 1970; Norbert Wiener Award, IEEE, 1980; IEEE Centennial Medal, 1984. Other Awards: Frederick Emmonds Terman Award for Excellence in Engineering Education, American Society for Engineering Education, 1970; Fellow, American Association for the Advancement of Science, 1983; Listed in: "Who's Who in America," "Who's Who in the World," "Who's Who in Engineering."

SAH, CHIH-TANG, Fellow 1969. Born: November 10, 1932, Peking, China. Degrees: B.S.(Engr.Physics), 1953, B.S.E.E., 1953, University of Illinois; M.S., 1954, Ph.D., 1956, Stanford University; Doctor honoris causa, K.U. Leuven, 1975.

Fellow Award "For contributions to the theory, understanding and development of solid-state electronic devices"; Browder J. Thompson Paper Award, IRE, 1962; J.J. Ebers Award, IEEE Electron Devices Society, 1981. Other Awards: Fellow, American Physical Society, 1970; Franklin Institute Award, 1975; Life Fellow, Franklin Institute of Philadelphia, 1975; 1000 Contemporary Scientists Most-Cited, 1965-1978, Institute of Scientific Information, 1981.

SAIN, MICHAEL KENT, Fellow 1978. Born: March 22, 1937, St. Louis, MO. Degrees: B.S.E.E., 1959, M.S.E.E., 1962, St. Louis University; Ph.D., 1965, University of Illinois.

Fellow Award "For contributions to the theory of multivariable control systems"; Distinguished Member, IEEE Control Systems Society, 1983; IEEE Centennial Medal, 1984. Other Awards: Bendix Energy Controls Division Recognition and Commendation Award; Fellow, National Science Foundation; Fellow, National Electronics Conference; Paul Galvin Fellow, Motorola Corporation; General Motors Scholar; Frank M. Freimann Chair in Electrical Engineering, University of Notre Dame, 1982.

SAITO, SHIGEBUMI, Fellow 1974. Born: September 17, 1919, Tokyo, Japan. Degrees: B.E., 1941, D.E., 1951, University of Tokyo.

Fellow Award "For contributions to microwave and laser technology." Other Awards: Awards for Distinguished Paper, Institute of Electronics and Communications Engineers of Japan, 1953, 1959, 1961, 1962; Award for Distinguished Achievement, Institute

of Electronics and Communications Engineers of Japan, 1966; Award for Distinguished Paper, Institute of Electrical Engineers of Japan, 1969; Commendation Award, Minister of Post and Telecommunication of Japan, 1971; Imperial Award for Marked Inventions, Japan Institute of Invention and Innovation, 1975; Distinguished Service Award, Institute of Electronics and Communications Engineers of Japan, 1976; Prime Minister's Award of Communication Year, 1983.

SAITO, YUKIO, Fellow 1978. Born: July 22, 1904. Degrees: B.S., 1931, Tohuku University; Ph.D., 1939, Tokyo Institute of Technology.

Fellow Award "For contributions to the understanding of electrical insulating materials."

SAKAI, HIROSHI, Fellow 1981. Born: November 9, 1921, Japan. Degrees: B.E., 1944, Port Arthur Technical College; Ph.D.E., 1953, Kyoto University.

Fellow Award "For contributions to vehicular communications in a critical high-speed environment"; Chairman, Tokyo Chapter, IEEE, Vehicular Technology Society; Secretary and Treasurer, Tokyo Chapter, IEEE, Vehicular Technology Society, 1979. Other Awards: Expert Member, Examination Committee of Consulting Engineers, Science and Technology Agency, Japanese Government, 1975-1977, 1981-1982; Chairman, Science and Technology Exchange Committee, Sophia University, 1978-1979.

SAKAMOTO, TOSHIFUSA, Fellow 1972. Born: July 16, 1906, Japan. Degrees: B.Eng., 1929, D.Eng., 1940, University of Tokyo.

Fellow Award "For contributions to research and development in the field of communication and biomedical engineering, and to engineering education"; IEEE Centennial Medal, 1984. Other Awards: Distinguished Service Award, Institute of Electronics and Communication Engineers of Japan, 1958; Distinguished Service Award, Institute of Electrical Engineers of Japan, 1964.

SAKSHAUG, EUGENE C., Fellow 1980. Born: October 18, 1923. Degrees: B.E.E., 1952, North Carolina State University.

Fellow Award "For contributions to the development and standardization of surge arresters"; Prize Paper Award, Power Engineering Society, 1971, 1973, 1978; W.R.G. Baker Prize Award, IEEE, 1978; Prize Paper Award, IEEE Surge Protection Devices Committee, 1981, 1983; IEEE Surge Protective Devices Committee Award.

SALATI, OCTAVIO M., Fellow 1973. Born: December 12, 1914, Philadelphia, Pa. Degrees: B.S.E.E., 1936, M.S.E.E., 1939, Ph.D., 1963, University of Pennsylvania.

Fellow Award "For contributions to electromagnetic compatibility technology, and for leadership in engineering education"; Past Chairman Award, Philadelphia Section, 1972; Section Award, Philadelphia Section, 1972, "For leadership in the advancement of engineering education and professional contributions to electromagnetic compatibility technology." Other Awards: Listed in "Who's Who in the East"; Invited by Republic of China to lecture in Xian, Nanging and Shanghai from 1 Sept. thru 15 Dec., 1982.

SALINE, LINDON E., Fellow 1983. Born: March 16, 1924, Minneapolis, Minnesota. Degrees: B.E.E., 1945, Marquette University; M.S.E.E., 1948, Ph.D.(E.E.), 1950, University of Wisconsin.

Fellow Award "For contributions to the professional development of engineers and for leadership in national programs for the entrance of minorities into the engineering profession"; Prize Paper, AIEE, 1955; Achievement Award, IEEE Education Society, 1981. Other Awards: HKN Outstanding Electrical Engineer (Runner-up), 1955; Distinguished Service Citation, University of Wisconsin, 1973; Honorary Doctor of Science, Western N.E. College, 1973; Distinguished Service Award for CPDD, ASEE, 1980; Distinguished Engineering Alumnus Award, Marquette University, 1981.

SALISBURY, WINFIELD W., AIEE Fellow 1947. Born: December 27, 1903, Carthage, III. Degrees: B.A., 1926, University of Iowa; Sc.D.(Hon.), 1950, Cornell College, Iowa.

Other Awards: Harvard Astronomical Annual, 1949; Presidential Certificate of Merit, 1951; Certificate of Merit, Massachusetts Institute of Technology, 1951; Certificate of Merit, Harvard University, 1951.

SALTER, ERNEST H., AIEE Fellow 1951. Born: September 7, 1898, Pikesville, Md. Degrees: B.E., 1922, The Johns Hopkins University.

Other Awards: Fellow, Illuminating Engineering Society, 1950; Tau Beta Pi; Omicron Delta Kappa; Scabbard and Blade.

SALTZBERG, BURTON R., Fellow 1976. Born: June 20, 1933, New York, N.Y. Degrees: B.E.E., 1954, New York University; M.S., 1955, University of Wisconsin; Sc.D., 1964, New York University.

Fellow Award "For contributions to data communication."

SALTZER, JEROME H., Fellow 1983. Born: October 9, 1939, Nampa, Idaho. Degrees: S.B., 1961, S.M., 1963, Sc.D., 1966, Mass. Inst. of Tech.

Fellow Award "For contributions to the design of large-scale computer operating systems."

SALZBERG, BERNARD, IRE Fellow 1952. Born: July 22, 1907, New York, N.Y. Degrees: E.E., 1929, M.E.E., 1933, D.E.E., 1941, Polytechnic Institute of Brooklyn.

Fellow Award "For his contributions in the fields of electron tube development, circuit design, and military electronic systems"; Harry Diamond Award, IRE, 1955, "For his contributions in the fields of electron tubes, circuits, and military electronics."

SALZER, JOHN M., IRE Fellow 1962. Born: September 12, 1917, Vienna, Austria. Degrees: B.S., 1947, M.S., 1948, Case Institute of Technology; Sc.D., 1951, Massachusetts Institute of Technology.

Fellow Award "For contributions to the technology of the electronic digital computers"; Distinguished Member, IEEE Control Systems Society, 1983; IEEE Centennial Medal, 1984. Other Awards: Tau Beta Pi, 1946; Eta Kappa Nu, 1946; Sigma Xi, 1947; Fellow, Institute for the Advancement of Engineering (IAE), 1980.

SAMMIS, WALTER H., AIEE Fellow 1961. Born: June 28, 1896, Hempstead, N.Y. Degrees: E.E., 1917, Columbia University; LL.D.(Hon.), 1958, University of Akron; LL.D.(Hon.), 1958, Theil College.

Fellow Award "For contributions to electric utility engineering, development and management."

SAMMON, JOHN W., JR., Fellow 1982. Born: February 19, 1939, Buffalo, NY. Degrees: B.S.E.E., 1960, U.S. Naval Academy, Annapolis; M.S.(Aeronaut. & Astronaut.), 1962, Massachusetts Institute of Technology; Ph.D.(E.E.), 1966, Syracuse University.

Fellow Award "For technical leadership and direction in applying computer technology to pattern recognition."

SAMPSON, CLIFFORD L., AIEE Fellow 1957. Born: March 31, 1902, Minneapolis, Minn. Degrees: B.S., 1923, M.S., 1925, E.E., 1929, University of Minnesota.

Fellow Award "For excellent performance in directing engineering, construction, and operations in a large operating communications utility during a period of rapid expansion."

SAMSEL, RICHARD W., AIEE Fellow 1961. Born: July 2, 1920, Evanston, Ill. Degrees: B.S.E.E., 1941, Northwestern University.

Fellow Award "For contributions to the development of electroacoustic and electro-optic gear for submarine defense systems." Other Awards: Tau Beta Pi; Sigma Xi.

SAMUEL, ARTHUR L., IRE Fellow 1945, AIEE Fellow 1961. Born: December 5, 1901, Emporia, Kans. Degrees: A.B., 1923, College of Emporia; S.B., 1925, S.M., 1926, Massachusetts Institute of Technology; D.Sc.(Hon.), 1946, College of Emporia.

IRE Fellow Award "For his fundamental work in the field of electronic research and for development of electronic devices of particular value at high frequencies"; AIEE Fellow Award "For contributions to the development of electron discharge devices for operation in the ultra-high frequency range"; Paper Prize, IRE, 1937. Other Awards: Fellow, American Physical Society; American Association for the Advancement of Science; Sigma Xi.

SAMUELSON, ROBERT E., IRE Fellow 1957. Born: February 28, 1911, St. Paul, Minn. Degrees: B.S.M.E., 1933, University of Minnesota; M.S.E.E., 1947; Ph.D., 1949, Northwestern University.

Fellow Award "For leadership in research and development in the field of radio communication." Other Awards: Tau Beta Pi, 1931; Sigma Xi, 1949.

SAMULON, HENRY A., Fellow 1964. Born: December 26, 1915, Graudenz, Poland. Degrees: Dipl.Ing., 1939, Eid. Technische Hochschule, Zurich, Switzerland.

Fellow Award "For contributions to color television and electronic equipment for space vehicle guidance."

SANDBERG, IRWIN W., Fellow 1973. Born: January 23, 1934, Brooklyn, N.Y. Degrees: B.E.E., 1955, M.E.E., 1956, D.E.E., 1958, Polytechnic Institute of New York.

Fellow Award "For contributions to circuit and system theory, including the functional analysis of nonlinear and time-varying circuits and systems." Other Awards: Westinghouse Fellow, 1956; Bell Telephone Laboratories Fellow, 1957, 1958; Eta Kappa Nu; Tau Beta Pi; Sigma Xi; Best Paper Award, Asilomar Conference on Circuits, Systems and Computers, 1970; Distinguished Staff Award, Bell Laboratories, 1983; Member, National Academy of Engineering.

SANDERS, LON L., Fellow 1980. Born: June 13, 1927, New York, NY. Degrees: B.S.E.E., 1953, Newark College of Engineering.

Fellow Award "For pioneering research and continuing technical leadership in the development and standardization of the Microwave Landing System for use by civil and military aircraft." Other Awards: S.W. Gilfillan Professional Paper Award, 1971, 1973; Peter Reggia Omni Award, 1977; Engineer of the Year, San Fernando Valley, 1977.

SANDERS, ROYDEN C., JR., IRE Fellow 1961. Born: August 27, 1917, Camden, N.J.

Fellow Award "For contributions to the art of continuous-wave radar systems." Other Awards: Public Service Award, U.S. Dept. of Navy, 1958.

SANDRETTO, PETER C., IRE Fellow 1954. Born: April 14, 1907, Pont Canvese, Italy. Degrees: B.S.E.E., 1930, E.E., 1939, Purdue University.

Fellow Award "For his contributions to aeronautical communication and navigation." Other Awards: Bronze Star, 1945; Radio Technical Commission for Aeronautics, 1950; C.Eng., FIEE, 1963; Legion of Merit, 1967; Volare Award, Airlines Technical Institute, 1969.

SANGIOVANNI-VINCENTELLI, ALBERTO, Fellow 1983. Born: June 23, 1947, Milano, Italy. Degrees: Dr. Ing., Politecnico di Milano.

Fellow Award "For contributions to circuit simulation and computer aids for the design of integrated circuits"; Best Paper Award, 19th IEEE ACM Design Auto-

mation Conference, 1982; Best Presentation Award, 19th ACM-IEEE Design Automation Conference, 1982; Best Paper Award, 20th IEEE ACM Design Automation Conference, 1983; Cauer-Guillemin Award, Best Paper, IEEE Transactions on CAS and CAD, 1981-1982, 1983. Other Awards: Distinguished Teaching Award, University of California, 1981.

SARAZIN, ARMAND, Fellow 1975. Born: August 8, 1924, Rans-Jura, France. Degrees: Elec.Ing., 1949, ENSEM; Dr.Phys., 1953, University of Paris.

Fellow Award "For contributions to the advancement of nuclear instrumentation and particle acceleration techniques." Other Awards: Medaille Blondel, 1966.

SARD, EUGENE W., Fellow 1973. Born: December 21, 1923, Brooklyn, N.Y. Degrees: S.B.E.E, 1944, S.M.E.E., 1948, Massachusetts Institute of Technology.

Fellow Award "For contributions to the field of low-noise microwave, millimeter wave and infrared receivers."

SARIDIS, GEORGE N., Fellow 1978. Born: November, 17, 1931, Athens, Greece. Degrees: Dipl. of Mech. and Elec. Eng., 1955, National Technical University, Athens; M.S.E.E., 1952, Ph.D., 1965, Purdue University.

Fellow Award "For contributions to the theory of self-organizing control systems"; Interim President, IEEE Robotics and Automation Council, 1983.

SASAKI, HAJIME, Fellow 1983. Born: February 19, 1920, Sendai, Japan. Degrees: B.S., 1941, Ph.D., 1962, Tohoku University.

Fellow Award "For contributions to the development of hybrid circuits used to provide highly reliable communication systems." Other Awards: Technical Achievement Award, Agency of Industrial Science and Technology, Ministry of International Trade and Industry, Japan.

SASSCER, CLARENCE D., AIEE Fellow 1949. Born: March 9, 1900, Northkeys, Md. Degrees: B.S.E.E., 1922, University of Maryland.

SATO, GENTEI, Fellow 1984. Born: March 15, 1926, Sendai, Japan. Degrees: B.Eng., 1947, Ph.D.(Engineering), 1961, Tohoku Univ.

Fellow Award "For research and development in the field of specialized antenna design."

SATO, NORIAKI, Fellow 1982. Born: July 29, 1928, Tokyo, Japan. Degrees: B.Eng., 1953, Ph.D(Eng.), 1965, Tokyo Institute of Technology.

Fellow Award "For contributions to adjustable speed drive motor systems and static power converter technology"; Honorable Mention Award, IEEE Industry Applications Society, 1972. Other Awards: Authorship Prize, Teshima Industrial Education Society, Tokyo, 1981.

SATO, RISABURO, Fellow 1977. Born: September 23, 1921, Furukawa, Miyagi, Japan. Degrees: B.E., 1944, Ph.D., 1952, Tohoku University.

Fellow Award "For contributions to transmission theory and engineering education"; Chairman, IEEE Electromagnetic Compatibility Society, Tokyo Chapter; Certificate of Appreciation, IEEE Electromagnetic Compability Society, 1981; Microwave Prize, IEEE Microwave Theory and Techniques Society, 1982. Other Awards: Paper Award, IEE of Japan, 1952; Culture Award, Kahoku Press, 1963; Award, Japan Invention Association, 1966; Best Paper Award, EMC Symposium and Exhibition, Zurich, 1979; Paper Award, IECE of Japan, 1980; Award of Minister of Posts and Telecommunication, Japan, 1983; Paper Award, ITE of Japan, 1983.

SATTERLEE, WILLIAM W., AIEE Fellow 1950. Born: September 4, 1892, Mount Sinai, N.Y.

Other Awards: Silver W Award, Order of Merit, Westinghouse Electric, 1945.

SAUNDERS, ROBERT M., AIEE Fellow 1962. Born: September 12, 1915, Winnipeg, Manitoba, Canada. Degrees: B.E.E., 1938, M.S.E.E., 1942, University of Minnesota; D.Eng., 1971, Tokyo Institute of Technology.

Fellow Award "For contributions to engineering education and publications in the field of electrical engineering"; President, IEEE, 1977. Other Awards: Simon Fellow, Manchester University, England, 1960.

SAVAGE, CHARLES F., AIEE Fellow 1955. Born: February 24, 1906, Maywood, Ill. Degrees: B.S.E.E., 1928, Oregon State University.

Fellow Award "For contributions in the field of electrical measurement, especially in aircraft engine, guidance and control." Other Awards: Fellow, American Society of Mechanical Engineers, 1968; Honorary Member, C.E.S.S.E., 1971; Fellow, AAAS; Chrmn. Sect. M.; Fellow, AIAA; Prof. Engr., NY, MA, OR.

SAWAZAKI, NORIKAZU, Fellow 1974. Born: April 16,-1913. Degrees: B.E., 1938, D.Eng., 1952, Waseda University, Tokyo.

Fellow Award "For contributions to the development of video tape recorders." Other Awards: Prize Paper Award, Society of Motion Picture and Television Engineers, 1961; Prize Paper Award, Institute of Television Engineers of Japan, 1962; Commendation, Minister of Science and Technology Agency of Japan, 1963; Medal of Honor with Purple Ribbon, Japanese Government, 1968.

SAWYER, OGDEN E., Fellow 1969. Born: January 29, 1911, East Providence, R.I. Degrees: B.S.E.E., 1932, Brown University.

Fellow Award "For contributions to the arts of generation, transmission, and distribution in the expansion of a major power system."

SAWYER, R. TOM, AIEE Fellow 1951. Born: June 20, 1901, Schenectady, N.Y. Degrees: B.E.E., 1923, M.E., 1930, Ohio State University.

Other Awards: Sigma Xi, 1957; Honorary and Life Member, American Society of Mechanical Engineers, 1969; Life Member, S.A.E.; Life Member, American Institute of Aeronautics and Astronautics; Lamme Medal, Ohio State University, 1969; First R. Tom Sawyer Award, Gas Turbine Division, American Society of Mechanical Engineers, 1972; Listed in: "Who's Who in America," "Who's Who in the East (USA)," "World Who's Who in Finance and Industry (USA)," "The National Register of Prominent Americans and International Notables," "Dictionary of International Biography," "Who's Who in Engineering (USA)," "Leaders in American Sciences," "World Who's Who in Science," "Who's Who in Atoms," "Who's Who in Railroading." Honorary Member, Gas Turbine Society of Japan; Editor, Sawyers Gas Turbine Catalogs, 1963, and Gas Turbine International Magazine, 1960.

SAWYER, RALPH S., Fellow 1984. Born: January 9, 1921, Gray, ME. Degrees: B.S. (Electronics and Physics), 1944, Tufts University.

Fellow Award "For leadership in the communications and tracking systems for all U. S. manned spaceflight." Other Awards: Sigma Pi Sigma; NASA Medal for Exceptional Service, Jan., Sept. 1969; Television Academy Award, "Outstanding Achievements in Engineering Development--Color Television from Space", 1970; NASA Medal for Outstanding Leadership, 1981; Certificate of Merit, American Radio League, 1983.

SCANLAN, SEAN O., Fellow 1976. Born: September 20, 1937, Dublin, Ireland. Degrees: B.E.(E.E.), 1959, M.E., 1964, University College Dublin; Ph.D., 1966, University of Leeds; D.Sc., 1972, National University of Ireland.

Fellow Award "For contributions to distributed circuit theory, and for leadership in engineering education." Other Awards: J. J. Thompson Premium (with J. D. Rhodes), 1970, Institution of Electrical Engineers, U.K.

SCHAEFER, EDWARD J., AIEE Fellow 1957. Born: July 10, 1901, Baltimore, Md. Degrees: B.S.E.E., The Johns Hopkins University; D.Eng.(Hon.), Indiana Institute of Technology.

Fellow Award "For contributions to the design and overload protection of small motors." Other Awards: Board of Trustees (Emeritus), The Johns Hopkins University.

SCHAEFER, JACOB W., Fellow 1984. Born: June 27, 1919, Paullina, IA. Degrees: B.S.M.E., 1941, Dr.Sc.(Hon.); 1976, Ohio State University.

Fellow Award "For the invention of the command-guidance principle for ground-to-air/space intercept of an intruding aircraft/missile." Other Awards: U.S.

Army Commendation Medal; U.S. Army Outstanding Civilian Service Medal; Distinguished Alumnus Award from Ohio State University, 1966; The Centennial Award, The Bancroft School, 1983.

SCHAFER, RONALD W., Fellow 1977. Born: February 17, 1938, Tecumseh, Nebr. Degrees: B.S.E.E., 1961, M.S.E.E., 1962, University of Nebraska; Ph.D., 1968, Massachusetts Institute of Technology.

Fellow Award "For contributions to digital signal processing and speech communication"; Senior Award, IEEE Group on Audio and Electroacoustics, 1968 (with A.V. Oppenheim and T.G. Stockham, Jr.), for the paper "Nonlinear Filtering of Multiplied and Convolved Signals"; Achievement Award, IEEE Acoustics, Speech, and Signal Processing Society, 1980; Emanuel R. Piore Award (with L.R. Rabiner), IEEE, 1980. Other Awards: Fellow, Acoustical Society of America, 1975.

SCHALKWIJK, JOHAN P. M., Fellow 1983. Born: November 1, 1936, Rijswijk (Z.H.), The Netherlands. Degrees: M.Sc., 1959, Delft Institute of Technology; Ph.D., 1965, Stanford University.

Fellow Award "For contributions to the development of simple and effective coding schemes for data communication systems"; IEEE Information Theory Group Paper Award (with T. Kailath), 1967.

SCHANDA, ERWIN, Fellow 1984. Born: June 11, 1931, Kirchdorf, Austria. Degrees: Dipl. Ing., 1957, Technical University, Vienna, Austria; Ph.D. (Physics), 1968, University of Bern, Switzerland.

Fellow Award "For theoretical and experimental contributions to microwave emission from the earth and atmosphere."

SCHARFETTER, DONALD L., Fellow 1976. Born: February 21, 1934, Pittsburgh, Pa. Degrees: B.S.E.E., 1960, M.S.E.E., 1961, Ph.D.(E.E.), 1962, Carnegie-Mellon University.

Fellow Award "For contributions to computer modeling of solid-state microwave power sources and other semiconductor devices."

SCHAUFELBERGER, FRED G., Fellow 1980. Born: August 13, 1925. Degrees: B.S.E.E., 1948, Oregon State University.

Fellow Award "For contributions to the application of high-voltage circuit switching devices and their standardization."

SCHAWLOW, ARTHUR L., Fellow 1964. Born: May 5, 1921, Mt. Vernon, N.Y. Degrees: B.A., 1941, M.A., 1942, Ph.D., 1949, University of Toronto; Doctor honoris causa, University of Ghent, 1968; LL.D.(Hon.), 1970, University of Toronto; D.Sc.(Hon.), 1970, University of Bradford, England.

Fellow Award "For contributions to the achievement of coherent light and the concept of the optical maser"; Morris N. Liebmann Award, 1964, "For his pioneering and continuing contributions in the field of optical mas-

ers." Other Awards: Stuart Ballantine Medal, Franklin Institute, 1962; Thomas Young Medal and Prize, Institute of Physics and the Physical Society, U. K., 1963; California Scientist of the Year, 1973; Frederick Ives Medal, Optical Society of America, 1976; Honorary Member, Optical Society of America, 1982; Marconi International Fellowship, 1977; Nobel Prize for Physics, 1981; Schawlow Medal, Laser Institute of America, 1982.

SCHEER, GEORGE B., AIEE Fellow 1953. Born: April 2, 1898, San Francisco, Calif. Degrees: B.S., 1920, University of California at Berkeley.

Fellow Award "For outstanding achievements, ingenuity, and broad engineering ability in designing power supply and electrical equipment layout for aluminum reduction plants and steel mills, including a number of electrical applications new to the industry."

SCHELKUNOFF, SERGEI A., IRE Fellow 1944; AIEE Fellow 1951. Born: January 27, 1897, Samara, Russia. Degrees: B.A., M.A., 1923, State College of Washington; Ph.D., 1928, Columbia University; Dr. Sc. (Hon.), 1982, Mount Holyoke College.

IRE Fellow Award "In recognition of his mathematical contributions to electromagnetic theory"; Morris N. Liebmann Award, IRE, 1942. Other Awards: Stuart Ballantine Medal, Franklin Institute, 1949; Fellow, American Association for the Advancement of Science; John T. Bolljahn Memorial Award, 1969.

SCHELL, ALLAN C., Fellow 1973. Born: April 14, 1934, New Bedford, Mass. Degrees: S.B., S.M., 1956, Sc.D., 1961, Massachusetts Institute of Technology.

Fellow Award "For contributions to antenna pattern synthesis, spatial frequency analysis, and large antenna design"; John T. Bolljahn Award for Best Paper in IEEE TRANSACTIONS on Antennas and Propagation, 1966. Other Awards: Guenter Loeser Memorial Lecturer, AFCRL, 1965.

SCHELLENG, JOHN C., IRE Fellow 1928, AIEE Fellow 1951. Born: November 12, 1892, Freeport, Ill. Degrees: A.B., 1915, Cornell University.

Other Awards: Fellow, Acoustical Society of America.

SCHENKER, LEO, Fellow 1979. Born: January 3, 1922, Vienna, Austria. Degrees: B.Sc., 1942, University of London; M.A.Sc., 1950, University of Toronto; Ph.D., 1954, University of Michigan.

Fellow Award "For contributions to Touch-Tone telephone signaling systems." Other Awards: Duggan Medal, Canadian Institute of Steel Construction, 1951.

SCHERER, HAROLD N., JR., Fellow 1979. Born: April 5, 1929, Plainfield, N.J. Degrees: B.E.(Electrical Engineering), 1951, Yale University; M.B.A.(Finance), 1955, Rutgers University.

Fellow Award "For leadership in the design and operation of ultrahigh-voltage power transmission systems." Other Awards: Tau Beta Pi, 1950; Beta Gamma

Sigma, 1955.

SCHERR, ALLAN L., Fellow 1983. Born: November 18, 1940, Baltimore, Maryland. Degrees: B.S., M.S., 1962; Ph.D., 1965, Massachusetts Institute of Technology.

Fellow Award "For innovation in the field of operating system software for large computer systems and networks." Other Awards: Grace Murray Hopper Award, Association for Computer Machinery(ACM), 1975; Outstanding Contribution Award, IBM Corp., 1971.

SCHILLING, DONALD L., Fellow 1975. Born: June 11, 1935, Brooklyn, N.Y. Degrees: B.E.E., 1956, City College of the City University of New York; M.S.E.E., 1958, Columbia University; Ph.D., 1962, Polytechnic Institute of Brooklyn.

Fellow Award "For contributions to the design of communication systems and to engineering education"; President, IEEE Communications Society, 1980-1981; Director, Division III, IEEE, 1982-1983. Other Awards: Meritorious Service Award, IEEE Communications Society, 1978; H.G. Kayser Distinguished Professor of Electrical Engineering, City College of New York, 1980-present; Coauthored five textbooks in Telecommunications and in Electronics, 100 papers in Telecommunications.

SCHLEIF, FERBER R., Fellow 1972. Born: March 6, 1913, Oroville, Wash. Degrees: B.S., 1935, Washington State University.

Fellow Award "For his developments in power operations and contributions to stabilizing interconnected power systems." Other Awards: Award of Excellence, Dept. of the Interior, 1946; Special Act Award, Bureau of Reclamation, 1966; Engineering Achievement Award, Denver Federal Center Professional Engineers, 1968; Distinguished Service Award, Dept. of the Interior, 1970.

SCHLEIMANN-JENSEN, ARNE, IRE Fellow 1948. Born: July 27, 1906, Lindbjerg, Randers, Denmark.

Fellow Award "For his contributions as an engineer and executive in the electron tube industry in Sweden under adverse war conditions."

SCHLESINGER, PERRY S., Fellow 1982. Born: October 9, 1918, New York, NY. Degrees: B.A.(Math.), 1941, Michigan State; M.S.E., 1950, Union College; D.Eng., 1956, Johns Hopkins University.

Fellow Award "For contributions to the study of electromagnetic wave-plasma interaction and relativistic beam high-power millimeter-wave sources." Other Awards: Best Paper Award, Naval Research Laboratory, 1977, 1978.

SCHLICKE, HEINZ M., Fellow 1967. Born: December 13, 1912, Dresden, Germany. Degrees: Dr.Ing., 1939, Technische Hochschule Dresden, Germany.

Fellow Award "For his pioneering contributions to the understanding and applications of ferrites and high permittivity dielectrics"; Stoddart-Award, IEEE Electromagnetic Compatibility Society, 1982. Other Awards: Fellow, AAAS.

SCHLOEMANN, ERNST, F.R.A., Fellow 1978. Born: December 13, 1926. Degrees: Diplom Physiker, 1953, Ph.D., 1954, University of Gottingen; Post Doctoral Fulbright Fellow, 1955, Massachusetts Institute of Technology.

Fellow Award "For contributions to the theory and development of microwave ferrite materials and devices."

SCHLOSSBERG, VICTOR E., AIEE Fellow 1958. Born: May 28, 1905, New York, N.Y. Degrees: B.S.E.E., 1926, E.E., 1933, Rose Polytechnic Institute; AMP, 1954, Harvard Business School.

Fellow Award "For engineering and administrative leadership in a large steel company."

SCHMERLING, ERWIN ROBERT, Fellow 1971. Born: July 28, 1929, Vienna, Austria. Degrees: B.A., 1950, M.A., 1954, Ph.D., 1958, Cambridge University, England.

Fellow Award "For contributions to research and administration in ionospheric and radio physics." Other Awards: NASA, Exceptional Service Medal, 1979.

SCHMIDT, AUGUST, JR., AIEE Fellow 1951. Born: May 27, 1897, Ithaca, N.Y. Degrees: M.E.(E.E.), 1919, Cornell University.

SCHMIDT, RONALD V., Fellow 1984. Born: March 31, 1944, San Francisco, CA. Degrees: B.S., 1966, M.S., 1968, Ph.D., 1970, University of California, Berkeley (all in E.E.).

Fellow Award "For contributions to integrated optics, acoustic surface waves, and fiber optics."

SCHMIDT, W.C., AIEE Fellow 1962.

Fellow Award "For contributions to the development of wood pole high voltage transmission."

SCHMIDT, WILLIAM G., Fellow 1979. Born: November 19, 1930, New York, N.Y. Degrees: B.E.E., 1957, Manhattan College; M.S.E.E., 1962, Massachusetts Institute of Technology.

Fellow Award "For contributions to the development of time-division multiple-access satellite communications."

SCHMIDT-TIEDEMANN, KARL J., Fellow 1976. Born: July 20, 1929, Dresden, Germany. Degrees: Dipl.(Physics), 1954, Dr.rer.nat., 1957, Hamburg University.

Fellow Award "For contributions in management of industrial research." Other Awards: Honorary Prof. (physics), Univ. Hamburg, 1981; Mbr. of German Physical Soc.; European Physical Soc.; German Soc. of Electrical Engrs.

SCHMITT, HANS J., Fellow 1979. Born: August 3, 1930, Dortmund, Germany. Degrees: Diplom-

Physiker, 1954, Dr. rer. nat., 1955, University of Gottingen.

Fellow Award "For contributions to the field of microwave physics and techniques." Other Awards: Guggenheim Fellow, 1961-62.

SCHMITT, OTTO H., IRE Fellow 1955. Born: April 6, 1913, St. Louis, Mo. Degrees: B.A., 1934, Ph.D., 1937, Washington University.

Fellow Award "For his contributions to the application of electronics to the study of living organisms"; Morlock Award, IEEE, 1963; IEEE Centennial Medal, 1984. Other Awards: Award, Sigma Xi, 1935; Fellow, National Research Council, 1937; Sir Halley Stewart Fellow, 1938; Government Citation, 1950; Fellow, American Physical Society, 1953; Fellow, New York Academy of Sciences, 1959; Lovelace Award, 1960; Wetherill Medal, Franklin Institute, 1972; Minnesota Inventors Hall of Fame, 1978; National Academy of Engineering, 1979.

SCHMITT, ROLAND W., Fellow 1981. Born: July 24, 1923, Seguin, TX. Degrees: B.A., B.S., 1947, M.A., 1948, University of Texas; Ph.D., 1951, Rice University.

Fellow Award "For leadership in directing research and development in electronics, energy, materials, and chemistry"; Engineering Management Award, Region I, IEEE, 1977. Other Awards: Fellow, American Physical Society, 1963; National Academy of Engineering, 1978; Fellow, American Association for the Advancement of Science, 1981.

SCHNEIDER, DARREN B., Fellow 1971. Born: May 28, 1922, St. Francis, Kans. Degrees: B.S.E.E., 1944, Kansas State University.

Fellow Award "For contributions and leadership in developing numerical control equipment for machine tools, and in establishing national and international standards."

SCHNEIDER, HAROLD N., Fellow 1969. Born: March 25, 1920, New York, N.Y. Degrees: B.E.E., 1940, City College of the City University of New York.

Fellow Award "For outstanding creative ability and technical leadership in high voltage power circuit breaker development, design, and testing."

SCHNEIDER, MARTIN V., Fellow 1976. Born: October 20, 1930. Degrees: M.S., 1955, Ph.D., 1959, Swiss Federal Institute of Technology, Zurich, Switzerland.

Fellow Award "For contributions to millimeter-wave integrated circuits and devices"; Microwave Prize, IEEE Microwave Theory and Techniques Society, 1979; Region I Award, IEEE, 1984; IEEE Centennial Medal, 1984. Other Awards: Fellow, American Swiss Foundation for Scientific Exchange, 1955-1956; Member: American Physical Society, American Vacuum Society, American Association for the Advancement of Science, New York Academy of Sciences.

SCHNEIDER, SOL, Fellow 1973. Born: February 24, 1924, New York, N.Y. Degrees: B.A., 1946, Brooklyn College; M.S., 1949, New York University.

Fellow Award "For contributions to high-energy gaseous electron devices." Other Awards: U.S. Army Research and Development Award, 1963, 1978; Secretary of the Army Special Act Award, 1963; Army Science Conference Medallion for Outstanding Scientific Contributions, 1978.

SCHNEIDER, WILLIAM, AIEE Fellow 1962. Born: November 1, 1914, Germany. Degrees: B.S., Carnegie Institute of Technology.

Fellow Award "For contributions to the development of apparatus insulation systems." Other Awards: Member, American Society of Mechanical Engineers.

SCHOEFFLER, JAMES D., Fellow 1976. Born: February 9, 1933. Degrees: B.S.E.E., 1955, M.S.E.E., 1957, Case Institute of Technology; Sc.D.(E.E.), 1960, Massachusetts Institute of Technology.

Fellow Award "For contributions to applied networks, control theory, and computer science, and for leadership in engineering education." Other Awards: Teaching Fellowship, Case Institute of Technology, 1956-1957; General Electric Fellowship of MIT, 1957-1958; Member, Association for Computing Machines; Tau Beta Pi; Eta Kappa Nu; Sigma Xi.

SCHOENFELD, ROBERT L., Fellow 1977. Born: April 1, 1920, New York, N.Y. Degrees: B.A., 1942, Washington Square College, New York University; B.E.E., 1944, Columbia Engineering School; M.E.E., 1949, D.E.E., 1956, Polytechnic Institute of Brooklyn.

Fellow Award "For contributions to the development of instrumentation and techniques for automatic collection of real-time biological data." Other Awards: Postdoctoral Fellow, Sloan Kettering Cancer Research Institute, 1957.

SCHOFIELD, W. RICHISON, AIEE Fellow 1932. Born: October 20, 1895, Philadelphia, Pa.

SCHOLTZ, ROBERT A., Fellow 1980. Born: January 26, 1936, Lebanon, OH. Degrees: E.E., 1958, University of Cincinnati; M.S.E.E., 1960, University of Southern California; Ph.D., 1964, Stanford University.

Fellow Award "For contributions to the theory and design of synchronizable codes for digital communications and radar systems"; Donald G. Fink Prize Award, IEEE, 1984. Other Awards: Distinguished Engineering Alumnus Award, University of Cincinnati, 1982; Leonard G. Abraham Prize Paper Award, 1983.

SCHOLZ, H. J., AIEE Fellow 1946. Born: May 10, 1892, Campbell, Calif. Degrees: A.B., 1915, Stanford University.

SCHOOLEY, ALLEN H., IRE Fellow 1954. Degrees: B.S., 1931, Iowa State University; M.S., 1932, D.Eng.(-Hon.), 1966, Purdue University.

Fellow Award "For his pioneering development of

fire control radar, and his contributions to electronic measurements"; Harry Diamond Award, IRE, 1963, "For contributions in government service to radar and electronic research." Other Awards: Distinguished Civilian Service Award, U.S. Dept. of Navy, 1946; Knight, Legion of Naval Merit, Brazil, 1963; Citation, Supreme Allied Commander, Atlantic Theater, 1965; Iowa Honor Award, Buena Vista College, 1966; Fellow, American Association for the Advancement of Science; Applied Science Trophy, Scientific Research Society of America; American Institute of Physics; American Geophysical Union; Phi Kappa Phi; Sigma Xi; Distinguished Alumnus Citation, Iowa State University, 1969; Marston Gold Medal Engineering Award, Iowa State University, 1974; International Union of Geodesy and Geophysics; International Association for the Physical Sciences of the Ocean.

SCHOOLEY, CHARLES E., Fellow 1964. Born: September 18, 1905, Archie, Mo. Degrees: B.S., University of Missouri.

Fellow Award "For contributions to ground, undersea, radio and satellite communication systems." Other Awards: Engineering Achievement Award, University of Missouri, 1960.

SCHREIBER, WILLIAM F., Fellow 1976. Born: September 18, 1925, New York, N.Y. Degrees: B.S., 1945, M.S., 1947, Columbia University; Ph.D., 1953, Harvard University.

Fellow Award "For contributions to the understanding of image enhancement, and for the design of image processing and transmission systems"; Leonard G. Abraham Prize Paper Award, 1971. Other Awards: Pulitzer Free Scholarship, New York State Scholarship, Columbia University; Gordon McKay Fellowship, 1949-1951, Charles Coffin Fellowship (General Electric Co.), 1951-1953, Harvard University; Journal Award, Society of Motion Picture and Television Engineers, 1960; Bernard Gordon Professorship, MIT, 1980-83; Honors Award, Technical Ass'n of the Graphic Arts, 1983.

SCHRENK, MATTHEW H., IRE Fellow 1957. Born: September 6, 1902, Naperville, Ill. Degrees: B.A., 1922, North Central College; M.A., 1926, Syracuse University; Sc.D.(Hon.), 1959, North Central College.

Fellow Award "For pioneering achievements in naval aviation electronics;" Patron Award, Washington Section, IRE, 1963. Other Awards: Presidential Certificate of Merit, 1946; Edward Longstreth Medal, Franklin Institute, 1962; Superior Service Award, U.S. Dept. of Navy, 1965.

SCHROEDER, ALFRED C., IRE Fellow 1954. Born: February 28, 1915, West New Brighton, N.Y. Degrees: B.S., M.S., 1937, Massachusetts Institute of Technology.

Fellow Award "For his contributions to television re-ceiver circuitry, and his pioneering work in the development of color television"; Vladimir K. Zworykin Award, IEEE, 1971, "For his outstanding technical contributions to television and particularly his leadership in the development of color television." Other Awards: Awards, RCA, 1947, 1950, 1951, 1952, 1957, 1970; David Sarnoff Gold Medal, Society of Motion Picture and Television Engineers, 1965.

SCHROEDER, MANFRED R., Fellow 1971. Born: July 12, 1926, Ahlen, West Germany. Degrees: Diplom Physiker (Physics), 1951, Dr.rer.nat.(Physics), 1954, University of Goettingen.

Fellow Award "For contributions to speech communication, architectural acoustics, and electroacoustics"; W. R. G. Baker Prize Award, IEEE, 1977, for paper "Models of Hearing"; Senior Award (with B.S. Atal), IEEE Acoustics, Speech, and Signal Processing Society, 1979. Other Awards: Gold Medal, Audio Engineering Society, 1972; Foreign Scientific Member, Max Planck Society, 1975; Fellow, Acoustical Society of America; Fellow, Audio Engineering Society; Member, Academy of Sciences at Goettingen, 1973; Member, National Academy of Engineering, 1979; Member, New York Academy of Sciences, 1980; Companion of Western Europe Diploma, Who's Who in Western Europe, 1981.

SCHROEDER, THEODORE W., AIEE Fellow 1958. Born: December 17, 1913, Omaha, Nebr. Degrees: B.Sc., 1936, University of Nebraska.

Fellow Award "For contributions to planning and design of power systems." Other Awards: Eta Kappa Nu; Sigma Xi; Sigma Tau; Pi Mu Epsilon.

SCHUESSLER, HANS W., Fellow 1977. Born: February 28, 1928, Dortmund, West Germany. Degrees: Diplom-Ingenieur, 1954, Ph.D., 1958, Habilitation, 1961, Aachen University.

Fellow Award "For contributions to the theory of analog and digital filters"; Society Award, IEEE Acoustics, Speech, and Signal Processing Society, 1979; IEEE Centennial Medal, 1984. Other Awards: Borchers Medal, Aachen University; Silver Needle, German Society of Electrical Engineers (VDE), 1976, Bundesverdienstkreuz am Bande des Verdienstordens der Bundesrepublik Deutschland, 1980.

SCHULTZ, SOL E., AIEE Fellow 1947. Born: December 23, 1900, Oil City, Pa. Degrees: Certificate in Electrical Engineering, 1923, Drexel Institute of Technology.

Other Awards: Milton H. McGuire Award, Northwest Public Power Association, 1950; Outstanding Achievement Citation, Drexel Institute, 1961.

SCHULZ, ELMER H., AIEE Fellow 1950, IRE Fellow 1958. Born: October 30, 1913, Lockhart, Tex. Degrees: B.S. 1935, M.S., 1936, University of Texas; Ph.D., 1947, Illinois Institute of Technology.

IRE Fellow Award "For his contributions to electronic research."

SCHULZ, RICHARD B., Fellow 1970. Born: May 21, 1920, Philadelphia, Pa. Degrees: B.S.E.E., 1942, M.S.E.E., 1951, University of Pennsylvania.

Fellow Award "For leadership in the field of electromagnetic compatibility, and for technical contributions to RF shielding"; Laurence G. Cumming Award, IEEE Electromagnetic Compatibility Society, 1980; IEEE Centennial Medal, 1984.

SCHUTZ, HARALD, Fellow 1966. Born: June 24, 1907, Vienna, Austria. Degrees: Dipl.Ing., 1929, Dr.Tech.Sci., 1934, Technische Hochschule Vienna.

Fellow Award "For contributions to the performance improvement of military electronics and his contributions to the welfare of the engineering profession."

SCHWAN, HERMAN P., IRE Fellow 1959. Born: August 7, 1915, Aachen, Germany. Degrees: Ph.D., 1940, Dr.Habil., 1946, University of Frankfurt.

Fellow Award "For outstanding leadership in the medical electronic field"; Award, IRE, Philadelphia Section, 1963, "For his outstanding leadership in research and education in the electromedical field which has led to national recognition of Philadelphia as a center for such effort"; First Paper Prize, AIEE, Mideast District, 1952; W.J. Morlock Award, IEEE, 1967, "For contributions in education and research and for inspiration of others by his personal example and scientific achievement, for pioneering a formal academic program in bioengineering, and for research in the field of physical properties of biological materials, particularly with respect to the electrical properties of tissues and cells"; Edison Medal, IEEE, 1983. Other Awards: Fellow, American Association for the Advancement of Science, 1959; Foreign Scientific Member, Max Planck Society, 1961; Citation of Appreciation, U.S. Dept. of Health, Education and Welfare, 1966; Rajewsky Prize, 1974; Member, National Academy of Engineering, 1975; MacKay Visiting Professor, University of California, Berkeley, 1956; W.W. Clyde Visiting Professor, University of Utah, 1980; U.S. Senior Scientist Award, Alexander von Humboldt Foundation, West Germany, 1980-1981.

SCHWARTZ, JAMES W., Fellow 1980. Born: February 11, 1927. Degrees: B.E.P.(Eng. Physics), 1951, M.E.P.(Eng. Physics), 1952, Cornell University.

Fellow Award "For contributions to the theory and development of color television picture tubes and single-sideband aircraft radio equipment." Other Awards: Sigma Xi, 1951.

SCHWARTZ, LEONARD S., IRE Fellow 1961, AIEE Fellow 1961. Born: May 28, 1914, Pittsburgh, Pa. Degrees: B.S., 1936, M.S., 1939, University of Pittsburgh.

IRE Fellow Award "For applications of information and decision theory to communication systems"; AIEE Fellow Award "For contributions to the theory of communication systems." Other Awards: New York Academy of Sciences.

SCHWARTZ, MISCHA, Fellow 1966. Born: September 21, 1926, New York, N.Y. Degrees: B.E.E., 1947, Cooper Union; M.E.E., 1949, Polytechnic Institute of Brooklyn; Ph.D., 1951, Harvard University.

Fellow Award "For contributions to communication systems analysis and engineering education"; IEEE Education Medal, 1983; Director, IEEE, 1978, 1979; Chairman, Administrative Committee, IEEE Information Theory Group, 1964-1965; Vice President, IEEE Communications Society, 1982-1983, President, 1984-. Other Awards: Sperry Gyroscope Fellow, 1949-51; Science Faculty Fellow, National Science Foundation, 1965-66; Distinguished Visitor Award, Australian-American Education Foundation, 1975; Fellow, AAAS, 1977; Visiting Scientist Award, Nippon T & T, 1981; Great Teacher Award, Columbia University, 1983; Chairman, Commission C, USNC-URSI, 1978-1981.

SCHWARTZ, MORTON I., Fellow 1982. Born: May 31, 1934, New York, NY. B.E.E., 1956, City College of New York; M.E.E., 1959, Ph.D.(Eng.Sc.), 1964, New York University.

Fellow Award "For leadership and personal contributions to the practical realization of optical fiber technology"; Engineer of the Year Award, Region 3, IEEE, 1979.

SCHWARTZMAN, LEON, Fellow 1976. Born: February 6, 1931, Brooklyn, N.Y. Degrees: B.E.E., 1958, M.S.E.E., 1963, Polytechnic Institute of Brooklyn.

Fellow Award "For contributions to the theory and design of phased array radar antennas." Other Awards: Sigma Xi.

SCHWARZ, HARVEY F., Fellow 1978. Born: October 31, 1905, Edwardsville, IL. Degrees: B.Sc., 1926, Washington University.

Fellow Award "For contributions to electronic navigation systems"; Pioneer Award, IEEE Aerospace and Electronic Systems Group, 1969. Other Awards: Gold Medal, Institute of Navigation, 1956; First Pioneer Award, Airlines Avionics Institute, 1970; Honorary Commander of the Order of the British Empire, 1971; Founder, Member, Fellowship of Engineering; Past President, Fellow, Institute of Electronics and Radio Engineers, Great Britain.

SCHWARZ, RALPH J., IRE Fellow 1962. Born: June 13, 1922, Hamburg, Germany. Degrees: B.S., 1943, M.S., 1944, Ph.D., 1949, Columbia University.

Fellow Award "For contributions to engineering education and to system theory"; Chairman, IEEE Group on Circuit Theory, 1963-1965. Other Awards: Great Teacher Award, Columbia University, 1965; Thayer Lindsley Professor of Electrical Engineering, Columbia University, 1976.

SCHWEPPE, FRED C., Fellow 1977. Born: November 18, 1933. Degrees: B.S.E.E., 1955, M.S.E.E., 1957, University of Arizona; Ph.D., 1959, University of Wisconsin at Madison.

Fellow Award "For contributions to the application of state estimation, load forecasting, and system dynamics to the control of electric power systems." Other Awards: Tau Beta Pi; Sigma Xi; Sigma Pi Sigma; Pi Mu Epsilon.

SCIFRES, DONALD R., Fellow 1981. Born: September 10, 1946, Lafayette, IN. Degrees: B.S., 1968, Purdue University; M.S., 1970, Ph.D., 1972, University of Illinois.

Fellow Award "For contributions to the science and technology of diode lasers"; Technical Committees Chairman and elected Member, Administrative Committee, IEEE Quantum Electronics and Applications Society, 1980; Electro-Optics Technical Committee Chairman, Administrative Committee, IEEE Electron Devices Society. Other Awards: Fellow, University of Illinois, 1968-1969; Fellow, General Telephone and Electronics, 1970-1972; Fellow, Optical Society of America, 1978; Research Fellow, Xerox Corp., 1982.

SCLAR, NATHAN, Fellow 1982. Born: March 22, 1920, New York, NY. Degrees: B.S., 1948, New York University; M.S., 1952, Ph.D., 1967, Syracuse University.

Fellow Award "For contributions to infrared detector array technology." Other Awards: Principal Scientist, Rockwell International Science Center, 1978; Certificate of Recognition, NASA Infrared Astronomical Satellite Program (IRAS), 1980; Fellow, American Physical Society, 1981; American Association for the Advancement of Science, 1982; Member, Smithsonian Institution, 1982.

SCOTT, NORMAN R., Fellow 1964. Born: May 15, 1918, New York, N.Y. Degrees: B.S., M.S., 1941, Massachusetts Institute of Technology; Ph.D., 1950, University of Illinois.

Fellow Award "For contributions to engineering education and to analog and digital computer technology."

SCOTT, RONALD E., Fellow 1966. Born: March 25, 1921, Leslie, Saskatchewan, Canada. Degrees: B.A.Sc., 1943, M.A.Sc., 1944, University of Toronto; Sc.D., 1950, Massachusetts Institute of Technology.

Fellow Award "For contributions to engineering education and technical advances in the fields of computers and network theory." Other Awards: Tau Beta Pi; Sigma Xi; Eta Kappa Nu; Phi Kappa Phi; American Society for Engineering Education; Armed Forces Communications and Electronics Association; Best Paper Prize, American Society for Engineering Education, 1955.

SCOTT, W. MAXWELL, JR., Fellow 1966. Born: October 10, 1904, Philadelphia, Pa. Degrees:

Dr.Eng.(honoris causa), 1973, Drexel University.

Fellow Award "For contributions to electrical distribution and protection systems"; Award, Philadelphia Section, 1967, "For creativity, leadership and inspiration which have enriched the economic, scientific and educational growth of the Delaware Valley." Other Awards: Honorary Member, Eta Kappa Nu, 1958.

SCOVILLE, MERRITT E., AIEE Fellow 1949. Born: November 9, 1905, Valparaiso, Nebr. Degrees: B.S.E.E., 1930, University of Nebraska.

Fellow Award "For design, application, and development of power capacitors." Other Awards: Charles A. Coffin Award, General Electric, 1946; Elfun Award, General Electric, 1965; Charter Day Award, Glens Falls Rotary Club, 1982.

SEACAT, RUSSELL H., Fellow 1980. Born: March 31, 1924. Degrees: B.S.E.E., 1948, M.S.E.E., 1958, Ph.D.(E.E.), 1963, Texas A & M University.

Fellow Award "For innovative contributions to electrical engineering education."

SEAMAN, JOSEPH W., AIEE Fellow 1958. Born: May 14, 1905, Mishawaka, Ind. Degrees: B.S., 1929, Antioch College; A.M.P., 1948, Harvard University.

Fellow Award "For contributions to the design and manufacture of power circuit breakers."

SEARS, RAYMOND W., IRE Fellow 1961. Born: January 9, 1906, Cambridge, Ohio. Degrees: B.A., 1928, Ohio Wesleyan University; M.S., 1929, Ohio State University.

Fellow Award "For contributions in the field of coding and storage electron tubes"; The Haraden Pratt Award, IEEE, 1980.

SEDRA, ADEL S., Fellow 1984. Born: November 2, 1943, Assiout, Egypt. Degrees: B.Sc., 1964, Cairo University; M.A.Sc., 1968, University of Toronto; Ph.D., 1969, University of Toronto (all E.E.).

Fellow Award "For contributions to the theory and design of active-RC and switched-capacitor filters, and to engineering education."

SEELEN, HARRY R., Fellow 1965. Born: September 27 1907, New York, N.Y. Degrees: B.S., 1929, D.B.A.(-Hon.), 1968, Providence College.

Fellow Award "For contributions to the design and development of electronic receiving tubes and color kinescopes." Other Awards: Award of Merit, RCA, 1955.

SEELY, SAMUEL, IRE Fellow 1955. Born: May 8, 1909, New York, N.Y. Degrees: E.E., 1931, Polytechnic Institute of Brooklyn; M.S., 1932, Stevens Institute of Technology; Ph.D., 1936, Columbia University.

Fellow Award "For his contributions as an educator, author and as a director of research and development." Other Awards: Army-Navy Certificate of Appreciation, 1946; Fellow, American Physical Sociey, 1948; Certificate of Distinction, Polytechnic Institute of Brooklyn,

1956; Silver Cross, Royal Order of Phoenix, Greece, 1967.

SEIDEL, HAROLD, Fellow 1973. Born: October 7, 1922, Brooklyn, N.Y. Degrees: B.E.E., 1943, City College of the City University of New York; M.E.E., 1947, D.E.E., 1954, Polytechnic Institute of Brooklyn.

Fellow Award "For contributions to the field of active and passive microwave networks"; Microwave Prize, 1957, "For the most significant contribution to the field of endeavor of the professional group of microwave theory and techniques, for his paper 'Synthesis of a Class of Microwave Filters.'"

SEKI, HIDEO, Fellow 1976. Born: October 13, 1905, Yonezawa, Japan. Degrees: B.E., 1932, D.Eng., 1943, Tokyo Institute of Technology.

Fellow Award "For contributions to the noise theory in telecommunication systems and to engineering education." Other Awards: Purple Ribbon Medal, 1960; Third Order of Merit with the Order of the Sacred Treasure, 1975.

SEKIGUCHI, TADASHI, Fellow 1983. Born: May 3, 1926, Yonezawa, Japan. Degrees: B.Eng., 1949, Dr. Eng., 1956, University of Tokyo, Japan.

Fellow Award "For contributions to fundamental plasma science and applications and for electrical engineering education." Other Awards: Distinguished Technological Achievement Award, 1979; Excellent Technical Article Awards, 1968, 1970, The Inst. of Elec. Eng. of Japan.

SEKIMOTO, TADAHIRO, Fellow 1976. Born: November 14, 1926, Kobe-City, Japan. Degrees: B.S., 1948, Ph.D., 1962, University of Tokyo.

Fellow Award "For contributions and leadership in the field of communications." Other Awards: Japanese Government Prize, 1976.

SEKINE, YASUJI, Fellow 1981. Born: December 7, 1931, Tokyo, Japan. Degrees: B.Eng., 1954, M.S., 1956, Dr.Eng., 1959, University of Tokyo.

Fellow Award "For contributions to the methodology of power systems analysis, planning, control, and operation." Other Awards: Outstanding Paper Prize, 1960, 1970, 1975; Outstanding Book Prize, 1968, 1972; Distinguished Power Engineer Medal, 1974; Institute of Electrical Engineers, Japan.

SELLERS, JOHN F., SR., AIEE Fellow 1949. Born: July 16, 1903, Akron, Ohio. Degrees: B.S.E.E., 1923, Purdue University.

Fellow Award "For developments in d-c propulsion."

SELS, HOLLIS K., AIEE Fellow 1945. Born: June 12, 1897, Toledo, Iowa. Degrees: 1919, Iowa State University.

SELSTED, WALTER T., IRE Fellow 1960. Born: December 17, 1921, Berkeley, Calif. Degrees: B.S.E.E., 1944, University of California at Berkeley.

Fellow Award "For contributions to the art of magnetic recoding." Other Awards; Fellow, Audio Engineering Society, 1961.

SELVIDGE, HARNER, IRE Fellow 1958. Born: October 16, 1910, Columbia, Mo. Degrees: B.S., 1932, M.S., 1933, Massachusetts Institute of Technology; S.M., 1933, S.D., 1937, Harvard University.

Fellow Award "For his contributions to the field of military electronics."

SEMON, WARREN L., Fellow 1976. Born: January 17, 1921, Boise, Idaho. Degrees: S.B., 1944, University of Chicago; M.A., 1947, Ph.D., 1954, Harvard University.

Fellow Award "For contributions to the theory of switching circuits and techniques of computation."

SENIOR, THOMAS B. A., Fellow 1972. Born: June 26, 1928, Menston-in-Wharfedale, England. Degrees: B.Sc., 1949, M.Sc., 1950, Manchester University; Ph.D., 1954, Cambridge University.

Fellow Award "For his contributions and leadership in studies of diffraction and propagation of electromagnetic waves."

SENSIPER, SAMUEL, IRE Fellow 1960. Born: April 26, 1919, Elmira, N.Y. Degrees: S.B., 1939, Massachusetts Institute of Technology; E.E., 1941, Stanford University; Sc.D., 1951, Massachusetts Institute of Technology.

Fellow Award "For contributions in the fields of microwave instrumentation and radiation." Other awards: Certificate of Commendation, U.S. Dept. of Navy, Bureau of Ships; Industrial Electronics Fellow, Massachusetts Institute of Technology, 1947-48; Fellow, American Association for the Advancement of Science, 1959.

SEO, JUNG UCK, Fellow 1983. Born: November 14, 1934. Degrees: B.S.E.E., 1957, Seoul National University; M.S.E.E., 1963, Ph.D., 1969, Texas A&M University. Born: November 14, 1934, Seoul, Korea. Degrees: B.S., 1957, Seoul National University; M.E., 1963, A&M College of Texas; Ph.D., 1969, Texas A&M University.

Fellow Award "For leadership in building the technological base of defense electronics in Korea." Other Awards: Founder Member, Korea Amateur Radio League, 1954; Eta Kappa Nu, 1962; Ministry of National Defense Distinguished Service Award, 1972; Korean Institute of Electrical Engineers Distinguished Academic Achievement Award, 1974; Ministry of National Defense Distinguished Service Award, 1975; Agency for Defense Development Silver Medal Award for Distinguished Research and Development, 1975; Republic of Korea Iron Tower of Merit for Industrial Development, 1978; Ministry of National Defense Bounty for Distinguished Research and Development, 1979.

SEQUIN, CARLO H., Fellow 1982. Born: October 30, 1941, Winterthur, Switzerland. Degrees:

Diploma(Exp.Physics), 1965, Ph.D.(Exp. Physics), 1969, University of Basel, Switzerland.

Fellow Award "For developments in charge-coupled image sensors and signal processing devices."

SESSLER, GERHARD M., Fellow 1977. Born: February 15, 1931, Rosenfeld, Germany. Degrees: Vordiplom, 1953, University of Munich; Diplom, 1957, Dr.rer.nat., 1959, University of Goettingen.

Fellow Award "For contributions to the field of electroacoustic transducers, particularly the electret microphone"; Senior Award, IEEE Group on Audio and Electroacoustics, 1970, "For contributions to audio technology as documented in publications of The Institute of Electrical and Electronics Engineers." Other Awards: T. D. Callinan Award, Electrochemical Society, 1970; 120 scientific articles; 20 U.S. patents; Fellow Acoustical Society of America; Member, American Physical Society; German Physical Society; Nachrichtentechnische Gesellschaft (Germany).

SEVERIN, HANS K. F., Fellow 1971. Born: May 6, 1920, Grafenwoehr, Germany. Degrees: Dr.rer.nat, 1943, Universitaet Goettingen, Germany.

Fellow Award "For contributions to microwave theory and technology, and to engineering education." Other Awards: Full Professor of Electrical Engineering, 1965, Universitaet Bochum, Germany; International Union of Radio Science, Commission B; Society of German Electrical Engineers; Society of Communication Techniques; German Physical Society; Rotary Club, Hattingen, President, 1983-84.

SFORZINI, MARIO, Fellow 1976. Born: September 10, 1928, Bressana Bottarone (Pavia), Italy. Degrees: D.E.E., 1953, Politecnico of Milano, Italy; M.S.E.E., 1958, Illinois Institute of Technology.

Fellow Award "For contributions to the development of EHV and UHV transmission." Other Awards: Member, CIGRE.

SHAAD, PAUL E., AIEE Fellow 1958. Born: October 6, 1911, Lawrence, Kans. Degrees: B.S., 1933, University of Kansas.

Fellow Award "For contributions in reorganizing and planning an electric power district."

SHAFRITZ, ARNOLD B., Fellow 1973. Born: December 17, 1926, Atlantic City, N.J. Degrees: B.S.E.E., 1949, University of Pennsylvania.

Fellow Award "For contributions to computer systems architecture, logical design, and real-time data processing."

SHAHBENDER, RABAH, Fellow 1971. Born: July 23, 1924, Damascus, Syria. Degrees: B.E.E., 1946, Cairo University; M.S.E.E., 1949, Washington University; Ph.D., 1951, University of Illinois.

Fellow Award "For contributions to magnetic and ultrasonic devices and computer memories." Other Awards: Fellow, University of Illinois, 1949; Outstand-

ing Achievement Award, RCA Laboratories, 1960, 1963, 1974, 1982; Materials in Design Engineering, Awards Competition, 1963-64; AFIPS Best Paper Award, 1963; IR-100 Awards, 1964, 1969.

SHANK, ROBERT J., IRE Fellow 1958. Born: March 19, 1914, La Junta, Colo. Degrees: A.B., 1935, Goshen College; B.S.E.E., 1937, Purdue University.

Fellow Award "For his contributions to airborne military electronics."

SHANKLE, DERRILL F., Fellow 1966. Born: May 6, 1922, Kingsport, Tenn. Degrees: B.S., 1943, University of Pittsburgh.

Fellow Award "For contributions to the more effective use of insulation and to the design of field-test facilities for extra high voltage." Other Awards: Service Award, American National Standards Committee C63, Radio-Electrical Coordination, 1963-1973; Listed in "Who's Who in the East," 1968-69; Phi Eta Sigma, Sigma Tau, Eta Kappa Nu.

SHANKS, JOHN L., Fellow 1981. Born: February 20, 1934, Beeville, TX. Degrees: B.S.E.E., 1955, M.S.E.E., 1956, Ph.D., 1961, Texas A&M University.

Fellow Award "For contributions to the field of digital filter design and stability analysis." Other Awards: Best Presentation Award, Society of Exploration Geophysicists, 1966.

SHANNON, CLAUDE E., IRE Fellow 1950. Born: April 1916, Petoskey, Mich. Degrees: B.S., 1936, University of Michigan; S.M., 1940, Ph.D., 1940, Massachusetts Institute of Technology; S.M.(Hon.), 1954, Yale University; Sc.D.(Hon.), 1961, University of Michigan; 1962, Princeton University; 1964, University of Edinburgh; University of Pittsburgh; 1970, Northwestern University; 1978, University of Oxford; 1982, University of East Anglia.

Fellow Award "For his contributions to the philosophy of new pulse methods and to the basic theory of communications." Other Awards: Member: National Academy of Sciences; American Academy of Arts and Sciences; American Philosophical Society; Leopoldina Academy, DDR.; Royal Netherlands Academy of Arts and Sciences; Tau Beta Pi; Sigma Xi; Phi Kappa Phi; Eta Kappa Nu; Bolles Fellow, M.I.T. (1938-40); National Research Fellow, Institute for Advanced Study, Princeton (1940-41); Fellow of Center for Advanced Study in Behavioral Sciences (1957-58); Visiting Fellow at All Souls College, Oxford (1978); Honorary Fellow of Muir College of University of California; Alfred Noble Prize (1940); Morris Liebmann Memorial Award (1949); Stuart Ballantine Medal (1955); Research Corporation Award (1956); Rice University Medal of Honor (1962); Mervin J. Kelly Award (1962); IEEE Medal of Honor (1966); National Medal of Science (1966); Golden Plate Award (1967); Harvey Prize (1972); Jacquard Award (1978); Harold Pender Award

(1978); Vis. Prof. Elect. Communications, M.I.T. 1956; prof. comm. sci. and math 1957-; Donner Prof. of Science 1958-1979; Emeritus 1978-.

SHAPIRO, GUSTAVE, IRE Fellow 1961. Born: July 6, 1917, The Bronx, N.Y. Degrees: B.E.E., 1956, George Washington University.

Fellow Award "For contributions to the development of electronic miniaturization techniques and components"; Administrative Committee, IEEE Components, Hybrids, and Manufacturing Technology Group, 1952-1981; Organizer, Administrative Committee, Group on Component Parts; Organizer, Washington, D.C., Chapter, Group on Component Parts; 1953; Chairman, Washington, D.C. Chapter, Group on Component Parts, 1954-1955; Editor, IEEE Transactions on Components, Hybrids and Manufacturing Technology, 1955-1977; Member, IEEE Parts, Hybirds, and Packaging Awards Committee, 1956-1976; Administrative Committee, IEEE Microwave Theory and Techniques Group, 1957-1966; Newsletter Editor, IEEE Microwave Theory and Techniques Group, 1958-1967; Chairman, IEEE Parts, Hybrids, and Packaging Group Awards Committee, 1961-1962; Member, IEEE Standards Board, 1966-1976; Chairman, IEEE CHMT Standards Committee, IEEE, 1971-1981; Contribution Award, IEEE Parts, Hybrids, and Packaging Group, 1972; Fellow, Washington Academy of Sciences; 12 Patents; Registered Professional Engineer, District of Columbia.

SHARBAUGH, AMANDUS H., Fellow 1965. Born: March 28, 1919, Richmond, VA. Degrees: A.B., 1940, Western Reserve University; Ph.D., 1943, Brown University.

Fellow Award "For leadership and research in the field of electrical insulation"; First Paper Prize, AIEE, Middle Eastern District, 1962; Second Paper Prize, AIEE, Empire District, 1962; Dakin Award, IEEE Electrical Insulation Society, 1982. Other Awards: Potter Prize in Physical Chemistry, Brown University, 1943; World Record, Microwave Radio Communication, 1946-1982; "Who's Who in America"; "American Men in Science."

SHARP, SAMUEL M., AIEE Fellow 1948. Born: October 14, 1901, Alma, Ark. Degrees: B.E.E., 1923, University of Arkansas.

Other Awards: Bronze Star, U.S. Dept. of Army, 1945.

SHARPLESS, WILLIAM M., IRE Fellow 1958. Born: September 4, 1904, Minneapolis, Minn. Degrees: B.S., 1928, E.E., 1951, University of Minnesota.

Fellow Award "For research and development in the microwave field"; Outstanding Paper Prize, IRE, 1962, for "An Effective Advance in the Field of Ultra-High Speed Pulse Generation." Other Awards: Fellow, American Association for the Advancement of Science,

1967.

SHAW, HERBERT J., Fellow 1973. Born: 1918, Wash. Degrees: B.S.E.E., 1941, University of Washington; M.A., 1943, Ph.D., 1948, Stanford University.

Fellow Award "For contributions to the field of microwave acoustics"; Morris N. Liebmann Memorial Award, IEEE, 1976; Achievement Award, IEEE Group on Sonics and Ultrasonics, 1981. Other Awards: Member, American Physical Society; Tau Beta Pi; Sigma Xi.

SHAW, LEONARD G., Fellow 1984. Born: August 15, 1934, Toledo, OH. Degrees: B.S.E.E., 1956, Univ. of Pennsylvania; M.S.E.E., 1957, Ph.D., 1961, Stanford Univ.

Fellow Award "For contributions to modeling, estimation, and control of stochastic systems."

SHAW, THOMAS R., Fellow 1980. Born: August 7, 1915, Kansas City, KS. Degrees: B.S.(E.E.), Kansas State University.

Fellow Award "For contributions to the utilization of electrical power control and communications in the petroleum industry." Other Awards: Petroleum Industry, 1953; Petroleum Electric Power Association, 1953; Petroleum Industry Electrical Association, 1964; Chairman, Petroleum and Chemical Industry Committee, 1971-1972; Member at Large, 1971-1974; Vice Chairman, Technology Operations Dept., 1975-1976; Chairman, Petroleum and Chemical Technical Conference, 1976; Vice Chairman, Industry Operation Dept., 1974-1976; Certificate of Appreciation, Petroleum and Chemical Committee, 1978.

SHEA, RICHARD F., IRE Fellow 1954. Born: September 13, 1903, Boston, Mass. Degrees: B.S., 1924, Massachusetts Institute of Technology.

Fellow Award "For his contributions to FM receiver design, and his pioneering work in transistor applications"; Nuclear and Plasma Sciences Society Award, 1973, "In recognition of outstanding contributions in the field of nuclear science"; Editor, IEEE Transactions on Nuclear Science, 1958-1983; Editor-in-Chief, IEEE Nuclear and Plasma Sciences Society; IEEE Centennial Medal, 1984.

SHEALY, ALEXANDER N., SR., AIEE Fellow 1962. Born: March 18, 1903, Perry, S.C. Degrees: B.S., 1923, Clemson University.

Fellow Award "For contributions to extra high voltage conductor technology."

SHEARMAN, EDWIN D.R., Fellow 1982. Born: November 17, 1924, Cambridge, England. Degrees: B.Sc.(Eng.), 1945, Imperial College, England.

Fellow Award "For contributions to backscatter radio propagation studies." Other Awards: Leslie McMichael Premium, Institution of Electronics and Radio Engineers, 1957; Blumlein-Brown-Willans Premium, Institution of Electrical Engineers, London, 1963.

SHEEHAN, J. E., AIEE Fellow 1933. Born: March 9, 1895, Montrose, Colo. Degrees: B.A., 1913, Colorado College; LL.B., 1935, Houston Law School.

SHEINGOLD, LEONARD S., IRE Fellow 1962. Born: January 14, 1921, Boston, Mass. Degrees: B.S.E.E., 1947, M.S.E.E., 1949, Syracuse University; M.A., 1950, Ph.D., 1953, Harvard University.

Fellow Award "For contribution to microwave, and leadership in military electronics." Other Awards: Exceptional Civilian Service Award, U.S. Dept. of Air Force, 1962, 1968.

SHELDON, JOHN L., IRE Fellow 1956. Born: September 26, 1904, Battle Creek, Mich. Degrees: B.S., 1935, Battle Creek College; M.S., 1938, Ph.D., 1940, University of Michigan.

Fellow Award "For contributions in the field of glass technology of electron devices." Other Awards: Engineering Award of Excellence, Electronics Industries Association, 1973.

SHELTON, J. PAUL, Fellow 1976. Born: December 24, 1931, Detroit, Mich. Degrees: B.S.(Physics), 1953, Ohio State University.

Fellow Award "For contributions to the theory and design of microwave antennas and feed networks."

SHEN, DAVID W. C., Fellow 1965. Born: January 4, 1920, China. Degrees: B.S., 1938, National Tsing Hua University, Peking, China; Ph.D., 1948, University of London; M.A.(Hon.), 1972, University of Pennsylvania.

Fellow Award "For contributions to engineering education and the theory of adaptive control." Other Awards: Fellow, American Association for the Advancement of Science, 1968; Fellow, Institution of Electrical Engineers, U.K., 1970.

SHENOI, BELLE A., Fellow 1978. Born: December 23, 1929, Belle, India. B.Sc., 1951, University of Madras; D.I.I.Sc., 1955, Indian Institute of Science. M.S., 1958, Ph.D., 1962, University of Illinois.

Fellow Award "For contributions to the theory of active filters."

SHEPARD, B. R., AIEE Fellow 1960. Born: April 10, 1918, Denver, Colo. Degrees: B.S.E.E., 1940, University of Colorado.

Fellow Award "For contributions to guidance systems and associated computers."

SHEPARD, FRANCIS H., JR., Fellow 1969. Born: May 6, 1906, New York, N.Y. Degrees: B.S.M.E., 1929, Yale University.

Fellow Award "For contributions in the field of high-speed printers and electronic instrumentation." Other Awards: Armstrong Medal, The Radio Club of America; P.E., New Jersey; Past President, Radio Club of America, Present Secretary; Director, Armstrong Memorial Foundation.

SHEPHERD, JAMES E., IRE Fellow 1948. Born: May 29, 1910, Houston, Tex. Degrees: B.A., 1932, M.A.-

(E.E.), 1933, University of Missouri; M.S., 1935, D.Sc., 1940, Harvard University; D.Sc.(Hon.), 1959, Ohio Wesleyan University.

Fellow Award "For his contributions to the development of airborne radar armament and for his active participation and leadership in the functions of the Institute." Other Awards: Missouri Honor Award, Distinguished Service in Engineering, 1962; Fellow, American Association for the Advancement of Science, 1977.

SHEPHERD, MARK, JR., Fellow 1964. Born: January 18, 1923, Dallas, Tex. Degrees: B.S., 1942, Southern Methodist University; M.S., 1947, University of Illinois.

Fellow Award "For contributions to the evolution of the semiconductor art and the growth of the semiconductor industry."

SHEPHERD, NEAL H., Fellow 1970. Born: 1920, Texas. Degrees: B.S.E.E., 1942, Texas A & M.

Fellow Award "For contributions in the field of vehicular communication, particularly in effective spectrum utilization"; Honorable Mention, Awards Committee of IEEE Vehicular Technology Group (with co-author), 1961, 1965; Administrative Committee, IEEE Vehicular Technology Society, 1964-1979; Avant Garde, IEEE Vehicular Technology Society, 1980; Founder's Award, IEEE Electromagnetic Compatability Society, 1983. Other Awards: Associate Member, Research Society of America (RESA), 1958; Fellow, Radio Club of America; Life Member, Amateur Athletic Union, 1982.

SHEPHERD, WILLIAM G., IRE Fellow 1952. Born: August 28, 1911, Fort William, Ontario, Canada. Degrees: B.E.E., 1933, Ph.D., 1937, University of Minnesota.

Fellow Award "For his contributions to the development and design of electron tubes, particularly the reflex klystron." Other Awards: Sigma Xi, 1937; Citation, U.S. Dept. of Navy, Bureau of Ships, 1947; Medal of Honor, National Electronics Conference, 1965; Minnesota Engineer of the Year Award, 1966; Member, National Academy of Engineering, 1969; Regents Award, University of Minnesota, 1974; Outstanding Achievement Award, University of Minnesota, 1979.

SHERIDAN, THOMAS B., Fellow 1983. Born: December 23, 1929, Cincinnati, Ohio. Degrees: B.S.M.E., 1951, Purdue Univ., M.S.Eng., 1954, Univ. Calif., Los Angeles; Sc.D., 1959, Mass. Inst. of Technology.

Fellow Award "For contributions to engineering education and man-machine systems." Other Awards: Fellow, Human Factors Society; Paul M. Fitts Award (H.F.S.) "for contributions to education."

SHERMAN, HERBERT, Fellow 1971. Born: February 24, 1920, Brooklyn, N.Y. Degrees: B.E.E., 1940, City College of the City University of New York; M.E.E., 1949, D.E.E., 1955, Polytechnic Institute of Brooklyn.

Fellow Award "For contributions to satellite communications techniques and to medical electronics."

SHERMAN, SAMUEL M., Fellow 1976. Born: September 12, 1914, Camden, N.J. Degrees: B.A, 1934, M.A., 1939, Ph.D., 1965, University of Pennsylvania.

Fellow Award "For contributions to radar systems engineering and signal processing." Other Awards: Phi Beta Kappa; Sigma Xi; Pi Mu Epsilon.

SHERWIN, JAMES L., AIEE Fellow 1949. Born: April 20, 1899, London, Ontario, Canada.

SHEVEL, W. LEE, Fellow 1970. Born: October 26, 1932, Monessen, Pa. Degrees: B.S., 1954, M.S., 1955, Ph.D., 1960, Carnegie-Mellon University.

Fellow Award "For contributions to the development of computer memories and magnetic material." Other Awards: Outstanding Young Electrical Engineer, Eta Kappa Nu, 1961, 1964.

SHEW, E. BARRETT, Fellow 1964. Born: April 6, 1900, Philadelphia, Pa. Degrees: E.E., 1923, Drexel Institute of Technology.

Fellow Award "For contribution to the development of oilless circuit breakers and for leadership in the standardization of metal-clad switchgear."

SHILDNECK, LLOYD P., AIEE Fellow 1960. Born: April 25, 1899, Edgemont, S.D. Degrees: B.S., 1924, University of Nebraska; M.S., 1929, Union College.

Fellow Award "For contributions to the development and design of steam turbine generators."

SHIMA, SHIGEO, Fellow 1968. Born: August 29, 1905, Toyko, Japan. Degrees: B.S., 1933, Waseda University, Japan.

Fellow Award "For contributions to acoustics, broadcasting and studio analysis." Other Awards: Recognition Award, Association for Post and Telecommunications, 1942; Recognition Award, Association for Inventions Promotion, 1943; Recognition Award, Nippon Radio Wave Association, 1950; Recognition Award, Japan Broadcasting Corporation, 1956; Ranju Hosho, Medal of Honor with Blue Ribbon, H.M. Emperor of Japan, 1967; Niwa-Takayanagi Memorial Award, Institute of Television Engineers of Japan, 1972; Maejima Memorial Award, Society for Telecommunications, Japan, 1972; Recognition Award, Minister of Post and Telecommunications, Japan, 1974; Imperial Order of the Rising Sun, H.M. Emperor of Japan, 1975; Recognition Award, Institute of Acoustical Engineers of Japan, 1983.

SHINKAWA, HIROSHI, Fellow 1970. Born: November 13, 1909, Tokyo, Japan. Degrees: D.E., 1954, Waseda University.

Fellow Award "For contributions in the field of radio propagation and in the establishment of an earth station for satellite communications." Other Awards: Blue Ribbon Award, 1970; Award, Institute of Electronics and Communication Engineers of Japan, 1971; ITE Award, 1974; Imperial Order of the Rising Sun, 1981.

SHINNERS, STANLEY M., Fellow 1973. Born: May 9, 1933, New York, N.Y. Degrees: B.E.E., 1954, City College of the City University of New York; M.S.E.E., 1959, Columbia University.

Fellow Award "For contributions to the theory, design and development of control systems." Other Awards: Eta Kappa Nu, 1952; Tau Beta Pi, 1952; Career Achievement Medal, Engineering and Architecture Alumni of the City College, 1980.

SHIPLEY, R. BRUCE, AIEE Fellow 1957. Born: September 27, 1913, Milligan, Tenn. Degrees: B.S., 1936, M.S., 1945, University of Tennessee.

Fellow Award "For originality in the use of mathematical methods for the solution of complex electrical transmission system problems."

SHOBERT, ERLE I., AIEE Fellow 1962. Born: November 19, 1913, DuBois, Pa. Degrees: A.B., 1935, Susquenhanna University; M.A., 1939, Princeton University; D.Sc.(Hon.), 1957, Susquehanna University.

Fellow Award "For contribution to the development of electrical contacts and to the theory of commutation"; Ragnar Holm Scientific Achievement Award, 1974. Other Awards: Fellow, American Association for Advancement of Science, 1953; Award of Merit, American Society for Testing Materials, 1953; Achievement Award, Susquehanna University, 1969; President, American Society for Testing & Materials, 1971-72; 3 books; 40 technical papers; 20 patents.

SHOCKLEY, WILLIAM, IRE Fellow 1955. Born: February 13, 1910, London, England. Degrees: B.Sc., 1932, California Institute of Technology; Ph.D., 1936, Massachusetts Institute of Technology; Sc.D.(Hon.), 1955, University of Pennsylvania; Sc.D.(Hon.), 1956, Rutgers University; Sc.D.(Hon.), 1963, Gustavus Adolphus College.

Fellow Award "For his contributions to the development of the transistor"; Morris N. Liebmann Award, 1952, "In recognition of his contributions to the creation and development of the transistor"; IEEE Gold Medal, 1972; IEEE Medal of Honor, 1980. Other Awards: Medal for Merit, 1946; Oliver E. Buckley Prize, American Physical Society, 1953; Comstock Prize, National Academy of Sciences, 1954; Nobel Prize, 1956; Holley Medal, American Society of Mechanical Engineers, 1963; Wilhelm Exner Medaille, Osterreichischer Gewerbeverien, 1963; Distinguished Service Alumni Award, California Institute of Technology, 1966; National Inventors Hall of Fame, 1974; Fellowship Fund (in the names of Bardeen, Brattain, and Shockley), SEMI, Massachusetts Institute of Technology, 1977.

SHOHET, J. LEON, Fellow 1978. Born: June 26, 1937, Chicago, Ill. Degrees: B.S., 1958, Purdue University; M.S., 1960, Ph.D., 1961, Carnegie-Mellon University.

Fellow Award "For leadership in plasma science and

engineering"; Merit Award, IEEE Nuclear and Plasma Sciences Society, 1978. Other Awards: Frederick Emmons Terman Award, American Society for Engineering Education, 1977.

SHOOMAN, MARTIN L., Fellow 1979. Born: February 24, 1934, Trenton, N.J. Degrees: B.S., 1956, M.S., 1956 (both in Electrical Engineering), Massachusetts Institute of Technology; D.E.E., 1961, Polytechnic Institute of Brooklyn.

Fellow Award "For contributions to the field of reliability engineering"; P.K. McElroy Award for Best Technical Paper, IEEE Reliability Society, 1967, 1971, 1977, 1983; Annual Reliability Award, IEEE Reliability Society, 1977; Best Software Technical Paper, IEEE COMPCON Conference, 1977.

SHORES, RONALD B., Fellow 1966. Born: June 12, 1912, Spartanburg, S.C. Degrees: B.S., 1934, Clemson University.

Fellow Award "For contributions to the design of high voltage oil and air-blast circuit breakers, and related electric utility applications."

SHORT, H. DOUGLASS, AIEE Fellow 1959. Born: 1910, Thetford Mines, Quebec, Canada. Degrees: B.Sc., 1933, Queen's University, Kingston, Ontario.

Fellow Award "For contributions to the design and installation of high-voltage power cables of large capacity"; Second Paper Prize, AIEE, Toronto Section, 1951; First Paper Prize, AIEE, Toronto Section, 1953; First Paper Prize, AIEE Canadian Section, 1953; First Paper Prize, AIEE, Vancouver Section, 1954; First Paper Prize, AIEE, Montreal Section, 1958. Other Awards: Order of "U.E.," Royal Decree, 1972; Registered Professional Engineer; Certified Consulting Engineer.

SHOULTS, DAVID R., AIEE Fellow 1948. Born: June 23, 1903, Storms, Ohio. Degrees: B.S., 1925, Sc.D.(Hon.), 1953, University of Idaho.

SHOUPP, WILLIAM E., IRE Fellow 1951. Born: October 5, 1908, Troy, Ohio. Degrees: A.B., 1931, Miami University, Ohio; A.M., 1933, Ph.D., 1937, University of Illinois; D.Sc.(Hon), 1956, Miami University; D.Sc.(Hon), 1972, Indiana Institute of Technology.

Fellow Award "For his work in applying electronics to nuclear research"; Frederik Phillips Award, 1978. Other Awards: Man of the Year in Science Award, Pittsburgh Junior Chamber of Commerce, 1949; Order of Merit, Westinghouse Electric, 1953; National Academy of Engineering, 1967; Alumni Honor Award, University of Illinois, 1967; Fellow, American Nuclear Society; Fellow, American Physical Society; Fellow, American Society of Mechanical Engineers; Medal, Industrial Research Institute, 1973.

SHOWERS, RALPH M., Fellow 1964. Born: August 7, 1918, Plainfield, N.J. Degrees: B.S.E.E., 1939, M.S.E.E., 1941, Ph.D., 1950, University of Pennsyl-

vania.

Fellow Award "For leadership in engineering education and contributions to radio-frequency interference research"; Member, IEEE Committee on Communications Information Policy, 1978-1982; Chairman, IEC Committee of Action EMC Coordinating Working Group, 1979-; Richard R. Stoddard Award, IEEE EMC Society, 1979; Charles Proteus Steinmetz Award, IEEE, 1982; IEEE Centennial Medal, 1984. Other Awards: Chairman, Am. Nat. Standards Committee C63, Radio-Electrical Coordination, 1968-; Chairman, International Special Committee on Radio Interference (CISPR), 1979-; Vice-President, U.S. National Committee of the International Electrotechnical Commission, 1975-.

SHRADER, WILLIAM W., Fellow 1979. Born: October 17, 1930, Foochow, China. Degrees: B.S.E.E., 1953, University of Massachusetts; M.S.E.E., 1961, Northeastern University.

Fellow Award "For contributions to the theory of radar systems and moving-target indication techniques."

SHTRIKMAN, SHMUEL, Fellow 1973. Born: 1930, Poland. Degrees: B.Sc., 1953, E.E., 1954, D.Sc., 1958, Technion Institute of Technology, Haifa, Israel.

Fellow Award "For contributions to basic and applied magnetism, in particular, the theory of new magnetization processes and its impact on the development of new permanent magnets." Other Awards: Weizmann Prize for Science, Tel-Aviv Municipality, 1968; Michael Landau Prize, Mifal Hapyis Foundation, Israel, 1975; Fellow, American Physical Society, 1976.

SICHAK, WILLIAM, IRE Fellow 1960. Born: January 7, 1916, Lyndora, Pa. Degrees: B.A., 1942, Allegheny College.

Fellow Award "For contributions to the techniques of microwave transmission."

SIDWAY, CHARLES L., AIEE Fellow 1962. Born: December 19, 1902, Chicago, Ill. Degrees: A.B., 1924, Stanford University.

Fellow Award "For contributions to the development of electrical apparatus through standardization."

SIEBERT, WILLIAM M., Fellow 1964. Born: November 19, 1925, Pittsburgh, Pa. Degrees: S.B., 1946, Sc.D., 1952, Massachusetts Institute of Technology.

Fellow Award "For contributions to the theory and application of signal detection methods."

SIEGEL, ROBERT C., AIEE Fellow 1960. Born: January 18, 1898, Los Angeles, Calif. Degrees: B.S.E.E., 1921, University of Wisconsin.

Fellow Award "For contributions to direction of expansion of a state-wide communication system." Other Awards: Citation, University of Wisconsin, 1954.

SIEGMAN, ANTHONY E., Fellow 1966. Born: November 23, 1931, Detroit, Mich. Degrees: A.B., 1952, Harvard University; M.S., 1954, University of California at

Los Angeles; Ph.D., 1957, Stanford University.

Fellow Award "For contributions to the theory and application of masers and lasers to communications"; W.R.G. Baker Award (with D.J. Kuizenga), 1972. Other Awards: Fellow, Optical Society of America, 1967; Guggenheim Fellow, 1969-70; Member, National Academy of Engineering, 1973; J.J.Ebers Award, 1977; R.W. Wood Prize, Optical Society of America, 1980; Fellow, American Physical Society, 1980; Fellow, American Association for the Advancement of Science, 1982.

SIEWIOREK, DANIEL P., Fellow 1981. Born: June 2, 1946, Cleveland, OH. Degrees: B.S.E.E., 1968, University of Michigan; M.S.E.E., 1969, Ph.D., 1972, Stanford University.

Fellow Award "For contributions to the design of modular computer systems." Other Awards: Recognition of Outstanding Young Electrical Engineeers (Hon. Ment.), Eta Kappa Nu, 1977; Frederick Emmons Terman Award, American Society for Engineering Education, 1983.

SIFOROV, VLADIMIR IVANOVICH, Fellow 1967. Born: May 31, 1904, Moscow, Russia. Degrees: B.E.E., 1929, Institute of Electrical Engineering, Leningrad; D.Sc., 1937, Supreme Sertificate Commission of USSR.

Fellow Award "For his contributions and leadership in research in the fields of radio reception and information theory." Other Awards: Corresponding Member, USSR Academy of Sciences, 1953; Scroll of the Fifth International Television Symposium, Montreaux, Switzerland, 1967.

SILJAK, DRAGOSLAV D., Fellow 1981. Born: September 10, 1933, Beograd, Yugoslavia. Degrees: B.S.E.E., 1958, M.S.E.E., 1961, Dr.Sc., 1963, University of Beograd.

Fellow Award "For contributions to the theory of nonlinear control and large-scale systems."

SILLCOX, LEWIS K., AIEE Fellow 1931. Born: April 30, 1886, Germantown, Pa. Degrees: Dipl., 1903, University of Brussels, Belgium; D.Sc., 1932, Clarkson College of Technology; D.Eng., 1941, Cumberland University; LL.D., 1948, Syracuse University; D.Engr., 1951, Purdue University; LL.D., 1955, Queen's University (Canada), LH.D., 1963, Norwich College.

Other Awards: Gold Medal and Honorary Member, Institute of Locomotive Engineers, London, 1932; Gold Medal, American Society of Mechanical Engineers, 1943; Salzberg Medal, Syracuse University; Fellow, American Association for the Advancement of Science; Life Member, A.I.C.E.; Fellow, Institution of Electrical Engineers, U.K.; Fellow, Institute of Mechanical Engineers; Honorary Member, American Railing Engineering Assn.; Pi Tau Sigma; Tau Beta Pi; Sigma Xi; Who's Who in America.

SILVERMAN, DANIEL, IRE Fellow 1953. Born: June 24, 1905, Montreal, Quebec, Canada. Degrees: B.S., 1927, University of California at Berkeley; M.S., 1929, Sc.D., 1930, Massachusetts Institute of Technology.

Fellow Award "In recognition of his application of electronics to geophysical exploration." Other Awards: Medalist, University of California, 1927; Heller Traveling Fellow, University of California, 1928-30.

SILVERMAN, LEONARD M., Fellow 1979. Born: October 21, 1939, England. Degrees: A.B., 1961, B.S., 1962, M.S., 1963, Ph.D., 1966, Columbia University.

Fellow Award "For contributions to the theory of linear systems and its applications"; Prize Paper Award, IEEE Circuits and Systems Society, 1968.

SILVESTER, PETER P., Fellow 1983. Degrees: B.S., 1956, Carnegie Institute of Technology; M.A.Sc., 1958, University of Toronto; Ph.D., 1964, McGill University (all in Elec. Eng.).

Fellow Award "For contributions to the development of finite element methods and their application to electromagnetic field problems."

SIMMONS, ALAN J., Fellow 1979. Born: October 14, 1924, New York, N.Y. Degrees: B.S., 1945, Harvard University; M.S., 1948, Massachusetts Institute of Technology; Ph.D.(Electrical Engineering), 1957, University of Maryland.

Fellow Award "For contributions to the development of microwave and millimeter-wave components and antennas."

SIMON, MARVIN K., Fellow 1978. Born: September 10, 1939, New York, N.Y. Degrees: B.E.E., 1960, The City College of New York; M.S.E.E., 1961, Princeton University; Ph.D., 1966, New York University.

Fellow Award "For analytical contributions to communication system design"; IEEE Centennial Medal, 1984. Other Awards: NASA Paper Award, 1974; NASA Exceptional Service Medal, 1979.

SIMON, RALPH E., Fellow 1972. Born: October 20, 1930, Passaic, N.J. Degrees: B.A., 1952, Princeton University; Ph.D., 1959, Cornell University.

Fellow Award "For his contributions in the conception and practical development of electron emitters using the principle of negative electron affinity and of low-light level camera tubes"; Vladimir K. Zworykin Award, 1975, "For the invention and leadership in the development of the silicon target camera tube, and in the extension of electronic television into new applications." Other Awards: David Sarnoff Outstanding Achievement Award, 1970; Two RCA Achievement Awards.

SIMPSON, JOHN W., AIEE Fellow 1962. Born: September 25, 1914, Glenn Springs, S.C. Degrees: B.S., 1937, United States Naval Academy; M.S.(in E.E.), 1941, University of Pittsburgh; D.Sc.(Hon.), 1968, Seton Hill College; D.Sc.(Hon.), 1973, Wofford College.

Fellow Award "For contributions to the development of nuclear power for submarines, land based plants and spacecraft"; IEEE, Edison Medal, 1971, "For sustained contributions to society through the development and engineering design of nuclear power systems." Other Awards: Fellow, American Nuclear Society, 1960; Fellow, American Society of Mechanical Engineers, 1965; Member, National Academy of Engineering, 1966; Gold Medal for Advancement of Research, American Society of Metals, 1973; Honorary Member, American Society of Mechanical Engineers, 1974; Newcomen Gold Medal in Steam, 1974; George Westinghouse Gold Medal, American Society of Mechanical Engineers, 1976; Walter H. Zinn Award, American Nuclear Society, 1982; Energy Research Advisory Board, U.S. Dept. of Energy; Science Advisory Council, U. of Notre Dame.

SIMPSON, M., Fellow 1969. Born: July 27, 1921. Degrees: B.E.E., 1942, City College of the City University of New York; M.E.E., 1952, Polytechnic Institute of Brooklyn.

Fellow Award "For contributions to phased array radars and electronic countermeasures systems."

SIMPSON, WALTER L., AIEE Fellow 1946. Born: December 7, 1898, Elmvale, Ontario, Canada. Degrees: B.A.Sc., 1924, University of Toronto.

SIMRALL, HARRY C., AIEE Fellow 1962. Born: October 16, 1912, Memphis, Tenn. Degrees: B.S.(E.E.), 1934, B.S.(M.E.), 1935, Mississippi State University; M.S.(E.E.), 1939, University of Illinois.

Fellow Award "For contributions to engineering education." Other Awards: Engineer of the Year Award, Mississippi Engineering Society, 1963; Alumnus of the Year Award, Mississippi State University, 1963; Mississippi Engineering Leadership Award, 1968; Distinguished Service Certificate, National Council of Engineering Examiners, 1971; Distinguished Alumnus Award, Electrical Engineering Alumni Association, University of Illinois, 1978.

SINCLAIR, DONALD B., IRE Fellow 1943, AIEE Fellow 1957. Born: May 23, 1910, Winnipeg, Manitoba, Canada. Degrees: B.S., 1931, M.S., 1932, Sc.D., 1935, Massachusetts Institute of Technology.

IRE Fellow Award "For the development and application of various types of networks for high-frequency measurement of impedance." Other Awards: Sigma Xi, 1935; Presidential Certificate of Merit, 1948; Member, National Academy of Engineering, 1965; Fellow, American Association for the Advancement of Science, 1969.

SINCLAIR, GEORGE, IRE Fellow 1954. Born: November 5, 1912, Hamilton, Ontario, Canada. Degrees: B.Sc., 1933, M.Sc., 1935, University of Alberta; Ph.D., 1946, D.Sc.(Hon.), 1973, Ohio State University.

Fellow Award "For his contributions to the develop-ment of radiating systems and model techniques in antenna measurements"; General A.G.L. McNaughton Award, IEEE, Region 7, 1975. Other Awards: Army-Navy Certificate of Appreciation, 1948; Guggenheim Fellowship, 1958; Distinguished Alumnus Award, Ohio State University, 1966; Fellow, Royal Society of Canada, 1973; Fellow, Engineering Institute of Canada, 1974; Fellow, American Association for the Advancement of Science, 1978; Julian C. Smith Medal, Engineering Institute of Canada, 1980; Fellow Award, Ryerson Institute of Technology, Canada, 1980.

SINGH, AMARJIT, Fellow 1974. Born: November 19, 1924, Ramdas, India. Degrees: M.Sc., 1945, Punjab University; M.Eng.Sc., 1947, Ph.D., 1949, Harvard University; D.Sc.(honoris causa), 1975, Punjabi University.

Fellow Award "For contributions to the knowledge of microwave plasmas and devices, and for research leadership." Other Awards: S.K. Mitra Memorial Award, Institution of Telecommunication Engineers, India, 1969, 1976; Inventions Promotion Board Award, 1972; Distinguished Fellow, Institution of Electronics and Telecommunication Engineers, India, 1973; Fellow, Indian Academy of Sciences, 1974; Award, FICCI, 1979.

SIPRESS, JACK M., Fellow 1975. Born: April 9, 1935, Brooklyn, N.Y. Degrees: B.E.E., 1956, M.E.E., 1957, D.E.E., 1961, Polytechnic Institute of Brooklyn.

Fellow Award "For contributions to the development of pulse-code modulation systems."

SITTIG, ERHARD K., Fellow 1975. Born: June 3, 1928. Degrees: D.I.C. (E.E.), 1954, University of London; M.Sc., 1955, University of Tubingen; Ph.D., 1959, Technical University of Stuttgart.

Fellow Award "For contributions to acoustic wave devices."

SKELLETT, A. MELVIN, IRE Fellow 1956. Born: July 14, 1901, St Louis, Mo. Degrees: A.B., 1924, M.S, 1927, Washington University; Ph.D., 1933, Princeton University.

Fellow Award "For contributions to the physics of the upper atmosphere and to radial beam switching tubes."

SKILLING, HUGH H., AIEE Fellow 1945. Born: September 2, 1905, San Diego, Calif. Degrees: A.B., 1926, Eng., 1927, Stanford University; S.M., 1930, Massachusetts Institute of Technology; Ph.D., 1931, Stanford University.

Awards: Education Medal, IEEE, 1965, "For leadership in undergraduate education and, in particular, for innovation and lucid exposition of complex ideas in his textbooks."

SKIPPER, DONALD J., Fellow 1981. Born: December 3, 1924. Degrees: B.Sc.(Eng.) (first class honors), 1944, Queen Mary College, England.

Fellow Award "For contributions to EHV power cable technology."

SKLANSKY, JACK, Fellow 1981. Born: November 15, 1928, New York, NY. Degrees: B.E.E., 1950, City College of New York; M.S.E.E., 1952, Purdue University; Eng. Sc.D., 1955, Columbia University.

Fellow Award "For contributions to digital pattern classification and medical applications"; Certificate of Appreciation, IEEE Computer Society, 1978. Other Awards: Eta Kappa Nu, 1948; Tau Beta Pi, 1949; Sigma Xi; 1953; Engineer, California, 1977; Annual Award, Pattern Recognition Society, 1979.

SKOLNIK, MERRILL I., Fellow 1969. Born: November 6, 1927, Baltimore, Md. Degrees: B.E., 1947, M.S.E., 1949, Dr.Eng., 1951, The Johns Hopkins University.

Fellow Award "For outstanding contributions to gaseous electronics, antenna theory, and radar systems"; Harry Diamond Award, IEEE, 1983. Other Awards: Heinrich Hertz Premium, Institution of Electronic and Radio Engineers, U.K., 1964; Society of Scholars, The Johns Hopkins University, 1975; Distinguished Alumnus Award, The Johns Hopkins University, 1979; Navy Distinguished Civilian Service Award, 1982.

SKOMAL, EDWARD N., Fellow 1980. Born: April 15, 1926, Kansas City, MO. Degrees: B.A. (Physics), 1947, M.A. (Physics), 1949, Rice University.

Fellow Award "For contributions to the theory and measurement of man-made radio noise"; Paper Award IEEE Vehicle Technology Group, 1970; Certificate of Achievement, IEEE Electromagnetic Compatibility Society, 1971; Richard R. Stoddart Award, IEEE Electromagnetic Compatibility Society, 1980.

SKOOGLUND, CARL M., AIEE Fellow 1954. Born: January 18, 1894, Kane, Pa. Degrees: B.S.E.E., 1919, Pennsylvania State University.

Fellow Award "For contributions to electric system production and electrical design of power plants."

SKOOGLUND, JOHN W., Fellow 1979. Born: January 4, 1930, Pittsburgh, Pa. Degrees: B.S.E.E., 1951, Pennsylvania State University; M.S.E.E., 1958, University of Pittsburgh.

Fellow Award "For contributions to the analysis of power systems stability and the development of generator excitation systems."

SLATTERY, JOHN J., IRE Fellow 1956. Born: August 2, 1909, East Orange, N.J. Degrees: B.S.E.E., 1930, Villanova University; M.S., 1949, Stevens Institute of Technology.

Fellow Award "For contributions and administration in the radar and electronic counter-measure fields." Other Awards: Legion of Honor, U.S. Dept. of Army, 1945; Registered Professional Engineer, Maryland, Arizona, and Hawaii.

SLAUGHTER, JOHN BROOKS, Fellow 1978. Born: March 16, 1934. Degrees: B.S.E.E., 1956, Kansas State University; M.S.E., 1961, University of California at Los Angeles; Ph.D., 1971, University of California at San Diego.

Fellow Award "For contributions to the design of digital, sample-data control systems."

SLAYMAKER, FRANK H., Fellow 1971. Born: 1914, Nebr. Degrees: B.S.E.E., 1941, E.E., 1946, University of Nebraska.

Fellow Award "For contributions to the fields of acoustical and communication engineering." Other Awards: Fellow, Acoustical Society of America; Member, Optical Society of America; Member, Sigma Xi; Sigma Tau.

SLEMMER, WILBUR E., AIEE Fellow 1962. Born: July 2, 1908, Okeene, Okla. Degrees: B.S., 1930, B.S., 1932, M.S., 1934, Oklahoma State University.

Fellow Award "For contributions to the economic planning of domestic and foreign utility systems."

SLEMON, GORDON R., Fellow 1975. Born: August 15, 1924, Bowmanville, Ontario Canada. Degrees: B.A.Sc., 1946, M.A.Sc., 1948, University of Toronto; D.I.C., 1951, Imperial College, London; Ph.D., 1952, D.Sc., 1968, University of London.

Fellow Award "For contributions to the analysis and application of electrical machines and magnetic devices"; Canadian District Prize, AIEE, 1948, 1958; Montreal Section Prize, AIEE, 1955. Other Awards: Western Electric Award for Engineering Education, American Society for Engineering Education, 1965; Centennial Medal; 1967; Fellow, Engineering Institute of Canada, 1978; Ross Medal, Engineering Institute of Canada, 1978, 1983.

SLEPIAN, DAVID, IRE Fellow 1962. Born: June 30, 1923, Pittsburgh, Pa. Degrees: M.A., 1947, Ph.D., 1949, Harvard University.

Fellow Award "For contributions to information theory and noise analysis"; Alexander Graham Bell Award, IEEE, 1981. Other Awards: Fellow, Institute of Mathematical Statistics, 1968; Member, National Academy of Engineering, 1976; Member, National Academy of Sciences, 1977; SIAM von Neumann Lecturer, 1982.

SLETTEN, CARLYLE J., Fellow 1966. Born: 13, 1922, Chetek, Wis. Degrees: B.S., 1947, University of Wisconsin; M.A., 1949, Harvard University.

Fellow Award "In recognition of his contributions to antenna design and to the field of radar object scatter and diffraction." Other Awards: Research and Development Award, Air Research and Development Command, 1956; Fulbright Lecturer, College of Telecommunication, Madrid, Spain, 1963-1964; Decoration for Exceptional Civilian Service, Dept. of Air Force, 1970; Phi Beta Kappa, University of Wisconsin; United Nations Antenna Consultant, Catholic University, Brazil, 1976-1978; President, Solar Energy Technology, Inc., 1.

SLEVEN, ROBERT L., Fellow 1976. Born: April 25, 1932, New York, N.Y. Degrees: B.E.E., 1953, City College of the City University of New York; M.E.E., 1959, Polytechnic Institute of New York.

Fellow Award "For contributions to microwave filters and equalizers and to the development of low-noise amplifiers for satellite communications."

SLOAN, DAVID H., IRE Fellow 1959. Born: September 8, 1905, Lyman, Wash. Degrees: B.S., 1928, M.S., 1929, Washington State University; Ph.D., 1941, University of California.

Fellow Award "For contributions to high power electron tubes."

SLOAN, ROYAL D., AIEE Fellow 1956. Born: June 29, 1891, Mt. Vernon, Mo. Degrees: B.S., 1913, University of Montana; E.E., 1921, Montana State University; M.S., 1936, Massachusetts Institute of Technology.

Fellow Award "For his contributions as an educator and administrator"; Citation, Spokane Section, 1956.

SLOANE, N.J.A., Fellow 1978. Born: October 10, 1939, Beaumaris, Wales. Degrees: B.E.E., 1959, B.A.(-with honors), 1960, University of Melbourne; M.S., 1964, Ph.D., 1967, Cornell University.

Fellow Award "For contributions to the theory of error correction in communication systems"; Editor, IEEE Transactions on Information Theory, 1978-1981. Other Awards: Lester R. Ford Award, Mathematics Association of America, 1978; Chauvenet Prize, Mathematics Association of America, 1979.

SLOTNICK, DANIEL L., Fellow 1977. Born: November 12, 1931, New York, N.Y. Degrees: B.A., 1951, Columbia College; M.A., 1952, Columbia University; Ph.D., 1956, New York University Institute of Mathematical Science.

Fellow Award "For contributions to the development of centrally controlled array computers"; W.W. McDowell Award, IEEE Computer Society, 1983. Other Awards: Prize, American Federation of Information Processing Societies, 1962; Mellon Award, Carnegie-Mellon University, 1967.

SMITH, C. PRICE, Fellow 1971. Born: June 28, 1920, Agency, Mo. Degrees: B.S.E.E., University of Missouri.

Fellow Award "For contributions to the processing and fabrication techniques for large-scale manufacture of shadow-mask color television picture tubes."

SMITH, CARL E., IRE Fellow 1955, AIEE Fellow 1956. Born: November 18, 1906, Eldon, Iowa. Degrees: B.S., 1930, Iowa State University; M.S., 1932, E.E., 1936, Ohio State University.

IRE Fellow Award "For his contributions to broadcast engineering and for his training activities"; AIEE Fellow Award "For his contributions to broadcasting antenna design, military electronics, and electrical engineering

education." Other Awards: Exceptional Civilian Service Award, War Dept., 1946; Distinguished Alumnus Award, Ohio State University, 1974; Distinguished Achievement Citation, Iowa State University, 1980; Honor Leadership Award, Cleveland Institute of Electronics, 1981.

SMITH, DAVID B., IRE Fellow 1948, AIEE Fellow 1955. Born: December 3, 1911, Newton, N.J. Degrees: S.B., 1933, S.M., 1934, Massachusetts Institute of Technology.

IRE Fellow Award "For outstanding leadership in standardization activities, particularly in the field of television"; AIEE Fellow Award "For his work in establishing national color television standards and for his contributions to the radio and television arts." Other Awards: Sigma Xi; Tau Beta Pi; Eta Kappa Nu; American Association for the Advancement of Science.

SMITH, EDWARD P., Fellow 1964. Born: December 11, 1917, Morrill, Kans. Degrees: B.S.E.E., 1939, Kansas State University.

Fellow Award "For contributions to the development of large direct-current machinery for industrial, marine, and excitation applications"; Committees of IEEE, IEC, & NSPE.

SMITH, ERNEST K., Fellow 1965. Born: May 31, 1922, Peking, China. Degrees: B.A.(Physics), 1944, Swarthmore College; M.Sc.(E.E.), 1951, Ph.D.(E.E.), 1956, Cornell University.

Fellow Award "For research on sporadic-E radio propagation." Other Awards: Member, Sigma Xi, 1956; Fellow, American Association for the Advancement of Science, 1966; CCIR 50th Anniversary Diplome d'Honneur, International Telecommunication Union, 1978; Member, Commissions C, E, F, G, International Scientific Radio Union (URSI), U.S. National Committee.

SMITH, FRANK V., AIEE Fellow 1948. Born: April 29, 1900, Kilmore, Australia. Degrees: B.E.E., 1922, M.E.E., 1925, University of Melbourne, Australia.

SMITH, GEORGE E., Fellow 1975. Born: 1930, White Plains, N.Y. Degrees: B.A., 1955, University of Pennsylvania; M.S., 1956, Ph.D., 1959, University of Chicago.

Fellow Award "For contributions to the development of charge-coupled devices"; Morris N. Liebmann Award, 1974, "For invention of the charge-coupled device (CCD) and leadership in the field of metal oxide semiconductor (MOS) device physics." Other Awards: Fellow, American Physical Society, 1963; Stuart Ballantine Medal, 1973; Member, National Academy of Engineering, 1983.

SMITH, GEORGE F., Fellow 1965. Born: May 9, 1922, Franklin, Ind. Degrees: B.S., 1944, M.S., 1948, Ph.D., 1952, California Institute of Technology.

Fellow Award "For contributions in the areas of electron emission, storage tubes and lasers." Other Awards: Fellow, American Physical Society, 1963.

SMITH, GEORGE S., AIEE Fellow 1951. Born: 1889, Hay Springs, Nebr. Degrees: B.S., 1916, E.E., 1924, University of Washington.

Other Awards: Engineer of the Year Award, National Society of Professional Engineers, 1957; Professor Emeritus, University of Washington, 1961.

SMITH, HAROLD A., Fellow 1968. Born: March 17, 1919, Lucknow, Ontario, Canada. Degrees: B.Sc., 1940, D.Sc.(Hon.), 1977, Queen's University, Canada; D.Sc.(Hon.), 1978, McMaster University, Canada.

Fellow Award: "For contributions to the application of nuclear energy to power generation"; McNaughton Medal, IEEE, Region 7, 1978. Other Awards: Engineering Medal of the Professional Engineers of the Province of Ontario, 1966; Fellow, The Royal Society of Canada, 1972; First Recipient of W.B. Lewis Medal, Canadian Nuclear Assoc., 1974; Foreign Associate, National Academy of Engineering (U.S.), 1978; Gold Medal, Association of Professional Engineers of the Province of Ontario, 1981.

SMITH, HAROLD W., Fellow 1970. Born: February 8, 1923, Brookfield, Mo. Degrees: B.S.E.E., 1944, M.S.E.E., 1949, Ph.D., 1954, University of Texas at Austin.

Fellow Award "For contributions to the field of geoscience electronics." Other Awards: General Dynamics Award, 1960.

SMITH, HARRY B., Fellow 1965. Born: August 11, 1921, Baltimore, Md. Degrees: B.S.E.E., 1942, University of Missouri; M.S., 1949, University of Maryland; A.M.P., 1964, Harvard University; P.E.(Hon.), "Management Engineer," 1977, University of Missouri.

Fellow Award "For contributions to the development and application of pulse Doppler radar"; David Sarnoff Award, AIEE, 1962, "For achievements in pulse Doppler radar and other areas of applied electronics." Other Awards: Citation, War Dept., 1945; Special Patent Award, Westinghouse Electric, 1960; Golden Plate Award, Academy of Achievement, 1962; Order of Merit, Westinghouse Electric, 1969; Meritorious Service Award, U.S. Air Force, 1974.

SMITH, HENRY W., AIEE Fellow 1960. Born: January 6, 1902.

Fellow Award "For contributions to development of a large electric power system."

SMITH, J. HERBERT, AIEE Fellow 1962. Born: November 21, 1909, Fredericton, New Brunswick, Canada. Degrees: B.Sc., 1932, M.Sc., 1942, D.Sc., 1958, University of New Brunswick.

Fellow Award "For contributions to electrical system design and project administration." Other Awards: Engineering Institute of Canada; Association of Professional Engineers, Ontario; Canada Confederation Medal, 1967.

SMITH, JOSEPH S., Fellow 1967. Born: July 16, 1925, New York, N.Y. Degrees: B.S., 1946, Iowa State University; M.E.E., 1950, D.Sc.Eng., 1955, New York University.

Fellow Award "For his contribution to system engineering and technical leadership of major defense systems."

SMITH, KENNETH C., Fellow 1978. Born: May 8, 1932, Toronto, Ontario, Canada. Degrees: B.A.Sc.(Eng.Phys.), 1954, M.A.Sc.(E.E.), 1956, Ph.D.(Phys.), 1960, Toronto.

Fellow Award "For contributions to digital circuit design."

SMITH, MARION P., Fellow 1973. Born: December 12, 1925, Waycross, Ga. Degrees: B.S.E.E., 1949, Louisiana State University.

Fellow Award "For contributions to the development and application of reliability engineering technology, and leadership of humanitarian organizations"; Florida West Coast Engineer of the Year Award, IEEE and Florida Engineering Society, 1970, "For outstanding service to society and to the engineering profession"; Annual Reliability Award, IEEE Reliability Society, 1978; IEEE Centennial Medal, 1984. Other Awards: Outstanding Service Award, ASQC, A&M Division, 1968-69.

SMITH, MARVIN W., AIEE Fellow 1942. Born: October 8, 1893, Overton, Tex. Degrees: B.S., 1915, E.E., 1938, D.Eng.(Hon.), 1950, Texas A.&M. University.

SMITH, MERLIN G., Fellow 1978. Born: May 12, 1928, Germantown, Ky. Degrees: B.S.E.E., 1950, University of Cincinnati; M.S.E.E., 1957, Columbia University.

Fellow Award "For contributions to the development of large-scale integration"; Distinguished Service Award, IEEE Computer Society, 1977-78; Richard E. Merwin Award for Distinguished Service, 1983.

SMITH, NEWBERN, IRE Fellow 1952. Born: January 21, 1909, Philadelphia, Pa. Degrees: B.S., 1930, M.S., 1931, Ph.D., 1935, University of Pennsylvania.

Fellow Award "For his contributions to the theory and measurement of ionospheric radio-wave propagation and the development of methods of predicting ionospheric transmission on a world-wide scale"; Harry Diamond Award, IRE, 1952. Other Awards: Eta Kappa Nu; Sigma Xi; Tau Beta Pi, Sigma Tau; Fellow, American Association for the Advancement of Science, 1968; NDRC Certificate; Special Achievement Award, U.S. Department of Commerce, 1974.

SMITH, OTTO J.M., IRE Fellow 1959. Born: August 6, 1917, Urbana, Ill. Degrees: B.S.(Chemistry), 1938, Oklahoma State University; B.S.E.E., 1938, University of Oklahoma; Ph.D., 1941, Stanford University.

Fellow Award "For contributions in the fields of electrical measurement and feedback control systems";

First Prize Paper, AIEE, 1940, 1953, 1959. Other Awards: Fellow, AAAS, 1954; Second Prize, National Editorial Award, Scientific Apparatus Makers Association, 1959; Guggenheim Fellow, 1960; Senior Research Fellow in Economics and Engineering, Monash University, Melbourne, Australia, 1966-1967.

SMITH, P. C., AIEE Fellow 1958. Born: January 16, 1900, Capron, Va. Degrees: B.S., 1921, University of North Carolina; A.M.P., 1952, Harvard University.

Fellow Award "For contributions to design and development of large motors and generators."

SMITH, PETER W., Fellow 1978. Born: November 3, 1937, London, England. Degrees: B.Sc., M.Sc., Ph.D., McGill University, Montreal.

Fellow Award "For contributions to tunable gas lasers"; Treasurer, QEAS, 1981, 1982; Vice President, QEAS, 1983; President, QEAS, 1984. Other Awards: Fellow, Optical Society of America; NATO Senior Scientist Award, 1979; Distinguished Technical Staff Award, Bell Laboratories, 1982; Bell System Technical Journal Prize Paper Award, 1983.

SMITH, PHILLIP H., IRE Fellow 1952. Born: April 29, 1905, Lexington, Mass. Degrees: B.S.E.E., 1928, Tufts University.

Fellow Award "For his contributions to the development of antennas and graphical analysis of transmission line characteristics"; Microwave Theory and Techniques Society, Special Recognition Award, "For the application of microwave theory and techniques to the practical realization of a circular transmission line chart for analyzing microwave circuits, the Smith Chart." Other Awards: Tau Beta Pi.

SMITH, RALPH J., Fellow 1969. Born: June 5, 1916, Herman, Nebr. Degrees: B.S.E.E., 1938, M.S.E.E., 1940, E.E., 1942, University of California; Ph.D., 1945, Stanford University.

Fellow Award "For eminent work in the advancement of engineering education, including excellent teaching, highly responsible administration, and effective writing that has guided, inspired, and aided scores of thousands of young men."

SMITH, RICHARD G., Fellow 1981. Born: January 19, 1937, Flint, MI. Degrees: B.S.(E.E.), 1958, M.S.(E.E.), 1959, Ph.D.(E.E.), 1963, Stanford University.

Fellow Award "For contributions to the development of Nd:YAG solid-state lasers and nonlinear optical devices"; Outstanding Electrical Engineering Student, Stanford University, 1958; President, IEEE Quantum Electronics and Applications Society, 1981. Other Awards: Tau Beta Pi, 1957; Phi Beta Kappa, 1958; Chairman CLEOS, 1980; Chairman, Steering Committee, CLEO, 1981.

SMITH, SIDNEY R., JR., Fellow 1957. Born: January 2, 1906, Gainesville, Tex. Degrees: B.S., 1929, Georgia Tech.

Fellow Award "For proficiency in development and design of electrical equipment."

SMITH, SIDNEY T., IRE Fellow 1962. Born: May 27, 1918, Montezuma, Ga. Degrees: B.S.E.E., 1939, Georgia Institute of Technology; D.Eng., 1942, Yale University.

Fellow Award "For contributions to electron tube technology."

SMITH, THEODORE A., IRE Fellow 1961. Born: February 17, 1905, New York, N.Y. Degrees: M.E., 1925, Stevens Institute of Technology.

Fellow Award "For contributions in many fields of radio engineering." Other Awards: Honorary Member, Electronic Industries Association.

SMITH, WARREN L., Fellow 1973. Born: July 6, 1924, Wayne, Nebr. Degrees: B.S.E.E., 1945, University of Wisconsin.

Fellow Award "For the application of quartz crystal devices to precision frequency generators of high special purity." Other Awards: C.B. Sawyer Memorial Award, 1975; Technical Advisor to Technical Committee 49, International Electrotechnical Commission, 1973 to present.

SMITS, FRIEDOLF M., Fellow 1972. Born: November 10, 1924, Stuttgart, Germany. Degrees: Ph.D., 1950, University of Freiburg.

Fellow Award "For contributions and leadership in semiconductor device development."

SMOLINSKI, ADAM K., Fellow 1975. Born: October 1, 1910, Radziechow, Poland. Degrees: Dipl.Eng., 1933, Dr.Eng., 1945, Technical University, Warszawa.

Fellow Award "For achievements in research, development, and teaching of electronic engineering." Other Awards: Golden Cross of Merit, 1956; Member, Polish Academy of Sciences, 1962; Officer's Cross, 1964; Polish State Prize, 2nd Grade, 1964; Golden Honor Mark of Distinction, Association of Polish Electrical Engineers, 1969; Commander's Cross of the Order of Polonia Restituta, 1973; Golden Honor Mark of Distinction, Polish Federation of Engineering Associations, 1977; Vice President, International Union of Radio Sciences, 1978; Order of the Banner of Labor (second class), 1980; Meritorious (Honorary) Teacher of the Polish People Republic, 1981.

SNIVELY, HOWARD D., AIEE Fellow 1957. Born: January 30, 1911, Canton, Ohio. Degrees: B.S.E.E., 1934, University of Cincinnati.

Fellow Award "For his contributions to the design and development of ac and dc machinery." Other Awards: Fellow, American Association for Advancement of Science; Distinguished Alumnus Award, University of Cincinnati, 1969; Sigma Xi; Tau Beta Pi; Eta Kappa Nu; Fellow, American Society of Mechanical Engineers.

SNODGRASS, JAMES MARION, Fellow 1970. Born: May 3, 1908, Marysville, Ohio. Degrees: A.B., 1931, Oberlin College.

Fellow Award "For contributions to biomedical and oceanographic instrumentation." Other Awards: Admiral of the Fleet Award, Pacific Asiatic Area, 1945; Telemetering Man of the Year, NTC, 1966; Fellow, ISA, 1967; Compass Distinguished Achievement Award, MTS, 1968; Distinguished Public Service Award, U.S. Navy, 1968; Oceanographer of the U.S. Navy, 1974.

SNOW, EDWARD H., Fellow 1979. Born: June 26, 1936, St. George, Utah. Degrees: B.A.(Physics), 1958, Ph.D.(Physics), 1963, University of Utah.

Fellow Award "For contributions to the theoretical understanding and analysis of MOS instabilities." Other Awards: Certificate of Merit, Franklin Institute, 1975.

SNOW, HAROLD A., IRE Fellow 1956. Born: October 1898, Santa Barbara, Calif. Degrees: B.S., 1921, George Washington University.

Fellow Award "For contributions to electronic circuitry, measurement techniques, and the development of the variable mu tube."

SNYDER, ALBERT W., Fellow 1977. Born: December 28, 1925, Halifax, Pa. Degrees: B.S.(Physics), 1951, Franklin and Marshall College; M.S. (Physics), 1953, Iowa State University.

Fellow Award "For contributions to and management of research in the fields of nuclear measurements and reactor safety."

SNYDER, DONALD LEE, Fellow 1981. Born: January 8, 1935. Degrees: B.S.E.E., 1961, University of Southern California; M.S.E.E., 1963, Ph.D., 1966, Massachusetts Institute of Technology.

Fellow Award "For contributions to estimation theory and applications to communications and medicine."

SNYDER, EDWIN H., AIEE Fellow 1949. Born: March 17, 1901, Washington, D.C. Degrees: E.E., 1923, Lehigh University; D.Eng.(Hon.), 1968, Lehigh University.

Other Awards: Licensed Professional Engineer, New Jersey, 1942; Chairman, Engg. Div. General Committee, EEI, 1952-54; Chairman, Research Projects Committee, Edison Electric Institute, 1958-1959; L-in-Life Award, Lehigh Club of New York, 1967; Outstanding Leadership, American Society of Mechanical Engineers, 1967; President, Association of Edison Illuminating Companies, 1969-1970.

SNYDER, JAMES N., Fellow 1984. Born: February 17, 1923, Akron, OH. Degrees: B.S.(Physics), 1945, Harvard College; M.A.(Physics), 1947, Ph.D.(Physics), 1949, Harvard University.

Fellow Award "For contributions to pioneering curricula in computer science and to computational physics"; IEEE Centennial Medal, 1984. Other Awards: Fellow, American Physical Society; Fellow, American As-

sociation for the Advancement of Science.

SNYDER, RICHARD L., IRE Fellow 1962. Born: November 1, 1911, Pittsburgh, Pa. Degrees: B.S., 1934, Lehigh University.

Fellow Award "For contributions to the development of electron tubes and of digital computers."

SNYDER, WILLIAM R., Fellow 1968. Born: January 7, 1903, Stony Creek Mills, Pa. Degrees: B.S.E.E., 1923, Pennsylvania State University; D.Sc., 1967, Albright College.

Fellow Award "For contributions to the growth and effective operation of the Manila Utility System."

SOBOL, HAROLD, Fellow 1973. Born: June 21, 1930, Brooklyn, N.Y. Degrees: B.S.E.E., 1952, City College of the City University of New York; M.S.E.E., 1955, Ph.D., 1959, University of Michigan.

Fellow Award "For contributions in the field of microwave techniques"; MTT National Lecturer, 1970; Dallas Section Outstanding Achievements Award, 1975; IEEE Centennial Medal, 1984. Other Awards: IR-100 Award, 1969.

SODERMAN, ROBERT A., Fellow 1974. Born: January 16, 1919. Degrees: A.B., 1940, E.E., 1942, Stanford University.

Fellow Award "For contributions to the development of instrumentation and measurement methods, and for leadership in the establishment and acceptance of associated standards." Other Awards: OSRD Certificate of Merit, 1945.

SOLOMON, DAVID L., Fellow 1967. Born: June 20, 1914, Chicago, Ill. Degrees: B.S.E.E., 1936, Purdue University.

Fellow Award "For contributions to the technical planning and direction of national communications networks." Other Awards: Distinguished Civilian Service Medal, Dept. of Defense, 1972.

SOLOMON, JAMES E., Fellow 1978. Born: July 20, 1936. Degrees: A.A., 1954, Boise College; B.S.E.E., 1958, M.S.E.E., 1960, University of California.

Fellow Award "For contributions in research, design, and application of analog integrated circuits."

SOMERS, RICHARD M., IRE Fellow 1961. Born: December 22, 1904, Orange, N.J. Degrees: E.E., 1926, Rensselaer Polytechnic Institute.

Fellow Award "For contributions to phonographic recording and dictating machines"; Annual Science Achievement Award, IRE, Kansas City Section, 1961, "For outstanding service contributed toward the advancement of science and technology." Other Awards: Audio Engineering Award, 1953; Tau Alpha Phi, 1961; Tau Beta Pi, 1962.

SOMEYA, ISAO, Fellow 1978. Born: March 23, 1915. Degrees: Bachelor of Engineering, 1938, Doctor of Engineering, 1950, University of Tokyo.

Fellow Award "For contributions in the use of sampling theory in development of microwave communications systems."

SOMMER, ALFRED H., IRE Fellow 1960. Born: November 19, 1909, Frankfurt, Germany. Degrees: Dr.Phil., 1934, Berlin University.

Fellow Award "For contributions in the field of photoemissive surfaces." Other Awards: Gaede-Langmuir Award, American Vacuum Society, 1982.

SOMMERMAN, GEORGE M. L., AIEE Fellow 1951. Born: July 2, 1909, Baltimore, Md. Degrees: B.E., 1929, D.Eng., 1933, The Johns Hopkins University.

Other Awards: Alfred Noble Prize, Founder Engineering Societies, 1938; Eta Kappa Nu.

SOMMERS, OTTO W., AIEE Fellow 1962. Born: November 20, 1906, San Antonio, Tex. Degrees: B.S., 1929, Texas A and M University.

Fellow Award "For contributions to operation and management of an electric utility." Other Awards: Tau Beta Pi, 1929.

SOMOS, ISTVAN, Fellow 1971. Born: January 1, 1911, Budapest, Hungary. Degrees: Dipl.Eng., 1933, Karlsruhe Technological University, Karlsruhe, Germany.

Fellow Award "For contributions to the field of high-power semiconductor devices." Other Awards: Managerial Award, General Electric, 1963.

SONNEMANN, WILLIAM K., AIEE Fellow 1956. Born: December 15, 1902, Port Lavaca, Tex. Degrees: B.S., 1924, University of Texas.

Fellow Award "For his contributions to design, development and application of protective relays for electric power systems"; Paper Prize, AIEE, Power Division, 1950, for "A study of directional elements connections for phase relays"; Paper Prize, AIEE, 1958, for "Magnetizing inrush phenomena in transformer banks"; Distinguished Service Award, Power System Relaying Committee, Power Engineering Society, 1981. Other Awards: Eta Kappa Nu, 1967.

SONNENFELDT, RICHARD W., IRE Fellow 1962. Born: July 3, 1923, Berlin, Germany. Degrees: B.S.E.E., 1949, The Johns Hopkins University.

Fellow Award "For contributions to color television and digital techniques"; District Branch Competition Prize, AIEE. Other Awards: Tau Beta Pi; Omicron Delta Kappa; Award of Merit, RCA Victor; Council on Foreign Relations; Conference Board; American Management Association; National Association of Corporate Directors; Lecturer, Harvard Business School, M.I.T. Sloan School.

SOOHOO, RONALD F., Fellow 1970. Born: September 1, 1928, Canton, China. Degrees: S.B., 1948, Massachusetts Institute of Technology; M.S., 1952, Ph.D., 1956, Stanford University.

Fellow Award "For contributions in the fields of magnetics, ferrites, and thin films." Other Awards: NATO

Fellowship to CNRS, France, 1970; National Science Foundation Faculty Research Participation; IBM Fellow, 1974; NATO Fellowship, Centre National de la Recherche Scientifique, France, 1970; "Outstanding Educators of America," 1972; "Who's Who in America," 1972; National Science Foundation Faculty Participation Fellow, IBM, 1974; Outstanding Teaching Award, Tau Beta Pi, 1976.

SORENSON, HAROLD WAYNE, Fellow 1978. Born: August 28, 1936. Degrees: B.S.(Aeronautical Engineering), 1957, Iowa State University; M.S.(Control Systems Engineerng), 1963, Ph.D.(Control Systems Engineering), 1966, University of California at Los Angeles.

Fellow Award "For contributions to control, estimation, and optimization of stochastic dynamic systems."

SORGER, GUNTHER U., Fellow 1984. Born: April 4, 1925, Riedlingen, W. Germany. Degrees: B.Sc., M.Sc., Ph.D.(Physics), 1948, 1950 and 1953 resp., Institute of Technology, Stuttgart, W. Germany.

Fellow Award "For innovative contributions to the development of precision, electronic measuring instruments, and standards."

SORIA, RODOLFO M., IRE Fellow 1958. Born: May 16, 1917, Berlin, Germany. Degrees: B.S., 1939, M.S., 1940, Massachusetts Institute of Technology; Ph.D., 1947, Illinois Institute of Technology.

Fellow Award "For his contributions to high frequency transmission and to engineering management." Other Awards: Award of Merit, Chicago Technical Societies Council, 1969.

SPANDORFER, LESTER M., Fellow 1971. Born: 1925, Virginia. Degrees: B.S.E.E., 1947, M.S.E.E., 1948, University of Michigan; Ph.D., 1956, University of Pennsylvania.

Fellow Award "For contributions and technical leadership in the field of computer technology." Other Awards: Sigma Xi; Delta Epsilon Pi.

SPANG, H. AUSTIN, III, Fellow 1981. Born: July 16, 1934. Degrees: B.Eng.(E.E.), 1956, M.Eng.(E.E.), 1958, D.Eng.(E.E.), 1960, Yale University.

Fellow Award "For contributions to the development and application of digital control systems."

SPARKS, MORGAN, Fellow 1966. Born: July 6, 1916, Pagosa Springs, Colo. Degrees: B.A., 1938, M.A., 1940, Rice University; Ph.D., 1943, University of Illinois.

Fellow Award "For contributions to semiconductor science and technology, and for leadership in development of solid-state electronics"; Jack A. Morton Award, IEEE, 1977. Other Awards: Fellow, American Physical Society; Fellow, American Institute of Chemists; Member, National Academy of Engineering.

SPARLING, THOMAS E., Fellow 1983. Born: July 6, 1917, Stanley, Wisconsin. Degrees: B.S.E.E., 1939, Montana State University.

Fellow Award "For innovation and leadership in the electrical design of industrial plants and commercial buildings"; Achievement Award, Industrial Power Systems Department, IEEE, 1979; Standards Medallion, Standards Department, IEEE, 1982. Other Awards: Engineer of the Year Awards, Seattle Section, Washington Society of Professional Engineers (NSPE), 1977; Consulting Engineers Council of Washington (ACEC), 1982.

SPEAKMAN, EDWIN A., IRE Fellow 1954. Born: August 14, 1909, Gratz, Pa. Degrees: B.S., 1931, Haverford College.

Fellow Award "In recognition of his leadership in administration of electronic research and development"; Special Award; IEEE, 1958-59, "For his leadership and guidance as National Chairman of Professional Group on Military Electronics." Other Awards: Associate Fellow, American Institute of Aeronautics and Astronautics; Meritorious Award, U.S. Navy; Special Commendation, U.S. Navy, 1948; Life Member, 1980, and National Secretary, 1976-81, Association of Old Crows; Science Advisor, 1968 to date, U.S. Army Intelligence & Security Command, Arlington, Va.

SPECHT, THEODORE R., Fellow 1973. Born: July 8, 1918, Felton, Minn. Degrees: B.E.E., 1940, University of Minnesota; M.S., 1951, University of Pittsburgh.

Fellow Award "For contributions to the theory and performance of electromagnetic apparatus"; Second Prize, Science and Electronic Divison (with R.N.Wagner), AIEE, 1949-50, for paper "The Theory of the Current Transductor and Its Application in the Aluminum Industry"; Prize Winning Paper Award, Transformers Committee, IEEE Transactions on Power Apparatus and Systems, 1983.

SPEISER, AMBROS P., Fellow 1967. Born: November 13, 1922, Basle, Switzerland. Degrees: Dipl.Ing., 1948, D.Tech.Sci., 1951, Eidg. Technische Hochschule, Zurich, Switzerland.

Fellow Award "For outstanding work in the design of digital computers and computer components." Other Awards: Secretary-Treasurer, IFIP, 1960-1965; President, IFIP, 1965-1968; Honorary Member, Zurich Physical Society, 1968; Silver Core Award, IFIP, 1974; Prognos Prize (Basle, Switzerland) for innovation.

SPENCE, ROBERT, Fellow 1977. Born: July 11, 1933, Hull, England. Degrees: B.Sc.(Eng), 1954, D.I.C.(Diploma of Imperial College), 1955, Ph.D.(Eng.), 1959, Imperial College; D.Sc.(Eng.), 1983, University of London.

Fellow Award "For contributions to the theory of and education in the algorithmic, linguistic, and perceptual aspects of computer-aided circuit design."

SPENCER, NED A., Fellow 1969. Born: August 28, 1925, Omaha, Nebr. Degrees: B.S.E.E., 1946, Massachusetts Institute of Technology; M.S. (Computer

Science), 1982, George Washington University.

Fellow Award "For outstanding technical and managerial leadership in the development of antennas for radar and communications."

SPENCER, WILLIAM J., Fellow 1972. Born: September 25, 1930, Raytown, Mo. Degrees: A.B., 1952, William Jewell College, Liberty, Mo.; M.S., 1956, Ph.D., 1959, Kansas State University.

Fellow Award "For contributions to the theory and development of piezoelectric devices for communication systems"; C.B. Sawyer Award, 1972. Other Awards: Citation for Achievement, William Jewell College, 1969; Regents Meritorious Service Medal, University of New Mexico, 1981.

SPICER, WILLIAM E., Fellow 1984. Born: September 7, 1929, Baton Rouge, LA. Degrees: B.S., College of William and Mary; B.S., M.I.T.; M.S., 1953, Ph.D., 1955, University of Missouri; D.Tech.(Hon.), Linkoping U., Sweden.

Fellow Award "For development of photoemission as a key technique for the investigation of the physical properties of solids and of surfaces." Other Awards: Guggenheim Fellowship, 1978-79; Overseas Fellow, 1979, Churchill College, Cambridge (reappointed for 1984); American Physical Society Oliver E. Buckley Solid State Physics Prize, 1980; Research and Development Magazine (Dun & Bradstreet) Scientist of the Year, 1981; Visiting Scholar, Chinese University Development Project--World Bank--to Fudan University, Shanghai, 1983; Fellow, American Physical Society; Who's Who (since 1966).

SPILKER, JAMES J., JR., Fellow 1982. Born: August 4, 1933, Philadelphia, PA. Degrees: B.S.E.E., 1955, M.S.E.E., 1956, Ph.D., 1958, Stanford University.

Fellow Award "For contributions to the development of digital satellite communications and navigation systems"; Chairman, IEEE Technical Appraisal Committee, 1980-1982. Other Awards: Rotary Club Scholarship, 1954; Hewlett-Packard Fellowship, 1956; Lockheed Publications Award, 1960, 1961, 1962; Outstanding Achievement Award, Ford Aerospace and Communications Corporation, 1969, 1970, 1972; Member, U.S. Congressional Advisory Panel on Civil Space Stations, 1983 to present; Recipient, Premio Prize, International Institute of Communications, 1983.

SPITZER, EDWIN E., IRE Fellow 1956. Born: February 22, 1905, Fitchburg, Mass. Degrees: B.S., M.S., 1927, Massachusetts Institute of Technology.

Fellow Award "For contributions and leadership in the design of power tubes."

SPRAGUE, ROBERT C., AIEE Fellow 1949. Born: August 3, 1900, New York, N.Y. Degrees: B.S., 1922, United States Naval Post Graduate School; S.M., 1924, Massachusetts Institute of Technology; D.Eng.(Hon.), 1953, Northeastern University; D.Sc.(Hon.), 1954,

Williams College; LL.D.(Hon.), 1959, Tufts University; D.Sc.(Hon.), 1959, Lowell Technological Institute; LL.D.(Hon.), 1967, University of New Hampshire; D.Sc.(Hon.), 1972, North Adams State College; D.Sc.(Hon.), 1975, University of Massachusetts.

Other Awards: Medal of Honor, Radio-Electronics-Television Manufacturers Association, 1954; Distinguished Citizenship Award, Bates College, 1958; Man of the Year Award, Hotchkiss Alumni Association, 1958; Gold Knight Award, National Management Association, 1961; Man of the Year Award, New England Council, 1965; Fellow, American Academy of Arts and Sciences; Medal of Honor, Electronic Industries Association, 1979.

SPROUSE, MCBETH N., Fellow 1981. Born: November 24, 1920, Gastonia, NC. Degree: B.E.E., 1942, Clemson University.

Fellow Award "For leadership in standardizing design processes for power generation stations." Other Awards: Knoxville Area Outstanding Achievement Award, 1974; Eminent Engineer, Tennessee Alpha Chapter, Tau Beta Pi, 1980.

SPROWL, P. R., AIEE Fellow 1962. Born: January 12, 1917, Indiana Co., Pa. Degrees: B.S.E.E., 1940, Carnegie Institute of Technology; 1952, University of Pittsburgh.

Fellow Award "For contributions to organized engineering development of electric power equipment." Other Awards: Member, National Society of Professional Engineers; Member, American Society of Naval Engineers.

SRINIVASAN, A., Fellow 1968. Born: October 15, 1908, Madura, India. Degrees: B.S., 1932, M.Sc., 1943, University of Madras; M.S., 1947, Illinois Institute of Technology.

Fellow Award "For contribution to education and research in electrical power engineering." Other Awards: Sigma Xi; Eta Kappa Nu; Institute of Engineers, India.

STAATS, GUSTAV, Fellow 1978. Born: November 30, 1919, Forest Park, Ill. Degrees: B.S., 1942, M.S., 1948, Ph.D., 1956, Illinois Institute of Technology.

Fellow Award "For contributions to the development, design, and construction of very large conductor-cooled steam turbine generators"; Memorial Award, Milwaukee Section, IEEE, 1975. Other Awards: Eta Kappa Nu Recognition of Outstanding Young Electrical Engineers, 1952.

STACY, JOHN D., AIEE Fellow 1953. Born: September 14, 1895, Hume, N.Y. Degrees: E.E., 1917, Rensselaer Polytechnic Institute.

Fellow Award "For leadership in development, design and standardization in the manufacture of capacitors." Other Awards: Charles A. Coffin Award, General Electric Co., 1952.

STADLIN, WALTER OLAF, Fellow 1982. Born: February 15, 1929. Degrees: B.S.E.E., 1952, Newark College of Engineering; M.S.E.E., 1961, University of Pennsylvania.

Fellow Award "For contributions to digital computer techniques for automated control of interconnected power distribution networks."

STADTFELD, NICHOLAS, AIEE Fellow 1960. Born: December 2, 1907, New York, N.Y. Degrees: B.S., 1929, E.E., 1930, City College of the City University of New York.

Fellow Award "For contributions to application of the principles of power system protection and control."

STAELIN, DAVID HUDSON, Fellow 1979. Born: May 25, 1938, Toledo, Ohio. Degrees: S.B., 1960, S.M., 1961, Sc.D., 1965, Massachusetts Institute of Technology.

Fellow Award "For advances in radio astronomy and the development of microwave radiometric probes."

STAMPFL, RUDOLF A., Fellow 1971. Born: January 21, 1926, Vienna, Austria. Degrees: B.S.E.E., 1948, M.S.E.E., 1950, Ph.D., 1953, Institute of Technology, Vienna.

Fellow Award "For contributions to space programs in the fields of telecommunication and system design"; Harry Diamond Memorial Prize Award, 1967, "For his outstanding technical contribution and his able direction of a highly complex engineering organization that has contributed to the exploration of space"; President, IEEE Aerospace and Electronic Systems Society, 1979-1980. Other Awards: U.S. Army Project Score, NASA Tiros, Nimbus, Orbiting Astronomical Observatory Group Awards; Civilian Executive Citation.

STANLEY, C. MAXWELL, Fellow 1965. Born: June 16, 1904, Corning, Iowa. Degrees: B.S., 1926, M.S., 1930, State University of Iowa; LHD(Hon.), 1961, Iowa Wesleyan College; HHD(Hon.), 1970, University of Manila; LHD(Hon.), Augustana College; Honorary Rector, University of Dubuque, 1983.

Fellow Award "For contributions to the planning and design of electric power systems." Other Awards: Alfred Noble Prize, 1933, and Collingwood Prize, 1935, American Society of Civil Engineers; John Dunlap Prize, 1943, Anson Marston Award, 1947, Distinguished Service Award, 1962, and Hoover Humanitarian Award, 1979, Iowa Engineering Society; Engineer of the Year Award, 1965, and PEPP Award, 1975, National Society of Professional Engineers; Fellow, American Society of Civil Engineers; American Society of Mechanical Engineers; American Consulting Engineers Council; Distinguished Service Award, 1967, Hancher-Finkbine Medallion, 1971, and Iowa Business Leadership Award, 1980, University of Iowa; Past Presidents's Award, 1983, American Consulting Engineer's Council.

STARK, LAWRENCE, Fellow 1970. Born: 1926, New York. Degrees: A.B., 1945, Columbia University; M.D., 1948, Albany Medical College.

Fellow Award "For contributions to neurological control systems, to bioengineering education, and to the use of computers in medical instrumentation"; Morlock Award, IEEE Engineering in Medicine and Biology Society. Other Awards: Guggenheim Fellow, 1968-69; Sigma Xi; Eta Kappa Nu; Member, American Physiological Society; Member, Optical Society of America; Member, American Academy of Neurology; Member, Boston Society of Psychiatry and Neurology; Member, International Brain Research Organization; Member, Biophysical Society; Member, Instrument Society of America; Member, American Institute of Aeronautics and Astronautics; Member, Association for Computing Machinery; Member, American Society of Engineering Education; Member, American Society of University Professors.

STARK, LOUIS, Fellow 1971. Born: April 5, 1926, Detroit, Mich. Degrees: B.S., 1950, M.S., 1952, Massachusetts Institute of Technology.

Fellow Award "For inventions and developments in phased-array antenna technology." Other Awards: Eta Kappa Nu, 1950; Sigma Xi, 1952; L.A. Hyland Patent Award, 1969.

STARR, EUGENE C., AIEE Fellow 1941. Born: August 6, 1901, Falls City, Oreg. Degrees: B.S.E.E., 1923, E.E., 1938, Oregon State University.

Awards: First Paper Prize, AIEE, 1941, for "High-Voltage D-C Point Discharges" and "Aircraft Precipitation Static Radio Interference"; First Paper Prize, AIEE, Northwest District, 1953, for "Shunt Capacitors in Large Transmission Networks"; William M. Habirshaw Award, 1968, "For meritorious achievement in the field of electrical transmission and distribution"; Lamme Medal, IEEE, 1980; IEEE Centennial Medal, 1984. Other Awards: Distinguished Service Award, Gold Medal, U.S. Dept. of Interior, 1958; Engineer of the Year Award, Professional Engineers of Oregon, 1965; Distinguished Service Award, Oregon State University, 1976; Administrator's Award for Distinguished Service, Bonneville Power Administration, 1982; National Academy of Engineering, 1977; National Society of Professional Engineers; Fellow, American Association for the Advancement of Science; American Nuclear Society; Eta Kappa Nu; Sigma Tau; Tau Beta Pi; Sigma Xi; Phi Kappa Phi.

STATZ, HERMANN, Fellow 1980. Born: January 9, 1928, Herrenberg, Wurttemberg, Germany. Degrees: Abiturum, Kepler Oberrealschule, 1946, Tubingen, Germany; Vordiplom (Physics), 1948, Diplom Physiker, 1949, Dr. rer.nat., 1951, Technische Hochschule Stuttgart.

Fellow Award "For contributions to semiconductor devices and lasers"; IEEE Elmer A. Sperry Board of Awards. Other Awards: Fellow, American Physical Society; Board of Editors, Journal of Applied Physics and Applied Physics Letters, 1969-1971.

STAUNTON, JOHN J.J., Fellow 1970. Born: July 4, 1911, Binghamton, N.Y. Degrees: B.S., 1932, M.S., 1934, E.E., 1941, University of Notre Dame; D.Eng.(Hon.), 1969, Midwest College of Engineering.

Fellow Award "For contributions in the field of electronic and optical instrumentation related to chemical and clinical analyses." Other Awards: Award of Merit, Chicago Technical Societies Council, 1970.

STAVIS, GUS, Fellow 1970. Born: June 5, 1921, New York. Degrees: B.E.E., 1941, City College of the City University of New York; B.A., 1975, William Paterson College; M.B.A., 1980, Fairleigh Dickinson University.

Fellow Award "For contributions in the development of aircraft navigation and landing systems." Other Awards: Award for Outstanding Invention, New Jersey Research and Development Council, 1967; Tau Beta Pi; Eta Kappa Nu.

STEAR, EDWIN B., Fellow 1981. Born: December 8, 1932, Peoria, IL. Degrees: B.S. (Mech. Eng.), 1954, Bradley University; M.S. (Mech.Eng.), 1956, University of Southern California; Ph.D.(E.E.), 1961, University of California at Los Angeles.

Fellow Award "For contributions to linear and nonlinear filter theory and their application to biological systems." Other Awards: Distinguished Alumnus, Bradley University, 1980; Exceptional Civilian Service Award, U.S. Air Force, 1982.

STEARNS, HORACE MYRL, IRE Fellow 1959. Born: April 24, 1916, Kiesling, Wash. Degrees: B.S., 1937, University of Idaho; E.E., 1939, Stanford University; D.Sc.(Hon.), 1960, University of Idaho.

Fellow Award "For contributions in the fields of microwave tubes and Doppler radar." Other Awards: Certificate of Commendation, Secretary of Navy, 1946; Sigma Xi; Fellow, American Association for the Advancement of Science, 1960.

STEELE, EARL L., IRE Fellow 1962. Born: September 24, 1923, Denver, Colo. Degrees: B.S., 1945, University of Utah; Ph.D., 1952, Cornell University.

Fellow Award "For contributions to semiconductor devices." Other Awards: Outstanding Teacher Award, Engineering College, University of Kentucky, 1971; University of Kentucky Alumni Great Teacher Award, 1975.

STEELE, MARTIN C., Fellow 1981. Born: December 25, 1919, New York, NY. Degrees: B.Ch.E., 1940, Cooper Union Institute of Technology; M.S. (Physics), 1949, Ph.D. (Physics), 1952, University of Maryland.

Fellow Award "For contributions to solid-state electronics, particularly plasma effects in solids."

STEEN, JEROME R., IRE Fellow 1950. Born: June 29, 1901, West Bloomfield, N.Y. Degrees: B.S.E.E., 1923,

University of Wisconsin.

Fellow Award "For his work in the introduction and development of statistical quality control techniques in electron tube manufacturing." Other Awards: Fellow, American Society for Quality Control, 1947; Fellow, Radio Club of America, 1959.

STEEVES, CECIL M., AIEE Fellow 1961. Born: December 16, 1906, Moncton, New Brunswick, Canada. Degrees: B.Sc., 1927, University of New Brunswick.

Fellow Award "For contributions to the design and operation of a major electric power system in Brazil."

STEIER, WILLIAM H., Fellow 1976. Born: May 25, 1933, Kendallville, Ind. Degrees: B.S.E.E., 1955, Evansville College; M.S.E.E., 1957, Ph.D.(E.E.), 1960, University of Illinois at Urbana.

Fellow Award "For contributions in the fields of optical and microwave devices."

STEIGLITZ, KENNETH, Fellow 1981. Born: January 30, 1939. Degrees: B.E.E., 1959, M.E.E., 1960, Eng. Sc.D., 1963, New York University.

Fellow Award "For contributions to the theory and application of digital signal processing"; Technical Achievement Award, IEEE Acoustics, Speech, and Signal Processing Society, 1981.

STEIN, GERHARD M., AIEE Fellow 1962. Born: February 28, 1902, Breslau, Germany. Degrees: Dipl.Ing., 1925, Dr.Ing., 1927, Technische Hochschule Breslau.

Fellow Award "For contributions to the theory and practice of transformer design"; Paper Prizes, AIEE, Sharon Section, 1945, for "A Relationship Between Leakage Reactance and Winding Resistance of Transformers"; 1946, for "A Contribution to the Analytical Design of Transformers"; 1947, for "The Influence of the Core Form upon the Iron Losses of Transformers"; 1948, for "Methods for Testing Overload Temperatures in Transformers"; 1949, for "Temperature Characteristics of Oil-Insulated Transformers Protected by Internal Circuit Breakers"; 1950, for "Life Studies of Oil-Insulated Transformers and Their Application to Small Distribution Units"; 1951, for "Hot Spot Temperature in the Winding of a Small Distribution Transformer"; 1952, for "Determination of Temperature Peaks in Transformer Load Cycles"; 1954, for "Influence of Unbalanced Turn Ratios upon the Parallel Operation of Transformers Windings"; 1958, for "Asymmetrical Short Circuit Currents in Transformers"; 1960, for "The Magnetizing Power Supplied to Three-Phase Core Type Transformer Cores"; 1961, for "A Study of the Initial Surge Distribution in Transformer Coils." Other Awards: Special Patent Award, Westinghouse Electric, 1965, 1968.

STEIN, KARL-ULRICH, Fellow 1984. Born: November 2, 1936, Neuruppin. Degrees: Dipl.-Ing.(E.E.), 1961, Dr.-Ing.(E.E.), 1965, Techn. Hochschule Stuttgart.

Fellow Award "For contributions to the development

of widely used MOS memory cells." Other Awards: Award of the NTG, 1966.

STEIN, SEYMOUR, Fellow 1975. Born: April 4, 1928, Brooklyn, N.Y. Degrees: B.E.E., 1949, City College of the City University of New York; S.M., 1950, Ph.D., 1955, Harvard University.

Fellow Award "For technical leadership in the development of advanced communication and signal processing systems"; Paper Award, Information Theory Group, 1964.

STEINBERG, BERNARD D., Fellow 1966. Born: October 19, 1924, Brooklyn, N.Y. Degrees: B.S.E.E., 1949, M.S.E.E., 1949, Massachusetts Institute of Technology; Ph.D., 1971, University of Pennsylvania.

Fellow Award "For contributions toward advancing the theory of radar, sonar, and communication systems, and for reducing advanced theories to practical forms for engineering applications."

STEINER, HARRY C., AIEE Fellow 1957. Born: October 5, 1902, Larned, Kans. Degrees: B.S., 1926, University of Kansas; P.E., 1949, New York University.

Fellow Award "For his contributions to the development and application of mercury vapor rectifier tubes." Other Awards: Sigma Tau, 1924; Tau Beta Pi, 1925; "Who's Who in Engineering," 1977; "Leaders in Electronics, 1979."

STEINHOFF, ERNST A., Fellow 1968. Born: February 11, 1908, Treysa, West Germany. Degrees: B.S., 1931, M.S., 1933, Ph.D., 1940, Darmstadt Institute of Technology, Germany.

Fellow Award "For his contributions to aerospace science and technology, missile systems, range testing facilities, and the exploration of space." Other Awards: German Civilian Meritorious Service Awards, 1st and 2nd Class, 1942, 1944; Honorary Member and Golden Honor Pin, Hermann Oberith Society, 1957; Exceptional Civilian Service Award, U.S. Air Force, 1958; Sigma Xi, 1969; Fellow, American Astronautical Society, 1969; Golden Honor Pin, Deutsche Gesellschaft fuer Ortung and Navigation, 1969; Member, International Academy of Astronautics, 1970; Hermann Oberth Honor Ring, 1973; Nicolaus Copernicus Medal in Silver, Kuratorium Der Mensch und der Weltraum e.V., 1973.

STEPHENSON, W. BOYD, AIEE Fellow 1944. Born: January 16, 1891. Degrees: B.S., 1913, Purdue University; M.L.A., 1970, Southern Methodist University.

Other Awards: Certificate of Achievement, Headquarters European Command, Paris, France.

STERN, ARTHUR P., IRE Fellow 1962. Born: July 20, 1925, Budapest, Hungary. Degrees: Dipl.Ing., 1948, Swiss Federal Institute of Technology, Zurich, Switzerland; M.S.E.E., 1955, Syracuse University.

Fellow Award "For contributions and leadership in the field of solid-state circuits." Other Awards: Fellow

Award, American Association for the Advancement of Science, 1982.

STERN, ERNEST, Fellow 1980. Born: June 5, 1928. Degrees: B.S., 1953, Columbia University.

Fellow Award "For leadership in the development of surface-acoustic-wave devices for signal processing in radar and communications systems."

STERN, THOMAS E., Fellow 1972. Born: 1930, New York. Degrees: B.S., 1953, M.S., 1953, Sc.D., 1956, Massachusetts Institute of Technology.

Fellow Award "For contributions to nonlinear network theory and engineering education"; Prize Paper Award, IEEE Communications Society, 1978. Other Awards: Fulbright Research Grant, 1965-66; IBM Post-Doctoral Fellowship, 1971-72; Member, American Association of University Professors; Sigma Xi; Eta Kappa Nu.

STERZER, FRED, Fellow 1969. Born: November 18, 1929, Vienna, Austria. Degrees: B.S., 1951, City College of The City University of New York; M.S., 1952; Ph.D., 1955, New York University.

Fellow Award: "For contributions in the field of microwave solid-state energy sources and microwave modulation and demodulation of light."

STEVENS, FREDERICK, Fellow 1970. Born: June 13, 1922, Burbank, Calif. Degrees: A.B., 1944, Whitman College; M.S.E.E., 1947, California Institute of Technology.

Fellow Award "For contributions to the inertial and astroinertial guidance art." Other Awards: Meritorious Civilian Award, U.S. Air Force, 1968.

STEVENS, KENNETH N., Fellow 1965. Born: March 23, 1924, Toronto, Ontario, Canada. Degrees: B.A.Sc., 1945, M.A.Sc., 1948, University of Toronto; Sc.D., 1952, Massachusetts Institute of Technology.

Fellow Award "For fundamental research in speech synthesis and analysis." Other Awards: Fellow, Acoustical Society of America; J.S. Guggenheim Fellowship, 1962-63; National Institute of Health Special Fellowship, 1969-70; Silver Medal in Speech Communication, Acoustical Society of America, 1983.

STEVENS, RICHARD F., AIEE Fellow 1958. Born: June 5, 1902, Columbus, Ohio. Degrees: B.S.E.E., 1924, E.E., 1952, University of Washington.

Fellow Award "For contributions to the development of major high voltage power systems in the Pacific Northwest"; Distinguished Service Award, Substations Committee, 1978. Other Awards: Tau Beta Pi, 1924; Sigma Xi, 1953; Gold Medal for Distinguished Service, U.S. Dept. of Interior, 1964.

STEVENS, ROBERTSON, Fellow 1973. Born: August 7, 1922, San Diego, Calif. Degrees: B.S., 1945, U.S. Naval Academy; M.S.E.E., 1949, University of California at Berkeley.

Fellow Award "For leadership in development and effective use of the Deep Space Tracking Network in lunar and planetary exploration missions, and the application of the stations for scientific investigations." Other Awards: NASA Exceptional Service Medal.

STEVENSON, WILLIAM D., JR., AIEE Fellow 1962. Born: July 21, 1912, Pittsburgh, Pa. Degrees: B.S.E., 1934, Princeton University; B.S.E.E., 1939, Carnegie Institute of Technology; M.S., 1942, University of Michigan.

Fellow Award "For contributions to electrical engineering education and literature." Other Awards: Naval Ordnance Development Award, 1946; Phi Beta Kappa; Sigma Xi; Tau Beta Pi; Eta Kappa Nu; Special Citation, Edison Electric Institute, 1977.

STEWART, HERBERT R., AIEE Fellow 1946. Born: June 30, 1902, London, England. Degrees: S.B., 1924, S.M., 1925, Massachusetts Institute of Technology.

Fellow Award "For contributions to the design, protection, and control of electric power generation, transmission, and distribution systems"; William M. Habirshaw Award, 1974, "For meritorious achievements in the field of electrical transmission and distribution." Other Awards: Tau Beta Pi, 1924; Distinguished New England Engineer Award, Engineering Societies of New England, Inc., 1975.

STICHER, JOSEPH, AIEE Fellow 1954. Born: August 13, 1903, Bochum-Linden, Germany.

Fellow Award "For his contributions to the knowledge of the behavior of dielectrics in high-voltage cables." Other Awards: Attwood Associate Award in the U.S. National Committee Cigre "For having made notable contributions to the Conference Internationale des Grands Reseaux Electriques a Haute Tension (CIGRE) over an extended period of time."

STICKLEY, C. MARTIN, Fellow 1980. Born: October 30, 1933, Washington, D.C. Degrees: E.E., 1957, University of Cincinnati; M.S. (E.E.), 1958, Massachusetts Institute of Technology; Ph.D. (E.E.), 1964, Northeastern University.

Fellow Award "For contributions to the development of optically pumped solid-state lasers, and for initiation and management of research for laser fusion and solid-state electronics." Other Awards: Outstanding Senior Engineer, University of Cincinnati, 1957; Marcus O'-Day Award, ARCRL, 1966; Secretary of Defense Meritorious Civilian Service Medal, 1976.

STIFFLER, JACK J., Fellow 1975. Born: May 22, 1934, Mitchellville, Iowa. Degrees: A.B., 1956, Harvard University; M.S., 1957, Ph.D., 1962, California Institute of Technology.

Fellow Award "For contributions to the field of synchronous communications."

STILLMAN, GREGORY E., Fellow 1977. Born: February 15, 1936, Scotia, Nebr. Degrees: B.S., 1958, University of Nebraska; M.S., 1965, Ph.D., 1967 University of Illinois.

Fellow Award "For contributions to the characterization of ultra-pure gallium arsenide, to the extension of the long wavelength range of extrinsic photoconductors, and to the development of avalanche photodetectors."

STITCH, MALCOLM L., Fellow 1971. Born: April 23, 1923, Elizabeth, N.J. Degrees: B.S., 1947, B.A., 1947, Southern Methodist University; Ph.D., 1953, Columbia University.

Fellow Award "For contributions to laser technology and applications, particularly in the development of the laser range finder." Other Awards: Sigma Xi, 1951; Fellow, Institute for Advancement of Engineering (Southern California), 1971; Fellow, Society of Photo-Optical Instrumentation Engineers, 1981.

STOCKHAM, THOMAS G., JR., Fellow 1977. Born: December 22, 1933. Degrees: S.B., 1955, S.M., 1956, Sc.D., 1959, Massachusetts Institute of Technology.

Fellow Award "For contributions to engineering education and to digital signal processing"; Senior Award (with A. V. Oppenheim and R.W. Schafer), IEEE Group on Audio and Electroacoustics, 1969; Award for Outstanding Technical Achievement, IEEE, Utah Chapter, 1972; Engineering Excellence Award, Utah Chapter, IEEE, 1978. Other Awards: Goodwin Medal, Massachusetts Institute of Technology, 1957; T.V. Shares Management Award, Massachusetts Institute of Technology, 1957; Speech Winner, Science Symposium, Air Force Special Weapons Center, 1960; Member, Association for Computing Machinery; Member, Audio Engineering Society; Sigma Xi; Eta Kappa Nu; Tau Beta Pi; Fellow Award, Audio Engineering Society, 1978; President, Audio Engineering Society, 1982-1983.

STODOLA, E. KING, IRE Fellow 1961. Born: October, 1914, Brooklyn, N.Y. Degrees: B.E.E., 1936, E.E., 1947, Cooper Union.

Fellow Award "For contributions to the development of extended-range radar systems." Other Awards: 20 patents (radar and radio).

STOKER, WARREN C., AIEE Fellow 1961. Born: January 30, 1912, Union Springs, N.Y. Degrees: E.E., 1933, M.E.E., 1934, Ph.D., 1938, Rensselaer Polytechnic Institute.

Fellow Award "For contributions to electrical engineering education." Other Awards: Engineer of the Year Award, Connecticut Society of Professional Engineers, 1958; Founders Award, Rensselaer Polytechnic Institute of Connecticut, 1971.

STOKES, STANLEY, AIEE Fellow 1929. Born: 1890, Columbia, Mo. Degrees: E.E., 1912, University of Missouri.

Other Awards: Distinguished Service Award, University of Missouri.

STONE, LOUIS N., AIEE Fellow 1960. Born: June 18, 1914, Wasco, Oreg. Degrees: B.S., 1939, Oregon State University.

Fellow Award "For contributions to undergraduate and graduate electrical engineering education and to research in the area of high-voltage phenomena." Other Awards: Carter Teaching Award, 1955; Engineer of the Year Award, Professional Engineers of Oregon, 1968.

STONE, ROBERT R., JR., Fellow 1977 Born: December 14, 1921. Degrees: A.B.(Physics), 1942, Western Maryland College.

Fellow Award "For contributions to the field of precision frequency and time control." Other Awards: Commendation, U.S. Navy, 1966; Commendation, Commander Naval Electronic Systems Command, 1967; Commendation, Chief of Naval Research, 1967; Member, Sigma Xi, Naval Research Laboratory Chapter; Navy Distinguished Civilian Service Award.

STORER, JAMES E., Fellow 1967. Born: October 26, 1927, Buffalo, N.Y. Degrees: A.B., 1947, Cornell University; A.M., 1948, Ph.D., 1951, Harvard University.

Fellow Award "For contributions in the application of mathematics to antennas, information processing, and computers." Other Awards: John Simon Guggenheim Fellow, 1951.

STORM, HERBERT F., AIEE Fellow 1962. Born: June 28, 1909, Vienna, Austria. Degrees: E.E., 1932, Dr.Ing.Sci., 1933, Technische Universitat Vienna.

Fellow Award "For his contributions to the theory and application of magnetic amplifiers"; Magnetics Group Award, IEEE, "In recognition of outstanding contributions to the IEEE Magnetics Group", 1968; Honorary Chairman, IEEE-INTERMAG Conference, Stuttgart, 1966; Honorary Life Member, IEEE Magnetic Society, 1975; "Herbert F. Storm Best Paper Award" for Students was established by the Magnetics Society to honor Herbert F. Storm's contributions, 1968; IEEE Magnetics Society Achievement Award, "For founding the IEEE Magnetics Society and the IEEE INTERMAG Conference, and for outstanding contributions to Power Magnetics," 1982; IEEE Centennial Medal, 1984. Other Awards: Industrial Research 100 Award, 1973.

STOTT, BRIAN, Fellow 1983. Born: August 5, 1941, Manchester, England. Degrees: B.Sc., 1962, M.Sc., 1963, Ph.D., 1971, UMIST, University of Manchester, England.

Fellow Award "For contributions to the advancement of computational methods for solutions to power system network problems."

STOUT, G. PHILIP, AIEE Fellow 1961. Born: August 21, 1904, Knoxville, Tenn. Degrees: B.S.E.E., B.S.M.E., 1928, University of Tennessee.

Fellow Award "For contributions to telemetering, fire

control systems, servo systems and coordinated research programs."

STOUT, MELVILLE B., AIEE Fellow 1950. Born: October 17, 1895, Pittsburgh, Pa. Degrees: B.S.E.E., 1920, M.S., 1924, University of Michigan.

Other Awards: Fellow, Instrument Society of America, 1963; American Association for the Advancement of Science, 1981.

STRAITON, ARCHIE W., IRE Fellow 1953. Born: August 27, 1907, Arlington, Tex. Degrees: B.S., 1929, M.A., 1931, Ph.D., 1939, University of Texas.

Fellow Award "For his many technical contributions in the field of radio propagation." Other Awards: National Academy of Engineering, 1976.

STRANDBERG, MALCOM W. P., IRE Fellow 1959. Born: March 9, 1919, Box Elder, Mont. Degrees: B.S., 1941, Harvard University; Ph.D., 1948, Massachusetts Institute of Technology.

Fellow Award "For research in the field of microwave spectroscopy." Other Awards: Fellow, American Physical Society, 1955; Fellow, American Academy of Arts and Sciences, 1957; Member, New York Academy of Sciences, 1957; Fellow, American Association for the Advancement of Science, 1959.

STRANG, HAROLD E., AIEE Fellow 1945. Born: April 11, 1901, Mechanicville, N.Y. Degrees: E.E., 1922, Rensselaer Polytechnic Institute.

STRATFORD, J. P., AIEE Fellow 1948. Born: December 22, 1900, Cincinnati, Ohio. Degrees: E.E., 1924, Cornell University.

STRATFORD, RAY P., Fellow 1982. Born: February 26, 1925, Pocatello, ID. Degrees: B.S.E.E., 1950, Stanford University.

Fellow Award "For contributions to the understanding and control of nonsinusoidal currents in electric power systems"; Honorable Mention Paper Award, IEEE Industry Applications Society, 1980; Best Paper Award, IEEE Industry Applications Society, 1981. Other Awards: Registered Professional Engineer, New York, 1963; Chairman, Subcommittee on Reactive Compensation and Harmonic Control of Static Power Converters, Static Power Converter Committee, IEEE-IAS.

STRATTON, JULIUS A., IRE Fellow 1945. Born: May 18, 1901, Seattle, Wash. Degrees: S.B., 1923, S.M., 1926, Massachusetts Institute of Technology; Sc.D., 1928, Eid. Technische Hochschule, Zurich, Switzerland; D.Eng.(Hon.), 1955, New York University; LL.D.(Hon.), 1957, Northeastern University; D.Sc.(Hon.), 1957, St. Francis Xavier University, Canada; LL.D.(Hon.), 1958, Union College; LL.D.(Hon.), 1959, Brandeis University; LL.D.(Hon.), 1959, Harvard University; LL.D.(Hon.), 1960, Carleton College; LL.D.(Hon.), 1961, University of Notre Dame; LH.D.(Hon.), 1962, Hebrew Union College, Jewish Institute of Religion; LL.D.(Hon.), 1962, The Johns Hopkins University; LH.D.(Hon.), 1963, Oklahoma City University; D.Sc.(Hon.), 1964, College of William and Mary; D.Sc.(Hon.), 1965, Carnegie Institute of Technology; LH.D.(Hon.), 1965, Jewish Theological Seminary of America; D.Sc.(Hon.), 1967, University of Leeds, England; D.Sc.(Hon.), 1971, Heriot-Watt University (Edinburgh); D.Sc.(Hon.), 1972, Cambridge University.

Fellow Award "For his work as a teacher and author in the field of fundamental research and the application of his knowledge to improve radio communication." Other Awards: Medal of Honor, IRE, 1957, "For his inspiring leadership and outstanding contributions to the development of radio engineering, as teacher, physicist, engineer, author and administrator." Other Awards: Medal for Merit, 1946; Distinguished Public Service Award, U.S. Dept. of Navy, 1957; Member, Naval Research Advisory Committee, 1954-59 (Chairman, 1956-57); Member, National Science Board, 1956-62; 1964-67; Faraday Medal, Institution of Electrical Engineers, U.K., 1961; Officer, French Legion of Honor, 1961; Honorary Fellow, Manchester College of Science and Technology, England, 1963; Orden de Boyaca, Colombia, 1964; Member, Governing Board, National Research Council, 1961-65; Boston Medal for Distinguished Achievement, 1966; Honorary Member of Senate, Technical University of Berlin, 1966; Knight Commander, Order of Merit, Federal Republic of Germany, 1966; Fellow, American Academy of Arts and Sciences; Fellow, American Association for the Advancement of Science; Member, American Philosophical Society; Fellow, American Physical Society; Member, Bostonian Society; Member, Council on Foreign Relations; Eminent Member, Eta Kappa Nu; Member, National Academy of Sciences (Vice President, 1961-65); Founding Member, National Academy of Engineering; Sigma Xi; Tau Beta Pi; Zeta Psi; Silver Stein Award, M.I.T. Alumni Center of New York, 1967; Chairman, Commission on Marine Science, Engineering and Resources, 1967-69; Bronze Beaver, M.I.T. Alumni Association, 1968; Marine Technology Society Citation, 1969; Man of the Year, National Fisheries Institute, 1969; Individual Distinguished Achievement Award, Offshore Technology Conference, 1971; Member, National Advisory Committee on Oceans and Atmosphere, 1971-73; Neptune Award, American Oceanic Organization, 1979.

STRATTON, ROBERT, Fellow 1974. Born: August 14, 1928, Vienna, Austria. Degrees: B.Sc., 1949, Ph.D., 1952, Manchester University, England.

Fellow Award "For contributions to the theory of semiconductor devices and to technical management of semiconductor research and development." Other Awards: Fellow, American Physical Society; Fellow, Institute of Physics (Great Britain).

STREETMAN, BEN G., Fellow 1980. Born: June 24, 1939, Cooper, TX. Degrees: B.S., 1961, M.S., 1963, Ph.D., 1966, University of Texas at Austin.

Fellow Award "For contributions to the understanding of ion implantation and radiation damage of compound semiconductors." Other Awards, Frederick Emmons Terman Award, American Society for Engineerng Education, 1981; Janet S. Cockrell Centennial Chair in Engineering, University of Texas, 1983.

STREIFER, WILLIAM, Fellow 1979. Born: September 13, 1936, Poland. Degrees: B.E.E., 1957, City University of New York; M.S.E.E., 1959, Columbia University; Ph.D., 1962, Brown University.

Fellow Award "For contributions to distributed feedback lasers and integrated optics"; Editor, IEEE Journal of Quantum Electronics; Treasurer, IEEE Quantum Electronics and Applications Society, 1980. Other Awards: Fellow, Optical Society of America.

STRINGER, LOREN FRANK, Fellow 1979. Born: September 28, 1925. Degrees: B.S.E.E., 1946, University of Texas; M.S.E.E., 1947, California Institute of Technology; Ph.D., 1963, University of Pittsburgh.

Fellow Award "For contributions to the development and application of solid-state power electronics systems." Other Awards: Westinghouse Order of Merit, 1966.

STRINGHAM, L. K., AIEE Fellow 1949. Born: September 30, 1913, Fishkill Plains, N.Y. Degrees: E.E., 1933, Cornell University.

STROKE, GEORGE W., Fellow 1975. Born: July 29, 1924, Yugoslavia. Degrees: B.Sc., 1942, University of Montpellier, France; Ing.Dipl., 1942, Institut D'Optique Theorique et Appliquee, Paris; Dr.Es.Sc.(Physics), 1960, University of Paris.

Fellow Award "For contributions to holography, coherent optics, and application to optical computing." Other Awards: Allan Gordon Memorial Award, Society of Photo-Optical Instrumentation Engineers, 1971. Fellow, American Physical Society; Fellow, Optical Society of America; Member, American Society of Physicists in Medicine; Member, Astronomical Society of America; Member, Biomedical Engineering Society; Member, European Physical Society; Member, Societe Francaise de Physique; Sigma Xi; "Blue-Ribbon" Task-Force on Ultrasonic Diagnostics (reporting to Science Advisor to President), National Science Foundation, 1973; Member, U.S. Congress House "Select Committee" on President John F. Kennedy's Assasination, 1977-1979; Humboldt Prize, Alexander von Humboldt Foundation, 1978; Visiting Professor, Harvard University (Medical School), 1970-1973; Visiting Professor, Technical University, Munich, 1978-1979.

STROM, CHARLES A., JR., Fellow 1966. Born: August 10, 1917, St. Paul, Minn. Degrees: B.E.E., 1940, University of Minnesota.

Fellow Award "For contributions to military global communications systems and military electronic computers for command and control."

STRONG, CHARLES E., IRE Fellow 1949. Born: November 1898, Omagh, Tyrone County, Northern Ireland. Degrees: B.A., B.A.I., 1922, M.A., 1963, Trinity College, Dublin.

Fellow Award "For his pioneering work in the radio equipment design and development field, particularly broadcasting transmitter, both medium and high frequency, and his many wartime contributions in England." Other Awards: Officer of the Order of the British Empire, 1947.

STRONG, EVERETT M., AIEE Fellow 1958. Born: January 23, 1900, Portland, Me. Degrees: B.S., 1922, Massachusetts Institute of Technology.

Fellow Award "For contributions to cooperative engineering education and illuminating engineering." Other Awards: Fellow, Illuminating Engineering Society, 1952; Gold Medal Award, Illuminating Engineering Society, 1966; Distinguished Service Award, Illuminating Engineering Society, 1967.

STRULL, GENE, Fellow 1967. Born: May 15, 1929, Chicago, Ill. Degrees: B.S.E.E., 1951, Purdue University; M.S.E.E., 1952, Ph.D., 1954, Northwestern University.

Fellow Award "For pioneering contributions to solid state and microsystems electronics." Other Awards: Fellow, National Science Foundation, 1952-54.

STRUTT, MAX J.O., IRE Fellow 1956. Born: October 2, 1903, Soerakarta, Java. Degrees: M.Sc., 1926, Dr.Techn.Sc., 1927, Institute of Technology, Delft, Holland; Dr.Ing.(Hon.), 1950, Technische Hochschule Karlsruhe, West Germany.

Fellow Award "For contributions to the knowledge of electron tubes and associated circuits, particularly at high frequencies." Other Awards: C. F. Gauss Medal, Society of Sciences, Brunswick, West Germany, 1954; Honorary Member, Society of Sciences, Brunswick, West Germany, 1955; Honorary Member, International Television Committee, 1956; Gordon McKay Professor of Electrical Engineering, University of California at Berkeley, 1961-1963; Honorary Member, Electronics Association of Japan, 1965; Senior Foreign Scientist Fellowship, National Science Foundation, 1966; Honorary Member, Institute of Electronics and Communication Engineers, Japan, 1967.

STRYKER, CLINTON E., AIEE Fellow 1935. Born: February 27, 1897, Chicago, Ill. Degrees: B.S., 1917, E.E., 1924, Armour Institute of Technology.

STUBBERUD, ALLEN R., Fellow 1977. Born: August 14, 1934, Glendive, Mont. Degrees: B.S.E.E., 1956, University of Idaho; M.S.E., 1958, Ph.D.(Eng.), 1962, University of California at Los Angeles.

Fellow Award "For contributions to the theory and

application of time variable systems and to engineering education."

STUETZER, OTMAR M., Fellow 1967. Born: April 30, 1912, Nuernberg, Germany. Degrees: Dr.Rer.Tech., 1938, Dr.Habil., 1943, Technische Hochschule Munich, Germany.

Fellow Award "For contributions to semiconductor and electrohydrodynamic devices, microwave systems, and management of applied science." Other Awards: Award, German Lilienthal Society, 1941.

STUMPERS, F. LOUIS H.M., IRE Fellow 1962. Born: August 30, 1911, Eindhoven, The Netherlands. Degrees: M.Sc., 1937, Utrecht University, Netherlands; Dr.Tech.Sci., 1946, Institute of Technology, Delft.

Fellow Award "For contributions to communication theory"; Certificate of Recognition, IEEE Electromagnetic Compatiblity Group, 1973; International Award, IEEE Electromagnetic Compatibility Group, 1975; IEEE Award in International Communications in Honor of Hernand and Sosthenes Behn, 1978. Other Awards: Veder Radio Prize, Veder Foundation, Netherlands, 1956, 1972; Member, Royal Academy of Science and Letters, Netherlands, 1968; Columbus Gold Medal, City of Genoa, Italy, 1976; CCIR Honor Award, Kyoto, Japan, 1978; Honorary Member, Hungarian Academy of Sciences, Budapest, 1979; Honorary Member, Dutch Radio Electronics Society, NERG, 1980; Honorary Chairman, Technical Programme Committee, Forum 83, ITU, Geneva.

STURLEY, KENNETH R., Fellow 1967. Born: February 19, 1908, Smethwick, England. Degrees: B.Sc., 1928, Ph.D., 1930.

Fellow Award "For outstanding direction and leadership in the training of engineers for broadcasting and television in both government and industry." Other Awards: Student Premium, Institution of Electrical Engineers, U.K., 1934; Fellow, Institution of Electrical Engineers, U.K., 1945; Ambrose Fleming Premiums, Institution of Electrical Engineers, U.K., 1941, 1945; Honorary Associate, University of Aston, 1956.

SU, KENDALL L., Fellow 1977. Born: July 10, 1926, Nanping, Fujian, China. Degrees: B.S., 1947, Amoy University, China; M.S., 1949, Ph.D., 1954 Georgia Institute of Technology.

Fellow Award "For contributions to the theory of active and distributed-parameter networks and to engineering education." Other Awards: Faculty Research Award, Georgia Tech Chapter of Sigma Xi, 1957.

SUDAN, RAVINDRA N., Fellow 1979. Born: June 8, 1931, Kashmir, India. Degrees: B.A.(with honors), 1948, Panjab University; D.I.I. Sc., 1952, Indian Institute of Science; D.I.C., 1955, Imperial College, London, England; Ph.D., 1955, University of London.

Fellow Award "For contributions to plasma theory and to graduate engineering education in plasma science." Other Awards: Best Paper Award, Naval Research Laboratory, 1974; Fellow, American Physical Society; IBM Professor of Engineering, and Director, Laboratory of Plasma Studies, Cornell University; Member, Editorial Boards of Nuclear Fusion, Physics of Fluids; 2 U.S. patents, 125 papers.

SUEMATSU, YASUHARU, Fellow 1980. Born: September 22, 1932, Japan. Degrees: B.S., 1955, M.S., 1957, Ph.D., 1960, Tokyo Institute of Technology.

Fellow Award "For contributions to semiconductor lasers, integrated optical circuits, and optical waveguides;" Quantum Electronics Award, IEEE Quantum Electronics and Application Society, 1982. Other Awards: Inada Memorial Award, IECE, Japan, 1961; Paper Awards, IECE, Japan, 1965, 1976, and 1977; Book Award, IECE, Japan, 1978; Achievement Award, IECE, Japan, 1978; Valdemar Poulsen Gold Medal, Danish Academy of Technical Science, 1983.

SUETAKE, KUNIHIRO, Fellow 1984. Born: May 14, 1920, Hokkaido, Japan. Degrees: B.E.(E.E.), 1944, Ph.D., 1957, Tokyo Inst. of Technology.

Fellow Award "For contributions to the theory and techniques of microwave absorbers and the introduction of educational technology." Other Awards: Paper Prize Award, IECEJ, 1960; Tokyo Metropolitan Technical Award, Tokyo Metropolitan Government, 1980.

SUGANO, TAKUO, Fellow 1983. Born: August 25, 1931, Tokyo, Japan. Degrees: B.Sc.(E.E.), 1954, M.S.E.E., 1956, Ph.D., 1959, The University of Tokyo.

Fellow Award "For contributions to semiconductor technology and devices and to engineering education." Other Awards: Research on MOS Devices, Inst. of Electronics & Comm. Engineers, Japan, 1974; Research on Surface Devices, Matsunaga Award, 1974; Academy of Science, DDR, 1983.

SUGATA, EIZI P., Fellow 1973. Born: February 16, 1908, Tottori Prefecture, Japan. Degrees: B.S., 1932, Dr.Eng., 1944, Osaka University.

Fellow Award "For contributions to the development of the electron microscope and to engineering education." Other Awards: Japanese Society of Electron Microscopy, 1958; Medal of Honor with Purple Ribbon, Japanese Government, 1971; Institute of Electronics and Communication Engineers of Japan, 1972; Medal of the Second Class, Order of the Rising Sun, Japanese Emperor, 1978.

SUGDEN, ARTHUR C., AIEE Fellow 1960. Born: February 11, 1908, Wilkes Barre, Pa. Degrees: B.S., 1930, Pennsylvania State University; M.S., 1931, Massachusetts Institute of Technology.

Fellow Award "For contribution in the field of power transmission and distribution."

SUGIYAMA, TAKASHI, Fellow 1983. Born: August 10, 1924, Osaka, Japan. Degrees: B.Eng., 1947, Ph.D.(Eng.), 1969, University of Tokyo.

Fellow Award "For developments in precision power measurement techniques and for leadership in electronic measurement and control industries." Other Awards: Prize of Japanese Minister, Japan Institute of Invention and Innovation, 1972; Invention Research Merit Award, Tokyo Governor, 1974; Purple Ribbon Medal, Japanese Government, 1975.

SUITS, C. G., AIEE Fellow 1947, IRE Fellow 1956. Born: March 12, 1905, Oshkosh, Wis. Degrees: B.A., University of Wisconsin; D.Sc., Eid. Technische Hochschule, Zurich, Switzerland; D.Sc.(Hon.), 1944, Union College; D.Sc.(Hon.), 1946, Hamilton College; D.Eng.(-Hon.), 1950, Rensselaer Polytechnic Institute; D.Sc.(Hon.), 1955, Drexel Institute of Technology; D.Sc.(Hon.), 1959, Marquette University.

IRE Fellow Award "For leadership in industrial research." Other Awards: Frederik Philips Award, IEEE, 1974. Other Awards: Outstanding Young Electrical Engineer Award, Eta Kappa Nu, 1937; The King's Medal for Service in the Cause of Freedom, Great Britain, 1948; U.S. Presidential Medal for Merit, 1948; William Proctor Prize, Scientific Research Society of America, 1958; Distinguished Service Award, American Management Association, 1959; Medal, Industrial Research Institute, 1962; Charter Member, National Academy of Engineering, 1964; Advancement of Research Medal, American Society for Metals, 1966; National Academy of Sciences; American Philosophical Society; American Academy of Arts and Sciences; Mem. Div. 14, Chief Div. 15 Nat, Def. Res. Comm., 1942-46; Mem. N.Y.S.Sci. Tech. Found. 1967-81; V. Chm. 1969-81; Naval Res. Adv. Comm. 1956-64, Chm. 1958-61;

SULLIVAN, A. H., JR., Fellow 1969. Born: Lincoln, Nebr. Degrees: E.E., Cornell University.

Fellow Award "For leadership in the field of electromagnetic compatibility and in radio spectrum utilization planning"; Certificate of Appreciation, IEEE, Group on Electromagnetic Compatibility, 1966; Honorary Life Member, IEEE Electromagnetic Compatibility Society, 1976; Chairman, IEEE EMC Society, 1966-67; Editor, IEEE Transactions on Electromagnetic Compatibility, 1962-1968; Certificate of Acknowledgement, IEEE Society on Electromagnetic Compatibility, 1983. Other Awards: Dept. of the Navy, Superior Civilian Service Award, 1971; Dept. of the Navy, Meritorious Civilian Service Award, 1975; Dept. of the Navy, Outstanding Performance Awards, 1968, 69, 70.

SULZER, PETER G., IRE Fellow 1958. Born: August 3, 1922, Media, Pa. Degrees: B.S.E.E., 1948, M.S.E.E., 1949, Pennsylvania State University.

Fellow Award "For his contributions to the development and application of ionosphere measuring equipment."

SUMERLIN, WILLIAM T., Fellow 1968. Born: March 29, 1917, Philadelphia, Pa. Degrees: B.A., 1937, E.E., 1939, Stanford University.

Fellow Award "For contribution to the introduction, implementation and advancement of reliability technology in industrial and defense programs."

SUMMERS, CLAUDE M., AIEE Fellow 1949. Born: September 1, 1903, Boone Co., Mo. Degrees: B.S., 1927, E.E., 1933, University of Colorado.

Other Awards: Distinguished Engineering Alumnus Award, University of Colorado, 1977.

SUMNER, ERIC E., Fellow 1970. Born: December 17, 1924, Vienna, Austria. Degrees: B.M.E., 1948, Cooper Union; M.S., 1953, E.E., 1960, Columbia University.

Fellow Award "For contributions to pulse-code modulation and signal processing systems"; Alexander Graham Bell Medal, 1978.

SUNAHARA, YOSHIFUMI, Fellow 1984. Born: July 16, 1927, Mie-Prefecture, Japan. Degrees: B.E., 1953, Osaka Prefectural University; MS.E., 1955, Ph.D., 1962, Kyoto University.

Fellow Award "For advancement of stochastic and nonlinear control system theories and their engineering applications." Other Awards: Contributed Paper Award, Japan Society of Mechanical Engineers (JSME), 1963; Contributed Paper Award, Society of Instrument and Control Engineers (SICE), Japan, 1982.

SUNDE, ERLING D., Fellow 1964. Born: December 22, 1902, Haugesund, Norway. Degrees: Dipl.Ing., 1926, Technische Hochschule Darmstadt, Germany.

Fellow Award "For contributions to knowledge of earth conduction effects."

SUOZZI, JOSEPH J., Fellow 1973. Born: March 2, 1926, New York, N.Y. Degrees: B.E.E., 1949, M.E.E., 1954, Catholic University; Ph.D., 1958, Carnegie-Mellon University.

Fellow Award "For contributions to power magnetics and computer magnetics, and for educational programs." Other Awards: Alumni Achievement Award, Catholic University, 1973.

SURAN, JEROME J., Fellow 1964. Born: January 11, 1926, New York, N.Y. Degrees: B.S.E.E., 1949, Columbia University; Dr.Eng.(honoris causa), 1976, Syracuse University.

Fellow Award "For contributions to the field of solid-state devices and circuits"; Paper Prize, International Solid State Circuits Conference, 1964, "Use of circuit redundancy to increase system reliability." Other Awards: Eminent Engineer, Tau Beta Pi, 1978.

SUSSKIND, CHARLES, IRE Fellow 1962. Born: August 19, 1921, Prague, Czechoslovakia. Degrees: B.S., 1948, California Institute of Technology; M.Eng., 1949, Ph.D., 1951, Yale University.

Fellow Award "For contributions to the theory of electron beams and electrical engineering education."

Other Awards: Clerk-Maxwell Premium, Institution of Electronic and Radio Engineers; Senior Faculty Fellow, National Science Foundation.

SUTTON, CHARLES T. W., Fellow 1966. Born: December 16, 1903, West Ham, Essex, England. Degrees: B.Sc.Eng., 1932, M.Sc.Eng., 1948, University of London.

Fellow Award "For development of high voltage underground cables and contribution to standardization of cable design and installation." Other Awards: Sebastian de Ferranti Premium, Institution of Electrical Engineers, U.K., 1956; Fellow, Institute of Electrical Engineers, U.K.; Member, Institute of Civil Engineers, U.K.; Member, Institution of Mechanical Engineers, U.K.; Member, British Standards Institution; Member, British Electrical and Allied Industries Research Association.

SUTTON, HOWARD J., AIEE Fellow 1962. Born: September 28, 1908, Garnett, Kans. Degrees: B.S.E.E., 1931, University of Kansas.

Fellow Award "For contributions to microwave communications and relaying." Other Awards: Joint Power Generation Sponsor's Committee, 1971-1972.

SUZUKI, KEIJI, Fellow 1969. Born: June 20, 1911, Hamamatsu City, Japan. Degrees: B.S., 1932, Hamamatsu Technical College; D.Eng., 1950, Toyko University, Japan.

Fellow Award "For contributions and leadership in the relaying of television signals via satellite"; Vladimir K. Zworykin Prize Award, IEEE, "For his outstanding technical contribution in the field of television and engineering leadership in the application of the relaying of television signals via satellite," 1967. Other Awards: Awards, Institute of Electrical Communication Engineers of Japan, 1944, 1954, 1956; Scroll of Appreciation, Communications Satellite Corp., 1964; Purple Ribbon Award, Medal of Honor, Japanese Government, 1965; Maejima Award, 1961; Institute of Television Engineers Award, Japan, 1963, 1965; Award for Excellence, Japan Invention Association, 1969; Honorary Member, Institute of Television, Japan, 1977.

SUZUKI, KOUZO, Fellow 1984. Born: September 8, 1933, Sendai City, Japan. Degrees: B.S.E.E., 1956, Tohoku University.

Fellow Award "For development of advanced mobile telephone systems and advancement of consumer communication service." Other Awards: Maeshima Prize, Post and Telecommunications Association of Japan, 1982.

SVALA, C. GUNNAR, Fellow 1970. Born: October 16, 1918, Stockholm, Sweden. Degrees: M.S.E.E., Royal Institute of Technology, Stockholm.

Fellow Award "For research in system theory fundamental electronic switching and saturation signaling." Other Awards: Gold Medal, Royal Swedish Academy of Engineering Sciences, 1967.

SWAMY, SRIKANTA M.N., Fellow 1980. Born: April 7, 1935, Bangalore, India. Degrees: B.Sc.(Hons.)(Math.), 1954, Mysore University; Diploma (Elect. Comm. Eng.), 1957, Indian Institute of Science; M.Sc., 1961, Ph.D.(E.E.), 1963, University of Saskatchewan, Canada.

Fellow Award "For leadership in engineering education and contributions to circuit theory." Other Awards: Fellow, Institution of Engineers, India, 1979; Eta Kappa Nu, 1979; Fellow, Institution of Electronics and Telecommunication Engineers, India, 1980; Fellow, Engineering Institute of Canada, 1981; Fellow, Institution of Engineers (U.K.), 1983; Fellow, Dictionary of International Biographical Association, 1977; Fellow, American Biographical Association, 1979.

SWANK, ARTHUR J., AIEE Fellow 1960. Born: September 28, 1896, Colusa, Calif. Degrees: B.S.E.E.(cum laude), 1918, University of California.

Fellow Award "For contributions to construction and operation of a large power system."

SWENSON, GEORGE W., JR., Fellow 1965. Born: September 22, 1922, Minneapolis, Minn. Degrees: B.S., 1944, Michigan College of Mining and Technology; S.M., 1948, Massachusetts Institute of Technology; E.E., 1950, Michigan College of Mining and Technology; Ph.D., 1951, University of Wisconsin.

Fellow Award "For contributions to radio astronomy, ionospheric physics, and engineering education." Other Awards: Fellow, American Association for the Advancement of Science, 1966; National Academy of Engineering, 1978.

SWERDLOW, NATHAN, Fellow 1977. Born: June 5, 1907, Philadelphia, Pa. Degrees: B.S.E.E., 1930, Drexel University.

Fellow Award "For contributions to the development of the high-current isolated phase bus"; First Prize Paper, AIEE, Lehigh Valley Section Meeting, 1960, for "Practical Solutions of Inductive Heating Problems Resulting from High Current Busses." Other Awards: Managerial Award, General Electric Co., 1950, 1959, 1968.

SWERLING, PETER, Fellow 1968. Born: March 4, 1929. Degrees: B.S., 1947, California Institute of Technology; B.A., 1949, Cornell University; M.A., 1951, Ph.D., 1955, University of California at Los Angeles.

Fellow Award "For contributions to signal theory as applied to errors in tracking and trajectory prediction of missiles by radar."

SWERN, LEONARD, Fellow 1969. Born: February 12, 1925, New York, N.Y. Degrees: A.B., 1945, Columbia College, M.A., 1947, Columbia University.

Fellow Award "For contribution to the application of ferrites to microwave devices, for the advancement of microwave radiometry, and for exceptional staff work in planning complex research and development programs."

SWIFT, CALVIN T., Fellow 1983. Born: February 6, 1937, Prince William County, Virginia. Degrees: S.B., 1959, Massachusetts Institute of Technology; M.S., 1965, Virginia Polytechnic Institute; Ph.D., 1969, College of William and Mary.

Fellow Award "For contributions to the area of microwave remote sensing of the oceans"; Distinguished Service Award, IEEE Council on Oceanic Engineering, 1977.

SWINYARD, WILLIAM O., IRE Fellow 1945. Born: July 17, 1904, Logan, Utah. Degrees: B.S., 1927, Utah State University.

Fellow Award "In recognition of his work in promoting electronics and the affairs of the Institute, particulary in his district"; IEEE Consumer Electronics Outstanding Contributions Award, 1973. Other Awards: Eta Kappa Nu, 1958.

SYKES, ROGER A., IRE Fellow 1957. Born: February 17, 1908, Windsor, Vt. Degrees: B.S., 1929, M.S., 1930, Massachusetts Institute of Technology.

Fellow Award "For contributions to the development of quartz crystal units for filter networks and frequency control"; Mervin J. Kelly Award, 1972. Other Awards: Distinguished Engineer Award, Pennsylvania Society of Professional Engineers, 1967; C.B. Sawyer Memorial Award, 1981.

SYLVAN, TAGE P., Fellow 1967. Born: July 5, 1928, Montclair, N.J. Degrees: B.A., 1952, Bowdoin College.

Fellow Award "For creativity in the design, evaluation and application of the unijunction transistor, tunnel diode and other semiconductors." Other Awards: Phi Beta Kappa, 1951; Swope Fellow, 1954; Semi Award, 1983.

SYMONS, ROBERT S., Fellow 1972. Born: July 3, 1925, San Francisco, Calif. Degrees: B.S.E.E., 1946, M.S.E.E., 1948, Stanford University.

Fellow Award "For development of super-power klystrons and hybrid amplifiers."

SZABLYA, JOHN F., Fellow 1977 Born: June 25, 1924, Budapest, Hungary. Degrees: Dipl.Eng, 1947, Dipl.Econ., 1948, Dipl.Educ., 1948, Ph.D.(Econ.), 1948, Jozsef Nador University, Budapest.

Fellow Award "For contributions to the design and theory of electric machines and for leadership in engineering education." Other Awards: Zipernowsky Medal, Institute of Hungarian Electrical Engineers, 1954.

SZE, SIMON M., Fellow 1977. Born: March 21, 1936, Nanking, China. Degrees: B.S., 1957, National Taiwan University; M.S., 1960, University of Washington; Ph.D., 1963, Stanford University.

Fellow Award "For contributions to semiconductor device research and to education."

SZEGHO, CONSTANTIN S., IRE Fellow 1952. Born: March 15, 1905, Nagybocsko, Hungary. Degrees: Dipl.Ing., 1928, Technische Hochschule Munich, Germany; Dr.Ing., 1932, Technische Hochschule Aachen, Germany.

Fellow Award "In recognition of his contributions to the development of cathode-ray devices." Other Awards: 70 patents, cathode ray device field.

SZENTIRMAI, GEORGE, Fellow 1972. Born: October 8, 1928, Budapest, Hungary. Degrees: Dipl.E.E., 1951, Cand.Tech.Sci.(EE.), 1955, Technical University of Budapest; Ph.D., 1963, Polytechnic Institute of Brooklyn.

Fellow Award "For contributions to filter design techniques."

SZIKLAI, GEORGE C., IRE Fellow 1955. Born: July 9, 1909, Budapest, Hungary. Degrees: Dipl.Chem.Ing., Technische Hochschule Munich, Germany; Absolutorium, 1930, University of Budapest, Hungary.

Fellow Award "For his contributions to television circuits and systems." Other Awards: Radio Corporation of America, 1947, 1951, 1952.

T

TAGG, GEORGE F., AIEE Fellow 1935. Born: February 16, 1903, Brighton, Sussex, England. Degrees: B.Sc., Ph.D., University of London.

TAGGART, HAROLD E., Fellow 1983. Born: January 16, 1927, Madison, Kansas. Degrees: B.S.(Chemistry), 1950, University of Denver.

Fellow Award "For developing standard field-strength measurement techniques." Other Awards: Fellow, Radio Club of America, 1980.

TAI, CHEN T., IRE Fellow 1962. Born: December 30, 1915, Soochow, China. Degrees: B.S., 1937, Tsing Hua University; D.Sc., 1947, Harvard University.

Fellow Award "For contributions to electromagnetic and antenna theory"; IEEE Centennial Medal, 1984. Other Awards: Walker-Ames Professor, University of Washington, 1973; Distinguished Faculty Achievement Award, University of Michigan, 1975; Honorary Professor, Shanghai Normal University, 1979.

TAKAGI, NOBORU, Fellow 1981. Born: June 19, 1908. Tokyo, Japan. Degrees: B.E., 1931, Dr. Eng., 1942, University of Tokyo.

Fellow Award "For leadership in space technology and related advances in reliability engineering." Other Awards: Paper Award, Institution of Electrical Engineers, Japan, 1959; Awards for Distinguished Services, Institute of Electronics and Communication Engineers, Japan, 1965; Institution of Electrical Engineers, Japan, 1968; Japan Broadcasting Corporation, 1970; Purple Ribbon Medal (Emperor), 1972; Second Order of Merit with the Rising Sun (Emperor), 1978.

TAKEDA, YUKIMATSU, Fellow 1972. Born: April 24, 1911, Yamagata, Japan. Degrees: B.Eng., 1934, D.Eng., 1945, Ryojun Engineering College, Port Arthur, Manchuria.

Fellow Award "For his inventive and educational contributions in manufacture, research, and development in the electronics industry." Other Awards: The Fourth Order of the Sacred Treasure, Emperor of Japan, 1981.

TAKEUCHI, HIKOTARO, Fellow 1975. Born: October 16, 1904, Tokyo, Japan. Degrees: D.Eng., 1948, Waseda University.

Fellow Award "For contributions to the development of antenna systems and overseas telecommunication facilities." Other Awards: Distinguished Services Award, Science and Technology Agency of Japan, 1962; Distinguished Services Award, The Institute of Electronics and Communication Engineers of Japan, 1968; Honorary Member, The Institute of Electronics and Communication Engineers of Japan, 1970.

TAKI, YASUO, Fellow 1973. Born: March 17, 1919, Nagoya, Japan. Degrees: B.E. 1942, D.Eng., 1953, University of Toyko.

Fellow Award "For contribiubtions in the field of Communication, and to engineering education." Other Awards: Paper Awards, Institute of Electronics and Communication Engineers of Japan, 1951, 1970, 1981; Award for Publication, Institute of Electronics and Communication Engineers of Japan. 1961, 1972; Paper Award, Institute of Television Engineers of Japan, 1962; Distinguished Services Award, Institute of Television Engineers of Japan, 1975; Distinguished Services Award, Institute of Electronics and Communication Engineers of Japan, 1980; Minister's Prize, Ministry of Post and Telecommunications, Japan, 1980; Broadcasting Cultural Award, Japan Broadcasting Corporation (NHK), 1981.

TAMIR, THEODOR, Fellow 1977. Born : September 17, 1927, Bucharest, Roumania. Degrees: B.S.E.E., 1952, Dipl.Ing.(E.E.), 1953, M.S.E.E., 1956, Technion, Israel Institute of Technology; Ph.D.(Electrophysics), 1962, Polytechic Institute of Brooklyn.

Fellow Award "For contributions to guided wave propagation, with application to electromagnetics, optics and acoustics"; Special Recognition, IEEE Antennas and Propagation Society, 1968, for the paper "On Radio Wave Propagation in Forest Environments"; Other Awards: Institution Premium, IEE (London), 1964; Electronics Section Premium, IEE (London), 1967; Citation, Distinguished Research, Polytechnic Chapter, Sigma Xi, 1978.

TAMURA, YASUO, Fellow 1982. Born: September 21, 1931. Degrees: B.Eng., 1953, M.Eng., 1955, Dr.Eng., 1961, Waseda University, Tokyo, Japan.

Fellow Award "For contributions to the control, oper-

ation, and planning of power systems and for leadership in power system engineering education." Other Awards: KURATA Memorial Prize, Hitachi Works, Japan, 1971; ISHIKAWA Memorial Industrial Prize, Japan, 1978.

TANAKA, IKUO, Fellow 1979. Born: June 14, 1911, Mie Prefecture, Japan. Degrees: Ph.D., 1951, Tokyo University.

Fellow Award "For contributions to computer-aided analysis of electric field effects and the development of low skin-effect stranded conductors." Other Awards: President Award, Sumitomo Mining and Sumitomo Kyodo Electric Power Co., 1951; President Award, Sumitomo Electric Industries Ltd., 1953; I.C.C. Award, 1979.

TANAKA, RICHARD I., Fellow 1973. Born: December 17, 1928, Sacramento, Calif. Degrees: B.S.E.E., 1950, M.S.E.E., 1951, University of California at Berkeley; Ph.D., 1958, California Institute of Technology.

Fellow Award "For contribution to the technology and applications of computers and leadership in the computer profession." Other Awards: Honorary Member, Information Processing Society of Japan, 1970; Honorary Member, International Federation for Information Processing, 1979; Distinguished Service Award, American Federation of Information Processing Societies, 1983.

TANENBAUM, MORRIS, Fellow 1970. Born: November 10, 1928, Huntington, W.Va. Degrees: B.A.(Chemistry), 1949, The Johns Hopkins University; M.A.(Chemistry), 1950, Ph.D.(Physical Chemistry), 1952, Princeton University.

Fellow Award "For contributions to solid-state junction devices, and for leadership in the development of advanced manufacturing technology." Other Awards: ASM Campbell Lecture, 1975; Fellow, American Physical Society; National Academy of Engineering.

TANG, CHUNG L., Fellow 1977. Born: May 14, 1934, Shanghai, China. Degrees: B.S.E.E., 1955, University of Washington; M.S.E.E., 1956, California Institute of Technology; Ph.D.(Applied Physics), 1960, Harvard University.

Fellow Award "For contributions to the development of lasers and nonlinear optical devices and processes." Other Awards: Fellow, American Physical Society, 1974

TANG, DONALD T., Fellow 1981. Born: May 9, 1932, China. Degrees: B.S. (E.E.), 1953, Taiwan University; Ph.D (E.E.), 1960, University of Illinois at Urbana.

Fellow Award "For contributions to the design and analysis of computer communication networks." Other Awards: Invention Achievement Awards, IBM, 1968, 1969, 1973, 1975, and 1981; Outstanding Contribution Award, IBM Research, 1971; Research Division Awards, IBM, 1975 and 1979.

TANK, FRANZ, IRE Fellow 1949. Born: March 6, 1890, Zurich, Switzerland. Degrees: Dr.Phil., 1916, Zurich University; Dr.Tech.Sc.(Hon.), 1954, Ecole Polytechnique, Lausanne, Switzerland; Laurea, 1960, Politecnico, Turin, Italy.

Fellow Award "For his contributions to the field of radio education in Switzerland, and his accomplishments in ultra short-wave communications."

TANNER, R. L., Fellow 1964. Born: December 4, 1921, Idaho Falls, Idaho. Degrees: B.S., 1944, M.S., 1947, Ph.D., 1953, Stanford University.

Fellow Award "For contributions leading to the elimination of precipitation static on aircraft." Other Awards: Phi Beta Kappa; Tau Beta Pi; Sigma Xi; Scientific Research Society of America.

TANNER, ROBERT H., IRE Fellow 1958. Born: July 22, 1915, London, England. Degrees: B.Sc., 1936, M.Sc., 1961, University of London.

Fellow Award "For his contributions to the design and application of audio equipment in the broadcast field"; McNaughton Gold Medal, IEEE, Canadian Region, 1972, "For outstanding and sustained contributions to the professional, technical and industrial aspects of electrical and electronics engineering both nationally and transnationally"; Haraden Pratt Award, IEEE, 1981; "For contributions towards professionalism, and dedicated services to the Canadian Region, to the IEEE, and to the profession over many years;" President, IEEE, 1972.

TARZIAN, SARKES, Fellow 1965. Born: October 5, 1900, Kharpoot, Armenia. Degrees: B.S., 1924, M.S., 1927, University of Pennsylvania; Ph.D.(Hon.), 1969, Indiana Institute of Technology; LL.D.(Hon.), 1974, University of Pennsylvania; LL.D.(Hon.), 1975, Indiana University.

Fellow Award "For pioneering work in broadcasting and for the development of electronic components and equipment"; Consumer Electronics Award, IEEE, 1979. Other Awards: Eta Kappa Nu; Sigma Delta Chi; Tau Beta Pi; Charter Member, Society of Television Pioneers; Member, Armenian Students Association, 1961; Indiana Penn Man of the Year Award, University of Pennsylvania, 1962; Member, Friars Senior Society, 1966; Certificate of Appreciation, Radio Free Europe, 1969; D. Robert Yarnall Award, Engineering Alumni Society, University of Pennsylvania, 1969; Sagamore of the Wabash, State of Indiana, 1969; Man of the Year Award, Ararat Square Club of Delaware Valley, 1973; Distinguished American Award, Eisenhower Memorial Scholarship, 1975; National Horatio Alger Award, 1977.

TASCH, AL F., JR., Fellow 1984. Born: May 12, 1941, Corpus Christi, TX. Degrees: B.S.(Physics), 1963, U. of Texas, Austin; M.S.(Physics), 1965, Ph.D.(Physics), 1969, University of Illinois, Urbana.

Fellow Award "For the advancement of high-density MOS dynamic memory technology."

TAUB, JESSE J., Fellow 1967. Born: April 27, 1927, New York, N.Y. Degrees: B.E.E., 1948, City College of the City University of New York; M.E.E., 1949, Polytechnic Institute of Brooklyn.

Fellow Award "For contributions to microwave networks and millimeter quasi-optic techniques"; IEEE Centennial Medal, 1984.

TAUSWORTHE, ROBERT C., Fellow 1978. Born: September 28, 1934, Asherton, Tex. Degrees: B.S.E.E., 1957, New Mexico State University; M.S.E.E., 1958, Ph.D., 1963, California Institute of Technology.

Fellow Award "For contributions to communication theory and development of deep space communication and tracking systems." Other Awards: Exceptional Service Medal, NASA, 1977; Exceptional Engineering Achievement Medal, NASA, 1983.

TAYLOR, ARCHER S., Fellow 1978. Born: February 14, 1916. Degrees: B.S.(Physics), 1938, Antioch College.

Fellow Award "For leadership in the development of professional engineering procedures and standards for the cable television industry."

TAYLOR, BARRY N., Fellow 1977. Born: March 27, 1936, Philadelphia, Pa. Degrees: A.B. 1957, Temple University; M.S., 1960, Ph.D., 1963, University of Pennsylvania.

Fellow Award "For contributions to the understanding of electron tunneling in superconductors and for leadership in the development of Josephson junction voltage standards." Other Awards: Outstanding Achievement Award in Science, RCA, 1969; Fellow, American Physical Society, 1972; Fellow, Washington Academy of Sciences, 1974; John Price Wetherill Medal, Franklin Institute, 1975; Silver Medal Award for Meritorious Federal Service, U.S. Dept. of Commerce, 1975.

TAYLOR, DONALD W., AIEE Fellow 1945. Born: October 19, 1893, Cedar Rapids, Iowa. Degrees: E.E., 1916, Columbia University.

Other Awards: Professional Engineer, New Jersey; former Chairman, New York Section, AIEE; former Secretary of District 1, AIEE; Chairman, AIEE Automatic Stations Committee; Chairman, Electrical Equipment Committee, Edison Electric Institute.

TAYLOR, EDGAR RIVES, JR., Fellow 1974. Born: March 14, 1929, Jacksonville, Fla. Degrees: B.E.E., 1952, Cornell University.

Fellow Award "For contributions in the field of extra-high-voltage transmission as pertains to radio and audible noise, and to the study of transients and surges on EHV systems"; District Prize Award, 1962; Working Group Prize Paper Award, 1972; P.E.S. Prize Paper Award, 1981. Other Awards: Tau Beta Pi, 1952; Eta

Kappa Nu, 1952.

TAYLOR, EDMUND R., Fellow 1972. Born: 1898, Virginia. Degrees: B.E., 1921, The Johns Hopkins University.

Fellow Award "For pioneering and invention in the fields of radio transmission and telephone switching and signaling."

TAYLOR, GEORGE J., AIEE Fellow 1958. Born: May 13, 1903, Chicago, Ill. Degrees: B.S., 1925, Illinois Institute of Technology; S.M., 1926, Massachusetts Institute of Technology; E.E., 1933, Illinois Institute of Technology.

Fellow Award "For contributions to the application of electrical lighting to industry." Other Awards: Fellow, Illuminating Engineering Society, 1953; Gold Medal, Illuminating Engineering Society, 1971.

TAYLOR, LEONARD S., Fellow 1979. Born: December 28, 1928, New York, N.Y. Degrees: A.B., 1951, Harvard College; M.S., 1956, Ph.D., 1960, New Mexico State University.

Fellow Award "For contributions to the theory and application of electromagnetic scattering by turbulent media." Other Awards: Distinguished Alumnus Award, New Mexico State University, 1971.

TAYLOR, RALPH E., Fellow 1982. Born: November 28, 1923, Hickory, NC. Degrees: B.A.,(Physics), 1951, George Washington University.

Fellow Award "For contributions to electromagnetic interference predictions"; Certificate of Achievement, IEEE Electromagnetic Compatibility Society, 1977; Richard R. Stoddart Award, IEEE Electromagnetic Compatibility Society, 1981. Other Awards: Recognition Award, U.S. Patent Invention, 1950, 1963, 1967, 1969, 1970, 1981; Sustained Superior Performance Award, Diamond Ordnance Fuze Labs, U.S. Army, 1958; Special Achievement Award, National Aeronautics and Space Administration, Goddard Space Flight Center, 1975.

TAYLOR, THOMAS T., Fellow 1972. Born: April 18, 1921, Montpelier, Ind. Degrees: B.S., 1942, Purdue University; M.S., 1953, Ph.D., 1958, California Institute of Technology.

Fellow Award "For contributions to the theory of antenna aperture synthesis"; Best Paper Prize, P.G.A.P., 1960, for "Design of Circular Apertures for Narrow Beamwidth and Low Side Lobes." Other Awards: Sigma Xi, 1955; Tau Beta Pi, 1978.

TAYLOR, WADE H., AIEE Fellow 1951. Born: May 6, 1908, Ambia, Ind. Degrees: B.S., 1934, M.S., 1935, University of Colorado.

Other Awards: Distinguished Service Award, U.S. Dept. of the Interior, 1966; Distinguished Engineering Alumnus, University of Colorado, 1974.

TEAL, GORDON K., Fellow 1956. Born: January 10, 1907, Dallas, Tex. Degrees: A.B.(Math.and Chem.), 1927, Baylor University; Sc.M.(Chem.), 1928, Ph.D.(Phys.Inorg.Chem.), 1931, Brown University; LL.D(Hon.), 1969, Baylor University; Sc.D.(Hon.), 1969, Brown University.

Fellow Award "For advances in electronics, particularly in the field of semiconductors"; Medal of Honor, 1968, IEEE, "For his contributions to single crystal germanium and silicon technology and the single crystal grown junction transistor"; IEEE Centennial Medal, 1984. Other Awards: Fellow, American Association for the Advancement of Science, 1959; Honorary Member, Sigma Pi Sigma Physics Honor Society, 1959, "In Recognition of Achievement in Physics"; Honorary Life Fellow, Texas Academy of Science, 1960; Distinguished Alumnus Award, Baylor University, 1964; Chartered Electrical Engineer, Institution of Electrical Engineers, U.K., 1965; Inventor of the Year Award, The Patent, Trademark and Copyright Research Institute, George Washington University, 1966; Fellow, Institution of Electrical Engineers, U.K., 1966; Golden Plate Award, American Academy of Achievement, 1967; Certificate of Appreciation and Honor Scroll, National Bureau of Standards, 1967; Fellow, American Institute of Chemists; Member, National Academy of Engineering, 1969, "For essential contributions to the development of the semiconductor and electronic industries"; Medal for Creative Invention, American Chemical Society, 1970, "For essential contributions to the achievement of the junction transistor and for his germanium and silicon single crystal developments vital to semi-conductor science and technology"; Honor Scroll Recipient and 34th Fellows Lecturer, 50th Annual Meeting, American Institute of Chemists, 1973; The Wilfred T. Doherty Award, Dallas-Fort Worth Chapter, American Chemical Society, 1974; Omicron Delta Kappa Outstanding Baylor University Alumnus, 1978; Brown University Graduate School 50th Anniversary Commemoration Citation, 1978; Citation, Texas Instruments, 1980; Sigma Xi.

TEARE, B. RICHARD, JR., AIEE Fellow 1942. Born: January 12, 1907, Menomonie, Wis. Degrees: B.S., 1927, M.S., 1928, University of Wisconsin; D.Eng., 1937, Yale University; D.S.(Hon.), 1965, Tri-State College; D.Eng.(Hon.), 1970, Cleveland State University; D.Eng.(Hon.), 1977, Carnegie-Mellon University.

Awards: Vice President, IEEE, 1963; Active in merger of AIEE & IRE; IEEE Education Medal, 1964. Other Awards: George Westinghouse Award, American Society for Engineering Education, 1947; Lamme Medal, American Society for Engineering Education, 1963; Tasker H. Bliss Medal, SAME, 1977; President, ASEE, 1959-60; Fellow, ASEE, 1983.

TEBO, JULIAN D., AIEE Fellow 1951. Born: July 5, 1903, Harpers Ferry, W.Va. Degrees: B.E.E., 1924, D.Eng., 1928, The Johns Hopkins University.

Other Awards: Sigma Xi, 1928; Fellow, New York

Academy of Sciences, 1960; Life Member, Montclair Society of Engineers, 1971; Eta Kappa Nu, 1975.

TEER, KEES, Fellow 1974. Born: June 6, 1925, Haarlem, The Netherlands. Degrees: M.S., 1949, Ph.D., 1959, Technical University of Delft, The Netherlands.

Fellow Award "For contributions to television, acoustics, and electronic systems, and for leadership in research." Other Awards: "Veder" Award, 1955; Member, Royal Dutch Academy of Sciences, 1977.

TEGOPOULOS, JOHN A., Fellow 1975. Born: September 30, 1924, Trikala, Greece. Degrees: M.E.E., 1948, Technical University of Athens; M.S.E.E., 1954, Ph.D., 1956, Purdue University.

Fellow Award "For contributions to electric machine design and in the investigation of eddy currents in conducting media."

TEHON, STEPHEN W., Fellow 1972. Born: October 20, 1920, Shenandoah, Iowa. Degrees: B.S., 1942, M.S., 1947, Ph.D., 1958, University of Illinois.

Fellow Award "For contributions to the understanding and application of electroacoustic devices."

TELLIER, JOSEPH C., IRE Fellow 1959. Born: June 24, 1914, Des Moines, Iowa. Degrees: B.S., 1935, University of Pennsylvania.

Fellow Award "For contributions to receiver design and transistorization."

TELLIER, ROGER A., Fellow 1967. Born: February 3, 1926, St. Prix, France. Degrees: Ing.E.S.E., 1946, Ecole Superieure D'Electricite, France.

Fellow Award "For his outstanding contributions to the development of underground and submarine power transmission"; Paper Prize, AIEE, 1960, for "Two Examples of Industrial Research in France Relating to the Transmission of Electrical Energy." Other Awards: Laureate Member, Societe Francaise des Electriciens, 1952; Fellow, Institution of Electrical Engineers, U.K., 1965.

TEMES, GABOR C., Fellow 1972. Born: October 14, 1929, Budapest, Hungary. Degrees: Dipl.Ing., 1952, Technical University of Budapest; Dipl.Phys., 1954, Eotvos University, Budapest; Ph.D., 1961, University of Ottawa.

Fellow Award "For contributions to filter theory and computer-aided circuit design"; Best Paper Award (Darlington Award) with corecipient, IEEE Circuits and Systems Society, 1968; IEEE Centennial Medal, 1984. Other Awards: Outstanding Engineer Merit Award, Institute for the Advancement of Engineering, 1981; Best Paper (Darlington Prize) Award (co-recipient), 1982; Western Electric Fund Award, 1982.

TEMOSHOK, MICHAEL, Fellow 1974. Born: February 9, 1920. Degrees: B.S., 1941, Lehigh University.

Fellow Award "For contributions to the continuing development and application of excitation systems for large alternating current generators." Other Awards:

Member, C.I.G.R.E.; Tau Beta Pi; Eta Kappa Nu; Sigma Xi.

TENNEY, GEORGE C., Fellow 1957. Born: July 13, 1898, Provo, Utah. Degrees: LL.D.(Hon.), 1970, University of California at Berkeley.

Fellow Award "For important contributions to the advancement of engineering knowledge."

TERHUNE, ROBERT W., Fellow 1984. Born: February 7, 1926, Detroit, MI. Degrees: B.S.(Physics), 1947, University of Michigan; M.A.(Physics), 1948, Dartmouth College; Ph.D.(Physics), 1957, University of Michigan.

Fellow Award "For contributions to nonlinear optics and quantum electronics." Other Awards: Science and Engineering Award, Drexel Institute of Technology, 1964; Sesquicentennial Award, University of Michigan, 1967.

TERMAN, LEWIS M., Fellow 1975. Born: August 26, 1935, San Francisco, Calif. Degrees: B.S., 1956, M.S., 1958, Ph.D., 1961, Stanford University.

Fellow Award "For contributions to the design and development of semiconductor computer memory and logic circuitry." Other Awards: IBM Outstanding Contribution Award, 1968; IBM Outstanding Invention Awards, 1971, 1974, 1977; Fellow, American Association for the Advancement of Science, 1981.

TERRY, IRA A., AIEE Fellow 1937. Born: October 31, 1903, Ogden, Utah. Degrees: B.S.E.E., 1925, University of Utah; M.S.E.E., 1929, Union College.

Awards: First Paper Prize, Summer Meeting, 1929; VP, Region VI, 1948. Other Awards: Distinguished Alumni Award, Weber State College, 1982.

THALER, GEORGE J., Fellow 1967. Born: March 15, 1918, Baltimore, Md. Degrees: B.E., 1940, D.Eng., 1947, The Johns Hopkins University.

Fellow Award "For leadership in the field of automatic control and contributions to engineering education"; Harry Diamond Award, IEEE, 1965. Other Awards: Sigma Xi; American Society for Engineering Education.

THATCHER, JOHN W., Fellow 1973. Born: December 7, 1905, Hatch, Idaho. Degrees: B.S.(with honor), 1928, M.S., 1930, California Institute of Technology.

Fellow Award "For contributions to space exploration through his efforts in establishment and operation of the Deep Space Network." Other Awards: Tau Beta Pi, 1928; Sigma Xi, 1929; Achievement Award, NASA Lunar Orbiter Team, 1967; Scroll of Appreciation, U.S. Department of State, 1969; Fellow, Institute for Advancement of Engineering, 1971.

THAXTON, GUY WERTER, AIEE Fellow 1943. Born: September 30, 1894, Lake, Miss. Degrees: B.Sc.E.E.(with special honors), 1916, Mississippi State University.

Awards: Tau Beta Pi, 1934; Phi Kappa Phi, 1956; Registered Professional Electrical Engineer, Missouri.

THAYER, GORDON N., IRE Fellow 1951. Born: October 6, 1908, Delta, Colo. Degrees: M.E., 1930, D.Eng.(Hon.), 1960, Stevens Institute of Technology.

Fellow Award "For his leadership in the engineering of long distance microwave links for network television and multiplex telephony."

THIELE, GARY A., Fellow 1982. Born: May 5, 1938, Cleveland, OH. Degrees: B.S.E.E., 1960, Purdue University; M.Sc., 1964, Ph.D., 1968, Ohio State University.

Fellow Award "For contributions to computational methods in electromagnetic theory"; Best Paper Award, IEEE Antennas and Propagation Society, 1976.

THIEMANN, HUGO E., Fellow 1972. Born: February 2, 1917, St. Gall, Switzerland. Degrees: Elec.Ing., 1939, D.Sc., 1947, Federal Institute of Technology, Zurich; D.Sc.(Hon.), 1965, University of Geneva.

Fellow Award "For contributions to projection television and leadership in electronic research and development."

THOMAS, DAVID G., Fellow 1977. Born: August 4, 1928, London, England. Degrees: B.A.(Chemistry), 1949, M.A.(Chemistry), 1950, Oxford University, Oriel College; D.Phil.(Chemistry), 1952, Oxford University, Merton College.

Fellow Award "For contributions to the understanding of luminescence in semiconductors and to the development of light-emitting diodes." Other Awards: Oliver E. Buckley Solid State Physics Prize (corecipient J. J. Hopfield), American Physical Society, 1969.

THOMAS, EVERETT J., AIEE Fellow 1951. Born: November 11, 1901, Ipswich, S.D. Degrees: B.S.E.E., 1924, University of Wisconsin.

Other Awards: National Electrical Manufacturers Association; American Standards Association.

THOMAS, J. EARL, Fellow 1965. Born: September 7, 1918, Seattle, Wash. Degrees: A.B., 1939, The Johns Hopkins University; Ph.D., 1943, California Institute of Technology.

Fellow Award "For contributions to the field of semiconductor devices." Other Awards: Phi Beta Kappa, 1939; Sigma Xi, 1942; Award, U.S. Dept. of Navy, Bureau of Ordnance, 1944; Award for Effective Service, Office of Scientific Research and Development, 1946; Honorary Member, Tau Beta Pi, 1959; Fellow, American Physical Society, 1959.

THOMAS, JOHN B., Fellow 1965. Born: July 14, 1925, New Kensington, Pa. Degrees: A.B., 1944, Gettysburg College; B.S., 1952, The Johns Hopkins University; M.S., 1953, Ph.D., 1955, Stanford University.

Fellow Award "For contributions to communication theory and engineering education." Other Awards: Senior Postdoctoral National Science Foundation Fellow, 1967-1968.

THOMAS, JULIAN B., AIEE Fellow 1943. Born: July 19, 1891, San Marcos, Tex. Degrees B.S., 1911, M.E., 1931, Texas A.&M. University.

Other Awards: Fellow, American Society of Mechanical Engineers, 1958; Distinguished Alumnus Award, Texas A.&M. University, 1974; 50-Year Pin Award, American Society of Mechanical Engineers, 1977.

THOMAS, LEONARD W., SR., Fellow 1978. Born: May 11, 1909, Birmingham, Ala. Degrees: B.S.(E.E.), 1931, Auburn University.

Fellow Award "For leadership in electromagnetic compatibility and development of interference measurement instrumentation and standards"; Debt of Gratitude, IEEE, 1968; Certificate of Appreciation, IEEE Electromagnetic Compatibility Society, 1965; Laurence G. Cumming Award, IEEE Electromagnetic Compatibility Society, 1979. Other Awards: Certificate, American Standards Association, 1946; Certificate of Appreciation, Society of Automotive Engineers, 1970; Chartered Electrical Engineer, Institution of Electrical Engineers (UK); Corporate Member, Institution of Electrical Engineers (UK), 1956; Member Washington Society of Engineers; National Society of Professional Engineers; District of Columbia Society of Professional Engineers; American Society of Naval Engineers; Society of Automotive Engineers; Registered Professional Engineer, District of Columbia, 1952; Listed in "Who's Who in the South and Southwest"; "Who's Who in Technology Today"; and "Who's Who in Engineering."

THOMAS, MELVIN A., AIEE Fellow 1948. Born: January 5, 1903, New Castle, Nebr. Degrees: B.S., 1925, University of Wisconsin; M.S., 1931, University of Texas.

THOMPSON, DAVID A., Fellow 1983. Born: December 17, 1940, Devils Lake, North Dakota. Degrees: B.S., 1962, M.S., 1963, Ph.D., 1966,(Elec. Eng.) Carnegie Institute of Technology.

Fellow Award "For developing techniques for thin-film magnetic recording heads." Other Awards: IBM Fellow, 1980; IBM Awards: Outstanding Contribution, Outstanding Invention, and Outstanding Innovation awards.

THOMPSON, FRANCIS T., Fellow 1975. Born: November 22, 1930, New York, N.Y. Degrees: B.S.E.E., 1952, Rensselaer Polytechnic Institute; M.S.E.E., 1955, Ph.D., 1964, University of Pittsburgh.

Fellow Award "For leadership in the application of solid-state electronics." Other Awards: B.G. Lamme Scholarship Award, Westinghouse Electric, 1961; Order of Merit Award, Westinghouse Electric, 1976. Eta Kappa Nu; Sigma Xi; Tau Beta Pi; Instrument Society of America.

THOMPSON, LELAND E., IRE Fellow 1957. Born: September 18, 1905, Creighton, Nebr. Degrees: B.S.,

1929, University of South Dakota.

Fellow Award "For contributions to microwave communication systems and development of special purpose radio receivers."

THOMSON, J. M., AIEE Fellow 1947. Born: June 3, 1898, Couva, Trinidad, British West Indies. Degrees: B.A.Sc., 1924, M.A.Sc., 1933, Ph.D., 1937, University of Toronto, Ontario.

THOREN, BERTIL H., Fellow 1982. Born: March 28, 1921, Vanersborg, Sweden. Degrees: M.Sc., 1944, D.Tech., 1950, Royal Institute of Technology, Sweden,

Fellow Award "For advancing the understanding of the design and operation of equipment for ultrahigh-voltage power transmission systems."

THORNTON, CLARENCE G., Fellow 1966. Born: August 3, 1925, Detroit, Mich. Degrees: B.S., 1949, M.S., 1950, Ph.D., 1952, University of Michigan.

Fellow Award "For contributions in research and development in the field of semiconductor devices"; IEEE Centennial Medal, 1984. Other Awards: Science Conference Award, Dept. of the Army, 1976; Research and Development Achievement Award, Dept. of the Army, 1976; Honors Award, University of Michigan; Sigma Xi; Phi Kappa Phi; Phi Lambda Upsilon; Senior Executive Service Awards, Department of Army, 1980, 1982, 1983; AFCEA Gold Medal, 1983.

THORNTON, RALPH E., AIEE Fellow 1959. Born: April 29, 1903, Keokuk Falls, Okla. Degrees: B.S., 1925, University of Oklahoma.

Fellow Award "For contributions to planned expansion of a utility power system."

THUN, RUDOLF E., Fellow 1971. Born: January 30, 1921, Berlin, Germany. Degrees: Dipl.Phy., 1954, Ph.D., 1955, University of Frankfurt.

Fellow Award "For contributions to the science and technology of thin films, and solid-state structures." Other Awards: Award for Technological Achievement, U.S. Army Corps of Engineers, 1953; Fellow, American Physical Society, 1972.

THURSTON, MARLIN O., Fellow 1969. Born: September 20, 1918, Denver, Colo. Degrees: B.A., 1940, M.Sc., 1946, University of Colorado; Ph.D., 1955, The Ohio State University.

Fellow Award "For contributions to electron devices, and for ability as both a teacher and as an administrator." Other Awards: Robert M. Critchfield Professor, Emeritus, Ohio State University, Columbus, Ohio.

TICE, THOMAS E., IRE Fellow 1961. Born: January 24, 1924, Florence, Ala. Degrees: B.E.E., 1947, M.Sc., 1948, Ph.D., 1951, Ohio State University.

Fellow Award "For contributions to radome theory and techniques." Other Awards: Chi Beta Phi, 1943; Eta Kappa Nu, 1947; Tau Beta Pi, 1947; Sigma Xi, 1948; Commendations, U.S. Dept. of Air Force, 1956, 1957, 1966.

TIEMANN, JEROME J., Fellow 1976. Born: February 21, 1932. Degrees: Sc.B., 1953, Massachusetts Institute of Technology; Ph.D., 1960, Stanford University.

Fellow Award "For clarifying the understanding of interband tunneling and surface charge transport, and for applying these phenomena to new devices." Other Awards: Miniaturization Award, Miniature Precision Bearings, Inc., 1959; I-R 100 Award, Industrial Research Magazine, 1971, 1974. Coolidge Fellow, Research and Development Center, General Electric.

TIEN, P. K., Fellow 1966. Born: August 2, 1919, Checkiang, China. Degrees: B.S., 1942, National Central University, China; M.S., 1948, Ph.D., 1951, Stanford University.

Fellow Award "For creativity and leadership in advanced electronics"; Morris N. Liebmann Award, 1979. Other Awards: Sigma Xi; Achievement Award, Chinese Institute of Engineers, 1966; National Academy of Engineers, 1975; Fellow, Optical Society of America, 1976; National Academy of Sciences, 1978; Bell Laboratories Fellow, 1983.

TILLES, ABE, AIEE Fellow 1951. Born: March 9, 1907, New York, N.Y. Degrees: B.S., 1928, M.S., 1932, Ph.D., 1934, University of California at Berkeley.

Awards: Alfred Noble Prize, 1936. Other Awards: Honorary Member, Eta Kappa Nu, 1931; Full Member, Sigma Xi, 1935. Life Member, American Society for Engineering Education; Founder and Editor, San Francisco Engineer, 1948; Chairman San Francisco Bay Area Engineering Council, 1952.

TIMASCHEFF, ANDREW S., Fellow 1967. Born: May 10, 1899, St. Petersburg, Russia. Degrees: Dipl.Ing., 1924, Technische Hochschule Karlsruhe; Dr.Ing., 1942, Dr.Ing.Habil., 1943, Technische Hochschule Munich.

Fellow Award "For notable contributions to the design of electrical apparatus and transmission systems," Other Awards: Ross Medal, Engineering Institute of Canada, 1963; Fellow, Engineering Institute of Canada, 1970; Honorary Director, Research Institute of Hydro-Que (IREQ), 1979.

TIMOSHENKO, GREGORY S., AIEE Fellow 1954. Born: November 1, 1904, St. Petersburg, Russia. Degrees: Dipl.Ing.(Applied Physics), 1929, Technical University, Berlin, Germany; Ph.D.(Electrical Engineering), 1932, University of Michigan.

Fellow Award "For his contributions to the advancement of engineering education." Other Awards: Fellow, American Physical Society, 1941.

TINNEY, WILLIAM F., Fellow 1976. Born: May 5, 1921, Portland, Oreg. Degrees: B.S., 1948, M.S., 1949, Stanford Unversity.

Fellow Award "For contributions to the application of digital computers to solve large power network problems"; Prize Paper Award, IEEE Power Engineering Society, 1970, 1975. Other Awards: Gold Medal, U.S. Dept. of the Interior, 1976.

TIPTON, EARL W., AIEE Fellow 1956. Born: March 23, 1902, Kansas City, Kans. Degrees: B.S., 1925, University of Kansas.

Fellow Award "For his many contributions to the fundamental knowledge and the development of transformer design."

TISCHER, FREDERICK J., IRE Fellow 1962. Born: March 14, 1913, Plan, Austria. Degrees: M.S., Ph.D., 1938, Prague, Czechoslovakia.

Fellow Award "For technical contributions in the field of microwaves"; Distinguished Service Awards, MTT/AP PGT, and Huntsville, Alabama Section, 1964 and 1965. Other Awards: Distinguished Service Award, NASA, 1963; Tau Beta Pi; Sigma Xi; Distinguished Professor, NPS, 1978-1979.

TITUS, CHARLES H., AIEE Fellow 1958. Degrees: B.S., 1938, M.S., 1940, Lehigh University.

Fellow Award "For contributions to design and development of industrial and high voltage circuit interrupting devices."

TODD, ZANE G., Fellow 1977. Born: February 3, 1924, Hanson, Ky. Degrees: B.S.E.E., 1951, D.Eng.(-Hon.), 1979, Purdue University.

Fellow Award "For leadership in power system operations and engineering." Other Awards: Distinguished Engineering Alumnus, Purdue University, 1976.

TOHMA, YOSHIHIRO, Fellow 1980. Born: August 22, 1933, Kawasaki, Japan. Degrees: B.E., 1956, M.E., 1958, Dr. Eng., 1961, Tokyo Institute of Technology.

Fellow Award "For contributions to the theory and design of fault-tolerant digital systems, and to engineering education." Other Awards: Okabe Memorial Award, Institute of Electronics and Communiction Engineers of Japan, 1965; Excellent Book Award, Institute of Electronics and Communication Engineers of Japan, 1976.

TOLER, JAMES C., Fellow 1981. Born: January 31, 1936, Carthage, AR. Degrees: B.S.E.E., 1957, University of Arkansas; M.S.E.E., 1970, Georgia Tech.

Fellow Award "For contributions to electromagnetic compatibility testing which led to the development of reliable cardiac pacemakers"; Certificate of Recognition, Certificate of Appreciation, Certificate of Acknowledgement, Certificate of Achievement, IEEE Electromagnetic Compatibility Society, 1976, 1978, 1980; IEEE Centennial Medal, 1984. Other Awards: Electromagnetic Compatiblity Symposium and Exhibition Citation, Montreux, Switzerland, 1977; University of Arkansas Academy of Electrical Engineering, 1981.

TOLLES, WALTER E., IRE Fellow 1960. Born: February 1, 1916, Moline, Ill. Degrees: B.S., 1939, Antioch College; M.S., 1942, University of Minnesota; Ph.D., 1969, State University of New York.

Fellow Award "For the application of electronics to the field of medicine." Other Awards: Fellow, Association for the Advancement of Science, 1943; Fellow, New York Academy of Sciences, 1957.

TOMIYASU, KIYO, IRE Fellow 1962. Born: September 25, 1919, Las Vegas, Nev. Degrees: B.S., 1940, California Institute of Technology; M.S., 1941, Columbia University; Ph.D., 1948, Harvard University.

Fellow Award "For contributions to microwave theory"; Honorary Life Member, MTT-S, IEEE, 1973, "For his continued outstanding contributions and services to the IEEE Society on Microwave Theory and Techniques"; Career Award, IEEE Microwave Theory and Techniques Society, "For a career of meritorious achievement and outstanding technical contributions in the field of microwave theory and techniques"; 1980; IEEE Centennial Medal, 1984. Other Awards: Steinmetz Award, General Electric, 1977.

TOMOTA, MIYAJI, IRE Fellow 1961. Born: August 22, 1905, Hiroshima, Japan. Degrees: Ph.D., 1952, Tokyo University.

Fellow Award "For development of electronic measuring instruments and automatic control devices." Other Awards: Blue Ribbon, Japanese Government, 1958; Honorary Member, SICE, Japan, 1972; Sunshine Medal, Japanese Government, 1975.

TOMPKINS, JOEL S., AIEE Fellow 1962. Born: March 11, 1905, Swampscott, Mass. Degrees: S.B., 1926, S.M., 1927, Massachusetts Institute of Technology.

Fellow Award "For contributions to the development of transmission conductor and hydroelectric power plant design."

TORGOW, EUGENE N., Fellow 1968. Born: November 26, 1925, Bronx, N.Y. Degrees: B.S.E.E., 1946, Cooper Union; M.E.E., 1949, Polytechnic Institute of Brooklyn; Engineer's Degree, 1980, Polytechnic Institute of N.Y.; Executive Program, 1983, UCLA Graduate School of Management.

Fellow Award "For contributions to the theory and design of microwave components"; Distinguished Service Award, IEEE Microwave Theory and Techniques Society, 1978. Other Awards: Fellow, Institute for Advancement of Engineering, 1971.

TOU, JULIUS T., Fellow 1966. Born: August 15, 1926, Wusih, China. Degrees: B.S.E.E., 1947, National Chiao Tung University, China; M.S., 1950, Harvard University; D.Eng., 1952, Yale University.

Fellow Award "For contributions to discrete systems theory, computer control theory, and engineering education." Other Awards: Distinguished Alumni Award, Yale University, 1970; Fellow, Academia Sinica, 1971.

TOWLSON, HAROLD G., Fellow 1968. Born: June 17, 1908, Gouverneur, N.Y. Degrees: B.E.E., 1929, Clarkson College of Technology; M.E.E., 1954, Syracuse University.

Fellow Award "For leadership in the introduction of high power levels in short-wave radio and in TV broadcasting, and in setting of standards for TV broadcast equipment."

TOWNER, ORRIN W., IRE Fellow 1958. Born: March 29, 1903, Peterson, Iowa. Degrees: A.A., 1924, Junior College of Kansas City; B.S., 1927, E.E., 1933, University of Kansas.

Fellow Award "For his contributions to war-time electronics"; Louisville Section IEEE, "Engineer of the Year," 1981; Member, AIEE Committee on TV and Aural Broadcast Systems, 1956-1957; Publicity Chairman, AIEE Winter Covention Committee, 1954; Chairman Louisville Section, IRE, 1948-1949; 1955-1957; Chairman, Louisville Section, IEEE, 1964-1965; Representative for Louisville Section, IEEE, to Louisville Engineering & Scientific Societies Council, 1965-1966. Other Awards: Presidential Certificate of Merit, 1948; Vice-Chairman, FCC National Industry Advisory Committee Field Test of Emergency Alerting Ad Hoc Committee, 1964; Co-Chairman, FCC National Industry Advisory Committee, Sub-Committee on Emergency Alerting System Test and Evaluation Ad Hoc Committee, 1966-1968.

TOWNES, CHARLES H., IRE Fellow 1962. Born: July 28, 1915, Greenville, S.C. Degrees: B.A., B.S., 1935, Furman University; M.A., 1937, Duke University; Ph.D., 1939, California Institute of Technology.

Fellow Award "For fundamental contributions to the maser"; Morris N. Liebmann Award, IRE, 1958; David Sarnoff Award, AIEE, 1961; Medal of Honor, 1967. Other Awards: Annual Award, Research Corporation, 1958; Comstock Prize, National Academy of Sciences, 1959; Stuart Ballantine Medals, Franklin Institute, 1959, 1962; Exceptional Service Award, U.S. Dept. of Air Force, 1959; Rumford Premium, American Academy of Arts and Sciences, 1961; Arnold O. Beckman Award, Instrument Society of America, 1961; John J. Carty Medal, National Academy of Sciences, 1962; Thomas Young Medal and Prize, Institute of Physics and the Physical Society, U.K., 1963; John Scott Award, Philadelphia, 1963; Nobel Prize, 1964; Alumni Distinguished Service Award, California Institute of Technology, 1966; C.E.K. M.E.E.S. Medal, Optical Society of America, 1968; Distinguished Public Service Medal, NASA, 1969; Wilhelm Exner Award (Austria), 1970; Foreign Member, Royal Society of London, 1976; Earle K. Plyler Prize, American Physical Society, 1977; Niels Bohr International Gold Medal, 1979; LeConte Medal, 1980; National Medal of Science, 1982; Engineering and Science Hall of Fame Enshrinee, 1983.

TOWNSEND, MARJORIE R., Fellow 1980. Born: March 12, 1930, Washington, D.C. Degree: B.E.E., 1951, George Washington University.

Fellow Award "For management and technical con-tributions in the space exploration program"; IEEE Centennial Medal, 1984. Other Awards: TIROS Project Group Achievement Award, 1963; NIMBUS Project Group Achievement Award, 1969; Volunteers for Education Award, 1970; NASA Exceptional Service Medal, 1971; Knight of Italian Republic Order, 1972; Special Citation for Exceptional Achievement in Engineering Management, D.C. Council of Engineering and Architectural Societies, 1973; Federal Woman's Award, 1973; Award for Culture, Association EUR, Rome, 1974; Engineer Alumni Achievement Award, George Washington University, 1975; (General Alumni Association Achievement Award, George Washington University, 1976; SAS Project Group Achievement Award, 1975; 1976;) Goddard Space Flight Center Quality Increase for Exceptional Service, 1978; Heat Capacity Mapping Mission Project Team Group Achievement Award, 1978; SES Bonus, 1980; NASA Outstanding Leadership Medal, 1980; Listed in "Who's Who of American Women," "Who's Who in the East," "Who's Who in Aviation," "Who's Who in Government," "Two Thousand Women of Achievement," "International Biographical Centre-World Who's Who of Women," "Men and Women of Distinction," "American Biographical Institute," "American Men and Women of Science"; "Personalities of America," "Who's Who in Aviation and Aerospace," "Who's Who in Frontier Science and Technology," "Who's Who in Technology Today," "International Who's Who in Engineering."

TOY, WING N., Fellow 1981. Born: February 3, 1926, China. Degrees: B.S.E.E., 1950, M.S.E.E., 1952, University of Illinois; Ph.D.E.E., 1969, University of Pennsylvania.

Fellow Award "For contributions to the conception, design, and development of fault-tolerant computers for electronic telephone switching systems and telecommunication systems."

TRAINOR, JAMES H., Fellow 1984. Born: August 22, 1935, Lancaster, NH. Degrees: B.S.(Physics), 1958, M.S.(Physics), 1959, Ph.D.(Physics), 1964, University of New Hampshire.

Fellow Award "For contributions to the development of unique long-life spacecraft scientific instruments." Other Awards: NASA Medal for Exceptional Scientific Achievement, 1974; NASA Medal for Exceptional Service, 1977; Phi Kappa Phi; NASA Research and Study Fellowship and Visiting Associate, Caltech, 1979.

TRAUTMAN, DEFOREST L. (WOODY), Fellow 1972. Born: June 14, 1920, Canal Zone, Panama. Degrees: B.S., 1942, M.S., 1943, Carnegie Institute of Technology; Ph.D., 1949, Stanford University.

Fellow Award "For his contributions and leadership in engineering education." Other Awards: Engineer of the Year, Toledo Technical Society (Ohio), 1978.

TRAVIS, IRVEN, IRE Fellow 1953. Born: March 30, 1904, McConnellsville, Ohio. Degrees: B.S., 1926, Drexel Institute of Technology; M.S., 1928, Sc.D., 1938, University of Pennsylvania; D.Eng.(Hon.), 1962, Drexel Institute of Technology.

Fellow Award "For his contributions to the fields of computing devices and of gun-fire control equipment, and to engineering education." Other Awards: Alumni Citation, Drexel Institute of Technology; Citation, U.S. Dept. of Navy, 1946; Eta Kappa Nu; Tau Beta Pi; Sigma Xi; 50th Anniversary Gold Medal, Moore School of Electrical Engineering, University of Pennsylvania, 1974.

TREITEL, SVEN, Fellow 1983. Born: March 5, 1929, Freiburg i/Breisgau, Germany. Degrees: B.S.(Geophysics), 1953; M.S.(Geophysics), 1955; Ph.D.(Geophysics), 1958, MIT.

Fellow Award "For contributions to the development of digital signal processing techniques and their application to geophysical data." Other Awards: SEG Fessenden Medal, 1969; EAEG Conrad Schlumberger Award, 1969; SEG Honorary Member, 1983.

TREVES, DAVID, Fellow 1976. Born: June 28, 1930, Italy. Degrees: B.Sc., 1953, M.Sc., 1956, D.Sc., 1958, Technion, Israel Institute of Technology.

Fellow Award "For contributions to magnetooptic memories, and for leadership in the field of applied physics."

TRICK, TIMOTHY N., Fellow 1977. Born: July 14, 1939, Dayton, Ohio. Degrees: B.E.E., 1961, University of Dayton; M.S.E.E., 1962, Ph.D., 1966, Purdue University.

Fellow Award "For contributions to the analysis of communication circuits and to engineering education"; Guillemin-Cauer Award, IEEE, 1976, for the paper "Computation of Capacitor Voltage and Inductor Current Sensitivities with Respect to Initial Conditions for the Steady State Analysis of Nonlinear Periodic Circuits." Other Awards: ASEE-NASA Summer Faculty Fellowship, 1970, 1971; Best Paper Award, Asilomar Conference on Circuits and Systems, 1972, Member: Sigma Xi, Eta Kappa Nu, Tau Beta Pi, Pi Mu Epsilon.

TRICKEY, PHILIP H., AIEE Fellow 1947. Born: March 29, 1906, Charleston, Me. Degrees: B.S., 1928, M.S., 1931, E.E., 1934, University of Maine.

Awards: Nikola Tesla Award, IEEE, 1980.

TRIEBWASSER, SOL, Fellow 1973. Born: August 16, 1921, New York, N.Y. Degrees: B.A., 1941, Brooklyn College; M.A., 1948, Ph.D., 1952, Columbia University.

Fellow Award "For leadership in the research and development of large-scale integration for computer circuits."

TRIVELPIECE, ALVIN WILLIAM, Fellow 1981. Born: March 15, 1931, Stockton, CA. Degrees: B.S., 1953, California State Polytechnic College; M.S., 1955, Ph.D.,

1958, California Institute of Technology.

Fellow Award "For contributions to the understanding of wave-plasma interactions and to plasma education." Other Awards: Fulbright Scholar, The Netherlands, 1958; Guggenheim Fellowship, 1966; Distinguished Alumnus Award, California State Polytechnic College, 1978; Fellow, American Assocation for the Advancement of Science; Fellow, American Physical Society.

TROXEL, FRANKLIN D., AIEE Fellow 1950. Born: September 15, 1900, Arcanum, Ohio.

Other Awards: Life Member, Western Society of Engineers; Fellow, A.A.A.S.; Eta Kappa Nu.

TRUMP, JOHN G., AIEE Fellow 1959. Born: August 21, 1907, New York, N.Y. Degrees: B.S.E.E., 1929, Polytechnic Institute of Brooklyn; M.S.(Physics), 1931, Columbia University; Sc.D.(E.E.), 1933, Massachusetts Institute of Technology.

Fellow Award "For contributions in the field of high voltage engineering, particularly super voltage X rays"; Lamme Medal, AIEE, 1960; Power Life Award, 1973. Other Awards: The King's Medal for Freedom, 1946; Presidential Citation, 1946; Holmes Lecturer, New England Roentgen Ray Society, 1963; Public Service Award, Polytechnic Institute of Brooklyn Alumni Association, 1963; New England Award, Engineering Societies of New England, 1967; Honorary Member, Alumni Association, Lahey Clinic Foundation, 1970; Honorary Fellow, American College of Radiology, 1974; Member, National Academy of Engineering, 1977; Gold Medal of American College of Radiology, 1982.

TRUXAL, JOHN G., IRE Fellow 1959. Born: February 19, 1924, Lancaster, Pa. Degrees: A.B., 1944, Dartmouth College; B.S., 1947, Sc.D., 1950, Massachusetts Institute of Technology.

Fellow Award "For fundamental contributions to the theory of feedback control systems"; Education Medal, IEEE, 1974. Other Awards: Phi Beta Kappa; Sigma Xi; Tau Beta Pi; Eta Kappa Nu; American Association for the Advancement of Science; American Society for Engineering Education; National Academy of Engineering; Education Award, ISA, 1963; Westinghouse Award, American Society for Engineering Education, 1965.

TSAI, CHEN SHUI, Fellow 1983. Born: Nov. 3, 1935, Taiwan. Degrees: B.Sc.(E.E.), 1957, National Taiwan Univ.; M.Sc.(E.E.), 1961, Utah State Univ.; Ph.D.(E.E.), 1965, Stanford University.

Fellow Award "For contributions to acoustooptic devices for wideband real-time signal processing and for acoustic microscopy"; Best Paper Award, IEEE Circuits & Systems Society, 1980; Best Paper Award, IEEE Reliability & Electron Devices Societies, 1980. Other Awards: Fellow, Optical Society of America, 1983; Fellow, Institute for Advancement of Engineering, 1983.

TSAO, TSEN C., AIEE Fellow 1961. Born: October 16, 1901, Shanghai, China. Degrees: B.S.E.E., 1924, National Chiao Tung University; M.S.E.E., 1929, Harvard University; D.Eng.(Hon.), 1973, China Academy; Ph.D.(Hon.), 1983, National Chiao Tung University.

Fellow Award "For rehabilitation of electrical utilities in Greater Shanghai after WW II, and contributions to electrical communications." Other Awards: First Class Ching Sing Meritorious Medal, Republic of China, 1945; Professional Achievement Award, Chinese Institute of Engineers, 1964; Distinguished Service Award, Phi Tau Phi, 1973; Award of Honor, Chinese Culture Society, 1974; Fellow, Institution of Electrical Engineers (U.K.); Fellow, AAAS; Fellow, Radio Club of America; Who's Who, Marquis 1979-1980; Director, Chinese Institute of Engineers-U.S.A., 1980-1983; Chairman of Technical Seminar Committee, CIE-USA, 1981-present.

TUDBURY, CHESTER A., Fellow 1980. Born: January 3, 1913, Warwick, RI. Degrees: B.S.(E.E.), M.S.(E.E.), 1934, Massachusetts Institute of Technology.

Fellow Award "For contributions to the theory and industrial application of induction heating"; Past Chairman, Electric Process Heating Committee, Induction and Dielectric Heating Subcommittee, IEEE Industry Applications Society.

TUFTS, DONALD W., Fellow 1982. Born: March 5, 1933, Yonkers, NY. Degrees: B.A.(Math.), 1955, Williams College; B.S.E.E., 1957, M.S.E.E., 1958, D.Sc.(E.E.), 1960, Massachusetts Institute of Technology.

Fellow Award "For contributions to digital communications and signal processing."

TULL, WILLIAM J., Fellow 1972. Born: 1918, Ontario, Canada. Degrees: B.S.E.E., 1942, University of Michigan.

Fellow Award "For his contributions to the development of electronic aids to aircraft navigation"; Pioneer Award, IEEE Aerospace and Electronics Systems Group, 1968. Other Awards: Member, Institute of Navigation, 1946; Thurlow Award, Institute of Navigation, 1959; Association of U.S. Army; American Ordnance Association; Air Force Association; American Helicopter Society; Flight Safety Foundation; Navy League of the United States; Aerospace Industries Association; NSIA; President, Institute of Navigation, 1968-1969; Distinguished Service Award, Institute of Navigation, 1969.

TUNIS, CYRIL J., Fellow 1982. Born: July 31, 1932, Montreal, Canada. Degrees: B.(Eng. and Physics), 1954, M.Sc.(Physics), 1956, Mcgill University; Ph.D.(E.E.), 1958, Manchester University, England.

Fellow Award "For contributions to pattern recognition and read-only memories;" Treasurer, IEEE, 1983-84. Other Awards: Charles Babbage Award, Institution

of Electronics and Radio Engineers, Great Britain, 1964; IBM Outstanding Contribution Award, 1965; IBM First Level Invention Achievement Award, 1966; IBM Second Level Invention Achievement Award, 1968; IBM Division Award, 1980.

TUOMENOKSA, LEE S., Fellow 1980. Born: July 17, 1928, Helsinki, Finland. Degrees: B.S., 1952, Worcester Polytechnic Institute; S.M., 1954, Massachusetts Institute of Technology.

Fellow Award "For contributions to the development of telephone electronic switching systems."

TURIN, GEORGE L., Fellow 1971. Born: January 27, 1930, New York, N.Y. Degrees: S.B., 1952, S.M., 1952, Sc.D., 1956, Massachusetts Institute of Technology.

Fellow Award "For contributions to statistical communication theory and its applications." Other Awards: Guggenheim Fellowship, 1966-67; British Science and Engineering Council Senior Fellowship, 1983.

TURNER, GEORGE S., IRE Fellow 1957. Born: June 25, 1900, Independence, Mo. Degrees: LL.B., 1936, LL.M., 1939, Atlanta Law School.

Fellow Award "For achievements in telecommunications and in their international regulation."

TURNER, WILLIAM O., AIEE Fellow 1962. Born: March 19, 1897, Lonoke, Ark. Degrees: B.E.E., 1919, University of Arkansas.

Fellow Award "In recognition of his contributions to the engineering development and management of a large electric utility."

TUSKA, CLARENCE D., IRE Fellow 1957. Born: August 15, 1896, New York, N.Y. Degrees: B.S., 1919, Trinity College; LL.B., 1934, LaSalle Extension University.

Fellow Award "For pioneering services to radio communications." Other Awards: Hall of Fame, American Radio Relay League; Honorary Member, Antique Wireless Association, 1981.

TUTTLE, ALBERT D., Fellow 1971. Born: January 11, 1917, Elmira, N.Y. Degrees: B.E.E., 1938, Rensselaer Polytechnic Institute.

Fellow Award "For contributions to the design, construction, and operation of large electrical power systems, including computer applications to such systems."

TUTTLE, DAVID F., JR., AIEE Fellow 1959, IRE Fellow 1960. Born: July 5, 1914, Briarcliff Manor, N.Y. Degrees: B.A., 1934, Amherst College; S.B., S.M., 1938, Sc.D., 1948, Massachusetts Institute of Technology.

AIEE Fellow Award "For contributions to the theory and solution of electrical networks"; IRE Fellow Award "For contributions to network theory education." Other Awards: Western Electric Fund Award, 1977.

TUTTLE, W. NORRIS, IRE Fellow 1949. Born: March 29, 1902, Croton-on-Hudson, N.Y. Degrees: A.B., 1924, S.M., 1926, Ph.D., 1929, Harvard University.

Fellow Award "For his application of sound theoretical principles to the design of commercial measuring equipment." Other Awards: Medal of Freedom, 1946; Fellow, American Association for the Advancement of Science, 1971.

TWERSKY, VICTOR, IRE Fellow 1962. Born: August 10, 1923, Lublin, Poland. Degrees: B.S., 1947, City College of the City University of New York; A.M., 1948, Columbia University; Ph.D., 1950, New York University.

Fellow Award "For contributions to electromagnetic scattering theory." Other Awards: Fellow, American Physical Society; Fellow, Optical Society of America; Fellow, Acoustical Society of America; Fellow, American Association for the Advancement of Science; John Simon Guggenheim Fellow, 1972-73, 1979-80.

TYSON, BENJAMIN F., IRE Fellow 1958. Born: November 18, 1913, Chatham, N.J. Degrees: M.E., 1935, M.S., 1937, Stevens Institute of Technology.

Fellow Award "For his contributions to radar and color television systems." Other Awards: Certificate of Commendation, U.S. Dept. of Navy, 1947.

U

UDA, SHINTARO, Fellow 1970. Born: June 1, 1896, Toyama Ken, Japan. Degrees: B.S., 1924, D.Eng., 1931, Tohoku University.

Fellow Award "For contributions to the theory and design of antennas, for pioneering work in the field of microwaves." Other Awards: The Imperial Academy Prize of Japan, 1932.

UDO, MUNEYUKI, Fellow 1984. Born: July 26, 1928, Hiroshima, Japan. Degrees: B.Eng., 1951, Dr.Eng., 1962, University of Tokyo.

Fellow Award "For contributions to the methodology of computer control of power systems." Other Awards: Progress Award, IEE of Japan, 1972; Prize, Remarkable Invention, Ministry of Science & Technology of Japan, 1975.

UDO, TATSUO, Fellow 1980. Born: September 15, 1925, Tokyo, Japan. Degrees: B.E., 1949, Dr.Eng., 1960, Tokyo University.

Fellow Award "For contributions to the technology of insulation and electrical breakdown in high-voltage power systems." Other Awards: Prize for Technical Progress, 1954, 1962; Prize for Technical Thesis, 1964; and Prize for Electric Power Engineering, 1969; all from the Institute of Electrical Engineers, Japan; Prize for Contribution to Research from the Minister of Science and Technology, Japan, 1982.

UENOHARA, MICHIYUKI, Fellow 1971. Born: September 5, 1925, Kagoshima, Japan. Degrees: B.S., 1949, Nihon University, Tokyo; M.S., 1953, Ph.D., 1956, Ohio State University; D.Eng., 1958, Tohoku University, Sendai.

Fellow Award "For technical contributions in the field of parametric devices, and for leadership in electron devices research and development." Other Awards: INADA Award, Institute of Electronic and Communication Engineers of Japan, 1956; Distinguised Paper Award, National Electronic Convention, 1967.

UHLIR, ARTHUR, JR., Fellow 1967. Born: February 2, 1926, Chicago, Ill. Degrees: B.S., 1945, M.S., 1948, Illinois Institute of Technology; S.M., 1950, Ph.D., 1952, University of Chicago.

Fellow Award "For original contributions to the theory, development, and application of varactor diodes in parametric amplifiers."

ULABY, FAWWAZ T., Fellow 1980. Born: February 4, 1943, Damascus, Syria. Degrees: B.S.(Physics), 1964, American University of Beirut, Lebanon; M.S.E.E., 1966, Ph.D.(E.E.), 1968, University of Texas at Austin.

Fellow Award "For contributions to the application of radar to remote sensing for agriculture and hydrology"; Past President, IEEE Geoscience and Remote Sensing Society; IEEE Geoscience and Remote Sensing Society, Outstanding Service Award, 1982; IEEE Geoscience and Remote Sensing Society, Distinguished Achievement Award, 1983; IEEE Centennial Medal, 1984; Executive Editor, IEEE Transactions on Geoscience and Remote Sensing, 1983-1985; Distinguished Lecturer, IEEE Geoscience and Remote Sensing Society, 1984. Other Awards: J.L. Constant Distinguished Professor of Electrical Engineering-KU; Henry E. Gould Award, University of Kansas, 1973; Eta Kappa Nu Association C. Holmes MacDonald Award, 1975; University of Kansas Chancellor's Award for Excellence in Teaching, 1979-1980; Eta Kappa Nu; Tau Beta Pi; Sigma Xi.

ULRICH, WERNER, Fellow 1979. Born: March 12, 1931, Munich, West Germany. Degrees: B.S., 1952, M.S., 1953, Dr.Eng.Sc., 1957, Columbia University, School of Engineering; M.B.A., 1975, University of Chicago.

Fellow Award "For contributions to the development of telephone electronic switching systems."

UMEZU, TERUHIRO, Fellow 1979. Born: September 23, 1925. Degrees: Bachelor of Engineering, 1948, Ph.D. of Engineering, 1961, Tokyo University.

Fellow Award "For contributions to the development of analytical methods and their application to electric power systems."

UNDRILL, JOHN M., Fellow 1978. Born: September 22, 1940, New Zealand. Degrees: B.E.(with honors), 1963, Ph.D, 1965, University of Canterbury, New Zealand.

Fellow Award "For development of interactive simulation methods and analysis techniques for electric power systems."

UNGER, HANS-GEORG, Fellow 1974. Born: September 14, 1926, Braunschweig, Germany. Degrees: Dipl.-Ing., 1951, Dr.Ing., 1954, Technical University, Braunschweig.

Fellow Award "For contributions to the theory of multimode millimeter waveguides."

UNGER, STEPHEN H., Fellow 1976. Born: July 7, 1931, New York, N.Y. Degrees: B.E.E., 1952, Polytechnic Institute of Brooklyn; S.M. 1953, Sc.D., 1957, Massachusetts Institute of Technology.

Fellow Award "For contributions to switching circuit theory and computer science education." Other Awards: Guggenheim Fellow, 1967.

URBAN, LOUIS J., Fellow 1981. Born: June 26, 1932, Canton, OH. Degrees: B.S.(E.E.), 1956, University of Cincinnati; M.S., 1970, Massachusetts Institute of Technology.

Fellow Award "For leadership in avionics research and development and contributions to guidance and control for missiles and aircraft"; Service Award, Life Member, IEEE Aerospace and Electronic Systems Society, 1977; IEEE Centennial Medal, 1984. Other Awards: Sigma Xi; Tau Beta Phi; Meritorious Civilian Service Medal 1970; NATO Advisory Group for Aerospace Research and Development Guidance and Control Panel; U.S. Air Force Representative/Consultant to NASA, Aerospace Safety Advisory Council; Distinguished Alumnus Award, University of Cincinnati, 1982.

URKOWITZ, HARRY, Fellow 1973. Born: October 1, 1921, Philadelphia, Pa. Degrees: B.S.E.E., 1948, Drexel University; M.S.E.E., 1954, Ph.D., 1972, University of Pennsylvania.

Fellow Award "For contributions to radar signal processing and to graduate education."

UTLAUT, WILLIAM F., Fellow 1970. Born: July 26, 1922, Sterling, Colo. Degrees: B.S., 1944, M.S., 1950, Ph.D., 1966, University of Colorado.

Fellow Award "For leadership in radiowave propagation research." Other Awards: Department of Commerce Gold Medal, 1972; Letter of Commendation from the President, 1972; Distinguished Engineering Alumnus Award, University of Colorado, 1973.

UTSUNOMIYA, TOSHIO, Fellow 1982. Born: November 20, 1921, Kagawa, Japan. Degrees: B.Eng., 1943, D.Eng., 1961, University of Tokyo.

Fellow Award "For leadership in the field of biomedical engineering and education." Other Awards: Tokyo Metropolitan Governor's Award, 1980; Niwa-Takayanagi Prize, Institute of Television Engineers of Japan, 1982.

V

VADASZ, LESLIE L., Fellow 1977. Born: September 12, 1936, Budapest, Hungary. Degrees: B.S.E.E.,

1961, McGill University.

Fellow Award "For leadership in the development of semiconductor memories and microcomputer components."

VAHAVIOLOS, SOTIRIOS J., Fellow 1983. Born: April 16, 1946, Mistras, Sparta, Greece. Degrees: B.S., 1970, Fairleigh Dickinson University; M.S.E.E., 1972, M.Phi., 1975, Ph.D., 1976, Columbia University.

Fellow Award "For contributions to the techniques of acoustic-emission measurements in industrial process monitoring"; 2nd Prize Student Paper Award, IEEE, 1970; Meritorious Award, IE Society, 1979; Editor IEEE Trans on IE. Other Awards: Sigma Xi.

VAIL, CHARLES R., AIEE Fellow 1960. Born: October 16, 1915, Glens Falls, N.Y. Degrees: B.S.E.E., 1937, Duke University; M.S., 1946, Ph.D., 1956, University of Michigan.

Fellow Award "For contributions to electrical engineering teaching and to synthetic insulations." Other Awards: Member: Phi Eta Sigma, Omicron Delta Kappa, Phi Beta Kappa, Sigma Xi; Sigma Tau; Tau Beta Pi; Eta Kappa Nu; Pi Mu Epsilon; Distinguished Alumnus Award, Duke University School of Engineering, 1967; Engineer of the Year in Education Atlanta Chapter, Georgia Society of Professional Engineers, 1975; Engineer of the Year in Education Award, State, Georgia Society of Professional Engineers, 1976; Engineer of the Year in Georgia Award, State, Georgia Society of Professional Engineers, 1979. Life Member, Georgia Engineering Foundation Board of Directors, 1978; ASEE 1979 College Industry Education Conference Best Paper Award; Listed in: "Who's Who in Engineering, 1964", "Who's Who in the South and Southwest, 1965-66", "American Men and Women of Science, 1973", "Who's Who in America, 1980-81", "Who's Who in the World, 1980-81."

VALLESE, LUCIO, Fellow 1964. Born: 1915, Italy, Degrees: D.E.E., 1937, University of Naples; D.Sc., 1948, Carnegie-Mellon University.

Fellow Award "For contributions to the theory and applications of solid-state electronic devices."

VALLEY, GEORGE E., JR., IRE Fellow 1960. Born: September 5, 1913, New York, N.Y. Degrees: S.B., 1935, Massachusetts Institute of Technology; Ph.D., 1939, University of Rochester.

Fellow Award "For contributions to military systems engineering." Other Awards: Letter of Appreciation, U.S. Dept. of Army, 1945; Presidential Certificate of Merit, 1948; Distinguished Service Award, Air Force Association, 1951; Exceptional Civilian Service Medals, 1956, 1958, 1964; Fellow, American Physical Society.

VAN ALLEN, JAMES A., IRE Fellow 1960. Born: September 7, 1914, Mt. Pleasant, Iowa. Degrees: B.S., 1935, Iowa Wesleyan College; M.S., 1936, Ph.D.,

1939, University of Iowa; D.Sc.(Hon.), 1951, Iowa Wesleyan College; D.Sc.(Hon.), 1957, Grinnell College; D.Sc.(Hon.), 1958, Coe College; D.Sc.(Hon.), 1959, Cornell College; D.Sc.(Hon.), 1960, University of Dubuque; D.Sc.(Hon.), 1961, University of Michigan; D.Sc.(Hon.), 1961, Northwestern University; D.Sc.(Hon.), 1963, Illinois College; D.Sc.(Hon.), 1966, Butler University; D.Sc.(Hon.), 1966, Boston College; D.Sc.(Hon.), 1967, Southampton College; D.Sc.(Hon.), 1969, Augustana College; D.Sc.(Hon.), 1982, St. Ambrose College.

Fellow Award "For the experimental discovery and exploration of radiation belts around the earth." Other Awards: C. N. Hickman Medal, American Rocket Society, 1949; Physics Award, Washington Academy of Science, 1949; Research Fellow, Guggenheim Memorial Foundation, 1951; Space Flight Award, American Astronautical Society, 1958; Distinguished Civilian Service Medal, U.S. Dept. of Army, 1959; Louis W. Hill Space Transportation Award, 1960; First Iowa Award in Science, 1961; First Annual Research Award, American Rocket Society, 1961; Elliott Cresson Medal, Franklin Institute, 1961; Space Flight Award, International Academy of Astronautics, 1961; David and Florence Guggenheim International Astronautics Award, 1962; John A. Fleming Award, American Geophysical Union, 1963; Golden Omega Award, Electrical Insulation Conference, 1963; Commander, Order du Merite pour la Recherche et l'Invention, 1964; Award, Iowa Broadcasters Association, 1964; Carver Distinguished Professor, University of Iowa; Medal for Exceptional Scientific Achievement, NASA, 1974; Distinguished Fellow, Iowa Academy of Science, 1975; Distinguished Civilian Service Award, U.S. Navy, 1976; William Bowie Medal, American Geophysical Union, 1977; Fellow, American Physical Society; Fellow, American Geophysical Union; National Academy of Sciences; American Astronomical Society; Fellow, American Astronautical Society; Sigma Xi; Gold Medal, Royal Astronomical Society, 1978; Award of Merit, American Consulting Engineers Council, 1978; Fellow, American Rocket Society; Fellow, American Academy of Arts and Sciences; Regents Fellow, Smithsonian Institution, 1980; Space Science Award, American Institute of Aeronautics and Astronautics, 1982; Governor's Science Medal, 1982.

VAN ATTA, LESTER C., IRE Fellow 1952. Born: April 18, 1905, Portland, Oreg. Degrees: B.A., 1927, Reed College; M.S., 1929, Ph.D., 1931, Washington University.

Fellow Award "In recognition of his contributions in the field of microwave antenna theory and design"; IEEE Centennial Medal, 1984. Other Awards: NDRC Citation, Office of Scientific Research and Development, 1945; Presidential Certificate of Merit, 1948; Sigma Xi Citation, 1957; Certificate of Appreciation, Los Angeles City Schools, 1959; S.C.I.E.C. Award,

1960; Fellow, American Physical Society; Associate Fellow, American Institute of Aeronautics and Astronautics.

VAN BLADEL, JEAN G., Fellow 1975. Born: July 24, 1922, Antwerp, Belgium. Degrees: E.M.E., 1947, Radio Engr., 1948, Brussels University; Ph.D., 1950, University of Wisconsin.

Fellow Award "For contributions to electromagnetic theory." Other Awards: International Montefiore Prize, 1965; Fellow, IEE.

VANCE, PAUL A., Fellow 1964. Born: November 17, 1902, Glencoe, Minn. Degrees: B.S., 1923, University of Illinois.

Fellow Award "For contributions to the design and application of saturable reactors, nonlinear magnetic devices, and transformers."

VAN DER ZIEL, ALDERT, IRE Fellow 1956. Born: December 12, 1910, Zandeweer Groningen, The Netherlands. Degrees: B.A., 1930, M.A., 1933, Ph.D., 1934, University of Groningen; Hon. Degree, 1975, Universite Paul Sabatier, France; Hon. Degree, 1981, Eindhoven University of Technology, The Netherlands.

Fellow Award "For research leadership and for studies of fluctuation phenomena in electron devices"; IEEE Education Medal, 1980. Other Awards: Invited Paper Award, National Electronics Conference, 1961; Western Electric Award, 1967; Vincent Bendix Award, 1975; National Academy of Engineering, 1978.

VAN DUZER, THEODORE, Fellow 1977. Born: December 27, 1927, Piscataway Tsp., N.J. Degrees: B.S., 1954, Rutgers University; M.S., 1957, University of California at Los Angeles; Ph.D., 1960, University of California at Berkeley.

Fellow Award "For contributions to superconducting devices and to engineering education." Other Awards: Outstanding Engineering Alumnus, Rutgers University, 1975.

VANIER, JACQUES, Fellow 1983. Born: January 4, 1934, Dorion, Quebec, Canada. Degrees: B.Sc., 1958, Universite de Montreal; M.Sc., 1960, Ph.D., 1963, McGill University.

Fellow Award "For contributions to the theory and development of atomic-resonance frequency standards."

VAN LINT, VICTOR A. J., Fellow 1977. Born: May 10, 1928, Indonesia. Degrees: B.S.(Physics), 1950, Ph.D.(Physics), 1954, California Institute of Technology.

Fellow Award "For contributions to the understanding of radiation effects and to the applications of this knowledge to improve the survivability of military and space systems."

VAN NESS, JAMES E., Fellow 1982. Born: June 24, 1926, Omaha, NB. Degrees: B.S., 1949, Iowa State University; M.S., 1951, Ph.D., 1954, Northwestern

University.

Fellow Award "For contributions to research and education in computer analysis of power systems."

VAN NORDEN, R.W., AIEE Fellow 1913.

VAN OVERSTRAETEN, ROGER J., Fellow 1984. Born: December 7, 1937, Vlezenbeek, Belgium. Degrees: Engineer's Degree (E.E. and M.E.), 1960, Katholieke Universiteit, Leuven, Belgium; Ph.D. (Electronics), 1963, Stanford University.

Fellow Award "For contributions to semiconductor device theory and solar-cell development and for leadership in academic research." Other Awards: Project Leader, Research Program on Photovoltaics, European Economic Community; C.R.B. Fellowship, 1960, Advanced C.R.B. Fellowship, 1972, Belgian-American Educational Foundation; Member, American Physical Society; European Physical Society; Koninklyke Vlaamse Ingenieurs Vereniging (K.V.I.V.).

VAN SICKLE, ROSWELL C., AIEE Fellow 1945. Born: August 25, 1900, Buffalo, N.Y. Degrees: M.E., 1923, E.E., 1924, Cornell University; M.S., 1928, University of Pittsburgh.

VAN TASSEL, E. KENNETH, Fellow 1970. Born: September 28, 1904, Paw Paw, Mich. Degrees: B.S., 1926, M.S., M.S.C., 1928, Michigan State University.

Fellow Award "For contributions to telephone transmission, radar circuitry, and applications of digital techniques to data processing."

VAN TREES, HARRY L., Fellow 1974. Born: June 27, 1930. Degrees: B.S., 1952, U.S. Military Academy; M.S., 1958, University of Maryland; Sc.D., 1961, Massachusetts Institute of Technology.

Fellow Award "For contributions to teaching and research in the detection, estimation and modulation theory area, and the design of military communications systems."

VAN VALKENBURG, M.E., IRE Fellow 1962. Born: October 5, 1921, Union, Utah. Degrees: B.S.E.E., 1943, University of Utah; M.S., 1946, Massachusetts Institute of Technology; Ph.D., 1952, Stanford University.

Fellow Award "For contributions to circuit theory"; Education Medal, IEEE, 1972, "For outstanding textbooks in circuit theory, innovations in undergraduate teaching, inspired guidance of students, and professional leadership in electrical engineering." Other Awards: George Westinghouse Award, American Society for Engineering Education, 1963; Member, National Academy of Engineering, 1973; Benjamin Garver Lamme Award, American Society for Engineering Education, 1978; Ernst Guillemin Prize, The Guillemin Foundation, 1978; Halliburton Engineering Education Leadership Award, 1979; W. W. Grainger Chair, University of Illinois, 1982.

VAN VESSEM, JAN C., Fellow 1965. Born: November 8, 1920, Den Helder, Netherlands. Degrees: M.Sc., 1942, Ph.D., 1947, Utrecht University.

Fellow Award "For technical contributions and leadership in the semiconductor industry."

VAN ZEELAND, FRED J., AIEE Fellow 1957. Born: September 30, 1906, Kimberly, Wis. Degrees: B.S., 1928, Milwaukee School of Engineering.

Fellow Award "For his contributions as an educator and an administrator in close contact with industrial research."

VANZETTI, RICCARDO, Fellow 1982. Born: May 1, 1910, Milano, Italy. Degrees: D.Eng., 1932, University of Rome, Italy.

Fellow Award "For the introduction of new concepts and products in infrared electronics and industrial instrumentation." Other Awards: Fellow, American Society for Nondestructive Testing, 1976.

VARAIYA, PRAVIN P., Fellow 1980. Born: October 29, 1940. Degrees: B.S., 1960, V.J. Technical University, India; M.S., 1962, Ph.D., 1966, University of California at Berkeley.

Fellow Award "For fundamental contributions to the theory and control of large-scale stochastic systems."

VARNERIN, LAWRENCE J., JR., Fellow 1974. Born: July 10, 1923, Boston, Mass. Degrees: S.B., 1944, Ph.D., 1949, Massachusetts Institute of Technology.

Fellow Award "For contributions to electronic and magnetic devices and materials."

VASSELL, GREGORY S., Fellow 1978. Born: December 24, 1921, Moscow, U.S.S.R. Degrees: Dipl. Ing., 1951, Technical University, Berlin, Germany; M.B.A., 1954, New York University.

Fellow Award "For contributions to the planning of reliable and economic electric power systems." Other Awards: National Academy of Engineering, 1980.

VAUGHAN, HAROLD R., AIEE Fellow 1954. Born: August 25, 1903, Independence, Kans. Degrees: B.S., 1928, University of Colorado.

Fellow Award "For his contributions in the field of turbine generator design and in the development of transmission systems"; National Award, Best Paper on Application, AIEE, 1946. Other Awards: American Society of Mechanical Engineers; National Society of Professional Engineers; International Executive Service Corps.

VAUGHAN, VIRGINIUS N., JR., Fellow 1969. Born: May 22, 1915, Ashland, Va. Degrees: B.S., 1936, Randolph-Macon College; B.S.E.E., M.S.E.E., 1938, Massachusetts Institute of Technology; D.Sc.(Hon.), 1980, Randolph-Macon College.

Fellow Award "For pioneering work and subsequent leadership in data communications engineering including systems planning, project management, and standards activities." Other Awards: Tau Beta Pi, 1936; "Who's Who in Engineering," 1977.

VEINOTT, CYRIL G., AIEE Fellow 1948. Born: February 15, 1905, Somerville, Mass. Degrees: B.S., 1926, E.E., 1938, D.Eng., 1951, University of Vermont.

Awards: Citation, IEEE Rotating Machinery Committee, 1970; Nikola Tesla Award, IEEE, 1977. Other Awards: Phi Beta Kappa, 1926; Eta Kappa Nu Recognition of Young Engineers, 1936; Silver W Order of Merit, Westinghouse Electric, 1945; Eta Kappa Nu, 1957; Tau Beta Pi, 1959; Medal, Research Institute for Rotating Electrical Machines, Brno, Czechoslovakia, 1968; Nikola Tesla Medal and Golden Plaque, Yugoslav Society for the Promotion of Scientific Knowledge, 1977.

VERHAGEN, CORNELIS J. D. M., Fellow 1969. Born: April 28, 1915, Hertogenbosch, The Netherlands. Degrees: Dipl.Ing., 1939, Dr.Ing., 1942, Technical University of Delft; D.Sc.(honoris causa), 1976, The City University, London.

Fellow Award "For contributions to the development of interest and education in the field of instrumentation and control."

VER PLANCK, DENNISTOUN W., AIEE Fellow 1951. Born: January 29, 1906, Swampscott, Mass. Degrees: B.S., 1928, M.S., 1929, Massachusetts Institute of Technology; D.Eng., 1940, Yale University.

Other Awards: Order of the British Empire, Military Division, 1946; Commendation Ribbon, U.S. Dept. of Navy, 1946; Distinguished Civilian Service Award, U.S. Dept. of Navy, 1946; Fellow, American Society of Mechanical Engineers, 1966; Fellow, American Nuclear Society, 1969.

VICK, CHARLES R., Fellow 1981. Born: October 7, 1934, Steele, MO. Degreees: B.A. (Math.), 1965, Oklahoma City University; Ph.D.(E.E.), 1979, Auburn University.

Fellow Award "For contributions to software engineering theory and applications"; Member, Governing Board, IEEE Computer Society; Chairman, Technical Committee on Distributed Data Processing, IEEE Computer Society. Other Awards: Dept. of Army Research and Development Achievement Award, 1972; Senior Executive Honorable Mention, 1979.

VICTOR, WALTER K., Fellow 1974. Born: December 18, 1922, The Bronx, N.Y. Degrees: B.S.M.E., 1942, University of Texas.

Fellow Award "For contributions to the design of radio systems for space communications and tracking." Other Awards: Tau Beta Pi, 1942; Sigma Xi, 1962; Space Act Award, NASA, 1963.

VIDYASAGAR, MATHUKUMALLI, Fellow 1983.

Fellow Award "For contributions in the stability analysis of linear and nonlinear distributed systems."

VIGILANTE, FRANK S., Fellow 1975. Born: March 15, 1930, Brooklyn, N.Y. Degrees: B.S.E.E., 1957, University of California; M.S.E.E., 1959, New York University.

Fellow Award "For contributions and leadership in the development and deployment of electronic switching systems." Other Awards: Outstanding Young American Engineer, Eta Kappa Nu, 1964.

VIJH, ASHOK K., Fellow 1982. Born: March 15, 1938, Multan, India. Degrees: B.Sc.(Hons.), 1960, M.Sc.(Hons.), 1961, Panjab University, India; Ph.D., 1966, University of Ottawa, Canada.

Fellow Award "For contributions to the theory of electrochemical reactions involved in electrical and electronic products." Other Awards: Fellow, Chemical Institute of Canada, 1973; Fellow, Royal Society of Chemistry, London, 1973; Lash Miller Award, Electrochemical Society, Quebec, 1973; NORANDA Lecture Award, Chemical Institute of Canada, 1979; Maitre de recherche, Institut de recherche d'Hydro- Quebec, 1973; Member, Editorial Board, several international research journals; Invited Professor, Institut National de la Recherche Scientifique-Energie, 1970-.

VILLARD, OSWALD G., JR., IRE Fellow 1957. Born: September 17, 1916. Degrees: A.B., 1938, Yale University; E.E., 1943, Ph.D., 1949, Stanford University.

Fellow Award "For contributions to knowledge of the ionosphere and its role in the propagation of radio waves"; Morris N. Liebmann Award, IRE, 1957, "For his contributions in the fields of meteor astronomy and ionospheric physics which led to the solution of outstanding problems in radio propagation." Other Awards: Outstanding Young Bay Area Engineer Award, 1955; Member, National Academy of Science, 1961; Member, National Academy of Engineering, 1966; Meritorious Civilian Service Award, Dept. of the Air Force, 1975; Medal for Outstanding Public Service, Dept. of Defense, 1981; Fellow, American Academy of Arts and Sciences, 1982.

VISWANATHAN, CHAND R., Fellow 1981. Degrees: B.Sc., 1948, M.A., 1949, University of Madras, India; M.S., 1959, Ph.D., 1964, University of California.

Fellow Award "For leadership in engineering education and contributions to the theory of metal-oxide-semiconductor devices."

VITERBI, ANDREW J., Fellow 1973. Born: March 9, 1935, Bergamo, Italy. Degrees: S.B., 1957, S.M., 1957, Massachusetts Institute of Technology; Ph.D., 1962, University of Southern California.

Fellow Award "For contributions to information and communication theory"; Outstanding Paper Award, IEEE Information Theory Group, 1968; Alexander Graham Bell Medal, IEEE, 1984. Other Awards: Valuable Contribution to Telemetry, PGSET, 1962; Best Original Paper, National Electronics Conference, 1962; Christopher Columbus International Communication Award, 1975; National Academy of Engineering, 1978; Aerospace Communication Award, American Institute of Aeronautics and Astronautics, 1980.

VITHAYATHIL, JOHN J., Fellow 1984. Born: February 17, 1937, Kerala, India. Degrees: B.Sc.(Eng.), University of Kerala; M.Sc.(Eng.), 1962, Ph.D., 1967, Indian Institute of Science, Bangalore.

Fellow Award "For contributions to the analysis and design of high-voltage dc power systems."

VLACH, JIRI, Fellow 1982. Born: October 5, 1922, Prague, Czechoslovakia. Degrees: Dipl.Eng., 1947, CsC, 1958, Technical University of Prague.

Fellow Award "For contributions to computer-aided analysis and design of electrical networks."

VOELCKER, HERBERT B., Fellow 1973. Born: January 7, 1930, Tonawanda, N.Y. Degrees: S.B., 1951, S.M., 1954, Massachusetts Institute of Technology; D.I.C., 1961, Ph.D., 1961, Imperial College of Science and Technology.

Fellow Award "For contributions to modulation theory and digital signal processing, and for the teaching of electrical engineering"; Best Paper Prize, IEEE Communication Society, 1967. Other Awards: Fulbright Predoctoral Fellow, 1958-60; American Society for Engineering Education Regional Award, Western Electric, 1967; Postdoctoral Fellowship, NATO, 1967-68; Curtis Prize, University of Rochester, 1969; Senior Visiting Fellow, Science Research Council of Great Britain, 1981-1983; Conference Presentation Prize, Society of Automotive Engineers, 1981.

VOGE, JEAN P., Fellow 1968. Born: February 21, 1921, Casablanca, Morocco. Degrees: B.S., 1937, College Saint-Chamond; 1942, Ancien Eleve de l'Ecole Polytechnique; 1944, Ingenieur de l'Ecole Nationale Superieure des Telecommunications.

Fellow Award "For outstanding work and teaching on UHF tubes, wave propagation, and space communication"; Chairman, France Section, IEEE, 1972-1981. Other Awards: Chairman, Institut pour le Developpement et l'Amenagement des Telecommunications et de l'Economie (IDATE); Deputy Director and Special Advisor, Direction des Affaires Industrielles et Internationales, Ministry of Post and Telecommunications, Paris; President, International Union of Radio Science (URSI), 1975-1978.

VOGEL, FRED J., AIEE Fellow 1949. Born: April 15, 1893, Bangor, Me. Degrees: B.S., 1915, Massachusetts Institute of Technology.

Awards: William Martin Habirshaw Medal and Award, 1970, "For meritorious achievement in the field of electrical transmission and distribution." Other Awards: Certificate of Commendation, U.S. Dept. of Navy, 1946; awarded several patents; Honorary Sigma Xi; Tau Beta Pi; Eta Kappa Nu; Chairman, Sharon Section, Transformer Committee and Dielectric Tests Subcommittee; Coodinating Committee #4 on Insulation Life; ANSI Transformer Standards.

VOGELMAN, JOSEPH H., IRE Fellow 1959. Born: August 18, 1920, New York, N.Y. Degrees: B.S., 1940, City College of the City University of New York; M.E.E., 1948, D.E.E., 1957, Polytechnic Institute of Brooklyn.

Fellow Award "For contributions to military electronics." Other Awards: Outstanding Performance Award, U.S. Air Force, 1957; Fellow, American Association for the Advancement of Science, 1959.

VOLLMER, JAMES, Fellow 1971. Born: April 19, 1924, Philadelphia, Pa. Degrees: B.S., 1945, Union College; M.A., 1951, Ph.D., 1956, Temple University.

Fellow Award "For contributions to radiation, physics, and quantum electronics, for applied research, and for leadership in engineering education"; IEEE Centennial Medal, 1984. Other Awards: Fellow, AAAS, 1967.

VOLLUM, HOWARD, IRE Fellow 1955. Born: May 31, 1913, Portland, Oreg. Degrees: B.A., 1936, Reed College; D.Sc.(Hon.), 1955, University of Portland; LL.D.(Hon.), 1967, Lewis and Clark College.

Fellow Award "For his contributions to the development and manufacture of electronic laboratory instruments." Other Awards: Legion of Merit, 1945; Oak Leaf Cluster, 1946; Medal of Achievement, Western Electrical Manufacturers Association, 1964.

VON AULOCK, WILHELM H., Fellow 1974. Born: 1915, Pirna, Germany. Degrees: Dipl.Ing., 1937, Institute of Technology, Berlin; Dr.Ing., 1953, Institute of Technology, Stuttgart.

Fellow Award "For leadership in the codification of the theory and application of microwave ferrites, and for contributions to the theory of phased array antennas."

VON BAEYER, HANS J., IRE Fellow 1959.

Fellow Award "For contributions to development of radio communication techniques and systems."

VON RECKLINGHAUSEN, DANIEL R., Fellow 1966. Born: January 22, 1925, New York, N.Y. Degrees: S.B., 1951, Massachusetts Institute of Technology.

Fellow Award "For contributions to FM multiplex broadcasting and reception in the field of high fidelity music reproduction"; Achievement Award, Group on Audio and Electroacoustics, 1967. Other Awards: Eta Kappa Nu, 1950; Tau Beta Pi, 1951; Sigma Xi, 1951; Fellow, Audio Engineering Society, 1962; Gold Medal, Audio Engineering Society, 1978.

VON ROESCHLAUB, FRANK, Fellow 1975. Born: May 11, 1912, Denver, Colo. Degrees: B.S., 1933, D.Eng., 1936, Yale University.

Fellow Award "For contributions to the development of international standards for power system protection and control"; Distinguished Service Award, Power System Relaying Committee, IEEE, 1979.

W

WACHTER, ROBERT V., Fellow 1977. Born: March 17, 1920, Portland, Oreg. Degrees: B.S.E.E., 1942, Washington State University.

Fellow Award "For leadership in the development and application of silicon power conversion equipment"; Outstanding Achievement Award, IEEE Industry Applications Society, 1973.

WADA, SHIGENOBU, Fellow 1983. Born: April 9, 1910, Tokyo, Japan. Degrees: B.Eng., 1934, Ph.D.(Eng.), 1950, Tokyo University.

Fellow Award "For development of magnetic materials for electric power apparatus and for research and development leadership in advanced energy conversion." Other Awards: Prize for the Promotion of Science, IEE Japan, 1949.

WADE, GLEN, IRE Fellow 1962. Born: March 19, 1921, Ogden, Utah. Degrees: B.S., 1948, M.S., 1949, University of Utah; Ph.D., 1954, Stanford University.

Fellow Award "For contributions to parametric amplification"; IEEE Centennial Medal, 1984. Other Awards: Outstanding Young Electrical Engineer, Eta Kappa Nu, 1955; Annual Award, National Electronics Conference, 1959; Visiting Professorship at Tokyo University, Japan Society for the Advancement of Science, 1971; Fulbright-Hays Lectureship in Spain, 1972; Distinguished Teaching Award, UCSB Academic Senate, 1977; "Special Chair" Visiting Professorship at National Taiwan University, National Science Council, Republic of China, 1980.

WAGERS, ROBERT S., Fellow 1984. Born: January 25, 1943, Covington, KY. Degrees: B.S.E.E., 1966, M.S.E.E., 1967, Arizona State Univ.; Ph.D., 1972, Stanford University.

Fellow Award "For contributions to the theory and technology of surface-acoustic-wave devices"; Best Paper Award, IEEE Sonics and Ultrasonics Group, 1976. Other Awards: Rhodes Scholar, 1966.

WAGHORNE, JOHN H., Fellow 1973. Born: June 25, 1911, Lynn Valley, British Columbia, Canada. Degrees: B.S., 1939, M.Sc., 1940, Queen's University.

Fellow Award "For leadership in the management of utility research programs."

WAGNER, CHARLES L., Fellow 1971. Born: November 23, 1925, Pittsburgh, Pa. Degrees: B.S.E.E., 1945, Bucknell University; M.S.E.E., 1949, University of Pittsburgh.

Fellow Award "For contributions in the field of extra-high-voltage transmission, protective relaying, and power circuit-breaker application"; Prize Paper Award, Power Engineering Society, 1973; IEEE Standards Medallion, 1980; IEEE Switchgear Committee Distinguished Service Award, 1982; IEEE Centennial Medal, 1984; President, IEEE Power Engineering Society, 1984. Other Awards: Tau Beta Pi; Pi Mu Epsilon.

WAGNER, ROBERT N., AIEE Fellow 1962. Born: April 1, 1910, Springville, N.Y. Degrees: B.S., 1931, Clarkson College of Technology.

Fellow Award "For contributions to the development, design and installation of aluminum reduction plants and their associated power conversion equipment." Other Awards: Silver Beaver Award, 1961; St. George Award, 1964; Distinguished Service Award, Clarkson College of Technology, 1971; Silver Antelope Award, 1983.

WAIDELICH, DONALD L., AIEE Fellow 1951, IRE Fellow 1958. Born: May 3, 1915, Allentown, Pa. Degrees: B.S., 1936, M.S., 1938, Lehigh University; Ph.D., 1946, Iowa State University.

IRE Fellow Award "For his distinguished teaching in the engineering field and for furtherance of international understanding through technical services in the Middle East." Other Awards: Fulbright Award, Egypt, 1951-52; Fulbright Award, Australia, 1961-62; Research Award, Sigma Xi, 1977, Teaching Award, Halliburton, 1983, University of Missouri.

WAIT, JAMES R., IRE Fellow 1962. Born: January 23, 1924, Ottawa, Ontario, Canada. Degrees: B.A.Sc., 1948, M.A.Sc., 1949, Ph.D., 1951, University of Toronto.

Fellow Award "For contributions in electromagnetic theory and radio propagation"; Harry Diamond Award, 1964, "For outstanding contributions in the field of electromagnetic wave theory"; Founder's Award, IEEE Electromagnetic Compatibility Society, 1983; IEEE Centennial Medal, 1984. Other Awards: Gold Medal, U.S. Dept. of Commerce, 1958; Boulder Scientist Award, Scientific Research Society of America, 1960; Samuel Wesley Stratton Award, National Bureau of Standards, 1962; Arthur S. Flemming Award, Washington, D.C. Chamber of Commerce, 1964; Publication Award, Office of Telecommunications, 1972; Research and Achievement Award, National Oceanic and Atmospheric Administration, 1973; Special Achievement Award, Office of Telecommunications, 1975; Member, National Academy of Engineering, 1977; Fellow, Institution of Electrical Engineers (London), 1977; Balth Van der Pol Gold Medal, International Union of Radio Science, 1978.

WALDHAUER, FRED D., Fellow 1977, Born: December 6, 1927, Brooklyn, N.Y. Degrees: B.E.E., 1948, Cornell University; M.S.E.E., 1960, Columbia University.

Fellow Award "For contributions to the development of pulse code modulation systems and of design techniques for feedback amplifiers."

WALKER, A. PROSE, Fellow 1964. Born: February 25, 1910, Bird's Run, Ohio. Degrees: B.A., 1932, Denison University.

Fellow Award "For contributions to international standards in the utilization of the radio spectrum." Other Awards: Alumni Citation, Denison University; Diplome d'Honneur, International Telecommunications Union; Licensed Radio Amateur, Licensee of Experimental-Research Station KM2XK0 for investigation of ionospheric effects.

WALKER, ERIC A., AIEE Fellow 1947, IRE Fellow 1960. Born: April 29, 1910, Long Eaton, Derby, England. Degrees: B.S., 1932, M.B.A., 1933, Sc.D., 1935, Harvard University; LL.D. (Hon.), 1957, Temple University; LL.D.(Hon.), 1957, Lehigh University; LH.D.(Hon.), 1958, Elizabethtown College; LL.D(Hon.), 1960, Hofstra University; LL.D.(Hon.), 1960, Lafayette College; LL.D.(Hon.), 1960, University of Pennsylvania; Litt.D.(Hon.), 1960, Jefferson Medical College; LL.D.(Hon.), 1962, University of Rhode Island; D.Sc.(Hon.), 1965, Wayne State University; D.Sc.(Hon.), 1966, Thiel College; LH.D.(Hon.), 1968, St. Vincent College; D.Sc.(Hon.), 1968, University of Notre Dame.

IRE Fellow Award "For services as an engineering teacher and administrator." Other Awards: Fellow, American Physical Society, 1938; Fellow, American Acoustical Society, 1944; Presidential Certificate of Merit, 1948; Distinguished Service Medal, Pennsylvania Department, American Legion, 1957; Distinguished Public Service Award, U.S. Dept. of Navy, 1958; Horatio Alger Award, 1959; Tasket H. Bliss Award, American Society of Military Engineers, 1959; Golden Omega Award, Electrical Insulation Industry, 1962; National Science Foundation Board, 1962; National Academy of Engineering, 1964; Lamme Award, Honorary Member, American Society for Engineering Education, 1965; American Academy of Arts and Sciences, 1966; Benjamin Franklin, Royal Society of Arts, London, 1969; President's Certificate of Merit, 1970.

WALKER, JOHN R., AIEE Fellow 1948. Born: November 30, 1901, Erie, Pa. Degrees: E.E., 1924, Rensselaer Polytechnic Institute. Awards: Vice President, AIEE, 1954-1956.

Other Awards: Listed in "Who's Who in America," 1961; "Who's Who in Engineering."

WALLACE, JAMES D., IRE Fellow 1959. Born: March 6, 1904, Gloster, Miss. Degrees: B.A., 1925, M.A., 1927, University of Mississippi.

Fellow Award "For contributions in the fields of radio transmitters, antennas and propagation."

WALLACE, PAUL G., AIEE Fellow 1961. Born: January 15, 1903, Omaha, Tex.

Fellow Award "For contributions to the electrical design and construction of a rapidly expanding utility system." Other Awards: Electrical Engineer of the Year Award, North Texas, 1963.

WALLACE, ROBERT L., JR., IRE Fellow 1956. Born: February 21, 1916, Callina, Tex. Degrees: B.A., 1936,

M.A., 1939, University of Texas; M.A., 1941, Harvard University.

Fellow Award "For contributions in the field of transistor technology and applications." Other Awards: Fellow, Audio Engineering Society.

WALLENSTEIN, GERD D., Fellow 1973. Born: January 2, 1913, Berlin, Germany. Degrees: M.S.(Cybernetic Systems), 1972, San Jose State University; Ph.D.(International Planning), 1976, Stanford University.

Fellow Award "For pioneering leadership in planning and implementing worldwide telecommunications." Other Awards: "El Capitan Juan Bautista de Anza" Award, Public Relations Society of America, Peninsula Chapter, 1963.

WALLIS, CLIFFORD M., AIEE Fellow 1948. Born: March 7, 1904, Waitsfield, Vt. Degrees: B.S., 1926, University of Vermont; M.S., 1928, Massachusetts Institute of Technology; D.Sc., 1941, Harvard University.

Other Awards: Fulbright Lecturer, Ankara University, Turkey, 1960-61; Fulbright Lecturer, National Taiwan University, Taipei, 1967-68; Professor Emeritus of Electrical Engineering, University of Missouri, 1970; Honor Award, Distinguished Service in Engineering, University of Missouri, 1973.

WALLMARK, J. TORKEL, Fellow 1964. Born: June 4, 1919, Stockholm, Sweden. Degrees: Civilin.E.E., 1944, Tekn. Licentiat, 1947, Dr.Tekn., 1953, Royal Institute of Technology, Stockholm.

Fellow Award "For contributions to the concepts of integrated electronic devices and field-effect transistors." Other Awards: L. J. Wallmark Award, Royal Academy of Science, Sweden, 1954; David Sarnoff Outstanding Team Award, 1964; Fellow, American Association for the Advancement of Science; Member, Royal Academy of Engineering Sciences, Sweden; Polhem Award, Swedish Engineer Association, 1982.

WALSH, GEORGE W., Fellow 1974. Born: March 22, 1923, Felt, Idaho. Degrees: B.S.E.E., 1947, University of Idaho; M.E.E., 1960, Rensselaer Polytechnic Institute.

Fellow Award "For leadership in the engineering and protection of industrial power systems, and for contributions to the education of power systems engineers"; IEEE Centennial Medal, 1984. Other Awards: Sigma Xi; Sigma Tau.

WALSH, JOHN B., Fellow 1972. Born: August 20, 1927, Brooklyn, N.Y. Degrees: B.E.E., 1948, Manhattan College; M.S., 1950, Columbia University.

Fellow Award "For contributions to radar theory, missile guidance, and electronics research." Other Awards: Exceptional Civilian Service Award, U.S. Air Force, 1969; Meritorious Civilian Service Award, Dept. of Defense, 1971; Citation of Honor as Outstanding Air Force Civilian Employee of the Year, Air Force Association, 1971; Distinguished Civilian Service Award, Dept.

of Defense, 1977; Theodore von Karman Award for Science and Engineering, Air Force Association, 1977.

WALTER, CARLTON H., Fellow 1971. Born: July 22, 1924, Willard, Ohio. Degrees: B.E.E., 1948, M.S., 1951, Ph.D., 1957, Ohio State University.

Fellow Award "For contributions to traveling-wave and Luneberg lens antennas, and to graduate education." Other Awards: Robert M. Critchfield Award, Ohio State University; Ronald W. Thompson Award, Ohio State University.

WALTER, F. JOHN, Fellow 1980. Born: March 31, 1931. Degrees: B.S.(Chem.Eng.), 1953, Kansas State University; M.S.(Physics), 1958, University of Tennessee.

Fellow Award "For contributions and leadership in the development and application of semiconductor radiation spectrometers."

WALTERS, THEODORE R., AIEE Fellow 1952. Born: May 10, 1906, New York, N.Y. Degrees: I.E.E., 1926, Pratt Institute.

Fellow Award "For outstanding contributions in the development of measuring instruments, improved methods of measuring dielectric phenomena, and for inventions in the field of insulation, particularly for high-temperature applications." Other Awards: Charles A. Coffin Award, General Electric, 1950; Life Member, NSPE and MSPE, 1972; Senior Member, FSPE, 1977.

WANAMAKER, ROBERT L., AIEE Fellow 1962. Born: April 9, 1917, Melrose, Mass. Degrees: B.E.E.E., 1939, Yale University.

Fellow Award "For contributions to aircraft flight control and missile guidance technology." Other Awards: Member, Tau Beta Pi, associate of Sigma Xi, Registered Professional Engineer in N.Y. State; Yale Engr. Assoc., Mass; National Soc. of Professional Engrs.

WANG, AN, Fellow 1971. Born: February 7, 1920, Shanghai, China. Degrees: B.S.E.E., 1940, Chiao Tung University; M.S., 1946, Ph.D., 1948, Harvard University; Honorary Doctoral Degrees: Lowell Technological Institute, Suffolk University, Southeastern Massachusetts University, Syracuse University, Emmanuel College, Bryant College, Fairleigh Dickinson University, Tufts University, Polytechnic Institute of New York, University of Hartford, Boston College.

Fellow Award "For leadership and contributions in developing electronic desktop calculators and magnetic-core memories." Other Awards: Achievement Award, Chinese Institute of Engineers (in America), 1969; Fellow, American Academy of Arts & Sciences, 1981; Newcomen Society in America, 1981; Member, Babson College's Academy of Distinguished Entrepreneurs, 1981; Member, National Academy of Engineering, 1982; The Golden Door Award (The International Institute of Boston), 1982; New England Inven-

tors Award, 1983.

WANG, CHAO C., IRE Fellow 1957. Born: October 20, 1914, Changchow, Kiansu, China. Degrees: B.S., 1936, Chiao Tung University, Shanghai; M.S., 1938, Sc.D., 1940, Harvard University.

Fellow Award "For basic contributions in the field of microwave tubes." Other Awards: Victor Emmanuel Distinguished Professor, Cornell University, 1960; Achievement Award, Seven Long Island Colleges, Special Convocation, 1960; Achievement Award, Chinese Institute of Engineers, 1961; Fellow, Academia Sinica, 1968.

WANG, SHYH, Fellow 1978. Born: June 15, 1925. Degrees: B.S.E.E., 1945, Chiao-Tung University; M.A., 1949, Ph.D., 1951 (both in Applied Physics), Harvard University.

Fellow Award "For contributions to the theory and technique of integrated optics devices."

WARD, HAROLD R., Fellow 1976. Born: July 31, 1931, Lancaster, N.Y. Degrees: B.S.E.E., 1953, Clarkson College; M.S.E.E., 1957, University of Southern California.

Fellow Award "For contributions to the design, analysis, and evaluation of radar systems."

WARD, JAMES B., AIEE Fellow 1961. Born: April 29, 1917, Monte Vista, Colo. Degrees: B.S.E.E., 1939, Colorado State University; M.S.E.E., 1945, Ph.D., 1949, Purdue University.

Fellow Award "For contributions to electrical engineering through teaching and research."

WARD, JOHN E., Fellow 1968. Born: January 4, 1920, Toledo, Ohio. Degrees: B.S.E.E., 1943, M.S.E.E., 1947, Massachusetts Institute of Technology.

Fellow Award "For his outstanding contributions to computer controlled systems"; Distinguished Member, IEEE Control Systems Society, 1983; IEEE Centennial Medal, 1984.

WARE, LAWRENCE, IRE Fellow 1962. Born: May 21, 1901, Bonapart, Iowa. Degrees: B.E., 1926, M.S., 1927, Ph.D., 1930, E.E., 1935, University of Iowa.

Fellow Award "For contributions to electrical engineering education."

WARE, WILLIS H., IRE Fellow 1962. Born: August 31, 1920, Atlantic City, N.J. Degrees: B.S., 1941, University of Pennsylvania; S.M., 1942, Massachusetts Institute of Technology; Ph.D., 1951, Princeton University.

Fellow Award "For contributions in the early development of digital computers"; Achievement Award, Los Angeles Section, IRE, 1957; IEEE Centennial Medal, 1984. Other Awards: Atwater Kent Prize, Moore School of Electrical Engineering, University of Pennsylvania, 1941; Tau Beta Pi National Fellowship, 1941-42; Award, American Federation of Information Processing Societies, 1963; Computer Sciences Man of the Year, Data Processing Management Association,

1975; Air Force Exceptional Civilian Service Medal, 1979.

WARFIELD, JOHN N., Fellow 1977 Born: November 21, 1925, Sullivan, Mo. Degrees: B.A., 1948, B.S.E.E., 1948, M.S.E.E., 1949, University of Missouri at Columbia; Ph.D., 1952, Purdue University.

Fellow Award "For contributions to systems engineering, automatic computation, and engineering education"; Certificate Outstanding Service, IEEE, 1973; Outstanding Contribution Award, IEEE Systems, Man, and Cybernetics Society, 1977. Other Awards: Western Electric Fund Award for Excellence in Instruction of Engineering Students, American Society for Engineering Education, Midwest Section, 1966.

WARING, MOWTON L., AIEE Fellow 1950. Born: June 14, 1906, Chicago, Ill. Degrees: B.S., 1927, Virginia Military Institute; M.S., 1932, Union College.

WARNER, NORWOOD A., AIEE Fellow 1962. Born: August 23, 1902, South Yarmouth, Mass. Degrees: B.S., 1922, University of New Hampshire; LL.B., 1928, New Jersey Law School.

Fellow Award "For contributions to the development and application of nationwide switching."

WARNER, RAYMOND M., JR., Fellow 1977. Born: March 21, 1922, Barberton, Ohio. Degrees: B.S.(Physics), 1947, Carnegie-Mellon University; M.S.(Physics), 1950, Ph.D.(Physics), 1952, Case Western Reserve University.

Fellow Award "For continued contributions to the field of semiconductor devices."

WARREN, CLIFFORD A., Fellow 1971. Born: November 6, 1913, Plainfield, N.J. Degrees: B.S.E.E., 1936, Cooper Union; M.S.E.E., 1949, Stevens Institute of Technology.

Fellow Award "For contributions to the development of radar and large-scale guided missile systems." Other Awards: U.S. Army, Outstanding Civilian Service Award, 1969, 1972; U.S. Army Decoration for Distinguished Civilian Service, 1976.

WARREN, S. REID, JR., AIEE Fellow 1953. Born: January 31, 1908, Philadelphia, Pa. Degrees: B.S., 1928, M.S., 1929, Sc.D., 1937, University of Pennsylvania.

Fellow Award "For outstanding leadership in the application of electrical engineering principles in the medical field, and for leadership in the promotion of appreciation of such work among both electrical engineers and physicians." Other Awards: Fellow, American Association for the Advancement of Science; Associate Fellow, American College of Radiology; Tau Beta Pi; Eta Kappa Nu; Sigma Xi.

WARTERS, WILLIAM D., Fellow 1976. Born: March 22, 1928, Des Moines, Iowa. Degrees: A.B., 1949, Harvard University; M.S., 1950, Ph.D., 1953, California Institute of Technology.

Fellow Award "For contributions to the understand-

ing of wave propagation in multimode media and to the development of millimeter waveguide transmission systems."

WATANABE, HITOSHI, Fellow 1972. Born: December 26, 1930, Shimane, Japan. Degrees: B.E.E., 1953, Dr.Eng., 1961, Kyoto University, Japan.

Fellow Award "For contributions to filter design, computer-aided circuit theory, and application." Other Awards: Inata Memorial Award, 1960; Best Paper Awards, Institute of Electronics and Communications Engineers, Japan, 1961, 1968, 1969; Best Book Award, Institute of Electronics and Communications Engineers, Japan, 1969.

WATCHORN, CARL W., AIEE Fellow 1962. Born: April 25, 1900, Waltham, Mass. Degrees: B.S., 1923, E.E., 1936, Worcester Polytechnic Institute; LLB., 1936; J.D., 1969, University of Maryland; M.S., 1962, Lehigh University.

Fellow Award "For contributions to economic solutions of planning and operating problems in the electric utility industry." Other Awards: Tau Beta Pi, 1921; Sigma Xi, 1923.

WATERMAN, ALAN T., JR., Fellow 1966. Born: July 8, 1918, Northampton, Mass. Degrees: A.B., 1939, Princeton University; B.S., 1940, California Institute of Technology; M.A., 1949, Ph.D., 1952, Harvard University.

Fellow Award "For contributions to the science and practice of transhorizon radio propagation."

WATKINS, DEAN A., IRE Fellow 1958. Born: October 23, 1922, Omaha, Nebr. Degrees: B.S., 1944, Iowa State University; M.S., 1947, California Institute of Technology; Ph.D., 1951, Stanford University.

Fellow Award "For his contributions to the development of microwave tubes"; Achievement Award, IRE, Region 7, 1957; Frederik Philips Award, IEEE, 1981. Other Awards: Member, National Academy of Engineering, 1968; Fellow, American Association for the Advancement of Science, 1981.

WATSON, GEORGE O., AIEE Fellow 1955. Born: May 14, 1891, London, England. Degrees: Diploma, 1910, Battersea Polytechnic, England.

Fellow Award "For his accomplishments in the application of electricity to ship propulsion and in the formulation of the regulations governing electric installations on shipboard." Other Awards: Fellow, Institute of Marine Engineers, 1934; Fellow, Institution of Electrical Engineers, U.K., 1940.

WAY, WILLIAM R., AIEE Fellow 1953. Born: May 29, 1896, North Bay, Ontario, Canada. Degrees: B.Sc.(E.E.), 1918, McGill University.

Fellow Award "For contributions to the electrical industry in the administrative and technical development of utility engineering and operating practices and intersystem collaboration."

WEAVER, ALFRED BRADLEY, AIEE Fellow 1951. Born: October 6, 1894, Texas. Degrees: B.S.E.E., 1923, Texas A & M University.

Other Awards: Tau Beta Pi; Member, Texas Society of Professional Engineers.

WEAVER, CHARLES HADLEY, Fellow 1970. Born: January 27, 1920, Murfreesboro, Tenn. Degrees: B.S.E.E., 1943, M.S.E.E., 1948, University of Tennessee; Ph.D., 1956, University of Wisconsin.

Fellow Award "For leadership in engineering education and for contributions to the theory and application of control systems." Other Awards: Engineer of the Year, East Tennessee Chapter, Tennessee Society of Professional Engineers, 1969; Outstanding Service Award, Air Force ROTC, 1969; Tasker H. Bliss Medal, Society of American Military Engineers, 1970; Nathan W. Dougherty Engineering Award, 1976; First Recipient, "University Professor" Chair, University of Tennessee, 1982.

WEBB, RICHARD C., IRE Fellow 1958. Born: September 2, 1915, Omaha, Nebr. Degrees: B.S.E.E., 1937, University of Denver; M.S.E.E., 1944, Ph.D., 1951, Purdue University.

Fellow Award "For his contributions to the development of color television." Other Awards: Outstanding Work in Research Award, RCA Laboratories, 1947,-1949; Distinguished Engineering Alumnus Award, Purdue University, 1970; "Outstanding Professional Achievement Award," University of Denver Alumni Association, 1983.

WEBB, ROY L., AIEE Fellow 1951. Born: March 17, 1904, Felsenthal, Ark. Degrees: B.S.E.E., 1926, Rice University.

Awards: Chairman, Committee on Membership, AIEE, 1938-1940; Chairman, Committee on Switchgear, AIEE, 1948-1952. Other Awards: Life Member, National Society of Professional Engineers, 1934-1982; Life Member, NSPE, 1983; Chairman, AEIC Committee on Electric Switching and Switchgear, 1962-1964; Chairman EEI-AEIC-NEMA Joint Committee on Power Circuit Breakers, 1966-1968.

WEBBER, HUGH E., Fellow 1965. Born: July 11, 1914, Ludlow, Ill. Degrees: B.A., 1937, Ohio State University.

Fellow Award "For pioneering work in fundamental microwave measurements"; Engineer of the Year Award, Central Florida Section, 1963. Other Awards: Associate Fellow, American Institute of Aeronautics and Astronautics, 1960; "Who's Who in America," 1963; Citation, Florida Engineering Society, Central Florida Chapter, 1964; "Who's Who in Engineering," Engineers Joint Council, 1977; "Who's Who in the South," 1984.

WEBBER, STANLEY E., IRE Fellow 1959. Born: June 8, 1919, Boston, Mass. Degrees: B.S.E.E., 1941, M.S.E.E., 1942, Massachusetts Institute of Technology.

Fellow Award "For contributions to high power klystron amplifiers." Other Awards: Sigma Xi.

WEBER, ERNST, AIEE Fellow 1934, IRE Fellow 1951. Born: September 6, 1901, Vienna, Austria. Degrees: Dipl.Ing., 1924, Technische Hochschule Vienna; Ph.D., 1926, University of Vienna; D.Sc., 1927, Technische Hochschule Vienna.

Awards: Education Medal, 1960, "For excellence as a teacher in science and electrical engineering, for creative contributions in research and development, for broad professional and administrative leadership, and in all, for a considerate approach to human relations"; Founders Award, IEEE, 1971, "For leadership of great value to the profession"; Microwave Career Award, IEEE Microwave Theory and Techniques Society, 1977. Other Awards: Fellow, American Physical Society, 1946; Presidential Certificate of Merit, 1948; Honorary Member, Institute of Electrical Engineers of Japan and Institute of Radio Engineers of Japan, 1963; Member, National Academy of Engineering, 1964; Member, National Academy of Sciences, 1965; Howard Coonley Medal, ASA, 1966; Honorary Member, American Society for Engineering Education, 1972; Distinguished Service Award, Transportation Research Board, National Research Council, 1976; Eminent Member, Eta Kappa Nu, 1962.

WEBER, HEINRICH E., Fellow 1971. Born: March 22, 1907, Switzerland. Degrees: Dipl.El.Ing., 1929, Swiss Federal Institute of Technology.

Fellow Award "For contributions to theory and practice in electroacoustical transducers and engineering education." Other Awards: Honorary Member, Swiss Association for Electrical Engineers, 1969; Naturforschende Gesellschaft Zurich, 1964, 1977.

WEBER, HOWARD H., SR., AIEE Fellow 1940. Born: March 22, 1896, Allentown, Pa. Degrees: B.S.E.E., 1920, Bliss Electrical School; M.Sc., 1924, Lehigh University.

Other Awards: Life Member, International Association of Electrical Inspectors, 1961; Life Member, International Municipal Signal Association, 1961.

WEBER, JOSEPH, IRE Fellow 1958. Born: May 17, 1919, Paterson, N.J. Degrees: B.S., 1940, United States Naval Academy; Ph.D., 1951, Catholic University of America.

Fellow Award "For his early recognition of concepts leading to the development of the maser." Other Awards: Guggenheim Fellow, 1955, 1962; Scientific Achievement Award, Washington Academy of Sciences, 1958; First Prize, Gravity Research Foundation, 1959; Fellow, American Physical Society; Babson Award, Gravity Research Foundation, 1970; Award, Sigma Xi, 1970; Boris Pregel Award, 1973.

WEBSTER, EDWARD M., IRE Fellow 1944. Born: February 28, 1889, Washington, D.C. Degrees: B.A., 1912, United States Coast Guard Academy.

Fellow Award "For his contributions to the development of the maritime mobile services and his leadership in promoting measures for enhancing the safety of life and property at sea." Other Awards: Marconi Medal, 1950.

WEBSTER, ROGER R., Fellow 1968. Born: May 4, 1920, Holtville, Calif. Degrees: B.S.E.E., 1943, University of California.

Fellow Award "For outstanding personal contributions and leadership in microwave device and integrated circuit technology development"; Best Paper Award, IRE, 1961. Other Awards: Citation, AVIATION WEEK AND SPACE TECHNOLOGY, 1967.

WEBSTER, WILLIAM M., IRE Fellow 1960. Born: June 13, 1925, Warsaw, N.Y. Degrees: B.S., 1945, Union College; Ph.D., 1954, Princeton University.

Fellow Award "For contributions to gaseous electronic and solid-state devices"; Editor's Award, IRE, 1952; Frederik Philips Award, IEEE, 1980.

WEED, LESLIE J., AIEE Fellow 1951. Born: April 17, 1906, Sandwich, N.H. Degrees: B.S.E.E., 1927, M.S.E.E., 1928, Massachusetts Institute of Technology.

Awards: Chairman, Boston Section, AIEE, 1954-1955; Vice President, District 12, AIEE, 1962. Other Awards: Tau Beta Pi, 1960; Eta Kappa Nu, 1961; Lawrence F. Cleveland Award, 1979; President, Engineering Societies of New England, 1959.

WEEKS, PAUL T., IRE Fellow 1954. Born: November 1890, Clarksfield, Ohio. Degrees: B.A., 1913, Oberlin College; Ph.D., 1917, Cornell University.

Fellow Award "For contributions to electron tube research, engineering, and manufacture." Other Awards: Fellow, American Association for the Advancement of Science, 1925.

WEGE, HARRY R., Fellow 1966. Born: January 28, 1903, Bangor, Wis. Degrees: B.S.E.E., 1925, Kansas State University.

Fellow Award "For contributions to radar and weapons systems." Other Awards: Award of Merit, RCA Victor, 1954; Achievement Award, Kansas State University, 1964.

WEHNER, GOTTFRIED K., Fellow 1974. Born: September 23, 1910, Baerenwalde, DDR. Degrees: Dipl.Ing., 1937, Dr.Ing., 1939, Technical University, Munich, Germany.

Fellow Award "For contributions to the understanding of the sputtering process." Other Awards: 2nd Welch Award, American Vacuum Society, 1971; Honorary Member, American Vacuum Society, 1981.

WEIDENHAMMER, JAMES A., Fellow 1975. Born: March 1, 1918, Allentown, Pa. Degrees: B.S.M.E.,

1938, Lehigh University.

Fellow Award "For contributions to the development of computer tape drives."

WEIHE, VERNON I., IRE Fellow 1958. Born: March 3, 1909, Louisville, Ky. Degrees: B.S.E.E., 1931, University of Louisville.

Fellow Award "For contributions to systems planning in air navigation and traffic control." Other Awards: Collier Award, 1948; Superior Achievement Award, Institute of Navigation, 1967; Honorary Life Member, Institute of Navigation, 1981.

WEIL, THOMAS A., Fellow 1975. Born: January 22, 1930, New York, N.Y. Degrees: B.S.E.E., 1951, Massachusetts Institute of Technology.

Fellow Award "For contributions to radar transmitter and system technology." Other Awards: Eta Kappa Nu, 1950; Tau Beta Pi, 1951; Sigma Xi, 1951.

WEILL, JACKY, Fellow 1981. Born: July 27, 1924, Strasbourg, France. Degrees: Engineer Institute Electrotechnique Grenoble, 1947; Dr.Eng., 1953, Paris.

Fellow Award "For contributions as a research manager and leader in the field of electronic control and protection systems for nuclear reactors." Other Awards: Chairman, Technical Committee 45 Nuclear Instrumentation International Electrotechnical Commission; Chevalier de la Legion d'Honneur, Medaille de la Reconnaissance Francaise; Officier de l'Ordre National du Merite, 1982.

WEIMER, PAUL K., IRE Fellow 1955. Born: November 5, 1914, Wabash, Ind. Degrees: A.B., 1936, Manchester College; M.A., 1938, University of Kansas; Ph.D., 1942, Ohio State University; D.Sc.(Hon.), 1968, Manchester College.

Fellow Award "For his contributions to the development of television pickup tubes"; Vladimir K. Zworykin Award, IRE, 1959, "For contributions to photoconductive-type pickup tubes"; Morris N. Liebmann Award, 1966, "For invention, development and applications of the thin-film transistor." Other Awards: Television Broadcasters Award, 1946; David Sarnoff Outstanding Achievement Award, Radio Corporation of America, 1963; Outstanding Paper Award Plaques, International Solid State Circuit Conferences, 1963, 1965; National Academy of Engineering, 1981.

WEINBERG, LOUIS, IRE Fellow 1960. Born: July 15, 1919, Brooklyn, N.Y. Degrees: A.B., 1941, Brooklyn College; M.S., 1947, Harvard University; Sc.D., 1951, Massachusetts Institute of Technology.

Fellow Award "For contributions to the field of network theory." Other Awards: Dean Bildersee Award, Brooklyn College, 1941; Alumnus of the Year Award, Brooklyn College, 1964; Fellow, A.A.A.S., 1960; Alumnus of the Year, Thomas Jefferson High School, 1971; Research Fellowship, Japan Society for Promotion of Science.

WEINBERGER, ARNOLD, Fellow 1984. Born: October 23, 1924, Czechoslovakia. Degrees: B.S.E.E., 1950, CCNY.

Fellow Award "For contributions to the theory of computer arithmetic logic, and to the layout of large-scale integrated circuits." Other Awards: IBM Corporate Invention Award, 1981.

WEINBERGER, JULIUS, IRE Fellow 1925. Born: July 22, 1893, New York, N.Y. Degrees: B.S., 1913, City College of the City University of New York.

Other Awards: Modern Pioneer Award, National Association of Manufacturers, 1940.

WEINREB, SANDER, Fellow 1978. Born: December 9, 1936, New York, N.Y. Degrees: B.S.E.E., 1958, Ph.D., 1963, Massachusetts Institute of Technology.

Fellow Award "For contributions to instrumentation in radio astronomy"; IRE Award, Boston Section, IEEE, 1958.

WEINSCHEL, BRUNO, Fellow 1966. Born: May 26, 1919, Stuttgart, Germany. Degrees: Dr.Ing., 1966, Technische Hochschule Munich, Germany.

Fellow Award "For contributions in the field of precision microwave measurements and advancement of attenuation measurements." Other Awards: Fellow, Institution of Electrical Engineers, 1977.

WEINSTEIN, STEPHEN B., Fellow 1984. Born: November 25, 1938, New York, N.Y. Degrees: S.B., 1960, M.I.T.; M.S., 1962, Univ. of Michigan; Ph.D., 1966, Univ. of Calif., Berkeley (all in E.E.).

Fellow Award "For contributions to the theory and practice of voiceband data communications, and to IEEE publications activities"; Donald W. McLellan Meritorious Service Award, IEEE Communications Society, 1983.

WEINSTOCK, WALTER W., Fellow 1978. Born: August 18, 1925, Philadelphia, Pa. Degrees: B.S.E.E., 1946, M.S.E.E., 1954, Ph.D., 1964, University of Pennsylvania.

Fellow Award "For contributions to radar systems and for leadership in development of modern air defense systems." Other Awards: David Sarnoff Outstanding Achievement Award and Medal in Engineering, 1972.

WEISBERG, LEONARD R., Fellow, 1978. Born: October 17, 1929, New York, N.Y. Degrees: B.A., 1950, Clark University; M.A., 1952, Columbia University.

Fellow Award "For contributions in semiconductor compound device research." Other Awards: Outstanding Achievement Award, RCA, 1959; Secretary of Defense Meritorious Civilian Service Medal, 1979.

WEISS, MAX T., Fellow, 1967. Born: December 29, 1922, Hungary. Degrees: B.S.E.E., 1943, City College of the City University of New York; M.S.E.E., 1947, Ph.D., 1950, Massachusetts Institute of Technology.

Fellow Award "For contributions to microwave fer-

rites and in the management of research." Other Awards: Fellow, American Physical Society, 1964; Fellow, American Institute of Engineering.

WEISZ, WILLIAM J., Fellow 1966. Born: January 8, 1927, Chicago, Ill. Degrees: B.S.E.E., 1948, Massachusetts Institute of Technology; Doctor of Business Administration (Hon.), 1976, St. Ambrose College.

Fellow Award "For contributions in the field of effective radio spectrum utilization by the Land Mobile Radio Services." Other Awards: Award of Merit, National Electronics Conference, 1970; Corporate Leadership Award, Massachusetts Institute of Technology; 1976; Freedom Foundation of Valley Forge Award, 1974; Electronic Industries Association Medal of Honor, 1981.

WELBER, IRWIN, Fellow 1973. Born: 1924, New York. Degrees: B.S.E.E., 1948, Union College; M.E.E., 1950, Rensselaer Polytechnic Institute.

Fellow Award "For contributions and leadership in the development of communications by submarine cable and satellites." Other Awards: Honorary RPI, New York, 1950; Sigma Xi.

WELCH, A. U., JR., AIEE Fellow 1960. Born: July 25, 1905. Degrees: B.S.E.E., Rutgers University.

Fellow Award "For contributions to the development of reactors, transformers and arc welding apparatus." Other Awards: Charles A. Coffin Award, General Electric.

WELCH, H. WILLIAM, JR., IRE Fellow 1958. Born: October 21, 1920, Beardstown, Ill. Degrees: B.A., 1942, DePauw University; M.S., 1948, Ph.D., 1952, University of Michigan.

Fellow Award "For his contributions to research on microwave tubes and solid state devices"; Paper Prize, IRE, 1952, for "Effects of Space Charge on Frequency Characteristics of Magnetrons." Other Awards: Phi Beta Kappa, 1942; Sigma Pi Sigma, 1946; Sigma Xi, 1950; Eta Kappa Nu, 1956; Tau Beta Pi, 1957; Fellow, American Association for the Advancement of Science, 1978.

WELDON, EDWARD J., Fellow 1980. Born: April 8, 1938. Degrees: B.S.E.E., 1958, Manhattan College; M.S.(E.E.), 1960, Ph.D.(E.E.), 1963, University of Florida.

Fellow Award "For contributions to the development of error-correcting codes."

WELDON, JAMES O., IRE Fellow 1954. Born: March 15, 1905, Canton, Mo.

Fellow Award "For his contributions to the design of high power transmitters and their use in international broadcasting."

WELKOWITZ, WALTER, Fellow 1976. Born: August 3, 1926, Brooklyn, N.Y. Degrees: B.S., 1948, The Cooper Union; M.S., 1949, Ph.D., 1954, University of Illinois at Urbana.

Fellow Award "For contributions to the analysis of cardiovascular systems and to the development of heart-assist devices." Other Awards: Rutgers Research Council Fellow, 1974-1975.

WELLER, EDWARD F., Fellow 1977, Born: November 30, 1919, Baltimore, Md. Degrees: E.E., 1943, University of Cincinnati.

Fellow Award "For leadership in automotive electronics and instrumentation and for electronic control of automotive emissions." Other Awards: Distinguished Alumnus, University of Cincinnati, 1973.

WELLS, CLARENCE A., AIEE Fellow 1960. Born: February 5, 1895, New Haven, Conn. Degrees: B.S., 1917, Occidental College.

Fellow Award "For technical and educational contributions in the field of communication transmission engineering." Other Awards: Phi Beta Kappa, 1917; Eta Kappa Nu, 1955.

WELLS, FRANK H., IRE Fellow 1958. Born: March 1915, London, England. Degrees: M.Sc.(Eng.), 1936, University of London.

Fellow Award "For his contributions to pulse techniques."

WELSH, J. W., AIEE Fellow 1914. Born: October 20, 1880, Springfield, Ohio. Degrees: A.B., 1900, Wittenberg University; A.B., 1901, Harvard University; B.S., 1903, Massachusetts Institute of Technology.

WELSH, JAMES P., Fellow 1968. Born: December 27, 1917, Buffalo, N.Y. Degrees: B.S.E.E., 1938, Carnegie Institute of Technology.

Fellow Award "For his contributions to thermal measurement and cooling of electronic equipment." Other Awards: Buffalo's Man of the Week, 1969.

WENDT, KARL R., IRE Fellow 1954. Born: January 3, 1906, Coshocton, Ohio.

Fellow Award "For his contributions to the development of television equipment and circuits." Other Awards: Award, RCA, 1947.

WENGER, FLOYD E., IRE Fellow 1962. Born: May 13, 1898, Orrville, Ohio.

Fellow Award "For contributions to reliability and performance of military components"; Commendation, IRE; Award, IRE, Professional Group on Reliability; Contribution Award, IEEE, 1967. Other Awards: Certificate of Merit, Certificate of Service, Air Force System Command; Commendation for Meritorious Civilian Service, Superior Performance Rating, Department of Air Force; Certificate of Merit, National Security Industrial Association; Certificate of Appreciation, Dept. of Defense; Ohio Senior Citizens Hall of Fame, 1979.

WENTZ, EDWARD C., AIEE Fellow 1956. Born: July 3, 1905, St. Paul, Minn. Degrees: B.S., 1926, University of Minnesota; M.S., 1936, University of Pittsburgh; E.E., 1939, University of Minnesota.

Fellow Award "For his design advances on high-capacity current transformers."

WENZEL, ROBERT J., Fellow 1983. Born: September 11, 1939, Milwaukee, Wisconsin. Degrees: B.S., 1961, Marquette University; M.S., 1962, Massachusetts Institute of Technology.

Fellow Award "For contributions to the theory and synthesis of microwave filter and multiplexer networks"; G-MTT Microwave Prize, 1967.

WEPPLER, H. EDWARD, Fellow 1967. Born: September 4, 1915, Elizabeth, N.J. Degrees: B.S.E.E., 1937, Purdue University.

Fellow Award "For his contributions to radio communications and communication satellite systems."

WERTH, ANDREW M., Fellow 1982. Born: March 2, 1934, Saarbruck, Germany. Degrees: B.S. 1955, M.S., 1961, Columbia University.

Fellow Award "For leadership in the design and development of digital satellite communications systems."

WERTS, ROBERT W., Fellow 1978. Born: November 28, 1916, Reading, Pa. Degrees: B.S.(Electrical Engineering), 1938, Penn State University.

Fellow Award "For contributions in the planning and developing of large-scale electric power systems."

WESTBYE, JOHN B., AIEE Fellow 1945. Born: January 15, 1898, Norway. Degrees: E.E., University of Darmstadt, Germany.

Other Awards: American Society for Testing and Materials; CIGRE.

WESTENDORP, WILLEM F., AIEE Fellow 1949. Born: May 7, 1905, Amsterdam, Netherlands. Degrees: E.E., 1928, Technical University of Delft, Netherlands; D.Eng.(Hon.), 1947, Rensselaer Polytechnic Institute.

Fellow Award "For research, development and design of one and two million volt resonant transformer X-ray machines and the 100 MeV. Betatron and the 70 MeV. Synchrotron electron accelerators." Other Awards: John Price Wetherill Medal, Franklin Institute, 1944.

WESTRATE, MILLARD C., AIEE Fellow 1958. Born: August 16, 1908, Allegan, Mich. Degrees: B.S., 1930, University of Michigan.

Fellow Award "For contributions to the design and economic operations of large power systems."

WETHERILL, LYNN, SR., AIEE Fellow 1948. Born: July 12, 1904, Philadelphia, Pa. Degrees: B.S.E.E., 1926, M.S.E.E., 1926, Massachusetts Institute of Technology.

WEYGANDT, CORNELIUS N., Fellow 1981. Born: August 13, 1904, Degrees: B.S.E.E., 1928, University of Pennsylvania; M.S.E.E., 1933, Massachusetts, Institute of Technology; Ph.D., 1947, University of Pennsylvania.

Fellow Award "For contributions to differential

analyzer solutions of rotating machinery and power system problems."

WHARTON, CHARLES B., Fellow 1977, Born: March 29, 1926, Gold Hill, Oreg. Degrees: B.S.E.E., 1950, M.S.E.E., 1952, University of California at Berkeley.

Fellow Award "For contributions to the understanding of plasmas and to the development of plasma diagnostic techniques." Other Awards: Humboldt-Preis, Alexander von Humboldt Foundation, Germany, 1973; Fellow, American Physical Society, 1973; ORC Visiting Professor, Occidental Research Corp., 1979-1981.

WHEELER, HAROLD A., IRE Fellow 1935, AIEE Fellow 1946. Born: May 10, 1903, St. Paul, Minn. Degrees: B.S. 1925; D.Sc.(Hon.), 1972, George Washington University; D.Eng.(Hon.), 1978, Stevens Institute of Technology.

Awards: Morris N. Liebmann Award, IRE, 1940, "For his contribution to the analysis of wideband high-frequency circuits particularly suitable for television"; Medal of Honor, 1964, "For his analyses of the fundamental limitations on the resolution in television systems and on wide-band amplifiers, and for his basic contributions to the theory and development of antennas, microwave elements, circuits, and receivers"; Microwave Career Award, IEEE, Society of Microwave Theory and Techniques, 1975. Other Awards: Modern Pioneer Award, National Association of Manufacturers, 1940; Certificate of Commendation, U.S. Dept. of Navy, 1947; Armstrong Medal, Radio Club of America, 1964.

WHEELON, ALBERT D., Fellow 1970. Born: January 18, 1929, Moline, Ill. Degrees: B.S., 1949, Stanford University; Ph.D., 1952, Massachusetts Institute of Technology.

Fellow Award "For contributions to electromagnetic propagation in turbulent media, and for basic exploration in missile guidance." Other Awards: Member, National Academy of Engineering.

WHICKER, LAWRENCE R., Fellow 1978. Born: October 3, 1934, Bristol, Va. Degrees: B.S.(Electrical Engineering), 1957, M.S.E.E., 1958, University of Tennessee; Ph.D., 1964, Purdue University.

Fellow Award "For contributions to the development of microwave and millimeter-wave nonreciprocal components."

WHINNERY, JOHN R., IRE Fellow 1952. Born: July 26, 1916, Read, Colo. Degrees: B.S., 1937, Ph.D., 1948, University of California at Berkeley.

Fellow Award "For his contributions to the knowledge of electromagnetic theory and the application of that theory to microwave problems"; IEEE Education Medal, 1967; Microwave Career Award, IEEE Microwave Theory and Techniques Society, 1976, "For a career of meritorious achievement and outstanding technical contributions in the field of microwave theory and techniques"; IEEE Centennial Medal, 1984. Other Awards: Guggenheim Fellow, 1959; Member, National Academy of Engineering, 1965; Member, National Academy of Sciences, 1972; Lamme Award, American Society for Engineering Education, 1975; Fellow, Optical Society of America, 1977; Distinguished Alumnus Award, University of California at Berkeley, 1980; Appointed University Professor, University of California, 1980; American Academy of Arts and Sciences, 1980.

WHITAKER, H. BARON, Fellow 1969. Born: May 25, 1913, Durham, N.C. Degrees: B.S.E.E., 1936, North Carolina State College; LL.D.(Hon.), 1968, Illinois Institute of Technology.

Fellow Award "For contributions to the development of procedures and standards to bring about safety in the use of electrical equipment." Other Awards: Distinguished Service Award, 1965, Honorary Life Member, 1978, National Fire Protection Association; Joseph B. Finnegan Award, Society of Fire Protection Engineers, 1968; Honorary Life Fellow, Standards Engineers Society, 1976; Honorary Life Member, International Association of Electrical Engineers, 1978; Howard Coonley Medal, American National Standards Institute, 1978.

WHITE, ALAN D., Fellow 1976. Born: July 6, 1923, Rahway, N.J. Degrees: A.B., 1949, Rutgers University; M.S., 1951, Syracuse University.

Fellow Award "For development and subsequent improvements of the visible light helium-neon laser"; David Sarnoff Award, IEEE, 1984.

WHITE, CHARLES E., Fellow 1969. Born: January 22, 1911, Hoboken, N.J. Degrees: B.S.E.E., 1934, New York University; M.S.E.E., 1955, University of Connecticut.

Fellow Award "For contributions to the organization and development of standardization activities in support of the national measurement system." Other Awards: Letter of Commendation for Outstanding Services as a Civilian Engineer, U.S. Navy, 1946; Contributions to the Literature of Standards, American Society for Testing and Materials, 1963; Outstanding Achievement in Metrology and Calibration, U.S. Air Force Systems Command, 1964; Outstanding Services Award, National Conference of Standards Laboratories, 1970; Life Member, National Society of Professional Engineers, 1975.

WHITE, DAVID C., Fellow 1965. Born: February 18, 1922, Sunnyside, Wash. Degrees: B.S., 1946, M.S., 1947, Ph.D., 1949, Stanford University.

Fellow Award "For contributions to engineering education." Other Awards: George Westinghouse Award, American Society for Engineering Education, 1961; Fellow, American Academy of Arts and Sciences, 1963; Member, National Academy of Engineering, 1975; Chairman, Electric Power Research Institute Advisory Council, 1984-1986; Ford Professor of Engineering and Director of the Energy Laboratory at M.I.T.

WHITE, EDWIN LEE, AIEE Fellow 1947, IRE Fellow 1955. Born: July 5, 1896, Valley City, N.D. Degrees: A.B., 1922, M.S., 1925, George Washington University.

IRE Fellow Award "For his leadership in advancing the use of radio in the interest of safety and efficiency in industry"; IEEE Centennial Medal, 1984. Other Awards: Bronze Star, 1943; Presidential Unit Citation, 1944; China-Burma-India Theater Citation, 1944; Alumni Service Award, George Washington University, 1973.

WHITE, EUGENE L., Fellow 1966. Born: January 7, 1900, Harford, N.Y. Degrees: B.S., 1924, University of Nebraska.

Fellow Award "For achievement as an engineering executive in the field of high-voltage transmission of electric energy." Other Awards: Distinguished Service Award, U.S. Dept. of Interior, 1962.

WHITE, GIFFORD E., Fellow 1964. Born: February 17, 1912, San Saba, Tex. Degrees: B.A., M.A., 1939, University of Texas.

Fellow Award "For contributions to radar systems and physical instrumentation."

WHITE, H. BRIAN, Fellow 1978. Born: September 18, 1922, Toronto, Canada. Degrees: Bachelor of Applied Science (Structural, Honours), 1944, University of Toronto.

Fellow Award "For contributions to the development of extra- high-voltage transmission line structures."

WHITE, HARRY J., AIEE Fellow 1961. Born: July 29, 1905, Fremont, Nebr. Degrees: B.S.E.E., 1928, M.S., 1931, Ph.D. (Physics), 1934, University of California at Berkeley.

Fellow Award "For contributions to the theory and practice of electrostatic precipitation." Other Awards: Sigma Xi, 1933; Fellow, American Physical Society, 1937; Fellow, American Association for the Advancement of Science, 1946; Frank A. Chambers Award, Air Pollution Control Association, 1971; Fellow, Institute of Electrostatics, Japan, 1977; Larry Faith Technical Achievement Award, Air Pollution Control Association, 1979; Award of Merit, Institute of Electrostatics, Japan, 1982.

WHITE, J. COLEMAN, Fellow 1975. Born: August 27, 1923, Fort Dodge, Iowa. Degrees: B.E.E., 1947, Cornell University.

Fellow Award "For contributions to the development and digital computer analysis of large ac and dc rotating machines."

WHITE, JAMES A., AIEE Fellow 1962. Born: September 21, 1908, Mount Vernon, Wash. Degrees: B.S., 1931, University of Washington; M.S., 1961, Stanford University.

Fellow Award "For contributions to wind-tunnel electrical instrumentation."

WHITE, JOSEPH F., Fellow 1979. Born: June 5, 1938, Cleveland, Ohio. Degrees: B.S.E.E., 1960, Case Institute of Technology; M.S.E.E., 1965, Northeastern University; Ph.D., 1968, Rensselaer Polytechnic Institute.

Fellow Award "For contributions to the development of diode phase shifters for microwave array antennas"; Application Award, IEEE Microwave Theory and Techniques Society, 1975. Other Awards: Member, Eta Kappa Nu; Sigma Xi.

WHITE, MARVIN H., Fellow 1974. Born: September 6, 1937, The Bronx, N.Y. Degrees: A.S., 1957, Henry Ford Community College; B.S.E., 1960, M.S., 1961, University of Michigan; Ph.D., 1969, The Ohio State University.

Fellow Award "For contributions to the theory and development of solid-state electronic devices, especially memory transistors and charge-coupled imaging arrays"; IEEE Electron Devices National Newsletter Editor, 1973-1976; IEEE Electron Devices National Lecturer, 1982. Other Awards: Patent Awards, Westinghouse Electric, 1973-1975; Fellow, Washington Academy of Sciences, 1976; Sigma Xi; Eta Kappa Nu; Fulbright Professorship, 1978-1979; Sherman Fairchild Professor of Electrical and Computer Engineering, Lehigh University, Bethlehem, PA, 1981-.

WHITE, RICHARD M., Fellow 1972. Born: April 25, 1930, Denver, Colo. Degrees: A.B., 1951, A.M., 1952, Ph.D., 1956, Harvard University.

Fellow Award "For contributions to the discovery and applications of surface elasic waves." Other Awards: Guggenheim Fellowship, 1968-69.

WHITE, ROBERT L., Fellow 1977. Born: February 14, 1927, Plainfield, N.J. Degrees: B.S., 1949, Columbia College; M.A., 1951, Ph.D., 1954, Columbia University.

Fellow Award "For teaching and research in the fields of magnetic and optical properties of materials and of biomedical engineering." Other Awards: Fellow, American Physical Society, 1962; Guggenheim Fellowship, 1969-1960, 1977-1978; Fellowship, Japan Society for Promotion of Science, 1975.

WHITE, STANLEY A., Fellow 1982. Born: September 25, 1931, Providence, RI. Degrees: B.S.E.E., 1957, M.S.E.E., 1959, Ph.D., 1965, Purdue University.

Fellow Award "For contributions to digital signal processing techniques and applications"; Commendation for Outstanding Service, Orange County Section, IEEE, 1976, 1977; Los Angeles Council, 1981-1982; IEEE Centennial Medal, 1984. Other Awards: Eta Kappa Nu, 1957; National Career Guidance Award, 1958; Tau Beta Pi, 1959; Letter of Commendation, Purdue University, 1963; North American Aviation Science Engineering Fellow, 1963-1965; Sigma Xi, 1965; RESA,

1966; "International Who's Who in Medical Engineering," 1968; "Distinguished Lecturer Award," National Electronics Conference, 1973; Fellow, Institute for the Advancement of Engineering, 1981; "Who's Who in Engineering," "Who's Who in America," and "Who's Who in the World," 1982.

WHITE, WARREN D., IRE Fellow 1957. Born: July 7, 1915, Springfield, Mo. Degrees: B.S., 1936, Drury College; B.S., 1938, Missouri School of Mines.

Fellow Award "For achievements in the fields of air traffic control, information theory, countermeasures, and radar."

WHITEHEAD, DANIEL L., AIEE Fellow 1962. Born: December 25, 1915, Walland, Tenn. Degrees: B.S., 1939, University of Tennessee; M.S., 1940, Cornell University.

Fellow Award "For contributions in the fields of power transmission and high voltage testing."

WHITEHEAD, EDWIN R., AIEE Fellow 1945. Born: October 8, 1906, Ash Grove, Mo. Degrees: B.S.E.E., 1928, University of Colorado; M.S.E.E., 1935, Ph.D., 1944, University of Pittsburgh.

Other Awards: Distinguished Service Award, National Council of Engineering Examiners, 1968; Western Electric Fund Award, American Society for Engineering Education, 1970; Distinguished Engineering Alumnus Award, University of Colorado, 1972.

WHITELOCK, LELAND D., Fellow 1971. Born: September 11, 1907, Petersburg, Ind. Degrees: B.S.E.E., 1931, Carnegie Institute of Technology.

Fellow Award "For contributions to communications and computer systems, and to methodology of systems effectiveness." Other Awards: Meritorious Civilian Service Award, U.S. Navy, 1946.

WHITMAN, LAWRENCE C., AIEE Fellow 1956. Born: October 31, 1903, Franklin, Vt. Degrees: B.S., 1927, M.S., 1933, University of Vermont.

Fellow Award "For his contributions to the improved design of dry-type insulating structures and to a better knowledge of their thermal aging"; Second Paper Prize, AIEE, 1954, for "Calculation of Life Characteristics of Insulation." Other Awards: Phi Beta Kappa, 1927; Managerial Award, General Electric, 1948; Sigma Tau, 1965; Eta Kappa Nu, 1966; 8 Patents; Full Professor, South Dakota State University; Listed in Who's Who in Engineering; American Men of Science.

WHITMAN, WEBSTER C., Fellow 1971. Born: December 27, 1907, Auburn, Me. Degrees: B.S., 1930, Brown University.

Fellow Award "For contributions to the design of generation, transmission, substation, and distribution facilities of electric power systems."

WHITNEY, EUGENE C., AIEE Fellow 1962. Born: August 26, 1913, Columbus, Ohio. Degrees: B.S., 1935, University of Michigan.

Fellow Award "For contributions to the theory and design of large rotating electrical machines." Other Awards: Silver W Award, Westinghouse Electric, 1965.

WHITTEMORE, LAURENS E., IRE Fellow 1927. Born: August 20, 1892, Topeka, Kans. Degrees: A.B., 1914, Washburn College; M.A., 1915, University of Kansas.

Awards: Vice President, IRE, 1928. Other Awards: Sigma Xi, 1915.

WHITTINGTON, BERNARD W., Fellow 1982. Born: July 19, 1920, Charleston, WV. Degrees: B.S.E.E., 1951, West Virginia University.

Fellow Award "For leadership in electrical standards formulation and implementation"; Achievement Award, Industrial and Commercial Power Systems Department, IEEE Industry Applications Society, 1978.

WICKIZER, GILBERT S., IRE Fellow 1961. Born: Augusrt 20, 1904, Warren, Pa. Degrees: B.S., 1926, Pennsylvania State University.

Fellow Award "For contributions to experimental wave propagation research." Other Awards: Award, RCA, 1954.

WIDROW, BERNARD, Fellow 1976. Born: December 24, 1929, Norwich, Conn. Degrees: S.B., 1951, S.M., 1953, Sc.D., 1956, Massachusetts Institute of Technology.

Fellow Award "For contributions to adaptive antenna systems and to the theory of quantization error"; First Prize Paper, AIEE, Eighth District, 1960-1961, for "Statistical Analysis of Amplitude-Quantized Sampled-Data Systems"; IEEE Centennial Medal, 1984. Other Awards: Francqui Chair, University of Louvain, Belgium, 1967; Tau Beta Pi; Fellow, American Association for the Advancement of Science, 1980.

WIESNER, JEROME B., IRE Fellow 1952. Born: May 30, 1915, Detroit, Mich. Degrees: B.S.E.E., 1937, M.S., 1938, Ph.D., 1950, University of Michigan; D.Eng.(Hon.), 1961, Polytechnic Institute of Brooklyn; D.Sc.(Hon.), 1962, Lowell Technological Institute; D.Sc.(Hon.), 1962, University of Michigan; D.Eng.(Hon.), 1962, Rensselaer Polytechnic Institute; D.Sc.(Hon.), 1965, Lehigh University; D.Sc.(Hon.), 1965, Brandeis University; D.Sc.(Hon.), 1965, University of Massachusetts; D.Sc.(Hon.), 1966, Northwestern University; Ph.D.(Hon.), 1968, Williams College; D.Sc.(Hon.), 1969, Oklahoma City University; D.Sc.(Hon.), 1970, Yeshiva University; Dr. of Law, 1974, Harvard University; Hon. Doctorate, University of Notre Dame.

Fellow Award "For his contributions in the field of information theory and administration of research on advanced techniques and concepts"; Founders Medal, IEEE, 1977. Other Awards: Eta Kappa Nu, 1937; Phi Kappa Phi, 1937; Sigma Xi, 1937; Fellow, American Academy of Arts and Sciences, 1950; Outstanding Engineering Graduate, University of Michigan, 1954; Alfred P. Sloan Award, Electronic Industries Association, 1956; National Academy of Sciences, 1958; Medal of Honor, 1961; Star of Pakistan, 1963; National Academy of Engineering, 1966; Migel Medal, American Foundation for the Blind, 1971; First Class Order of the Sacred Treasure, Emperor of Japan for Japanese Government, 1983.

WILD, EARLE, AIEE Fellow 1951. Born: April 28, 1902, Boston, Mass. Degrees: S.B., 1924, Massachusetts Institute of Technology.

WILDER, HAROLD F., Fellow 1967. Born: November 29, 1907, Quincy, Mass. Degrees: B.E.E., 1929, Northeastern University.

Fellow Award "For significant contributions to the art of submarine cable telegraphy." Other Awards: F.E. d'Humy Award, Western Union, 1959.

WILDES, KARL L., AIEE Fellow 1940. Born: September 25, 1895, Belmont, N.H. Degrees: B.S., 1920, University of New Hampshire; S.M., 1922, Massachusetts Institute of Technology.

WILEY, CARL ATWOOD, Fellow 1979. Born: December 30, 1918. Degrees: B.S.(Mathematics), 1944, Antioch College.

Fellow Award "For contributions to the development of high-resolution synthetic aperture radar."

WILEY, WILLIAM C., Fellow 1984. Born: August 7, 1924, Monmouth, IL. Degrees: B.S.(Engineering Physics), University of Illinois.

Fellow Award "For technical leadership and creativity in the development of scientific instruments." Other Awards: ISA Fellow Award, 1976; University of Illinois Alumni Award, 1975; Michigan Patent Law Association Award, 1963.

WILHELM, GEORGE RAYMOND, AIEE Fellow 1948. Born: August 7, 1896, Monkton, Md.

Awards: Section Chairman, Washington Section, AIEE, 1942-1943; Chairman, National AIEE Transfers Committee, 1959-1961; Patron Award, Washington Section, "For substantial contributions to the welfare of the Section." Other Awards: Certificate of Service, National Production Authority, and member of the National Defense Executive Reserve, U.S. Dept. of Commerce, 1953; Executive Reservist Emeritus, Secretary of Commerce, 1967; Award of Merit, Bureau of Domestic Commerce, U.S. Dept. of Commerce, 1976; Registered Professional Engineer, 1952, District of Columbia.

WILKINS, ARNOLD F., IRE Fellow 1958. Born: February 20, 1907, Manchester, Lancashire, England. Degrees: M.Sc., 1928, Manchester University.

Fellow Award "For his contributions to research, to short wave direction finding, and to the early development of radar." Other Awards: Radio Section Premium, Institution of Electrical Engineers, U. K., 1932; Officer of the Order of the British Empire, 1942; Leslie McMi-

chael Premium, British Institution of Radio Engineers, 1957; Fellow, Institution of Electrical Engineers.

WILKINSON, ROGER I., Fellow 1968. Born: March 18, 1903, Mason City, Iowa. Degrees: B.S.E.E., 1924, E.E., 1950, Iowa State University.

Fellow Award "For contributions to the application of probability and statistics in the engineering of communication systems." Other Awards: Medal for Merit, 1946; Eta Kappa Nu, honored as founder of the Award for the Recognition of Outstanding Young Electrical Engineers, 1962; Iowa State Alumni Distinguished Service Award, 1973; Honorary Member, International Advisory Committee, International Teletraffic Congress.

WILLCUTT, FREDERICK W., AIEE Fellow 1948. Born: January 9, 1906, Lawrence, Mass. Degrees: S.B., 1927, S.M., 1928, Massachusetts Institute of Technology.

WILLEMS, JAN C., Fellow 1980. Born: September 18, 1939, Brugge, Belgium. Degrees: E.E., 1963, University of Gent, Belgium; M.Sc.(E.E.), 1965, University of Rhode Island; Ph.D.(E.E.), 1968, Massachusetts Institute of Technology.

Fellow Award "For contributions to the theory of dynamical systems"; Associate Editor, IEEE Transactions on Automatic Control, 1971-1973. Other Awards: Fulbright Fellow, 1963-1965; Member, Society for Industrial and Applied Mathematics; Member Dutch General Systems Society; Member, Dutch Wiskundig Genootschap; Associate Editor, SIAM T. on Control and Optimization, RAIRO Automatique; Managing Editor, Systems and Control Letters.

WILLENBROCK, F. KARL, IRE Fellow 1962. Born: July 19, 1920, New York, N.Y. Degrees: Sc.B., 1942, Brown University; A.M., 1947, Ph.D., 1950, Harvard University.

Fellow Award "For contributions to electrical engineering education"; IEEE Centennial Medal, 1984. Other Awards: Distinguished Engineering Service Award, Brown University, 1962; Member, National Academy of Engineering, 1975; Gold Medal, Dept. of Commerce, 1975; Fellow, American Association for the Advancement of Science, 1978; Cecil H. Green Professor of Engineering, Southern Methodist University.

WILLHEIM, RAOUL, Fellow 1969. Degrees: D.Sc.tech., 1921, Vienna.

Fellow Award "For extensive contributions, major achievements, and leadership in all areas of high-capacity and high-voltage power engineering."

WILLIAMS, BROWN F, Fellow 1983. Born: December 22, 1940, Evanston, Illinois. Degrees: B.A.(Math and Physics), 1962, M.A.(Physics), 1963, Ph.D.(Physics), 1966, University of California at Riverside.

Fellow Award "For technical contributions and innovative leadership in research and development of electron devices." Other Awards: RCA Laboratories Outstanding Achievement Award, 1967; RCA Laboratories David Sarnoff Award for Outstanding Technical Achievement, 1970.

WILLIAMS, CARL H., Fellow 1965. Born: October 22, 1915, Mansfield, Ga. Degrees: B.S., 1939, Georgia Institute of Technology.

Fellow Award "For the development of low-cost power distribution systems." Other Awards: Engineer of the Year, Hawaii Society of Professional Engineers (NSPE), 1972.

WILLIAMS, CHARLES E., IRE Fellow 1952. Born: March 15, 1896; Seattle, WA.

Fellow Award "For his contributions to electronic and radio engineering, and for planning and development of Naval radio installations." Other Awards: Meritorious Civilian Service Award, U.S. Navy, 1947.

WILLIAMS, CHARLES W., Fellow 1982. Born: February 18, 1931, Palestine, AR. Degrees: B.S.E.E., 1959, M.S., 1963, University of Tennessee.

Fellow Award "For contributions to the development and application of nuclear instrumentation"; Administrative Committee, IEEE Nuclear Plasma Science Society, 1976-1979; Secretary, Ad Com, 1977; Vice President, IEEE Nuclear Plasma Science Society, 1978. Other Awards: Registered Professional Engineer; National Society of Professional Engineers; Tau Beta Pi; Eta Kappa Nu.

WILLIAMS, FREDERICK C., Fellow 1980. Born: March 9, 1927, Monticello, NY. Degrees: A.B.(Math. and Physics), 1950, University of California at Berkeley; M.S.E.E., 1963, University of California at Los Angeles.

Fellow Award "For contributions to the development and application of pulse Doppler and imaging radars." Other Awards: L.A. Hyland Award; Associate Fellow, American Institute of Aeronautics and Astronautics.

WILLIAMS, LOYD T., AIEE Fellow 1961. Born: January 22, 1901, Flagstaff, Ariz. Degrees: B.S., 1925, Texas A and M University.

Fellow Award "For contributions to the design and development of oil field electric supply systems."

WILLIAMS, ROBERT C. G., AIEE Fellow 1948. Born: December 28, 1907, London, England. Degrees: B.Sc., 1929, D.I.C., 1930, Ph.D., 1931, University of London.

Other Awards: Clothworkers Scholar, 1926; Royal Scholar, 1926; Henrici Medal, 1928; Siemens Medal, 1928; Unwin Scholar, 1929; Life Fellow, Royal Television Society, 1948; Fellow, Institution of Electrical Engineers, 1966; Fellow, City and Guilds of London Institute, 1967; Fellow, IMechE, 1968; Officer of the Order of the British Empire, 1969; Fellow, Institution of Electronic and Radio Engineers, 1971; Fellow, InstP, 1971; Honorary Fellow, Institution of Electrical and El-

ectronics Incorporated Engineers, 1976.

WILLIAMSON, RAYMOND H., IRE Fellow 1951. Born: April 6, 1907, Eagle Grove, Iowa. Degrees: B.S., 1928, Iowa State University; M.S., 1935, Union College.

Fellow Award "For his accomplishments in the design of very high power radio transmitters and for his contributions to transmitter industrial standards." Other Awards: Southwestern Regional Community Service Award, Elfun Society, General Electric, 1965; U.S. Assay Commission, 1968.

WILLIAMSON, RICHARD CARDINAL, Fellow 1982. Born: September 10, 1939. Degrees: B.S.(Physics), 1961, Ph.D.(Physics), 1966, Massachusetts Institute of Technology.

Fellow Award "For contributing reflective grating devices to the field of surface-acoustic-wave filters."

WILLSON, ALAN N., JR., Fellow 1978. Born: October 16, 1939, Baltimore, Md. Degrees: B.E.E., 1961, Georgia Institute of Technology; M.S.E.E., 1965, Ph.D., 1967, Syracuse University.

Fellow Award "For contributions to circuit and system theory in the area of nonlinear circuits"; Guillemin-Cauer Best Paper Award, IEEE Circuits and Systems Society, 1978. Other Awards: Distinguished Faculty Member Award, UCLA Engineering Alumni Association, 1982; George Westinghouse Award, American Society for Engineering Education, 1982.

WILLYOUNG, DAVID M., Fellow 1978. Born: May 7, 1924, Ridgewood, N.J. Degrees: Mechanical Engineer, 1945, Stevens Institute of Technology; M.S. M.E., 1967, Union College.

Fellow Award "For contributions to the design of large steam turbine-driven generators." Other Awards: Fellow, American Society of Mechanical Engineers, 1978; Power Systems Sector Engineering Award, General Electric Company, 1981.

WILMOTTE, RAYMOND M., IRE Fellow 1938. Born: August 13, 1901, Paris, France. Degrees: B.A., 1921, M.A., 1923, Sc.D., 1958, University of Cambridge.

Awards: Antenna Research Award, 1929; Premium, Institution of Electrical Engineers, U. K., 1930; Bureau of Ordnance Development Award, 1945.

WILSON, DELANO D., Fellow 1979. Born: April 15, 1934, Great Falls, Mont. Degrees: B.S.E.E., 1959, Montana State University.

Fellow Award "For contributions to the development of compact transmission-line technology in the United States."

WILSON, GERALD L., Fellow 1977. Born: April 29, 1939, Springfield, Mass. Degrees: B.S.E.E., 1961, M.S.E.E., 1963, Sc.D.(M.E.), 1965, Massachusetts Institute of Technology.

Fellow Award "For contributions to electric power engineering education and to the understanding of arc and noise phenomena in power systems"; Hickernell

Award, 1971, "For inspiring Roger Chang to present the 1970 Hickernell Prize Winning Paper in Electrical Power Engineering."

WILSON, MYRON S., AIEE Fellow 1958. Born: July 21, 1899, Spencer, Mass. Degrees: Dipl., Lowell Technological Institute; Massachusetts Institute of Technology.

Fellow Award "For contributions to the design of electrical measuring instruments especially by the improvement in accuracy of current transformers."

WILSON, T. LAMONT, Fellow 1977. Born: June 4, 1914, Salt Lake City, Utah. Degrees: B.S.E.E., 1940, University of Utah; M.S.(Physics), 1963, University of Louisville.

Fellow Award "For leadership in the development of high-powered dielectric heating equipment"; Outstanding Engineer Region III, IEEE, 1977; Achievement Award, IEEE Industrial Electronics and Control Instrumentation Society, 1979; IEEE Centennial Medal, 1984. Other Awards: Eminent Engineer, Tau Beta Pi, 1977; U.S. Delegate to CISPR and IWP 1/4 of CCIR.

WILSON, WALTER R., AIEE Fellow 1959. Born: May 3, 1919, South Bend, Ind. Degrees: B.S. (Eng. Physics) 1941, University of Michigan.

Fellow Award "For contributions to development of high voltage switchgear"; Alfred Noble Paper Prize, AIEE, 1944, for "Corona in Aircraft Electric Systems at High Altitude"; IEEE Standards Medallion, 1977. Other Awards: Cordiner Award, General Electric, 1961; Sesquicentennial Award, University of Michigan, 1967.

WILTSE, JAMES C., JR., Fellow 1974. Born: March 16, 1926, Tannersville, N.Y. Degrees: B.E.E., 1947, M.E.E., 1952, Rensselaer Polytechnic Institute; D.Eng., 1959, The Johns Hopkins University.

Fellow Award "For contributions to microwave and millimeter wave technology in the areas of radar, radiometry, and transmission line research."; Engineer of the Year Award, Orlando Section, IEEE, 1968; Outstanding Engineer, IEEE, Region 3, 1975; Popov Society Congress Delegate, 1979; MTT National Lecturer, 1979-80. Other Awards: Author of the Year, Martin Marietta Corporation, 1967; Engineer of the Year, Martin Marietta Corporation, 1970; Listed in: "American Men and Women of Science," 1973, 1979, "Who's Who in the South and Southwest," 1974, "Men of Achievement," 1974, "Dictionary of International Biography," 1974, 1978-79, "Who's Who in Engineering," 1977; Sigma Xi; Tau Beta Pi; Eta Kappa Nu.

WINCHESTER, ROBERT L., Fellow 1976. Born: February 1, 1926, St. Louis, Mo. Degrees: B.S.E.E., 1948, California Institute of Technology.

Fellow Award "For contributions to the design and performance analysis of large steam-turbine generators and to the advancement of their excitation sys-

tems." Other Awards: Tau Beta Pi, 1946.

WINDER, ROBERT O., Fellow 1977. Born: October 9, 1934, Boston, Mass. Degrees: A.B., 1954, University of Chicago; B.S., 1956, University of Michigan; M.A., 1958, Ph.D., 1962, Princeton University.

Fellow Award "For contributions in switching theory, computer structure, and microprocessor design and applications." Other Awards: Fellowship, National Science Foundation, 1956-1957; David Sarnoff Award for Outstanding Technical Achievement, RCA, 1976.

WING, ARTHUR K., JR., IRE Fellow 1957. Born: July 1, 1908, New York, N.Y. Degrees: B.S., 1930, Yale University; M.S., 1931, Massachusetts Institute of Technology.

Fellow Award "For contribution to the advance of vacuum tube techniques."

WING, OMAR, Fellow 1973. Born: March 2, 1928, Detroit, Mich. Degrees: B.S., 1950, University of Tennessee; M.S., 1952, Massachusetts Institute of Technology; Sc.D., 1959, Columbia University.

Fellow Award "For contributions to circuit theory and engineering education." Other Awards: Great Teacher Award, Society of Older Graduates of Columbia; Fulbright Lecturer, Taiwan, 1961; Ford Foundation Engineering Resident(IBM), 1966; Fulbright-Hays Senior Lecturer, The Netherlands, 1979.

WINJE, SEVERT W., AIEE Fellow 1949. Born: May 30, 1896, Lidgerwood, N.Dak. Degrees: B.S.E.E., 1922, University of North Dakota.

Fellow Award "For assistance in development and application of AC network systems of underground electric power distribution"; Past Section Chairman of I.E.E.E., Fort Wayne, Ind. Section. Other Awards: Past President of Fort Wayne Engineers Club.

WINN, OLIVER H., Fellow 1970. Born: August 11, 1920, Detroit, Mich. Degrees: B.S.E.E., 1942, Michigan Technological University; Ph.D.(E.E.), 1976, Syracuse University.

Fellow Award "For contributions and leadership in the development of radar and high-power capacitors."

WINOGRAD, SHMUEL, Fellow 1974. Born: January 4, 1936, Tel-Aviv, Israel. Degrees: B.S.E.E., 1959, M.S.E.E., 1959, Massachusetts Institute of Technology; Ph.D., 1968, New York University.

Fellow Award "For analysis of computational complexity which established bounds on the time required to compute certain mathematical functions"; W. Wallace McDowell Award, IEEE Computer Society, 1974. Other Awards: IBM Fellow, 1972.

WINTER, DAVID F., IRE Fellow 1958. Born: November 9, 1920, St. Louis, Mo. Degrees: B.S.E.E., 1942, Washington University; M.S., 1948, Massachusetts Institute of Technology.

Fellow Award "For his electronic research at ultra-high-frequencies." Other Awards: Tau Beta Pi, 1942;

Sigma Xi, 1948; Eta Kappa Nu, 1958.

WISCHMEYER, CARL R., AIEE Fellow 1951, IRE Fellow 1961. Born: October 2, 1916, Terre Haute, Ind. Degrees: B.S., 1937, Rose Polytechnic Institute; M.E., 1939, Yale University; E.E., 1942; D.Sc.(Hon.), 1970, Rose Polytechnic Institute (Rose-Hulman Institute of Technology.)

IRE Fellow Award "For contributions to engineering education and instrumentation"; Paper Prize, AIEE, Houston Section, 1959. Other Awards: Special Science Education Grant, National Science Foundation, Technische Hogeschool te Eindhoven, Netherlands; Honor Alumnus, Rose-Hulman Institute of Technology.

WISEMAN, ROBERT SWERN, Fellow 1970. Born: February 27, 1924, Robinson, Illinois. Degrees: B.S., 1948, M.S., 1950, Ph.D., 1954, University of Illinois.

Fellow Award "For scientific leadership and research and development in devices for image detection in darkness." Other Awards: Fellow, Illuminating Engineering Society; Honors Bronze Tablet, University of Illinois, 1948; Outstanding and Sustained Performance Awards, Dept. of Army, 1959 through 1981; Meritorious Civilian Award, Dept. of Army, 1965; Research and Development Achievement Award, Dept. of Army, 1965; Exceptional Civilian Service Award, 1968; Dept. of Defense Distinguished Civilian Service Award, 1969; Presidential Senior Executive Service Distinguished Executive Award, 1980; Distinguished E.E. Alumni Award, University of Illinois, 1980.

WISSEMAN, WILLIAM R., Fellow 1984. Born: November 2, 1932, Halletsville, TX. Degrees: Bachelor of Nuclear Engineering, 1954, North Carolina State College; Ph.D.(Physics), 1959, Duke University.

Fellow Award "For technical leadership in the development of gallium arsenide power field-effect transistors and integrated circuits." Other Awards: Phi Eta Sigma, Tau Beta Pi, Phi Kappa Phi, Sigma Xi, Phi Beta Kappa.

WITT, VICTOR R., Fellow 1974. Born: May 17, 1920, New York, N.Y. Degrees: B.S.E.E., 1950, New York University.

Fellow Award "For contributions to magnetic recording on tapes and disks and to secondary storage devices for computers." Other Awards: IBM Invention Achievement Award, 1961; IBM Outstanding Invention Award, 1962; Master Design Award, Product Engineering, 1963, 1965; IBM Fellow, 1970.

WITZKE, RAYMOND L., AIEE Fellow 1951. Born: November 5, 1911, Chester, Iowa. Degrees: B.S.(Elec. Eng.), 1934, M.S., 1936, State University of Iowa.

Other Awards: Westinghouse Order of Merit, 1968.

WOHL, JOSEPH G., Fellow 1984. Born: May 8, 1927, Chicago, IL. Degrees: B.Sc.(Physics-Math), 1949, University of Wisconsin.

Fellow Award "For contributions to the development of predictive models of human performance in decision making." Other Awards: NASA New Technology Award, 1968.

WOHLFARTH, ERICH P., Fellow 1982. Born: December 7, 1924, Gleiwitz, Germany. Degrees: B.Sc., 1945, Ph.D., 1948, D.Sc., 1957, University of Leeds.

Fellow Award "For contributions to the theories of fine-particle ferromagnetism and the properties of electrons in metals"; Distinguished Lecturer Award, IEEE Magnetics Society, 1979; IEEE Centennial Medal, 1984.

WOLF, EDWARD D., Fellow 1977. Born: May 30, 1935, Quinter, Kans. Degrees: B.S.(Chemistry), McPherson College; Ph.D.(Physical Chemistry), Iowa State University.

Fellow Award "For contributions to scanning electron beam diagnostic and microfabrication techniques." Other Awards: Fellow of the American Institute of Chemists, 1971; Listed in: "Who's Who in American Universities and Colleges," 1957, "Outstanding Young Men of America," 1966, "American Men and Women of Science," 1973, "Who's Who in America," 1983; Bohmische Physical Society.

WOLF, HERMAN B., AIEE Fellow 1945. Born: September 10, 1896, Dallas, N.C.

Other Awards: Tau Beta Pi; Robinson Award and Citation, Duke Power Co.; Distinguished Service Citation, Charlotte Engineers Club.

WOLF, JACK K., Fellow 1973. Born: March 14, 1935, Newark, N.J. Degrees: B.S.E.E., 1956, University of Pennsylvania; M.S.E., 1957, M.A., 1958, Ph.D., 1960, Princeton University.

Fellow Award "For contributions to coding theory"; Prize Paper Award (corecipient), IEEE Information Theory Group, 1975. Other Awards: Senior Postdoctoral Fellowship, National Science Foundation, 1971-1972; Guggenheim Fellow, 1979-80.

WOLF, MARTIN, Fellow 1980. Born: August 22, 1922. Degrees: Vordiplomer-Physiker, 1948, Diplom-Physiker, 1952, Georg August University, Germany.

Fellow Award "For contributions to the development of silicon solar cells and their applications in spacecraft power systems and for terrestrial energy supply; Marconi Award, IRE, 1958.

WOLF, WERNER P., Fellow 1984. Born: April 22, 1930, Vienna, Austria. Degrees: B.A.(Physics), 1951, D.Phil.(Physics), M.A., 1954, Oxford University; M.A.(-Hon.), 1965, Yale University.

Fellow Award "For contributions to fundamental understanding of magnetic materials and leadership in engineering education." Other Awards: Scholar, New College, Oxford University, 1948; Fulbright Fellowship, 1956; Fellow, American Physical Society, 1963; Senior Visiting Fellow, Oxford University, 1980, 1984; Member, Connecticut Academy of Science and Engineering, 1982; Humboldt Senior U.S. Scientist Award, 1983.

WOLFE, CHARLES M., Fellow 1978. Born: December 21, 1935, Morgantown, W.Va. Degrees: B.S.E.E., 1961, M.S.E.E., 1962, West Virginia University; Ph.D., 1965, University of Illinois.

Fellow Award "For contributions to the development of high-purity gallium-arsenide for microwave and optical device applications." Other Awards: Electronic Division Award, Electrochemical Society, 1978; Samuel C. Sachs Professor, Washington University, 1982.

WOLFF, EDWARD A., Fellow 1973. Born: October 31, 1929, Chicago, Ill. Degrees: B.S.E.E., 1951, University of Illinois; M.S., 1953, Ph.D., 1961, University of Maryland.

Fellow Award "For contributions to antennas, geoscience instrumentation, and management of scientific projects." Other Awards: Fellow, Washington Academy of Sciences; Phi Eta Sigma; Sigma Tau; Eta Kappa Nu.

WOLFF, HANNS H., Fellow 1972. Born: December 19, 1903, Berlin, Germany. Degrees: Cand.Ing., 1926, Dipl.Ing., 1928, Dr.Ing., 1945, Dr.Ing.habil., 1946, Technical University, Berlin.

Fellow Award "For contributions and leadership in the field of simulation systems for training." Other Awards: Radio Amateur; Adjunct Professor, Polytechnic Institute of Brooklyn.

WOLFF, S. S., AIEE Fellow 1950. Born: October 5, 1900, Chicago, Ill. Degrees: B.S., 1922, University of Illinois.

WOLL, HARRY J., IRE Fellow 1962. Born: August 25, 1920, Farmington, Minn. Degrees: B.S.E.E., 1940, North Dakota State University; Ph.D., 1953, University of Pennsylvania.

Fellow Award "For contributions to tube and transistor circuitry"; IEEE Philadelphia Section Award, 1971. Other Awards: David Sarnoff Employee Fellow, 1947; Gold Medal, Moore School, University of Pennsylvania, 1973; Yarnall Award, Engineering Alumni Society, University of Pennsylvania, 1982.

WOLL, RICHARD F., Fellow 1971. Born: August 15, 1915, Pittsburgh, Pa. Degrees: B.S.E.E., 1937, University of Pittsburgh.

Fellow Award "For contributions to the design and understanding of ac machines." Other Awards: Sigma Tau, 1937.

WOLLASTON, FRANCIS O., AIEE Fellow 1960. Born: March 26, 1903, Victoria, British Columbia, Canada. Degrees: B.S., 1925, E.E., 1931, University of Washington.

Fellow Award "For contributions to design of transmission at extra high voltages; overhead, underground and underwater"; AIEE National Prize (with K.W. Miller), 1932, for "Best Paper in Engineering Practise."

WOLOVICH, WILLIAM A., Fellow 1984.

Fellow Award "For contributions to the algebraic theory for multivariable control systems."

WONG, EUGENE, Fellow 1974. Born: December 24, 1934. Degrees: B.S., 1955, A.M., 1958, Ph.D., 1959, Princeton University.

Fellow Award "For contributions to the theory of random processes and its engineering applications, and to engineering education." Other Awards: John Simon Guggenheim Fellow, 1968.

WONHAM, W. MURRAY, Fellow 1977 Born: November 1, 1934. Degrees: B.Eng., 1956, McGill University; Ph.D., 1961, University of Cambridge.

Fellow Award "For contributions to multivariable control system theory and design." Other Awards: National Academy of Sciences Senior Postdoctoral Resident Research Associateship, held at NASA, 1967-1969.

WOOD, ALAN B., Fellow 1980. Born: March 4, 1920, Middlesbrough, United Kingdom. Degrees: B.S. (First Class Honours), University of Durham.

Fellow Award "For leadership in the design of high-voltage transmission lines." Other Awards: Past Chairman, CIGRE Study Committee 22; Fellow, Institution of Electrical Engineers.

WOOD, ALLEN J., Fellow 1974. Born: October 1, 1925. Degrees: B.E.E., 1949, Marquette University; M.S.E.E., 1951, Illinois Institute of Technology; Ph.D., 1959, Rensselaer Polytechnic Institute.

Fellow Award "For contributions to engineering and economic analyses of large scale electric power systems"; First Prize Paper, AIEE, District 1, 1960; Vice Chairman, Power Eng. Education Committee, IEEE PES, 1982-1983.

WOOD, HARRIS O., Fellow 1967. Born: October 14, 1918, Canada. Degrees: B.S.E.E., 1942, Michigan State University.

Fellow Award "For leadership in the standardization of all channel television receivers by industry and government and in the development of commercial television receivers." Other Awards: Engineering Department Outstanding Accomplishment Award, Electronic Industries Association, 1971; Consumer Electronics Engineering Panel, Electronic Industries Association, 1967; Consumer Electronics Group, Electronic Industries Association, 1971.

WOODBURY, ERIC J., Fellow 1973. Born: February 9, 1925, Washington, D.C. Degrees: B.S., 1947, Ph.D., 1951, California Institute of Technology.

Fellow Award "For contributions to simulated Raman Scattering and laser engineering."

WOODROW, CHARLES A., AIEE Fellow 1953. Born: April 26, 1903, Des Moines, Iowa. Degrees: B.S., 1926, University of Caifornia at Berkeley.

Fellow Award "For contributions in the electrical

utility fields of transmission, power system analysis, and standardization of ratings of large power circuit breakers."

WOODSON, HERBERT H., Fellow 1970. Born: April 5, 1925, Stamford, Tex. Degrees: S.B., 1952, S.M., 1952, Sc.D., 1956, Massachusetts Institute of Technology.

Fellow Award "For contributions to teaching and research in the areas of energy conversion, electric machinery, and power-systems technology"; Nikola Tesla Award, IEEE, 1984. Other Awards: Member: American Society for Engineering Education; American Association for the Advancement of Science; National Academy of Engineering; Eta Kappa Nu; Tau Beta Pi; Sigma Xi; Phi Kappa Phi; Edison Electric Institute Power Engineering Educator Award, 1978; Fellow, American Association for the Advancement of Science.

WOODWARD, J. GUY, Fellow 1970. Born: November 19, 1914, Carleton, Mich. Degrees: B.A., 1936, North Central College; M.S., 1939, Michigan State College; Ph.D., 1942, Ohio State University.

Fellow Award "For contributions in magnetic tape and disk recording." Other Awards: Emile Berliner Award, Audio Engineering Society, 1963; Honorary Membership, Audio Engineering Society, 1973.

WOOLDRIDGE, DEAN E., IRE Fellow 1954. Born: May 30, 1913, Chickasha, Okla. Degrees: A.B., 1932, M.S., 1933, University of Oklahoma; Ph.D., 1936, California Institute of Technology.

Fellow Award "For his contributions to physics and electronics research, and his leadership in development efforts for national defense." Other Awards: Citation of Honor, Air Force Association, 1955; Distinguished Service Citation, University of Oklahoma, 1960; Westinghouse A.A.A.S. Award for Science Writing, 1963; Fellow, American Physical Society; Fellow, American Institute of Aeronautics and Astronautics; Fellow, American Academy of Arts and Sciences; Member, National Academy of Sciences, National Academy of Engineering.

WOOLEY, BRUCE A., Fellow 1982. Born: October 14, 1943, Milwaukee, WI. Degrees: B.S., 1966, M.S., 1968, Ph.D., 1970, University of California at Berkeley.

Fellow Award "For contributions to the design of integrated circuits for communication systems"; IEEE Fortescue Fellow, 1966. Other Awards: University Medalist, University of California, Berkeley, 1966.

WOZENCRAFT, JOHN M., Fellow 1965. Born: September 30, 1925, Dallas, Tex. Degrees: B.S., 1946, United States Military Academy; S.M., 1951, Sc.D., 1957, Massachusetts Institute of Technology.

Fellow Award "For contributions to coding theory and practice."

WRAY, JAMES Q., JR., AIEE Fellow 1961. Born: June 10, 1904, York, S.C. Degrees: B.S.E.E., 1926, Clemson University.

Fellow Award "For contributions to the design of economical electric power generating plants."

WRIGHT, JAY W., IRE Fellow 1958. Born: November 1909, Salt Lake City, Utah. Degrees: A.B., 1935, M.Sc., 1936, University of Utah.

Fellow Award "For his contributions to electronic devices for the Armed Services." Other Awards: Certificate of Appreciation, U.S. Dept. of Navy, 1948.

WRIGHT, SHERWIN H., AIEE Fellow 1950. Born: October 31, 1906, San Francisco, Calif. Degrees: B.S., 1927, University of California at Berkeley; M.S., 1940, University of Pittsburgh.

Other Awards: National Society of Professional Engineers; Member, American Society of Mechanical Engineers, 1958; Institute of Engineers, Australia, 1958.

WYLIE, CHARLES J., Fellow 1972. Born: 1926, N.C. Degrees: B.S.E.E., 1950, University of South Carolina.

Fellow Award "For contributions to the design of economical power generation and to formulation of nuclear standards for the electric power industry"; Nominee, Outstanding Engineer Award, IEEE, Region III, 1971; Recognition Award, Charlotte Section, IEEE, 1971; Outstanding Service Award, Power Generation Committee, IEEE Power Engineering Society, 1978. Other Awards: Outstanding Service Award, Charlotte Engineers Club, 1978; Tau Beta Pi.

WYMAN, BRYCE W., Fellow 1970. Born: 1915, Nebr. Degrees: B.S.E.E., 1937, University of Nebraska.

Fellow Award "For leadership in the development of land transportation including higher powered diesel-electric locomotives, commercial frequency electrification, and high speed rapid transit."

WYNDRUM, RALPH W., JR., Fellow 1975. Born: April 20, 1937, New York, N.Y. Degrees: B.S., 1959, M.S., 1960, Columbia University; Sc.D., 1963, New York University; M.S., (Bus. Ad.), 1978, Columbia University.

Fellow Award "For contributions to hybrid integrated circuit development and application"; Best Paper Award, Eascon, 1962, 1963; Best Paper Award, ECC, 1968; Chairman, Globecom/ICC Conference Board; Secretary, IEEE Communications Society Board of Governors. Other Awards: Outstanding Young Electrical Engineer Award, Honorable Mention, Eta Kappa Nu, 1968.

WYNER, AARON D., Fellow 1975. Born: March 17, 1939, New York, N.Y. Degrees: B.S., 1960, Queens College; B.S.E.E., 1960, M.S., 1961, Ph.D., 1963, Columbia University.

Fellow Award "For basic contributions to information theory"; Prize Paper Award, IEEE Information Theory Group, 1977. Other Awards: Guggenheim Fellowship,

1966.

X

XENIS, CONSTANTINE P., AIEE Fellow 1949. Born: Aidin, Turkey.

Other Awards: Distribution Achievement Award, American Gas Association, 1960.

Y

YAMAMOTO, MITSUYOSHI, Fellow 1982. Born: January 17, 1923, Japan. Degrees: B.Eng., 1946, D.Eng., 1961, Tokyo University.

Fellow Award "For contributions to advanced electrical devices for energy generation and transmission." Other Awards: Achievement Award, Institution of Electrical Engineers, Japan, 1964, 1975.

YAMAMURA, SAKAE, Fellow 1974. Born: February 26, 1918, Tokyo, Japan. Degrees: B.E., 1941, University of Tokyo; M.S., 1951, Michigan State University; Ph.D., 1953, Ohio State University; Dr.E., 1955, University of Tokyo.

Fellow Award "For contributions to the theory of high-speed linear induction motors and development of the numerical control technique"; Nikola Tasla Award, IEEE, 1982. Other Awards: Technical Paper Award, Institute of Electrical Engineers of Japan, 1958; Technology Advancement Award, Institute of Electrical Engineers of Japan, 1968; Electric Power Technology Award, Institute of Electrical Engineers of Japan, 1970; Technical Book Award, Institute of Electrical Engineers of Japan, 1974; Award of Honor, Institute of Electrical Engineers of Japan, 1975; Blue Ribbon Award, Government of Japan, 1978; Japan Academy Award, 1983; Professor Emeritus, Faculty of Engr., Bunkyo-ku, Tokyo, Japan.

YAMANAKA, CHIYOE C., Fellow 1983. Born: December 14, 1923, Osaka, Japan. Degrees: B.Eng., 1948, Ph.D.(Eng.), 1960, Osaka University.

Fellow Award "For contributions to high power laser systems for material processing and nuclear fusion and for leadership in education." Other Awards: Award, Toray Science Foundation, 1968; Award, IEE of Japan, 1972.

YAMASHITA, EIKICHI, Fellow 1984. Born: February 4, 1933, Tokyo, Japan. Degrees: B.S.E.E., 1956, University of Electro-Communications; M.S.E.E., 1963, Ph.D., 1966, University of Illinois.

Fellow Award "For contributions to the analysis and design of microstrip networks."

YANAI, HISAYOSHI, Fellow 1977. Born: May 19, 1920, Okayama, Japan, Degrees: B.Eng., 1942, D.Eng., 1953, University of Tokyo.

Fellow Award "For contributions to research and development in the fields of semiconductor devices and microwave technology, and to engineering education."

Other Awards: Excellent Paper Awards, 1957, 1971, Distinguished Services Award, 1970, Excellent Book Award, 1982, Institute of Electronics and Communication Engineers of Japan; Minister's Award for Standardization, MITI, Japan, 1973; Award for Research and Invention, Governor of Tokyo Metropolis, 1974; International GaAs Symposium Award (Heinrich Welker Medal), 1980.

YANG, RICHARD F. H., Fellow 1967. Born: November 5, 1917, Nanking, China. Degrees: B.S., 1942, National Wu-Han University; M.S., 1948, Ph.D., 1951, University of Illinois.

Fellow Award "For his contributions to antenna design, industry-government frequency utilization, and engineering education"; Annual Paper Award, IRE, Professional Group on Vehicular Communications, 1960.

YARIV, AMNON, Fellow 1970. Born: April 13, 1930, Tel-Aviv, Israel. Degrees: B.S., 1954, M.S., 1956, Ph.D., 1958, University of California at Berkeley.

Fellow Award "For basic contributions to research and education in quantum electronics and solid-state devices"; Quantum Electronics Award, IEEE, 1980. Other Awards: Fellow, Optical Society of America; Member, National Academy of Engineering; American Academy of Arts and Sciences.

YAU, STEPHEN S., Fellow 1973. Born: August 6, 1935, Wusei, Kiangsu, China. Degrees: B.S.E.E., 1958, National Taiwan University; M.S.E.E., 1959, Ph.D.(E.E.), 1961, University of Illinois at Urbana.

Fellow Award "For contributions to switching theory, the reliability of computing systems, and engineering education"; First Richard E. Merwin Award, IEEE Computer Society, 1981. Other Awards: Louis E. Levy Medal, Franklin Institute, 1963; Golden Plate Award, American Academy of Achievement, 1964; Life Fellow, Franklin Institute, 1971; Fellow, American Association for the Advancement of Science, 1983; Eta Kappa Nu; Sigma Xi; Tau Beta Pi; Pi Mu Epsilon.

YEE, SINCLAIR S., Fellow 1980. Born: January 20, 1937, China. Degrees: B.S., 1959, M.S., 1961, Ph.D., 1965, University of California at Berkeley.

Fellow Award "For contributions to the development and application of biotransducers using hybrid technology." Other Awards: Special Fellow, National Institutes of Health, 1972-1974; Listed in "Who's Who in America."

YEH, KUNG C., Fellow 1973. Born: August 4, 1930, Hangchow, China. Degrees: B.S., 1953, University of Illinois; M.S., 1954, Ph.D., 1958, Stanford University.

Felllow Award "For contributions to theory and observation in ionospheric research and to engineering education"; Certificate of Achievement Award for the paper, "Displacement of Rays in a Turbulent Medium," IEEE TRANSACTIONS on Antenna and Propagation,

1968.

YEH, RAYMOND T., Fellow 1983. Born: Nov. 5, 1937, Hunan, China. Degrees: B.S.E.E., M.A.(Math), Ph.D.(Math).

Fellow Award "For research and leadership in software engineering"; IEEE Centennial Medal, 1984. Other Awards: Special Award, IEEE Computer Society, Founding Editor of IEEE Transactions on Software Engineering.

YEH, YU-SHUAN, Fellow 1984. Born: September 9, 1939, Kiangsu, China. Degrees: B.S.E.E., 1961, Taiwan University, China; M.S.E.E., 1964; Ph.D., 1966, U.C. Berkeley.

Fellow Award "For contributions to advanced communication satellites and high-capacity mobile radio systems." Other Awards: Best Paper Awards, IEEE Transaction on Antennas & Propagation, 1968; IEEE Transactions on Vehicular Technology, 1974.

YERGER, LLOYD K., AIEE Fellow 1950. Born: March 17, 1902, Souderton, Pa.

YETTER, JOHN W., Fellow 1968. Born: September 7, 1917, Canandaigua, N.Y. Degrees: E.E., 1939, Cornell University.

Fellow Award "For his contributions to the principles, design, and economic application of electrical equipment to power systems and for investigations and technological studies leading to the use of EHV transmission."

YOKELSON, BERNARD J., Fellow 1976. Born: September 14, 1924, Brooklyn, N.Y. Degrees: B.S.E.E., 1948, Columbia University; M.E.E., 1954, Polytechnic Institute of Brooklyn.

Fellow Award "For contributions to the development of telephone electronic switching systems and operator service systems."

YONEZAWA, SHIGERU, Fellow 1969. Born: Feburary 1, 1911, Toyama Prefecture, Japan. Degrees: B.S.E.E., 1933, D.Eng., 1942, Tokyo University, Japan.

Fellow Award "For outstanding contribution to the development of very-high-frequency microwave multichannel radio relay systems and leadership to the rapid growth of the telecommunication industry in Japan"; IEEE Founders Medal Award, 1982. Other Awards: 13th Mainichi Industrial Engineering Prize, Mainichi, 1961; Merit Prize, Institute of Electrical Communication Engineers of Japan, 1962; Merit Prize, Ministry of Post and Telecommunications, 1977; Testimonial, Nippon Telegraph and Telephone Public Corporation, 1978; The Special Maejima Prize, Association of Communications, 1979; First order of merit decorated in recognition of Communications from the Emperor, 1982.

YORK, RAYMOND A., Fellow 1966. Born: May 15, 1917, Baldwin, Kans. Degrees: B.S.E.E., 1941, University of Kansas.

Fellow Award "For contributions to power semiconductor device development and standardization."

YOSHIDA, SUSUMU, Fellow 1981. Born: May 30, 1923, Toyama Prefecture, Japan. Degrees: B.E.E., 1945, Dr.(E.E.), 1978, Tohoku University.

Fellow Award "For invention of the Trinitron tube and the development of CRT production techniques"; Outstanding Paper Awards, Chicago Spring Conference, BTR, 1968, 1973, 1974. Other Awards: 11th Niwa and Takayanagi Award, Institute of Television Engineers, Japan, 1971; Award for Contribution to Science and Technology, Science and Technology Agency, Japan, 1973; Okochi Memorial Prize, 1973; Medal of Honor with Purple Ribbon, The Emperor of Japan, 1973.

YOST, ARNOLD G., Fellow 1975. Born: January 10, 1927, Toledo, Ohio. Degrees: B.S.E.E., 1949, Purdue University.

Fellow Award "For contributions in the field of surge arrester design and standardization in the electric power industry." Other Awards: Eta Kappa Nu, 1947; Tau Beta Pi, 1948.

YOULA, DANTE C., Fellow 1966. Born: October 17, 1925, Brooklyn, N.Y. Degrees: B.E.E., 1947, City College of the City University of New York; M.S., 1950, New York University.

Fellow Award "For contributions to network synthesis techniques"; W.R.G. Baker Prize Award, 1965; Guillemin-Cauer Circuit Award, 1973. Other Awards: Air Force Systems Command Award, 1965; Special Award, School of Engineering, City College of the City University of New York, 1969.

YOUNG, CHARLES J., IRE Fellow 1953. Born: December 17, 1899, Cambridge, Mass. Degrees: B.A., 1921, Harvard University.

Fellow Award "In recognition of his many contributions in the fields of facsimile and electronic timing equipment." Other Awards: Modern Pioneer Award, National Association of Manufacturers, 1946; Sigma Xi.

YOUNG, CLIFFORD C., JR., Fellow 1981. Born: September 19, 1926, Houston, TX. Degrees: B.S.E.E., 1949, University of Texas at Austin; M.S.E., 1971, Union College.

Fellow Award "For the development and application of digital computer techniques for the dynamic analysis of power systems."

YOUNG, FRANK S., Fellow 1977. Born: April 26, 1933, Salt Lake City, Utah. Degrees: B.S.E.E., 1955, Stanford University; M.S.E.E., 1962, University of Pittsburgh.

Fellow Award "For contributions to the engineering and management of laboratory facilities for underground transmission technology"; First Prize Paper Award, IEEE Power Group, 1963, for "Shielding of

Transmission Lines." Other Awards: Tau Beta Pi.

YOUNG, FREDERIC C., AIEE Fellow 1945. Born: January 4, 1899, N.Y. Degrees: 1922, Rensselaer Polytechnic Institute.

Other Awards: American Society for Testing Materials.

YOUNG, JOHN A. I., Fellow 1983. Born: November 12, 1915, Creston, British Columbia, Canada. Degrees: B.Sc.(Mathematics & Physics), 1950, University of Western Ontario.

Fellow Award "For contributions to industrial power conversion systems." Other Awards: Engineering Medal, Association of Professional Engineers of Ontario, 1980; General Electric Steinmetz Award, 1981.

YOUNG, LAURENCE RETMAN, Fellow 1979. Born: December 19, 1935. Degrees: A.B.(Physics), 1957, Amherst College; S.B.(E.E.), 1957, S.M.(E.E.), 1959, Sc.D.(Instrumentation), 1962, Massachusetts Institute of Technology; Cert. de License (Math), 1958, Sorbonne.

Fellow Award "For contributions to biomedical instrumentation and biomedical engineering education"; Franklin Taylor Award, IEEE, 1965. Other Awards: Member, National Academy of Engineering, 1980; Fellow, Explorers Club, 1980; Dryden Lecturer, American Institute of Aeronautics and Astronautics, 1982; Member, U.S. Air Force Scientific Advisory Board, 1979; NASA Life Sciences Advisory Board, 1980; National Research Council-Aerospace Engineering Board, Exec. Council of Committee on Hearing, Audition and Biomechanics, Air Force Study Board; Principal Investigator, Spacelabs 1,4&D-1.

YOUNG, LEO, Fellow 1968. Born: August 18, 1926, Vienna, Austria. Degrees: B.A.(Math.), 1945, B.A.(Physics), 1947, M.A, 1950, Cambridge University; M.S.E.E., 1956, D.Eng., 1959, The Johns Hopkins University.

Fellow Award "For his contributions in the field of microwaves"; Microwave Prize, IEEE, 1963; National Lecturer, IEEE, 1968; Citation of Honor, United States Activities Board, 1978; Distinguished Service Award, IEEE Microwave Theory and Techniques Society, 1979; Certificate of Recognition, IEEE Electromagnetic Compatibility Society, 1980; President, IEEE, 1980; IEEE Centennial Medal, 1984. Other Awards: Major Open Scholarship, Cambridge University, 1943; Sigma Xi, 1956; Benjamin Garver Lamme Graduate Scholarship, Westinghouse Electric Corp., 1958; Fellow, American Association for the Advancement of Science, 1981; Life Member, IEEE Microwave Theory and Techniques Society, 1982.

YOUNG, W. RAE, Fellow 1964. Born: October, 30, 1915, Lawton, Mich. Degrees: B.S.(E.E.), 1937, University of Michigan.

Fellow Award "For contributions to mobile radio and

data communication systems"; Technical Program Chairman, VTS/79. Other Awards: Professional Engineer, State of New Jersey.

YOVITS, MARSHALL C., Fellow 1983. Born: May 16, 1923, New York, New York. Degrees: B.S.(Physics), 1944, M.S., 1948, Union College, Schenectady, N.Y.; M.S., 1950, Ph.D., 1951, Yale University.

Fellow Award "For leadership in the field of computer science and education." Other Awards: AEC Predoctoral Fellowship, 1950; Navy Outstanding Performance Award, 1961; Navy Superior Civilian Service Award, 1964; Fellow, American Association for the Advancement of Science, 1982.

YU, YAO-NAN, Fellow 1978. Born: October 25, 1909. Degrees: B.S., 1936, Dr. Sc.Engg., 1962, Tokyo Institute of Technology.

Fellow Award "For contribution to the development of analysis and testing techniques applied to stability in large electric power systems."

Z

ZABORSZKY, JOHN, AIEE Fellow 1962. Born: May 13, 1914, Budapest, Hungary. Degrees: Dipl.Eng., 1937, D.Sc., 1942, Royal Hungarian Technical University, Budapest.

Fellow Award "For teaching and research in the field of automatic control."

ZACHARIAS, JERROLD R., Fellow 1964. Born: January 23, 1905. Degrees: A.B., 1926, M.A., 1927, Ph.D., 1932, Columbia University; D.(Hum.)(Hon.), 1964, Tufts University; D.Sc.(Hon.), 1964, Oklahoma City University; D.Sc.(Hon.), 1965, St. Lawrence University; D.Sc.(Hon.), 1967, Western Reserve University; L.L.D.(Hon.), 1968, Jacksonville University.

Fellow Award "For contributions to defense systems, atomic frequency standards, and education." Other Awards: President's Certificate of Merit, 1948; Certificate of Appreciation, Dept. of Defense, 1955; Oersted Medal, American Association of Physics Teachers, 1961; National Science Teachers Citation, 1969; Fellow, American Association for the Advancement of Science; National Academy of Sciences; American Academy of Arts and Sciences; American Physical Society; American Association of Physics Teachers.

ZADEH, LOTFI A., IRE Fellow 1958. Born: February 4, 1921, Baku, Russia. Degrees: B.S., 1942, University of Teheran; M.S., 1946, Massachusetts Institute of Technology; Ph.D., 1949, Columbia University.

Fellow Award "For his contributions to theory and teaching of time-varying networks and filters"; IEEE Education Medal, 1973. Other Awards: Guggenheim Fellow, 1967-1968; National Academy of Engineering, 1973; Fellow, American Association for the Advancement of Science, 1980.

ZAFFANELLA, LUCIANO E., Fellow 1981. Born: November 22, 1937. Degrees: Dr.(E.E.), 1960, Politecnico of Milan, Italy.

Fellow Award "For contributions to high-voltage transmission research and development."

ZAININGER, KARL H., Fellow 1975. Born August 3, 1929, Endorf, Bavaria. Degrees: B.S.E.E.(magna cum laude) 1959, CCNY; M.S.E., 1969, M.A., 1962, Ph.D., 1964, Princeton University.

Fellow Award "For contributions to metal-oxide-semiconductor interface and radiation damage theory and technology." Other Awards: David Sarnoff Fellow; two RCA Laboratories Achievement Awards; Eta Kappa Nu; Tau Beta Pi; Department of the Army commendation for contributions to "... the development of the DoD and Army Very High Speed Integrated Circuit (VHSIC) Program", 1980; IEEE 'Outstanding Contributions to the ICCD, 1983.'

ZAKAI, MOSHE, Fellow 1974. Born: 1926, Poland. Degrees: B.S.E.E., 1951, Dipl.Ing., 1952, Technion, Israel; Ph.D., 1958, University of Illinois.

Fellow Award "For contributions in the field of statistical communication theory."

ZAMES, GEORGE, Fellow 1979. Born: January, 7, 1934, Poland. Degrees: B.Eng., 1954, McGill University; Sc.D., 1960, Massachusetts Institute of Technology.

Fellow Award "For contributions to the stability theory of nonlinear feedback systems"; Outstanding Paper Award, IEEE Transactions on Automatic Control, 1978 (Hon. Mention), 1980, 1982. Other Awards: Athlone Fellowship, 1954-56; British Association Medal (McGill), 1954; Guggenheim Fellowship, 1966; R.R. Associateship, National Academy of Sciences, 1967; Best Paper Award, American Automatic Control Council, 1968; Classic Paper Selection, Institute for Scientific Information, 1981.

ZAREM, ABE M., AIEE Fellow 1950, IRE Fellow 1961. Born: March 7, 1917, Chicago, Ill. Degrees: B.S.E.E., 1939, Illinois Institute of Technology; M.S., 1940, Ph.D., 1943, California Institute of Technology; LL.D.(Hon.), 1967, University of California; LL.D.(Hon.), 1968, Illinois Institute of Technology.

IRE Fellow Award "For contributions in the application of millimicrosecond electronic instrumentation techniques." Other Awards: Eta Kappa Nu, 1937; Outstanding Young Electrical Engineer Award, Eta Kappa Nu, 1948; America's Ten Outstanding Young Men Award, Junior Chamber of Commerce, 1950; Fellow, American Institute of Aeronautics and Astronautics, 1968; Albert F. Sperry Medal Award, Instrument Society of America, 1969; Illinois Institute of Technology Hall of Fame, 1982.

ZEBROWITZ, STANLEY, Fellow 1974. Born: November 28, 1927, New York, N.Y. Degrees: B.E.E., 1949, City College of the City University of New York; M.S.E.E., 1954, University of Pennsylvania; M.B.A., 1981, Temple University.

Fellow Award "For contributions to the introduction of microwave systems in developing nations."

ZEGERS, LEO E., Fellow 1979. Born: July 11, 1935, Kerkrade, The Netherlands. Degrees: Master's Degree (Electrical Engineering), 1959, Technical University, Delft; Doctor's Degree (Electrical Engineering), 1972, Technical University, Twente.

Fellow Award "For contributions to error correction, jitter reduction, and synchronization in data transmission."

ZEMANEK, HEINZ, Fellow 1970. Born: January 1, 1920, Vienna, Austria. Degrees: Dipl.Ing., 1944, Dr.Ing., 1951, University of Technology, Vienna; Honorary doctorate, Johannes Kepler University, Linz, 1982.

Fellow Award "For contributions to the theory of programming languages, particularly the formal descriptions of syntax and semantics, and for work in computer language design." Other Awards: Prize of the Nachrichtentechnische Gesellschaft im VDE, 1960; Goldene Stefan Ehrenmedaille of Oesterreichischer Verband fuer Elektrotechnik, 1969; Fellow, British Computer Society; Wilhelm Exner-Medaille, 1972; Honorary Member, Computer Society of South Africa, 1972; Grosses Ehrenzeichen fuer Verdienste um die Republik Oesterreich, 1974; Honorary Member, Information Processing Society of Japan, 1975; IBM Fellow, 1976; Silver-Core and Honorary Member, International Federation for Information Processing, 1977; Prechtl-Medaille, University of Technology, Vienna, 1978; Corresponding Member, Austrian Academy of Sciences, Vienna, 1979; Merit Medal, Bulgarian Academy of Sciences, 1980.

ZEMANIAN, ARMEN H., Fellow 1970. Born: April 16, 1925, Bridgewater, Mass. Degrees: B.E.E., 1947, City College of the City University of New York; M.E., 1949, Sc.D., 1953, New York University.

Fellow Award "For the application of generalized functions and integral transforms to network theory." Other Awards: National Science Foundation Faculty Fellow in Science, 1975-1976; Science Award, Armenian Students Association of America, 1982; Leading Professor of Electrical Engineering, State University of New York at Stony Brook.

ZEMEL, JAY N., Fellow 1977. Born: June 26, 1928, New York, N.Y. Degrees: B.S., 1949, M.S., 1952, Ph.D., 1956, Syracuse University; M.A.(honoris causa), 1972, University of Pennsylvania.

Fellow Award "For contributions to solid-state electronics and the development of IV-VI compound semiconductors for infrared photoconductive applications."

ZENNER, RAYMOND E., Fellow 1971. Born: April 16, 1910, Chicago, Ill. Degrees: B.S., 1933, University of Chicago.

Fellow Award "For contributions to magnetic recording and printing telegraphy."

ZENNER, WALTER J., Fellow 1969. Born: February 21, 1904, Chicago, Ill. Degrees: B.S.E.E., 1928, Illinois Institute of Technology.

Fellow Award "For outstanding contributions to the development of printing telegraph and data communication apparatus and systems." Other Awards: Certificate of Commendation (Civilian), U.S. Navy, 1947; Life Member, Western Society of Engineers; Inventor, Communication Products, 100 U.S. patents.

ZETTERBERG, LARS H., Fellow 1977. Born: January 6, 1925, Uppsala, Sweden. Degrees: E.E., 1949, Lic.-Tech., 1954, Dr.Tech., 1961, The Royal Institute of Technology, KTH, Stockholm.

Fellow Award "For contributions to communication theory and medical signal processing and to engineering education." Other Awards: Member, Swedish Academy of Engineering Sciences, 1976.

ZIEGLER, HANS K., IRE Fellow 1961. Born: March 1, 1911, Munich, Germany. Degrees: B.S., 1932, M.S., 1934, Ph.D., 1936, Technische Hochschule Munich, Germany.

Fellow Award "For guidance and leadership in military electronics." Other Awards: Siemens Ring Stiftung, 1934; Fellow, American Astronautical Society, 1960; Meritorious Civil Service Award, 1963, 1970; Meritorious Service Award, Armed Forces Communications and Electronics Association, 64; Anarctica Service Medal, 1964; Gold Medal for Meritorious Service, Armed Forces Communications and Electronics Association, 1974; Secretary of the Army's Exceptional Civil Service Award, 1976.

ZIEMER, RODGER E., Fellow 1983. Born: August 22, 1937, Sargeant, MN. Degrees: B.S., 1960, M.S., 1962, Ph.D., 1965, University of Minnesota.

Fellow Award "For contributions to digital communications systems and to engineering education"; Education Award, St. Louis Section, IEEE, 1980. Other Awards: Halliburton Award of Excellence, Halliburton Education Foundation, 1983.

ZIERDT, CONRAD H., JR., Fellow 1965. Born: July 15, 1916, Edgewood, Pa. Degrees: B.S.E.E., 1936, Pennsylvania State University.

Fellow Award "For contributions to the reliability and standardization of semiconductor devices." Other Awards: Engineering Award of Excellence, Electronic Industries Association, 1976.

ZIERLER, NEAL, Fellow 1979. Born: September 17, 1926, Baltimore, Md. Degrees: A.B., 1947, The Johns Hopkins University; A.M., 1950, Ph.D., 1959, Harvard University.

Fellow Award "For the application of finite mathematics to the design and analysis of communication systems, including error-correcting codes and cryptology."

ZIMMERER, CECIL W., AIEE Fellow 1951. Born: April 8, 1902, Brooklyn, N.Y. Degrees: 1922, Pratt Institute.

Other Awards: Technical Service Award, National Fire Protection Association, 1966.

ZIMMERMAN, STANLEY W., AIEE Fellow 1962. Born: July 30, 1907, Detroit, Mich. Degrees: B.S., 1930, M.S., 1930, University of Michigan.

Fellow Award "For contributions in the field of high voltage engineering." Other Awards: Professor Emeritus, Cornell University, 1973.

ZIMMERMANN, HENRY J., Fellow 1969. Born: May 11, 1916, St. Louis, Mo. Degrees: B.S., 1938, Washington University; S.M., 1942, Massachusetts Institute of Technology.

Fellow Award "For contributions as an educator and director of research and graduate training in the broad field of electronics." Other Awards: Alumni Achievement Award, Washington University Engineering School, 1975.

ZIPPLER, W. N., AIEE Fellow 1949. Born: March 19, 1896, Philadelphia, Pa. Degrees: B.S., 1920, E.E., 1927, University of Pennsylvania.

Fellow Award "For outstanding work in the development of electric plants and electric propulsion on large seagoing vessels."

ZITELLI, LOUIS T., Fellow 1980. Born: October 2, 1922. Degrees: B.A., 1944, San Jose State College; M.S.(E.E.), 1946, E.E., 1948, Ph.D.(E.E.), 1950, Stanford University.

Fellow Award "For contributions to space communications and space radar by leadership in the design and fabrication of high-power microwave tubes."

ZIV, JACOB, Fellow 1973. Born: 1931, Israel. Degrees: B.Sc., 1954, Dipl.Ing., 1955, M.Sc., 1957, Technion, Israel; D.Sc., 1962, Massachusetts Institute of Technology.

Fellow Award "For contributions in the fields of information theory and engineering education, and for leadership in establishing engineering research and development in Israel."

ZOBEL, EUGENE S., Fellow 1979. Born: December 6, 1918, Charleston, S.C. Degrees: B.S.C.E., 1940, The Citadel, Charleston, S.C.

Fellow Award "For contributions to the research and development of transmission-line engineering"; Outstanding Engineer Award, North Carolina Council, IEEE, 1980; Outstanding Engineer Award, Charlotte Section, IEEE, 1980.

ZRAKET, CHARLES A., Fellow 1978. Born: January 9, 1924, Lawrence, Mass. Degrees: B.S.(Electrical Engi-

...gna cum laude), 1951, Northeastern University, M.S.(Electrical Engineering, cum laude), 1953, ...ssachusetts Institute of Technology.

Fellow Award "For technical management and contributions in the application of systems engineering to large military and civilian problems." Other Awards: Associate Fellow and Corporate Representative, American Institute of Aeronautics and Astronautics; The New York Academy of Sciences; Fellow, The American Association for the Advancement of Science; Tau Beta Pi; Eta Kappa Nu; Sigma Xi.

ZVEREV, ANATOL I., Fellow 1968. Born: November 20, 1913. Degrees: Dip.Ing., 1938, Leningrad Electrotechnic Institute; D.Eng., 1940, Academy of Transportation.

Fellow Award "For inspiring leadership in network synthesis theory in communications, radar, weapon control, and navigation systems."